Discrete Mathematics

Fifth Edition

John A. Dossey
Albert D. Otto
Lawrence E. Spence
Charles Vanden Eynden
Illinois State University

PEARSON
Addison
Wesley

Boston San Francisco New York
London Toronto Sydney Tokyo Singapore Madrid
Mexico City Munich Paris Cape Town Hong Kong Montreal

Publisher Greg Tobin
Senior Acquisitions Editor Bill Hoffman
Editorial Assistants Lianna Kong and Emily Portwood
Managing Editor Karen Wernholm
Senior Production Supervisor Kathleen A. Manley
Senior Designer/Cover Designer Barbara T. Atkinson
Photo Researcher Beth Anderson
Marketing Manager Phyllis Hubbard
Marketing Coordinator Celena Carr
Senior Author Support/Technology Specialist Joe Vetere
Senior Prepress Supervisor Caroline Fell
Rights and Permissions Advisor Dana Weightman
Senior Manufacturing Buyer Evelyn Beaton
Text Design and Production Coordination WestWords, Inc.
Composition and Illustrations Techsetters, Inc.

Cover photo: Shell Fish Farming © The Boston Globe/David L. Ryan

Photo credits on p. 658

Library of Congress Cataloging-in-Publication Data

Discrete mathematics / John A. Dossey ... [et al.].–5th ed.
p. cm.
Includes bibliographical references and index.
ISBN 0-321-30515-9 (alk. paper)
1. Mathematics. 2. Computer science–Mathematics. I. Dossey, John A.

QA39.3.D58 2006
511'.1–dc22

2005045880

6 7 8 9 10—CRW—11

To our families

Contents

*Sections denoted by an asterisk are optional.

Preface

Today an increasing proportion of the applications of mathematics involves discrete rather than continuous models. The main reason for this trend is the integration of the computer into more and more of modern society. This book is intended for a one-semester introductory course in discrete mathematics.

Prerequisites Even though a course taught from this book requires few formal mathematical prerequisites, students are assumed to have the mathematical maturity ordinarily obtained by taking at least two years of high school mathematics, including problem-solving and algorithmic skills, and the ability to think abstractly.

Approach This book has a strong algorithmic emphasis that serves to unify the material. Algorithms are presented in English so that knowledge of a particular programming language is not required.

Choice of Topics The choice of topics is based upon the recommendations of various professional organizations, including those of the MAA's Panel on Discrete Mathematics in the First Two Years, the NCTM's *Principles and Standards for School Mathematics*, and the CBMS's recommendations for the mathematical education of teachers.

Flexibility Although designed for a one-semester course, the book contains more material than can be covered in either one semester or two quarters. Consequently, instructors will have considerable freedom to choose topics tailored to the particular needs and interests of their students. Users of previous editions have reported considerable success in courses ranging from freshman-level courses for computer science students to upper-level courses for mathematics majors. The present edition continues to allow instructors the flexibility to devise a course that is appropriate for a variety of different types of students.

Changes in the Fifth Edition The principal change in this fifth edition is the inclusion of a new chapter (Chapter 3, entitled Coding Theory) that treats congruences; the Euclidean algorithm and related topics from number theory; the RSA method of public key cryptography; and error-detecting and error-correcting codes, including matrix codes. The remainder of the book is independent of this new chapter, however, and so it may be included at the discretion of the instructor. The material on matrix codes requires familiarity with matrices, and so students who are not familiar with matrices will need to read Appendix B before beginning the coding theory chapter. (See the subsequent items on Chapter Independence and Possible Courses for details.)

Additional changes in this new edition have been made to improve the clarity of the text and to note recent developments in discrete mathematics.

Exercises The exercise sets in this book have been designed for flexibility. Many straightforward computational and algorithmic exercises are included after each section. These exercises give students hands-on practice with the concepts and algorithms of discrete mathematics and are especially important for students whose mathematical backgrounds are weak. Other exercises extend the material in the text or introduce new concepts not treated there. Exercise numbers in color indicate the more challenging problems. An instructor should choose those exercises appropriate to his or her course and students. Answers to odd-numbered computational exercises appear at the end of the book. At the end of each chapter, a set of **Supplementary Exercises** is provided. These reprise the most important concepts and techniques of the chapter and also explore new ideas not covered elsewhere.

Chapter Independence The sequence of chapters allows considerable flexibility in teaching a course from this book. The accompanying diagram shows the logical dependence of the chapters. The dashed line indicates that only the beginning sections of Chapter 4 are needed for Chapter 6. Although this book assumes only the familiarity with logic and proof ordinarily gained in high-school geometry, an appendix (Appendix A) is provided for those who prefer a more formal treatment. If Appendix A is covered, it may be taught at any time as an independent unit or in combination with Chapter 9. Familiarity with matrices (Appendix B) is required only for Sections 3.5–3.6 and for discussions of adjacency matrices in Chapter 4. The brackets in the diagram on page xiii denote these dependencies.

Chapters 1 and 2 are introductory in nature. Chapter 1, which should be covered fairly quickly, gives a sampling of the sort of discrete problems the course treats. Some questions are raised that will not be answered until later in the book. Section 1.4 contains a discussion of complexity that some instructors may want to omit or delay until students have had more experience with algorithms. An instructor may wish to cover only the illustrative algorithms in this section that are most relevant to his or her students.

Chapter 2 reviews various basic topics, including sets, relations, functions, and mathematical induction. It can be taught more or less rapidly, depending on the mathematical backgrounds of the students and the level of the course. It should be possible for students with good mathematics backgrounds to be able to read much of Chapter 2 on their own. The remaining chapters are, as the

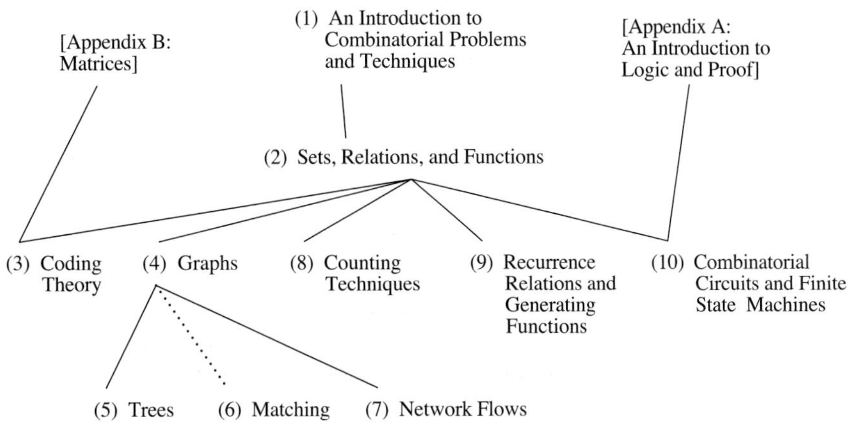

preceding diagram shows, independent, except that Chapters 5 and 7 depend on Chapter 4, and the beginning concepts of Chapter 4 are needed in Chapter 6.

To accommodate the new Chapter 3, the material on congruences has been moved from Chapter 2. As in the fourth edition, however, this material (Section 3.1 of the fifth edition) can be studied at any time after Section 2.2 (Equivalence Relations).

Possible Courses Three sample three-semester-hour courses are suggested in the table that follows. Course A emphasizes graph theory and its applications; it covers most of Chapters 4–7. Course B includes less graph theory, but more counting techniques. Course C emphasizes topics of interest to computer science students.

Course A		*Course B*		*Course C*	
Chapter	*Hours*	*Chapter*	*Hours*	*Chapter*	*Hours*
1	4	1 (skip 1.4)	3	1	4
2	5	2	5	2	5
4	6	4	6	Appendix B	1
5	7	5	6	3	6
6	6	8	8	4	6
7	4	9	5	5	6
8	8	Appendix A	3	9	5
		9	4	Appendix A	3
				10	4

Courses of various levels of sophistication can be taught from this book. For example, the topic of computational complexity is of great importance, and so attention is given to the complexity of many algorithms in this text. Yet it is a difficult topic, and the detail with which it is treated should correspond to the intended level of the course and the preparation of students.

Computer Projects Each chapter ends with a set of computer projects related to its content, algorithmic and otherwise. These are purposely stated in general terms, so as to be appropriate for students using various computing systems and languages.

Supplements A *Student's Solutions Manual*, available for purchase by students, provides detailed, worked-out solutions to the odd-numbered exercises. To order, use ISBN 0-321-30516-7. An *Instructor's Solutions Manual*, containing answers to all even-numbered computational exercises, is also available (ISBN 0-321-30517-5).

Acknowledgments We would like to thank the following mathematics professionals whose reviews guided the preparation of this text: Dorothee Blum, Millersville University; Richard Brualdi, University of Wisconsin, Madison; John L. Bryant, Florida State University; Richard Crittenden, Portland State University; Klaus Fischer, George Mason University; Dennis Grantham, East Texas State University; William R. Hare, Clemson University; Christopher Hee, Eastern Michigan University; Frederick Hoffman, Florida Atlantic University; Julian L. Hook, Florida International University; Carmelita Keyes, Broome Community College; Richard K. Molnar, Macalester College; Catherine Murphy, Purdue University, Calumet; Charles Nelson, University of Florida; Fred Schuurmann, Miami University; Karen Sharp, Charles S. Mott Community College; and Donovan H. Van Osdol, University of New Hampshire. The second edition was improved by the helpful comments of our colleagues Saad El-Zanati, Michael Plantholt, and Shailesh Tipnis, and other users of the text throughout the country, as well as by Elaine Bohanon, Bemidji State University; George Dimitroff, Evergreen State College; Richard Enstad, University of Wisconsin-Whitewater; Donald Goldsmith, Western Michigan University; Thomas R. Graviss, Kentucky Educational Network; Gary Klatt, University of Wisconsin-Whitewater; Mark Michael, Kings College; Peter Morris, Shepherd College; Dix H. Pettey, University of Missouri; Matt Pickard, University of Puget Sound; Terry Walters, University of Tennessee; Porter Webster, University of Southern Mississippi; Richard Weimer, Frostburg State University; Thomas Weininger, University of Wisconsin-Eau Claire; and Mark Woodard, Furman University. The third edition was further improved by the comments of users of the text throughout the country, as well as by the following list of colleague reviewers: Veena Chadha, University of Wisconsin-Eau Claire; Gary Chartrand, West Michigan University; Tilak de Alwis, Southeastern Louisiana University; Ron Dalla, Eastern Washington University; George Dimitroff, Evergreen State College; Gayla S. Domke, Georgia State University; Jerome Eisenfeld, University of Texas at Arlington; Kathleen Elder, Frostburg State University; Klaus Fischer, George Mason University; Donald A. Goral, Northern Virginia Community College; Natasa Jonoska, University of South Florida; Thomas Kiley, George Mason University; Theresa D. Magnus, St. Mary's College; Chris Mawata, University of Tennessee; Robert C. Mers, North Carolina A&T State University; Catherine M. Murphy, Purdue University, Calumet; Anne Quinn, Edinboro University of Pennsylvania; Steen Pedersen, Wright State University; Dix H. Pettey, University of Missouri, Columbia; James L. Richards, Bemidji State University; A. Allan Riveland, Washburn University; Mohan Shrikhande, Central Michigan University; Allan Schwenk, Western Michigan University; and our colleague, Roger Eggleton.

We wish to offer special thanks to Michael Plantholt and Dean Sanders of Illinois State University, who independently checked all the algorithms in the third edition for both correctness and readability. Their suggestions led to substantial revisions and improvements of the algorithms.

Changes made for the fourth edition were guided by the comments of the following reviewers: Mark Ferris, Midwestern State University; Johanne Hattingh, Georgia State University; Colleen Hoover, St. Mary's College; Jason Miller, Truman State University; and Richard Rockwell, Pacific Union College.

For their comments that led to this fifth edition, we would like to thank the following reviewers: Glen Aguiar, City College of San Francisco; Stephen Comer, The Citadel; Lowell Doerder, Black Hawk College; Mark Hall, Hastings College; George Hilton, Pacific Union College; Kenneth Myers, Bloomfield College; Charles Parry, Virginia Polytechnic Institute and State University; Richard Schoyen, University of Great Falls; Allen Schwenk, Western Michigan University; and Fereja Tahir, Illinois Central College.

We also appreciate the excellent editorial work of Jami Darby, Emily Portwood, and Bill Hoffman and the fine technical assistance provided by Joe Vetere, Senior Author Support/Technology Specialist, during the production process.

John A. Dossey
Albert D. Otto
Lawrence E. Spence
Charles Vanden Eynden

To the Student

This book is concerned with the *discrete*—that is, finite processes and sets of elements that can be listed. This contrasts with calculus, which has to do with infinite processes and intervals of real numbers.

Although discrete mathematics has been around for a long time, it has enjoyed a recent rapid expansion, paralleling the growth in the importance of computers. A digital computer is a complicated, but essentially finite, machine. At any given time it can be described by a large, but finite, sequence of 0s and 1s corresponding to the internal states of its electronic components. Thus discrete mathematics is essential in understanding computers and how they can be applied.

An important part of discrete mathematics has to do with *algorithms*, which are explicit instructions for performing certain computations. You first learned algorithms in elementary school, for arithmetic is full of them. For example, there is the *long division algorithm*, which might cause an elementary school student to write down something like the following tableau.

$$\begin{array}{r} 32 \\ 13\,\overline{)\,425} \\ \underline{39} \\ 35 \\ \underline{26} \\ 9 \end{array}$$

Internally, the student is applying certain memorized procedures: *There are three 13s in 42, 3 times 13 is 39, 42 minus 39 is 3, bring down the 5*, etc. These procedures constitute the algorithm.

Another example of an algorithm is a computer program. Suppose a small business wants to identify all customers who owe it more than $100 and have been delinquent in payments for at least three months. Even though the company's computer files contain this information, it constitutes only a small portion of their data. Thus a program must be written to sift out exactly what the company wants to know. This program consists of a precise set of instructions to the computer, covering all possibilities, that causes it to isolate the desired list of customers.

Our two examples of algorithms are similar in that the entity executing the algorithm does not have to understand why it works. Students in elementary school generally do not know why the long division algorithm gives the correct answer, only what the proper steps are. Of course, a computer doesn't understand anything; it just follows orders (and if its orders are incorrect, so that the program is wrong, the computer will dutifully produce the wrong answer).

If you are taking a course using this book, however, you are no longer in elementary school and you are a human being, not a computer. Thus you will be expected to know not only *how* our algorithms work, but *why*.

We will investigate some algorithms you probably have never seen before. For example, suppose you are planning to drive from Miami, Florida, to Seattle, Washington. Even if you stick to the interstate highways, there are hundreds of ways to go. Which way is the shortest? You might get out a map and, after playing around, find a route you *thought* was shortest; but could you be sure?

There is an algorithm that you could apply to this problem that would give you the correct answer. Better yet, you could program the algorithm into a computer, and let it find the shortest route. That algorithm is explained in this book.

We will be interested not only in the how and why of algorithms, but also in the *how long*. Computer time can be expensive, so before we give a computer a job to do we may want an estimate of how long it will take. Sometimes the surprising answer is that the computation will take so long as to make a computer solution impractical, even if we use the largest and fastest existing machines. It is a popular, but incorrect, idea that computers can do any computation. No computer can take the data from the world's weather stations and use it to predict future weather accurately more than a few days in advance. The fact that no one knows how to do certain computations efficiently can actually be useful. For example, if n is the product of two primes of about 200 decimal digits, then to factor n takes hundreds of years (even using the best methods and computers known), and this is the basis of an important system of cryptography.

You have probably already heard a number of times that mathematics is not a spectator sport, and that the only way to learn mathematics is by doing it. There is an important reason we are repeating this advice here: *It's true!* Moreover, it's the best thing we know to tell you. You can't learn to play the guitar or shoot free throws just by watching someone else do these things, and you can't learn discrete mathematics just by reading this book or attending lectures. The mind must be in gear and active. When reading a mathematics book, you should always have paper and pencil handy to work out examples and the details of computations. When attending a mathematics lecture, it is best if you have read the material already. Then you can concentrate on seeing if your understanding of the content agrees with that of the professor, and you can ask questions about any difficult points.

Of course, one of the best ways to be active in learning mathematics is by doing exercises. There are many of these in this book. Some are purely computational, others test understanding of concepts, and some require constructing proofs. Answers to odd-numbered computational exercises are in the back of the book, but *don't look before you have determined your own answer*. If your work consistently gives the same answer as in the back of the book, then you can have confidence that you are on the right track.

Some exercises are harder than others. The more time you spend on such exercises, the more you learn. There is a common notion (reinforced by some courses) that if you can't figure out how to do a problem in five minutes, you should go on to the next problem. This attitude becomes less and less relevant the more skillful you become. Very few accomplishments of any importance can be done in five minutes.

Many students do not realize the importance of learning the technical language of what they are studying. It is traditional in mathematics to assign special meaning to short, common words such as *set, function, relation, graph, tree, network.* These words have precise definitions that you must learn. Otherwise, how can you understand what you read in this book or what your professor is saying? These technical words are necessary for efficient communication. How would you like to explain a baseball game to someone if you were not allowed to use the particular language of that sport? Every time you wanted to say that a pitch was a *ball*, you would have to say that it was a pitch that was not in the strike zone and that the batter didn't swing at it. For that matter, *strike zone* is a technical term that would also need an explanation in each instance. Communication on such a basis would be almost impossible.

Finally, proper terminology is necessary to share information in a useful way with others. Mathematics is a human endeavor, and human cooperation depends on communication. In the real world, it is seldom sufficient simply to figure something out. You must be able to explain it to other people and to convince them that your solution is correct.

We hope that your study of discrete mathematics is successful and that you get from it techniques and attitudes that you will find useful in many contexts.

CHAPTER 1 An Introduction to Combinatorial Problems and Techniques

Combinatorial analysis is an area of mathematics concerned with solving problems for which the number of possibilities is finite (though possibly quite large). These problems may be broken into three main categories: determining existence, counting, and optimization. Sometimes it is not clear whether a certain problem has a solution or not. This is a question of **existence**. In other cases, solutions are known to exist, but we want to know how many there are. This is a **counting** problem. Or we may desire a solution that is "best" in some sense. This is an **optimization** problem. We will give a simple example of each type.

Determining Existence

Four married couples play mixed doubles tennis on two courts each Sunday night. They play for two hours, but switch partners and opponents after each half-hour period. Does a schedule exist so that each man plays with and against each woman exactly once and plays against each other man at least once?

Counting

A six-person investment club decides to rotate the positions of president and treasurer each year. How many years can pass before they will have to repeat the same people in the same offices?

Optimization

An employer has three employees, Pat, Quentin, and Robin, who are paid \$6, \$7, and \$8 per hour, respectively. The employer has three jobs to assign. The following table shows how much time each employee requires to do each job:

	Pat	*Quentin*	*Robin*
Job 1	7.5 hr	6 hr	6.5 hr
Job 2	8 hr	8.5 hr	7 hr
Job 3	5 hr	6.5 hr	5.5 hr

How should the employer assign one job to each person to get the work done as cheaply as possible?

Often, the solution we develop for a combinatorial problem will involve an **algorithm**, that is, an explicit step-by-step procedure for solving the problem. Many algorithms lend themselves well to implementation by a computer, and the importance of combinatorial mathematics has increased because of the wide use of these machines. However, even with a large computer, solving a combinatorial problem by simply running through all possible cases may be impossible. More sophisticated methods of attack are needed. In this chapter we will present more complicated examples of combinatorial problems and some analysis of how they might be solved.

1.1 THE TIME TO COMPLETE A PROJECT

The Problem

A large department store is having a Fourth of July sale (which will actually start July 2) and plans to send out an eight-page advertisement for it. This advertisement must be mailed out at least 10 days before July 2 to be effective, but various tasks must be done and decisions made first. The department managers decide which items in stock to put on sale, and buyers decide what merchandise should be brought in for the sale. Then a management committee decides which items to put in the advertisement and sets their sale prices.

The art department prepares pictures of the sale items, and a writer provides copy describing them. Then the final design of the advertisement, integrating words and pictures, is put together.

A mailing list for the advertisement is compiled from several sources, depending on the items put on sale. Then the mailing labels are printed. After the advertisement itself is printed, labels are attached, and the finished product, sorted by zip code, is taken to the post office.

Of course, all these operations take time. Unfortunately, it is already June 2, so only 30 days are available for the whole operation, including delivery. There is some concern whether the advertisements can be gotten out in time, and so estimates are made for the number of days needed for each task, based on past experience. These times are listed in the following table:

Task	*Time in Days*
Choose items (department managers)	3
Choose items (buyers)	2
Choose and price items for ad	2
Prepare art	4
Prepare copy	3
Design advertisement	2
Compile mailing list	3
Print labels	1
Print advertisement	5
Affix labels	2
Deliver advertisements	10

If the time needed for all the jobs is added up, we get 37 days, which is more time than is available. Some tasks can be done simultaneously, however. For example, the department managers and the buyers can be working on what they want to put on sale at the same time. On the other hand, many tasks cannot even be started until others are completed. For example, the mailing list cannot be compiled until it is decided exactly what items will be advertised.

In order to examine which jobs need be done before which other jobs, we label them A, B, . . . , K and list after each job any task that must immediately precede it:

Task		*Preceding Tasks*
A	Choose items (department managers)	None
B	Choose items (buyer)	None
C	Choose and price items for ad	A, B
D	Prepare art	C
E	Prepare copy	C
F	Design advertisement	D, E
G	Compile mailing list	C
H	Print labels	G
I	Print advertisement	F
J	Affix labels	H, I
K	Deliver advertisements	J

For example, the letters A and B are listed after task C because the items to be advertised and their prices cannot be determined until the department managers and buyers decide what they want to put on sale. Likewise, the letter C is listed after task D because the art cannot be prepared until the items to be advertised are identified. Notice that tasks A and B must also precede the preparation of the art, but this information is omitted because it is implied by what is given. That is, since A and B must precede C, and since C must precede D, logically A and B must go before D also, so this need not be said explicitly.

Let us assume that workers are available to start on each task as soon as it is possible to do so. Even so, it is not clear whether the advertisement can be prepared in time, although we have all the relevant information. Here we have a problem of *existence*. Does a schedule exist that will allow the advertisement to be sent out in time for the sale?

Analysis

Sometimes a body of information can be understood more easily if it is presented in graphical form. Let us represent each task by a point and draw an arrow from one point to another if the task represented by the first point must immediately precede the task represented by the second. For example, tasks A and B must precede task C, and C must precede D, so we start as in Figure 1.1.

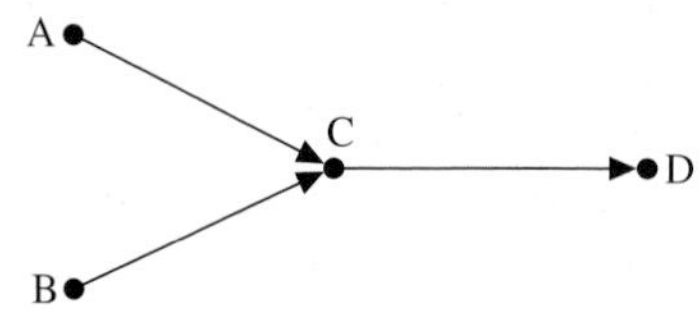

FIGURE 1.1

Continuing in this way produces the diagram of Figure 1.2(a). Note that the appearance of the diagram is not uniquely determined. For example, Figure 1.2(b) is consistent with the same information.

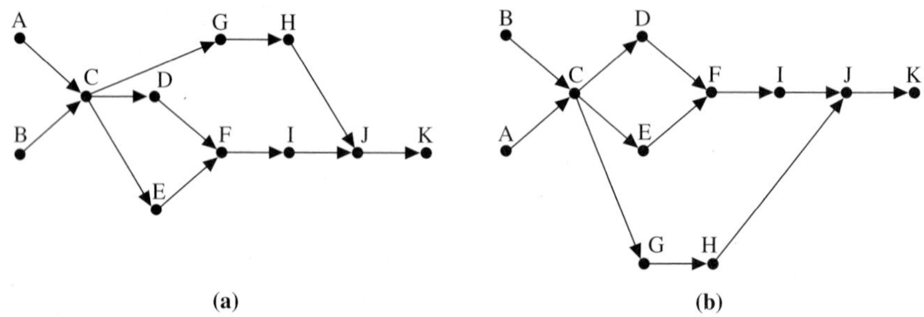

FIGURE 1.2

If we agree that all arrows go from left to right, we can omit the arrowheads, which we will do from now on.

This picture makes the whole project seem somewhat more comprehensible, but we must still take into account the time needed to do each task. Let us introduce these times into our diagram in Figure 1.3 by replacing each point with a circle containing the time in days needed for the corresponding task.

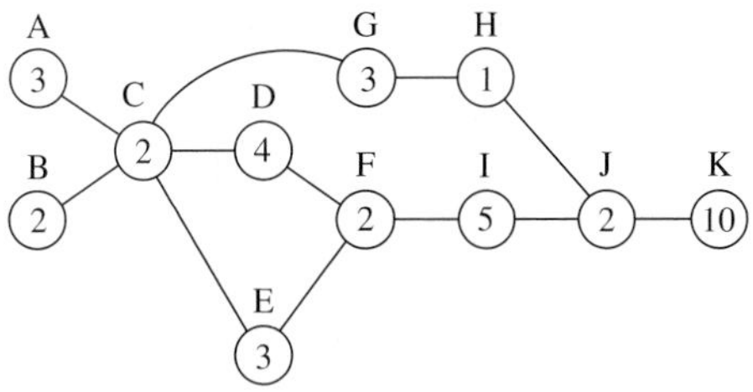

FIGURE 1.3

Now we will determine the smallest number of days after the start of the whole project in which each task can be finished. For example, task A can be started at once, so it will be done after 3 days. We will write the number 3 by the corresponding circle to indicate this. Likewise, we write a 2 by circle B.

How we treat task C is the key to the whole algorithm we will develop. This task cannot be started until both A and B are done. This will be after 3 days, since that is the time needed for A. Then task C will take 2 days. Thus 5 days are needed until C can be completed, and this is the number we write by the circle for C. So far our diagram looks like Figure 1.4.

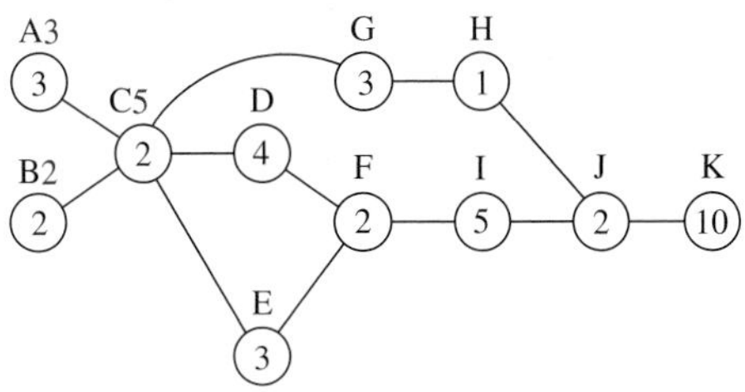

FIGURE 1.4

We carry on in the same fashion. Notice that if more than one line comes into a point from the left, then we add to the time for that point the *maximum* of all the incoming times to determine when it can first be completed. For example, it will take 9 days until D is finished and 8 days until E is finished. Since task F must wait for both of these, it will not be done for

$$(\text{maximum of 8 and 9}) + 2 = 11 \text{ days}.$$

The reader should check the numbers on the completed diagram in Figure 1.5.

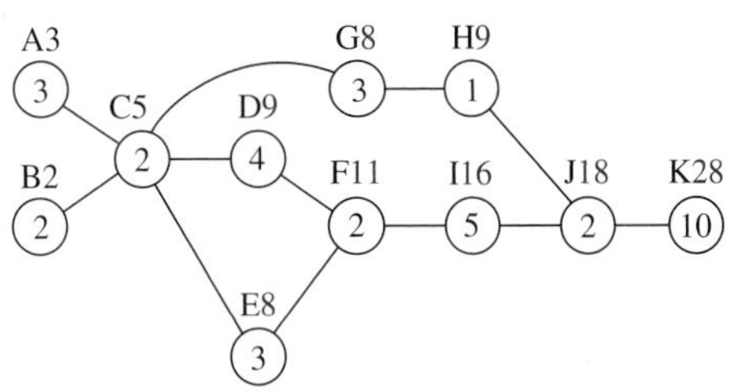

FIGURE 1.5

We see that the advertisement can be produced and delivered in 28 days, in time for the sale!

Critical Path Analysis

The method just described is called **PERT**, which stands for Program Evaluation and Review Technique. The PERT method (in a somewhat more complicated form) was developed in 1958 for the U.S. Navy Polaris submarine and missile project, although similar techniques were invented at about the same time at the E.I. du Pont de Nemours chemical company and in England, France, and Germany. Its usefulness in scheduling and estimating completion times for large projects, involving hundreds of steps and subcontracts, is obvious, and in various forms it has become a standard industrial technique. Any large library will contain dozens of books on the subject (look under PERT, Critical Path Analysis, or Network Analysis).

More information may be gleaned from the diagram we have just created. Let us work backward, starting from task K, and see what makes the project take all of 28 days. Clearly, it takes 28 days to finish K because there are 18 days until J is completed. Tasks H and I lead into J, but it is the 16 days needed to finish I that is important. Of course, task I cannot be completed before F is finished. So far we have traced a path back from K to F, as shown in color in Figure 1.6.

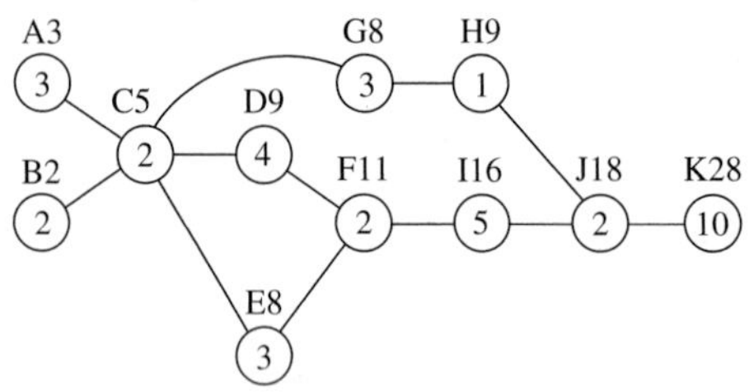

FIGURE 1.6

In the same way, we work back from F to D (since the reason it takes 11 days to finish F is that it cannot be started until the 9 days are up that it takes to complete D), then C, and finally A. The path A-C-D-F-I-J-K (in color in Figure 1.7) is called a **critical path**. The method of identifying the path is called the **critical path method**.

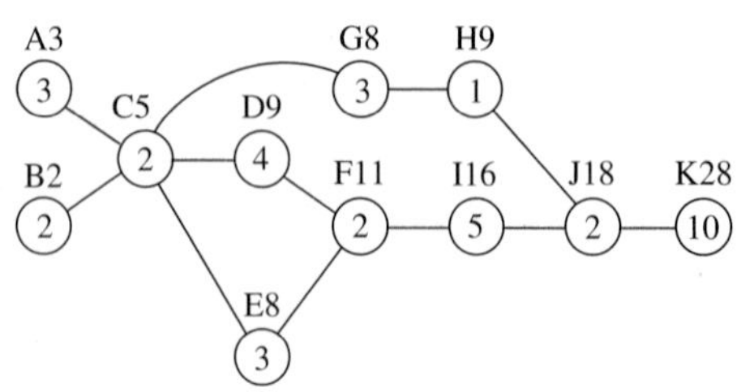

FIGURE 1.7

A critical path is important because the tasks on it are those that determine the total project time. If this time is to be reduced, then some task on a critical path must be done faster. For example, if the mailing list is compiled in 2 days instead of 3, it will still take 28 days to prepare and deliver the advertisement, since compiling the list (task G) is not on a critical path. Shortening the printing time (task I) by a day, however, would reduce the total time to 27 days; I is on the critical path. (Note, however, that changing the time for one task may change the critical path, altering whether or not the other tasks are on it.)

A Construction Example

The following table gives the steps necessary in building a house, the number of days needed for each step, and the immediately preceding steps:

Task		*Time in Days*	*Preceding Steps*
A	Site preparation	4	None
B	Foundation	6	A
C	Drains and services	3	A
D	Framing	10	B
E	Roof	5	D
F	Windows	2	E
G	Plumbing	4	C, E
H	Electrical work	3	E
I	Insulation	2	G, H
J	Shell	6	F
K	Drywall	5	I, J
L	Cleanup and paint	3	K
M	Floors and trim	4	L
N	Inspection	10	I

We prepare the diagram in Figure 1.8 showing times and precedences. Working first from left to right, then from right to left, we find the total times to complete each task and determine the critical path, which is marked with color in Figure 1.9. The only decision to be made in finding the critical path comes in working back

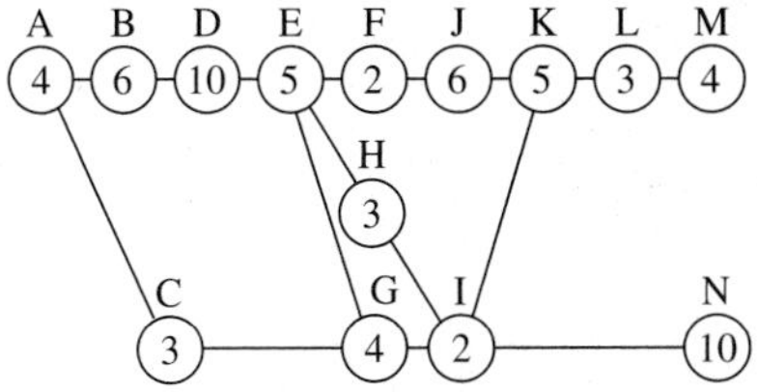

FIGURE 1.8

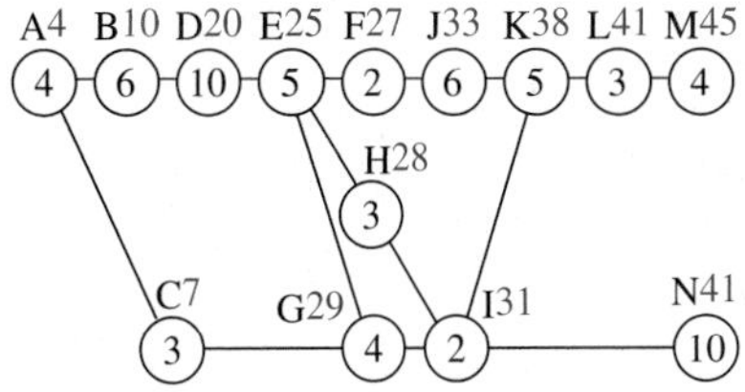

FIGURE 1.9

from K, where it is the 33 days needed to complete J that is important. We see that a total of 45 days are needed to build the house, and the critical path is A-B-D-E-F-J-K-L-M.

The technique of representing a problem by a diagram of points, with lines between some of them, is useful in many other situations and will be used throughout this book. The formal study of such diagrams will begin in Chapter 4.

EXERCISES 1.1

In Exercises 1–8, use the PERT method to determine the total project time and all the critical paths.

1.

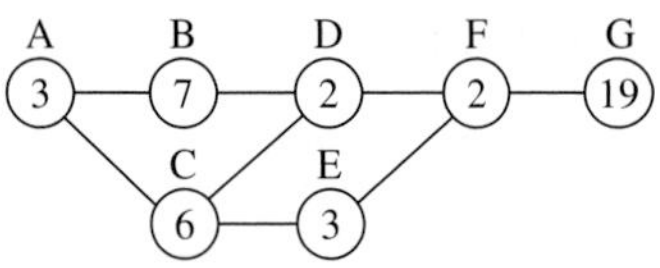

2.

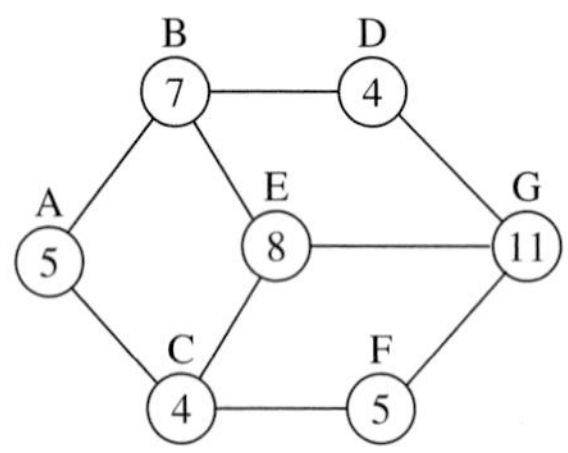

3.

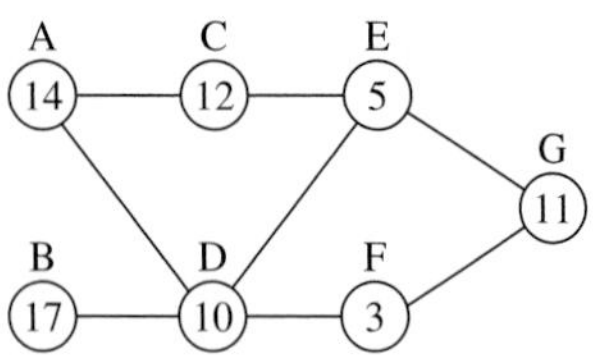

4.

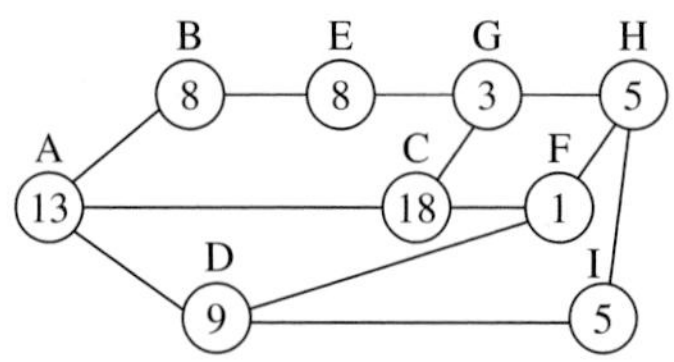

5.

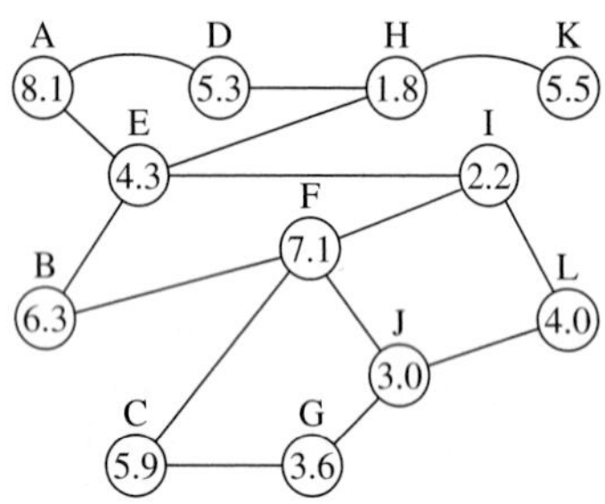

6.

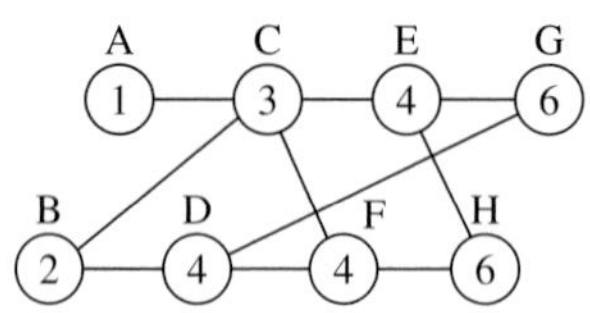

7.

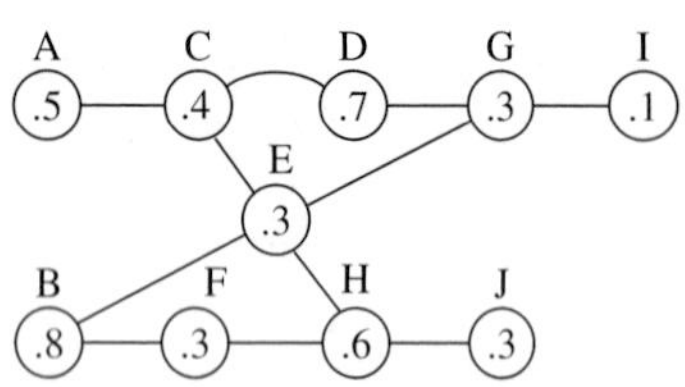

8.

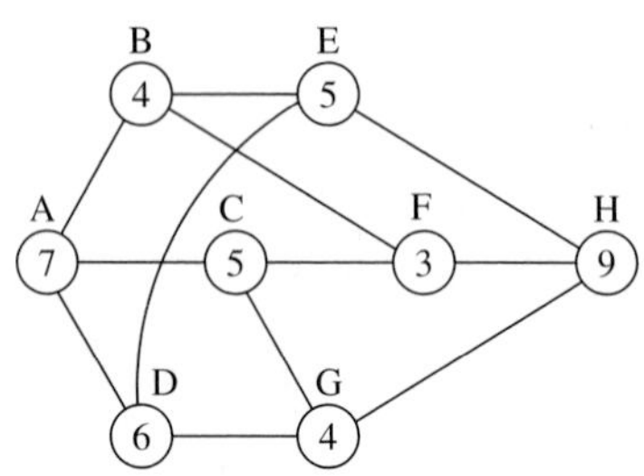

In Exercises 9–16, a table is given telling the time needed for each of a number of tasks and which tasks must immediately precede them. Make a PERT diagram for each problem, and determine the project time and critical path.

9.

Task	*Time*	*Preceding Tasks*
A	5	None
B	2	A
C	3	B
D	6	A
E	1	B, D
F	8	B, D
G	4	C, E, F

10.

Task	*Time*	*Preceding Tasks*
A	5	None
B	6	A
C	7	A
D	10	B
E	8	B, C
F	7	C
G	6	D, E, F

11.

Task	*Time*	*Preceding Tasks*
A	3	None
B	5	None
C	4	A, B
D	2	A, B
E	6	C, D
F	7	C, D
G	8	E, F

12.

Task	*Time*	*Preceding Tasks*
A	10	None
B	12	None
C	15	None
D	6	A, C
E	3	A, B
F	5	B, C
G	7	D, F
H	6	D, E
I	9	E, F

13.

Task	*Time*	*Preceding Tasks*
A	3.3	None
B	2.1	None
C	4.6	None
D	7.2	None
E	6.1	None
F	4.1	A, B
G	1.3	B, C
H	2.0	F, G
I	8.5	D, E, G
J	6.2	E, H

14.

Task	*Time*	*Preceding Tasks*
A	6	None
B	9	A, D
C	10	B, I
D	8	None
E	9	B
F	13	I
G	5	C, E, F
H	9	None
I	6	D, H

15.

Task	*Time*	*Preceding Tasks*
A	.05	None
B	.09	A
C	.10	A, F
D	.07	B, C
E	.02	None
F	.04	E
G	.11	E
H	.09	F, G
I	.06	D, H

16.

Task	*Time*	*Preceding Tasks*
A	11	None
B	13	None
C	12	None
D	14	None
E	8	A, C, D
F	6	A, B, D
G	10	A, B, C
H	5	B, C, D
I	9	E, F, H
J	7	F, G, H

17. A small purse manufacturer has a single machine that makes the metal parts of a purse. This takes 2 minutes. Another single machine makes the cloth parts in 3 minutes. Then it takes a worker 4 minutes to sew the cloth and metal parts together. Only one worker has the skill to do this. How long will it take to make 6 purses?

18. What is the answer to the previous problem if the worker can do the sewing in 2 minutes?

19. A survey is to be made of grocery shoppers in Los Angeles, Omaha, and Miami. First, a preliminary telephone survey is made in each city to identify consumers in certain economic and ethnic groups willing to cooperate, and also to determine what supermarket characteristics they deem important. This will take 5 days in Los Angeles, 4 days in Miami, and 3 days in Omaha. After the telephone survey in each city, a list of shoppers to be personally visited is prepared for that city. This takes 6 days for Miami and 4 days each for Omaha and Los Angeles. After all three telephone surveys are made, a standard questionnaire is prepared. This takes 3 days. When the list of consumers to be visited has been prepared and the questionnaire is ready, the questionnaire is administered in each city. This takes 5 days in Los Angeles and Miami and 6 days in Omaha. How long will it take until all three cities are surveyed?

1.2 A MATCHING PROBLEM

The Problem

An airline flying out of New York has seven long flights on its Monday morning schedule: Los Angeles, Seattle, London, Frankfurt, Paris, Madrid, and Dublin. Fortunately, seven capable pilots are available: Alfors, Timmack, Jelinek, Tang, Washington, Rupp, and Ramirez. There is a complication, however. Pilots are allowed to request particular destinations, and these requests are to be honored if possible. The pilots and the cities they requested are as follows:

Los Angeles: Timmack, Jelinek, Rupp
Seattle: Alfors, Timmack, Tang, Washington
London: Timmack, Tang, Washington
Frankfurt: Alfors, Tang, Rupp, Ramirez
Paris: Jelinek, Washington, Rupp
Madrid: Jelinek, Ramirez
Dublin: Timmack, Rupp, Ramirez

This information could also be represented by a diagram (Figure 1.10(a)), where we draw a black line between a city and a pilot if the former is on the pilot's request list.

The person assigning the flights would like to please all the pilots if this can be done, and if not, would like to accommodate as many as possible. This may be thought of as an *optimization* problem. We desire a matching of pilots with flights such that the number of pilots who get flights they have requested is as large as possible.

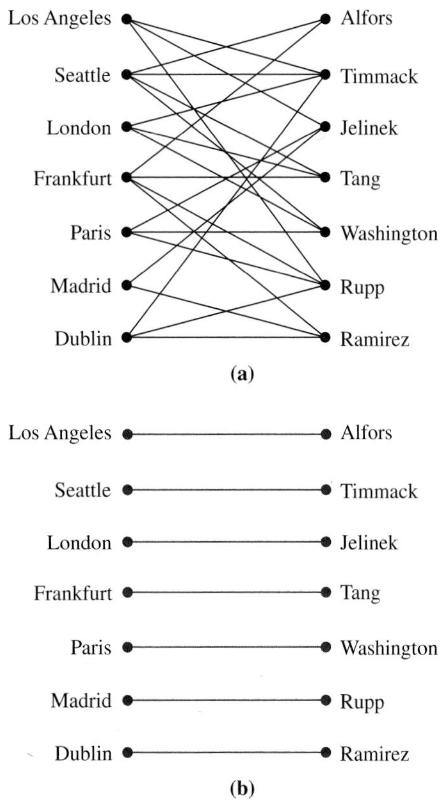

FIGURE 1.10

Analysis

Let us start with a very crude attack on our matching problem. We could simply list all possible ways of assigning one pilot to each flight and count for each the number of pilots who are assigned to flights they requested. For example, one matching would be to take the flights and pilots in the order they were listed:

Flight	*Pilot*	*Requested?*
Los Angeles	Alfors	No
Seattle	Timmack	Yes
London	Jelinek	No
Frankfurt	Tang	Yes
Paris	Washington	Yes
Madrid	Rupp	No
Dublin	Ramirez	Yes

This matching is indicated by the colored lines in Figure 1.10(b). Here, four of the pilots would get flights they want, but perhaps a different matching would do even better.

If we agree always to list the flights in the same order—say, that of our original list—then an assignment will be determined by some arrangement of the seven pilots' names. For example, the arrangement

Timmack, Alfors, Jelinek, Tang, Washington, Rupp, Ramirez

would send Timmack to Los Angeles and Alfors to Seattle, while assigning the same pilots to the other flights as previously. Likewise, the arrangement

Ramirez, Rupp, Washington, Tang, Jelinek, Timmack, Alfors

would send Ramirez to Los Angeles, Rupp to Seattle, etc. The reader should check that this matching will accommodate only three pilots' wishes.

Several questions come to mind concerning our plan for solving this problem:

(1) How much work will this be? In particular, how many arrangements will we have to check?
(2) How can we generate all possible arrangements so that we are sure we have not missed any?

The second question is somewhat special, and we will not answer it until Chapter 8, but the first question is easier. (Note that it is a *counting* problem.) In order to make the count, we will invoke a simple principle that will be useful many times in this book.

The Multiplication Principle Consider a procedure that is composed of a sequence of k steps. Suppose that the first step can be performed in n_1 ways; and for each of these, the second step can be performed in n_2 ways; and, in general, no matter how the preceding steps are performed, the ith step can be performed in n_i ways ($i = 2, 3, \ldots, k$). Then the number of different ways in which the entire procedure can be performed is $n_1 \cdot n_2 \cdots n_k$.

Example 1.1 A certain Japanese car is available in 6 colors, with 3 different engines and either a manual or automatic transmission. What is the total number of ways the car can be ordered?

We can apply the multiplication principle with $k = 3$, $n_1 = 6$, $n_2 = 3$, and $n_3 = 2$. The number of ways is $(6)(3)(2) = 36$. ■

Now we return to the problem of counting the number of ways the 7 flights can be assigned. Let us start with the Los Angeles flight. There are 7 pilots who can be assigned to it. We pick one, and turn to the Seattle flight. Now only 6 pilots are left. Choosing one of these leaves 5 from which to pick for the London flight. We can continue in this manner all the way to the Dublin flight, at which time only 1 pilot will be left. Thus, the total number of matchings we can devise will be $7 \cdot 6 \cdot 5 \cdot 4 \cdot 3 \cdot 2 \cdot 1$.

The same argument will work whenever we have the same number of flights and available pilots, producing

$$n(n-1)(n-2)\cdots 3 \cdot 2 \cdot 1$$

possible matchings if there are n flights and n pilots.

Permutations

The reader is probably aware that there is a shorter notation for a product of the type we just developed. If n is any nonnegative integer, we define ***n* factorial**, which is denoted by $n!$, as follows:

$$0! = 1, \quad 1! = 1, \quad 2! = 1 \cdot 2, \quad \text{and, in general,} \quad n! = 1 \cdot 2 \cdots (n-1)n.$$

Notice that if $n > 1$, then $n!$ is just the product of the integers from 1 to n.

By a **permutation** of a set of objects, we mean any ordering of those objects. For example, the permutations of the letters a, b, and c are

$$abc, \quad acb, \quad bac, \quad bca, \quad cab, \quad cba.$$

The analysis of the number of matchings when there are n flights and n pilots can be modified to prove the following result:

Theorem 1.1 There are exactly $n!$ permutations of a set of n objects.

There is a generalization of the idea of a permutation that often arises. Suppose that in the flight assignment problem the flights to Madrid and Dublin are cancelled because of bad weather. Now 7 pilots are available for the 5 remaining flights. There are 7 ways to choose a pilot for the Los Angeles flight, then 6 pilots to choose from for the Seattle flight, etc. Since only 5 pilots need to be chosen, there are

$$7 \cdot 6 \cdot 5 \cdot 4 \cdot 3$$

possible ways to make the assignments. (Notice that this product has 5 factors.) The same argument works in general.

Theorem 1.2 The number of ways an ordered list of r objects can be chosen without repetition from n objects is

$$n(n-1)\cdots(n-r+1) = \frac{n!}{(n-r)!}.$$

Proof. The first object can be chosen in n ways, the second in $n-1$ ways, etc. Since it is easy to check that the product on the left has r factors, by the multiplication principle it counts the total number of arrangements. As for the second expression, note that

$$\begin{aligned} n(n-1)\cdots(n-r+1) \\ &= \frac{n(n-1)\cdots(n-r+1)(n-r)(n-r-1)\cdots 2\cdot 1}{(n-r)(n-r-1)\cdots 2\cdot 1} \\ &= \frac{n!}{(n-r)!}. \end{aligned}$$ ■

Example 1.2 The junior class at Taylor High School is to elect a president, vice-president, and secretary from among its 30 members. How many different choices are possible?

We are to choose an ordered list of 3 officers from 30 students. The number of possibilities is

$$30\cdot 29\cdot 28 = 24{,}360.$$ ■

The number of ordered lists of r objects that can be chosen without repetition from n objects is denoted by $P(n,r)$. These lists are called **permutations of n objects, taken r at a time**. For example, we have just seen that $P(30,3) = 24{,}360$, and in general, according to the last theorem,

$$P(n,r) = \frac{n!}{(n-r)!}.$$

The Practicality of Our Solution to the Airline Problem

We were going to run through all the ways of assigning a pilot to each of the 7 flights to see which would please the most pilots. We now know the number of possible assignments is $7! = 5040$. This number is large enough to discourage us from trying this method by hand. If a computer were available, however, the method would look more promising. We would need a way to tell the computer how to generate these 5040 permutations—that is, an algorithm. This would amount to an explicit answer to question (2) asked earlier.

Of course, 7 flights and 7 pilots are really unrealistically small numbers. For example, at O'Hare field in Chicago, an average of more than 1100 airplanes take off every day. Let us consider a small airline with 20 flights and 20 pilots to assign to them, and consider the practicality of running through all possible assignments. The number of these is $20!$, which we calculate on a hand calculator to be about $2.4\cdot 10^{18}$. This is a number of 19 digits, and a computer is apparently required. Let us suppose that our computer can generate one million assignments per second and, for each of them, check how many pilots get their requested flights. How long would it take to run through them all?

The answer is not hard to calculate. Doing $2.4 \cdot 10^{18}$ calculations at 1,000,000 per second would take

$$\frac{2.4 \cdot 10^{18}}{1{,}000{,}000} = 2.4 \cdot 10^{12} \text{ seconds,}$$

$$\text{or} \quad \frac{2.4 \cdot 10^{12}}{60} = 4 \cdot 10^{10} \text{ minutes,}$$

$$\text{or} \quad \frac{4 \cdot 10^{10}}{60} \approx 6.7 \cdot 10^{8} \text{ hours,}$$

$$\text{or} \quad \frac{6.7 \cdot 10^{8}}{24} \approx 2.8 \cdot 10^{7} \text{ days,}$$

$$\text{or} \quad \frac{2.8 \cdot 10^{7}}{365} \approx 7.6 \cdot 10^{4} \text{ years.}$$

The calculation would take about 76,000 years, just for 20 flights and 20 pilots.

The point of this calculation is that even with a computer you sometimes have to be clever. In Chapter 5 we will explain a much more efficient way to solve our matching problem. This method will allow a person to handle 7 flights and 7 pilots in a few minutes and enable a computer to deal with hundreds of flights and pilots in a reasonable time.

EXERCISES 1.2

In Exercises 1–16, calculate the number shown.

1. $5!$

2. $6!$

3. $\frac{8!}{3!}$

4. $\frac{7!}{4!}$

5. $\frac{8!}{2!6!}$

6. $\frac{9!}{3!6!}$

7. $P(7, 4)$

8. $P(8, 4)$

9. $P(10, 7)$

10. $P(11, 9)$

11. $\frac{P(9, 4)}{5!}$

12. $\frac{P(10, 2)}{3!5!}$

13. $P(6, 6)$

14. $P(7, 7)$

15. $\frac{P(8, 3)}{P(3, 3)}$

16. $\frac{P(9, 5)}{P(5, 5)}$

17. A baseball manager has decided who his 9 starting hitters are to be, but not the order in which they will bat. How many possibilities are there?

18. A president, vice president, and treasurer are to be chosen from a club with 7 members. In how many ways can this be done?

19. A music company executive must decide the order in which to present 6 selections on a compact disk. How many choices does she have?

20. A Halloween makeup kit contains 3 different moustaches, 2 different sets of eyebrows, 4 different noses, and a set of ears. (It is not necessary to use any moustache, etc.) How many disguises using at least one of these items are possible?

21. A man has 5 sport coats, 4 pairs of slacks, 6 shirts, and 1 tie. How many combinations of these can he wear, if he must wear at least slacks and a shirt?

22. Different prizes for first place, second place, and third place are to be awarded to 3 of the 12 finalists in a beauty contest. How many ways is this possible?

23. Seven actresses have auditioned for the parts of the three daughters of King Lear: Goneril, Regan, and Cordelia. In how many ways can the roles be filled?

24. A farmer with 7 cows likes to milk them in a different order each morning. How many days can he do this without repeating?

25. A busy summer resort motel has 5 empty rooms and 3 travelers who want rooms. In how many ways can the motel manager assign a single room to each guest?

26. An Alaskan doctor visits each of 5 isolated settlements by plane once a month. He can use either of two planes, but once he starts out he visits all 5 settlements in some order before returning home. How many possibilities are there?

27. A tennis coach must pick her top 6 varsity and top 6 junior varsity players in order from among 9 varsity and 11 junior varsity players. In how many ways is this possible?

28. A dinner special for 4 diners at a Chinese restaurant allows one shrimp dish (from 3 selections), one beef dish (from 5), one chicken dish (from 4), and one pork dish (from 4). Each diner can also choose either soup or an egg roll. How many different orders might be sent to the kitchen?

29. How many ways can 6 keys be placed on a circular key ring? Both sides of the ring are the same, and there is no way to tell which is the "first" key on the ring.

30. Show that if $n > 1$, then $P(n, 2) = n^2 - n$.

31. Show that if $n > 0$, then $P(n, n-1) = n!$.

32. Show that if $0 \le 2r \le n$, then $\dfrac{P(n, 2r)}{P(n, r)} = P(n-r, r)$.

1.3 A KNAPSACK PROBLEM

The Problem

A U.S. shuttle is to be sent to a space station in orbit around the earth, and 700 kilograms of its payload are allotted to experiments designed by scientists. Researchers from around the country apply for the inclusion of their experiments. They must specify the weight of the equipment they want taken into orbit. A panel of reviewers then decides which proposals are reasonable. These proposals are then rated from 1 (the lowest score) to 10 (the highest) on their potential importance to science. The ratings are as follows:

	Experiment	Weight in Kilograms	Rating
1	Cloud patterns	36	5
2	Speed of light	264	9
3	Solar power	188	6
4	Binary stars	203	8
5	Relativity	104	8
6	Seed viability	7	6
7	Sun spots	92	2
8	Mice tumors	65	8
9	Weightless vines	25	3
10	Space dust	170	6
11	Cosmic rays	80	7
12	Yeast fermentation	22	4

It is decided to choose experiments so that the total of all their ratings is as large as possible. Since there is also the limitation that the total weight cannot exceed 700 kilograms, it is not clear how to do this. If we just start down the list, for example, experiments 1, 2, 3, and 4 have a total weight of 691 kilograms. Now we cannot take experiment 5, since its 104 kilograms would put us over the 700-kilogram limit. We could include experiment 6, however, which would bring us up to 698 kilograms. The following table shows how we might go down the list this way, keeping a running total of the weight and putting in whichever experiments do not put us over 700 kilograms:

Experiment	*Weight*	*Include?*	*Total Weight*	*Rating*
1	36	Yes	36	5
2	264	Yes	300	9
3	188	Yes	488	6
4	203	Yes	691	8
5	104	No	691	—
6	7	Yes	698	6
7	92	No	698	—
8	65	No	698	—
9	25	No	698	—
10	170	No	698	—
11	80	No	698	—
12	22	No	698	—

Note that the ratings total of the experiments chosen this way is

$$5 + 9 + 6 + 8 + 6 = 34.$$

The question is whether we can do better than this. Since we just took the experiments as they came, without paying any attention to their ratings, it seems likely that we can. Perhaps it would be better to start with the experiments with the highest rating and include as many of them as we can, then go on to those with the next highest rating, and so on. If two experiments have the same rating, we would naturally choose the lighter one first. The following table shows how such a tactic would work:

Experiment	*Rating*	*Weight*	*Include?*	*Total Weight*
2	9	264	Yes	264
8	8	65	Yes	329
5	8	104	Yes	433
4	8	203	Yes	636
11	7	80	No	636
6	6	7	Yes	643
10	6	170	No	643
3	6	188	No	643
1	5	36	Yes	679
12	4	22	No	679
9	3	25	No	679
7	2	92	No	679

Using this method, we choose experiments 2, 8, 5, 4, 6, and 1, giving a rating total of

$$9 + 8 + 8 + 8 + 6 + 5 = 44.$$

This is 10 better than our previous total, but perhaps it can be improved further.

Another idea would be to start with the experiment of smallest weight (number 6), then include the next lightest (number 12), and so on, continuing until we reach the 700-kilogram limit. The reader should check that this would mean including experiments 6, 12, 9, 1, 8, 11, 7, 5, and 10 for a rating total of 49, which is still better.

Yet another idea would be to compute a rating points-per-kilogram ratio for each experiment and to include, whenever possible, the experiments for which this ratio is highest. To illustrate this idea, consider an example where only three experiments are submitted, with the limit of 700 kilograms still in effect, as in the following table:

Experiment	*Weight*	*Rating*	*Ratio*
1	390	8	$\frac{8}{390} \approx .0205$
2	350	6	$\frac{6}{350} \approx .0171$
3	340	5	$\frac{5}{340} \approx .0147$

Using our new scheme, we would choose experiment 1, since it has the highest ratio, and then not be able to include either of the other two. The total rating would be 8. But this is not as good as choosing experiments 2 and 3 for a rating total of 11. The ratio method does not assure us of the best selection either. It turns out that if this method is applied to our original 12 experiments, it yields a subset of 9 experiments with a total rating of 51. (See Exercise 19.) Even this is not the optimal subset, however.

Analysis

We could play around with this problem, taking experiments out and putting experiments in, and perhaps find a collection of experiments with a higher rating total than 51. Even then, it would be hard for us to be sure that we could not do even better somehow. Notice that this is another *optimization* problem. We want to find a selection of experiments from the 12 given whose total weight is no more than 700 kilograms and whose rating total is as large as possible.

As in the case of the matching problem of the previous section, we will turn to the tedious method of trying all the possibilities. Getting a computer to do the calculations, even if there are many experiments, might be a practical way to attack the problem. Since the experiments are numbered from 1 to 12, we will save time by simply using the numbers. We will introduce some language (with which the reader is probably already familiar) in order to state the problem in a compact way.

We need the idea of a **set**. Although we cannot give a formal definition of a set in terms of simpler ideas, we can think of a set as a collection of objects of some sort such that, given any object, we can tell whether that object is in the set or not. If the object x is in the set S, we write $x \in S$, and if not, we write $x \notin S$.

Example 1.3 Let P be the set of all presidents of the United States. Then

$$\text{George Washington} \in P,$$

but

$$\text{Benjamin Franklin} \notin P.$$

If U is the set of all integers from 1 to 12, then

$$5 \in U, \text{ but } 15 \notin U. \quad \blacksquare$$

If a set has only a finite number of objects, one way to define it is simply to list them all between curly braces. For example, the set U of the example could also be defined by

$$U = \{1, 2, 3, 4, 5, 6, 7, 8, 9, 10, 11, 12\}.$$

If the set has more elements than we care to list, we may use three dots to indicate some elements. For example, we could also write

$$U = \{1, 2, 3, \ldots, 11, 12\}.$$

Another way to express a set is to enclose inside curly braces a variable standing for a typical element of the set, followed by a colon, followed by a description of what condition or conditions the variable must satisfy in order to be in the set—for example,

$$U = \{x : x \text{ is an integer and } 0 < x < 13\}$$

and

$$P = \{x : x \text{ is a president of the U.S.}\}.$$

The latter expression is read "the set of all x such that x is a president of the United States." In these two examples the use of x for the variable is arbitrary; any other letter having no previous meaning could be used just as well.

Let A and B be sets. We say that A is a **subset** of B and write

$$A \subseteq B$$

if every element of A is also in B. In this case, we also say that A **is contained in** B and that B **contains** A. An equivalent notation is

$$B \supseteq A.$$

Example 1.4 If U is the set that we just defined and

$$T = \{1, 2, 3, 4, 6\},$$

then $T \subseteq U$. Likewise, if

$$C = \{\text{Lincoln, A. Johnson, Grant}\},$$

then $C \subseteq P$, where, as before, P is the set of all American presidents. On the other hand, $P \subseteq C$ is false. $\blacksquare$

If A is a finite set, we will denote by $|A|$ the number of elements in A. For example, if C, T, U, and P are as previously defined, we have $|C| = 3$, $|T| = 5$, $|U| = 12$, and (in 2005) $|P| = 42$. (Although George W. Bush is often listed as the 43rd president of the United States, this number is achieved by counting Grover Cleveland twice, because Benjamin Harrison was president between Cleveland's two terms. But an element is either in a set or is not; it cannot be in the set more than once.) The **empty set** is the set that has no elements at all. We denote it by $\emptyset$. Thus, if A is a set, then $A = \emptyset$ if and only if $|A| = 0$.

We say that two sets are **equal** if every element in the first is also in the second and, conversely, every element in the second is also in the first. Thus $A = B$ if and only if $A \subseteq B$ and $B \subseteq A$.

The Problem Revisited

Armed with the language of sets, we return to the question of selecting experiments. The set of all experiments corresponds to the set

$$U = \{1, 2, \ldots, 11, 12\},$$

and each selection corresponds to some subset of U. For example, the choice of experiments 1, 2, 3, 4, and 6 corresponds to the subset

$$T = \{1, 2, 3, 4, 6\}.$$

This happens to be the selection of our first attempt to solve the problem, with a rating total of 34.

Of course, some subsets of U are unacceptable because their total weight exceeds 700 kilograms. An example of such a subset is

$$\{2, 3, 4, 10\},$$

with a total weight of 825 kilograms.

We could simply go through all the subsets of U, computing for each its total weight. If this does not exceed 700, then we will add up the ratings of the corresponding experiments. Eventually we will find which subset (or subsets) has the maximal rating total.

As in the last section, two questions arise:

(1) How many subsets are there (another counting problem)?
(2) How can we list all the subsets without missing any?

We will start with problem (1), saving problem (2) for Section 1.4. Let us start with some smaller sets to get the idea:

Set	*Subsets*	*Number of Subsets*
$\{1\}$	$\emptyset, \{1\}$	2
$\{1, 2\}$	$\emptyset, \{1\}, \{2\}, \{1, 2\}$	4
$\{1, 2, 3\}$	$\emptyset, \{1\}, \{2\}, \{1, 2\}, \{3\}, \{1, 3\}, \{2, 3\}, \{1, 2, 3\}$	8

We see that a set with 1 element has 2 subsets, a set with 2 elements has 4 subsets, and a set with 3 elements has 8 subsets. This suggests the following theorem, which will be proved in Section 2.6:

Theorem 1.3 A set with n elements has exactly 2^n subsets.

The set U has 12 elements, so, by Theorem 1.3, it has exactly $2^{12} = 4096$ subsets. This is more than we would like to run through by hand, but it would be easy enough for a computer. In fact, as n gets large, the quantity 2^n does not grow as fast as the quantity $n!$ that arose in the previous section. For example, 2^{20} is only about a million. Our hypothetical computer that could check one million subsets per second could run through the possible selections from 20 experiments in about a second, which is considerably less than the 76,000 years we found it would take to check the 20! ways of assigning 20 pilots to 20 flights. Even so, 2^n can get unreasonably large for modest values of n. For example, 2^{50} is about $1.13 \cdot 10^{15}$, and our computer would take about 36 years to run through this number of subsets.

Our problem of choosing experiments is an example of the **knapsack problem**. The name comes from the idea of a hiker who has only so much room in his knapsack, and must choose which items—food, first aid kit, water, tools, etc.—to include. Each item takes up a certain amount of space and has a certain value to the hiker, and the idea is to choose items that fit with the greatest total value.

In contrast to the matching problem of the previous section, there is no efficient way known to solve the knapsack problem. What exactly is meant by an "efficient way" will be made clearer when complexity theory is discussed in the next section.

EXERCISES 1.3

In Exercises 1–14, let $A = \{1, 2\}$, $B = \{2, 3, 4\}$, $C = \{2\}$, $D = \{x : x \text{ is an odd positive integer}\}$, *and* $E = \{3, 4\}$. *Tell whether each statement is true or false.*

1. $A \subseteq B$

2. $C \subseteq A$

3. $2 \subseteq A$

4. $B \subseteq D$

5. $10^6 \in D$

6. $C \in B$

7. $A \subseteq A$

8. $B = \{C, E\}$

9. $2 \in \{C, E\}$

10. $|\{C, E\}| = 2$

11. $|\{2, 3, 4, 3, 2\}| = 5$

12. $|\{\emptyset, \emptyset\}| = 1$

13. $\emptyset \in A$

14. $\emptyset \subseteq \{C, E\}$

In Exercises 15–18, the given sets represent a selection of space shuttle experiments from among the 12 given in the text. Determine whether each selection is acceptable (i.e., not over 700 kilograms). If it is, then find the total rating.

15. $\{2, 3, 9, 10, 12\}$

16. $\{2, 3, 9, 10, 11\}$

17. $\{2, 4, 6, 7, 9, 11\}$

18. $\{2, 3, 4, 6, 9\}$

19. Suppose that the rating/kilogram ratio is computed for each of the 12 proposed space shuttle experiments. Experiments are chosen by including those with the highest ratio that do not push the total weight over 700 kilograms. What set of experiments does this produce, and what is its total rating?

20. List the subsets of $\{1, 2, 3, 4\}$. How many are there?

21. How many subsets does {Sunday, Monday, . . . , Saturday} have?

22. How many subsets does {Dopey, Happy, ..., Doc} have?

23. How many subsets does {Chico, Harpo, Groucho, Zeppo, Gummo} have?

24. How many subsets does $\{13, 14, \ldots, 22\}$ have?

25. Suppose m and n are positive integers with $m < n$. How many elements does $\{m, m+1, \ldots, n\}$ have?

26. How many subsets does $\{2, 4, 8, \ldots, 256\}$ have?

27. A draw poker player may discard some of his 5 cards and be dealt new ones. The rules say he cannot discard all 5. How many sets of cards can be discarded?

28. Suppose that in the previous problem no more than 3 cards may be discarded. How many choices does a player have?

29. How long would it take a computer that can check one million subsets per second to run through the subsets of a set of 40 elements?

30. Find a subset of the 12 experiments with a total weight of 700 kilograms and a total rating of 49.

1.4 ALGORITHMS AND THEIR EFFICIENCY

Comparing Algorithms

In previous sections we developed algorithms for solving certain practical combinatorial problems. We also saw that in some cases solving a problem of reasonable size, even using a high-speed computer, can take an unreasonable amount of time. Obviously, an algorithm is not practical if its use will cost more than we are prepared to pay, or if it will provide a solution too late to be of value. In this section we will examine in more detail the construction and efficiency of algorithms.

We assume that a digital computer will handle the actual implementation of the algorithms we develop. This means a precise set of instructions (a "program") must be prepared telling the computer what to do. When an algorithm is presented to human beings, it is explained in an informal way, with examples and references to familiar techniques. (Think of how the operation of long division is explained to children in grade school.) Telling a computer how to do something requires a more organized and precise presentation.

Most computer programs are written in some specific higher-level computer language, such as FORTRAN, BASIC, COBOL, or C. In this book we will not write our algorithms in any particular computer language, but rather use English in a form that is sufficiently organized and precise that a program could easily be written from it. Usually, this will mean a numbered sequence of steps, with precise instructions on how to proceed from one step to the next. See Appendix C for an explanation of the technical terms in our algorithms.

Of course, a big, complicated problem will require a big, complicated solution, no matter how good an algorithm we find for it. For the problem of choosing an optimal set of experiments to place in a space shuttle, introduced in Section 1.3, choosing from among 12 submitted experiments requires looking at about 4000 subsets, while if there are 20 experiments, then there are about 1,000,000 subsets, 250 times as many. A reasonable measure of the "size" of the problem in this example would be n, the number of experiments. For each type of problem, we may be able to identify some number n that measures the amount of information upon which a solution must be based. Admittedly, choosing the quantity to be labeled n is often somewhat arbitrary. The precise choice may not be important, however, for the purpose of comparing two algorithms that do the same job, so long as n represents the same quantity in both algorithms.

We will also try to measure the amount of work done in computing a solution to a problem. Of course, this will depend on n, and for a desirable algorithm, it will not grow too quickly as n gets larger. We need some unit by which to measure the size of an algorithmic solution. In the space shuttle problem, for example, we saw that a set of n experiments has 2^n subsets that need to be checked. "Checking" a subset itself involves certain computations, however. The weights of the experiments in the subset must be totaled to see if the 700-kilogram limit is exceeded. If the limit is not exceeded, then the ratings of the subset must be added and compared with the previous best total. How much work this will be for a particular subset will also depend on the number of experiments in it.

We will take the conventional course of measuring the size of an algorithm by counting the total number of elementary operations it involves, where an elementary operation is defined as the addition, subtraction, multiplication, division, or comparison of two numbers. For example, adding up the k numbers $a_1, a_2, \ldots, a_k$ involves $k - 1$ additions, as we compute

$$a_1 + a_2, (a_1 + a_2) + a_3, \ldots, (a_1 + \cdots + a_{k-1}) + a_k.$$

We will call the total number of elementary operations required the **complexity** of the algorithm.

There are two disadvantages to measuring the complexity of an algorithm this way:

(1) This method essentially tries to measure the time it will take to implement an algorithm, assuming that each elementary operation takes the same time. But computers are also limited by their memory. An algorithm may require storage of more data than a given computer can hold. Or additional slower memory may have to be used, thereby slowing down the process. In any case, computer storage itself has a monetary value that our simple counting of elementary operations does not take into account.

(2) It may be that not all operations take the same amount of computer time; for example, division may take longer than addition. Also, the time an elementary operation takes may depend on the size of the numbers involved; computations with larger numbers take longer. Indeed, just the assignment of a value to a variable takes computer time—time that we are not taking into account.

In spite of the criticisms that can be made of our proposed method of measuring the complexity of an algorithm, we will use it for simplicity and to avoid considering the internal operations of particular computers.

Evaluating Polynomials

We will consider some examples of algorithms and their complexity. Let us start with the problem of evaluating x^n, where x is some number and n is a positive integer. To break this down into elementary operations, we could compute $x^2 = x \cdot x$, then $x^3 = (x^2)x, \ldots,$ until we get to x^n. Since computing x^2 takes 1 multiplication, computing x^3 takes 2 multiplications, and so on, a total of $n - 1$ multiplications is necessary. An algorithm for this process might be as follows:

Algorithm for Evaluating x^n

Given a real number x and a positive integer n, this algorithm computes $P = x^n$.

Step 1 (initialization)
Set $P = x$ and $k = 1$.
Step 2 (next power)
while $k < n$
(a) Replace P with Px.
(b) Replace k with $k + 1$.
endwhile
Step 3 (output $P = x^n$)
Print P.

Notice that step 2 entails $n - 1$ multiplications and $n - 1$ additions. There are also n comparisons, since we only exit step 2 when $k = n$. Thus we see that computing x^n with this algorithm involves a total of $3n - 2$ elementary operations.

Since our method of estimating the number of operations in an algorithm involves various inaccuracies anyway, usually we are not interested in an exact count. Knowing that computing x^n takes about $3n$ operations, or even a number of operations that is less than some constant multiple of n, may satisfy us. We are mainly interested in avoiding, when possible, an algorithm whose number of operations grows very quickly as n gets large.

Later in this book we will often write algorithms in a less formal way that may make impractical an exact count of the elementary operations involved. For example, in the last algorithm, instead of incrementing k by 1 at each step and then comparing its new value to that of n, we might simply have said something like "for $k = 1$ to $n - 1$, replace P by Px." High-level computer languages usually allow loops to be defined by some such language. If the number of operations required for each value of k does not exceed some constant C, then the complexity of the algorithm does not exceed Cn. (We could actually take $C = 3$ for the algorithm just presented.) Often, knowing the precise value of C is not important to us. We will indicate why this is true later in this section, after we have more examples of algorithms.

Now we will give an example of two algorithms with the same purpose, one of which is more efficient than the other. By a **polynomial of degree *n* in *x***, we mean an expression of the form

$$P(x) = a_n x^n + a_{n-1} x^{n-1} + \cdots + a_1 x + a_0,$$

where $a_n, a_{n-1}, \ldots, a_0$ are constants and $a_n \neq 0$. Thus a polynomial in x is a sum of terms, each of which is either a constant or else a constant times a positive integral power of x. We will consider two algorithms for computing the value of a polynomial. The first one, which may seem the more natural, will start with a_0, then add $a_1 x$ to that, then add $a_2 x^2$ to that, etc.

Polynomial Evaluation Algorithm

This algorithm computes $P(x) = a_n x^n + a_{n-1} x^{n-1} + \cdots + a_0$, given the nonnegative integer n and real numbers $x, a_0, a_1, \ldots, a_n$.

Step 1 (initialization)
  Set $S = a_0$ and $k = 1$.
Step 2 (add next term)
  while $k \leq n$
    (a) Replace S with $S + a_k x^k$.
    (b) Replace k with $k + 1$.
  endwhile
Step 3 (output $P(x) = S$)
  Print S.

In this algorithm we will check whether $k \leq n$ in step 2 a total of $n + 1$ times, with $k = 1, 2, \ldots, n + 1$. For a particular value of $k \leq n$, this will entail 1 comparison, 2 additions (in computing the new values of S and k), and 1 multiplication (multiplying a_k by x^k), a total of 4 operations. But this assumes that we know what x^k is. We just saw that this number takes $3k - 2$ operations to compute. Thus, for a given value of $k \leq n$, we will use $4 + (3k - 2) = 3k + 2$ operations. Letting $k = 1, 2, \ldots, n$ accounts for a total of

$$5 + 8 + 11 + \cdots + (3n + 2)$$

operations.

It will be proved in Section 2.5 that the value of this sum is $\frac{1}{2}(3n^2 + 7n)$. The extra comparison when $k = n + 1$ gives a total of $\frac{1}{2}(3n^2 + 7n) + 1$ operations.

Notice that here the complexity of our algorithm is itself a polynomial in n, namely $1.5n^2 + 3.5n + 1$. For a specific polynomial in n of degree k—say, $a_k n^k + a_{k-1}n^{k-1} + \cdots + a_0$—the term $a_k n^k$ will exceed the sum of all the other terms in absolute value if n is sufficiently large. Thus, if the complexity of an algorithm is a polynomial in n, we are interested mainly in the degree of that polynomial. Even the coefficient of the highest power of n appearing is of secondary importance. That the complexity of the algorithm just presented is a polynomial with n^2 as its highest power of n (instead of n or n^3, for example) is more interesting to us than the fact that the coefficient of n^2 is 1.5.

We will say that an algorithm has **order at most $f(n)$**, where $f(n)$ is some nonnegative expression in n, in case the complexity of the algorithm does not exceed $Cf(n)$ for some constant C. Recall that the polynomial evaluation algorithm has complexity $1.5n^2 + 3.5n + 1$. It is not hard to see that $3.5n + 1 \le 4.5n^2$ for all positive integers n. Then

$$1.5n^2 + 3.5n + 1 \le 1.5n^2 + 4.5n^2 = 6n^2$$

for all positive integers n. Thus (taking $C = 6$) we see that the polynomial evaluation algorithm has order at most n^2. A similar proof shows that in general an algorithm whose complexity is no more than a polynomial of degree k has order no more than n^k.

Now we will present a more efficient algorithm for polynomial evaluation. It was first published in 1819 by W.G. Horner, an English schoolmaster. The idea behind it is illustrated by the following computation with $n = 3$:

$$\begin{aligned} a_3x^3 + a_2x^2 + a_1x + a_0 &= x(a_3x^2 + a_2x + a_1) + a_0 \\ &= x(x(a_3x + a_2) + a_1) + a_0 = x(x(x(a_3) + a_2) + a_1) + a_0. \end{aligned}$$

Horner's Polynomial Evaluation Algorithm

This algorithm computes $P(x) = a_n x^n + a_{n-1}x^{n-1} + \cdots + a_0$, given the nonnegative integer n and real numbers $x, a_0, a_1, \ldots, a_n$.

Step 1 (initialization) Set $S = a_n$ and $k = 1$.
Step 2 (compute next expression)
while $k \le n$
(a) Replace S with $xS + a_{n-k}$.
(b) Replace k with $k + 1$.
endwhile
Step 3 (output $P(x) = S$)
Print S.

The following table shows how this algorithm works for $n = 3$, the polynomial $P(x) = 5x^3 - 2x^2 + 3x + 4$ (so $a_3 = 5$, $a_2 = -2$, $a_1 = 3$, and $a_0 = 4$), and $x = 2$:

	S	k
Start:	$a_3 = 5$	1
	$xS + a_2 = 2(5) + (-2) = 8$	2
	$xS + a_1 = 2(8) + 3 = 19$	3
	$xS + a_0 = 2(19) + 4 = 42$	4

We check whether $k \le n$ in step 2 for $k = 1, 2, \ldots, n + 1$, a total of $n + 1$ times, and each time except the last requires just 1 comparison, 1 multiplication, 2 additions, and 1 subtraction. Thus the algorithm evaluates a polynomial of degree n by using just $5n + 1$ operations, as opposed to $\frac{1}{2}(3n^2 + 7n) + 1$ for our first version. If n were 10, the second algorithm would take 51 operations while the first would take 186; and for larger values of n, the difference would be even more marked. In broader terms, Horner's polynomial evaluation algorithm is superior to the previous polynomial evaluation algorithm because it has order no more than n, while we could say only that the first algorithm had order no more than n^2.

The algorithms we have presented so far are simple enough that we can compute their complexities exactly. Often, however, the exact number of operations necessary may not depend only on n. An algorithm to sort n numbers into numerical order, for example, may entail more or fewer steps, depending on how the numbers are arranged initially. Here we might count the number of operations in the worst possible case. The actual number of operations will then be less than or equal to this number.

Later we may present more complicated algorithms in an informal way. An exact analysis of their complexity would entail a more detailed description revealing each elementary operation, akin to an actual program in some computer language. In this case the statement that the algorithm has order no more than $f(n)$ is to be interpreted as saying that a computer implementation exists for which the number of elementary operations does not exceed $Cf(n)$ for some constant C.

A Subset-Generating Algorithm

Now we will consider an algorithm for generating subsets of a set, as our solution of the space shuttle problem requires. If a set S consists of the n elements $x_1, x_2, \ldots, x_n$, a compact way of representing a subset of S is as a string of 0s and 1s, where the kth entry in the string is 1 if $x_k \in S$ and 0 otherwise. If $n = 3$, for example, the $8 = 2^3$ strings and the subsets to which they correspond are as follows:

000	$\emptyset$
001	$\{x_3\}$
010	$\{x_2\}$
011	$\{x_2, x_3\}$
100	$\{x_1\}$
101	$\{x_1, x_3\}$
110	$\{x_1, x_2\}$
111	$\{x_1, x_2, x_3\}$

By examining this list, we see how these digits might be generated in the order given. We look for the rightmost 0, change it to a 1, and then change any digits still further to the right to 0s. If we let

$$a_1a_2\ldots a_n$$

be a given string with n 0s and 1s, the following algorithm generates the next string:

Next Subset Algorithm

Given a positive integer n and the string $a_1a_2\ldots a_n$ of 0s and 1s corresponding to a subset of a set with n elements, this algorithm computes the string corresponding to the next subset.

Step 1 (initialization)
  Set $k = n$.
Step 2 (look for rightmost 0)
  while $k \geq 1$ and $a_k = 1$
    Replace k with $k - 1$.
  endwhile
Step 3 (if there is a zero, form the next string)
  if $k \geq 1$
    Step 3.1 (change the rightmost 0 to 1)
      Replace a_k with 1.
    Step 3.2 (change succeeding 1s to 0s)
      for $j = k + 1$ to n
        Replace a_j with 0.
      endfor
    Step 3.3 (output)
      Print $a_1a_2\ldots a_n$.
  otherwise
    Step 3.4 (no successor)
      Print "This string contains all 1s."
  endif

In this algorithm, step 2 finds the rightmost string digit a_k that is 0. (If all digits are 1, k reaches 0 and we stop at step 3.4.) Then a_k is changed to 1 in step 3.1, and the digits to its right are changed to 0s in step 3.2. The actual number of arithmetic operations required depends on the string $a_1a_2 \ldots a_n$ that we start with, although since the replacements in steps 2 and 3 each can be repeated at most n times, the number of operations will be no more than some constant multiple of n.

Let us consider how this algorithm might be applied to the space shuttle problem. We will restrict our attention to deciding whether the subset we have generated has a total weight of no more than 700 kilograms. Let W_i be the weight of the ith experiment. Then we need to compute

$$a_1W_1 + a_2W_2 + \cdots + a_nW_n$$

and see if this exceeds 700 or not. If we are including experiment i when a_i is 1 and excluding it when a_i is 0, then this sum gives the total weight of the included experiments. The reader should check that the sum may be computed by using n multiplications and $n - 1$ additions, for a total of $2n - 1$ operations, not counting any comparisons or index changes.

Since, for n experiments, there are 2^n subsets of experiments, and since generating and checking each subset takes a multiple of n operations, the complexity of this method of finding the best choice of experiments is $Cn{\cdot}2^n$, where C is some constant. This is an expression that gets large quite quickly as n increases. The following table shows how long a computer that executes one million operations per second would take to do $n{\cdot}2^n$ operations for various values of n (for comparison purposes, we also show how long $1000n^2$ operations would take):

n:	10	20	30	40	50
$n2^n$ operations	0.01 sec	21 sec	9 hr	1.4 years	1785 years
$1000n^2$ operations	0.1 sec	0.4 sec	0.9 sec	1.6 sec	2.5 sec

This table indicates that simply increasing computer speed may not make an algorithm practical, even for modest values of n. If, for example, our computer were capable of one *billion* operations per second instead of one million (making it 1000 times faster), performing $n2^n$ operations for $n = 50$ would still require 1.785 years.

In general, an algorithm is considered "good" if its complexity is no more than some polynomial in n. Of course, in practice, a nonpolynomial complexity may be acceptable if only small values of n arise.

Expressions depending on n increase at different rates as n gets large. In general, expressions with n as the exponent of a number greater than 1 grow faster than any polynomial in n, and $n!$ grows even faster. On the other hand, although $\log_2 n$ (to be explained in Section 2.4) increases without bound as n increases, it grows more slowly than any positive power of n. The mathematical comparison of these expressions entails analytic techniques not appropriate for this course, but we offer the following table to give an idea of how fast various expressions grow (the time given is that for a computer executing one million operations per second to run through $f(n)$ operations):

n:	10	20	30	40	50	60
$f(n)$						
$\log_2 n$	.0000033 S	.0000043 S	.0000049 S	.0000053 S	.0000056 S	.0000059 S
$n^{1/2}$	.0000032 S	.0000045 S	.0000055 S	.0000063 S	.0000071 S	.0000077 S
n	.00001 S	.00002 S	.00003 S	.00004 S	.00005 S	.00006 S
n^2	.0001 S	.0004 S	.0009 S	.0016 S	.0025 S	.0036 S
$n^2 + 10n$	.0002 S	.0006 S	.0012 S	.0020 S	.003 S	.0042 S
n^{10}	2.8 H	119 D	19 Y	333 Y	3097 Y	19174 Y
2^n	.001 S	1 S	18 M	13 D	36 Y	36559 Y
$n!$	3.6 S	77147 Y	$8.4{\cdot}10^{18}$ Y	$2.6{\cdot}10^{34}$ Y	$9.6{\cdot}10^{50}$ Y	$2.6{\cdot}10^{68}$ Y

The Bubble Sort

Now we will give an algorithm for sorting a list of items possessing a natural order, such as scores in a golf tournament (where the order is numerical) or names in a list (where the order is alphabetical). The most widely known such algorithm, called the **bubble sort**, is so called because of the similarity between its actions and the movement of a bubble to the surface in a glass of water. The smaller items "bubble" to the beginning of the list. To keep the discussion simple, we will assume that the items $a_1, a_2, \ldots, a_n$ in the list are real numbers.

We first consider the last two items in the list, a_{n-1} and a_n, exchanging their values if a_n is less than a_{n-1}. We next consider the values in the $n-2$ and $n-1$ positions. Again, we exchange them if the $n-1$ item is less than the $n-2$ item. This process of comparing two adjacent items continues until the values in the first two positions are compared and exchanged, if necessary.

At this point, the smallest value in the list has been brought to the first position. We now start over again, this time operating on the smaller list consisting of the elements in the second through the nth positions. This will bring the smallest of the items in the second through the nth positions into the second position in the list. This process continues until all of the elements in the original list have been arranged in nondecreasing order.

Bubble Sort Algorithm

This algorithm places the numbers in the list $a_1, a_2, \ldots, a_n$ in nondecreasing order.

Step 1 (set beginning of sublist)
for $j = 1$ to $n - 1$

Step 1.1 (find smallest element of sublist)
for $k = n - 1$ to j by -1
Step 1.1.1 (interchange if necessary)
if $a_{k+1} < a_k$
Interchange the values of a_k and a_{k+1}.
endif
endfor
endfor

Step 2 (output list in nondecreasing order)
Print $a_1, a_2, \ldots, a_n$.

Example 1.5 We will use the bubble sort to order the list 7, 6, 14, 2. The following chart shows the positions of the numbers in the list as step 1.1.1 of the algorithm is performed (the circled numbers are those being compared):

a_1	a_2	a_3	a_4	j	k
7	6	14	2	1	3
7	6	2	14		2
7	2	6	14		1
2	7	6	14	2	3
2	7	6	14		2
2	6	7	14	3	3
2	6	7	14		

Thus 6 comparisons are required to sort the given list into nondecreasing order. ■

To measure how efficient the bubble sort is, we will count the number of comparisons required to get the n items in order. Notice that each comparison is accompanied by a bounded number of other operations, so that the complexity of the algorithm will not exceed some constant times the number of comparisons.

The first pass through the list requires $n - 1$ comparisons; this moves the smallest element to the front of the list. The second pass, using the items in the second through last positions, requires $n - 2$ comparisons. This pattern continues until the final pass, which compares only the items in the last two positions of the altered list. In all, there are

$$(n-1) + (n-2) + (n-3) + \cdots + 3 + 2 + 1$$

comparisons. By a formula given in the next chapter, the preceding expression has the value $\frac{n^2-n}{2}$. Thus the bubble sort algorithm has order at most n^2.

EXERCISES 1.4

In Exercises 1–6, tell whether the given expression is a polynomial in x or not, and if so, give its degree.

1. $5x^2 - 3x + \frac{1}{2}$

2. 16

3. $x^3 - \frac{1}{x^2}$

4. $2^x + 3x$

5. $\frac{1}{2x^2 + 7x + 1}$

6. $2x + 3x^{1/2} + 4$

In Exercises 7–10, compute the various values S takes on when the polynomial evaluation algorithm is used to compute $P(x)$. Then do the same thing, using Horner's polynomial evaluation algorithm.

7. $P(x) = 5x + 3,\ x = 2$

8. $P(x) = 3x^2 + 2x - 1,\ x = 5$

9. $P(x) = -x^3 + 2x^2 + 5x - 7,\ x = 2$

10. $P(x) = 2x^3 + 5x^2 - 4,\ x = 3$

In Exercises 11–14, tell what next string will be produced by the next subset algorithm.

11. 110101

12. 110111

13. 001101

14. 001001

In Exercises 15–18, make a table listing the values of k, j, and $a_1, a_2, \ldots, a_n$ after each step when the next subset algorithm is applied to the given string.

15. 101

16. 111

17. 1101

18. 1110

In Exercises 19–22, illustrate as in Example 1.5 the use of the bubble sort algorithm to sort each given list of numbers.

19. 13, 56, 87, 42

20. 23, 5, 17, 12

21. 6, 33, 20, 200, 9

22. 88, 2, 75, 10, 48

In Exercises 23–26, estimate how long a computer doing one million operations per second would take to do 3^n and $100n^3$ operations.

23. $n = 20$

24. $n = 30$

25. $n = 40$

26. $n = 50$

In Exercises 27–30, tell how many elementary operations the given algorithm uses. (It depends on n.)

27. Algorithm for evaluating $n!$.

Step 1 Set $k = 0$ and $P = 1$.
Step 2 **while** $k < n$
  (a) Replace k with $k + 1$.
  (b) Replace P with kP.
endwhile
Step 3 Print P.

---

28. Algorithm for computing the sum of an arithmetic progression of n terms with first term a and common difference d.

Step 1 Set $S = a$, $k = 1$, and $t = a$.
Step 2 **while** $k < n$
(a) Replace t with $t + d$.
(b) Replace S with $S + t$.
(c) Replace k with $k + 1$.
endwhile
Step 3 Print S.

---

29. Algorithm for computing the sum of a geometric progression of n terms with first term a and common ratio r.

Step 1 Set $S = a$, $P = ar$, and $k = 1$.
Step 2 **while** $k < n$
(a) Replace S with $S + P$.
(b) Replace P with Pr.
(c) Replace k with $k + 1$.
endwhile
Step 3 Print S.

---

30. Algorithm for computing F_n, the nth Fibonacci number (defined in Section 2.5).

Step 1 Set $a = 1$, $b = 1$, $c = 2$, and $k = 1$.
Step 2 **while** $k < n$
(a) Replace c with $a + b$.
(b) Replace a with b.
(c) Replace b with c.
(d) Replace k with $k + 1$.
endwhile
Step 3 Print b.

The polynomial evaluation algorithm is inefficient because it computes x^k anew for each value of k. The following revision corrects this shortcoming.

Revised Polynomial Evaluation Algorithm

This algorithm computes $P(x) = a_n x^n + a_{n-1} x^{n-1} + \cdots + a_0$, given the nonnegative integer n and real numbers $x, a_0, a_1, \ldots, a_n$.

Step 1 (initialization) Set $S = a_0$, $y = 1$, and $k = 1$.
Step 2 (add next term)
while $k \le n$
(a) Replace y with xy.
(b) Replace S with $S + ya_k$.
(c) Replace k with $k + 1$.
endwhile
Step 3 (output $P(x) = S$)
Print S.

In Exercises 31–32, compute the various values S takes on when the revised polynomial evaluation algorithm is used to compute $P(x)$, where $P(x)$ and x are as in the given exercise.

31. Exercise 9

32. Exercise 10

33. Show that the complexity of the revised polynomial evaluation algorithm is $5n + 1$.

HISTORICAL NOTES

Leonardo of Pisa (Fibonacci)

Chapter 1 features the organizing aspects of algorithms. From PERT charts to matching and knapsack problems, the emphasis is on thinking about orderly procedures for confronting and solving problems. Algorithms have been part of mathematics since its beginning. The contents of the Rhind papyrus (c. 1650 B.C.) and the many cuneiform tablets of the Babylonians (c. 1750 B.C.) give evidence of the attempt to generalize the solutions to problems into computational formulas [73, 74].[1] Such formulas are forms of algorithms.

The Greeks provided directions on how to perform various geometric constructions and analyze elementary problems in number theory. Perhaps the most famous among the latter are the sieve of Eratosthenes for developing a listing of the first n primes, Euclid's algorithm for finding the greatest common divisor for a pair of positive integers, and Diophantus's methods for finding solutions to algebraic equations. It was with the publication of the *Liber Abaci* of Leonardo of Pisa (Fibonacci) in 1202 that Europeans had their first organized exposure to Arabic numerals and algorithms for operating with them [78, 79, 85].

[1]Numbers in brackets refer to "References for the Historical Notes" on pages 616–617.

Fibonacci stated his algorithms by referring to the quantities held by the first person and second person and giving a verbal explanation of how to craft an answer for the situation at hand. While the Greeks had used letters to refer to points in earlier times, it was not until the work of François Viète (1540–1603) that algorithms for algebraic operations began to be represented in present-day symbolic forms.

The very name "algorithm" has a historical trail of its own. One of the earliest texts, if not the first, dealing with the arithmetic of the Hindu–Arabic numerals was written in Arabic by Mohammad ibn-Musa al-Khwarizmi (c. 783–850). His book was entitled *Al-jabr wa'l muqbalah.* Although the original text has been lost and we have only portions that were transcribed in Latin in the 1100s, the author and text had a large influence on what we say and do today. The word "algorithm" is an Anglicized version of al-Khwarizmi's name, and the title of his book has given way to our present-day word "algebra." Al-Khwarizmi's name, in the form "algorithmicians," was used in the Middle Ages to separate calculating by Hindu–Arabic algorithms from the work of the "abacists" who calculated with counting tables and abacus-related methods based on the Roman numerals [73, 74, 85].

Forms of the word "algorithm" disappeared from common usage after the Middle Ages, to reappear around 1850 as a result of a reanalysis of the work of al-Khwarizmi. Since 1900, the notion of carefully specifying steps in carrying out a procedure has grown in importance as mathematicians and computer scientists have struggled to understand the efficiency of various procedures.

Augusta Ada Byron

Augusta Ada Byron (1815–1852), the Countess of Lovelace, was one of the first to see the importance that algorithms could play in the development of computing devices. The only child of the English poet Lord Byron, Ada Byron had the great opportunity to be tutored by Augustus De Morgan (1806–1871) and to work with Charles Babbage (1791–1871), the inventor of the first automated computing machine. While the machine never worked flawlessly, the ideas behind it were as much the work of Byron as they were the work of Babbage. Her work in describing the programming of Babbage's Analytical Engine to carry out computations and derivations has been memorialized through the naming of the programming language ADA in her honor [83].

The program evaluation and review technique and the critical path method, as mentioned earlier, were first developed and used by the U.S. Navy and Westinghouse engineers in 1958 to organize the building of the first nuclear powered submarines. The process was essentially discovered simultaneously by chemical engineers at du Pont and operations researchers in England, France, and Germany. As graph theoretic methods became better known, mathematical scientists around the world began to represent and study problems regarding orderings via graph theoretic representations.

SUPPLEMENTARY EXERCISES

1. Use the PERT method to determine the total project time and critical path for the following diagram:

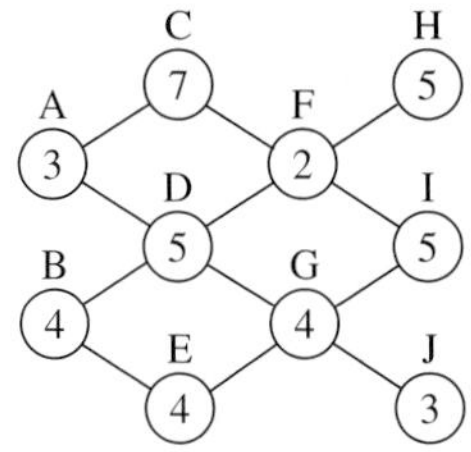

2. The following table tells the time needed for a number of tasks and which tasks precede them:

Task	*Time*	*Preceding Tasks*
A	3	None
B	5	None
C	2	A
D	4	A, B
E	6	A, B
F	6	C, D, E
G	3	D, E
H	3	F, G
I	2	F, G

Make a PERT diagram, and determine the project time and critical path.

3. In manufacturing a certain toy, Machine 1 makes part A in 3 minutes, Machine 2 makes part B in 5 minutes, Machine 3 makes the box in 2 minutes, Machine 4 assembles the two parts in 2 minutes, and Machine 5 puts an assembled toy into a box in 1 minute. How long will it take to make and box 5 toys?

4. Calculate $\frac{8!}{5!}$.

5. Calculate $P(11, 6)$.

6. A baseball team has 2 catchers, 4 starting pitchers, and 5 relief pitchers. A catcher, starting pitcher, and relief pitcher must be chosen for an all-star game. In how many ways can this be done?

7. A basketball team has 3 centers, 4 guards, and 4 forwards. A most-valuable player, captain, and most-improved player are to be chosen from the team. If all three players are to be different, in how many ways can this be done?

8. A student honor society has 10 juniors and 13 seniors. According to its bylaws, the president and treasurer must be seniors, and the vice president and secretary must be juniors. In how many ways can the four offices be filled by four different students?

Let $A = \{1, 3, 5\}$, $B = \{2, 6, 10\}$, and $C = \{x : x$ is an integer and $0 < x < 10\}$. In Exercises 9–16, tell whether each statement is true or false.

9. $B \subseteq C$ **10.** $A \subseteq C$ **11.** $6 \subseteq B$ **12.** $A \in C$

13. $\{1\} \subseteq A$ **14.** $6 \in C$ **15.** $\emptyset \subseteq B$ **16.** $|\{\emptyset\}| \in A$

17. What is $|\{X : X \subseteq \{2, 4, 6, 8, 10\}|$?

18. In Cincinnati, chili consists of spaghetti topped by any (or none) of meat sauce, cheese, chopped onions, and beans. In how many ways can chili be ordered?

19. Five students decide to send a delegation to a professor to ask her to delay a test. The delegation is to have a spokesperson and perhaps some accompanying members. In how many ways can it be chosen?

In Exercises 20–23, tell whether each expression is a polynomial in x, and if so, give its degree.

20. $2x + 3 + 4x^{-1}$ **21.** $x^{100} - 3$ **22.** $\log_2 10$ **23.** $3x^{1.5} + 1.5x^3$

24. Let $P(x) = 3x^3 + 4x - 5$. Compute the various values S takes on when the polynomial evaluation algorithm is used to compute $P(x)$ for $x = 3$.

25. Repeat the previous problem, using Horner's polynomial evaluation algorithm.

26. Let $S = \{1, 2, 3, 4\}$. Find the ordered sequence of all subsets of S as produced by the next subset algorithm, starting with $\emptyset$.

27. Illustrate the use of the bubble sort algorithm to sort the following list of numbers, as in Example 1.5: 44, 5, 13, 11, 35.

28. How long would it take a computer to do 25! operations if it can do one billion operations per second?

29. Apply the following algorithm to $n = 18$.

Step 1 Set $d = 1$ and $s = 0$.
Step 2 **while** $d \leq n$
  Step 2.1 **if** $\frac{n}{d}$ is an integer
    Replace s with $s + d$.
   endif
  Step 2.2 Replace d with $d + 1$.
 endwhile

What is the value of s when the algorithm stops?

30. Apply the following algorithm to $n = 100$.

Step 1 Set $s = n$.
Step 2 **repeat**
  Step 2.1 Set $t = s$.

Step 2.2 **while** s is even
Replace s with $\frac{s}{2}$.
endwhile
Step 2.3 Replace s with $3s + 1$.
until $s = t$

What is the value of s when the algorithm stops?

31. Determine how many elementary operations the following algorithm for finding the sum of the first n squares uses (the answer depends on n).

Step 1 Set $S = 1$ and $k = 1$.
Step 2 **while** $k < n$
(a) Replace S with $S + k \cdot k$.
(b) Replace k with $k + 1$.
endwhile
Step 3 Print S.

---

32. Determine how many elementary operations the following algorithm for finding $P(n, r)$ uses (the answer depends on n and r).

Step 1 Set $k = 1$ and $Q = n$.
Step 2 **while** $k < r$
(a) Replace Q with $(n - k)Q$.
(b) Replace k with $k + 1$.
endwhile
Step 3 Print Q.

---

33. Write a formal algorithm for performing PERT to determine the project time and critical path for a project. Assume that the input data is given in tabular form, as in Exercise 2.

COMPUTER PROJECTS

Write a computer program having the specified input and output.

1. Given n, compute n factorial.
2. Given n and r, with $0 \leq r \leq n$, compute $P(n, r)$.

3. Use the algorithm for evaluating x^n to compute x^n, given a real number x and a positive integer n.
4. Use the polynomial evaluation algorithm to evaluate

$$P(x) = 35x^4 - 17x^3 + 5x^2 + 41x - 29,$$

given a real number x. Have the program call the program of the previous problem. Time the program in evaluating $P(x)$ for $x = 1, 2, \ldots, 100$.
5. Using Horner's polynomial evaluation algorithm, evaluate $P(x)$ for any real number x, where $P(x)$ is the same as in the previous exercise. Time the program in evaluating $P(x)$ for $x = 1, 2, \ldots, 100$.
6. Given a string of twelve 0s and 1s, use the next subset algorithm to compute the next string.
7. Use the previous exercise to output all possible strings of twelve 0s and 1s.
8. Examine all subsets of the set of 12 space shuttle experiments of Section 1.3 to find an optimal subset having a total rating of 53. What experiments are used?
9. Given n, output all possible strings of n 0s and 1s.
10. Given a list of 10 numbers, output the list in nondecreasing order. Use the bubble sort algorithm.

SUGGESTED READINGS

1. Graham, Ronald L. "The Combinatorial Mathematics of Scheduling." *Scientific American* (March 1978): 124–132.
2. Lawler, Eugene L. *Combinatorial Optimization: Networks and Matroids.* New York: Holt, Rinehart, and Winston, 1976.
3. Lewis, Harry R. and Christos H. Papadimitriou. "The Efficiency of Algorithms." *Scientific American* (March 1978): 96–109.
4. Lockyer, K.G. *An Introduction to Critical Path Analysis.* 3d ed. London: Pitman and Sons, 1969.
5. Niven, Ivan. *Mathematics of Choice or How to Count Without Counting.* New York: L.W. Singer, 1965.
6. Pipenger, Nicholas. "Complexity Theory." *Scientific American* (June 1978): 114–124.

CHAPTER

2

Sets, Relations, and Functions

As we saw in Chapter 1, discrete mathematics is concerned with solving problems in which the number of possibilities is finite. Often, as in the analysis of the knapsack problem in Section 1.3, the discussion of a problem requires consideration of all the possibilities for a solution. Such an approach can be made easier by the use of sets. In other situations we may need to consider relationships between the elements of sets. Such relationships frequently can be expressed using the mathematical ideas of relations and functions. In this chapter we will study these basic concepts, as well as the principle of mathematical induction, an important method of proof in discrete mathematics.

2.1 SET OPERATIONS

In Section 1.3 we presented some of the basic ideas about sets. In this section we will discuss several ways in which sets can be combined to produce new sets.

Suppose that in the example discussed in Section 1.3 it is decided that the space shuttle will carry experiments on two successive trips. If $S = \{1, 5, 6, 8\}$ is the set of experiments carried on the first trip and $T = \{2, 4, 5, 8, 9\}$ is the set of experiments carried on the second trip, then $\{1, 2, 4, 5, 6, 8, 9\}$ is the set of experiments carried on the first or second trip or both. This set is called the union of S and T.

More generally, by the **union** of sets A and B we mean the set consisting of all the elements that are in A or in B. Note that, as always in mathematics, the word "or" in this definition is used in the inclusive sense. Thus an element x is in the union of sets A and B in each of the following cases:

(1) $x \in A$ and $x \notin B$,
(2) $x \notin A$ and $x \in B$, or
(3) $x \in A$ and $x \in B$.

The union of sets A and B is denoted $A \cup B$. Thus

$$A \cup B = \{x : x \in A \text{ or } x \in B\}.$$

Another set of interest in the space shuttle example is the set $\{5, 8\}$ of experiments carried on both trips. This set is called the intersection of S and T. In general, the **intersection** of sets A and B is the set consisting of all the elements that are in both A and B. This set is denoted $A \cap B$, so

$$A \cap B = \{x : x \in A \text{ and } x \in B\}.$$

If the intersection of two sets is the empty set, then these sets are said to be **disjoint**.

Example 2.1 If $A = \{1, 2, 4\}$, $B = \{2, 4, 6, 8\}$, and $C = \{3, 6\}$, then

$$A \cup B = \{1, 2, 4, 6, 8\} \quad \text{and} \quad A \cap B = \{2, 4\},$$
$$A \cup C = \{1, 2, 3, 4, 6\} \quad \text{and} \quad A \cap C = \emptyset,$$

and

$$B \cup C = \{2, 3, 4, 6, 8\} \quad \text{and} \quad B \cap C = \{6\}.$$

Thus A and C are disjoint sets. ■

The set $\{1, 6\}$ of elements carried on the first space shuttle trip, but not on the second trip, is called the difference of S and T. More generally, the **difference** of sets A and B, denoted $A - B$, is the set consisting of the elements in A that are not in B. Thus

$$A - B = \{x : x \in A \text{ and } x \notin B\}.$$

Note that, as the following example shows, the sets $A - B$ and $B - A$ are not usually equal.

Example 2.2 If A and B are as in Example 2.1, then $A - B = \{1\}$ and $B - A = \{6, 8\}$. ■

In many situations all of the sets under consideration are subsets of a set U. For instance, in our discussion of the space shuttle example in Section 1.3, all of the sets were subsets of the set of experiments

$$U = \{1, 2, 3, 4, 5, 6, 7, 8, 9, 10, 11, 12\}.$$

Such a set containing all of the elements of interest in a particular situation is called a **universal set**. Since there are many different sets that could be used as a universal set, the particular universal set being considered must always be described explicitly. Given a universal set U and a subset A of U, the set $U - A$ is called the **complement** of A and is denoted $\overline{A}$.

Example 2.3 If $A = \{1, 2, 4\}$, $B = \{2, 4, 6, 8\}$, and $C = \{3, 6\}$ are the sets in Example 2.1 and

$$U = \{1, 2, 3, 4, 5, 6, 7, 8\}$$

is the universal set, then

$$\overline{A} = \{3, 5, 6, 7, 8\}, \quad \overline{B} = \{1, 3, 5, 7\}, \quad \text{and} \quad \overline{C} = \{1, 2, 4, 5, 7, 8\}. \quad \blacksquare$$

Theorem 2.1 lists some elementary properties of set operations. These properties follow immediately from the definitions previously given by using the fact that $A = B$ if and only if $A \subseteq B$ and $B \subseteq A$.

Theorem 2.1 Let U be a universal set. For any subsets A, B, and C of U, the following are true:

(a) $A \cup B = B \cup A$ and $A \cap B = B \cap A$ (commutative laws)

(b) $(A \cup B) \cup C = A \cup (B \cup C)$ and $(A \cap B) \cap C = A \cap (B \cap C)$ (associative laws)

(c) $A \cup (B \cap C) = (A \cup B) \cap (A \cup C)$ and $A \cap (B \cup C) = (A \cap B) \cup (A \cap C)$ (distributive laws)

(d) $\overline{\overline{A}} = A$

(e) $A \cup \overline{A} = U$

(f) $A \cap \overline{A} = \emptyset$

(g) $A \subseteq A \cup B$ and $B \subseteq A \cup B$

(h) $A \cap B \subseteq A$ and $A \cap B \subseteq B$

(i) $A - B = A \cap \overline{B}$

Relationships among sets can be pictured in **Venn diagrams**, which are named after the English logician John Venn (1834–1923). In a Venn diagram, the universal set is represented by a rectangular region, and subsets of the universal set are usually represented by circular disks drawn within the rectangular region. Sets that are not known to be disjoint should be represented by overlapping circles, as in Figure 2.1.

Figure 2.2 contains Venn diagrams for the four set operations defined earlier. In each diagram the colored region depicts the set being represented.

Venn diagrams depicting more complicated sets can be constructed by combining the basic diagrams found in Figure 2.2. For example, Figure 2.3 shows how to construct a Venn diagram for $(\overline{A \cup B})$. (See also Example 2.4.)

The next theorem enables us to determine the complement of a union or intersection of sets. This result will be needed in Section 8.6 to help us find the number of elements in the union of several sets.

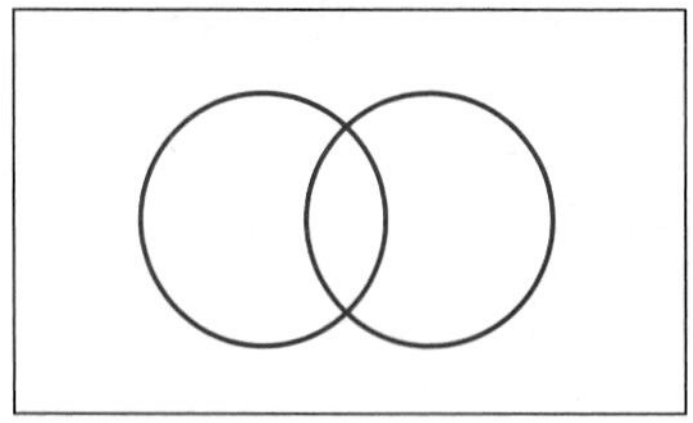

FIGURE 2.1

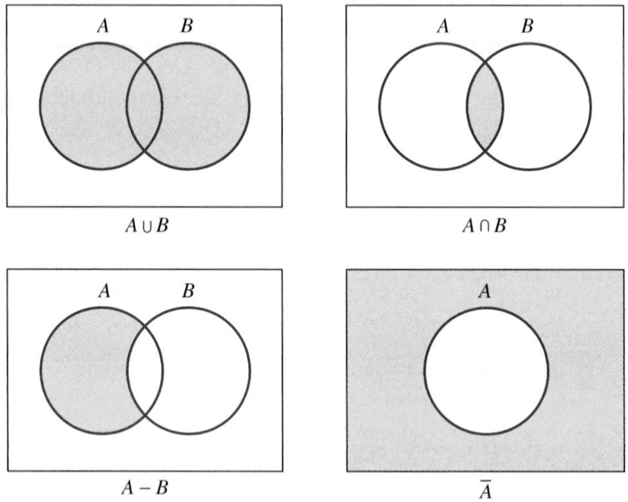

FIGURE 2.2

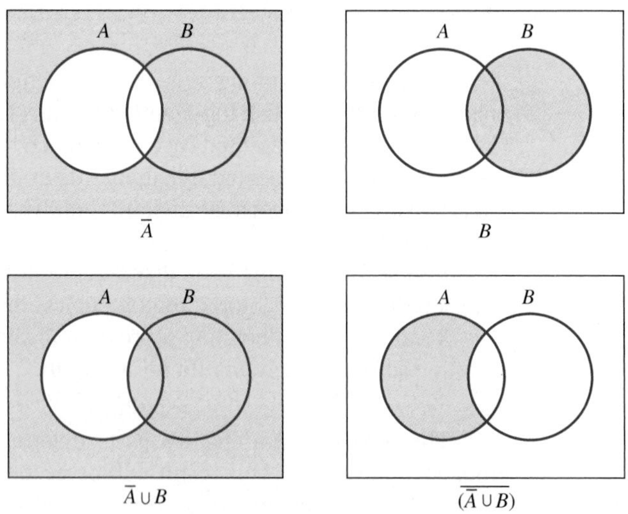

FIGURE 2.3

Theorem 2.2 ***De Morgan's Laws*** For any subsets A and B of a universal set U, the following are true:

(a) $\overline{(A \cup B)} = \overline{A} \cap \overline{B}$

(b) $\overline{(A \cap B)} = \overline{A} \cup \overline{B}$

Proof. To prove that $\overline{(A \cup B)} = \overline{A} \cap \overline{B}$, we will show that each of the sets $\overline{(A \cup B)}$ and $\overline{A} \cap \overline{B}$ is a subset of the other.

First suppose that $x \in \overline{(A \cup B)}$. Then $x \notin A \cup B$. But since this is true, $x \notin A$ and $x \notin B$. So $x \in \overline{A}$ and $x \in \overline{B}$. It follows that $x \in \overline{A} \cap \overline{B}$. Therefore $\overline{(A \cup B)} \subseteq \overline{A} \cap \overline{B}$.

Now suppose that $x \in \overline{A} \cap \overline{B}$. Then $x \in \overline{A}$ and $x \in \overline{B}$. Hence $x \notin A$ and $x \notin B$. It follows that $x \notin A \cup B$. So $x \in \overline{(A \cup B)}$. Thus $\overline{A} \cap \overline{B} \subseteq \overline{(A \cup B)}$.

Because we have $\overline{(A \cup B)} \subseteq \overline{A} \cap \overline{B}$ and $\overline{A} \cap \overline{B} \subseteq \overline{(A \cup B)}$, it follows that $\overline{(A \cup B)} = \overline{A} \cap \overline{B}$. This proves part (a).

The proof of part (b) is similar and will be left as an exercise. ■

Example 2.4 We can compute $\overline{(\overline{A} \cup B)}$ by using Theorems 2.2(a), 2.1(d), and 2.1(i):

$$\overline{(\overline{A} \cup B)} = \overline{\overline{A}} \cap \overline{B} = A \cap \overline{B} = A - B.$$

This equality is illustrated by Figure 2.3. ■

Example 2.5 According to the U.S. customs laws, a person is not allowed to bring liquor into the United States duty-free if he or she is not over 21, or if he or she has brought duty-free liquor into this country in the previous 30 days. Who is allowed to bring duty-free liquor into the United States?

Let A denote the set of people aged 21 or over, and let B denote the set of people who have brought duty-free liquor into the United States in the previous 30 days. Then the persons who are not allowed to bring duty-free liquor into the United States are those in the set $\overline{A} \cup B$. This means that those who are eligible to bring duty-free liquor into this country are those in the set $\overline{(\overline{A} \cup B)}$. Example 2.4 shows that the persons who can bring duty-free liquor into the United States are those in the set $A - B$, the set of people aged 21 or over who have not brought duty-free liquor into the United States in the previous 30 days. ■

When listing the elements of a set, the order in which the elements are written is immaterial. Thus, for example, $\{1, 2, 3\} = \{2, 3, 1\} = \{3, 1, 2\}$. Often, however, we need to be able to distinguish the order in which two elements are listed. In an **ordered pair** of elements a and b, denoted (a, b), the order in which the entries are written is taken into account. Thus $(1, 2) \neq (2, 1)$, and $(a, b) = (c, d)$ if and only if $a = c$ and $b = d$.

The final set operation that we will consider is the Cartesian product, which arises in connection with relations (to be studied in Section 2.2). Given sets A and

B, the **Cartesian product** of A and B is the set consisting of all the ordered pairs (a, b), where $a \in A$ and $b \in B$. The Cartesian product of A and B is denoted $A \times B$. Thus

$$A \times B = \{(a, b) : a \in A \text{ and } b \in B\}.$$

The Cartesian product is often encountered in discussions of the Euclidean plane, for if R denotes the set of all real numbers, then $R \times R$ is the set of all ordered pairs of real numbers, which can be pictured as the Euclidean plane.

Example 2.6 Let $A = \{1, 2, 3\}$ and $B = \{3, 4\}$. Then

$$A \times B = \{(1, 3), (1, 4), (2, 3), (2, 4), (3, 3), (3, 4)\} \text{ and}$$
$$B \times A = \{(3, 1), (3, 2), (3, 3), (4, 1), (4, 2), (4, 3)\}. \quad \blacksquare$$

As Example 2.6 shows, usually $A \times B \neq B \times A$.

Example 2.7 A public opinion poll was taken to see how several of the leading Democratic presidential candidates would fare in the 1976 presidential election against the leading Republican candidates. The set of Democratic candidates considered was

$$D = \{\text{Brown, Carter, Humphrey, Udall}\},$$

and the set of Republican candidates considered was

$$R = \{\text{Ford, Reagan}\}.$$

How many pairings of a Democratic and a Republican candidate are there?

The set of all possible pairs of a Democratic and a Republican candidate is $D \times R$. The elements of this set are the ordered pairs

(Brown, Ford), (Brown, Reagan), (Carter, Ford), (Carter, Reagan),
(Humphrey, Ford), (Humphrey, Reagan), (Udall, Ford), (Udall, Reagan).

Thus there are eight different pairings of a Democratic and a Republican candidate. ■

EXERCISES 2.1

In Exercises 1–4, evaluate $A \cup B$, $A \cap B$, $A - B$, $\overline{A}$, and $\overline{B}$ for each of the given sets A and B. In each case, assume that the universal set is $U = \{1, 2, \ldots, 9\}$.

1. $A = \{2, 3, 5, 7, 8\}$ and $B = \{1, 3, 4, 5, 6, 9\}$

2. $A = \{1, 4, 6, 9\}$ and $B = \{1, 2, 4, 5, 6, 7, 9\}$

3. $A = \{1, 2, 4, 8, 9\}$ and $B = \{3, 7\}$

4. $A = \{3, 4, 6, 7, 8, 9\}$ and $B = \{2, 5, 7, 9\}$

In Exercises 5–8, compute $A \times B$ for each of the given sets A and B.

5. $A = \{1, 2, 3, 4\}$ and $B = \{7, 8\}$

6. $A = \{3, 4, 5\}$ and $B = \{1, 2, 3\}$

7. $A = \{a, e\}$ and $B = \{x, y, z\}$

8. $A = \{p, q, r, s\}$ and $B = \{a, c, e\}$

Draw Venn diagrams representing the sets in Exercises 9–12.

9. $\overline{(A \cap \overline{B})}$ **10.** $\overline{A} - \overline{B}$ **11.** $\overline{A} \cap (B \cup C)$ **12.** $A \cup (B - C)$

13. Give an example of sets for which $A \cup C = B \cup C$, but $A \neq B$.

14. Give an example of sets for which $A \cap C = B \cap C$, but $A \neq B$.

15. Give an example of sets for which $A - C = B - C$, but $A \neq B$.

16. Give an example of sets for which $(A - B) - C \neq A - (B - C)$.

Use Theorems 2.1 and 2.2 as in Example 2.4 to simplify the sets in Exercises 17–24.

17. $A \cap (B - A)$ **18.** $(A - B) \cup (A \cap B)$ **19.** $(A - B) \cap (A \cup B)$ **20.** $\overline{A} \cap (A \cap B)$

21. $\overline{A} \cap (A \cup B)$ **22.** $\overline{(A - B)} \cap A$ **23.** $A \cap \overline{(A \cap B)}$ **24.** $A \cup \overline{(A \cup B)}$

25. If A is a set containing m elements and B is a set containing n elements, how many elements are there in $A \times B$?

26. Under what conditions is $A - B = B - A$?

27. Under what conditions is $A \cup B = A$?

28. Under what conditions is $A \cap B = A$?

29. Prove parts (c) and (i) of Theorem 2.1.

30. Prove part (b) of Theorem 2.2 by using an argument similar to that in the proof of part (a).

31. Note that if $A = B$, then $\overline{A} = \overline{B}$. Use this fact to prove part (b) of Theorem 2.2 from part (a).

32. Let A and B be subsets of a universal set U. Prove that if $A \subseteq B$, then $\overline{B} \subseteq \overline{A}$.

Prove the set equalities in Exercises 33–38.

33. $(A - B) \cup (B - A) = (A \cup B) - (A \cap B)$

34. $(A - B) \cup (A - C) = A - (B \cap C)$

35. $(A - B) - C = (A - C) - (B - C)$

36. $(A \times C) \cap (B \times D) = (A \cap B) \times (C \cap D)$

37. $(A - B) \cap (A - C) = A - (B \cup C)$

38. $(A - C) \cap (B - C) = (A \cap B) - C$

39. Give an example where $(A \times C) \cup (B \times D) \neq (A \cup B) \times (C \cup D)$.

40. Prove that $(A \times C) \cup (B \times D) \subseteq (A \cup B) \times (C \cup D)$.

2.2 EQUIVALENCE RELATIONS

In Section 1.2 we considered a problem involving the matching of pilots to flights having different destinations. Recall that the seven destinations and the pilots who requested them are as follows:

Los Angeles: Timmack, Jelinek, Rupp
Seattle: Alfors, Timmack, Tang, Washington
London: Timmack, Tang, Washington
Frankfurt: Alfors, Tang, Rupp, Ramirez
Paris: Jelinek, Washington, Rupp
Madrid: Jelinek, Ramirez
Dublin: Timmack, Rupp, Ramirez

This list establishes a relation between the set of destinations and the set of pilots, where a pilot is related to one of the seven destinations whenever that destination was requested by the pilot.

From this list, we can construct a set of ordered pairs in which the first entry of each ordered pair is a destination and the second entry is a pilot who requested that destination. For example, the pairs

(Los Angeles, Timmack), (Los Angeles, Jelinek), and (Los Angeles, Rupp)

correspond to the three pilots who requested the flight to Los Angeles. Let

$$S = \{(\text{Los Angeles, Timmack}), (\text{Los Angeles, Jelinek}), \ldots, (\text{Dublin, Ramirez})\}$$

denote the set of all 22 ordered pairs of destinations and the pilots who requested them. This set contains exactly the same information as the original list of pilots and their requested destinations. Thus the relation between the destinations and the pilots who requested them can be completely described by a set of ordered pairs. Notice that S is a subset of $A \times B$, where A is the set of destinations and B is the set of pilots.

Generalizing from the previous example, we define a **relation from a set *A* to a set *B*** to be any subset of the Cartesian product $A \times B$. If R is a relation from set A to set B and (x, y) is an element of R, we will say that ***x* is related to *y* by *R*** and write $x \ R \ y$ instead of $(x, y) \in R$.

Example 2.8 Among three college professors, suppose that Lopez speaks Dutch and French, Parr speaks German and Russian, and Zak speaks Dutch. Let

$$A = \{\text{Lopez, Parr, Zak}\}$$

denote this set of professors and

$$B = \{\text{Dutch, French, German, Russian}\}$$

denote the set of foreign languages they speak. Then

$$R = \{(\text{Lopez, Dutch}), (\text{Lopez, French}), (\text{Parr, German}), (\text{Parr, Russian}), (\text{Zak, Dutch})\}$$

is a relation from A to B in which x is related to y by R whenever Professor x speaks language y. So, for instance, Lopez R French and Parr R German are both true, but Zak R Russian is false. ■

Often, we need to consider a relation between the elements of some set. In Section 1.1, for instance, we considered the problem of determining how long it would take to produce and deliver the advertisements for a department store sale. In analyzing this problem, we listed all the tasks that needed to be done and represented each task by a letter. Then we determined which tasks immediately preceded each task. This listing establishes a relation on the set of tasks

$$S = \{\text{A, B, C, D, E, F, G, H, I, J, K}\}$$

in which task X is related to task Y if X immediately precedes Y. The resulting relation

$$\{(\text{A, C}), (\text{B, C}), (\text{C, D}), (\text{C, E}), (\text{D, F}), (\text{E, F}), (\text{C, G}), (\text{G, H}), (\text{F, I}), (\text{H, J}), (\text{I, J}), (\text{J, K})\}$$

is a relation from set S to itself, that is, a subset of $S \times S$.

A relation from a set S to itself is called a **relation on S**.

Example 2.9 Let $S = \{1, 2, 3, 4\}$. Define a relation R on S by letting $x \ R \ y$ mean $x < y$. Then 1 is related to 4, but 4 is not related to 2. Likewise $2 \ R \ 3$ is true, but $3 \ R \ 2$ is false. ■

A relation R on a set S may have any of the following special properties:

(1) If for each x in S, $x \ R \ x$ is true, then R is called **reflexive**.
(2) If $y \ R \ x$ is true whenever $x \ R \ y$ is true, then R is called **symmetric**.
(3) If $x \ R \ z$ is true whenever $x \ R \ y$ and $y \ R \ z$ are both true, then R is called **transitive**.

The relation R in Example 2.9 is not reflexive, since $1 \ R \ 1$ is false. Likewise, it is not symmetric, because $1 \ R \ 4$ is true but $4 \ R \ 1$ is false. However, R is transitive because if x is less than y and y is less than z, then x is less than z.

Example 2.10 Let S be the set of positive integers, and define $x \ R \ y$ to mean that x divides y (that is, $y = kx$ for some integer k). Thus $3 \ R \ 6$ and $7 \ R \ 35$ are true, but $8 \ R \ 4$ and $6 \ R \ 9$ are false. Then R is a relation on S. Furthermore, R is reflexive, since every positive integer divides itself, and R is transitive, since if x divides y and y divides z, then x divides z. (To see why this is so, note that, if $y = mx$ and $z = ny$ for some integers m and n, then $z = n(mx) = (mn)x$ for the integer mn. However, R is not symmetric, because $2 \ R \ 8$ is true but $8 \ R \ 2$ is false. ■

Example 2.11 Let S denote the set of all *nonempty* subsets of $\{1, 2, 3, 4, 5\}$, and define $A \ R \ B$ to mean that $A \cap B \neq \emptyset$. Then R is clearly reflexive and symmetric. However, R is not transitive, since $\{1, 2\} \ R \ \{2, 3\}$ and $\{2, 3\} \ R \ \{3, 4\}$ are true, but $\{1, 2\} \ R \ \{3, 4\}$ is false. ■

A relation that is reflexive, symmetric, and transitive is called an **equivalence relation**. The most familiar example of an equivalence relation is the relation of equality. Two more examples follow.

Example 2.12 On the set of students attending a particular university, define one student to be related to another whenever their surnames begin with the same letter. This relation is easily seen to be an equivalence relation on the set of students at this university. ■

Example 2.13 An integer greater than 1 is called **prime** if its only positive integer divisors are itself and 1. The first few prime numbers are 2, 3, 5, 7, 11, 13, 17, 19, and 23. We will see in Theorem 2.11 that every integer greater than 1 is either prime or a product of primes. For example, 67 is prime and $65 = 5 \cdot 13$ is a product of primes. On the set S of integers greater than 1, define $x\ R\ y$ to mean that x has the same number of *distinct* prime divisors as y. Thus, for example, $12\ R\ 55$ since $12 = 2 \cdot 2 \cdot 3$ and $55 = 5 \cdot 11$ both have two distinct prime divisors. Then R is an equivalence relation on S. ■

Example 2.14 Let S denote the set of all people in the United States. Define a relation R on S by letting $x\ R\ y$ mean that x has the same mother or father as y. Then R is easily seen to be reflexive and symmetric. But R is not transitive, for x and y may have the same mother and y and z may have the same father, but x and z may have no parent in common. Hence R is not an equivalence relation on S. ■

If R is an equivalence relation on a set S and $x \in S$, the set of elements of S that are related to x is called the **equivalence class** containing x and is denoted $[x]$. Thus

$$[x] = \{y \in S\colon y\ R\ x\}.$$

Note that by the reflexive property of R, $x \in [x]$ for each element x in S. In the equivalence relation described in Example 2.12, there are 26 possible equivalence classes; namely, the set of students whose surnames begin with A, the set of students whose surnames begin with B, and so forth.

Example 2.15 Let S denote the set of integers greater than 1. For x and y in S, define $x\ R\ y$ to mean that the largest prime divisor of x equals the largest prime divisor of y. Then R is an equivalence relation on S.

The equivalence class of R containing 2 consists of all elements in S that are related to 2—that is, all the elements in S whose largest prime divisor is 2. Such integers must be powers of 2; hence

$$[2] = \{2^k\colon k = 1, 2, 3, \ldots\}.$$

Similarly, [3] consists of all the elements in S whose largest prime divisor is 3. Such integers must have 3 as a divisor and may or may not have 2 as a divisor. Thus

$$[3] = \{2^i 3^j\colon i = 0, 1, 2, \ldots \text{ and } j = 1, 2, 3, \ldots\}.$$

Note that because the largest prime divisor of 4 is 2, we have 4 R 2. Since R is an equivalence relation on S, any element of S that is related to 2 must also be related to 4. Hence

$$[4] = [2] = \{2^k\colon k = 1, 2, 3, \ldots\}. \quad \blacksquare$$

The fact that $[2] = [4]$ in Example 2.15 illustrates part of the following theorem:

Theorem 2.3 Let R be an equivalence relation on a set S.

(a) If x and y are elements of S, then x is related to y by R if and only if $[x] = [y]$.

(b) Two equivalence classes of R are either equal or disjoint.

Proof. (a) Let x and y be elements of S such that $x \; R \; y$. We will prove that $[x] = [y]$ by showing that $[x] \subseteq [y]$ and $[y] \subseteq [x]$.

If $z \in [x]$, then z is related to x, that is, $z \; R \; x$. But if $z \; R \; x$ and $x \; R \; y$, then $z \; R \; y$ by the transitive property of R. So $z \in [y]$. This proves that $[x] \subseteq [y]$.

If $z \in [y]$, then $z \; R \; y$ as in the previous paragraph. By assumption, $x \; R \; y$, and so $y \; R \; x$ is true by the symmetric property of R. But then $z \; R \; y$ and $y \; R \; x$ imply $z \; R \; x$ by the transitive property. Hence $z \in [x]$, proving that $[y] \subseteq [x]$. Since we have both $[x] \subseteq [y]$ and $[y] \subseteq [x]$, it follows that $[x] = [y]$.

Conversely, suppose that $[x] = [y]$. Now $x \in [x]$ by the reflexive property, and so $x \in [y]$ because $[x] = [y]$. But if $x \in [y]$, then $x \; R \; y$. This completes the proof of part (a).

(b) Let $[u]$ and $[v]$ be any two equivalence classes of R. If $[u]$ and $[v]$ are not disjoint, then they contain a common element w. Since $w \in [u]$, part (a) shows that $[w] = [u]$. Likewise, $[w] = [v]$. It follows that $[u] = [v]$. Hence $[u]$ and $[v]$ are either disjoint or equal. $\blacksquare$

Because of part (b) of Theorem 2.3, the equivalence classes of an equivalence relation R on set S divide S into disjoint subsets. This family of subsets has the following properties:

(1) No subset is empty.
(2) Each element of S belongs to some subset.
(3) Two distinct subsets are disjoint.

Such a family of subsets of S is called a **partition** of S.

Example 2.16 Let $A = \{1, 3, 4\}$, $B = \{2, 6\}$, and $C = \{5\}$. Then $\mathcal{P} = \{A, B, C\}$ is a partition of $S = \{1, 2, 3, 4, 5, 6\}$. (See Figure 2.4.) $\blacksquare$

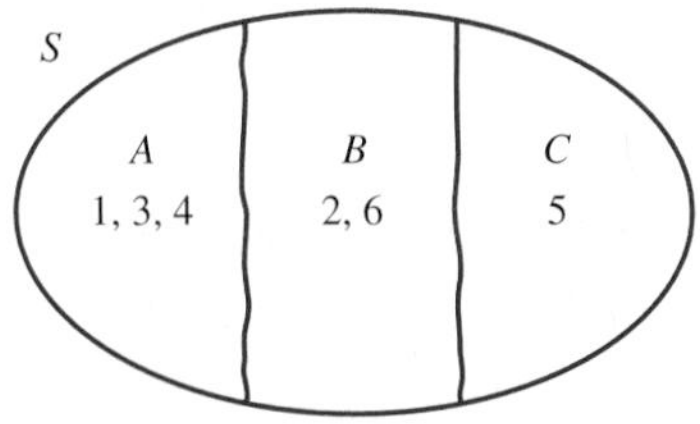

FIGURE 2.4

We have seen that every equivalence relation on S gives rise to a partition of S by taking the family of subsets in the partition to be the equivalence classes of the equivalence relation. Conversely, if $\mathcal{P}$ is a partition of S, we can define a relation R on S by letting $x\ R\ y$ mean that x and y lie in the same member of $\mathcal{P}$. Using the partition in Example 2.16, for instance, we obtain the relation

$$\{(1,1), (1,3), (1,4), (3,1), (3,3), (3,4), (4,1), (4,3), (4,4), \\ (2,2), (2,6), (6,2), (6,6), (5,5)\}.$$

Then R is clearly an equivalence relation on S, and the equivalence classes of R are precisely the members of $\mathcal{P}$. We will state these facts formally as our next theorem.

Theorem 2.4 (a) An equivalence relation R gives rise to a partition $\mathcal{P}$ in which the members of $\mathcal{P}$ are the equivalence classes of R.

(b) Conversely, a partition $\mathcal{P}$ induces an equivalence relation R in which two elements are related by R whenever they lie in the same member of $\mathcal{P}$. Moreover, the equivalence classes of this relation are the members of $\mathcal{P}$.

Although the definitions of an equivalence relation and a partition appear to be quite different, as a result of Theorem 2.4 we see that these two concepts are actually just different ways of describing the same situation.

EXERCISES 2.2

In Exercises 1–12, determine which of the reflexive, symmetric, and transitive properties are satisfied by the given relation R defined on set S.

1. $S = \{1, 2, 3\}$ and $R = \{(1,1), (1,2), (2,1), (2,2)\}$.
2. $S = \{1, 2, 3\}$ and $R = \{(1,1), (1,3), (2,2), (2,3), (3,1), (3,2), (3,3)\}$.
3. S is the set of all Illinois residents and $x\ R\ y$ means that x has the same mother as y.
4. S is the set of all citizens of the United States and $x\ R\ y$ means that x has the same weight as y.
5. S is the set of all students at Illinois State University and $x\ R\ y$ means that the height of x differs from the height of y by no more than one inch.
6. S is the set of all teenagers and $x\ R\ y$ means that x has a grandfather in common with y.

7. S is the set of all graduates of Michigan State University and $x \ R \ y$ means that x first attended Michigan State University in the same year as y.
8. S is the set of all residents of Los Angeles and $x \ R \ y$ means that x is a brother of y.
9. S is the set of all real numbers and $x \ R \ y$ means that $x^2 = y^2$.
10. S is the set of positive integers and $x \ R \ y$ means that x divides y or y divides x.
11. S is the set of all subsets of $\{1, 2, 3, 4\}$ and $X \ R \ Y$ means that $X \subseteq Y$.
12. S is the set of ordered pairs of real numbers and $(x_1, x_2) \ R \ (y_1, y_2)$ means that $x_1 = y_1$ and $x_2 \le y_2$.

In Exercises 13–18, show that the given relation R is an equivalence relation on set S. Then describe the equivalence class containing the given element z in S, and determine the number of distinct equivalence classes of R.

13. Let S be the set of integers, let $z = 7$, and define $x \ R \ y$ to mean that $x - y$ is even.
14. Let S be the set of all possible strings of 3 or 4 letters, let $z =$ ABCD, and define $x \ R \ y$ to mean that x has the same first letter as y and also the same third letter as y.
15. Let S be the set of integers greater than 1, let $z = 60$, and define $x \ R \ y$ to mean that the largest prime divisor of x equals the largest prime divisor of y.
16. Let S be the set of all subsets of $\{1, 2, 3, 4, 5\}$, let $z = \{1, 2, 3\}$, and define $X \ R \ Y$ to mean that $X \cap \{1, 3, 5\} = Y \cap \{1, 3, 5\}$.
17. Let S be the set of ordered pairs of real numbers, let $z = (3, -4)$, and define $(x_1, x_2) \ R \ (y_1, y_2)$ to mean that $x_1^2 + x_2^2 = y_1^2 + y_2^2$.
18. Let S be the set of ordered pairs of positive integers, let $z = (5, 8)$, and define R so that $(x_1, x_2) \ R \ (y_1, y_2)$ means that $x_1 + y_2 = y_1 + x_2$.

19. Write the equivalence relation on $\{1, 2, 3, 4, 5\}$ that is induced by the partition with $\{1, 5\}$, $\{2, 4\}$, and $\{3\}$ as its partitioning subsets.
20. Write the equivalence relation on $\{1, 2, 3, 4, 5, 6\}$ that is induced by the partition with $\{1, 3, 6\}$, $\{2, 5\}$, and $\{4\}$ as its partitioning subsets.
21. Let R be an equivalence relation on a set S. Prove that if x and y are any elements in S, then $x \ R \ y$ is false if and only if $[x] \cap [y] = \emptyset$.
22. Let R be an equivalence relation on a set S, and let x and y be elements of S. Prove that if $a \in [x]$, $b \in [y]$, and $[x] \ne [y]$, then $a \ R \ b$ is false.
23. What is wrong with the following argument that attempts to show that if R is a relation on set S that is both symmetric and transitive, then R is also reflexive?

    Since $x \ R \ y$ implies $y \ R \ x$ by the symmetric property, $x \ R \ y$ and $y \ R \ x$ imply $x \ R \ x$ by the transitive property. Thus $x \ R \ x$ is true for each $x \in S$, so R is reflexive.

24. Let R_1 and R_2 be equivalence relations on sets S_1 and S_2, respectively. Define a relation R on $S_1 \times S_2$ by letting $(x_1, x_2) \ R \ (y_1, y_2)$ mean that $x_1 \ R_1 \ y_1$ and $x_2 \ R_2 \ y_2$. Prove that R is an equivalence relation on $S_1 \times S_2$, and describe the equivalence classes of R.
25. Determine the number of relations on a set S containing n elements.
26. Call a relation R "circular" if $x \ R \ y$ and $y \ R \ z$ imply $z \ R \ x$. Prove that R is an equivalence relation if and only if R is both reflexive and circular.
27. Let S be a set containing n elements, where n is a positive integer. How many partitions of S into two subsets are there?

28. How many partitions are there of a set containing three elements?

29. How many partitions are there of a set containing four elements?

30. Prove Theorem 2.4.

31. Let S be any nonempty set and f any function with domain S. Define $s_1 \; R \; s_2$ to mean that $f(s_1) = f(s_2)$. Prove that R is an equivalence relation on S.

32. State and prove a converse to Exercise 31.

33. Let $p_m(n)$ be the number of partitions of a set of n elements into m subsets. Show that for $1 \leq m \leq n$, $p_m(n+1) = mp_m(n) + p_{m-1}(n)$.

2.3* PARTIAL ORDERING RELATIONS

In Section 1.1, we discussed a construction example in which the tasks necessary to build a house, the number of days needed to complete each task, and the immediately preceding tasks are as given in the following table:

Task		*Time in Days*	*Preceding Steps*
A	Site preparation	4	None
B	Foundation	6	A
C	Drains and services	3	A
D	Framing	10	B
E	Roof	5	D
F	Windows	2	E
G	Plumbing	4	C, E
H	Electrical work	3	E
I	Insulation	2	G, H
J	Shell	6	F
K	Drywall	5	I, J
L	Cleanup and paint	3	K
M	Floors and trim	4	L
N	Inspection	10	I

In Section 1.1, we saw that by doing some of the tasks simultaneously, all the tasks could be finished (and so the house could be completely built) in 45 days. In this section we will consider a different question: In what sequence should the tasks be performed if all the tasks are to be carried out by a group of individuals who are capable of doing only one task at a time?

In this construction project, certain tasks cannot be started until others are completed. Task G (the plumbing), for example, cannot be started until both tasks C and E are completed. Recall that there are other requirements that are not as readily apparent from the table. In this case, task G cannot be started until each of A, B, C, D, and E are completed. This is because task E cannot be started until D is completed, D cannot be started until B is completed, and B cannot be started until A is completed. Seeing all of the dependencies among the tasks is easier in Figure 2.5, where an arrow from task X to task Y signifies that task Y cannot be started until task X and all of its predecessors have been completed. If we omit the arrowheads in the figure by agreeing, as in Section 1.1, that all arrows point from left to right, then the resulting diagram is shown in Figure 2.6. (It is essentially the same as the one in Figure 1.8.)

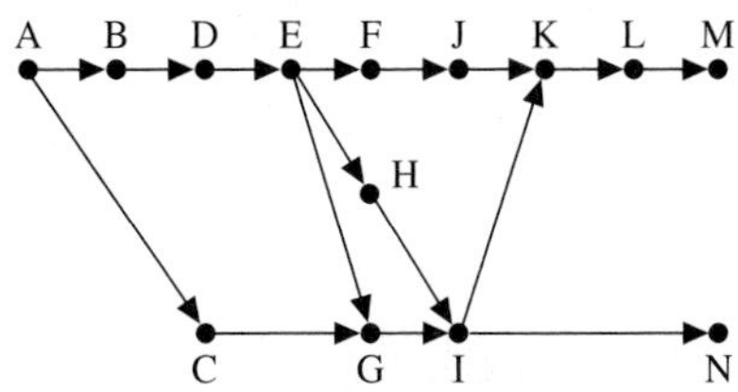

FIGURE 2.5

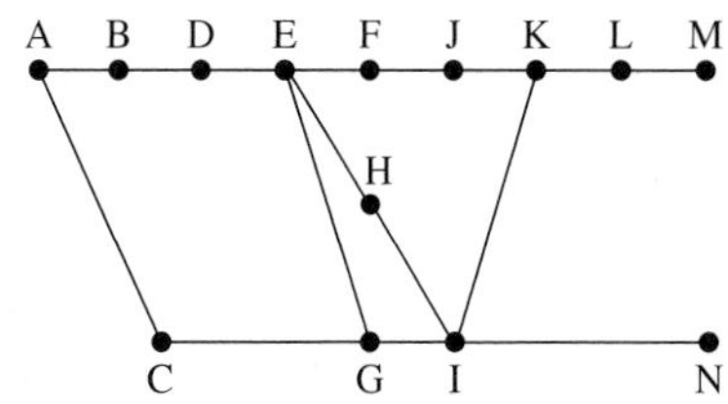

FIGURE 2.6

The sequencing of the tasks described in the original table and pictured in the diagram of Figure 2.6 creates a relation R on the set of tasks

$$S = \{\text{A, B, C, D, E, F, G, H, I, J, K, L, M, N}\},$$

where $X\ R\ Y$ means either that $X = Y$ or that Y cannot be started before X is completed. This relation has some of the special properties that we encountered in Section 2.2. First, this relation is obviously reflexive since each task in S is related to itself. Second, the relation is transitive. To see why, suppose that $X\ R\ Y$ and $Y\ R\ Z$. If $X = Y$ or $Y = Z$, then clearly $X\ R\ Z$. Otherwise, task Y cannot be started before task X is completed and task Z cannot be started before task Y is completed. It follows that task Z cannot be started before task X is completed. Thus in each case we have $X\ R\ Z$, proving that R is transitive.

However, this relation is *not* symmetric, because A R B is true but B R A is false. But the relation does have the following property: If both $X\ R\ Y$ and $Y\ R\ X$ hold, then necessarily $X = Y$. For if $X \neq Y$ and task Y cannot be started before task X is completed, and also task X cannot be started before task Y is completed, then neither X nor Y can be started, so the project cannot be finished!

A relation R on a set S is called **antisymmetric** if, whenever $x\ R\ y$ and $y\ R\ x$ are both true, then $x = y$. A relation R on a set S is called a **partial ordering relation** or, more simply, a **partial order**, if it has the following three properties:

(1) R is reflexive, that is, $x\ R\ x$ is true for every x in S.
(2) R is antisymmetric, that is, whenever both $x\ R\ y$ and $y\ R\ x$ are true, then $x = y$.
(3) R is transitive, that is, whenever both $x\ R\ y$ and $y\ R\ z$ are true, then $x\ R\ z$ is true.

Many familiar relations are partial orders, as Examples 2.17–2.20 show.

Example 2.17 The equality relation on any set is obviously antisymmetric. Since equality is also reflexive and transitive, it is a partial order on any set. ■

Example 2.18 Let S be any collection of sets. For $A, B \in S$ define $A\ R\ B$ to mean that $A \subseteq B$. Then R is an antisymmetric relation on S, for if $X\ R\ Y$ and $Y\ R\ X$ are both true, then $X \subseteq Y$ and also $Y \subseteq X$, from which it follows that $X = Y$.

Moreover, R is reflexive because $X \subseteq X$ for every $X \in S$, and R is transitive because $X \subseteq Y$ and $Y \subseteq Z$ imply that $X \subseteq Z$ for all $X, Y, Z \in S$. Therefore R is a partial order on S. ■

Example 2.19 Let S be a set of real numbers. The familiar less than or equal to relation ($\leq$) is an antisymmetric relation on S. Since this relation is also reflexive and transitive, it is a partial order on S. ■

Example 2.20 Let S be the set of positive integers. Define a relation R on S by letting $a\ R\ b$ mean that a divides b. Suppose that both $x\ R\ y$ and $y\ R\ x$ are true. Then

$$x = my \qquad \text{and} \qquad y = nx,$$

where m and n are positive integers. Thus $x = my = m(nx) = (mn)x$. Since m and n are positive integers such that $mn = 1$, it follows that $m = n = 1$. Hence $x = my = y$. Therefore R is an antisymmetric relation on S. Since this relation is also reflexive and transitive (see Example 2.10), it is a partial order on S. ■

Example 2.21 The equivalence relation R in Example 2.13 is not antisymmetric, because both $6\ R\ 10$ and $10\ R\ 6$ are true, but $6 \neq 10$. Hence R is not a partial order on the set of integers greater than 1. ■

Example 2.22 Today Mr. Webster is scheduled to interview three applicants for a summer internship at 9:00, 10:00, and 11:00, and Ms. Collins is to interview three other applicants at the same times. Unfortunately, both Mr. Webster and Ms. Collins have become ill, and so all six interviews are to be conducted by Ms. Herrera. She has decided to schedule the applicants to be interviewed by Ms. Collins in the order in which they were to appear, followed by the applicants to be interviewed by Mr. Webster in the order in which they were to appear. Thus the sequence in which the interviews are to be conducted is (Collins, 9:00), (Collins, 10:00), (Collins, 11:00), (Webster, 9:00), (Webster, 10:00), and (Webster, 11:00). The ordering of applicants used here by Ms. Herrera is an example of a *lexicographic order*. ■

In Example 2.22, there are two sets

$$S_1 = \{\text{Webster, Collins}\} \qquad \text{and} \qquad S_2 = \{9{:}00, 10{:}00, 11{:}00\}$$

and two partial orders R_1 and R_2 on those sets. In this case, the relation R_1 on S_1 is alphabetical order, and the relation R_2 on S_2 is numerical order (i.e., less than or equal to). The sequence in which the interviews are to be conducted is obtained by extending R_1 and R_2 to a partial order on $S_1 \times S_2$.

More generally, suppose that R_1 is a partial order on set S_1 and that R_2 is a partial order on set S_2. It is possible to use R_1 and R_2 to define a relation R on $S_1 \times S_2$. We define R by $(a_1, a_2)\ R\ (b_1, b_2)$ if and only if one of the following is true:

(1) $a_1 \neq b_1$ and $a_1\ R_1\ b_1$ or
(2) $a_1 = b_1$ and $a_2\ R_2\ b_2$.

This relation is called the **lexicographic order** on $S_1 \times S_2$. This ordering is also called "dictionary order" because it corresponds to the sequence in which words are listed in a dictionary.

Theorem 2.5 If R_1 is a partial order on set S_1 and R_2 is a partial order on set S_2, then the lexicographic order is a partial order on $S_1 \times S_2$.

Proof. Let R be the lexicographic order on $S_1 \times S_2$. For any $(a_1, a_2) \in S_1 \times S_2$, we have $(a_1, a_2)\ R\ (a_1, a_2)$ by condition (2) in the definition of the lexicographic order. Hence R is reflexive.

Next suppose that (a_1, a_2) and (b_1, b_2) are elements of $S_1 \times S_2$ such that both $(a_1, a_2)\ R\ (b_1, b_2)$ and $(b_1, b_2)\ R\ (a_1, a_2)$. If $a_1 \neq b_1$, then condition (1) in the definition of the lexicographic order applied to $(a_1, a_2)\ R\ (b_1, b_2)$ implies that $a_1\ R_1\ b_1$. Moreover, condition (1) applied to $(b_1, b_2)\ R\ (a_1, a_2)$ implies that $b_1\ R_1\ a_1$. Because R_1 is antisymmetric and both $a_1\ R_1\ b_1$ and $b_1\ R_1\ a_1$ are true, we must have $a_1 = b_1$, in contradiction of our assumption that $a_1 \neq b_1$. Thus $a_1 = b_1$ and condition (2) must apply. Applying (2) to $(a_1, a_2)\ R\ (b_1, b_2)$ implies that $a_2\ R_2\ b_2$, and applying condition (2) to $(b_1, b_2)\ R\ (a_1, a_2)$ implies that $b_2\ R_2\ a_2$. Since R_2 is antisymmetric, then $a_2 = b_2$. But if $a_1 = b_1$ and $a_2 = b_2$, then $(a_1, a_2) = (b_1, b_2)$, proving that R is antisymmetric.

An argument similar to that in the preceding paragraph shows that R is also transitive. Hence the lexicographic order on $S_1 \times S_2$ is a partial order. ■

Note that, despite the similarity in their names, the concepts of a symmetric relation and an antisymmetric relation are independent of each other. The relation in Example 2.17 is both symmetric and antisymmetric; the relations in Examples 2.18 through 2.20 are antisymmetric, but not symmetric; and the relation in Example 2.21 is symmetric, but not antisymmetric. Moreover, it is not difficult to find examples of relations on sets that are neither symmetric nor antisymmetric.

The reason for using the name *partial* order is that there may be elements in the underlying set that cannot be compared. For example, if we consider the partial order "is a subset of" ($\subseteq$) on the collection S of all subsets of $\{1, 2, 3, 4\}$, we see that the elements $A = \{1, 2\}$ and $B = \{1, 3\}$ of S cannot be compared; that is, neither $A \subseteq B$ nor $B \subseteq A$ is true.

If R is a partial order on S and S' is any subset of S, then R induces a partial order R' on S' by defining $x\ R'\ y$ if and only if $x\ R\ y$. (In other words, two elements of S' are related by R' if and only if the same two elements, regarded as elements of S, are related by R.) Using the ordered pair notation for a relation, this relation R' can be defined as $R' = R \cap (S' \times S')$. So, for instance, the "divides" relation in Example 2.20 induces a partial order on *any* subset of the set of positive integers.

A partial order R on set S is called a **total order** (or a **linear order**) on S if every pair of elements in S can be compared—that is, if for every $x, y \in S$, we have $x\ R\ y$ or $y\ R\ x$. Thus the "less than or equal to" relation ($\leq$) on the set of all real numbers is a total order. However, in Example 2.20 the "divides" relation is not a total order on the set of positive integers because neither 6 divides 15 nor 15 divides 6, and so the positive integers 6 and 15 cannot be compared.

Let R be a partial order on set S. An element x in S is called a **minimal element** of S (with respect to R) if the only element $s \in S$ satisfying $s\ R\ x$ is x itself, that is, if $s\ R\ x$ implies $s = x$. Likewise, an element z in S is called a **maximal element** of S (with respect to R) if the only element $s \in S$ satisfying $z\ R\ s$ is z itself, that is, if $z\ R\ s$ implies $s = z$. Minimal or maximal elements need not exist for a particular partial order.

Example 2.23 The "is a subset of" relation ($\subseteq$) is a partial order on the collection S of all subsets of $\{1, 2, 3\}$. Here $\emptyset$ (the empty set) is a minimal element of S because $A \subseteq \emptyset$ implies that $A = \emptyset$. Also, $\{1, 2, 3\}$ is a maximal element of S because $\{1, 2, 3\} \subseteq A$ implies that $A = \{1, 2, 3\}$. ■

Example 2.24 Let S denote the set of real numbers greater than or equal to 0 and less than or equal to 7. The "less than or equal to" relation ($\leq$) is a total order on S, and with respect to this relation 0 is a minimal element of S and 7 is a maximal element of S. ■

Example 2.25 The "less than or equal to" relation ($\leq$) is a total order on the set of real numbers, but here the set has neither minimal nor maximal elements. ■

Example 2.26 The "is a subset of" relation ($\subseteq$) is a partial order on the collection S of subsets of $\{1, 2, 3, 4\}$ that contain 1, 2, or 3 elements. In this setting, the one-element subsets of S are minimal elements of S and the three-element subsets of S are maximal elements of S. ■

In Example 2.25, we saw that an infinite set need not have minimal or maximal elements with respect to a particular partial order. This cannot happen if the set is finite.

Theorem 2.6 Let R be a partial order on a finite set S. Then S has both a minimal and a maximal element with respect to R.

Proof. Pick any element $s_1 \in S$. If there is no element $s \in S$ other than s_1 such that $s\ R\ s_1$, then s_1 is a minimal element of S. Otherwise, there exists an element

$s_2 \in S$ such that $s_2 \neq s_1$ and $s_2\ R\ s_1$. If there is no element $s \in S$ other than s_2 such that $s\ R\ s_2$, then s_2 is a minimal element of S. Otherwise, there exists an element $s_3 \in S$ such that $s_3 \neq s_2$ and $s_3\ R\ s_2$. Note that $s_3 \neq s_1$, because $s_3 = s_1$ would imply that $s_1\ R\ s_2$. Since R is antisymmetric and we have $s_2\ R\ s_1$, it would follow that $s_1 = s_2$, contrary to our choice of s_2. Because S is a finite set, continuing in this manner must produce a minimal element of S.

The proof that S contains a maximal element is similar. ▪

Hasse Diagrams

We have already seen that the diagram in Figure 2.6 is helpful in visualizing the precedence relations among the tasks in the construction example discussed at the beginning of this section. Similar diagrams can be constructed for any partial order R on a finite set S. Such diagrams are named **Hasse diagrams** after the German number theorist Helmut Hasse (1898–1979).

To construct a Hasse diagram for the partial order R on set S, we represent each element of S by a point, and for each pair of distinct elements x and y in S, we draw an arrow from the point representing x to the point representing y whenever $x\ R\ y$ and there is no $s \in S$ other than x and y such that $x\ R\ s$ and $s\ R\ y$. Finally, we arrange each arrow so that its initial point is below its terminal point and remove all the arrowheads. Thus, by convention, a Hasse diagram is read from bottom to top, so that all the line segments between points are regarded as pointing upward. For instance, a Hasse diagram for the construction example would be drawn as in Figure 2.7. (This is just the diagram in Figure 2.6 rotated by 90°.)

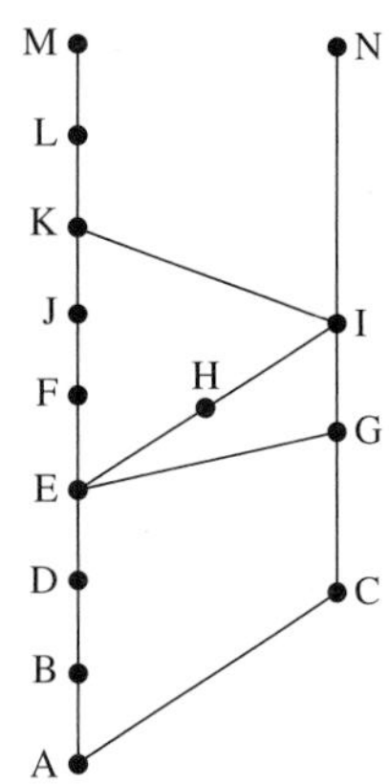

FIGURE 2.7

It is easy to detect minimal and maximal elements from a Hasse diagram. A minimal element is one that is joined by a segment to no lower point. Similarly, a maximal element is one that is joined by a segment to no higher point. Hence, in Figure 2.7, we see that A is the only minimal element and that M and N are the only maximal elements in the precedence relation for the tasks in the construction example.

Example 2.27 A Hasse diagram for the relation in Example 2.23 consists of eight points corresponding to the eight elements of S. A line segment is drawn upward from A to B whenever $A \subseteq B$ and there is no C in S other than A or B such that $A \subseteq C \subseteq B$. The resulting diagram is shown in Figure 2.8. ■

Example 2.28 Let R be the partial order "divides" on the set

$$S = \{2, 3, 4, 6, 8, 20, 24, 48, 100, 120\}.$$

A Hasse diagram for R and S is shown in Figure 2.9. Here 2 and 3 are minimal elements of S, and 48, 100, and 120 are maximal elements of S. ■

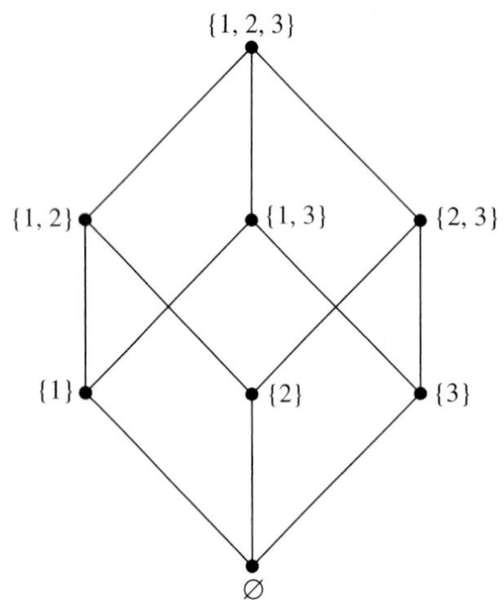

FIGURE 2.8

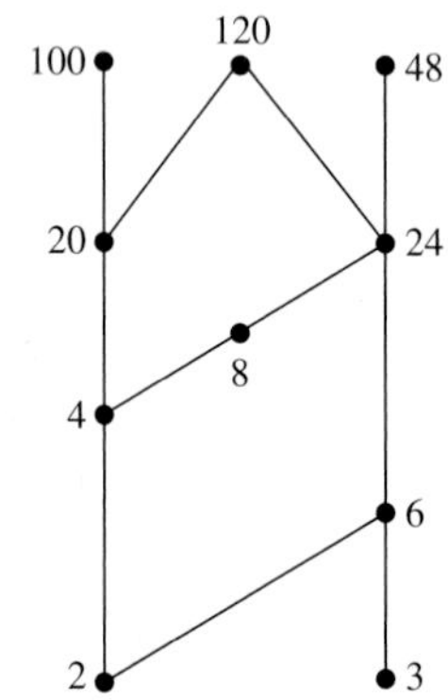

FIGURE 2.9

Example 2.29 Let R be the partial order "is a subset of" (defined in Example 2.18) on the set

$$S = \{\{1\}, \{2\}, \{3\}, \{4\}, \{1, 2\}, \{1, 5\}, \{3, 6\}, \{4, 6\}, \{0, 3, 6\}, \{1, 5, 8\}, \{0, 3, 4, 6\}\}.$$

A Hasse diagram for R and S is shown in Figure 2.10. Here $\{1\}$, $\{2\}$, $\{3\}$, and $\{4\}$ are minimal elements of S, and $\{1, 2\}$, $\{1, 5, 8\}$, and $\{0, 3, 4, 6\}$ are maximal elements of S. ■

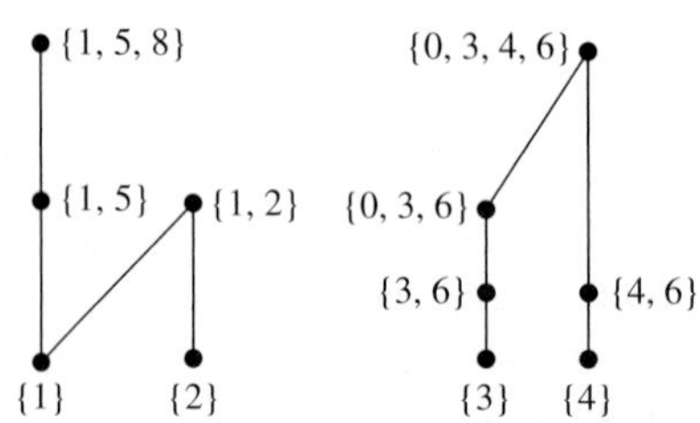

FIGURE 2.10

Topological Sorting

Now let us return to the question asked at the beginning of this section: In what sequence should the tasks in the construction example be performed if all the tasks are to be carried out by a group of individuals who are capable of doing only one task at a time? Since the tasks must be performed sequentially, we are seeking a total order T on the set of tasks that contains the original partial order R (the one depicted in the Hasse diagram in Figure 2.7). That is, we want a total order T such that $x\ R\ y$ implies $x\ T\ y$.

The process of constructing a total order that contains a given partial order is called **topological sorting**. The following algorithm will produce such a total order for any partial order on a finite set. It is based on the fact that R induces a partial order on every subset of S.

Topological Sorting Algorithm

Given a partial order R on a finite set S, this algorithm produces a total order T such that $x\ R\ y$ implies $x\ T\ y$.

Step 1 (initialization)
Set $k = 1$ and $S' = S$.

Step 2 (pick next element)
while S' is nonempty
(a) Choose any minimal element s_k of S' with respect to the order induced by R.
(b) Delete s_k from S'.
(c) Replace k with $k + 1$.
endwhile

Step 3 (define T)
Define the total order T on S by $s_i\ T\ s_j$ if and only if $i \leq j$.

To illustrate the use of the topological sorting algorithm, we will construct a total order for the set S of tasks in the construction example that contains the given partial order. From the Hasse diagram in Figure 2.7, we see that A is the only minimal element of S. Therefore we take $s_1 =$ A and delete A from S. The corresponding Hasse diagram for this new set is shown in Figure 2.11. Here both B and C are minimal elements, and we may arbitrarily select either one. Suppose we take $s_2 =$ C. A Hasse diagram for this new set is shown in Figure 2.12. Since B is the only minimal element in this new set, we take $s_3 =$ B.

Continuing in this manner, we choose $s_4 =$ D, $s_5 =$ E, $s_6 =$ G, $s_7 =$ H, $s_8 =$ I, $s_9 =$ F, $s_{10} =$ J, $s_{11} =$ N, $s_{12} =$ K, $s_{13} =$ L, and $s_{14} =$ M. Thus a sequence in which the tasks can be performed that contains the given partial order

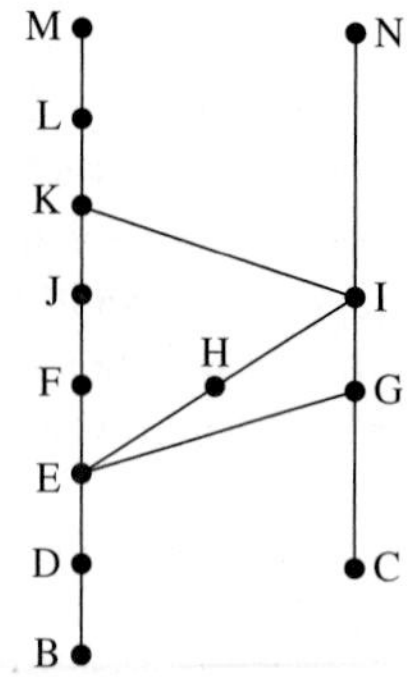

FIGURE 2.11

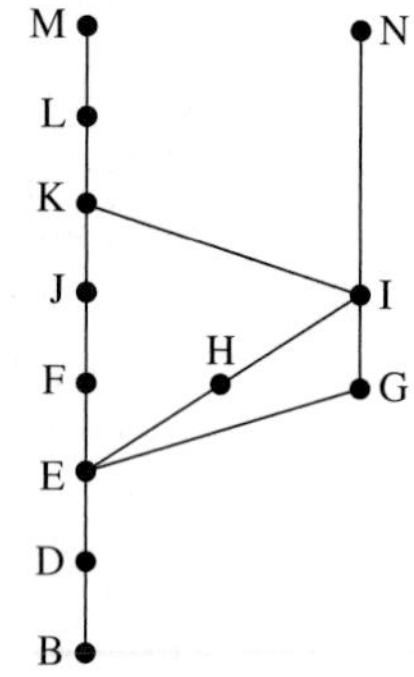

FIGURE 2.12

of the tasks is

A, C, B, D, E, G, H, I, F, J, N, K, L, and M.

Of course, other such sequences are possible because at various stages of the algorithm there were several possible minimal elements that could be chosen in step 2. Another possible sequence is

A, B, D, E, F, J, C, H, G, I, K, L, M, and N.

Each of these sequences corresponds to a total order of the tasks in S that contains the given partial order, namely, the order formed by defining X to be related to Y if and only if $X = Y$ or X occurs before Y in the sequence.

Example 2.30 To get the best seats at the basketball games of a large university, one must belong to the Basketball Booster Club. Occasionally, the athletic department receives complaints from members of the club who believe that their seats are inferior to those of another member. Experience shows that the complainer can be mollified by being assured that the other member either has been a member of the booster club longer or is giving a larger donation.

Let us define a relation R on the members of the Basketball Booster Club by $x\ R\ y$ in case both of the following are true:

(1) x has belonged to the booster club at least as many years as y.
(2) The current contribution of x is at least as great as that of y.

Suppose that the athletic department receives seating complaints from or about the following members of the booster club: Adams, Biaggi, Chow, Duda, El-Zanati, and Friedberg. It has agreed to examine the seats assigned to these six persons and make changes if the seating is found to be inequitable. The following table gives the number of years of membership in the Basketball Booster Club and the current contribution of these members:

Member x	*Years in the Club*	*Current Contribution*
Adams	6	\$150
Biaggi	3	\$200
Chow	5	\$600
Duda	8	\$500
El-Zanati	7	\$400
Friedberg	4	\$450

Because no two of these six persons have the same number of years of membership and the same current contribution, this relation is a partial order on the set

$$S = \{A, B, C, D, E, F\},$$

where each person is denoted by his or her initial. The Hasse diagram for this partial order on the set S is shown in Figure 2.13.

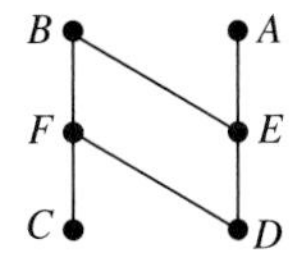

FIGURE 2.13

If we apply the topological sorting algorithm to this situation (choosing minimal elements in alphabetical order if there is a choice), we get the sequence

$$C, D, E, F, A, B.$$

If the athletic department assigns seats in this sequence (with Chow getting the best seats, and so on), then it can answer any complaint about the seats of these six persons. For example, Chow has better seats than Duda because Chow's contribution is larger, and El-Zanati's seats are better than Friedberg's because she has been a member of the booster club longer. ■

EXERCISES 2.3

In Exercises 1–8, determine whether the given relation R is a partial order on set S. Justify your answer.

1. $S = \{1, 2, 3\}$, and $R = \{(1, 1), (2, 3), (1, 3)\}$.

2. $S = \{1, 2, 3\}$, and $R = \{(1, 1), (1, 2), (2, 2), (2, 3), (3, 3), (3, 1), (1, 3), (3, 2)\}$.

3. $S = \{3, 4, 5\}$, and $R = \{(5, 5), (5, 3), (3, 3), (3, 4), (4, 4), (5, 4)\}$.

4. $S = \{\emptyset, \{1\}\}$, and $R = \{(\{1\}, \emptyset), (\{1\}, \{1\}), (\emptyset, \emptyset)\}$.

5. $S = \{1, 2, 3, 4\}$, and $x\ R\ y$ if y divides x.

6. S is the collection of all subsets of $\{1, 2, 3\}$, and $A\ R\ B$ if $A \subseteq B$ or $B \subseteq A$.

7. S is the set of integers, and $x\ R\ y$ if the remainder in the division of x by 6 equals the remainder in the division of y by 6.

8. S is the collection of all subsets of $\{1, 2, 3\}$, and $A\ R\ B$ if $|A| \leq |B|$.

In Exercises 9–12, determine a Hasse diagram for the given partial order R on set S.

9. $S = \{1, 2, 3\}$, and $R = \{(1, 1), (3, 1), (2, 1), (2, 2), (3, 3)\}$.

10. $S = \{1, 2, 3, 4\}$, and $R = \{(1, 1), (2, 2), (3, 3), (4, 4), (2, 4), (3, 1)\}$.

11. $S = \{1, 2, 3, 4, 5, 6, 7, 8\}$, and $x\ R\ y$ if x divides y.

12. S is the collection of all subsets of $\{1, 2, 3, 4\}$ with an even number of elements, and $A\ R\ B$ if $A \subseteq B$.

In Exercises 13–16, construct the partial order R on set S from the given Hasse diagram.

13.
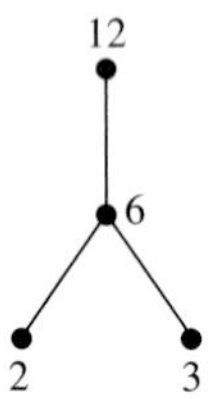

14.
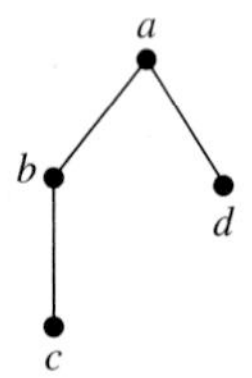

15.
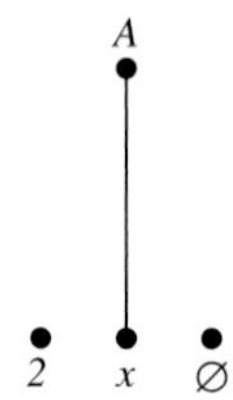

16.
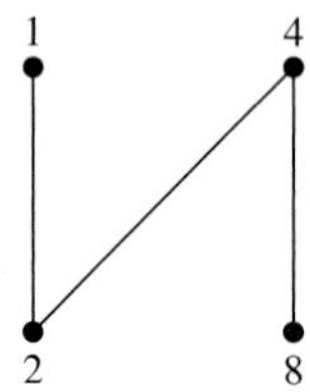

In Exercises 17–20, identify the minimal and maximal elements of S with respect to the given partial order R.

17. $S = \{1, 2, 3, 4, 5, 6\}$, and $x\ R\ y$ if x divides y.

18. S is the set of nonempty subsets of $\{1, 2, 3\}$, and $A\ R\ B$ if $B \subseteq A$.

19. $S = \{1, 2, 3, 4\}$, and $R = \{(1, 1), (2, 2), (3, 3), (4, 4), (4, 1), (3, 1)\}$.

20. S is the set of all real numbers x such that $0 \leq x < 1$, and $x\ R\ y$ if $x \leq y$.

In Exercises 21–24, apply the topological sorting algorithm to the given set S and partial order R, and give the sequence in which the elements of S are chosen.

21. $S = \{1, 2, 3, 4\}$, and $R = \{(1, 1), (2, 2), (3, 3), (4, 4), (1, 3), (2, 4), (4, 3), (2, 3)\}$.

22. $S = \{1, 2, 3, 4\}$, and $x\ R\ y$ if x divides y.

23. S is the collection of all subsets of $\{1, 2, 3\}$ such that the sum of their elements is less than 5, and $A\ R\ B$ if $A \subseteq B$.

24. $S = \{1, 2, 3, 4, 6, 12\}$, and $x\ R\ y$ if x divides y.

25. Let S be the set of nonzero integers. Define a relation R on S by letting $a\ R\ b$ mean that $\frac{b}{a}$ is an integer. Is R an antisymmetric relation on S?

26. Find a relation R on a set S that is neither symmetric nor antisymmetric.

27. Give an example of a subset S of the set of positive integers such that S has at least three elements and the "divides" relation is a total order on S.

28. Consider the "divides" relation on the set of positive integers greater than 1. Determine all the minimal and maximal elements of S.

29. Give an example of a set S and a partial order R on S such that S has exactly three minimal elements with respect to R and exactly four maximal elements with respect to R.

30. Let $S_1 = \{1, 2, 3\}$ and $S_2 = \{1, 2, 3, 4\}$, and let R_1 be the "less than or equal to" relation on S_1 and R_2 be the "less than or equal to" relation on S_2. The lexicographic order is a total order on $S_1 \times S_2$. List the elements of $S_1 \times S_2$ in the sequence given by the lexicographic order.

31. Let S be a set with five elements, and let R be a total order on S. Draw a Hasse diagram for R.

32. Let $S_1 = \{7, 8, 9\}$ and $S_2 = \{2, 3, 4, 6\}$. Let R_1 be the "less than or equal to" relation on S_1 and R_2 be the "divides" relation on S_2. Construct a Hasse diagram for the lexicographic order on $S_1 \times S_2$.

33. Let R be a partial order on S and S' be any subset of S. Prove that $R' = R \cap (S' \times S')$ is a partial order on S'.

34. Complete the proof of Theorem 2.6 by showing that if R is a partial order on a finite set S, then S has a maximal element with respect to R.

35. Let R be a total order on set S. Prove that if S has a minimal element, then the minimal element is unique.

36. Let R be a partial order on S, and suppose that x is a unique minimal element in S.

(a) Prove that if S is finite, then $x \; R \; s$ for all s in S.

(b) Show that the conclusion in (a) need not be true if S is infinite.

37. Let R_1 be a total order on set S_1 and R_2 a total order on set S_2. Prove or disprove that the lexicographic order is a total order on $S_1 \times S_2$.

38. Complete the proof of Theorem 2.5 by proving that the lexicographic order on $S_1 \times S_2$ is transitive.

39. Let R_1 be a partial order on set S_1 and R_2 a partial order on set S_2.

(a) Prove that if a is a minimal element of S_1 and b is a minimal element of S_2, then (a, b) is a minimal element of $S_1 \times S_2$ with respect to the lexicographic order.

(b) State and prove a result about a maximal element of $S_1 \times S_2$ that is similar to the result in (a).

40. Let S be a set containing exactly n elements. How many relations on S are both symmetric and antisymmetric?

41. Let S be a set containing exactly n elements. How many total orders on S are there?

42. Let S be a set containing exactly n elements. How many antisymmetric relations on S are there?

2.4 FUNCTIONS

In the matching problem described in Section 1.2, we are seeking an assignment of flights and pilots such that as many pilots as possible are assigned to flights that they requested. We saw in Section 2.2 that the list of destinations requested by the pilots gives rise to a relation between the set of destinations and the set of pilots. An assignment of flights and pilots can therefore be thought of as a special type of relation between the set of destinations and the set of pilots in which exactly one pilot is assigned to each destination. We will now study this special type of relation.

If X and Y are sets, a **function f from X to Y** is a relation from X to Y with the property that, for each element x in X, there is *exactly one* element y in Y such that $x\ f\ y$. Note that because a relation from X to Y is simply a subset of $X \times Y$, a function is a subset S of $X \times Y$ such that for each $x \in X$ there is a unique $y \in Y$ with (x, y) in S.

Example 2.31 Let $X = \{1, 2, 3, 4\}$, and $Y = \{5, 6, 7, 8, 9\}$. Then

$$f = \{(1, 5), (2, 8), (3, 7), (4, 5)\}$$

is a function from X to Y, because for each $x \in X$, there is exactly one $y \in Y$ with (x, y) in f. Note that in this case not every element of Y occurs as the second entry of an ordered pair in f (6 and 9 do not occur in any ordered pair in f), and some element of Y (namely, 5) occurs as the second entry of several ordered pairs in f.

On the other hand,

$$g = \{(1, 5), (1, 6), (2, 7), (3, 8), (4, 9)\}$$

is not a function from X to Y, because there is more than one $y \in Y$ (namely, 5 and 6) such that $(1, y)$ belongs to g. Also,

$$h = \{(1, 5), (2, 6), (4, 7)\}$$

is not a function from X to Y, because there is no element of Y associated with some element (namely, 3) of X. However, h is a function from $\{1, 2, 4\}$ to Y. ∎

We denote that f is a function from set X to set Y by writing $f\colon X \to Y$. The sets X and Y are called the **domain** and **codomain** of the function, respectively. The unique element of Y such that $x\ f\ y$ is called the **image of x under f** and is written $f(x)$—read "f of x." For the function f defined in Example 2.31, for instance, $f(1) = 5$, $f(2) = 8$, $f(3) = 7$, and $f(4) = 5$. Thus writing $y = f(x)$ is another way of expressing that (x, y) belongs to f.

It is often useful to regard a function $f\colon X \to Y$ as a pairing of each element x in X with a unique element $f(x)$ in Y. (See Figure 2.14.) In fact, functions are often defined by giving a formula that expresses $f(x)$ in terms of x; for example, $f(x) = 7x^2 - 5x + 4$.

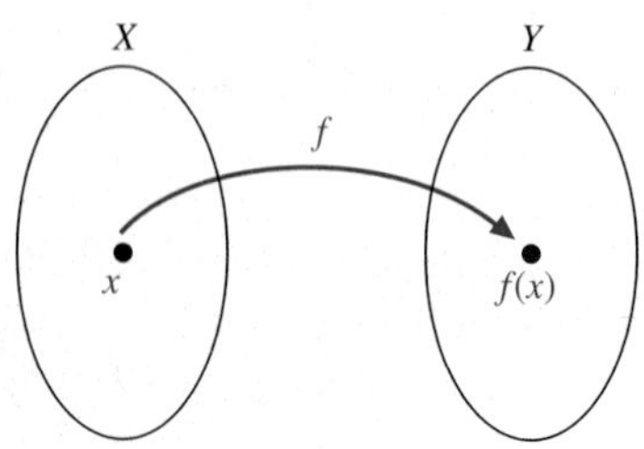

FIGURE 2.14

Note that in order for a set X to be the domain of a function g, it is necessary that $g(x)$ be defined for all x in X. Thus $g(x) = \sqrt{x}$ cannot have the set of all real numbers as its domain and codomain, for $g(x)$ is not a real number if $x < 0$. Likewise, $g(x) = 1/x$ cannot have the set of all nonnegative real numbers as its domain, because $g(x)$ is not defined if $x = 0$.

Example 2.32 Let $X = \{-1, 0, 1, 2\}$ and $Y = \{-4, -2, 0, 2\}$. Define a function $f\colon X \to Y$ by $f(x) = x^2 - x$. The images of f are as follows:

The image of -1 under f is the element $(-1)^2 - (-1) = 2$ in Y.
The image of 0 under f is the element $0^2 - 0 = 0$ in Y.
The image of 1 under f is the element $1^2 - 1 = 0$ in Y.
The image of 2 under f is the element $2^2 - 2 = 2$ in Y.

Thus $f(-1) = 2$, $f(0) = 0$, $f(1) = 0$, and $f(2) = 2$. (See Figure 2.15.) ■

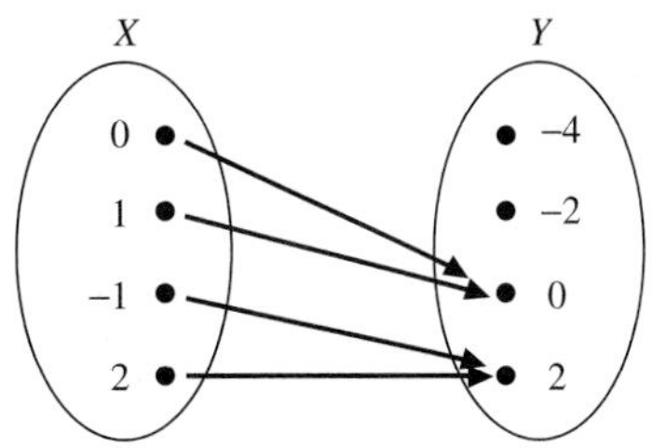

FIGURE 2.15

Example 2.33 Let X denote the set of all real numbers and Y denote the set of all nonnegative real numbers. The function $f\colon X \to Y$ defined by $g(x) = |x|$ assigns to each element x of X its absolute value $|x|$. The domain of g is X and the codomain is Y. ■

Example 2.34 Let X be the set of all real numbers between 0 and 100 inclusive, and let Y be the set of all real numbers between 32 and 212 inclusive. The function $F\colon X \to Y$ that assigns to each Celsius temperature c its corresponding Fahrenheit temperature $F(c)$ is defined by $F(c) = \frac{9}{5}c + 32$.

Unlike the situation in the preceding examples, here it is not immediately clear that the image under F of each element in X is an element of Y. To see that this is so, we must show that $32 \le F(c) \le 212$ if $0 \le c \le 100$. But if

$$0 \le c \le 100,$$

then

$$0 \le \frac{9}{5}c \le \frac{9}{5} \cdot 100 = 180.$$

So

$$32 \le \frac{9}{5}c + 32 \le 212.$$

Hence $F(c)$ is an element of Y, and so F is a function with domain X and codomain Y. ■

Example 2.35 Let Z denote the set of integers. The function $G: Z \to Z$ that assigns to each integer m the number $2m$ is defined by $G(m) = 2m$. The domain and the codomain of G are both equal to Z. ■

Example 2.36 Let $X = \{0, 1, 2, 3, 4, 5, 6, 7, 8, 9, 10, 11\}$ and Z denote the set of integers. A function $h: Z \to X$ can be defined by letting $h(x)$ equal the remainder when x is divided by 12. Here, the domain of h is Z and the codomain is X. ■

Example 2.37 On most scientific calculators, there is a built-in function named mod that behaves as follows: If m and n are positive integers, the value of the expression $\text{mod}(m, n)$ is the remainder in the division of m by n. Therefore we can regard mod as a function with $\{(m, n): m \text{ and } n \text{ are positive integers}\}$ as its domain and the set of nonnegative integers as its codomain. ■

Example 2.38 Let X denote the set of all subsets of $U = \{1, 2, 3, 4, 5\}$, and let Y be the set of nonnegative integers less than 20. If S is an element of X (that is, if S is a subset of U), define $H(S)$ to be the number of elements in S. Then $H: X \to Y$ is a function with domain X and codomain Y. ■

We have already noted that it is possible for a function to assign the same element of the codomain to different elements in the domain. The function $g(x) = |x|$ in Example 2.33, for instance, assigns to both -4 and 4 in the domain the element 4 in the codomain. If this does not occur—that is, if no two distinct elements of the domain are assigned the same element in the codomain—then the function is said to be **one-to-one**. Thus, to show that a function $f: X \to Y$ is one-to-one, we must show that $f(x_1) = f(x_2)$ implies $x_1 = x_2$.

It is also possible that one or more elements of the codomain are not paired by a function to any element in the domain. The function f in Example 2.32, for instance, does not pair the elements -4 and -2 in the codomain with any elements in the domain. Thus, in this case, only the elements 0 and 2 in the codomain are paired by f with elements in the domain. The subset of the codomain consisting of the elements that are paired with elements of the domain is called the **range** of the function. In Example 2.32 the range of f is $\{0, 2\}$. If the range and codomain of a function are equal, then the function is called **onto**. Hence, to show that a function $f: X \to Y$ is onto, we must show that if $y \in Y$, then there is an $x \in X$ such that $y = f(x)$.

A function that is both one-to-one and onto is called a **one-to-one correspondence**. Note that if $f: X \to Y$ is a one-to-one correspondence, then for each $y \in Y$ there is *exactly one* $x \in X$ such that $y = f(x)$.

For any set X, the function $I_X: X \to X$ defined by $I_X(x) = x$ is a one-to-one correspondence. This function is called the **identity function on X**.

Example 2.39 The function f in Example 2.32 is neither one-to-one nor onto. It is not one-to-one because f assigns the same element of the codomain (namely, 0) to both 0 and 1—that is, because 0 and 1 are distinct elements of the domain for which $f(0) = f(1)$. It is not onto because, as noted previously, the elements -4 and -2 in the codomain of f are not elements in the range of f. ■

Example 2.40 Let X be the set of real numbers. We will show that the function $f\colon X \to X$ defined by $f(x) = 2x - 3$ is both one-to-one and onto and, hence, is a one-to-one correspondence.

In order to show that f is one-to-one, we must show that if $f(x_1) = f(x_2)$, then $x_1 = x_2$. Let $f(x_1) = f(x_2)$. Then

$$\begin{aligned} 2x_1 - 3 &= 2x_2 - 3 \\ 2x_1 &= 2x_2 \\ x_1 &= x_2. \end{aligned}$$

Hence f is one-to-one.

In order to show that f is onto, we must show that if y is an element of the codomain of f, then there is an element x of the domain such that $y = f(x)$. Since the domain and codomain of f are both the set of real numbers, we need to show that for any real number y, there is a real number x such that $y = f(x)$. Take $x = \frac{1}{2}(y+3)$. (This value was found by solving $y = 2x - 3$ for x.) Then

$$\begin{aligned} f(x) &= f\left(\frac{1}{2}(y+3)\right) \\ &= 2\left[\frac{1}{2}(y+3)\right] - 3 \\ &= (y+3) - 3 \\ &= y. \end{aligned}$$

Thus f is onto and so is a one-to-one correspondence. ■

Note that whether or not a function is onto depends on the choice of domain and codomain. If, for instance, the codomain of the function f in Example 2.32 were changed from $\{-4, -2, 0, 2\}$ to $\{0, 2\}$, then f would be onto. Furthermore, if the set X in Example 2.40 were changed from the set of real numbers to the set of integers, then the function f in Example 2.40 would not be onto, because there would be no element $x \in X$ such that $f(x) = 0$. Likewise, whether or not a function is one-to-one depends on the choice of domain and codomain.

Example 2.41 The function $G\colon Z \to Z$ in Example 2.35 is one-to-one, for if $G(x_1) = G(x_2)$, then $2x_1 = 2x_2$; so $x_1 = x_2$. But G is not onto, because there is no element x in the domain Z for which $G(x) = 5$. In fact, it is easy to see that the range of G is the set of all even integers, and so the range and codomain of G are not equal. ■

Example 2.42 The function $h\colon Z \to X$ in Example 2.36 is easily seen to be onto. But h is not one-to-one, since $1 \neq 13$ but $h(1) = h(13) = 1$. ■

Example 2.43 Let $Y = \{x \in X\colon -1 < x < 1\}$ and X denote the set of real numbers. Define a function $f\colon X \to Y$ by

$$f(x) = \frac{x}{1 + |x|}.$$

We will show that f is a one-to-one correspondence.

As in Example 2.34, we will show first that if $x \in X$, then $f(x) \in Y$. For each $x \in X$, we have

$$-|x| \leq x \leq |x|.$$

Now $-1 - |x| < -|x|$ and $|x| < 1 + |x|$, so

$$-1 - |x| < x < 1 + |x|.$$

Dividing by $1 + |x|$ gives

$$-1 < \frac{x}{1 + |x|} < 1,$$

so that $f(x) \in Y$.

Next we will prove that f is one-to-one. If $x_1, x_2 \in X$ and $f(x_1) = f(x_2)$, then

$$\frac{x_1}{1 + |x_1|} = \frac{x_2}{1 + |x_2|}.$$

So

$$\begin{aligned}
\left|\frac{x_1}{1 + |x_1|}\right| &= \left|\frac{x_2}{1 + |x_2|}\right| \\
\frac{|x_1|}{1 + |x_1|} &= \frac{|x_2|}{1 + |x_2|} \\
|x_1|(1 + |x_2|) &= |x_2|(1 + |x_1|) \\
|x_1| + |x_1||x_2| &= |x_2| + |x_2||x_1| \\
|x_1| &= |x_2| \\
1 + |x_1| &= 1 + |x_2|.
\end{aligned}$$

Multiplying the original equation

$$\frac{x_1}{1 + |x_1|} = \frac{x_2}{1 + |x_2|}$$

by $1 + |x_1| = 1 + |x_2|$ gives $x_1 = x_2$. Hence f is one-to-one.

To show that f is onto, we solve $y = f(x)$ for x in terms of y as in Example 2.40. Then, by taking $\frac{y}{1-y} \in X$ if $0 \leq y < 1$, we have $f(\frac{y}{1-y}) = y$. And taking $\frac{y}{1+y} \in X$ if $-1 < y < 0$, we have $f(\frac{y}{1+y}) = y$. Thus f is onto. ■

Since functions are sets of ordered pairs, the definition of equality for functions follows from the definition of equality for sets. That is, $f\colon X \to Y$ and $g\colon V \to W$ are equal if $X = V$, $Y = W$, and

$$\{(x, f(x)) \colon x \in X\} = \{(v, g(v)) \colon v \in V\}.$$

It follows that $f = g$ if and only if $X = V$, $Y = W$, and $f(x) = g(x)$ for all x in X. It is possible for functions that appear different to be equal, as the next example shows.

Example 2.44 Let $X = \{-1, 0, 1, 2\}$ and $Y = \{-4, -2, 0, 2\}$. For each $x \in X$, define functions $f\colon X \to Y$ and $g\colon X \to Y$ by

$$f(x) = x^2 - x \qquad \text{and} \qquad g(x) = 2\left|x - \frac{1}{2}\right| - 1.$$

These functions are equal, since they have the same domains and codomains and $f(x) = g(x)$ for each $x \in X$:

$$f(-1) = 2 = g(-1), \qquad f(0) = 0 = g(0), \qquad f(1) = 0 = g(1),$$

and

$$f(2) = 2 = g(2). \quad \blacksquare$$

Composition and Inverses of Functions

If f is a function from X to Y and g is a function from Y to Z, then it is possible to combine them to obtain a function gf from X to Z. The function gf is called the **composition** of g and f and is defined by taking the image of x under gf to be $g(f(x))$. Thus $gf(x) = g(f(x))$ for all $x \in X$. The composition of g and f is therefore obtained by first applying f to x to obtain $f(x)$, an element of Y, and then applying g to $f(x)$ to obtain $g(f(x))$, an element of Z. (See Figure 2.16.) Note that, in evaluating $gf(x)$, we first apply f and then apply g. If $X = Z$, it is also possible to define the function fg; here we first apply g and then apply f. In general, however, the functions gf and fg are not equal.

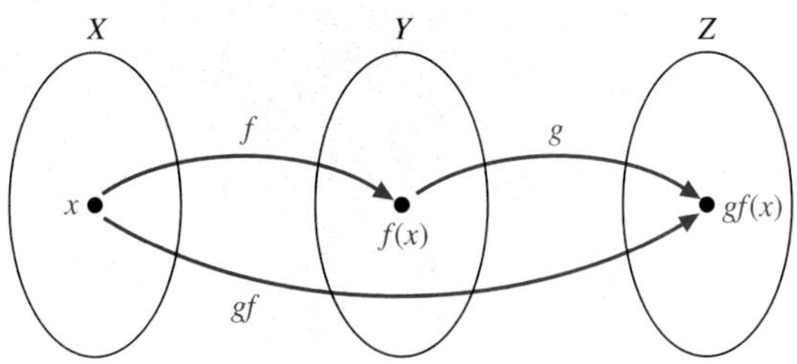

FIGURE 2.16

Example 2.45 Let X denote the set of subsets of $\{1, 2, 3, 4, 5\}$, Y the set of nonnegative integers less than 20, and Z the set of nonnegative integers. If S is an element of X, define

$f(S)$ to be the number of elements in set S; and if $y \in Y$, define $g(y) = 2y$. Then for $S = \{1, 3, 4\}$ we have $gf(S) = g(f(S)) = g(3) = 6$. In general, $gf(S) = g(f(S)) = 2 \cdot f(S)$; thus gf assigns to S the integer that is twice the number of elements in S. It follows that gf is a function with domain X and codomain Z. Note that in this case the function fg is not defined, because $g(y)$ does not lie in X, the domain of f. ■

Example 2.46 Let each of X, Y, and Z denote the set of real numbers. Define $f\colon X \to Y$ and $g\colon Y \to Z$ by $f(x) = |x|$ for all $x \in X$ and $g(y) = 3y + 2$ for all $y \in Y$. Then $gf\colon X \to Z$ is the function such that

$$gf(x) = g(f(x)) = g(|x|) = 3|x| + 2.$$

In this case we can also define the function fg, but

$$fg(x) = f(g(x)) = f(3x + 2) = |3x + 2|.$$

So $gf \neq fg$ because $gf(-1) = 5 \neq 1 = fg(-1)$. ■

Suppose that $f\colon X \to Y$ is a one-to-one correspondence. Then for each $y \in Y$ there is exactly one $x \in X$ such that $y = f(x)$. Hence we may define a function with domain Y and codomain X by associating to each $y \in Y$ the unique $x \in X$ such that $y = f(x)$. This function is denoted by f^{-1} and is called the **inverse** of function f. (See Figure 2.17.) The next theorem lists some properties that follow immediately from the definition of an inverse function.

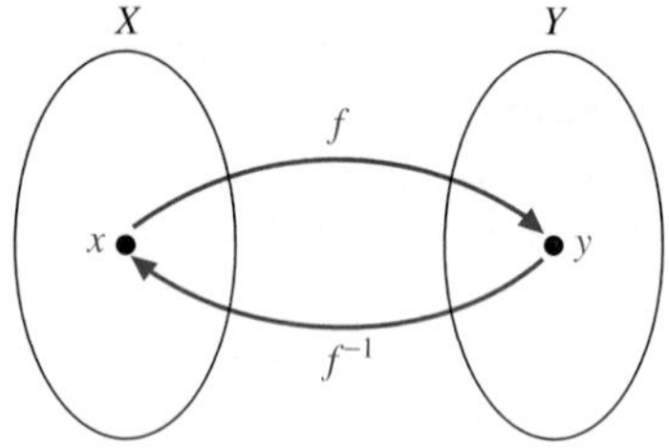

FIGURE 2.17

Theorem 2.7 Let $f\colon X \to Y$ be a one-to-one correspondence. Then

(a) $f^{-1}\colon Y \to X$ is a one-to-one correspondence.

(b) The inverse function of f^{-1} is f.

(c) For all $x \in X$, $f^{-1}f(x) = x$; and for all $y \in Y$, $ff^{-1}(y) = y$. That is, $f^{-1}f = I_X$ and $ff^{-1} = I_Y$.

Example 2.47 Theorem 2.7(c) can be used to compute the inverse of a given function. Suppose, for instance, that S is the set of real numbers,

$$X = \{x \in S\colon -1 < x \leq 3\}, \qquad Y = \{y \in S\colon 6 < y \leq 14\},$$

and $f\colon X \to Y$ is defined by $f(x) = 2x + 8$. It can be shown that f is a one-to-one correspondence and, hence, has an inverse.

If $y = f(x)$, then, by Theorem 2.7(c), $f^{-1}(y) = f^{-1}f(x) = x$. Thus, if we solve the equation $y = f(x)$ for x, we will obtain $f^{-1}(y)$. This calculation can be done as follows:

$$\begin{aligned} y &= 2x + 8 \\ y - 8 &= 2x \\ \frac{1}{2}(y - 8) &= x. \end{aligned}$$

Hence $f^{-1}(y) = \frac{1}{2}(y - 8)$, so $f^{-1}(x) = \frac{1}{2}(x - 8)$. ■

We will conclude this section by discussing an important inverse function that frequently arises in discussions about the complexity of algorithms. Recall that for any positive integer n, the number 2^n denotes the product of n factors of 2. Also,

$$2^0 = 1 \qquad \text{and} \qquad 2^{-n} = \frac{1}{2^n}.$$

It is possible to extend the definition of an exponent to include any real number in such a way that all of the familiar exponent properties hold. When this is done, the equation $f(x) = 2^x$ defines a function with the set of real numbers as its domain and the set of positive real numbers as its codomain. We call f the **exponential function with base 2**. The behavior of this function is shown in Figure 2.18.

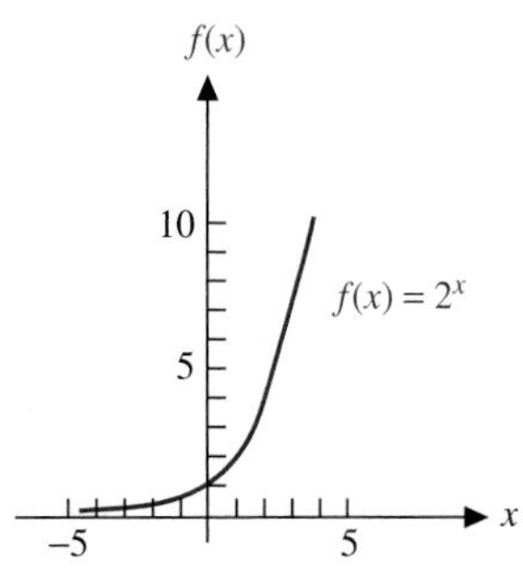

FIGURE 2.18

It can be seen in Figure 2.18 that the exponential function with base 2 is a one-to-one correspondence because each element of the codomain is associated with exactly one element of the domain. Hence this function has an inverse g called the **logarithmic function with base 2**. We denote this inverse function by $g(x) = \log_2 x$. Note that the definition of an inverse function implies that

$$y = \log_2 x \qquad \text{if and only if} \qquad x = 2^y.$$

Thus $\log_2 x$ is the exponent y such that $x = 2^y$. In particular, $\log_2 2^n = n$. So

$$\log_2 4 = \log_2 2^2 = 2, \quad \log_2 8 = \log_2 2^3 = 3, \quad \log_2 \frac{1}{2} = \log_2 2^{-1} = -1,$$

and so forth. Although $\log_2 x$ increases as x increases, the rate of growth of $\log_2 x$ is quite slow. For example,

$$\log_2 1000 < \log_2 1024 = \log_2 2^{10} = 10,$$

and similarly, $\log_2 1{,}000{,}000 < 20$. The behavior of the function $g(x) = \log_2 x$ is shown in Figure 2.19.

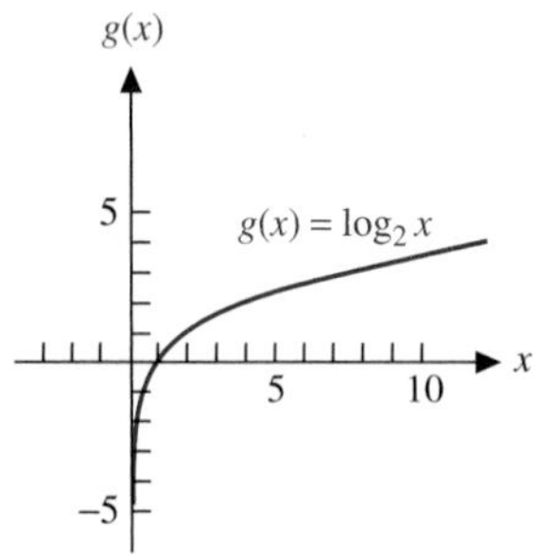

FIGURE 2.19

Scientific calculators usually contain a key marked LOG. This key can be used to find values of the logarithmic function with base 2, for

$$\log_2 x = \frac{\log x}{\log 2}.$$

Example 2.48 A swarm of killer bees escaped several years ago from South America. Suppose that the bees originally occupied a region with area of one square mile and the region occupied by the bees doubles in area each year. How long will it take for the bees to cover the entire surface of the earth, which is 197 million square miles?

Since the area of the region occupied by the bees doubles every year, after n years the bees will cover 2^n square miles. We must determine x such that $2^x = 197{,}000{,}000$. But then

$$x = \log_2 197{,}000{,}000 = \frac{\log 197{,}000{,}000}{\log 2} \approx 27.55.$$

Hence the bees will cover the entire surface of the earth in about 27.55 years. ■

EXERCISES 2.4

In Exercises 1–4, determine which of the given relations R are functions with domain X.

1. $X = \{1, 3, 5, 7, 8\}$, and $R = \{(1, 7), (3, 5), (5, 3), (7, 7), (8, 5)\}$.

2. $X = \{0, 1, 2, 3\}$, and $R = \{(0, 0), (1, 1), (1, -1), (2, 2), (3, -3)\}$.

3. $X = \{-2, -1, 0, 1\}$, and $R = \{(-2, 6), (0, 3), (1, -1)\}$.
4. $X = \{1, 3, 5\}$, and $R = \{(1, 9), (3, 9), (5, 9)\}$.

In Exercises 5–12, determine whether the given g is a function with domain X and some codomain Y.

5. X is the set of residents of Iowa and, for $x \in X$, $g(x)$ is the mother of x.
6. X is the set of computers currently in use on the Illinois State University campus and, for $x \in X$, $g(x)$ is the operating system that x is running.
7. X is the set of students at Illinois State University and, for $x \in X$, $g(x)$ is the oldest brother of x.
8. X is the set of presidents of the United States and, for $x \in X$, $g(x)$ is the year that x was first sworn into the office of president.
9. X is the set of real numbers and, for $x \in X$, $g(x) = \log_2 x$.
10. X is the set of real numbers and, for $x \in X$, $g(x) = x^2 + 3$.
11. X is the set of real numbers and, for $x \in X$, $g(x) = x2^x$.
12. X is the set of real numbers and, for $x \in X$, $g(x) = \frac{x}{|x|}$.

In Exercises 13–20, find the value of $f(a)$.

13. $f(x) = 5x - 7, a = 3$
14. $f(x) = 4, a = 8$
15. $f(x) = 2^x, a = -2$
16. $f(x) = 3|x| - 2, a = -5$
17. $f(x) = \sqrt{x - 5}, a = 9$
18. $f(x) = \frac{4}{x}, a = \frac{1}{2}$
19. $f(x) = -x^2, a = -3$
20. $f(x) = 2x^2 - x - 3, a = -2$

Evaluate the numbers in Exercises 21–28, using the fact that $\log_2 2^n = n$.

21. $\log_2 8$
22. $\log_2 \frac{1}{2}$
23. $\log_2 1$
24. $\log_2 64$
25. $\log_2 \frac{1}{16}$
26. $\log_2 \frac{1}{4}$
27. $\log_2 \frac{1}{32}$
28. $\log_2 1024$

Approximate the numbers in Exercises 29–36, using a calculator.

29. $\log_2 37$
30. $\log_2 1.72$
31. $\log_2 0.86$
32. $\log_2 100$
33. $\log_2 1.54$
34. $\log_2 9.31$
35. $\log_2 1000$
36. $\log_2 0.17$

Determine formulas for the functions gf and fg in Exercises 37–44.

37. $f(x) = 4x + 7$ and $g(x) = 2x - 3$
38. $f(x) = x^2 + 1$ and $g(x) = \sqrt{x}$
39. $f(x) = 2^x$ and $g(x) = 5x + 7$
40. $f(x) = 3x$ and $g(x) = \frac{1}{x}$
41. $f(x) = |x|$ and $g(x) = x \log_2 x$
42. $f(x) = 2^x$ and $g(x) = 5x - x^2$
43. $f(x) = x^2 - 2x$ and $g(x) = x + 1$
44. $f(x) = \dfrac{3x + 1}{2 - x}$ and $g(x) = x - 1$

In Exercises 45–52, Z denotes the set of integers. Determine whether each function g is one-to-one or onto.

45. $g: Z \to Z$ is defined by $g(x) = 3x$.
46. $g: Z \to Z$ is defined by $g(x) = x - 2$.

47. $g: Z \to Z$ is defined by $g(x) = 3 - x$.

48. $g: Z \to Z$ is defined by $g(x) = x^2$.

49. $g: Z \to Z$ is defined by $g(x) = \begin{cases} \frac{1}{2}(x+1) & \text{if } x \text{ is odd} \\ \frac{1}{2}x & \text{if } x \text{ is even.} \end{cases}$

50. $g: Z \to Z$ is defined by $g(x) = 3x - 5$.

51. $g: Z \to Z$ is defined by $g(x) = |x|$.

52. $g: Z \to Z$ is defined by $g(x) = \begin{cases} x - 1 & \text{if } x > 0 \\ x & \text{if } x \le 0. \end{cases}$

In Exercises 53–60, X denotes the set of real numbers. Compute the inverse of each function $f: X \to X$ if it exists.

53. $f(x) = 5x$

54. $f(x) = 3x - 2$

55. $f(x) = -x$

56. $f(x) = x^2 + 1$

57. $f(x) = \sqrt[3]{x}$

58. $f(x) = \frac{-1}{|x|+1}$

59. $f(x) = 3 \cdot 2^{x+1}$

60. $f(x) = x^3 - 1$

61. Find a subset Y of the set of real numbers X such that $g: X \to Y$ defined by $g(x) = 3 \cdot 2^{x+1}$ is a one-to-one correspondence. Then compute g^{-1}.

62. Find a subset Y of the set of real numbers X such that $g: Y \to Y$ defined by $g(x) = \frac{-1}{x}$ is a one-to-one correspondence. Then compute g^{-1}.

63. If X has m elements and Y has n elements, how many functions are there with domain X and codomain Y?

64. If X has m elements and Y has n elements, how many one-to-one functions are there with domain X and codomain Y?

65. Prove that if $f: X \to Y$ and $g: Y \to Z$ are both one-to-one functions, then $gf: X \to Z$ is also a one-to-one function.

66. Prove that if $f: X \to Y$ and $g: Y \to Z$ are both onto, then $gf: X \to Z$ is also onto.

67. Let $f: X \to Y$ and $g: Y \to Z$ be functions such that $gf: X \to Z$ is onto. Prove that g must be onto, and give an example to show that f need not be onto.

68. Let $f: X \to Y$ and $g: Y \to Z$ be functions such that $gf: X \to Z$ is one-to-one. Prove that f must be one-to-one, and give an example to show that g need not be one-to-one.

69. Let $f: X \to Y$ and $g: Y \to Z$ be one-to-one correspondences. Prove that gf is a one-to-one correspondence, and that $(gf)^{-1} = f^{-1}g^{-1}$.

70. Let $f: W \to X$, $g: X \to Y$, and $h: Y \to Z$ be functions. Prove that $h(gf) = (hg)f$.

2.5 MATHEMATICAL INDUCTION

In Section 1.4 we claimed that, for any positive integer n,

$$5 + 8 + 11 + \cdots + (3n+2) = \frac{1}{2}(3n^2 + 7n).$$

Since there are infinitely many positive integers, we cannot justify this assertion by verifying that this equation holds for each individual value of n. Fortunately, there is a formal scheme for proving that statements are true for all positive integers; this scheme is called the *principle of mathematical induction*.

The Principle of Mathematical Induction Let $S(n)$ be a statement involving the integer n. Suppose that, for some fixed integer n_0,

(1) $S(n_0)$ is true (that is, the statement is true if $n = n_0$) and
(2) whenever k is an integer such that $k \geq n_0$ and $S(k)$ is true, then $S(k+1)$ is true.

Then $S(n)$ is true for all integers $n \geq n_0$.

The induction principle is a basic property of the integers, and so we will give no proof of it. The principle seems quite reasonable, however, for if condition (1) in the principle holds, then we know that statement $S(n_0)$ is true. If condition (2) in the principle also holds, then we can use condition (2) with $k = n_0$ to conclude that $S(n_0 + 1)$ is true. Using condition (2) with $k = n_0 + 1$ now shows that $S(n_0 + 2)$ is true. If we then apply (2) with $k = n_0 + 2$, we see that $S(n_0 + 3)$ is true. A continuation of this argument makes it plausible that $S(n)$ is true for each integer $n \geq n_0$.

A proof by mathematical induction consists of two parts. Part (1) establishes a base for the induction by proving that some statement $S(n_0)$ is true. Part (2), called the **inductive step**, proves that if any statement $S(k)$ is true, then so is the next statement $S(k+1)$. In this section we will give several examples of the use of mathematical induction. In these examples n_0, the base for the induction, will usually be either 0 or 1.

The following example proves the result from Section 1.4 that was mentioned earlier:

Example 2.49 We will prove that $5 + 8 + 11 + \cdots + (3n+2) = \frac{1}{2}(3n^2 + 7n)$ for any positive integer n. The proof will be by induction on n, with $S(n)$ being the statement $5 + 8 + 11 + \cdots + (3n+2) = \frac{1}{2}(3n^2 + 7n)$. Since $S(n)$ is to be proved for all positive integers n, we will take the base of the induction to be $n_0 = 1$.

(1) For $n = 1$, the left side of $S(n)$ is 5 and the right side is

$$\frac{1}{2}[3(1)^2 + 7(1)] = \frac{1}{2}(3+7) = \frac{1}{2}(10) = 5.$$

Hence $S(1)$ is true.

(2) To perform the inductive step, we assume that $S(k)$ is true for some positive integer k and show that $S(k+1)$ is also true. Now $S(k)$ is the equation

$$5 + 8 + 11 + \cdots + (3k+2) = \frac{1}{2}(3k^2 + 7k).$$

To prove that $S(k+1)$ is true, we must show that

$$5 + 8 + 11 + \cdots + (3k+2) + [3(k+1)+2] = \frac{1}{2}[3(k+1)^2 + 7(k+1)].$$

But by using $S(k)$, we can evaluate the left side of the preceding equation as follows:

$$\begin{aligned}[5+8+11+\cdots+(3k+2)]+[3(k+1)+2] &= \frac{1}{2}(3k^2+7k)+[3(k+1)+2] \\ &= \left(\frac{3}{2}k^2+\frac{7}{2}k\right)+(3k+3+2) \\ &= \frac{3}{2}k^2+\frac{13}{2}k+5 \\ &= \frac{1}{2}(3k^2+13k+10).\end{aligned}$$

On the other hand, the right side of the equation to be proved is

$$\begin{aligned}\frac{1}{2}[3(k+1)^2+7(k+1)] &= \frac{1}{2}[3(k^2+2k+1)+7(k+1)] \\ &= \frac{1}{2}(3k^2+6k+3+7k+7) \\ &= \frac{1}{2}(3k^2+13k+10).\end{aligned}$$

Because the left and right sides are equal in the equation to be proved, $S(k+1)$ is true.

Since both (1) and (2) are true, the principle of mathematical induction guarantees that $S(n)$ is true for every integer $n \geq 1$, that is, for every positive integer n. ■

Example 2.50 Mathematical induction is often used to verify algorithms. To illustrate this, we will verify the polynomial evaluation algorithm stated in Section 1.4. Recall that this algorithm evaluates a polynomial

$$P(x) = a_m x^m + a_{m-1}x^{m-1} + \cdots + a_1 x + a_0$$

by the following steps:

Step 1 Let $T = a_0$ and $k = 1$.
Step 2 While $k \leq m$, replace T by $T + a_k x^k$ and k by $k+1$.
Step 3 $P(x) = T$.

Let $S(n)$ be the following statement: If the replacements in step 2 are executed exactly n times each, then $T = a_n x^n + a_{n-1}x^{n-1} + \cdots + a_1 x + a_0$. We will prove that $S(n)$ is true for all nonnegative integers n.

(1) If $n = 0$, then the replacements in step 2 are not performed, so the value of T is the value a_0 given in step 1. But the equality $T = a_0$ is the statement $S(0)$; so $S(0)$ is true.

(2) To perform the inductive step, we assume that $S(k)$ is true for some positive integer k and show that $S(k+1)$ is also true. For $S(k)$ to be true means that $T = a_k x^k + a_{k-1}x^{k-1} + \cdots + a_1 x + a_0$ when the replacements in step 2 are executed exactly k times. If the replacements in step 2 are executed one more time ($k+1$ times in all), then the value of T is

$$\begin{aligned} T + a_{k+1}x^{k+1} &= (a_k x^k + a_{k-1}x^{k-1} + \cdots + a_1 x + a_0) + a_{k+1}x^{k+1} \\ &= a_{k+1}x^{k+1} + a_k x^k + a_{k-1}x^{k-1} + \cdots + a_1 x + a_0. \end{aligned}$$

Thus $S(k+1)$ is true, completing the inductive step.

Since both (1) and (2) are true, the principle of mathematical induction guarantees that $S(n)$ is true for all nonnegative integers n. In particular, $S(m)$ is true. But $S(m)$ is the statement that $P(x) = T$.

The preceding proof shows that, after the replacements in step 2 are executed exactly k times each, the value of T is $T = a_k x^k + a_{k-1}x^{k-1} + \cdots + a_1 x + a_0$. Since this relationship holds for any number of repetitions of the while loop in step 2, it is called a **loop invariant**. ■

In our subsequent proofs by induction, we will follow the usual practice of not stating explicitly what the statement $S(n)$ is. Nevertheless, in every induction proof the reader should formulate this statement carefully.

Example 2.51 For any nonnegative integer n and any real number $x \neq 1$,

$$1 + x + x^2 + \cdots + x^n = \frac{x^{n+1} - 1}{x - 1}.$$

The proof will be by induction on n with 0 as the base of the induction. For $n = 0$, the right side of the equation is

$$\frac{x^{n+1} - 1}{x - 1} = \frac{x^1 - 1}{x - 1} = 1,$$

so the equation is true for $n = 0$.

Assume that the equation is true for some nonnegative integer k, that is,

$$1 + x + x^2 + \cdots + x^k = \frac{x^{k+1} - 1}{x - 1}.$$

Then

$$\begin{aligned}1+x+x^2+\cdots+x^k+x^{k+1} &= (1+x+x^2+\cdots+x^k)+x^{k+1}\\ &= \frac{x^{k+1}-1}{x-1}+x^{k+1}\\ &= \frac{x^{k+1}-1+x^{k+1}(x-1)}{x-1}\\ &= \frac{x^{k+1}-1+x^{k+2}-x^{k+1}}{x-1}\\ &= \frac{x^{k+2}-1}{x-1},\end{aligned}$$

proving that the equation is true for $k+1$. Thus the equation is true for all nonnegative integers n by the principle of mathematical induction. ■

In Examples 2.49 and 2.51, we used mathematical induction to prove that certain formulas are true. The principle of mathematical induction is not limited, however, to proving equations or inequalities. In the following example, induction is used to establish a geometric result.

Example 2.52 We will prove that, for any positive integer n, if *any* one square is removed from a $2^n \times 2^n$ checkerboard (one having 2^n squares in each row and column), then the remaining squares can be covered with L-shaped pieces (shown in Figure 2.20) that cover three squares.

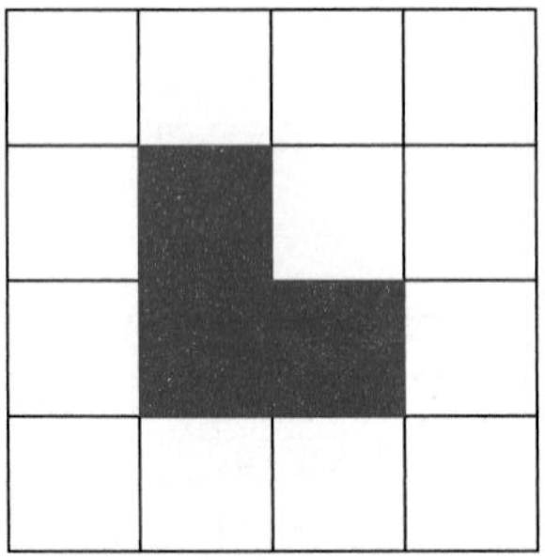

FIGURE 2.20

Figure 2.21 shows that every $2^1 \times 2^1$ checkerboard with one square removed can be covered by a single L-shaped piece. Hence the result is true for $n = 1$.

Now assume that the result is true for some positive integer k; that is, assume that every $2^k \times 2^k$ checkerboard with one square removed can be covered by L-shaped pieces. We must show that any $2^{k+1} \times 2^{k+1}$ checkerboard with one square removed can be covered by L-shaped pieces. If we divide the $2^{k+1} \times 2^{k+1}$ checkerboard in half both horizontally and vertically, we obtain four $2^k \times 2^k$

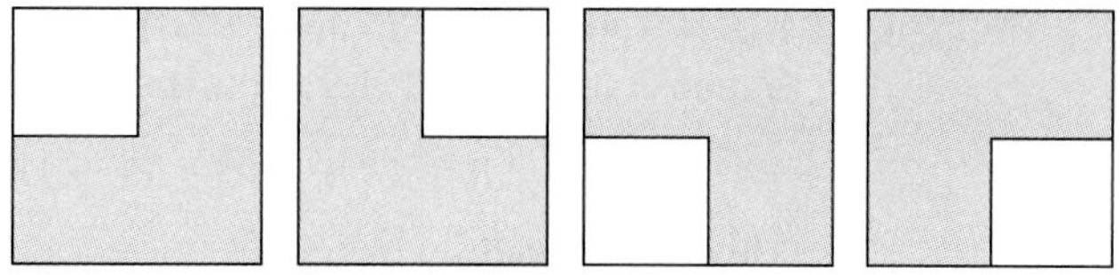

FIGURE 2.21

checkerboards. One of these $2^k \times 2^k$ checkerboards has a square removed, and the other three are complete. (See Figure 2.22.) From each of the complete $2^k \times 2^k$ checkerboards, remove the square that touches the center of the original $2^{k+1} \times 2^{k+1}$ checkerboard. (See Figure 2.23.) By the induction hypothesis, we know that all four of the $2^k \times 2^k$ checkerboards with one square removed in Figure 2.23 can be covered with L-shaped pieces. So, with one more L-shaped piece to cover the three squares touching the center of the $2^{k+1} \times 2^{k+1}$ checkerboard, we can cover with L-shaped pieces the original $2^{k+1} \times 2^{k+1}$ checkerboard with one square removed. This proves the result for $k + 1$. It now follows from the principle of mathematical induction that for every positive integer n, any $2^n \times 2^n$ checkerboard with one square removed can be covered by L-shaped pieces. ■

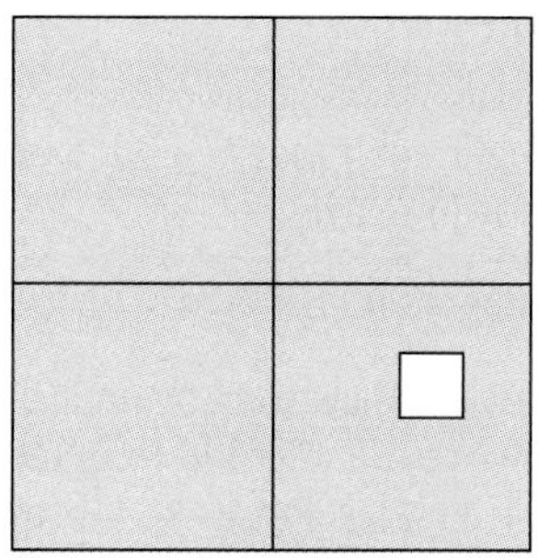

FIGURE 2.22

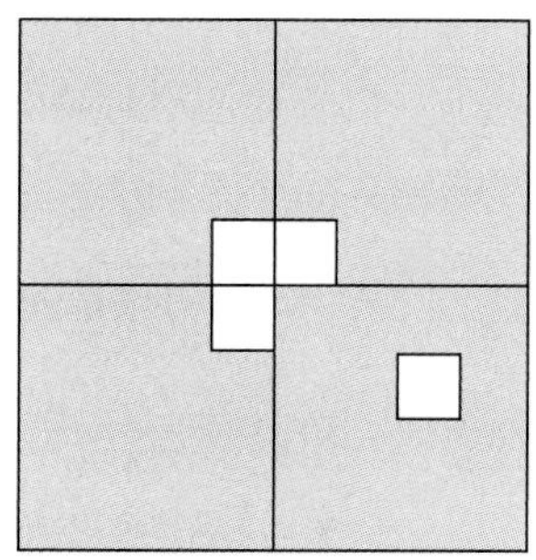

FIGURE 2.23

Closely related to the induction principle are what are known as **recursive definitions**. To define an expression recursively for integers $n \geq n_0$, we must give its value for n_0 and a method of computing its value for $k + 1$ whenever we know its value for $n_0, n_0 + 1, \ldots, k$. An example is the quantity $n!$, which was defined in Section 1.2. A recursive definition of $n!$ is

$$0! = 1 \qquad \text{and} \qquad \text{if } n > 0, \text{ then } n! = n(n-1)!.$$

By repeatedly using this definition, we can compute $n!$ for any nonnegative integer n. For example,

$$4! = (4)3! = (4)(3)2! = (4)(3)(2)1! = (4)(3)(2)(1)0! = (4)(3)(2)(1)1 = 24.$$

Example 2.53 We will prove that $n! > 2^n$ if $n \geq 4$, by applying the principle of mathematical induction with $n_0 = 4$.

(1) If $n = 4$, then $n! = 24$ and $2^n = 16$; so the statement holds.
(2) Suppose that $k! > 2^k$ for some integer $k \geq 4$. Then

$$(k+1)! = (k+1)k! \geq (4+1)k! > 2k! > 2(2^k) = 2^{k+1}.$$

This is the required inequality for $k + 1$.

Thus, by the induction principle, the statement holds for all $n \geq 4$. ■

Another example of a recursive definition is that of the **Fibonacci numbers** $F_1, F_2, \ldots$, which are defined by

$$F_1 = 1, \qquad F_2 = 1, \qquad \text{and} \qquad \text{if } n > 2, \text{ then } F_n = F_{n-1} + F_{n-2}.$$

For example, $F_3 = F_2 + F_1 = 1 + 1 = 2$, $F_4 = F_3 + F_2 = 2 + 1 = 3$, and $F_5 = F_4 + F_3 = 3 + 2 = 5$. Note that, since F_n depends on the two previous Fibonacci numbers, it is necessary to define both F_1 and F_2 at the start in order to have a meaningful definition.

In some circumstances, a slightly different form of the principle of mathematical induction is needed.

The Strong Principle of Mathematical Induction Let $S(n)$ be a statement involving the integer n. Suppose that, for some fixed integer n_0,

(1) $S(n_0)$ is true, and
(2) whenever k is an integer such that $k \geq n_0$ and $S(n_0), S(n_0+1), \ldots, S(k)$ are all true, then $S(k+1)$ is true.

Then $S(n)$ is true for all integers $n \geq n_0$.

The only difference between the strong principle of induction and the previous version is in (2), where now we are allowed to assume not only that $S(k)$ is true, but also that $S(n_0), S(n_0+1), \ldots, S(k-1)$ are true. Thus, from the point of view of logic, the strong principle should be easier to apply, since more can be assumed. It is more complicated than the previous form, however, and usually is not needed. In this book we will primarily use the strong principle to prove results about certain types of recurrence relations. To illustrate its use, we will establish a fact about the Fibonacci numbers.

Example 2.54 We will prove that $F_n \leq 2^n$ for every positive integer n. Since

$$F_1 = 1 \leq 2 = 2^1 \qquad \text{and} \qquad F_2 = 1 \leq 4 = 2^2,$$

the statement is true for $n = 1$ and $n = 2$. (We must verify the statement for both $n = 1$ and $n = 2$ because we need to assume that $k \geq 2$ in the inductive step in order to use the recursive definition of the Fibonacci numbers.)

Now suppose that, for some positive integer $k \geq 2$, the statement holds for $n = 1, n = 2, \ldots, n = k$. Then

$$F_{k+1} = F_k + F_{k-1} \leq 2^k + 2^{k-1} \leq 2^k + 2^k = 2 \cdot 2^k = 2^{k+1}.$$

So the statement is true for $n = k + 1$ if it holds for $n = 1, n = 2, \ldots, n = k$.

Thus, by the strong principle of mathematical induction, the statement is true for all positive integers n. ∎

EXERCISES 2.5

1. Compute the Fibonacci numbers F_1 through F_{10}.
2. Suppose that a number x_n is defined recursively by $x_1 = 7$ and $x_n = 2x_{n-1} - 5$ for $n \geq 2$. Compute x_1 through x_6.
3. Suppose that a number x_n is defined recursively by $x_1 = 3$, $x_2 = 4$, and $x_n = x_{n-1} + x_{n-2}$ for $n \geq 3$. Compute x_1 through x_8.
4. Give a recursive definition of x^n for any positive integer n.
5. Give a recursive definition of the nth even positive integer.
6. Give a recursive definition of the nth odd positive integer.

In Exercises 7–10, determine what is wrong with the given induction arguments.

7. We will prove that 5 divides $5n + 3$ for all positive integers n.

   Assume that, for some positive integer k, 5 divides $5k + 3$. Then there is a positive integer p such that $5k + 3 = 5p$. Now

   $$5(k + 1) + 3 = (5k + 5) + 3 = (5k + 3) + 5 = 5p + 5 = 5(p + 1).$$

   Since 5 divides $5(p + 1)$, it follows that 5 divides $5(k + 1) + 3$, which is the statement that we want to prove.

   Hence, by the principle of mathematical induction, 5 divides $5n + 3$ for all positive integers n.
8. We will prove that, in any set of n persons, all people have the same age.

   Clearly, all people in a set of 1 person have the same age, so the statement is true if $n = 1$.

   Now suppose that, in any set of k people, all persons have the same age. Let $S = \{x_1, x_2, \ldots, x_{k+1}\}$ be a set of $k + 1$ people. Then, by the induction hypothesis, all people in each of the sets $\{x_1, x_2, \ldots, x_k\}$ and $\{x_2, x_3, \ldots, x_{k+1}\}$ have the same age. But then $x_1, x_2, \ldots, x_k$ all have the same age, and likewise $x_2, x_3, \ldots, x_{k+1}$ all have the same age. It follows that $x_1, x_2, \ldots, x_{k+1}$ all have the same age. This completes the inductive step.

   The principle of mathematical induction therefore shows that, for any positive integer n, all people in any set of n persons have the same age.
9. We will prove that, for any positive integer n, if the maximum of two positive integers is n, then the integers are equal.

   If the maximum of any two positive integers is 1, then both of the integers must be 1. Hence the two integers are equal. This proves the result for $n = 1$.

   Assume that if the maximum of any two positive integers is k, then the integers are equal. Let x and y be two positive integers for which the maximum is $k + 1$. Then the maximum of $x - 1$ and $y - 1$ is k. So by the induction hypothesis, $x - 1 = y - 1$. But then $x = y$, proving the result for $n = k + 1$.

   It follows by the principle of mathematical induction that for any positive integer n, if the maximum of two positive integers is n, then the integers are equal. Hence any two positive integers are equal.

10. Let a be a nonzero real number. We will prove that, for any nonnegative integer n, $a^n = 1$.

Since $a^0 = 1$ by definition, the statement is true for $n = 0$.

Assume that, for some integer k, $a^m = 1$ for $0 \le m \le k$. Then

$$a^{k+1} = \frac{a^k a^k}{a^{k-1}} = \frac{1 \cdot 1}{1} = 1.$$

The strong principle of induction therefore implies that $a^n = 1$ for every nonnegative integer n.

In Exercises 11–26, prove each of the given statements by mathematical induction.

11. $1 + 2 + \cdots + n = \dfrac{n(n+1)}{2}$ for every positive integer n.

12. $1 + 4 + 9 + \cdots + n^2 = \dfrac{n(n+1)(2n+1)}{6}$ for every positive integer n.

13. $1 + 8 + 27 + \cdots + n^3 = \dfrac{n^2(n+1)^2}{4}$ for every positive integer n.

14. $\dfrac{1}{1 \cdot 2} + \dfrac{1}{2 \cdot 3} + \cdots + \dfrac{1}{n(n+1)} = \dfrac{n}{n+1}$ for every positive integer n.

15. $1(1!) + 2(2!) + \cdots + n(n!) = (n+1)! - 1$ for every positive integer n.

16. $\left(1 - \dfrac{1}{2}\right)\left(1 - \dfrac{1}{3}\right) \cdots \left(1 - \dfrac{1}{n+1}\right) = \dfrac{1}{n+1}$ for every positive integer n.

17. $1 \cdot 3 \cdots (2n-1) \ge 2 \cdot 4 \cdots (2n-2)$ for every integer $n \ge 2$.

18. $n^2 < 2^n$ for every integer $n \ge 5$.

19. $n! > 3^n$ for every integer $n \ge 7$.

20. $(2n)! < (n!)^2 4^{n-1}$ for every integer $n \ge 5$.

21. $F_n \le 2F_{n-1}$ for every integer $n \ge 2$.

22. $F_1 + F_2 + \cdots + F_n = F_{n+2} - 1$ for every positive integer n.

23. $F_2 + F_4 + \cdots + F_{2n} = F_{2n+1} - 1$ for every positive integer n.

24. $F_n \le \left(\dfrac{7}{4}\right)^n$ for every positive integer n.

25. $F_n \ge \left(\dfrac{5}{4}\right)^n$ for every integer $n \ge 3$.

26. For any integer $n \ge 2$, a $6 \times n$ checkerboard can be covered by L-shaped pieces of the form in Figure 2.20.

27. A sequence $s_0, s_1, s_2, \ldots$ is called a **geometric progression** with **common ratio** r if there is a constant r such that $s_n = s_0 r^n$ for all nonnegative integers n. If $s_0, s_1, s_2, \ldots$ is a geometric progression with common ratio r, find a formula for $s_0 + s_1 + \cdots + s_n$ as a function of s_0, r, and n. Then verify your formula by mathematical induction. (*Hint:* Use the equation in Example 2.51.)

28. A sequence, $s_0, s_1, s_2, \ldots$ is called an **arithmetic progression** with **common difference** d if there is a constant d such that $s_n = s_0 + nd$ for all nonnegative integers n. If $s_0, s_1, s_2, \ldots$ is an arithmetic progression with common difference d, find a formula for $s_0 + s_1 + \cdots + s_n$ as a function of s_0, d, and n. Then verify your formula by mathematical induction.

2.6 APPLICATIONS

In this section we will apply the two versions of the principle of mathematical induction stated in Section 2.5 to establish some facts that are needed elsewhere in the book. Our first two results give the maximum number of comparisons that are needed to search and sort lists of numbers; these facts will be used in our discussion of searching and sorting in Chapter 9.

Example 2.55 There is a common children's game in which one child thinks of an integer and another tries to discover what it is. After each guess, the person trying to determine the unknown integer is told if the last guess was too high or too low. Suppose, for instance, that we must identify an unknown integer between 1 and 64. One way to find the integer would be to guess the integers from 1 through 64 in order, but this method may require as many as 64 guesses to determine the unknown number. A much better way is to guess an integer close to the middle of the possible values, thereby dividing the number of possibilities in half with each guess. For example, the following sequence of guesses will discover that the unknown integer is 37:

Attempt	*Guess*	*Result*	*Conclusion*
1	32	Low	Integer is between 33 and 64.
2	48	High	Integer is between 33 and 47.
3	40	High	Integer is between 33 and 39.
4	36	Low	Integer is between 37 and 39.
5	38	High	Integer is between 37 and 37.
6	37	Correct	

■

If the strategy described here is used, it is not difficult to see that any unknown integer between 1 and 64 can be found with no more than 7 guesses. This simple game is related to the problem of searching a list of numbers by computer to see if a particular target value is in the list. Of course, this situation differs from the number-guessing game in that we do not know in advance what numbers are in the list being searched. But when the list of numbers is sorted in nondecreasing order, the most efficient searching technique is essentially the same as that used in the number-guessing game: Repeatedly compare the target value to a number in the list that is close to the middle of the range of values in which the target must occur. The following theorem describes the efficiency of this searching strategy:

Theorem 2.8 For any nonnegative integer n, at most $n + 1$ comparisons are required to determine whether a particular number is present in a list of 2^n numbers that are sorted in nondecreasing order.

Proof. The proof is by induction on n. For $n = 0$, we need to show that at most $n + 1 = 1$ comparison is required to see if a particular number m is in a list containing $2^0 = 1$ number. Since the list contains only one number, clearly only one comparison is needed to determine whether this number is m. This establishes the result when $n = 0$.

Now assume that the result is true for some nonnegative integer k; that is, assume that at most $k + 1$ comparisons are needed to determine whether a particular number is present in a sorted list of 2^k numbers. Suppose that we have a list of 2^{k+1} numbers in nondecreasing order. We must show that it is possible to determine whether a particular number m occurs in this list, using at most $(k+1)+1=k+2$ comparisons. To do so, we compare m with the number p in position 2^k of the list.

Case 1: $m \le p$ Since the list is in nondecreasing order, for m to be present in the list, it must lie in positions 1 through 2^k. But the numbers in positions 1 through 2^k are a list of 2^k numbers in nondecreasing order. Hence, by the induction hypothesis, we can determine whether m is present in this list by using at most $k+1$ comparisons. So in this case, at most $1+(k+1)=k+2$ comparisons are needed to determine whether m is present in the original list.

Case 2: $m > p$ Since the list is in nondecreasing order, for m to be present in the list, it must lie in positions 2^k+1 through 2^{k+1}. Again, the induction hypothesis tells us that we can determine whether m is present in this sorted list of 2^k numbers with at most $k+1$ comparisons. Hence, in this case also, at most $k+2$ comparisons are needed to determine whether m is present in the original list.

Thus, in each of the cases, we can determine whether m is present in the list of 2^{k+1} sorted numbers with at most $k+2$ comparisons. This completes the inductive step and therefore proves the theorem for all nonnegative integers n. ■

Although Theorem 2.8 is stated for lists of numbers in nondecreasing order, it is easy to see that the same conclusion is true for lists that are sorted in nonincreasing order. Moreover, the same conclusion is true for lists of words that are in alphabetical order. The next theorem is similar to Theorem 2.8; it gives an upper bound on the number of comparisons needed to merge two sorted lists of numbers into one sorted list. Before stating this result, we will illustrate the merging process to be used in proving Theorem 2.9.

Example 2.56 Consider the two lists of numbers in nondecreasing order:

$$2, 5, 7, 9 \qquad \text{and} \qquad 3, 4, 7.$$

Suppose that we want to merge them into a single list

$$2, 3, 4, 5, 7, 7, 9$$

in nondecreasing order. To combine the lists efficiently, first compare the numbers at the beginning of each list (2 and 3), and take the smaller one (2) as the first number in the combined list. (If the first number in one list is the same as the first number in the other list, choose either of the equal numbers.) Then delete this smaller number from the list that contains it to obtain the lists

$$5, 7, 9 \qquad \text{and} \qquad 3, 4, 7.$$

Second, compare the beginning numbers in each of these new lists (5 and 3), and take the smaller one (3) as the second number in the combined list. Delete this number from the list that contains it, and continue this process until all of the original numbers have been merged into a single list. Figure 2.24 illustrates this process. ■

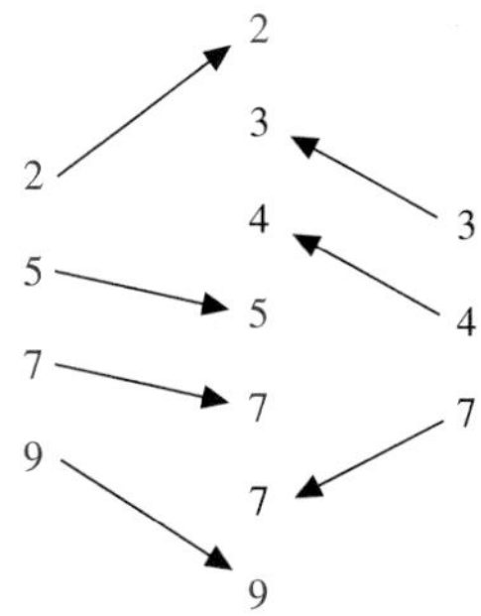

FIGURE 2.24

Theorem 2.9 Let A and B be two lists containing numbers sorted in nondecreasing order. Suppose that, for some positive integer n, there is a combined total of n numbers in the two lists. Then A and B can be merged into a single list of n numbers in nondecreasing order, using at most $n - 1$ comparisons.

Proof. The proof is by induction on n. If $n = 1$, then either A or B must be an empty list (and the other must contain one number). But then the list C obtained by adjoining list B to the end of list A will be in nondecreasing order, and C is obtained by making $0 = n - 1$ comparisons. This proves the theorem when $n = 1$.

Now suppose that the conclusion of the theorem holds for some positive integer k, and let A and B be sorted lists containing a total of $k + 1$ numbers. We must show that A and B can be merged into a sorted list C, using at most k comparisons. Compare a and b, the first elements of A and B, respectively.

Case 1: $a \leq b$ Let A' be the list obtained by deleting a from A. Then A' and B are sorted lists containing a total of k elements. So, by the induction hypothesis, A' and B can be merged into a single sorted list C', using at most $k - 1$ comparisons. Form the list C by adjoining a to C' as the first element. Then C is in nondecreasing order because C' is in nondecreasing order and a precedes all the other numbers in A and B. Moreover, C was formed by using 1 comparison to find that $a \leq b$ and at most $k - 1$ comparisons to form list C'; so C was formed using at most k comparisons.

Case 2: $a > b$ Delete b from B to form list B'. Then use the induction hypothesis as in case 1 to sort A and B' into a sorted list C', using at most $k - 1$ comparisons. The list C is then obtained by adjoining b to C' as the first element. As in case 1, C is in nondecreasing order and was formed by using at most k comparisons.

Thus in either case we can merge A and B into a sorted list by using at most k comparisons. This completes the proof of the inductive step, and so the conclusion is established for all positive integers n. ■

Our next two results involve the number of subsets of a set. These results arise in connection with the knapsack problem described in Section 1.3 and with counting techniques to be discussed in Chapter 8. Recall that we stated in Section 1.3 that the set $\{1, 2, \ldots, n\}$ has precisely 2^n subsets. If this result is true in general, then increasing n by 1 doubles the number of subsets. An example will indicate why this is true. Let us take $n = 2$ and consider the subsets of $\{1, 2\}$. They are

$$\emptyset, \qquad \{1\}, \qquad \{2\}, \qquad \text{and} \qquad \{1, 2\}.$$

Now consider the subsets of $\{1, 2, 3\}$. Of course, the four sets we have just listed are also subsets of this larger set; but there are other subsets, namely those containing 3. In fact, any subset of $\{1, 2, 3\}$ that is not a subset of $\{1, 2\}$ must contain the element 3. If we removed the 3, we would have a subset of $\{1, 2\}$ again. Thus the new subsets are just

$$\{3\}, \qquad \{1, 3\}, \qquad \{2, 3\}, \qquad \text{and} \qquad \{1, 2, 3\},$$

formed by including 3 in each of the previous four sets. The total number of subsets has doubled, as our formula indicates. This argument is the basis for a proof of Theorem 1.3.

Theorem 1.3 If n is any nonnegative integer, then a set with n elements has exactly 2^n subsets.

Proof. We prove this result by induction on n.

To establish a base for the induction, note that any set having 0 elements must be the empty set, whose only subset is $\emptyset$. Thus a set with 0 elements has exactly $2^0 = 1$ subset. This establishes the result when $n = 0$.

To perform the inductive step, we assume the result for some nonnegative integer k and prove it for $k + 1$. Thus we assume that any set with k elements has exactly 2^k subsets. Let S be a set with $k + 1$ elements, say, $a_1, a_2, \ldots, a_{k+1}$, and define a set R by

$$R = \{a_1, a_2, \ldots, a_k\}.$$

Since R has k elements, it has exactly 2^k subsets by our assumption. But each subset of S is either a subset of R or else a set formed by inserting a_{k+1} into a subset of R. Thus S has exactly

$$2^k + 2^k = 2(2^k) = 2^{k+1}$$

subsets, proving the result for $k + 1$.

It therefore follows from the principle of mathematical induction that the theorem is true for all nonnegative integers n. ■

Example 2.57 For many years, Wendy's Old Fashioned Hamburger Restaurants advertised that they serve hamburgers in 256 different ways. This claim can be justified by using Theorem 1.3, because hamburgers can be ordered at Wendy's with any combination of 8 toppings (cheese, ketchup, lettuce, mayonnaise, mustard, onions, pickles, and tomatoes). Since any selection of toppings can be regarded as a subset of the set of 8 toppings, the number of different toppings is the same as the number of subsets, which is $2^8 = 256$. ■

We can say even more about the number of subsets of a set containing n elements. The following theorem tells us how many of its 2^n subsets contain a specified number of elements:

Theorem 2.10 Let S be a set containing n elements, where n is a nonnegative integer. If r is an integer such that $0 \le r \le n$, then the number of subsets of S containing exactly r elements is

$$\frac{n!}{r!\,(n-r)!}.$$

Proof. The proof is by induction on n, starting with $n = 0$.

If $n = 0$, then S is the empty set and r must also be 0. But there is exactly one subset of Ø with no elements, namely, Ø itself. Also,

$$\frac{n!}{r!\,(n-r)!} = \frac{0!}{0!\,0!} = 1$$

because $0! = 1$ by definition. Thus the formula is correct for $n = 0$.

Now suppose that the formula is correct for some integer $k \ge 0$. Let S be a set containing $k+1$ elements, say, $S = \{a_1, a_2, \ldots, a_k, a_{k+1}\}$. We must count the subsets of S containing exactly r elements, where $0 \le r \le k+1$. Clearly, the only subset of S containing 0 elements is Ø. Likewise, there is only one subset of S containing $k+1$ elements, namely, S itself. In both of these cases the formula gives the correct value, since

$$\frac{(k+1)!}{0!\,(k+1-0)!} = 1 \quad \text{and} \quad \frac{(k+1)!}{(k+1)![(k+1)-(k+1)]!} = 1.$$

Let R be any subset of S containing exactly r elements, where $1 \le r \le k$. There are two cases to consider.

Case 1: $a_{k+1} \notin R$ Then R is a subset of $\{a_1, a_2, \ldots, a_k\}$ having r elements. By the induction hypothesis, there are

$$\frac{k!}{r!\,(k-r)!}$$

such subsets.

Case 2: $a_{k+1} \in R$ In this case, if we remove a_{k+1} from R, we have a subset of $\{a_1, a_2, \ldots, a_k\}$ containing $r-1$ elements. By the induction hypothesis, there are

$$\frac{k!}{(r-1)!\,[k-(r-1)]!}$$

sets like this.

Putting the two cases together, we see that S has a total of

$$\frac{k!}{r!\,(k-r)!} + \frac{k!}{(r-1)!\,(k-r+1)!}$$

subsets with r elements. But this number equals

$$\begin{aligned}
\frac{k!\,(k-r+1)}{r!\,(k-r)!\,(k-r+1)} + \frac{k!\,r}{r(r-1)!\,(k-r+1)!}
&= \frac{k!\,(k-r+1)}{r!\,(k-r+1)!} + \frac{k!\,r}{r!\,(k-r+1)!} \\
&= \frac{k!\,(k-r+1+r)}{r!\,(k-r+1)!} \\
&= \frac{(k+1)!}{r!\,(k+1-r)!}.
\end{aligned}$$

Since this is the number produced by the formula when $k+1$ is substituted for n, the formula is correct for $n = k+1$.

Thus, by the principle of mathematical induction, the formula is correct for all nonnegative integers n. ■

Many counting problems require knowing the number of r-element subsets of a set with n elements. We will denote[1] this number by $C(n, r)$. With this notation, Theorem 2.10 can be stated as

$$C(n, r) = \frac{n!}{r!\,(n-r)!}.$$

Example 2.58 How many 2-person committees can be chosen from a set of 5 people?

This is equivalent to asking how many subsets of $\{1, 2, 3, 4, 5\}$ have exactly 2 elements. Taking $n = 5$ and $r = 2$ in Theorem 2.10 gives the answer:

$$C(5, 2) = \frac{5!}{2!\,(5-2)!} = \frac{5!}{2!\,3!} = 10.$$

The actual subsets are $\{1, 2\}$, $\{1, 3\}$, $\{1, 4\}$, $\{1, 5\}$, $\{2, 3\}$, $\{2, 4\}$, $\{2, 5\}$, $\{3, 4\}$, $\{3, 5\}$, and $\{4, 5\}$. ■

The last result in this section proves a very basic fact about the positive integers. This fact was referred to in Example 2.13.

Theorem 2.11 Every integer greater than 1 is either prime or a product of primes.

---

[1]Another common notation is $\binom{n}{r}$.

Proof. Let n be an integer greater than 1. The proof is by induction on n, using the strong form of the principle of mathematical induction. Since 2 is a prime number, the statement is true for $n = 2$.

Assume that, for some integer $k > 1$, the statement is true for $n = 2, 3, \ldots, k$. We must prove that $k + 1$ is either prime or a product of primes. If $k + 1$ is prime, then there is nothing to prove; so suppose that $k + 1$ is not prime. Then there is a positive integer p other than 1 and $k + 1$ that divides $k + 1$. So $\frac{k+1}{p} = q$ is an integer. Now $q \neq 1$ (for otherwise $p = k + 1$) and $q \neq k + 1$ (for otherwise $p = 1$). Hence both p and q are integers between 2 and k inclusive. So the induction hypothesis can be applied to both p and q. It follows that each of p and q is either prime or a product of primes. But then $k + 1 = pq$ is a product of primes. This finishes the inductive step and therefore completes the proof of the theorem. ■

With the development of larger and faster computers, it is possible to discover huge prime numbers. In 1978, for instance, Laura Nickel and Curt Noll, two teenagers from Hayward, California, used 440 hours of computer time to find the prime number $2^{21701} - 1$. At that time this 6533-digit number was the largest known prime number. But factoring a particular positive integer into a product of primes remains a very difficult problem. Note that, although Theorem 2.11 tells us that positive integers greater than 1 are either prime or products of primes, it does not help determine which is the case. In particular, Theorem 2.11 is of no help in actually finding the prime factors of a specific positive integer.

Indeed, the difficulty of finding the prime factors of large numbers is the basis for an important method of cryptography (encoding of data or messages) called the RSA method. (The name comes from the initials of its discoverers, R.L. Rivest, A. Shamir, and L. Adleman.) For more information on the RSA method, see Sections 3.2 and 3.3.

EXERCISES 2.6

Evaluate the numbers in Exercises 1–12.

1. $C(7, 2)$	**2.** $C(8, 3)$	**3.** $C(10, 5)$	**4.** $C(12, 6)$
5. $C(11, 4)$	**6.** $C(10, 7)$	**7.** $C(11, 6)$	**8.** $C(13, 9)$
9. $C(n, 0)$	**10.** $C(n, 1)$	**11.** $C(n, 2)$	**12.** $\dfrac{P(n, r)}{C(n, r)}$

13. How many subsets of the set $\{1, 3, 4, 6, 7, 9\}$ are there?

14. How many nonempty subsets of the set {a, e, i, o, u} are there?

15. At Avanti's, a pizza can be ordered with any combination of the following ingredients: green pepper, ham, hamburger, mushrooms, onion, pepperoni, and sausage. How many different pizzas can be ordered?

16. If a test consists of 12 questions to be answered true or false, in how many ways can all 12 questions be answered?

17. A certain automobile can be ordered with any combination of the following options: air conditioning, automatic transmission, bucket seats, cruise control, power windows, rear window defogger, sun roof, and CD player. In how many ways can this car be ordered?

18. Jennifer's grandmother has told her that she can take as many of her 7 differently colored glass rings as she wants. How many choices are there?

19. How many subsets of $\{1, 3, 4, 5, 6, 8, 9\}$ contain exactly 5 elements?

20. How many subsets of {a, e, i, o, u, y} contain exactly 4 elements?

21. A basketball coach must choose a 5-person starting team from a roster of 12 players. In how many ways is this possible?

22. A beginning rock group must choose 2 songs to record from among the 9 they know. How many choices are possible?

23. A person ordering a complete dinner at a restaurant may choose 3 vegetables from among 6 offered. In how many ways can this be done?

24. A hearts player must pass 3 cards from his 13-card hand. How many choices of cards to pass does he have?

25. Three persons will be elected from among 10 candidates running for city council. How many sets of winning candidates are possible?

26. A sociologist intends to select 4 persons from a list of 9 people for interviewing. How many sets of persons to interview can be chosen?

27. How many 13-card bridge hands can be dealt from a 52-card deck? Leave your answer in factorial notation.

28. A racketeer is allowed to bring no more than 3 of the 7 lawyers representing him to a Senate hearing. How many choices does he have?

Prove each of the statements in Exercises 29–40 by mathematical induction.

29. For any distinct real numbers x and y and any nonnegative integer n,

$$x^n y^0 + x^{n-1} y^1 + \cdots + x^1 y^{n-1} + x^0 y^n = \frac{x^{n+1} - y^{n+1}}{x - y}.$$

30. $\dfrac{1}{1^2} + \dfrac{1}{2^2} + \cdots + \dfrac{1}{n^2} < 2 - \dfrac{1}{n}$ for all integers $n \geq 2$.

31. $\dfrac{(2n)!}{2^n}$ is an integer for all positive integers n.

32. $\dfrac{(n+1)(n+2)\cdots(2n)}{2^n}$ is an integer for all positive integers n.

33. For all positive integers n, 3 divides $2^{2n} - 1$.

34. For all positive integers n, 6 divides $n^3 + 5n$.

35. $\dfrac{(4n)!}{8^n}$ is an integer for all nonnegative integers n.

36. $\dfrac{(4n-2)!}{8^n}$ is an integer for all integers $n \geq 5$.

37. $(1 + 2 + \cdots + n)^2 = 1^3 + 2^3 + \cdots + n^3$ for all positive integers n.

38. $1^2 - 2^2 + \cdots + (-1)^{n+1} n^2 = \dfrac{(-1)^{n+1} n(n+1)}{2}$ for all positive integers n.

39. $\dfrac{1}{\sqrt{1}} + \dfrac{1}{\sqrt{2}} + \cdots + \dfrac{1}{\sqrt{n}} > \sqrt{n}$ for all integers $n \geq 2$.

40. $\dfrac{1(3)(5)\cdots(2n-1)}{2(4)(6)\cdots(2n)} \geq \dfrac{1}{2n}$ for all positive integers n.

41. Let n be a positive integer and $A_1, A_2, \ldots, A_n$ be subsets of a universal set U. Prove by mathematical induction that

$$\overline{(A_1 \cup A_2 \cup \cdots \cup A_n)} = \overline{A}_1 \cap \overline{A}_2 \cap \cdots \cap \overline{A}_n.$$

42. Let n be a positive integer and $A_1, A_2, \ldots, A_n$ be subsets of a universal set U. Prove by mathematical induction that

$$\overline{(A_1 \cap A_2 \cap \cdots \cap A_n)} = \overline{A}_1 \cup \overline{A}_2 \cup \cdots \cup \overline{A}_n.$$

43. If n is an integer larger than 3, determine the number of diagonals in a convex n-sided polygon. Then prove that your answer is correct by using mathematical induction.

44. Suppose that, for some positive integer n, n lines are drawn in the Euclidean plane such that no two are parallel and no three meet at the same point. Determine the number of regions into which the plane is divided by these n lines, and prove that your answer is correct by using mathematical induction.

45. Prove by mathematical induction that any list of 2^n numbers can be sorted into nondecreasing order with the use of at most $n \cdot 2^n$ comparisons.

46. Prove by mathematical induction that a uniform cake can be divided by n persons so that each person believes that the volume of cake he or she receives is at least $\frac{1}{n}$ of the total volume of the cake. Assume that each person is capable of dividing an object into parts that he or she considers to be of equal volume.

47. Mr. and Mrs. Lewis hosted a party for n married couples. As the guests arrived, some people shook hands. Later Mr. Lewis asked everyone else (including his wife) how many hands each had shaken. To his surprise, he found that no two people gave him the same answer. If no one shook his or her own hand, no spouses shook hands, and no two persons shook hands more than once, how many hands did Mrs. Lewis shake? Prove your answer by mathematical induction.

48. The **well-ordering principle** states that every nonempty set of positive integers contains a smallest element.

(a) Assume that the well-ordering principle holds, and use it to prove the principle of mathematical induction.

(b) Assume that the principle of mathematical induction holds, and use it to prove the well-ordering principle.

HISTORICAL NOTES

George Boole

The theory of sets, relations, functions, and mathematical induction share many common roots. Set theory, as a method of discussing classes of objects and their properties, got its start in the work of English mathematicians in the early to middle 1800s. George Boole's (1815–1864) publication of *Investigation of Laws of Thoughts* (1854) provided a basis for an algebra of sets and related logical forms. Boole recognized that, through the limiting of ordinary algebraic thought to the values of 0 and 1, one could develop a model for mathematical reasoning.

Boole's work amplified and clarified ideas developed earlier by George Peacock (1791–1858), Augustus De Morgan (1806–1871), and the Scottish philosopher Sir William Hamilton (1788–1856). This trio had been working to generalize the connections between arithmetic and algebra by reducing mathematical thought and argumentation to a series of symbolic forms involving generalized numbers and operations. In 1847, Boole published a work entitled *The Mathematical Analysis of Logic*. In it, he separated mathematical logic from the logics employed by the Greeks and scholastics. Boole's work elevated logic from its use in arguing particular cases to claim a role as a subdiscipline of the mathematical sciences in its own right.

In the development of the algebra of logic, Boole's work laid out an algebra of sets where the union and intersection of sets were denoted by the signs $+$ and $\times$, respectively. The empty set was denoted by 0. Our contemporary symbols $\cup$, $\cap$, and $\emptyset$ came later. The first two were developed from symbols used by the German algebraist Hermann Grassmann (1809–1877) in his 1844 work *Ausdehnungslehre*. These symbols were later popularized by the Italian Giuseppe Peano (1858–1932) in his 1894 work *Formulaire de Mathématiques*. In it, he added our present usage of $\in$ for set membership and $\subset$ for set containment. The origin of the symbol $\emptyset$ to denote the empty set is less clear, although it has been attributed to the Norwegian Niels Henrik Abel (1802–1829). Bertrand Russell (1872–1970) and Alfred North Whitehead (1861–1947) brought several other signs to common usage in their classic 1910–1913 multivolume *Principia Mathematica*. Among them were braces to denote sets, and bars above the symbols denoting sets to denote their complements [73, 74, 75, 80].

Venn diagrams are the work of the English logician John Venn (1834–1923). In his 1881 book *Symbolic Logic*, he used these diagrams to explain the ideas stated by Boole more than a quarter of a century earlier. Leonhard Euler (1707–1783) had earlier used a similar circle arrangement to make arguments about the relationships between logical classes. The use of such diagrams to represent sets, set operations, and set relationships provided a readily understandable way of reasoning about the properties of sets.

Carl Friedrich Gauss

The development of the concept of an equivalence relation is difficult to trace. However, the ideas central to the concept are found in the work of Joseph-Louis Lagrange (1736–1813) and Carl Friedrich Gauss (1777–1855) to develop congruence relations defined on the integers. The ideas are also present in Peano's 1889 work *I Principii di Geometria* [82].

The first mathematician to use the word "function" to describe a quantity associated with an algebraic relationship describing a curve was Gottfried Wilhelm Leibniz (1646–1716) in 1692. In 1748, Leonhard Euler (1707–1783) wrote in his *Introductio in Analysin Infinitorum*, that "a function of a variable quantity is an analytical expression composed in any manner from that variable quantity . . . ." It is from the work of Euler and Alexis Clairaut (1713–1765) that we have inherited the $f(x)$ notation that is still in use today.

LeJeune Dirichlet

In 1837, Peter Gustav LeJeune Dirichlet (1805–1859) set down a more rigorous formulation of the concepts of variable, function, and the correspondence between the independent variable x and the dependent variable y when $y = f(x)$. Dirichlet's definition did not depend on an algebraic relationship, but allowed for a more abstract relationship to define the connection between the entities. He stated that "y is a function of a variable x, defined on the interval $a < x < b$, if to every value of the variable x in this interval there corresponds a definite value of the variable y. Also, it is irrelevant in what way this correspondence is established." The modern set theoretic definition of a function as a subset of a Cartesian product is based on Dirichlet's work, but its formal development comes from a group of mathematicians writing under the pseudonym of Bourbaki in the late 1930s [81].

Mathematical induction was first used by the Italian mathematician and engineer Francesco Maurocyulus (1494–1575) in his 1575 book *Arithmetica* to prove that the sum of the first n positive odd integers is n^2. Blaise Pascal (1623–1662) used induction in his work on his arithmetic triangle, now called the Pascal triangle. In *Traité du triangle arithmétique* (1653), Pascal gave a clear explanation of induction in proving the fundamental property defining his triangle. The actual name "mathematical induction" was given to the principle by Augustus De Morgan in an article on the method of proof in 1838 [74].

SUPPLEMENTARY EXERCISES

Compute each of the sets in Exercises 1–8 if $A = \{1, 2, 3, 4\}$, $B = \{1, 4, 5\}$, $C = \{3, 5, 6\}$, *and the universal set is* $U = \{1, 2, 3, 4, 5, 6\}$.

1. $A \cap C$

2. $A \cup B$

3. $\overline{A}$

4. $\overline{C}$

5. $\overline{A} \cap \overline{B}$

6. $\overline{A} \cup \overline{C}$

7. $\overline{(B \cup C)}$

8. $A \cap (\overline{B} \cup C)$

Draw Venn diagrams depicting the sets in Exercises 9–12.

9. $\overline{(A - B)}$

10. $A - (B \cup C)$

11. $\overline{A} \cup (B - C)$

12. $A \cap (C - B)$

13. If $f(x) = 2x + 3$ and $g(x) = 1 - 4x^2$, determine gf and fg.

14. If $f(x) = x^3 + 1$ and $g(x) = 2x - 5$, determine gf and fg.

In Exercises 15–18, determine whether relation R is a function with domain $X = \{1, 2, 3, 4\}$.

15. $R = \{(1, 4), (2, 1), (3, 2), (4, 4)\}$

16. $R = \{(1, 3), (3, 4), (4, 1)\}$

17. $R = \{(1, -1), (2, -1), (3, 1), (4, 1)\}$

18. $R = \{(1, 2), (2, 3), (2, 1), (3, 0), (4, 1)\}$

Let Z denote the set of integers. Which of the functions $g : Z \to Z$ *in Exercises 19–22 are one-to-one? Which are onto?*

19. $g(x) = 2x - 7$

20. $g(x) = x^2 - 3$

21. $g(x) = \begin{cases} x - 2 \text{ if } x > 0 \\ x + 3 \text{ if } x \leq 0 \end{cases}$

22. $g(x) = 5 - x$

Let X denote the set of real numbers. In Exercises 23–26, compute the inverse of each function $f\colon X \to X$ *if it exists.*

23. $f(x) = |x| - 2$

24. $f(x) = 3^x + 1$

25. $f(x) = 3x - 6$

26. $f(x) = x^3 + 5$

27. At the local ice cream parlor, a sundae can be ordered with any combination of the following toppings: hot fudge, whipped cream, maraschino cherries, nuts, and marshmallows. How many different sundaes can be ordered?

28. In how many different ways can the Supreme Court render a 6-to-3 decision?

29. A grievance committee consisting of 6 persons is to be formed from 7 men and 8 women. How many different committees can be formed?

30. An investor is going to buy 100 shares of stock in each of 6 companies selected from a list of 10 companies prepared by her broker. How many different selections of 6 companies are available to the investor?

In Exercises 31–34, show that each relation R is an equivalence relation on set S. Then describe the distinct equivalence classes of R.

31. $S = \{1, 2, 3, 4, 5, 6, 7, 8\}$, and $x\ R\ y$ means that $x - y \in \{-4, 0, 4\}$.

32. $S = \{1, 2, 3, 4, 5, 6, 7, 8\}$, and $x\ R\ y$ means that $|4 - x| = |4 - y|$.

33. S is the set of integers, and $x\ R\ y$ means that either $x = y$ or $|x - y| = 1$ and the larger of x and y is even.

34. S is the set of nonzero real numbers, and $x\ R\ y$ means that $xy > 0$.

35. How many relations can be defined on $S = \{a, b, c\}$?

36. How many equivalence relations on $S = \{a, b, c\}$ are there?

37. How many functions $f\colon S \to S$ are there if $S = \{a, b, c\}$?

38. If $S = \{a, b, c\}$, how many functions $f\colon S \to S$ are one-to-one?

39. Suppose that R is an equivalence relation on set S and also a function with domain S. Describe R.

40. Let $g\colon Z \to Z$ be defined by $g(x) = ax + b$, where Z denotes the set of integers and $a, b \in Z$ with $a \neq 0$.

(a) Prove that g is one-to-one.

(b) What must be true about a and b if g is onto?

Exercises 41–43 give a relation R on a set S. Tell which of the reflexive, antisymmetric, and transitive properties R satisfies on S.

41. S is the set of all subsets of $\{1, 2, 3, 4\}$, and $A\ R\ B$ if and only if $A \subseteq B$ and $A \neq B$.

42. $S = \{\{1, 2, 3\}, \{2, 3, 4\}, \{3, 4, 5\}\}$, and $A\ R\ B$ if and only if $|A - B| \leq 1$.

43. S is the set of positive integers, and $x\ R\ y$ if and only if $y = n^2 x$ for some integer n.

44. Let S be a set of people. For $x, y \in S$, define $x\ R\ y$ to mean that $x = y$ or x is a descendant of y. Prove that R is a partial order on S.

45. Suppose that the advertisement for the Fourth of July sale (as described in Section 1.1) is to be created by a team of persons who perform only one task at a time. In what sequence should the tasks be performed?

46. Let R be a relation on set S. Define a relation R' on S by $x\ R'\ y$ if and only if $y\ R\ x$. Prove that if R is a partial order on S, then so is R'.

47. Suppose that R is a relation on set S that is both an equivalence relation and a partial order. Describe R.

Use the following information for Exercises 48–52. If R is a partial order on a set S and x, y, and z are in S, we call z the ***supremum*** *of x and y and write $z = x \vee y$ in case both*

(a) *$x\ R\ z$ and $y\ R\ z$, and*

(b) *if $w \in S$ and both $x\ R\ w$ and $y\ R\ w$, then $z\ R\ w$.*

48. Let $S = \{1, 2, 3, 4, 5, 6\}$ with $x\ R\ y$ if and only if x divides y. Compute $x \vee y$ for all pairs (x, y) in $S \times S$ for which it exists.

49. Let R be a partial order on set S, and let $x, y \in S$. Prove that, if $x \vee y$ exists, then so does $y \vee x$, and that $x \vee y = y \vee x$.

50. Let T be a set, and let S be the set of all subsets of T. For $A, B \in S$ define $A\ R\ B$ if and only if $A \subseteq B$. Prove that $A \vee B = A \cup B$ for all $A, B \in S$.

51. Let R be a partial order on set S, and let $x, y, z \in S$. Prove that if $x \vee y$, $y \vee z$, $(x \vee y) \vee z$, and $x \vee (y \vee z)$ all exist, then the latter two are equal.

52. Give an example of a partial order R on a set S where $x \vee y$, $y \vee z$, and $x \vee z$ all exist, but $(x \vee y) \vee z$ does not exist.

In Exercises 53–56, let Z denote the set of integers, $f\colon Z \to Z$ be a function, and let $x\ R\ y$ be defined to mean that $f(x) = f(y)$.

53. Prove that R is an equivalence relation on Z.

54. Determine $[n]$, the equivalence class containing $n \in Z$ with respect to R, if f is the function defined by $f(x) = x^2$.

55. What must be true about a function f if, for every $n \in Z$, the equivalence class containing $[n]$ consists of exactly one element?

56. Give an example of a function $f\colon Z \to Z$ for which every equivalence class $[n]$ contains exactly *three* elements.

Prove each of the set equalities in Exercises 57–62.

57. $A \times (B \cup C) = (A \times B) \cup (A \times C)$

58. $A \times (B \cap C) = (A \times B) \cap (A \times C)$

59. $A \times (B - C) = (A \times B) - (A \times C)$

60. $(A \cup B) - C = (A - C) \cup (B - C)$

61. $(A - B) - (A - C) = A \cap (C - B)$

62. $(A - B) \cup (A - C) = A - (B \cap C)$

Prove the results in Exercises 63–72 by mathematical induction.

63. For all positive integers n, $1^2 + 3^2 + \cdots + (2n-1)^2 = \dfrac{n(2n-1)(2n+1)}{3}$.

64. For all positive integers n, $\dfrac{1}{1 \cdot 3} + \dfrac{1}{3 \cdot 5} + \cdots + \dfrac{1}{(2n-1)(2n+1)} = \dfrac{n}{2n+1}$.

65. For all positive integers n, $\dfrac{5}{1 \cdot 2 \cdot 3} + \dfrac{6}{2 \cdot 3 \cdot 4} + \cdots + \dfrac{n+4}{n(n+1)(n+2)} = \dfrac{n(3n+7)}{2(n+1)(n+2)}$.

66. Any integer $n > 23$ can be written in the form $5r + 7s$ for some nonnegative integers r and s.

67. Any postage of 8 cents or more can be obtained using only 3-cent and 5-cent stamps.

68. For all positive integers n and any distinct real numbers x and y, $x - y$ divides $x^n - y^n$.

69. Choose any $n \geq 3$ distinct points on the circumference of a circle, and join consecutive points by line segments to form an n-sided polygon. Show that the sum of the interior angles of this polygon is $180n - 360$ degrees.

70. Prove that $F_{n+1} = F_{n-m}F_m + F_{n-m+1}F_{m+1}$ for any integers m and n such that $n > m \geq 1$. (*Hint:* Fix m, and use induction on n beginning with $n = m + 1$.)
71. Prove that F_m divides F_{mn} for all positive integers n.
72. Prove that if $n > 12$ is an even integer not divisible by 3, then an $n \times n$ checkerboard with one square removed can be covered by L-shaped pieces as in Figure 2.20. (*Hint:* Divide the $n \times n$ board into $(n-6) \times (n-6)$, $6 \times (n-6)$, $(n-6) \times 6$, and 6×6 subboards.)

COMPUTER PROJECTS

Write a computer program having the specified input and output.

1. Given a nonnegative integer n, list all the subsets of $\{1, 2, \ldots, n\}$.
2. Let U be a finite set of real numbers. Given lists of the elements in the universal set U and in subsets A and B, list the elements in the sets $A \cup B$, $A \cap B$, $A - B$, $\overline{A}$, and $\overline{B}$.
3. Given a finite set S of integers and a subset R of $S \times S$, determine which of the reflexive, symmetric, antisymmetric, and transitive properties are possessed by the relation R on S. Assume that the elements of the sets S and R are listed.
4. Given a finite set S of integers and a subset R of $S \times S$, determine if R is an equivalence relation on S. If so, list the distinct equivalence classes of R. Assume that the elements of the sets S and R are listed.
5. Given a partial order R on a finite set S, determine a total order on S that contains R. Assume that the elements of the sets S and R are listed.
6. Given a positive integer n, compute $1!, 2!, \ldots, n!$.
7. Given a positive integer n, compute $F_1, F_2, \ldots, F_n$.
8. Given sets $X = \{x_1, x_2, \ldots, x_m\}$ and $Y = \{y_1, y_2, \ldots, y_n\}$ containing m and n elements, respectively, list all the functions with domain X and codomain Y.
9. Given sets $X = \{x_1, x_2, \ldots, x_m\}$ and $Y = \{y_1, y_2, \ldots, y_n\}$ containing m and n elements, respectively, list all the one-to-one functions with domain X and codomain Y.
10. Given sets $X = \{x_1, x_2, \ldots, x_m\}$ and $Y = \{y_1, y_2, \ldots, y_n\}$ containing m and n elements, respectively, list all the onto functions with domain X and codomain Y.
11. Given sets $X = \{x_1, x_2, \ldots, x_n\}$ and $Y = \{y_1, y_2, \ldots, y_n\}$ containing n elements, list all the one-to-one correspondences with domain X and codomain Y.
12. Given two sorted lists of real numbers, merge them into a single sorted list, using the technique described in Example 2.56.

SUGGESTED READINGS

1. Buck, R.C. "Mathematical Induction and Recursive Definitions." *American Mathematical Monthly*, vol. 70, no. 2 (February 1963): 128– 135.
2. Halmos, Paul R. *Naive Set Theory*. New York: Springer, 1994.
3. Hayden, S. and J. Kennison. *Zermelo-Fraenkel Set Theory*. Columbus, Ohio: Charles Merrill, 1968.
4. Henken, L. "On Mathematical Induction." *American Mathematical Monthly*, vol. 67, no. 4 (April 1960): 323–337.

CHAPTER

3

Coding Theory

Modern life requires the ability to transmit large amounts of data. Banks send financial information over the telephone and the Internet, satellites transmit photographs to earth from outer space, and DVD players process audio and video signals for our entertainment. It is essential that data be sent both securely and accurately. For example, when a product is purchased over the Internet, it is crucial that a customer's credit card number be transmitted *safely,* so that it is not intercepted by an identity thief, and *correctly,* so that the transaction is not rejected by the issuer of the credit card.

This chapter concerns coding theory, which splits into two distinct parts. The first has the purpose of concealment, and this is what people usually think of when they hear the word "code." The type of concealment code we will consider is quite modern, however, and different from traditional codes used as recently as 30 years ago. It is called *public key cryptography* and is distinguished by the fact that it does not have to be prearranged between the sender and receiver of the message, as opposed to all previous methods. The public key method we will look at depends on the part of mathematics known as *number theory,* which is the study of the integers $\ldots, -3, -2, -1, 0, 1, 2, 3, 4, \ldots,$ and so the first two sections of this chapter will develop some background in that field. The security of public key methods depends as well on the complexity of algorithms, discussed in Section 1.4.

We will also consider *error-correcting* codes. These are ways of encoding a message with redundancy, so that even if part of the message is incorrectly transmitted, it still may be possible to reconstruct the original. One application of such codes is to allow a CD to sound good even when dust gets on the surface of the disc. The purpose of such a code is accuracy in an imperfect environment, not concealment.

3.1 CONGRUENCE

In this section we will discuss an important equivalence relation on the set of integers. This relation will lead to the study of number systems containing only a finite number of elements. Such number systems arise naturally in the study of computer arithmetic.

We will begin by discussing some ideas from arithmetic. If m and n are integers and $m \neq 0$, the *division algorithm* states that n can be expressed in the form

$$n = qm + r, \qquad \text{where} \qquad 0 \leq r < |m|,$$

for unique integers q and r. (Recall that $|m|$, the absolute value of m, is defined to be m if $m \geq 0$ and is defined to be $-m$ if $m < 0$.) These integers q and r are called the **quotient** and **remainder**, respectively, in the division of n by m and can be found by the process of long division. Thus, for instance, in the division of 34 by 9, the quotient is 3 and the remainder is 7, because

$$34 = 3 \cdot 9 + 7 \qquad \text{and} \qquad 0 \leq 7 < 9.$$

Note, however, that although

$$-34 = 3 \cdot (-9) + (-7),$$

3 is *not* the quotient in the division of 34 by -9, because -7 is not a possible remainder. (It does not lie between 0 and $|-9| = 9$.) In this case, we have

$$-34 = 4 \cdot (-9) + 2 \qquad \text{and} \qquad 0 \leq 2 < 9,$$

so that 4 is the quotient and 2 is the remainder in this division. If the remainder in the division of n by m is 0, then we say that n **is divisible by** m (or that m **divides** n). Thus to say that n is divisible by m means that $n = qm$ for some integer q.

Now let m be an integer greater than 1. If x and y are integers, we say that ***x* is congruent to *y* modulo *m*** if $x - y$ is divisible by m. If x is congruent to y modulo m, we write $x \equiv y \pmod{m}$; otherwise, we write $x \not\equiv y \pmod{m}$. We call this relation on the set of integers **congruence modulo *m***.

Example 3.1 Clearly, $3 \equiv 24 \pmod{7}$ because $3 - 24 = -21$ is divisible by 7. And similarly, $98 \equiv 43 \pmod{11}$ because $98 - 43 = 55$ is divisible by 11. But $42 \not\equiv 5 \pmod{8}$, since $42 - 5 = 37$ is not divisible by 8, and $4 \not\equiv 29 \pmod{6}$, since $4 - 29 = -25$ is not divisible by 6. ■

The most common situation in which congruence occurs is in connection with the telling of time. Standard clocks and watches keep track of time modulo 12. Because $7 + 15 \equiv 10 \pmod{12}$, we say that 15 hours after 7 o'clock is 10 o'clock. On the other hand, transportation schedules (such as train schedules) usually list times modulo 24 because there are 24 hours per day.

Example 3.2 Congruences often occur in applications involving error-detecting codes. In this example we will describe an application of such a code in the publishing industry.

Since 1972, a book published anywhere in the world has carried a 10-digit code number called an International Standard Book Number (ISBN). For instance, the ISBN for *Finite Mathematics* by Spence and Vanden Eynden is 0-673-38582-5. By providing a standard identifier for books, these numbers have allowed publishers and bookstores to computerize their inventories and billing procedures more easily than if each book had to be referred to by author, title, and edition.

An ISBN consists of four parts: a group code, a publisher code, an identifying number assigned by the publisher, and a check digit. In the ISBN 0-673-38582-5, the group code (0) denotes that the book was published in an English-speaking country (either Australia, Canada, New Zealand, South Africa, the United Kingdom, or the United States). The next group of digits (673) identifies the publisher, and the third group of digits (38582) designates this particular book among all those published by that publisher. The final digit of the ISBN (5) is the check digit, which is used to detect errors in copying or transmitting the ISBN. By using the check digit, publishers are often able to detect an incorrect ISBN and prevent the costly shipping charges that would result from filling an incorrect order.

The check digit has eleven possible values: 0, 1, 2, 3, 4, 5, 6, 7, 8, 9, or X. (A check digit of X represents the number 10.) This digit is determined in the following way: Multiply the first nine digits of the ISBN by 10, 9, 8, 7, 6, 5, 4, 3, and 2, respectively, and add these nine products to obtain a number y. The check digit d is then chosen so that $y + d \equiv 0 \pmod{11}$. For example, the check digit for *Finite Mathematics* is 5, because

$$\begin{aligned}&10(0) + 9(6) + 8(7) + 7(3) + 6(3) + 5(8) + 4(5) + 3(8) + 2(2)\\&\quad = 0 + 54 + 56 + 21 + 18 + 40 + 20 + 24 + 4 = 237\end{aligned}$$

and $237 + 5 = 242 \equiv 0 \pmod{11}$.

Likewise, the ISBN for this book (found on the back of the title page) is 0-321-30515-9. Here the check digit is 9, because

$$\begin{aligned}&10(0) + 9(3) + 8(2) + 7(1) + 6(3) + 5(0) + 4(5) + 3(1) + 2(5)\\&\quad = 0 + 27 + 16 + 7 + 18 + 0 + 20 + 3 + 10 = 101\end{aligned}$$

and $101 + 9 = 110 \equiv 0 \pmod{11}$. (For other uses of congruence in identification numbers, see suggested readings [1] and [2] at the end of this chapter.) ■

It can be shown that x is congruent to y modulo m precisely when $x = km + y$ for some integer k. In particular, x is congruent to the remainder in the division of x by m. Hence *x is congruent to y modulo m if and only if x and y have the same remainder when divided by m.* (See Exercise 51.) From this fact, the next theorem follows immediately.

Theorem 3.1 Congruence modulo m is an equivalence relation.

The equivalence classes for congruence modulo m are called **congruence classes modulo m**. The set of all the congruence classes modulo m will be denoted by Z_m. It follows from Theorem 2.3 that any two congruence classes modulo m are either equal or disjoint. Moreover, in Z_m, $[x] = [y]$ if and only if $x \equiv y \pmod{m}$. Thus, if r is the remainder in the division of x by m, then $[x] = [r]$ in Z_m. So there are m distinct congruence classes in Z_m, namely, $[0], [1], [2], \ldots, [m-1]$. These correspond to the m possible remainders when dividing by m.

Example 3.3 In Z_3 the distinct congruence classes are

$$[0] = \{\ldots, -6, -3, 0, 3, 6, 9, \ldots\},$$
$$[1] = \{\ldots, -5, -2, 1, 4, 7, 10, \ldots\}, \text{ and}$$
$$[2] = \{\ldots, -4, -1, 2, 5, 8, 11, \ldots\}.$$

Notice that each of the congruence classes in Z_3 has many possible representations. For instance, $[0] = [3] = [9] = [-12]$ and $[2] = [-4] = [11] = [32]$. ■

We would like to define addition and multiplication in Z_m. There is a natural way to do this, using the addition and multiplication of integers: Simply define

$$[x] + [y] = [x + y] \quad \text{and} \quad [x][y] = [xy].$$

In order for these definitions to make sense, however, we must be sure that they do not depend on the way in which congruence classes are represented. In other words, we must be certain that these definitions depend only on the congruence classes themselves. For example, in Z_3, we have $[0] = [9]$ and $[2] = [11]$; so we must be certain that the sums $[0] + [2]$ and $[9] + [11]$ give the same answer. The next result gives us that assurance.

Theorem 3.2 If $x \equiv x' \pmod{m}$ and $y \equiv y' \pmod{m}$, then

(a) $x + y \equiv x' + y' \pmod{m}$ and

(b) $xy \equiv x'y' \pmod{m}$.

Proof. If $x \equiv x' \pmod{m}$ and $y \equiv y' \pmod{m}$, then there are integers a and b such that $x = am + x'$ and $y = bm + y'$.

(a) Thus $x + y = (am + x') + (bm + y') = (a + b)m + (x' + y')$. Then

$$(x + y) - (x' + y') = (a + b)m,$$

so that $(x + y) - (x' + y')$ is divisible by m. This proves (a).

(b) Likewise, $xy = (am + x')(bm + y') = (amb + ay' + bx')m + x'y'$, so that $xy - x'y' = (amb + ay' + bx')m$. Hence $xy - x'y'$ is divisible by m, proving (b). ■

Notice that part (b) of Theorem 3.2 implies that if $x \equiv z \pmod{m}$, then $x^n \equiv z^n \pmod{m}$ for all positive integers n. Moreover, the definition of multiplication in Z_m shows that $[x]^n = [z]^n$.

Example 3.4 In Z_6, we have

$$[5] + [3] = [5 + 3] = [8] = [2],$$

since $8 \equiv 2 \pmod{6}$. Also,

$$[5][3] = [5 \cdot 3] = [15] = [3],$$

because $15 \equiv 3 \pmod{6}$; and

$$[8]^4 = [2]^4 = [2^4] = [16] = [4],$$

since $8 \equiv 2 \pmod{6}$ and $16 \equiv 4 \pmod{6}$. ■

Example 3.5 In Z_8, we have

$$[4] + [7] = [4 + 7] = [11] = [3],$$

since $11 \equiv 3 \pmod{8}$. Also,

$$[4][7] = [4 \cdot 7] = [28] = [4],$$

because $28 \equiv 4 \pmod{8}$; and

$$[7]^9 = [-1]^9 = [(-1)^9] = [-1] = [7],$$

since $7 \equiv -1 \pmod{8}$. ■

Example 3.6 A scientific recording instrument uses one foot of paper per hour. If a new roll of paper 100 feet long is installed at 11:00 A.M., at what hour of the day will the instrument run out of paper?

To answer this question, we will number the hours of a day with midnight being hour 0, 1:00 A.M. being hour 1, etc. Using arithmetic in Z_{24}, we see that the paper will run out at time $[11] + [100] = [111] = [15]$. Since hour 15 corresponds to 3:00 P.M., we see that the paper will run out at 3:00 P.M. ■

Example 3.7 Because there are seven days in a week, the day of the week on which a particular date occurs can be determined by using congruence modulo 7. For example, it can be shown that y, the day of the week on which January 1 falls in year x, satisfies

$$y \equiv x + f(x-1) - g(x-1) + h(x-1) \pmod 7,$$

where a Sunday is denoted by $y \equiv 0 \pmod 7$, a Monday by $y \equiv 1 \pmod 7$, a Tuesday by $y \equiv 2 \pmod 7$, etc. Here, $f(x)$, $g(x)$, and $h(x)$ are defined to be the quotients in the division of x by 400, 100, and 4, respectively.

To illustrate the use of this formula, we will determine the day of the week on which April 15, 2010, falls. Note that the quotients in the division of 2009 by 400, 100, and 4 are 5, 20, and 502, respectively, because

$$2009 = 5 \cdot 400 + 9, \quad 2009 = 20 \cdot 100 + 9, \quad \text{and} \quad 2009 = 502 \cdot 4 + 1.$$

Therefore

$$\begin{aligned} y &\equiv x + f(x-1) - g(x-1) + h(x-1) \\ &\equiv 2010 + f(2009) - g(2009) + h(2009) \\ &\equiv 2010 + 5 - 20 + 502 \\ &\equiv 2497 \\ &\equiv 5 \pmod 7, \end{aligned}$$

so January 1, 2010, falls on a Friday.

In the year 2010, which is not a leap year, there are 31 days in January, 28 days in February, and 31 days in March. Thus April 15, 2010, occurs

$$31 + 28 + 31 + (15 - 1) = 104$$

days after January 1, 2010. Now

$$5 + 104 \equiv 109 \equiv 4 \pmod 7,$$

so April 15, 2010, falls on a Thursday. Note that if the year 2010 had been a leap year, April 15 would have fallen on a Friday because February would have contained 29 days in this case. ■

Example 3.8 On a Sharp model EL-506S calculator, the value of 2^{30} is given as 1,073,741,820. If this value is correct, then the last digit of $2^{28} = \frac{2^{30}}{4}$ must be 5. But clearly, no power of 2 can be odd, so the last digit of 2^{30} must be wrong. What is the correct last digit of 2^{30}?

It is easy to see that two positive integers have the same last digit if and only if they are congruent modulo 10. But in Z_{10},

$$[2^{30}] = [2^5]^6 = [32]^6 = [2]^6 = [2^6] = [64] = [4].$$

Hence the last digit of 2^{30} is 4. (Actually, $2^{30} = 1{,}073{,}741{,}824$.) ■

EXERCISES 3.1

In Exercises 1–8, find the quotient and remainder in the division of n by m.

1. $n = 67$ and $m = 9$
2. $n = 39$ and $m = 13$
3. $n = 25$ and $m = 42$
4. $n = 103$ and $m = 8$
5. $n = -54$ and $m = 6$
6. $n = -75$ and $m = 23$
7. $n = -89$ and $m = -10$
8. $n = -57$ and $m = -11$

In Exercises 9–16, determine whether $p \equiv q \pmod{m}$.

9. $p = 29$, $q = -34$, and $m = 7$
10. $p = 47$, $q = 8$, and $m = 11$
11. $p = 96$, $q = 35$, and $m = 10$
12. $p = 21$, $q = 53$, and $m = 8$
13. $p = 39$, $q = -46$, and $m = 2$
14. $p = 75$, $q = -1$, and $m = 19$
15. $p = 91$, $q = 37$, and $m = 9$
16. $p = 83$, $q = -23$, and $m = 6$

In Exercises 17–36, perform the indicated calculations in Z_m. Write your answer in the form $[r]$ with $0 \le r < m$.

17. $[8] + [6]$ in Z_{12}
18. $[9] + [11]$ in Z_{15}
19. $[5] + [10]$ in Z_{11}
20. $[9] + [8]$ in Z_{13}
21. $[23] + [15]$ in Z_8
22. $[12] + [25]$ in Z_7
23. $[16] + [9]$ in Z_6
24. $[43] + [31]$ in Z_{22}
25. $[8][7]$ in Z_6
26. $[9][3]$ in Z_4
27. $[4][11]$ in Z_9
28. $[3][20]$ in Z_{11}
29. $[5][12]$ in Z_8
30. $[8][11]$ in Z_5
31. $[9][6]$ in Z_{10}
32. $[16][3]$ in Z_7
33. $[9]^7$ in Z_7
34. $[11]^8$ in Z_5
35. $[11]^9$ in Z_{12}
36. $[13]^6$ in Z_{15}

37. A newspaper teletypewriter that is in constant operation uses four feet of paper per hour. If a new roll of paper 200 feet long is installed at 6:00 P.M., at what hour of the day will the machine run out of paper?
38. A hospital heart monitoring device uses two feet of paper per hour. If it is attached to a patient at 8:00 A.M. with a supply of paper 150 feet long, at what hour of the day will the device run out of paper?
39. Use Example 3.2 to determine the correct check digit for the ISBN that has 3-540-90518 as its first nine digits.
40. Use Example 3.2 to determine the correct check digit for the ISBN that has 0-553-10310 as its first nine digits.
41. The Universal Product Code (UPC) is a 12-digit number found on products that enables them to be identified by electronic scanning devices. The first six digits identify the country of origin and the manufacturer, the next five digits indicate the product, and the last digit is a check digit. If the first eleven digits of a UPC are $a_1, a_2, \ldots, a_{11}$, then the check digit a_{12} is chosen so that $0 \le a_{12} < 10$ and $3a_1 + a_2 + 3a_3 + a_4 + \cdots + 3a_{11} + a_{12} \equiv 0 \pmod{10}$. Find the correct check digit for the product that has 0 70330 20118 as its first 11 digits.
42. Federal Express packages carry a 10-digit identification number n. Its last digit x is a check digit that equals the remainder in the division of $(n - x)/10$ by 7. Find the last digit of the Federal Express package tracking number with 903786299 as its first nine digits.
43. Use the formula in Example 3.7 to determine the day of the week on which the following dates fall:

    (a) February 12, 2020

    (b) August 8, 2020

44. Use the formula in Example 3.7 to determine all the months in the year 2015 in which the 13th of the month falls on a Friday.

45. Let A denote the equivalence class containing 4 in Z_6, and B denote the equivalence class containing 4 in Z_8. Is $A = B$?

46. In Z_8 which of the following congruence classes are equal: [2], [7], [10], [16], [39], [45], [−1], [−3], [−6], [−17], and [−23]?

47. Let R be the equivalence relation defined in Example 2.13. Give an example to show that it is possible that $p\ R\ x$ and $q\ R\ y$ are both true, yet $(p+q)\ R\ (x+y)$ and $pq\ R\ xy$ are both false. Thus the definitions $[p]+[q]=[p+q]$ and $[p][q]=[pq]$ do not define meaningful operations on the equivalence classes of R.

48. Show that there exist integers m, x, and y such that in Z_m we have $[x] \neq [0]$ and $[y] \neq [0]$, but $[x][y] = [0]$.

49. Let m and n be positive integers such that m divides n. Define a relation R on Z_n by $[x]\ R\ [y]$ in case $x \equiv y \pmod{m}$. Prove that R is an equivalence relation on Z_n. What can be said if m does not divide n?

50. A project has the nine tasks T_1, T_2, T_3, T_4, T_5, T_6, T_7, T_8, and T_9. Task T_i takes i days to complete for $i = 1, 2, \ldots, 9$. If i divides j and $i \neq j$, then task T_j cannot be started until task T_i is completed.

(a) Make a PERT diagram for this project. Include the task symbols $T_1, T_2, \ldots, T_9$ and the times in each circle.

(b) Apply the PERT method to assign to each task the shortest time until it can be completed. What is the shortest time to complete the whole project?

(c) What is the critical path?

51. **(a)** Prove that $x \equiv y \pmod{m}$ if and only if $x = km + y$ for some integer k.

(b) Prove that $x \equiv y \pmod{m}$ if and only if x and y have the same remainder when divided by m.

52. **(a)** Let a, b, and c be integers and $x \equiv y \pmod{m}$. Prove that $ax^2 + bx + c \equiv ay^2 + by + c \pmod{m}$.

(b) Show that the result in part (a) may be false if a, b, and c are not all integers, even if $ax^2 + bx + c$ and $ay^2 + by + c$ are both integers.

3.2 THE EUCLIDEAN ALGORITHM

Recall that, given integers d and n with $d \neq 0$, we say that d divides n if the remainder is zero when n is divided by d; that is, $n = qd$ for some integer q. Thus 3 divides 12, −4 divides 52, and 5 divides −35, but 7 does not divide 51. All integers divide 0, since $0 = 0 \cdot d$ for all d, but any nonzero integer will have a finite set of divisors. For example, the divisors of 12 are

$$-12, -6, -4, -3, -2, -1, 1, 2, 3, 4, 6, \text{ and } 12,$$

while the divisors of 18 are

$$-18, -9, -6, -3, -2, -1, 1, 2, 3, 6, 9, \text{ and } 18.$$

Note that the integers that divide *both* 12 and 18 are -6, -1, 1, and 6.

Given integers m and n, not both zero, we define the **greatest common divisor** of m and n to be the largest integer that divides both m and n, and we denote this by $\gcd(m, n)$. For example, the greatest common divisor of 12 and 18 is 6, and we write this as $\gcd(12, 18) = 6$. Note that if $m \neq 0$, then $\gcd(m, 0) = |m|$.

Computing the greatest common divisor of two integers by listing all their divisors is a very time-consuming process. Fortunately, there is a better way, based on the following theorem:

Theorem 3.3 Let a, b, c, and q be integers with $b > 0$. If $a = qb + c$, then $\gcd(a, b) = \gcd(b, c)$.

Proof. Let $d = \gcd(a, b)$ and $e = \gcd(b, c)$. Then there exist integers k_1, k_2, k_3, k_4 such that $a = dk_1$, $b = dk_2$, $b = ek_3$, and $c = ek_4$. Note that

$$a = qb + c = qek_3 + ek_4 = e(qk_3 + k_4),$$

so e divides a. Thus e is a common divisor of a and b, and so $e \leq d$, the *greatest* common divisor of a and b.

In the same way,

$$c = -qb + a = -qdk_2 + dk_1 = d(-qk_2 + k_1),$$

so d divides c. Thus d is a common divisor of b and c, and so $d \leq e$, the *greatest* common divisor of b and c. We conclude that $d = e$. ∎

To illustrate the usefulness of the preceding theorem, suppose we wish to know the greatest common divisor of 427 and 154. We use the division algorithm to divide 427 by 154:

$$427 = 2 \cdot 154 + 119.$$

Now, by Theorem 3.3, we know that $\gcd(427, 154) = \gcd(154, 119)$, and at least the numbers are smaller. But why stop here? We have

$$154 = 1 \cdot 119 + 35,$$

so $\gcd(154, 119) = \gcd(119, 35)$. Continuing this way, we have

$$\begin{aligned} 119 &= 3 \cdot 35 + 14, \\ 35 &= 2 \cdot 14 + 7, \\ 14 &= 2 \cdot 7 + 0. \end{aligned}$$

Thus we have $\gcd(427, 154) = \gcd(154, 119) = \gcd(119, 35) = \gcd(35, 14) = \gcd(14, 7) = \gcd(7, 0) = 7$.

Example 3.9 We will find the greatest common divisor of 804 and 654. We apply the division algorithm to 804 and 654, then to 654 and the remainder just computed, etc., as follows:

$$\begin{aligned}
804 &= 1 \cdot 654 + 150, && 0 \le 150 < 654,\\
654 &= 4 \cdot 150 + 54, && 0 \le 54 < 150,\\
150 &= 2 \cdot 54 + 42, && 0 \le 42 < 54,\\
54 &= 1 \cdot 42 + 12, && 0 \le 12 < 42,\\
42 &= 3 \cdot 12 + 6, && 0 \le 6 < 12,\\
12 &= 2 \cdot 6 + 0.
\end{aligned}$$

Then, as before, $\gcd(804, 654) = \gcd(654, 150) = \ldots = \gcd(6, 0) = 6$. ■

Notice that this process must produce a 0 remainder eventually because the remainders ($150 > 54 > 42 > 12 > 6 > 0$ in the last example) are nonnegative and decreasing.

This method of finding the greatest common divisor of two integers, which was known to the Greek geometer Euclid, is quite efficient. Note that m and $-m$ have the same set of divisors for any integer m. Thus, in computing $\gcd(m, n)$, we can assume that m and n are nonnegative, since $\gcd(m, n) = \gcd(|m|, |n|)$.

The Euclidean Algorithm

Given nonnegative integers m and n that are not both zero, this algorithm computes $\gcd(m, n)$.

Step 1 (initialization)
Set $r_{-1} = m$, $r_0 = n$, and $i = 0$.

Step 2 (apply division algorithm)
while $r_i \neq 0$
(a) Replace i with $i + 1$.
(b) Determine the quotient q_i and remainder r_i in the division of r_{i-2} by r_{i-1}.
endwhile

Step 3 (output greatest common divisor)
Print r_{i-1}.

The following table shows the working of the algorithm for $m = 804$ and $n = 654$:

i	r_i
−1	804
0	654
1	150
2	54
3	42
4	12
5	6
6	0
[Print 6]	

The Efficiency of the Euclidean Algorithm

Estimating the efficiency of the Euclidean algorithm, as we did for the algorithms in Section 1.4, is complicated by the fact that we do not know how many times the division algorithm will be used. In 1844, the French mathematician Gabriel Lamé proved that the number of such divisions is no more than five times the number of decimal digits in the smaller of m and n. Historically, this is the first instance of a complexity proof. We will prove a slightly weaker version of this theorem. We can arrange the integers m and n so that $m \geq n$.

Theorem 3.4 If the Euclidean algorithm is applied to the integers m and n with $m \geq n > 0$, then the number of divisions needed is no more than $2\log_2(n+1)$.

Proof. Let the division algorithm be applied k times until the remainder 0 is obtained, and let q_i, $1 \leq i \leq k$, and r_i, $-1 \leq i \leq k$, be as in the Euclidean algorithm. Then $r_{-1} \geq r_0 > r_1 > r_2 > \cdots > r_{k-1} > r_k = 0$. The ith division looks like $r_{i-2} = q_i r_{i-1} + r_i$. Since $r_{i-2} > r_i$, we have $q_i \geq 1$. Then

$$r_{i-2} = q_i r_{i-1} + r_i \geq r_{i-1} + r_i > r_i + r_i = 2r_i.$$

Thus $r_{i-2} \geq 2r_i + 1$, and $r_{i-2} + 1 \geq 2(r_i + 1)$. In particular,

$$n + 1 = r_0 + 1 \geq 2(r_2 + 1) \geq 2(2(r_4 + 1)) = 2^2(r_4 + 1) \geq \cdots \geq 2^i(r_{2i} + 1)$$

whenever $2i \leq k$.

Now define the integers s and t by $k - 1 = 2s + t, 0 \leq t < 2$. Then

$$n + 1 \geq 2^s(r_{2s} + 1) \geq 2^s(r_{2s+t} + 1) = 2^s(r_{k-1} + 1) \geq 2^{s+1},$$

because $r_{k-1} \geq 1$. Since the function $y = \log_2 x$ is increasing (see Section 2.4), we have $\log_2(n+1) \geq \log_2 2^{s+1} = s + 1$, so

$$2\log_2(n+1) \geq 2s + 2 \geq 2s + t + 1 = k,$$

as desired. ▪

The Extended Euclidean Algorithm

It is easy to see that if m, n, and d are integers such that d divides m and d divides n, then d divides $mx + ny$ for all integers x and y. In particular, $\gcd(m, n)$ divides $mx + ny$ for all integers x and y. It is a useful fact that the converse is also true, in the sense that whenever $\gcd(m, n)$ divides d, then there exist integers x and y such that $mx + ny = d$. In fact, a straightforward extension of the Euclidean algorithm produces integers x and y so that $mx + ny$ is precisely $\gcd(m, n)$.

The Extended Euclidean Algorithm

Given nonnegative integers m and n, not both zero, this algorithm computes $\gcd(m, n)$ and also integers x and y such that $mx + ny = \gcd(m, n)$.

Step 1 (initialization)
Set $r_{-1} = m$, $x_{-1} = 1$, $y_{-1} = 0$, $r_0 = n$, $x_0 = 0$, $y_0 = 1$, and $i = 0$.

Step 2 (apply division algorithm)
while $r_i \neq 0$
(a) Replace i with $i + 1$.
(b) Determine the quotient q_i and remainder r_i in the division of r_{i-2} by r_{i-1}.
(c) Set $x_i = x_{i-2} - q_i x_{i-1}$ and $y_i = y_{i-2} - q_i y_{i-1}$.
endwhile

Step 3 (output $\gcd(m, n)$, x, and y)
Print r_{i-1}, x_{i-1}, and y_{i-1}.

To apply this algorithm to $m = 66$ and $n = 51$, we set $r_{-1} = 66$, $x_{-1} = 1$, $y_{-1} = 0$, $r_0 = 51$, $x_0 = 0$, $y_0 = 1$, and $i = 0$. Since $r_0 \neq 0$, we set $i = 1$ and write $r_{-1} = q_1 r_0 + r_1$, $0 \leq r_1 < r_0$, that is, $66 = q_1 \cdot 51 + r_1$, $0 \leq r_1 < 51$. Therefore $q_1 = 1$ and $r_1 = 15$. Then $x_1 = x_{-1} - q_1 x_0 = 1 - 1 \cdot 0 = 1$, and $y_1 = y_{-1} - q_1 y_0 = 0 - 1 \cdot 1 = -1$.

Since $r_1 \neq 0$, we run through the while-loop again. The following table shows the whole calculation:

i	q_i	r_i	x_i	y_i
−1		66	1	0
0		51	0	1
1	1	15	1	−1
2	3	6	−3	4
3	2	3	7	−9
4	2	0	−17	22
[Print 3, 7, −9]				

Notice that $\gcd(66, 51) = 3 = 66 \cdot 7 + 51(-9)$. Since the sequence $r_1, r_2, \ldots$ is the same as in the original Euclidean algorithm, it is clear that the extended Euclidean algorithm also computes the greatest common divisor of m and n as r_{i-1} when r_i becomes 0. We leave the verification that then $r_{i-1} = mx_{i-1} + ny_{i-1}$ for the exercises.

Note that in the extended Euclidean algorithm each repetition of step 2 when $r_i \neq 0$ entails one comparison, one addition, one division, two subtractions, and two multiplications. Thus, if $r_k = 0$, the algorithm has $7k + 1$ elementary operations.

Example 3.10 Find, if possible, integers x and y such that

(a) $539x + 396y = 154$,

(b) $539x + 396y = 254$.

Applying the extended Euclidean algorithm to $m = 539$ and $n = 396$ gives the following table:

i	q_i	r_i	x_i	y_i
−1		539	1	0
0		396	0	1
1	1	143	1	−1
2	2	110	−2	3
3	1	33	3	−4
4	3	11	−11	15
5	3	0	36	49
[Print 11, −11, 15]				

We conclude that $\gcd(539, 396) = 11 = 539(-11) + 396 \cdot 15$.

What we want in (a), however, are integers x and y such that $539x + 396y = 154$. But since $154 = 11 \cdot 14$, these are easily found. We need only multiply the equation $539(-11) + 396 \cdot 15 = 11$ by 14, producing

$$539(14(-11)) + 396(14 \cdot 15) = 539(-154) + 396 \cdot 210 = 154.$$

Thus we can take $x = -154$ and $y = 210$.

On the other hand, a similar technique doesn't work in (b), because 254 is not divisible by 11. In fact, no solution is possible, because if there were integers x and y such that $539x + 396y = 254$, then 11, which divides both 539 and 396, would divide 254. ■

EXERCISES 3.2

1. List, in increasing order, the divisors of 45.

2. List, in increasing order, the divisors of 54.

3. List, in increasing order, the common divisors of 40 and 30.
4. List, in increasing order, the common divisors of 48 and 72.

In Exercises 5–10, make a table such as the one following the Euclidean algorithm for the given values of m and n.

5. $m = 715, n = 312$
6. $m = 341, n = 217$
7. $m = 247, n = 117$
8. $m = 451, n = 143$
9. $m = 76, n = 123$
10. $m = 89, n = 55$

11. Suppose the Euclidean algorithm is applied to m and n, where $m \geq n > 0$ and $n < 1000000$. What is the maximum number of divisions possible according to Theorem 3.4?
12. In Exercise 11, what is the maximum number of divisions possible according to Lamé's theorem?

In Exercises 13–18, make a table such as the one following the extended Euclidean algorithm for the given values of m and n.

13. $m = 1479, n = 272$
14. $m = 2030, n = 899$
15. $m = 4050, n = 1728$
16. $m = 231, n = 182$
17. $m = 546, n = 2022$
18. $m = 555, n = 2146$

In Exercises 19–22, use the Euclidean algorithm to decide whether each equation is solvable in integers x and y.

19. (a) $414x + 594y = 492$
    (b) $414x + 594y = 558$
20. (a) $637x + 259y = 357$
    (b) $637x + 259y = 408$
21. (a) $396x + 312y = 222$
    (b) $396x + 312y = 228$
22. (a) $638x + 165y = 451$
    (b) $638x + 165y = 583$

In Exercises 23–26, use the extended Euclidean algorithm to find, if possible, integers x and y satisfying the given equation.

23. $3157x + 656y = 2173$
24. $216x + 153y = 171$
25. $455x - 169y = 1157$
26. $1054x - 833y = 2277$

27. Prove that if a, b, and c are integers such that a divides b and b divides c, then a divides c.
28. Prove that if d, m, and n are integers, d divides m, and d divides n, then d divides $mx + ny$ for all integers x and y.
29. Let d and m be integers. Prove that d divides m if and only if d divides $-m$.
30. Prove that if the Euclidean algorithm is applied to integers m and n, not both 0, then the last nonzero remainder is $\gcd(m, n)$.

In Exercises 31–33, assume that the extended Euclidean algorithm is applied to integers m and n, not both 0, and that the last nonzero remainder is r_{k-1}.

31. Prove that $mx_i + ny_i = r_i$ for $i = -1$ and $i = 0$.
32. Show that if $j \leq k$ and $mx_i + ny_i = r_i$ for $i = j - 2$ and $i = j - 1$, then $mx_j + ny_j = r_j$.
33. Use the strong principle of mathematical induction to prove $mx_i + ny_i = r_i$ for $-1 \leq i \leq k$.

3.3 THE RSA METHOD

Two third-grade students might communicate secretly by using a code that replaces each letter in a message with a different letter. For example, they might replace each letter with the one just after it in the alphabet, with `Z` being replaced by `A`. With this system, `I HAVE A COOKIE` (called the *plaintext*) would become `J IBWF B DPPLJF` (the *ciphertext*). Disguising a message is called *enciphering*, and retrieving the original is called *deciphering*. Deciphering would be done by going backwards in the alphabet, so that `J XBOU B CJUF` would be deciphered to `I WANT A BITE`. Although this system is crude, it might be sufficient to baffle other third-graders who don't know the rules for enciphering and deciphering.

From the beginning of writing until about 30 years ago, all enciphering systems worked along the same general lines as just described. The plaintext was somehow changed to make it unreadable to someone not knowing the system, sent, and then changed back by reversing the original transformation. The rules of enciphering and deciphering, which came to involve complicated mathematical functions and were implemented with computers, had to be prearranged between sender and receiver. The system depended on a *key*, known only to the users of the code, for example, using the next letter of the alphabet. This key had to be kept secret, because if one knew how to encipher, one could easily figure out how to decipher and read the enciphered message.

You may have had the experience of buying something on the Internet and being told that you may safely input your credit card number because it will be transmitted in enciphered form. But how can you communicate secretly with someone if no method of enciphering and deciphering has been prearranged? The answer is to use a *public-key* system, one that allows anyone to encipher, but only someone knowing the system to decipher. This is possible because certain operations are easy to perform, but hard to reverse. Think of how easy it is to pull out of a parking space between two cars along a street, while backing into the same space requires much more skill.

In the method we will describe, the easy operation is multiplying together two large primes, and the hard reversal is factoring the resulting product to find what the two numbers are. To get an idea of how different the difficulties involved are, start by multiplying together, without using a calculator, the primes 71 and 59. Pretty easy, right? You should get 4189. Now figure out what two primes have the product 4161. Use a calculator to do the divisions if you want. It is a much harder problem.

Modular Exponentiation

The public-key method we will describe is called the RSA method, named for its inventors, R.L. **R**ivest, A. **S**hamir, and L. **A**dleman. In it, the message is first converted to a sequence of numbers in some standard way. As an example, we could use the following simple conversion:

Symbol	*Number*	*Symbol*	*Number*	*Symbol*	*Number*	*Symbol*	*Number*
space	00	G	07	N	14	U	21
A	01	H	08	O	15	V	22
B	02	I	09	P	16	W	23
C	03	J	10	Q	17	X	24
D	04	K	11	R	18	Y	25
E	05	L	12	S	19	Z	26
F	06	M	13	T	20		

For example, `I LOVE YOU` would be written as

I		L	O	V	E		Y	O	U
09	00	12	15	22	05	00	25	15	21

For the purposes of this example, we will group these digits into sets of 3, so that the message would be written as

$$090 \quad 012 \quad 152 \quad 205 \quad 002 \quad 515 \quad 210.$$

(An extra 0 was appended at the end to complete the last group.)

This transformation of a text message to numbers is not enciphering, but merely a way, known to all, to allow it to be treated mathematically. How to reverse the process is clear; one simply goes back to groups of two digits and uses the table in reverse. Thus

$$041 \quad 815 \quad 160 \quad 004 \quad 050 \quad 104$$

is regrouped as

$$04 \quad 18 \quad 15 \quad 16 \quad 00 \quad 04 \quad 05 \quad 01 \quad 04,$$

which deciphers as `DROP DEAD`.

In the RSA method, the actual enciphering consists of *modular exponentiation*, which can be thought of as raising to a power E (E for Enciphering) in Z_n. Thus, if the plaintext is $P_1, P_2, P_3, \ldots$, then the ciphertext is $C_1, C_2, C_3, \ldots$, where, for each i, we have

$$C_i \equiv P_i^E \pmod{n} \qquad \text{and} \qquad 0 \le C_i < n.$$

Example 3.11 Suppose that $n = 33$, $E = 3$, and the plaintext is 8, 7, 20, 3, 11, 13. Now $8^3 = 512$, and when 512 is divided by 33, the remainder is 17. Thus the ciphertext corresponding to 8 is 17. Likewise, $7^3 = 343 \equiv 13 \pmod{33}$, so 7 is enciphered as 13. Check that the entire enciphered message is 17, 13, 14, 27, 11, 19. ■

The preceding example used very small numbers to illustrate the idea of modular exponentiation. Notice that if $n = 33$, there are only 33 possibilities $(0, 1, 2, \ldots, 32)$ for plaintext numbers. This is not even enough for the lower and uppercase letters. For a more realistic example, let us take $n = 1189$. This allows three-digit numbers in our plaintext sequence, such as the 90, 12, 152, 205, 2, 515, 210 that corresponded to `I LOVE YOU`. We will take $E = 101$. Thus, to encipher, we need to compute the remainders when each of

$$90^{101}, \quad 12^{101}, \quad 152^{101}, \quad 205^{101}, \quad 2^{101}, \quad 515^{101}, \quad \text{and} \quad 210^{101}$$

is divided by 1189.

There is a problem. A hand calculator in floating point mode reveals that 90^{101} is approximately 2.39×10^{197}; that is, the integer 90^{101} has 198 digits. Even if we can compute this number and divide it by 1189, this entails an unnecessary amount of computation. Since all we need is the remainder, which will have at most four digits, it seems excessive to have to compute 198 digits. Furthermore, when the RSA method is implemented in the real world, security requires that the numbers be much larger than this example. For example, instead of $n = 1189$, n would be an integer of perhaps 400 decimal digits, and the powers of plaintext numbers would be too large for any computer to calculate with.

All is not lost, however, since, going back to our $n = 1189$ example, we can break the exponentiation up so as never to deal with numbers beyond exact manipulation with our hand calculator. For example, we can calculate 90^{101} in Z_{1189} a little at a time as follows:

$$\begin{aligned}
90^2 &= 90 \cdot 90 && = 8100 \equiv 966 \pmod{1189} \\
90^3 &= 90^2 \cdot 90 \equiv 966 \cdot 90 && = 86940 \equiv 143 \pmod{1189} \\
90^4 &= 90^3 \cdot 90 \equiv 143 \cdot 90 && = 12870 \equiv 980 \pmod{1189} \\
&\vdots
\end{aligned}$$

Although it would be tedious, we could continue this way, computing the remainders modulo 1189 of $90^4, 90^5, \ldots, 90^{101}$. This we will not do, since there is an even better way, based on squaring. Note that (all congruences are modulo 1189)

$$\begin{aligned}
90^2 &= && = 8100 \equiv 966 \\
90^4 &= (90^2)^2 &\equiv 966^2 &= 933156 \equiv 980 \\
90^8 &= (90^4)^2 &\equiv 980^2 &= 960400 \equiv 877 \\
90^{16} &= (90^8)^2 &\equiv 877^2 &= 769129 \equiv 1035 \\
90^{32} &= (90^{16})^2 &\equiv 1035^2 &= 1071225 \equiv 1125 \\
90^{64} &= (90^{32})^2 &\equiv 1125^2 &= 1265625 \equiv 529.
\end{aligned}$$

Now, $101 = 1 + 4 + 32 + 64$, so

$$\begin{aligned}
90^{101} &= 90^{1+4+32+64} = 90^1 \cdot 90^4 \cdot 90^{32} \cdot 90^{64} \equiv 90 \cdot 980 \cdot 1125 \cdot 529 \\
&= 88200 \cdot 1125 \cdot 529 \equiv 214 \cdot 1125 \cdot 529 = 240750 \cdot 529 \equiv 572 \cdot 529 \\
&= 302588 \equiv 582,
\end{aligned}$$

where again the congruences are modulo 1189.

Notice that here only six squarings are needed (with reduction modulo 1189), plus four more multiplications and reductions, to compute the final answer. This is much better than the 100 multiplications and divisions that our previous idea would have entailed. Our new method depends on the fact that every positive integer can be written as the sum of distinct powers of 2 (that is, has a binary representation), just as 101 can be written as $1 + 4 + 32 + 64$. It is embodied in the next algorithm, which performs the multiplication of the various squares as they are computed, instead of at the end as in our example.

The Modular Exponentiation Algorithm

Given positive integers P, E, and n, this algorithm computes the remainder when P^E is divided by n.

Step 1 (initialization)
Set $r_2 = 1$, $p = P$, and $e = E$.
Step 2 **while** $e \neq 0$
  Step 2.1 (determine parity of e)
    Determine the quotient Q and remainder R in the division of e by 2.
  Step 2.2 (square and reduce)
    Determine the remainder r_1 in the division of p^2 by n.
  Step 2.3 (multiply by new square if necessary)
    if $R = 1$
      Replace r_2 by the remainder in the division of $r_2 p$ by n.
    endif
  Step 2.4 (update variables)
    Replace p with r_1 and e with Q.
  endwhile
Step 3 (output remainder of P^E)
Print r_2.

Example 3.12 Suppose we want the remainder when 7^{11} is divided by 17, so that $P = 7$, $E = 11$, and $n = 17$. We set $r_2 = 1$, $p = 7$, and $e = 11$, and note in step 2.1 that $11 = 5 \cdot 2 + 1$; so $Q = 5$ and $R = 1$. In step 2.2, from $7^2 = 49 = 2 \cdot 17 + 15$, we see that $r_1 = 15$. Since $R = 1$ in step 2.3, from $7 \cdot 1 = 0 \cdot 17 + 7$, we conclude that $r_2 = 7$. Thus the updated values of p and e are 15 and 5, respectively. Continuing in this way gives the following table:

Q	R	r_1	r_2	p	e
			1	7	11
5	1	15	7	15	5
2	1	4	3	4	2
1	0	16		16	1
0	1	1	14	1	0
[Print 14]					

In the fourth line, notice that r_2 does not change, because $R = 0$ at this point. ∎

Let e_i, Q_i, and R_i be the values of e, Q, and R in the ith iteration of step 2 of this algorithm. Then $e_{i-1} = 2Q_i + R_i$, where R_i is 0 or 1, and so $e_{i-1} \geq 2Q_i = 2e_i$. Suppose e is first 0 in the kth full line, so that $e_{k-1} = 1$. Then

$$E = e_1 \geq 2e_2 \geq 2(2e_3) = 2^2 e_3 \geq 2^3 e_4 \geq \cdots \geq 2^{k-2} e_{k-1} = 2^{k-2}.$$

Taking the base 2 logarithm of both sides gives $\log_2 E \geq \log_2 2^{k-2} = k - 2$; so $k - 1 \leq \log_2 E + 1$.

Note that step 2.1 contains one division, step 2.2 contains one multiplication and one division, and step 2.3 contains one comparison and at most one multiplication and one division. Thus step 2 contains the comparison of e and 0, and, if $e \neq 0$, at most six more elementary operations. The complexity of the algorithm is thus at most $7(k-1)+1$, since only the comparison of e and 0 is done when $e = 0$. By the previous paragraph, this is no more than $7(\log_2 E + 1) + 1 = 7 \log_2 E + 8$.

The algorithm is very efficient. For example, even if E were a 400-digit number, so that $E < 10^{400}$, the number of elementary operations needed to perform the calculation would be less than $7 \log_2 10^{400} + 8 < 9310$. A computer could do this in less than one second.

Deciphering with the RSA Method

In the RSA method, we have seen that enciphering consists of modular exponentiation. So does deciphering, but with a different exponent. Of course, the new exponent must be carefully chosen.

Recall that in the RSA method the modulus n is chosen to be the product of prime numbers. Suppose $n = pq$, where p and q are distinct primes. In practice, p and q will be quite large—say, around 200 digits each—making n about 400 digits. First of all, the *E*nciphering exponent E is chosen so that $\gcd(E, b) = 1$, where $b = (p-1)(q-1)$. Such an E is easily found, since most odd integers will satisfy this condition. Thus we could choose a positive odd integer $E < n$ at random and use the Euclidean algorithm to check that $\gcd(E, b) = 1$. By Lamé's theorem, the number of divisions needed in the algorithm will be no more than $5 \cdot 400 = 2000$, and so this calculation is easily done by a computer.

The exponent D used for Deciphering is chosen as the smallest positive solution x to the congruence

$$Ex \equiv 1 \pmod{b}.$$

This congruence is always solvable, and in fact we already have an efficient method of finding a solution. Since $\gcd(E, b) = 1$, we can use the extended Euclidean algorithm to find integers x and y such that $Ex + by = 1$. It then follows that $Ex - 1 = -by \equiv 0 \pmod{b}$, and so x is a solution.

Notice that if $x' \equiv x \pmod{b}$, then x' will also be a solution, because

$$Ex' \equiv Ex \equiv 1 \pmod{b}$$

by Theorem 3.2. The solution x produced by the extended Euclidean algorithm may not be the smallest positive solution, but then its remainder when divided by b is.

To illustrate, let us return to Example 3.11. There we were enciphering with $E = 3$ and $n = 33 = 3 \cdot 11$. Thus $b = (3-1)(11-1) = 20$, and we need to solve the congruence

$$3x \equiv 1 \pmod{20}.$$

The numbers are so small that the smallest positive solution is almost obvious: $x = 7$. Our ciphertext was 17, 13, 14, 27, 11, 19, and we decipher by raising these numbers to the power $D = 7$ and finding their remainders when divided by 33. For example,

$$17^7 = 410338673 = 12434505 \cdot 33 + 8,$$

so the corresponding plaintext is 8. Likewise, $13^7 = 62748517 \equiv 7 \pmod{33}$, and so the next plaintext number is 7. Check that the whole deciphered sequence is 8, 7, 20, 3, 11, 13, which is the original plaintext.

Note that E and b are both less than n, so, by Lamé's theorem, the number k of divisions when the Euclidean algorithm is applied to E and b will be less than 5 times the number of decimal digits in n. We also saw that the complexity of the extended Euclidean algorithm is no more than $7k + 1$. Thus, if n is a 400-digit integer, we will have $k \leq 5 \cdot 400 = 2000$, and there will be no more than $7 \cdot 2000 + 1 = 14001$ elementary operations.

A larger example is to find D if $n = 1189$ and $E = 101$, as used previously. We need to find the smallest positive solution to the congruence $101x \equiv 1 \pmod{b}$, where $b = (p-1)(q-1)$. But what are p and q? We need to factor 1189. For a small number like 1189, the easiest way is to divide by the prime numbers in order until we find one that divides 1189 with remainder 0. Since 1189 is not even, 2 will not work; nor will 3 work, since $1189/3 \approx 396.3$. We need not try the nonprime $4 = 2 \cdot 2$, since, if 4 divided 1189, we would already have found that 2 did. Likewise, $1189/5 = 237.8$; so 5 doesn't work; nor does 7, since $1189/7 \approx 169.9$.

We might worry that 1189 is actually prime. How far must we carry this process until we know? The next theorem tells us.

Theorem 3.5 If the integer $n > 1$ is not prime, then n has a prime factor no larger than $\sqrt{n}$.

Proof. Suppose every prime divisor of n exceeds $\sqrt{n}$. Since n is not prime, we have $n = rs$, where r and s are integers greater than 1. Let r and s have prime divisors p and q, respectively. Then

$$n = rs \geq pq > \sqrt{n} \cdot \sqrt{n} = n,$$

which is a contradiction. ■

According to Theorem 3.5, we need only divide 1189 by primes less than $\sqrt{1189} \approx 34.5$. Actually, we find $1189/29 = 41$, so $1189 = 29 \cdot 41$.

Now we return to finding the deciphering exponent D when $n = 1189$ and $E = 101$. We can take $p = 29$ and $q = 41$, so $b = (29 - 1)(41 - 1) = 1120$. We need to solve the congruence

$$101x \equiv 1 \pmod{1120}.$$

Applying the extended Euclidean algorithm with $m = 101$ and $n = 1120$ produces the following table:

i	q_i	r_i	x_i	y_i
-1		101	1	0
0		1120	0	1
1	0	101	1	0
2	11	9	-11	1
3	11	2	122	-11
4	4	1	-499	45
5	2	0	1120	-101

[Print 1, −499, 45]

We check that $-499 \cdot 101 + 45 \cdot 1120 = 1$. Thus $101(-499) \equiv 1 \pmod{1120}$, and $x = -499$ is a solution to our congruence. The smallest positive solution will be the remainder when x is divided by 1120. But $-499 = (-1)1120 + 621$, $0 \leq 621 < 1120$, so $D = 621$.

Example 3.13 We previously found that the plaintext 90 corresponded to the ciphertext 582 when $n = 1189$ and $E = 101$. To decipher 582, we raise it to the power $D = 621$ and take the remainder modulo 1189. Using the modular exponentiation algorithm produces the following table:

Q	R	r_1	r_2	p	e
			1	582	621
310	1	1048	582	1048	310
155	0	857		857	155
77	1	836	583	836	77
38	1	953	1087	953	38
19	0	1002		1002	19
9	1	488	50	488	9
4	1	344	620	344	4
2	0	625		625	2
1	0	633		633	1
0	1	1185	90	1185	0
[Print 90]					

This is correct; 90 was the original plaintext. ■

There is an important lesson with the preceding computations. *In order to find the deciphering exponent* $D = 621$, *we had to be able to factor* $n = 1189$, *because the modulus of the congruence determining* D *is* b, *which is defined in terms of the primes* p *and* q.

Why the RSA Method Works

The first question to be answered is why raising the ciphertext numbers $C_1, C_2, \ldots$ to the power D previously defined produces the original plaintext numbers P_1, $P_2, \ldots$. Proving this entails a certain amount of elementary number theory, and we leave the details to Exercises 25–37 at the end of this section.

The security of the RSA method is another concern. The protocol of the RSA method is as follows: Suppose Adam has a store selling fancy candies and wants online buyers to be able to order from him securely. He chooses distinct large primes p and q, sets $n = pq$, and chooses an exponent E with $0 < E < n$ and $\gcd(E, b) = 1$, where $b = (p-1)(q-1)$. Then Adam makes the numbers n and E available on his website, keeping p and q secret. He also secretly computes the deciphering exponent D as we have explained.

Betty wants to buy various European licorice candies from Adam's website. Her computer translates her order and information, including her credit card number, to numbers $P_1, P_2, \ldots$ in some standard way, then uses modular exponentiation with exponent E and modulus n to produce the ciphertext sequence $C_1, C_2, \ldots$, which is sent over the Internet. Adam can retrieve her original message by applying modular exponentiation with exponent D.

We have seen previously that all these computations can be done very quickly by computers, even if p and q are quite large—say, about 200 digits each, making n about 400 digits.

The reason some third party, say, the nefarious Carl, cannot also compute D and steal Betty's credit card number is that he does not know p and q and thus cannot compute b and then D. But why doesn't Carl simply find p and q by factoring n?

The reason is that n is a 400-digit number, and factoring a 400-digit number is presently beyond the capability of even the fastest computer. According to Theorem 3.5, we need only divide n by all primes no larger than its square root, which is about 10^{200}. This is actually practical only for small values of n, since, as the possible divisors get larger, it takes too much time to decide whether they are prime or not. It is actually more efficient to try all odd divisors up to $\sqrt{n}$, of which there are $10^{200}/2$.

Suppose we had a computer which could perform one billion (that is, 10^9) divisions per second. Having it make $10^{200}/2$ divisions would take approximately 3.17×10^{183} years. Of course, by that time, the sun will be a cold cinder and Betty's credit card account long closed.

Number theorists have devised better methods for factoring large numbers than simply dividing by all odd integers up to the square root. None are really efficient, however, in the way that the Euclidean and modular exponentiation algorithms are. Even the best would take thousands of years to factor a 400-digit integer.

EXERCISES 3.3

In Exercises 1–4, change the given plaintext message to ciphertext, using the given values of E and n.

1. $E = 5, n = 35$, message $2, 5, 11, 8$

2. $E = 5, n = 21$, message $19, 3, 14$

3. $E = 3, n = 55$, message $40, 31, 9$

4. $E = 3, n = 51$, message $40, 31, 9$

In Exercises 5–10, apply the modular exponentiation algorithm, making a table as in Example 3.12.

5. $P = 19, E = 41, n = 91$

6. $P = 30, E = 29, n = 51$

7. $P = 11, E = 73, n = 187$

8. $P = 7, E = 53, n = 123$

9. $P = 90, E = 101, n = 1189$ (Note: The computations have all been done in the text following Example 3.11.)

10. $P = 12, E = 101, n = 1189$

In Exercises 11–14, find b corresponding to the given value of n, where b and n are as in the RSA method.

11. $n = 85$

12. $n = 143$

13. $n = 323$

14. $n = 299$

In Exercises 15–22, use the extended Euclidean algorithm to find the value of D corresponding to the constants in the given exercise.

15. Exercise 1

16. Exercise 2

17. Exercise 3

18. Exercise 4

19. Exercise 5

20. Exercise 6

21. Exercise 7

22. Exercise 8

23. Suppose $n = 55$, $E = 7$, and the ciphertext message is $C = 2$. Find the corresponding plaintext P.

24. Suppose $n = 93$, $E = 17$, and the ciphertext message is $C = 2$. Find the corresponding plaintext P.

25. Suppose a, b, and c are integers such that $\gcd(a, b) = 1$ and a divides bc. Prove that a divides c. (*Hint:* Use Exercises 30 and 33 of Section 3.2 to write $ax + by = 1$, and multiply by c.)

26. Prove that if p is a prime integer and p does not divide the integer a, then $\gcd(p, a) = 1$.

27. Prove that if p is a prime integer and a and b are integers such that p divides ab, then p divides a or p divides b.

28. Prove that if p and q are distinct prime integers such that p divides a and q divides a, then pq divides a. (*Hint:* Use Exercise 27.)

29. Prove that if a, b, x, and x' are integers such that $\gcd(a, b) = 1$ and $ax \equiv ax' \pmod{b}$, then $x \equiv x' \pmod{b}$.

30. Give an example to show that the conclusion of the previous exercise may not be true if the condition $\gcd(a, b) = 1$ does not hold.

31. Let p be a prime integer and a be an integer such that p does not divide a. Prove that the integers $a, 2a, 3a, \ldots, (p-1)a$ are congruent modulo p to the integers $1, 2, 3, \ldots, p-1$, in some order.

32. Prove *Fermat's theorem*: If p is a prime integer and a is an integer such that p does not divide a, then $a^{p-1} \equiv 1 \pmod{p}$. (*Hint:* Show that

$$a \cdot 2a \cdot 3a \cdots (p-1)a \equiv 1 \cdot 2 \cdot 3 \cdots (p-1) \pmod{p},$$

and use Exercise 29 to cancel $2, 3, \ldots, p-1$ from both sides.)

In Exercises 33–35, assume that p and q are distinct prime integers, $n = pq$, $b = (p-1)(q-1)$, E and D are positive integers with $ED \equiv 1 \pmod{b}$, and P and C are integers with $0 \le P < n$, $0 \le C < n$, and $C \equiv P^E \pmod{n}$. Write $ED = kb + 1$.

33. Prove that $C^D \equiv P \cdot P^{kb} \pmod{n}$.

34. Assume that p does not divide P. Use Fermat's theorem to show that $C^D \equiv P \pmod{p}$.

35. Assume that p divides P. Show that $C^D \equiv P \pmod{p}$.

36. Show that $C^D \equiv P \pmod{n}$. (*Hint:* Exercises 34 and 35 are also true if p is replaced by q.)

37. Prove that the remainder when C^D is divided by n is P.

3.4 ERROR-DETECTING AND ERROR-CORRECTING CODES

In Example 3.2 of Section 3.1, we discussed the International Standard Book Number (ISBN). Recall, for example, that the ISBN for this book is 0-321-30515-9 and that the last digit x is chosen so that

$$10(0) + 9(3) + 8(2) + 7(1) + 6(3) + 5(0) + 4(5) + 3(1) + 2(5) + x = 101 + x$$

is congruent to 0 modulo 11. Suppose that a bookstore wanting to order this book incorrectly orders copies of ISBN 0-321-35015-9 instead. By performing the calculation

$$10(0) + 9(3) + 8(2) + 7(1) + 6(3) + 5(5) + 4(0) + 3(1) + 2(5) + 9 = 115,$$

the publisher would know that an error has occurred because the number 115 is not congruent to 0 modulo 11. Thus the International Standard Book Number provides a means of detecting the occurrence of certain errors.

In Sections 3.4–3.6, we discuss algebraic coding theory, an area of applied mathematics that has made possible the development of devices such as compact disc players and fax machines, as well as the accurate transmission of data to and from spacecraft. This subject originated in the work of Claude Shannon, Marcel Golay, and Richard Hamming during the years 1948–1950.

For simplicity, we consider only codes composed of sequences of binary digits, that is, 0s and 1s. Such codes are called **binary codes**. This is the form in which information is stored in the memory of a computer, written on a compact disc or DVD, or transmitted between computers electronically. Blocks of the digits 0 and 1 may represent letters, numbers, punctuation marks, or mathematical symbols. One such representation scheme is the ASCII (American Standard Code for Information Interchange) code, in which letters, numerals, and certain symbols are represented by blocks of eight binary digits. For example, the blocks

$$01000001, \qquad 01100010, \qquad \text{and} \qquad 00110111$$

represent the capital letter A, the lower-case letter b, and the numeral 7, respectively. By combining blocks, we can transmit words or other information. For instance, the message BAT would be transmitted as the sequence

$$01000010 \quad 01000001 \quad 01010100.$$

(For legibility, we will display long sequences of binary digits with spaces. The preceding message would actually be transmitted as a sequence of 24 binary digits with no intervening spaces.) Unfortunately, when a message is sent, errors may occur that make the message received different from the one that was sent. For example, if an error occurs in the eighth digit, the received message is

$$01000011 \qquad 01000001 \qquad 01010100,$$

which would be decoded as CAT instead of BAT. By introducing some redundancy into the message, it is possible to detect, or even correct, such transmission errors.

Example 3.14 A simple scheme for detecting some errors is to append a **parity check digit** to each block of the message. For example, we might append to each eight-digit block a ninth digit chosen so that the total number of 1s in the block is even. Thus, for the message BAT, which is encoded as

$$01000010 \qquad 01000001 \qquad 01010100,$$

we append 0s to the first two blocks and 1 to the last. The message to be transmitted then becomes

$$010000100 \quad 010000010 \quad 010101001,$$

and the message CAT, which is encoded as

01000011 01000001 01010100,

would be transmitted as

010000111 010000010 010101001.

Suppose, as before, that the eighth digit of BAT is transmitted incorrectly. In this case, the message received would be

010000110 010000010 010101001.

Here, the first block must be wrong because it contains an odd number of 1s. Unfortunately, even if we know that only one digit was changed, the original block could have been

110000110, 000000110, 011000110, 010100110, 010010110, 010001110, 010000010, 010000100, or 010000111,

depending on which digit was changed.

Even though we do not know the proper interpretation of the first block of the received message, we know that it is not correct. Thus we would avoid taking any action based on this message until the first block can be corrected, perhaps by having the message (or just the first block) transmitted again. ■

Example 3.14 illustrates an *error-detecting* scheme. It is possible to detect that certain errors have occurred, but we cannot reconstruct the original message. Note that in Example 3.14, if both the sixth and eighth digits of the transmitted message

010000111 010000010 010101001

are changed, the received message

010001101 010000010 010101001

could be perceived as correct.

However, if the probability of an error in the transmission of any single digit is small, then the probability of two or more errors is considerably smaller. Throughout this discussion, we will make the following assumptions:

(1) The probability of changing a 0 to a 1 is the same as the probability of changing a 1 to a 0.
(2) The probability of an error in each digit is the same, and it is independent of whether there are errors in other digits. (That is, in the language of probability, the transmissions of any two digits are *independent events*.)
(3) The probability of an error in any digit is small, so that the probability of the correct transmission of each block is greater than the probability of a single error, and the probability of a single transmission error in any block is much greater than the probability of two or more errors.

It can be shown that the first two assumptions imply that if the probability of an error in the transmission of any single digit is p, then the probability that there are exactly k errors in the transmission of n digits is

$$C(n, k)p^k(1 - p)^{n-k}. \tag{3.1}$$

Here,

$$C(n, r) = \frac{n!}{r!\,(n - r)!}$$

is as defined in Section 2.6. So if $p = .01$, the probability of no errors in a nine-digit block is

$$C(9, 0)(.01)^0(.99)^9 \approx .9135,$$

and the probability of a single error is

$$C(9, 1)(.01)^1(.99)^8 \approx .0830.$$

In this situation, it can be shown that the probability of two or more errors is .0034. Thus, in this case, the probability of no errors is 11 times more likely than the probability of a single error, and the probability of a single error is 24 times more likely than the probability of two or more errors.

Example 3.15 Another error-detecting scheme is to transmit each block twice. For example, the first block, 01000010, in the message BAT is transmitted as

01000010 01000010.

The entire message is

01000010 01000010 01000001 01000001
01010100 01010100.

In this case, if the eighth digit of the message is transmitted incorrectly, the received message would be

01000011 01000010 01000001 01000001
01010100 01010100.

Because the first eight digits differ from the second eight digits, the receiver would know that there is an error. Because the probability of a single error is much greater than the probability of two or more errors, the first block is most likely to be either

01000010 or 01000011.

It is impossible, however, for the receiver to know which was intended. ■

The scheme used in Example 3.15 seems less efficient than that in Example 3.14 because it requires each letter to be represented by a block of 16 digits, compared with the block of 9 digits used in Example 3.14. In order to quantify the efficiency of a coding scheme, we must introduce some additional terminology.

Suppose that we wish to transmit information in the form of blocks of k binary digits. Each such block is called a **message word**, and k is called its **length**. What we transmit, however, are not k-digit words but longer strings called **codewords**. For instance, in Example 3.15, the 8-digit message word 01000010 was transmitted as the 16-digit codeword 0100001001000010. If each codeword has length n, then the coding scheme is called a **(k, n)-block code**. So the method used in Example 3.14 is an (8, 9)-block code, and the method used in Example 3.15 is an (8, 16)-block code. The **efficiency** of a (k, n)-block code is the ratio k/n. Thus the codes in Examples 3.14 and 3.15 have efficiencies of $8/9 \approx .89$ and $8/16 = .50$, respectively. The efficiency of a code measures the proportion of information in a transmission; so the code in Example 3.15 transmits one unit of information in two binary digits.

Because we will be considering only block codes, we often omit the word "block" and refer to a block code as simply a "code." The coding schemes described in Examples 3.14 and 3.15 formed their codewords by simply adjoining digits to the message words. Although other schemes are possible, in this book we will consider only schemes of this type.

All that is needed for a coding scheme is a one-to-one function E from the set of message words to the set of codewords. For a given message

$$w_1, w_2, \ldots, w_m,$$

we transmit

$$E(w_1), E(w_2), \ldots, E(w_m).$$

Because E is one-to-one, there is an inverse function D from the set of codewords to the set of message words. The received words are then decoded to obtain the original message:

$$w_1 = D(E(w_1)), w_2 = D(E(w_2)), \ldots, w_m = D(E(w_m)).$$

The functions E and D are called the **encoding** and **decoding** functions, respectively. Because of what is called *noise*, it is possible for an error to occur in transmission, so that the received word z is not a codeword. For example, sunspots could cause errors in satellite transmissions. In this case, $D(z)$ is not defined, and so the coding scheme indicates that an error has occurred. Figure 3.1 illustrates this type of coding scheme.

An important issue requiring special attention is how to interpret a received word that is not a codeword. *Throughout the rest of this chapter, we will assume that the receiver knows all the possible codewords and that if a word z is received which is not a codeword, then the receiver decodes z as the codeword that differs from z in the fewest digits.* If there is more than one such codeword, then z cannot

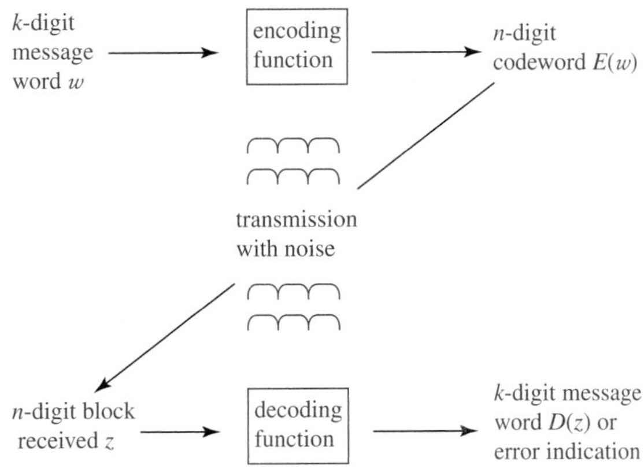

FIGURE 3.1

be decoded, and the receiver simply notes that an error occurred. This method of decoding is called **nearest-neighbor decoding**.

Example 3.16 Suppose that we modify Example 3.15 by transmitting each block three times. For example, the first block, 01000010, in the message BAT is transmitted as

$$01000010 \quad 01000010 \quad 01000010,$$

so the entire message is

$$01000010 \quad 01000010 \quad 01000010 \qquad 01000001 \quad 01000001 \quad 01000001$$
$$01010100 \quad 01010100 \quad 01010100.$$

The efficiency of this (8, 24)-code is only $8/24 = 1/3$. Although its efficiency is low, this code has a property not found in our previous codes: It is *error-correcting*. This means that in some cases it is possible to identify the intended codeword even if it is incorrectly transmitted.

Suppose, as before, that the eighth digit is incorrectly received, so that the first block is received as

$$01000011 \quad 01000010 \quad 01000010.$$

When this block is separated into three parts, we see that the first seven digits of each part are identical, but the eighth digits differ. Therefore we know that a transmission error has occurred, and an error must have occurred in digit 8, 16, or 24. Because the probability of one error in each block is greater than the probability of two or more errors, we assume that the error must be in digit 8. In general, the probability that one error is more likely than two or more errors implies that, when there is an error, the correct digit can be determined by majority rule. In this case, digits 16 and 24 are 0s and digit 8 is a 1, so we assume that the correct eighth digit is 0. Thus we assume that the original message word is 01000010 and that the original message is

$$01000010 \qquad 01000001 \qquad 01010100. \quad \blacksquare$$

In Example 3.16, we are able to correctly decode any codeword containing a single transmission error. For instance, suppose that the third block received is

$$01010100 \quad 01010100 \quad 01000100.$$

If we separate this block into three parts, majority rule shows that the correct message word is 01010100, and so the received block can be decoded correctly. Thus we are able to correct the most common errors that occur without needing to retransmit the original message. Here, the efficiency of the code is only $1/3$ (so that codewords take three times longer to transmit than message words); but when this code is compared with the alternative of sending the original message in uncoded form, we do not have the delays required to request and receive retransmissions if there is an error in the transmitted message.

The error-correcting property of the code in Example 3.16 requires more than that the encoding function be one-to-one. Consider, for example, the codewords

$$01001100 \qquad \text{and} \qquad 01001101.$$

If an error occurs in the transmission of the last digit of either of these codewords, then the other codeword will be received. In this case, we will have no reason to believe that the received codeword is incorrect, and so the message will be incorrectly decoded. More generally, if two codewords differ in a single digit and that digit of either word is changed during transmission, the received codeword will be decoded as the other word without our knowing that an error has occurred.

For two codewords c_1 and c_2 of the same length, the **Hamming distance** between c_1 and c_2 is defined to be the number of digits in which c_1 and c_2 differ. We denote this number by $d(c_1, c_2)$. For example,

$$d(010101, 011001) = 2 \qquad \text{and} \qquad d(11010100, 01111110) = 4.$$

Note that, for any codewords c_1 and c_2, we have $d(c_1, c_2) = d(c_2, c_1)$. The previous discussion shows that if a code is capable of detecting errors in a single digit, then the distance between distinct codewords must be at least 2. In order to obtain a condition that guarantees that errors in a single digit can be corrected, we need a property of the Hamming distance.

Theorem 3.6 *The Triangle Inequality* If c_1, c_2, and c_3 are any codewords of the same length, then

$$d(c_1, c_3) \le d(c_1, c_2) + d(c_2, c_3). \tag{3.2}$$

Proof. Suppose that c_1 and c_3 differ in their jth digits. Then either c_1 and c_2 differ in their jth digit, or c_2 and c_3 differ in their jth digit. Thus every difference that is counted on the left side of inequality (3.2) is also counted on the right side, and so the left side is less than or equal to the right. ■

Our general decoding scheme is as follows: If E is the coding function and C is the set of all codewords, then the decoding function $D = E^{-1}$ has domain C. When we receive a codeword c, we decode it as $D(c)$. If, however, we receive a block c' that is not a codeword, we decode it as $D(c)$, where c is the codeword nearest to c', that is, the codeword that differs from c' in the fewest digits. If there are two or more codewords nearest to c', then we record that an error has occurred.

Suppose that the distance between every pair of distinct codewords is at least 3 and that a received word c' is not a codeword. If there is a single error in transmission and $d(c', c_1) = 1$ for some codeword c_1, we will show that c_1 must be the codeword that was sent. Let c_2 be any codeword other than c_1. The triangle inequality implies that

$$3 \leq d(c_1, c_2) \leq d(c_1, c') + d(c', c_2) = 1 + d(c', c_2),$$

so $2 \leq d(c', c_2)$. Hence, if at most one digit in c' is incorrect, then c_2 cannot be the codeword that was sent. Consequently, c_1 must be the codeword that was sent.

The result stated in the preceding paragraph can be generalized as follows:

Theorem 3.7 Consider a block code in which m is the minimal Hamming distance between distinct codewords.

(a) This coding scheme can detect r or fewer errors if and only if $m \geq r + 1$.

(b) This coding scheme can correct r or fewer errors if and only if $m \geq 2r + 1$.

Proof. We will prove only (b), leaving (a) for the exercises.

Suppose first that $m \geq 2r + 1$. We will show that the coding scheme can correct r or fewer errors by proving that if codeword c is sent, c' is received, and $d(c, c') \leq r$, then codeword c' will be corrected to c. Suppose that c^* is a codeword such that $d(c', c^*) \leq r$. Then

$$d(c, c^*) \leq d(c, c') + d(c', c^*) \leq r + r = 2r < m.$$

Because the minimal Hamming distance between distinct codewords is m, we must have $c = c^*$. Thus codeword c' will be corrected to c.

Conversely, suppose that $m \geq 2r + 1$ is false, so that $m \leq 2r$. Then there exist distinct codewords c and c^* such that $d(c, c^*) = s \leq 2r$. Without loss of generality, let us suppose that c and c^* differ in their first s digits and agree in their last $n - s$ digits. Let t be the smaller of r and s, and let c' be the word formed by changing each of the first t digits of c. Then $d(c, c') = t \leq r$, and so c could be sent and c' received with no more than r errors. We will show that $d(c^*, c') \leq d(c, c')$.

Case 1: $s \leq r$ By the definition of t, we have $t = s$. Hence $c' = c^*$ in this case, and it follows that

$$d(c^*, c') = 0 \leq d(c, c').$$

Case 2: $s > r$ By the definition of t, we have $t = r$. Because c' agrees with c^* in precisely the first r and the last $n - s$ digits, we obtain

$$d(c^*, c') = n - (r + n - s) = s - r \leq 2r - r = r = t = d(c, c').$$

Thus, in either case, $d(c^*, c') \leq d(c, c') = t \leq r$. It follows that if either c or c^* were the intended codeword and no more than r errors occurred in transmission, then the received word could be c'. Hence c' cannot be corrected by this coding scheme, and so this scheme cannot correct r errors. ■

Theorem 3.7 shows that in order to have a good code—that is, one that corrects as many errors in a block as possible—the minimum distance between codewords must be as large as possible. (See Figure 3.2, where the large colored dots represent codewords.) In subsequent sections, we describe how to create such good codes.

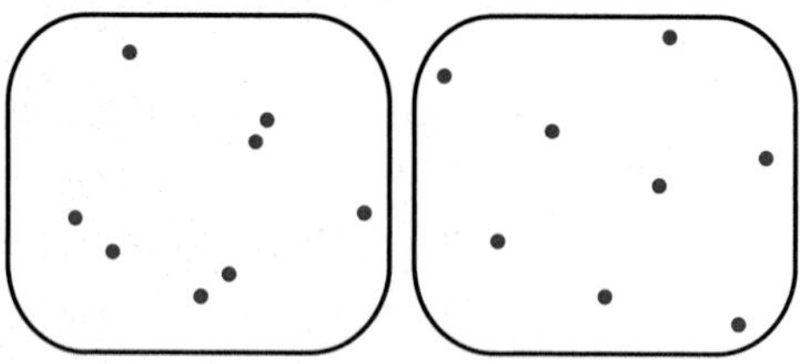

FIGURE 3.2

EXERCISES 3.4

In Exercises 1–8, determine the parity check digit that should be appended to each block so that the total number of 1s is even.

1. 01001010 **2.** 01101011 **3.** 00110011 **4.** 00110100

5. 01101010 **6.** 01010000 **7.** 00101010 **8.** 01110111

In Exercises 9–16, use formula (3.1) to determine the probability of exactly k errors in the transmission of n digits if the probability of an error in the transmission of a single digit is .01.

9. $k = 1, n = 5$ **10.** $k = 0, n = 6$ **11.** $k = 2, n = 7$ **12.** $k = 1, n = 8$

13. $k = 0, n = 8$ **14.** $k = 2, n = 10$ **15.** $k = 1, n = 10$ **16.** $k = 0, n = 10$

In Exercises 17–24, determine the Hamming distance between the given codewords.

17. $c_1 = 0110, c_2 = 1010$

18. $c_1 = 1001, c_2 = 1101$

19. $c_1 = 11001, c_2 = 01110$

20. $c_1 = 01010, c_2 = 10101$

21. $c_1 = 001100, c_2 = 111100$

22. $c_1 = 101010, c_2 = 001110$

23. $c_1 = 10011100, c_2 = 00111010$

24. $c_1 = 01101100, c_2 = 11000011$

In Exercises 25–32, add the given codewords using addition over Z_2 in each place. For example, $0110 + 0101 = 0011$.

25. $c_1 = 1110, c_2 = 0010$

26. $c_1 = 0101, c_2 = 1001$

27. $c_1 = 10111, c_2 = 10010$

28. $c_1 = 01011, c_2 = 11100$

29. $c_1 = 101011, c_2 = 001110$

30. $c_1 = 010111, c_2 = 110001$

31. $c_1 = 01100011, c_2 = 10110101$

32. $c_1 = 11010100, c_2 = 01001101$

In Exercises 33–36, suppose that the minimal Hamming distance between codewords in a certain block code is the given value m.

(a) *Determine the maximum number of errors that can be detected.*

(b) *Determine the maximum number of errors that can be corrected.*

33. $m = 8$ **34.** $m = 13$ **35.** $m = 15$ **36.** $m = 20$

37. Show that, for two codewords c_1 and c_2 of the same length, the Hamming distance between c_1 and c_2 equals the number of 1s in $c_1 + c_2$, where the addition is performed as described in Exercises 25–32.

38. Suppose that the minimal Hamming distance between codewords in a certain block code is m. Prove Theorem 3.7(a) by establishing each of the following results:

(a) If $m \geq r + 1$, then all errors in r or fewer digits can be detected.

(b) If all errors in r or fewer digits can be detected, then $m \geq r + 1$.

39. Consider the following table:

c_1	c_2	c_5
c_3	c_4	c_6
c_7	c_8	

Define a (4, 8)-block code as follows: Encode the message word $c_1c_2c_3c_4$ as the codeword $c_1c_2c_3c_4c_5c_6c_7c_8$, where c_5 is chosen so that the number of 1s in the first row is even, c_6 is chosen so that the number of 1s in the second row is even, c_7 is chosen so that the number of 1s in the first column is even, and c_8 is chosen so that the number of 1s in the second column is even. Show that this code corrects all errors in a single digit.

40. Generalize Exercise 39 to show that, for any positive integer s, there is an $(s^2, s^2 + 2s)$-block code that corrects all errors in a single digit.

41. Use Exercise 40 to show that, for any $\epsilon > 0$, there exists a code with efficiency greater than $1 - \epsilon$ that corrects all errors in a single digit.

3.5 MATRIX CODES

In this section, we describe a type of code for which the encoding and decoding functions are given by matrix multiplication (see Appendix B). To facilitate this discussion, for any positive integer m, we define $\mathcal{W}_m$ to be the set of all strings of 0s and 1s having length m. For example,

$$\mathcal{W}_3 = \{000, 001, 010, 011, 100, 101, 110, 111\}$$

is the set of strings of length 3. It follows from Theorem 1.3 that $|\mathcal{W}_m| = 2^m$.

With this notation, the encoding function for a (k, n)-block code is a one-to-one function $E\colon \mathcal{W}_k \to \mathcal{W}_n$. Suppose that A is a $k \times n$ matrix whose entries are 0s and 1s and x is a $1 \times k$ matrix of 0s and 1s. In this case, the product[1] xA over Z_2 of x and A is a $1 \times n$ matrix of 0s and 1s. Hence, if we regard an element in $\mathcal{W}_m$ as a $1 \times m$ matrix of 0s and 1s, then the product of an element x in $\mathcal{W}_k$ and matrix A is the element xA in $\mathcal{W}_n$.

Example 3.17 Let

$$A = \begin{bmatrix} 1 & 0 & 0 & 1 \\ 0 & 1 & 1 & 1 \\ 1 & 0 & 1 & 1 \end{bmatrix} \quad \text{and} \quad x = [1 \quad 1 \quad 0].$$

Because x is in $\mathcal{W}_3$ and A is a 3×4 matrix of 0s and 1s, the matrix product xA is a 1×4 matrix of 0s and 1s, and hence xA is an element of $\mathcal{W}_4$. In fact,

$$xA = [1 \quad 1 \quad 0] \begin{bmatrix} 1 & 0 & 0 & 1 \\ 0 & 1 & 1 & 1 \\ 1 & 0 & 1 & 1 \end{bmatrix} = [1 \quad 1 \quad 1 \quad 0].$$

---

[1]The multiplication of x and A is performed by using the arithmetic in Z_2, so that

$$0+0=1+1=0\cdot 0=0\cdot 1=1\cdot 0=0 \quad \text{and} \quad 0+1=1+0=1\cdot 1=1.$$

Likewise, if $y = [0 \quad 1 \quad 0]$, then

$$yA = [0 \quad 1 \quad 0]\begin{bmatrix} 1 & 0 & 0 & 1 \\ 0 & 1 & 1 & 1 \\ 1 & 0 & 1 & 1 \end{bmatrix} = [0 \quad 1 \quad 1 \quad 1].$$

Note that a product zA can be computed by adding (over Z_2) the rows of A that correspond to the nonzero entries of z. For example, since $x = [1 \quad 1 \quad 0]$, the product xA equals the sum of the first and second rows of A. Likewise, the only nonzero entry of y is the second entry, and so yA equals the second row of A. If $z = [0 \quad 0 \quad 0]$, then $zA = [0 \quad 0 \quad 0 \quad 0]$. ■

Consider the function $E\colon \mathcal{W}_k \to \mathcal{W}_n$ defined by $E(x) = xA$. When E is one-to-one, then we say that E defines a **matrix code**, and we call A its **generator matrix**. A simple way to ensure that E is one-to-one is to choose the $k \times n$ matrix A so that its first k columns form the $k \times k$ identity matrix I_k. We denote that A has this form by writing $A = [I_k|J]$, where J is the $k \times (n-k)$ matrix whose columns are the last $n-k$ columns of A. For such a matrix A, the first k entries of xA will be the same as those in x. Hence, if $x_1A = x_2A$, then $x_1 = x_2$, and it follows that E is one-to-one. For instance, if

$$A = \begin{bmatrix} 1 & 0 & 0 & 1 & 0 & 0 & 1 \\ 0 & 1 & 0 & 0 & 1 & 0 & 1 \\ 0 & 0 & 1 & 0 & 0 & 1 & 1 \end{bmatrix},$$

then $A = [I_3|J]$ for

$$J = \begin{bmatrix} 1 & 0 & 0 & 1 \\ 0 & 1 & 0 & 1 \\ 0 & 0 & 1 & 1 \end{bmatrix}.$$

Thus the function E defined by $E(x) = xA$ is one-to-one (and so produces a matrix code).

Because the preceding generator matrix A is a 3×7 matrix, the corresponding matrix code is a (3, 7)-block code. For any $[w_1 \quad w_2 \quad w_3]$ in $\mathcal{W}_3$, we have

$$[w_1 \quad w_2 \quad w_3]A = [w_1 \quad w_2 \quad w_3 \quad w_1 \quad w_2 \quad w_3 \quad w_1 + w_2 + w_3].$$

Note that the expression $w_1 + w_2 + w_3$ equals 0 if $[w_1 \quad w_2 \quad w_3]$ contains an even number of 1s and equals 1 if $[w_1 \quad w_2 \quad w_3]$ contains an odd number of 1s. Hence, for any message word w in $\mathcal{W}_3$, the corresponding codeword $E(w) = wA$ is obtained by repeating w twice and appending a parity check digit of 0 or 1, according to whether the number of 1s in w is even or odd, respectively.

The codewords for this code are as follows:

$$
\begin{aligned}
E([0\ \ 0\ \ 0]) &= [0\ \ 0\ \ 0\ \ 0\ \ 0\ \ 0\ \ 0]\\
E([0\ \ 0\ \ 1]) &= [0\ \ 0\ \ 1\ \ 0\ \ 0\ \ 1\ \ 1]\\
E([0\ \ 1\ \ 0]) &= [0\ \ 1\ \ 0\ \ 0\ \ 1\ \ 0\ \ 1]\\
E([0\ \ 1\ \ 1]) &= [0\ \ 1\ \ 1\ \ 0\ \ 1\ \ 1\ \ 0]\\
E([1\ \ 0\ \ 0]) &= [1\ \ 0\ \ 0\ \ 1\ \ 0\ \ 0\ \ 1]\\
E([1\ \ 0\ \ 1]) &= [1\ \ 0\ \ 1\ \ 1\ \ 0\ \ 1\ \ 0]\\
E([1\ \ 1\ \ 0]) &= [1\ \ 1\ \ 0\ \ 1\ \ 1\ \ 0\ \ 0]\\
E([1\ \ 1\ \ 1]) &= [1\ \ 1\ \ 1\ \ 1\ \ 1\ \ 1\ \ 1].
\end{aligned}
$$

It can be checked that the Hamming distance between any two of these codewords is at least 3, so, by Theorem 3.7, this code is capable of detecting errors in two digits and correcting errors in one digit.

The Check Matrix of a Code

When a transmission error occurs, the received word w' need not be a codeword. To decode w' correctly in such a case, we must determine which codeword is closest to w'. For a (k, n)-code, where k and n are large, an exhaustive search of all codewords could be a time-consuming process. Fortunately, a more efficient decoding method is available when A has the form $[I_k|J]$.

Suppose that $A = [I_k|J]$ is the $k \times n$ generator matrix of a (k, n)-code. Define the $n \times (n-k)$ matrix A^* so that its first k rows are the corresponding rows of J and its last $n-k$ rows are those of the $(n-k) \times (n-k)$ identity matrix. Symbolically, we write

$$A^* = \left[\frac{J}{I_{n-k}}\right]$$

and call A^* the **check matrix** associated with A.

Example 3.18 For the matrix

$$A = \begin{bmatrix} 1 & 0 & 0 & 1 & 0 & 1 & 0 \\ 0 & 1 & 0 & 1 & 1 & 0 & 1 \\ 0 & 0 & 1 & 0 & 1 & 1 & 1 \end{bmatrix},$$

we have $A = [I_3|J]$, where

$$J = \begin{bmatrix} 1 & 0 & 1 & 0 \\ 1 & 1 & 0 & 1 \\ 0 & 1 & 1 & 1 \end{bmatrix}.$$

The check matrix associated with A is

$$A^* = \left[\frac{J}{I_4}\right] = \begin{bmatrix} 1 & 0 & 1 & 0 \\ 1 & 1 & 0 & 1 \\ 0 & 1 & 1 & 1 \\ 1 & 0 & 0 & 0 \\ 0 & 1 & 0 & 0 \\ 0 & 0 & 1 & 0 \\ 0 & 0 & 0 & 1 \end{bmatrix}.$$

Note that

$$AA^* = \begin{bmatrix} 1 & 0 & 0 & 1 & 0 & 1 & 0 \\ 0 & 1 & 0 & 1 & 1 & 0 & 1 \\ 0 & 0 & 1 & 0 & 1 & 1 & 1 \end{bmatrix} \begin{bmatrix} 1 & 0 & 1 & 0 \\ 1 & 1 & 0 & 1 \\ 0 & 1 & 1 & 1 \\ 1 & 0 & 0 & 0 \\ 0 & 1 & 0 & 0 \\ 0 & 0 & 1 & 0 \\ 0 & 0 & 0 & 1 \end{bmatrix}$$

$$= \begin{bmatrix} 1+1 & 0 & 1+1 & 0 \\ 1+1 & 1+1 & 0 & 1+1 \\ 0 & 1+1 & 1+1 & 1+1 \end{bmatrix}$$

$$= \begin{bmatrix} 0 & 0 & 0 & 0 \\ 0 & 0 & 0 & 0 \\ 0 & 0 & 0 & 0 \end{bmatrix}. \quad \blacksquare$$

The calculation at the end of Example 3.18 is not a coincidence.

Theorem 3.8 Let $A = [I_k | J]$ be the generator matrix for a (k, n)-code, and let A^* be its check matrix.

(a) In the $k \times (n-k)$ matrix AA^*, each entry equals 0.

(b) A word c of length n is a codeword for this code if and only if $cA^* = O$, the $1 \times (n-k)$ zero matrix.

Proof. (a) Because A is the generator matrix for a (k, n)-code, A is a $k \times n$ matrix. Hence its check matrix A^* is an $n \times (n-k)$ matrix, and so the product AA^* is defined, and it is a $k \times (n-k)$ matrix. Let j_{rs} denote the entry of J in row r, column s. The entry in row r and column s of AA^* is the sum of the products (in Z_2) of corresponding entries from row r of A and column s of A^*, as shown in Figure 3.3. This entry is

$$1 \cdot j_{rs} + j_{rs} \cdot 1 = 0.$$

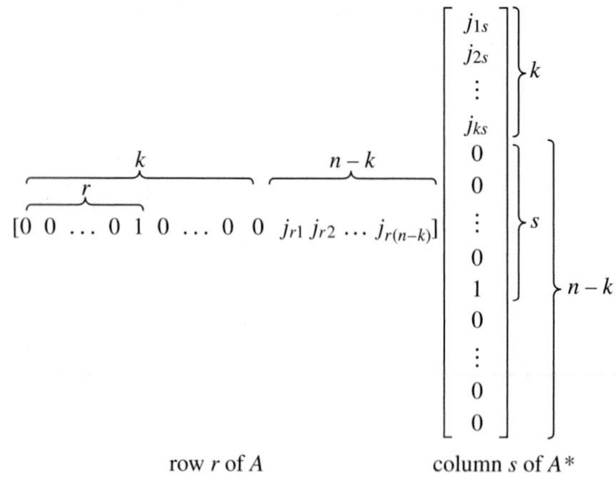

FIGURE 3.3

(b) If c is a codeword, then $c = xA$ for some message word x in $\mathcal{W}_k$. Thus, by (a), the associative property of matrix multiplication implies that

$$cA^* = (xA)A^* = x(AA^*) = O,$$

where O denotes the $1 \times (n-k)$ zero matrix.

Conversely, suppose c is a word of length n such that $cA^* = O$. Let x be the word of length k whose digits are the same as the first k digits of c. Then $c' = xA$ is a codeword whose first k digits agree with those of c. Thus the first k entries of $c - c'$ are 0. Because A^* has the form

$$A^* = \left[\frac{J}{I_{n-k}}\right],$$

the digits of $(c - c')A^*$ are just the last $n - k$ entries of $c - c'$. On the other hand, since both cA^* and $c'A^*$ are codewords, the distributive property of matrix multiplication and the first part of (b) imply that

$$(c - c')A^* = cA^* - c'A^* = O - O = O.$$

It follows that the last $n - k$ digits of $c - c'$ are all 0, and so all the digits of $c - c'$ are 0. Therefore $c = c'$, and hence c is a codeword. ■

Example 3.19 Consider the matrix

$$A = \begin{bmatrix} 1 & 0 & 0 & 1 & 0 & 0 & 1 \\ 0 & 1 & 0 & 0 & 1 & 0 & 1 \\ 0 & 0 & 1 & 0 & 0 & 1 & 1 \end{bmatrix},$$

discussed earlier in this section. Recall that its check matrix is

$$A^* = \begin{bmatrix} 1 & 0 & 0 & 1 \\ 0 & 1 & 0 & 1 \\ 0 & 0 & 1 & 1 \\ 1 & 0 & 0 & 0 \\ 0 & 1 & 0 & 0 \\ 0 & 0 & 1 & 0 \\ 0 & 0 & 0 & 1 \end{bmatrix}.$$

For $x = [1 \quad 0 \quad 1 \quad 0 \quad 0 \quad 1 \quad 1]$, we have

$$xA^* = [1 \quad 0 \quad 1 \quad 0 \quad 0 \quad 1 \quad 1] \begin{bmatrix} 1 & 0 & 0 & 1 \\ 0 & 1 & 0 & 1 \\ 0 & 0 & 1 & 1 \\ 1 & 0 & 0 & 0 \\ 0 & 1 & 0 & 0 \\ 0 & 0 & 1 & 0 \\ 0 & 0 & 0 & 1 \end{bmatrix} = [1 \quad 0 \quad 0 \quad 1].$$

Because $xA^* \neq O$, Theorem 3.8 implies that x is not a codeword. On the other hand, suppose that

$$[w_1 \quad w_2 \quad w_3 \quad w_4 \quad w_5 \quad w_6 \quad w_7]A^* = [0 \quad 0 \quad 0 \quad 0].$$

Using the matrix A^*, we see that

$$\begin{aligned} &[w_1 \quad w_2 \quad w_3 \quad w_4 \quad w_5 \quad w_6 \quad w_7]A^* \\ &= [w_1 \quad w_2 \quad w_3 \quad w_4 \quad w_5 \quad w_6 \quad w_7] \begin{bmatrix} 1 & 0 & 0 & 1 \\ 0 & 1 & 0 & 1 \\ 0 & 0 & 1 & 1 \\ 1 & 0 & 0 & 0 \\ 0 & 1 & 0 & 0 \\ 0 & 0 & 1 & 0 \\ 0 & 0 & 0 & 1 \end{bmatrix} \\ &= [w_1 + w_4 \quad w_2 + w_5 \quad w_3 + w_6 \quad w_1 + w_2 + w_3 + w_7]. \end{aligned}$$

If this product equals $[0 \quad 0 \quad 0 \quad 0]$, then

$$\begin{aligned} w_1 \qquad\qquad + w_4 \qquad\qquad &= 0 \\ w_2 \qquad\qquad + w_5 \qquad &= 0 \\ w_3 \qquad\qquad + w_6 &= 0 \\ w_1 + w_2 + w_3 \qquad\qquad\qquad + w_7 &= 0. \end{aligned}$$

The first of these equations implies that $w_1 = -w_4$, that is, $w_1 = w_4$ (because in Z_2, $-w_4 = w_4$). Similarly, the second and third equations imply that $w_2 = w_5$ and $w_3 = w_6$. Also, the equation $w_1 + w_2 + w_3 + w_7 = 0$ can be written as $w_1 + w_2 + w_3 = -w_7 = w_7$. Hence $w_7 = 0$ if $[w_1 \quad w_2 \quad w_3]$ contains an even number of 1s, and $w_7 = 1$ if $[w_1 \;\; w_2 \;\; w_3]$ contains an odd number of 1s. Thus $[w_1 \;\; w_2 \;\; w_3 \;\; w_4 \;\; w_5 \;\; w_6 \;\; w_7]A^* = [0 \;\; 0 \;\; 0 \;\; 0]$ when $[w_4 \;\; w_5 \;\; w_6]$ is identical to $[w_1 \;\; w_2 \;\; w_3]$ and w_7 is a check digit of 0 or 1, according to whether the number of 1s in $[w_1 \;\; w_2 \;\; w_3]$ is even or odd. But these are exactly the requirements that make xA^* a codeword for this code. ■

In Section 3.6, we will learn how to use the check matrix A^* to efficiently decode words that are encoded with the generator matrix A.

EXERCISES 3.5

In Exercises 1–4, determine the number of words in the given set.

1. $\mathcal{W}_5$ **2.** $\mathcal{W}_6$ **3.** $\mathcal{W}_8$ **4.** $\mathcal{W}_{10}$

In Exercises 5–8, suppose that the generator matrix for a (4, 8)*-code is*

$$\begin{bmatrix} 1 & 0 & 0 & 0 & 1 & 0 & 1 & 0 \\ 0 & 1 & 0 & 0 & 0 & 1 & 0 & 1 \\ 0 & 0 & 1 & 0 & 0 & 1 & 1 & 0 \\ 0 & 0 & 0 & 1 & 1 & 0 & 0 & 1 \end{bmatrix}.$$

Determine the codeword corresponding to each of the given message words.

5. 1001 **6.** 0011 **7.** 1101 **8.** 0111

In Exercises 9–12, determine the size of the check matrix for the given code.

9. a (3, 9)-block code **10.** a (4, 11)-block code

11. a (5, 10)-block code **12.** a (7, 13)-block code

13. If the check matrix of a matrix code is a 9×3 matrix, what is the efficiency of the code?

14. If the check matrix of a matrix code is an 11×4 matrix, what is the efficiency of the code?

In Exercises 15–20, determine all the codewords for the code having the given generator matrix.

15. $\begin{bmatrix} 1 & 0 & 1 & 0 & 1 \\ 0 & 1 & 1 & 1 & 0 \end{bmatrix}$

16. $\begin{bmatrix} 1 & 0 & 0 & 1 & 1 \\ 0 & 1 & 1 & 1 & 1 \end{bmatrix}$

17. $\begin{bmatrix} 1 & 0 & 0 & 0 & 0 & 1 \\ 0 & 1 & 0 & 0 & 1 & 1 \\ 0 & 0 & 1 & 1 & 1 & 1 \end{bmatrix}$

18. $\begin{bmatrix} 1 & 0 & 0 & 1 & 0 & 1 \\ 0 & 1 & 0 & 1 & 1 & 0 \\ 0 & 0 & 1 & 0 & 1 & 1 \end{bmatrix}$

19. $\begin{bmatrix} 1 & 0 & 0 & 1 & 0 & 0 & 1 \\ 0 & 1 & 0 & 0 & 1 & 1 & 0 \\ 0 & 0 & 1 & 0 & 1 & 0 & 1 \end{bmatrix}$

20. $\begin{bmatrix} 1 & 0 & 0 & 0 & 1 & 1 & 1 \\ 0 & 1 & 0 & 1 & 0 & 1 & 1 \\ 0 & 0 & 1 & 1 & 1 & 0 & 1 \end{bmatrix}$

In Exercises 21–28, determine the check matrix associated with the given generator matrix.

21. $\begin{bmatrix} 1 & 0 & 0 & 1 & 1 \\ 0 & 1 & 1 & 1 & 1 \end{bmatrix}$

22. $\begin{bmatrix} 1 & 0 & 1 & 0 & 1 \\ 0 & 1 & 1 & 1 & 0 \end{bmatrix}$

23. $\begin{bmatrix} 1 & 0 & 0 & 1 & 0 & 1 \\ 0 & 1 & 0 & 1 & 1 & 0 \\ 0 & 0 & 1 & 0 & 1 & 1 \end{bmatrix}$

24. $\begin{bmatrix} 1 & 0 & 0 & 0 & 0 & 1 \\ 0 & 1 & 0 & 0 & 1 & 1 \\ 0 & 0 & 1 & 1 & 1 & 1 \end{bmatrix}$

25. $\begin{bmatrix} 1 & 0 & 0 & 0 & 1 & 1 & 1 \\ 0 & 1 & 0 & 1 & 0 & 1 & 1 \\ 0 & 0 & 1 & 1 & 1 & 0 & 1 \end{bmatrix}$

26. $\begin{bmatrix} 1 & 0 & 0 & 1 & 0 & 0 & 1 \\ 0 & 1 & 0 & 0 & 1 & 1 & 0 \\ 0 & 0 & 1 & 0 & 1 & 0 & 1 \end{bmatrix}$

27. $\begin{bmatrix} 1 & 0 & 0 & 1 & 0 & 1 & 0 & 1 \\ 0 & 1 & 0 & 0 & 1 & 0 & 1 & 0 \\ 0 & 0 & 1 & 0 & 1 & 1 & 1 & 0 \end{bmatrix}$

28. $\begin{bmatrix} 1 & 0 & 0 & 1 & 1 & 1 & 1 & 0 \\ 0 & 1 & 0 & 1 & 1 & 0 & 1 & 1 \\ 0 & 0 & 1 & 0 & 1 & 1 & 1 & 1 \end{bmatrix}$

In Exercises 29 and 30, the check matrix A^* for a matrix code is given. Determine the generator matrix A.

29. $\begin{bmatrix} 1 & 0 & 1 \\ 1 & 1 & 1 \\ 0 & 1 & 1 \\ 1 & 1 & 0 \\ 1 & 0 & 0 \\ 0 & 1 & 0 \\ 0 & 0 & 1 \end{bmatrix}$

30. $\begin{bmatrix} 1 & 0 & 0 & 1 \\ 0 & 1 & 0 & 1 \\ 1 & 0 & 1 & 0 \\ 0 & 1 & 1 & 0 \\ 1 & 0 & 0 & 0 \\ 0 & 1 & 0 & 0 \\ 0 & 0 & 1 & 0 \\ 0 & 0 & 0 & 1 \end{bmatrix}$

In Exercises 31–38, use Theorem 3.8(b) to determine whether each of the given words is a codeword for the matrix code with the generator matrix

$$\begin{bmatrix} 1 & 0 & 0 & 1 & 0 & 1 & 1 \\ 0 & 1 & 0 & 0 & 1 & 1 & 0 \\ 0 & 0 & 1 & 0 & 0 & 1 & 1 \end{bmatrix}.$$

31. 0110110 **32.** 1101101 **33.** 1001011 **34.** 0011110

35. 0110101 **36.** 1111110 **37.** 1101010 **38.** 0111110

39. What proportion of the words in $\mathcal{W}_n$ are codewords for a particular (k, n)-code?

40. Consider the (3, 7)-code with generator matrix

$$\begin{bmatrix} 1 & 0 & 0 & 0 & 0 & 0 & 0 \\ 0 & 1 & 0 & 0 & 0 & 1 & 0 \\ 0 & 0 & 1 & 1 & 1 & 0 & 0 \end{bmatrix}.$$

Let $z = [0 \quad 0 \quad 0]$ and $w = [1 \quad 0 \quad 0]$.

(a) What is the Hamming distance between the codewords determined by z and w?

(b) Can this code detect all errors in a single digit?

41. Find the generator matrix of the code described in Exercise 39 of Section 3.4.

42. Find the generator matrix of the code that encodes the message word $c_1c_2c_3c_4$ as $c_1c_2c_3c_4c_5c_6c_7$, where the digits c_5, c_6, and c_7 are chosen to make the number of 1s even in disks A, B, and C in the following diagram:

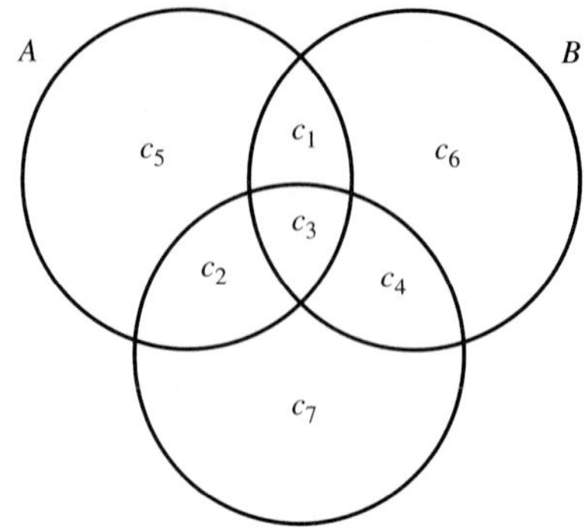

43. Suppose that A and A' are generator matrices for two (k, n)-codes having C and C', respectively, as their sets of codewords. Prove that if $C = C'$, then $A = A'$. Thus, in this case, the set of codewords completely determines the code.

3.6 MATRIX CODES THAT CORRECT ALL SINGLE-DIGIT ERRORS

In this section, we will learn how to create matrix codes capable of correcting all transmission errors in a single digit. The next theorem shows that the check matrix plays a prominent role in this process.

Theorem 3.9 Let A be the generator matrix of a (k, n)-block code. This code can correct all single-digit errors if and only if its check matrix A^* satisfies the following two conditions:
(i) No row of A^* consists entirely of zeros.
(ii) No two rows of A^* are identical.

Proof. Suppose that A^* satisfies conditions (i) and (ii). We first show that there are no codewords containing exactly one or exactly two 1s. Let e_r be the $1 \times n$ matrix in which all the entries are 0 except for entry r, which is 1. Note that the only nonzero words in $\mathcal{W}_n$ containing exactly one or exactly two 1s are those of the form e_r and $e_r + e_s$, where $r \neq s$. If e_r is a codeword, then $e_rA^* = O$ by Theorem 3.8. But the definition of matrix multiplication shows that e_rA^* equals row r of A^*, and hence row r of A^* consists entirely of zeros, contradicting (i). Therefore e_r is not a codeword. On the other hand, if $e_r + e_s$ is a codeword, then $(e_r + e_s)A^* = O$. Thus $e_rA^* + e_sA^* = O$, and so $e_rA^* = -e_sA^* = e_sA^*$. Therefore rows r and s of A^* are equal, contradicting (ii). Therefore $e_r + e_s$ is not a codeword. It follows that there are no codewords containing exactly one or exactly two 1s.

To prove that this code corrects all single-digit errors, it suffices by Theorem 3.7 to show that the minimal Hamming distance between distinct codewords is at least 3. Suppose that there are distinct message words x and y for which the distance between the corresponding codewords $E(x) = xA$ and $E(y) = yA$ is less than 3. Because E is one-to-one, $E(x) \neq E(y)$. Thus the Hamming distance between $E(x)$ and $E(y)$ must be either 1 or 2. But then, Exercise 37 in Section 3.4 shows that the number of 1s in $E(x) + E(y)$ must be 1 or 2. But $E(x) + E(y) = xA + yA = (x + y)A = E(x + y)$ is the codeword corresponding to the message word $x + y$. This contradicts the result proved in the preceding paragraph.

The proof of the converse is contained in Exercises 40 and 41. ■

Later in this section, we will use the two conditions in Theorem 3.9 to construct efficient error-correcting matrix codes that can correct all single-digit errors. For the present, we note that they guarantee that the matrix codes in Examples 3.18 and 3.19 correct all single-digit errors.

Check Matrix Row Decoding

Thus far in our discussion of matrix codes, we have focused on the process of encoding a message. Now we turn our attention to the process of decoding. For a matrix code generated by a matrix of the form $A = [I_k | J]$, the decoding of a codeword is easy: Simply take the first k digits of the codeword.

Example 3.20 For the code in Example 3.18, which is generated by the matrix

$$A = \begin{bmatrix} 1 & 0 & 0 & 1 & 0 & 1 & 0 \\ 0 & 1 & 0 & 1 & 1 & 0 & 1 \\ 0 & 0 & 1 & 0 & 1 & 1 & 1 \end{bmatrix},$$

the codeword corresponding to the message word $x = [1 \quad 0 \quad 1]$ is

$$xA = \begin{bmatrix} 1 & 0 & 1 \end{bmatrix} \begin{bmatrix} 1 & 0 & 0 & 1 & 0 & 1 & 0 \\ 0 & 1 & 0 & 1 & 1 & 0 & 1 \\ 0 & 0 & 1 & 0 & 1 & 1 & 1 \end{bmatrix} = \begin{bmatrix} 1 & 0 & 1 & 1 & 1 & 0 & 1 \end{bmatrix}.$$

Because this is a (3, 7)-code, the first three digits of the codeword are the original message word. Similarly, the codeword $[0 \quad 1 \quad 1 \quad 1 \quad 0 \quad 1 \quad 0]$ decodes as $[0 \quad 1 \quad 1]$. ■

When a received word c' is not a codeword, we must determine the codeword that is nearest to c'. As we have noted, this can be very time-consuming when

k and n are large. Fortunately, a simpler method is available for matrix codes whose check matrix satisfies the two conditions in Theorem 3.9.

Check Matrix Row Decoding

Suppose that a (k, n)-code has a check matrix A^* with distinct rows and no row of zeros.

1. If the word w is received, compute $s = wA^*$.
2. If $s = O$, the $1 \times n$ zero matrix, then w is a codeword, and we decode it as its first k digits.
3. If s equals row i of A^*, then form the word w' by changing digit i of w, and decode w' by taking its first k digits.
4. If s is nonzero, but is not a row of A^*, then more than one digit of w is incorrect. In this case, we record that an erroneous word was received.

The expression wA^* in step 1 is called the **syndrome** of w.

Example 3.21 Decode the message

0010111 1011100 0111010 1000111 1100010,

which was sent using the code in Example 3.18.

Recall from Example 3.18 that the check matrix for this code is

$$A^* = \begin{bmatrix} 1 & 0 & 1 & 0 \\ 1 & 1 & 0 & 1 \\ 0 & 1 & 1 & 1 \\ 1 & 0 & 0 & 0 \\ 0 & 1 & 0 & 0 \\ 0 & 0 & 1 & 0 \\ 0 & 0 & 0 & 1 \end{bmatrix}.$$

The decoding of this message, using the check matrix row decoding method, is shown in the following table:

Received Word w	*Syndrome wA^**	*Row of A^**	*Codeword*	*Message Word*
0010111	0000	—	0010111	001
1011100	0001	7	1011101	101
0111010	0000	—	0111010	011
1000111	1101	2	1100111	110
1100010	0101	not a row of A^*	unknown	unknown

Thus we decode the message as

$$001 \qquad 101 \qquad 011 \qquad 110 \qquad ??.$$

Because the syndrome of the last word is nonzero, but is not a row of A^*, there must have been two or more errors in the transmission of this word. Hence we have no way of decoding it reliably. ■

One final caution should be given. In Example 3.21, the minimal distance between codewords is 3. Thus this code can correct all single-digit errors and detect all two-digit errors. Consider the message word 011, which would be encoded as the codeword 0111010. If this were transmitted and errors occurred in the second, third, and sixth digits, then the received word would be 0001000. The syndrome of this received word is 1000, which is the fourth row of A^*. Thus the nearest codeword is 0000000, and we would assume that it is the correct codeword. Hence our decoded message would contain 000, instead of the original message word 011. Moreover, we would not be aware of this error, because it is an error involving three digits, and our code is capable of detecting only errors involving at most two digits.

Hamming Codes

Earlier, we noted that the $(3, 7)$-code in Example 3.19 is capable of correcting all single-digit errors. This code has efficiency $3/7$. It is natural to ask what is the most efficient $(3, n)$-code that can correct all single-digit errors. In other words, we are asking for the smallest integer $n \geq 3$ for which there is a $(3, n)$-code that is capable of correcting all single-digit errors. Equivalently, we are seeking the fewest digits that must be transmitted in order to be able to correct all single-digit errors in the transmission of message words of length 3. Recall that the conditions in Theorem 3.9 are necessary and sufficient for a matrix code to be capable of correcting all single-digit errors, and so we can use these conditions to answer our question.

Clearly, some additional digits will need to be transmitted. If $n = 3$, then every received word is a codeword, and we cannot detect any transmission errors. Suppose that $A = [I_3|J]$ is the generator matrix for a $(3, 4)$-code. Then A is a 3×4 matrix, so J is a 3×1 matrix. Necessarily, two rows of such a matrix must be identical, so

$$A^* = \left[\frac{J}{I_1}\right]$$

has two identical rows. Thus this code cannot correct all single-digit errors.

If $A = [I_3|J]$ is the generator matrix for a (3, 5)-code, then J is a 3×2 matrix. In this case, A^* has the form

$$A^* = \left[\frac{J}{I_2}\right] = \begin{bmatrix} ? & ? \\ ? & ? \\ ? & ? \\ 1 & 0 \\ 0 & 1 \end{bmatrix}.$$

Again, A^* must have two identical rows, and so no (3, 5)-code can correct all single-digit errors. However, for $n = 6$, J will be a 3×3 matrix, and it is easy to find a matrix A^* satisfying the conditions of Theorem 3.9. For example, we can choose

$$J = \begin{bmatrix} 1 & 1 & 0 \\ 1 & 0 & 1 \\ 0 & 1 & 1 \end{bmatrix},$$

so that

$$A^* = \left[\frac{J}{I_3}\right] = \begin{bmatrix} 1 & 1 & 0 \\ 1 & 0 & 1 \\ 0 & 1 & 1 \\ 1 & 0 & 0 \\ 0 & 1 & 0 \\ 0 & 0 & 1 \end{bmatrix}.$$

Thus the (3, 6)-matrix code with generator matrix

$$A = \begin{bmatrix} 1 & 0 & 0 & 1 & 1 & 0 \\ 0 & 1 & 0 & 1 & 0 & 1 \\ 0 & 0 & 1 & 0 & 1 & 1 \end{bmatrix}$$

is capable of correcting all single-digit errors.

The preceding argument is easily generalized. The check matrix of a (k, n)-code has size $n \times (n - k)$; so its rows have $n - k$ entries. But the number of distinct nonzero rows of length $n - k$ is $2^{n-k} - 1$. Since A^* has n rows, the conditions of Theorem 3.9 can be fulfilled if and only if $2^{n-k} - 1 \geq n$.

Theorem 3.10 A (k, n)-code exists that corrects all single-digit errors if and only if

$$2^{n-k} - 1 \geq n.$$

Codes for which $n = 2^{n-k} - 1$ are especially convenient. Setting $r = n - k$, we see that this condition is met when $n = 2^r - 1$, that is, when n is 1 less than a power of 2. Note that we must require $r > 1$ because $r = 1$ implies that $k = 0$, so there are no message words. The first five such values of k and n are given in the following table:

r	2	3	4	5	6
$k = 2^r - r - 1$	1	4	11	26	57
$n = 2^r - 1$	3	7	15	31	63

For these pairs k and n, we can construct a (k, n)-code that corrects all single-digit errors. In such cases, the check matrix, which is an $n \times (n - k) = n \times r$ matrix, has n rows, each of length r. But there are only 2^r such rows for A^*, so each nonzero row must be used. That is, the rows of A^* comprise all the nonzero words of length r. A matrix code for which the check matrix has this property is called a **Hamming code**.

Example 3.22 The preceding table shows that the Hamming code for $r = 2$ has $k = 1$ and $n = 3$. Thus the check matrix for this code has the form

$$A^* = \left[\frac{J}{I_2}\right] = \begin{bmatrix} ? & ? \\ 1 & 0 \\ 0 & 1 \end{bmatrix}.$$

Because the rows of A^* must be nonzero and distinct, we see that

$$A^* = \left[\frac{J}{I_2}\right] = \begin{bmatrix} 1 & 1 \\ 1 & 0 \\ 0 & 1 \end{bmatrix}.$$

Thus $J = [1 \quad 1]$, and $A = [I_1 | J] = [1 \quad 1 \quad 1]$. In this case, for every message word $[x]$, we have

$$[x]A = [x \quad x \quad x],$$

and so the code repeats each message word three times. The decoding of each possible word that can be received is given in the following table:

Word Received	*Syndrome*	*Row of A^**	*Codeword*	*Message Word*
000	00	—	000	0
001	01	3	000	0
010	10	2	000	0
011	11	1	111	1
100	11	1	000	0
101	10	2	111	1
110	01	3	111	1
111	00	—	111	1

■

Suppose that we have a Hamming code, and a word c' is received. The syndrome of c' is $s = c'A^*$. If $s = O$, then c' is a codeword, and we assume that there has been no error in transmission. But if $s \neq O$, then s appears as a row of A^* because every nonzero word of length r appears as a row of A^*. Let row i of A^* equal s. Then c' will be decoded as the first k digits of c, where c is the word obtained by changing the ith digit of c'. Thus, when check matrix row decoding is used with a Hamming code, *step 4 never happens, and every received word can be decoded.* The efficiency of the Hamming code for a particular value of r is

$$\frac{k}{n} = \frac{2^r - r - 1}{2^r - 1} = 1 - \frac{r}{2^r - 1}.$$

It can be shown that this quantity can be made arbitrarily close to 1 by choosing r to be sufficiently large. Thus there are Hamming codes with efficiencies that are arbitrarily close to 1.

EXERCISES 3.6

In Exercises 1–8, determine the syndrome of each word using the check matrix

$$A^* = \begin{bmatrix} 1 & 0 & 0 & 1 \\ 0 & 1 & 1 & 0 \\ 1 & 0 & 0 & 0 \\ 0 & 1 & 0 & 0 \\ 0 & 0 & 1 & 0 \\ 0 & 0 & 0 & 1 \end{bmatrix}.$$

Then decode the word if it is a codeword or if it differs from a codeword in a single digit.

1. 101111 **2.** 001001 **3.** 010110 **4.** 011010

5. 101010 **6.** 101001 **7.** 110110 **8.** 010000

In Exercises 9–28, the given word was received after being encoded using the generator matrix

$$A = \begin{bmatrix} 1 & 0 & 0 & 1 & 0 & 1 & 0 \\ 0 & 1 & 0 & 0 & 1 & 1 & 0 \\ 0 & 0 & 1 & 1 & 1 & 0 & 1 \end{bmatrix}$$

and transmitted. Determine the syndrome of each word, and decode it if it is a codeword or if it differs from a codeword in a single digit.

9. 1010101 **10.** 1110001 **11.** 1011001 **12.** 1001101

13. 0111011 **14.** 0110011 **15.** 1100110 **16.** 1100001

17. 0011101 **18.** 1010111 **19.** 1011100 **20.** 0011111

21. 1001100 **22.** 1011000 **23.** 1101100 **24.** 0010111

25. 1001011 **26.** 0100110 **27.** 0101011 **28.** 0000110

In Exercises 29 and 30, a check matrix and a list of received words is given. Decode the list.

29. $\begin{bmatrix} 1 & 1 & 0 \\ 0 & 1 & 1 \\ 1 & 0 & 0 \\ 0 & 1 & 0 \\ 0 & 0 & 1 \end{bmatrix}$, 11111, 11000, 10101, 10110

30. $\begin{bmatrix} 1 & 0 & 1 \\ 1 & 1 & 1 \\ 1 & 0 & 0 \\ 0 & 1 & 0 \\ 0 & 0 & 1 \end{bmatrix}$, 10101, 01110, 00101, 11100

In Exercises 31–34, determine the minimal value of n for which there exists a (k, n)-matrix code that corrects all errors in a single digit.

31. $k = 8$

32. $k = 10$

33. $k = 20$

34. $k = 25$

In Exercises 35–38, determine the smallest values of k and n for which there exists a (k, n)-Hamming code whose efficiency exceeds the given value.

35. .5

36. .75

37. .9

38. .99

39. Let z be the string of length n in which each digit equals 0. Show that z is a codeword in any (k, n)-block code.

40. Consider a (k, n)-code with check matrix A^*. Show that if some row of A^* consists entirely of zeros, then this code cannot detect all single-digit errors.

41. Consider a (k, n)-code with check matrix A^*. Show that if two rows of A^* are identical, then this code cannot detect all single-digit errors.

42. **(a)** Prove by mathematical induction that $r^2 + 1 \le 2^r$ for every integer $r \ge 5$.

(b) Use (a) to show that

$$1 - \frac{r}{2^r - 1} \ge 1 - \frac{1}{r}$$

for every integer $r \ge 5$.

43. Show that every Hamming code incorrectly decodes a received word having more than one error.

44. Consider a transmission method for which the probability of a transmission error in any digit is p.

(a) Compute the probability of no errors when an uncoded 4-digit message word is sent.

(b) Compute the probability that a word is correctly decoded by the (4, 7)-Hamming code.

(c) Show that the answer to (a) does not exceed the answer to (b) if p satisfies $0 \le p < .5$.

HISTORICAL NOTES

Coding theory has a long and rich history, reaching back to the times of the Greeks and Romans. Initially, codes were used to send secret messages and to encrypt state secrets. Today, codes are encountered everywhere, for example, when checking out at the local grocery and when ordering movies from a satellite television signal provider. In each of these instances, the mathematics of number theory, algebra, probability, and statistics provides a basis for the secure and accurate transmission of data.

Blaise de Vigenère

Historically, the first recorded uses of coding appeared in the works of Egyptian and Mesopotamian scribes. By 300 B.C., the Greeks were using various forms of substitution ciphers to code messages. Over the following centuries, various improvements were made. In 1412 A.D., al-Kalka-shandi published an encyclopedic work entitled *Subh al-a'sha*, which summarized previous systems and extended cryptanalysis to include both the use of letter-frequency techniques and combinations of substitution and transposition methods. In this same period, the authors Roger Bacon and Geoffrey Chaucer included encrypted materials in their writings. These and other findings were summarized in 1624 by Duke August of Brunswick in his work *Cryptomenytices et Cryptographiae Libri IX*, an encyclopedia of known coding theory.

Ronald L. Rivest

The next major advance in coding theory was the Vigenère tableau, invented in the 1580s by the Frenchman Blaise de Vigenère. This table-based method made coding systems in use at the time significantly more secure.

Several other advances appeared from the 1600s to the early 1900s, but it was World War I that cemented the importance of codes in military and political communications. New systems used advances based on Vigenère's work and also used mechanical means for encryption and decryption. Edward Hebern (1869–1952) of the United States, Hugo Koch (1870–1928) of the Netherlands, and Arthur Scherbius (1878–1929) of Germany each devised cipher machines employing connected rotors. The most famous of these was the Enigma machine used by German armed forces in World War II. Cracking the German cipher system was the work of a group of mathematicians at the secret Bletchley Manor project in England. Central in this group were Alan Turing (1912–1954) and Gordon Welchman (1906–1985). Their successful approach extended methods previously developed by the Polish mathematician Marian Rejewski (1905–1980).

Adi Shamir

Although these advances made decryption more difficult, ciphers were still based on some form of substitution. Moreover, all required communication of a shared secret key. The existence of such keys made the codes susceptible to espionage and outright theft of the key, as well as to mathematical analysis.

These problems were finally surmounted in the early 1970s by advances made by Whitfield Diffie (1944–), Martin Hellman (1945–), and Ralph Merkle (1952–), who provided a basis for the development of public key cryptosystems. Shortly after their breakthrough, other advances were proposed by Ronald L. Rivest (1947–), Adi Shamir (1952–), and Leonard Adleman (1945–), originators of the RSA method of public key cryptography.

Leonard Adleman

Parallel to these developments in cryptography, applications of coding theory became ubiquitous in daily life. Identification codes suddenly appeared on textbooks, grocery and other commercial items, and VIN numbers on vehicles, as well as in a myriad of other applications. The work of Bernard Silver (1924–1962) and Norman Woodland (1922–), who developed bar codes for the food industry in

Richard Hamming

1952 at IBM, ushered in a new industry resulting from the applications of coding theory to logistics and inventory control. Central to these applications was the concept of codes in which errors could be detected and, in some cases, corrected. These advances were based in large part on the seminal contributions in 1950 of Richard Hamming (1915–1998). Work in this and other areas of coding theory continues today.

SUPPLEMENTARY EXERCISES

Determine whether each statement in Exercises 1–4 is true or false.

1. $37 \equiv 18 \pmod 2$

2. $45 \equiv -21 \pmod{11}$

3. $-7 \equiv 53 \pmod{12}$

4. $-18 \equiv -64 \pmod 7$

In Exercises 5–10, perform the indicated operations in Z_m. Write your answer in the form $[r]$, where $0 \leq r < m$.

5. $[43] + [32]$ in Z_{11}

6. $[-12] + [95]$ in Z_{25}

7. $[5][11]$ in Z_9

8. $[-3][9]$ in Z_{15}

9. $[22]^7$ in Z_5

10. $[13]^6[23]^5$ in Z_{12}

11. If $x \equiv 4 \pmod{11}$ and $y \equiv 9 \pmod{11}$, what is the remainder when $x^2 + 3y$ is divided by 11?

12. When an integer is divided by 15, its remainder is 7. Find its remainders when this integer is divided by 5 and by 3.

13. If 100 is divided by a positive integer d, the remainder is 2. What is the remainder when 198 is divided by d?

14. When a positive integer n is divided by 7, the remainder is 5. What will the remainder be if $5n$ is divided by 7?

In Exercises 15–18, use the Euclidean algorithm to find the greatest common divisor of a and b.

15. $a = 770, b = 1764$

16. $a = 1320, b = 1575$

17. $a = -9798, b = 552$

18. $a = -7661, b = 4183$

In Exercises 19–22, use the extended Euclidean algorithm to write the greatest common divisor of a and b as a linear combination of a and b.

19. $a = 770, b = 1764$

20. $a = 2002, b = 1080$

21. $a = -9798, b = 552$

22. $a = -7661, b = 4183$

23. Use the extended Euclidean algorithm to find integer solutions x and y, if possible, to each of the following equations:

(a) $666x + 1414y = 30$

(b) $666x + 1414y = 55$

In Exercises 24 and 25, use the method of Example 3.11 with $E = 5$ and $n = 39$ to encipher the given plaintext message into ciphertext.

24. 4, 15, 17, 21

25. 18, 10, 6, 2

In Exercises 26 and 27, make a table, as in Example 3.12, to show the working of the modular exponentiation algorithm with the given values of P, E, and n.

26. $P = 25$, $E = 11$, and $n = 33$

27. $P = 18$, $E = 29$, and $n = 57$

In Exercises 28 and 29, determine the value of b corresponding to the given value of n, where b and n are as in the RSA method.

28. $n = 1763$

29. $n = 1829$

In Exercises 30 and 31, use the method of modular exponentiation with the given values of n and E to decipher the given ciphertext C.

30. $n = 35$, $E = 11$, $C = 4$

31. $n = 143$, $E = 11$, $C = 6$

In Exercises 32 and 33, append a parity check digit to the given block so that the number of 1s is even.

32. 01100110110101

33. 11011010101010

In Exercises 34 and 35, determine the probability of exactly k errors in the transmission of n digits if the probability of an error in the transmission of a single digit is .001.

34. $k = 3$, $n = 8$

35. $k = 4$, $n = 10$

In Exercises 36 and 37, determine the minimal Hamming distance between two codewords for the codewords in the given set.

36. {010101, 101010, 111111, 000000}

37. {1110110111, 010010010, 000000000, 111111111}

38. If a set of codewords contains the codeword in which each digit is 0, what can be said about the minimal Hamming distance between two codewords?

39. If a set of codewords contains the codeword in which each digit is 1, what can be said about the minimal Hamming distance between two codewords?

40. How many codewords are there in a (k, n)-block code?

41. Let z be an element of $\mathcal{W}_n$, and let s be an integer such that $0 \le s \le n$. How many elements of $\mathcal{W}_n$ are such that their Hamming distance from z equals s?

42. Let z be an element of $\mathcal{W}_n$, and let s be an integer such that $0 \le s \le n$. How many elements of $\mathcal{W}_n$ are such that their Hamming distance from z is at most s?

In Exercises 43 and 44, determine the codeword for the given message word in the (5, 9)*-code with the generator matrix*

$$\begin{bmatrix} 1 & 0 & 0 & 0 & 0 & 1 & 1 & 1 & 1 \\ 0 & 1 & 0 & 0 & 0 & 0 & 0 & 1 & 1 \\ 0 & 0 & 1 & 0 & 0 & 1 & 0 & 1 & 0 \\ 0 & 0 & 0 & 1 & 0 & 1 & 1 & 0 & 0 \\ 0 & 0 & 0 & 0 & 1 & 0 & 1 & 0 & 1 \end{bmatrix}.$$

43. 10101

44. 00101

45. Determine all the codewords for the code with generator matrix

$$\begin{bmatrix} 1 & 0 & 1 & 1 \\ 0 & 1 & 0 & 0 \end{bmatrix}.$$

46. Determine all the codewords for the code with generator matrix

$$\begin{bmatrix} 1 & 0 & 0 & 0 \\ 0 & 1 & 1 & 1 \end{bmatrix}.$$

In Exercises 47–50, determine the check matrix associated with the given generator matrix.

47. the generator matrix for Exercises 43–44

48. $[1 \ \ 1 \ \ 1 \ \ 1 \ \ 1 \ \ 1 \ \ 1]$

49. $\begin{bmatrix} 1 & 0 & 0 & 1 & 0 & 1 & 0 & 1 \\ 0 & 1 & 0 & 0 & 0 & 1 & 1 & 1 \\ 0 & 0 & 1 & 1 & 1 & 0 & 1 & 1 \end{bmatrix}$

50. $\begin{bmatrix} 1 & 0 & 0 & 0 & 1 & 0 & 0 & 0 & 1 \\ 0 & 1 & 0 & 0 & 0 & 1 & 1 & 1 & 0 \\ 0 & 0 & 1 & 0 & 0 & 0 & 0 & 1 & 1 \\ 0 & 0 & 0 & 1 & 1 & 0 & 0 & 0 & 1 \end{bmatrix}$

In Exercises 51 and 52, determine the generator matrix for the given check matrix.

51. $\begin{bmatrix} 1 & 0 & 1 \\ 0 & 1 & 1 \\ 1 & 0 & 0 \\ 0 & 1 & 0 \\ 0 & 0 & 1 \end{bmatrix}$

52. $\begin{bmatrix} 0 & 1 & 1 & 0 & 0 \\ 1 & 0 & 0 & 0 & 1 \\ 0 & 0 & 1 & 1 & 1 \\ 1 & 0 & 0 & 1 & 1 \\ 1 & 0 & 0 & 0 & 0 \\ 0 & 1 & 0 & 0 & 0 \\ 0 & 0 & 1 & 0 & 0 \\ 0 & 0 & 0 & 1 & 0 \\ 0 & 0 & 0 & 0 & 1 \end{bmatrix}$

53. Find the generator matrix for the code than encodes 0 as 000000 and 1 as 111111.

In Exercises 54–60, the given word was received after being encoded using the generator matrix

$$A = \begin{bmatrix} 1 & 0 & 0 & 1 & 0 & 1 & 1 \\ 0 & 1 & 0 & 0 & 1 & 1 & 1 \\ 0 & 0 & 1 & 1 & 0 & 0 & 1 \end{bmatrix}$$

and transmitted. Determine the syndrome of each word, and use check matrix row decoding to decode it if it is a codeword or if it differs from a codeword in a single digit.

54. 1001100 **55.** 0111110 **56.** 1010111 **57.** 1010110

58. 0101010 **59.** 1001011 **60.** 1100111

61. Consider the code having the following generator matrix:

$$\begin{bmatrix} 1 & 0 & 0 & 0 & 1 & 1 & 0 \\ 0 & 1 & 0 & 0 & 1 & 0 & 1 \\ 0 & 0 & 1 & 0 & 1 & 1 & 1 \\ 0 & 0 & 0 & 1 & 0 & 1 & 1 \end{bmatrix}.$$

(a) Show that this matrix generates the (4, 7)-Hamming code.

(b) Decode, if possible, the received words 0001110, 1110000, and 1001000.

(c) How many transmission errors can this code detect?

(d) How many transmission errors can this code correct?

62. (a) What is the size of the generator matrix for the (57, 63)-Hamming code?

(b) What is the size of the check matrix for the (57, 63)-Hamming code?

63. Consider a code in which each codeword has length 10 and the minimal Hamming distance between distinct codewords is 4. Determine an upper bound on the number of codewords for this code.

64. Consider a code that can correct two or fewer transmission errors. If each codeword has length 12, determine an upper bound on the number of codewords for this code.

65. Prove that if c and d are codewords for a matrix code, then $c + d$ is also a codeword for this code, where the addition is performed using addition over Z_2 in each place.

66. Suppose that a butcher wants to buy the fewest possible weights that will enable him to weigh integral amounts up to 100 pounds with a balance scale. What weights should he buy?

67. If a, b, and c are integers and a divides $b + c$, must a divide either b or c? Justify your answer with a proof or a counterexample.

68. If a, b, and c are integers, does $\gcd(a, bc) = \gcd(a, b) \cdot \gcd(a, c)$? Justify your answer with a proof or a counterexample.

69. If n is an integer greater than 1, prove that $n^3 + 1$ is not a prime.

70. Prove by mathematical induction that $2^n + 3^n \equiv 5^n \pmod 6$ for every positive integer n.

71. Prove by mathematical induction that $16^n \equiv 1 - 10n \pmod{25}$ for every positive integer n.

72. Prove by mathematical induction that $3^{2n+1} + 2(-1)^n \equiv 0 \pmod 5$ for all nonnegative integers n.

73. Prove by mathematical induction that $7^{n+2} + 8^{2n+1} \equiv 0 \pmod{57}$ for all nonnegative integers n.

COMPUTER PROJECTS

Write a computer program having the specified input and output.

1. Given integers x, y, and m with $m \geq 2$, compute $[x] + [y]$ and $[x][y]$ in Z_m. Write the answers in the form $[r]$, where $0 \leq r < m$.

2. Given integers m and n, not both zero, find $\gcd(m, n)$.
3. Given integers m and n, not both zero, find $\gcd(m, n)$ and also integers x and y such that $mx + ny = \gcd(m, n)$.
4. Given positive integers P, E, and n, compute the remainder when P^E is divided by n.
5. Given any integer n greater than 1, compute the smallest prime divisor of n. (*Hint:* If n is odd, divide n by the odd integers greater than 1 until a divisor is found or Theorem 3.5 applies.)
6. Write a computer program that will determine whether the sum of two given matrices A and B is defined. If defined, print the sum; otherwise, print a message saying that the sum is not defined.
7. Write a computer program that will determine whether the product of two given matrices A and B is defined. If defined, print the product; otherwise, print a message saying that the product is not defined.
8. Given a set of codewords, compute the minimal Hamming distance between distinct codewords.
9. Given a generator matrix A for a (k, n)-block code, determine all the codewords in $\mathcal{W}_n$.
10. Given a $k \times n$ generator matrix for a (k, n)-block code, determine its associated check matrix.
11. Given a $k \times n$ generator matrix for a (k, n)-block code and a received word w in $\mathcal{W}_n$, determine its syndrome.

SUGGESTED READINGS

1. Gallian, Joseph A., "Assigning Driver's License Numbers." *Mathematics Magazine*, vol. 64, no. 1 (February 1991): 13–22.
2. ________, "The Mathematics of Identification Numbers." *The College Mathematics Journal*, vol. 22, no. 3 (May 1991): 194–202.
3. Lehmer, D.H., "Computer Technology Applied to the Theory of Numbers," in *Studies in Number Theory*, MAA Studies in Mathematics, vol. 6, W.J. LeVeque, ed., Mathematical Association of America, 1969, 117–151.
4. Knuth, Donald E., *The Art of Computer Programming, vol. 2: Seminumerical Algorithms*, 3d ed., Reading, MA: Addison-Wesley, 1997.
5. Tuchinsky, Phillip M. "International Standard Book Numbers." *The UMAP Journal*, vol. 6, no. 1 (1985): 41–53.
6. Vanden Eynden, Charles. *Elementary Number Theory*, 2d ed. New York: McGraw–Hill, 2001.

CHAPTER

4

Graphs

Even though graphs have been studied for a long time, the increased use of computer technology has generated a new interest in them. Applications of graphs have been found not only in computer science, but also in many other areas, such as business and science. As a consequence, the study of graphs has become important to many.

4.1 GRAPHS AND THEIR REPRESENTATIONS

It is quite common to represent situations involving objects and their relationships by drawing a diagram of points, with segments joining those points that are related. Let us consider some specific examples of this idea.

Example 4.1 Consider an airline route map in which dots represent cities, and two dots are joined by a segment whenever there is a nonstop flight between the corresponding cities. A portion of such an airline map is shown in Figure 4.1. ■

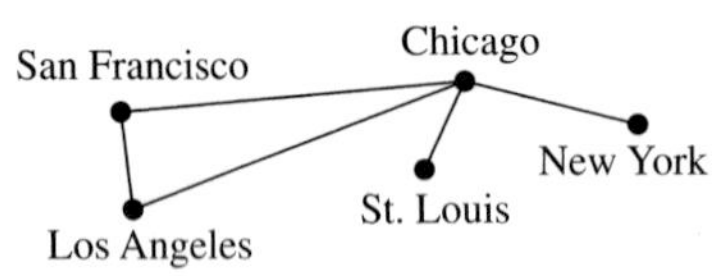

FIGURE 4.1

Example 4.2 Suppose we have four computers labeled A, B, C, and D, where there is a flow of information between computers A and B, C and D, and B and C. This situation can be represented by the diagram in Figure 4.2. This is usually referred to as a communication network. ■

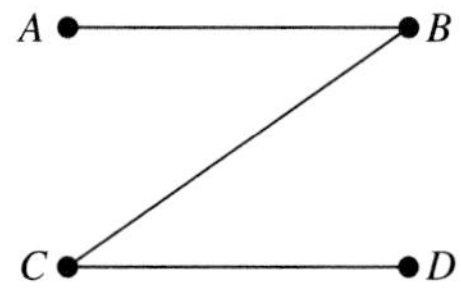

FIGURE 4.2

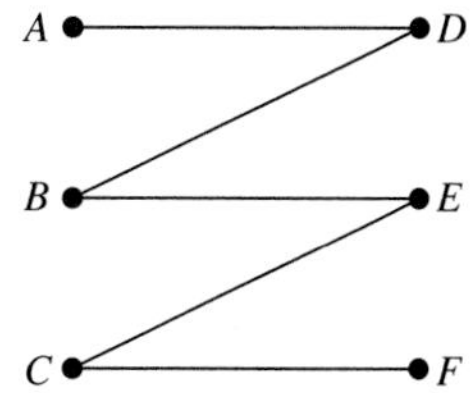

FIGURE 4.3

Example 4.3 Suppose that there is a group of people and a set of jobs that some of the people can do. For example, for individuals A, B, and C and jobs D, E, and F, suppose A can do only job D, B can do jobs D and E, and C can do jobs E and F. This type of situation can be represented by the diagram in Figure 4.3, where line segments are drawn between an individual and the jobs that person can do. ■

The general idea in the three examples is to represent by a picture a set of objects in which some pairs are related. We will now describe this type of representation more carefully.

A **graph** is a nonempty finite set $\mathcal{V}$ along with a set $\mathcal{E}$ of 2-element subsets of $\mathcal{V}$. The elements of $\mathcal{V}$ are called **vertices**, and the elements of $\mathcal{E}$ are called **edges**.

Figure 4.2 depicts a graph with vertices A, B, C, and D and edges $\{A, B\}$, $\{B, C\}$, and $\{C, D\}$. Thus a graph can be described either by the use of sets or by the use of a diagram, where segments between the vertices in $\mathcal{V}$ describe which 2-element subsets are being included. Figure 4.3 shows a graph with vertices A, B, C, D, E, and F and with edges $\{A, D\}$, $\{B, D\}$, $\{B, E\}$, $\{C, E\}$, and $\{C, F\}$.

We caution the reader that the use of terminology in graph theory is not consistent among users, and when consulting other books, definitions should always be checked to see how words are being used. In our definition of a graph, the set of vertices is required to be a finite set. Some authors do not make this restriction, but we find it convenient to do so. Also, our definition of a graph does not allow an edge from a vertex to itself, or different edges between the same two vertices. Some authors allow such edges, but we do not.

Whenever we have an edge $e = \{U, V\}$, we say that the edge e **joins** the vertices U and V and that U and V are **adjacent**. It is also said that edge e is **incident** with the vertex U and that the vertex U is **incident** with the edge e. For the graph in Figure 4.2, we see that vertices A and B are adjacent, whereas vertices A and C are not, because there is no segment between them (that is, the set $\{A, C\}$ is not an edge). In Figure 4.3 the edge $\{B, E\}$ is incident with the vertex B.

Note that the diagram in Figure 4.2 can be drawn differently and still represent the same graph. Another representation of this graph is given in Figure 4.4.

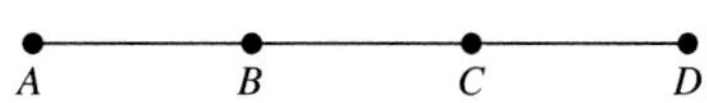

FIGURE 4.4

The way our picture is drawn is not important, although one picture may be much easier to understand than another. What is important in the picture is which vertices are joined by edges, for this describes what relationships exist between the vertices. In Figure 4.5 we have redrawn the graph from Figure 4.2 in such a way that the edges meet at a place other than a vertex. It is important not to be misled into believing that there is now a new vertex. Sometimes it is not possible to draw a picture of a graph without edges meeting in this way, and it is important to understand that such a crossing does not generate a new vertex of the graph. It is often very difficult to determine if a graph can be drawn without any edges crossing at points other than vertices.

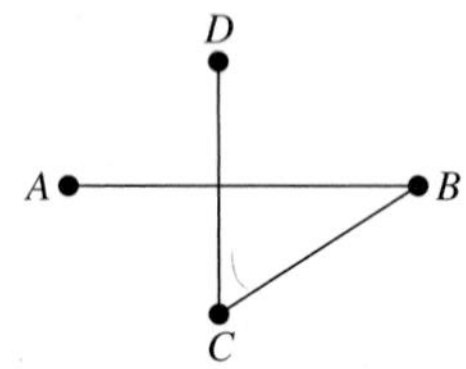

FIGURE 4.5

In a graph, the number of edges incident with a vertex V is called the **degree** of V and is denoted as $\deg(V)$. In Figure 4.6 we see that $\deg(A) = 1, \deg(B) = 3$, and $\deg(C) = 0$.

One special graph that is encountered frequently is the **complete graph** on n vertices, where every vertex is joined to every other vertex. This graph is denoted by $\mathcal{K}_n$. Figure 4.7 shows $\mathcal{K}_3$ and $\mathcal{K}_4$.

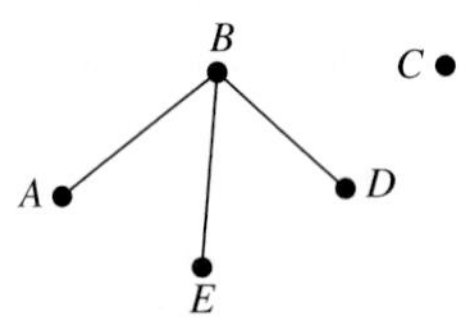

FIGURE 4.6

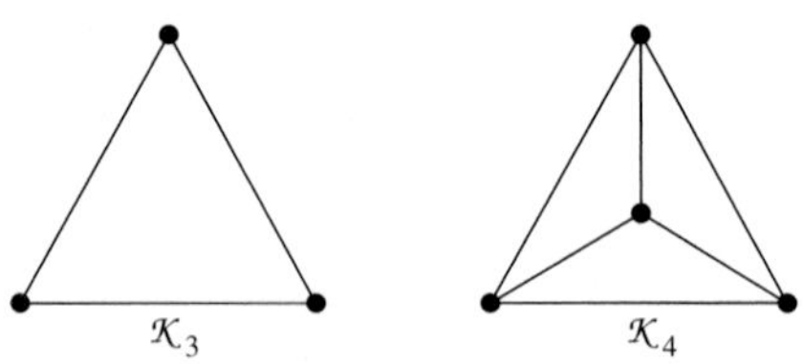

FIGURE 4.7

In Figures 4.6 and 4.7, notice that adding the degrees of the vertices in each graph yields a number that is twice the number of edges. This result is true in general.

Theorem 4.1 In a graph, the sum of the degrees of the vertices equals twice the number of edges.

Proof. The key to understanding why this theorem is true is to see that each edge is incident with two vertices. When we take the sum of the degrees of the vertices, each edge is counted twice in this sum. Thus the sum of the degrees is twice the number of the edges. Look again at Figures 4.5, 4.6, and 4.7 to see how this double counting of edges takes place. ■

Other Representations of Graphs

It is often necessary to analyze graphs and perform a variety of procedures and algorithms upon them. When a graph has many vertices and edges, it may be essential to use a computer to perform these algorithms. Thus it is necessary to communicate to the computer the vertices and edges of a graph. One way to do so is to represent a graph by means of matrices (discussed in Appendix B), which are easily manipulated with a computer.

Suppose we have a graph $\mathcal{G}$ with n vertices labeled $V_1, V_2, \ldots, V_n$. Such a graph is called a **labeled graph**. To represent the labeled graph $\mathcal{G}$ by a matrix, we form an $n \times n$ matrix in which the i, j entry is 1 if there is an edge between the vertices V_i and V_j and 0 if there is not. This matrix is called the **adjacency matrix** of $\mathcal{G}$ (with respect to the labeling) and is denoted by $A(\mathcal{G})$.

Example 4.4 Figure 4.8 contains two graphs and their adjacency matrices. For (a), the $1, 2$ entry is 1 because there is an edge between vertices V_1 and V_2, and the $3, 4$ entry is 0 because there is no edge between V_3 and V_4. For (b), we see that the $1, 2$ and $1, 3$ entries are 1 because of the edges between V_1 and V_2 and between V_1 and V_3.

Note that in $A(\mathcal{G}_1)$ the sum of the entries in row 1 is 1, which is the degree of V_1, and likewise the sum of the entries in row 2 is the degree of V_2. This illustrates a more general result. ■

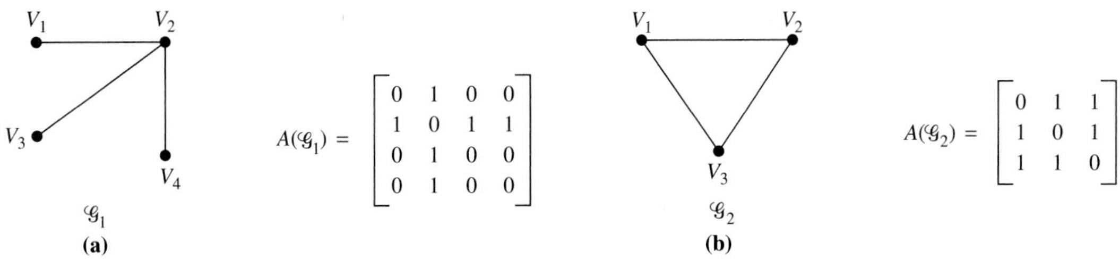

FIGURE 4.8

Theorem 4.2 The sum of the entries in row i of the adjacency matrix of a graph is the degree of the vertex V_i in the graph.

Proof. We recall that each 1 in row i corresponds to an edge on the vertex V_i. Thus the number of 1s in row i is the number of edges on V_i, which is the degree of V_i. ■

Matrices are not the only way to represent graphs in a computer. Although an adjacency matrix is easy to construct, this form of representation requires $n \cdot n = n^2$ units of storage for a graph with n vertices and so can be quite inefficient if the matrix contains many zeros. This means that if an algorithm to be performed on the graph requires a lot of searching of vertices and adjacent vertices, then the matrix representation can require much unnecessary time. A better representation for such a graph is an **adjacency list**.

The basic idea of an adjacency list is to list each vertex, followed by the vertices adjacent to it. This provides the basic information about a graph: the vertices and the edges. To form the adjacency list, we begin by labeling the vertices of the graph. Then we list the vertices in a vertical column, and after each one we write down the adjacent vertices. Thus we see that the 1s in a row of an adjacency matrix tell what vertices are listed in the corresponding row of an adjacency list.

Example 4.5 For the graph in Figure 4.9, there are six labeled vertices, and we list them in a vertical column, as in (b). Beside vertex V_1 we list the adjacent vertices, which are V_2 and V_3. Then, proceeding to the next vertex V_2, we list the vertices adjacent to it, V_1 and V_4. This process is continued until we get the adjacency list in (b). ■

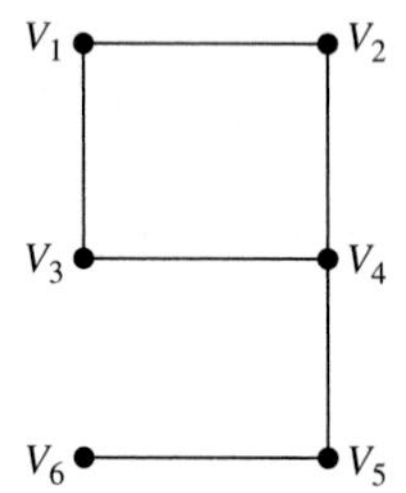

(a)

$V_1: V_2, V_3$

$V_2: V_1, V_4$

$V_3: V_1, V_4$

$V_4: V_2, V_3, V_5$

$V_5: V_4, V_6$

$V_6: V_5$

(b)

FIGURE 4.9

Isomorphism

In 1953 the CIA managed to photograph a KGB document listing their agents in a large third-world city, showing their past operations, duties, and contacts among themselves. Unfortunately, it listed the agents by the code designations D through L. The document shows that agent D's contacts were F and L; E's were J and K; F's were D, J, and L; G's were I and L; H's were I and J; I's were G, H, and K; J's were E, F, and H; K's were E and I; and L's were D, F, and G. Drawing an edge between agents if they are contacts produces the graph $\mathcal{A}$ shown in Figure 4.10. Unfortunately, the information in the document was of little use without the identities of the agents.

An inquiry to the CIA office in this city revealed that the suspected agents there were Telyanin, Rostov, Lavrushka, Kuragin, Ippolit, Willarski, Dolokhov, Balashev, and Kutuzov. Examining records of past meetings observed among them, the CIA created the contact graph $\mathcal{C}$ in Figure 4.11 by joining two individuals with an edge if they were known to have met together.

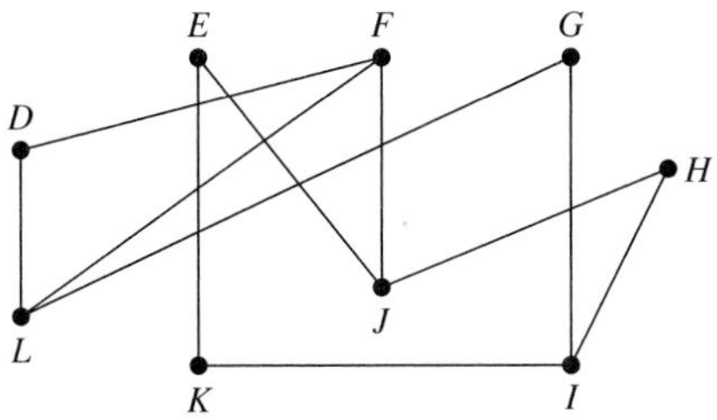

FIGURE 4.10

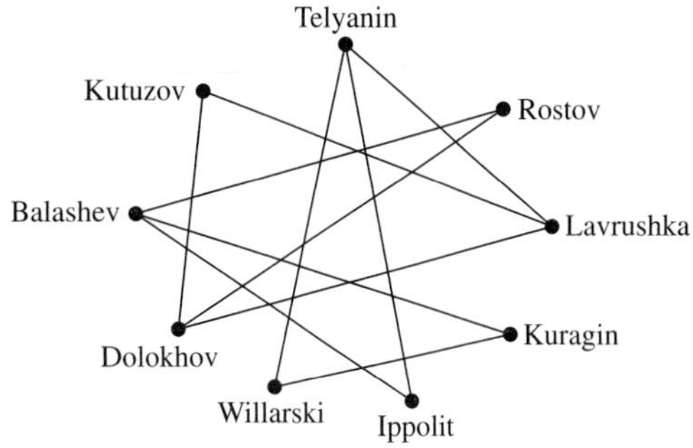

FIGURE 4.11

If these nine persons are the agents described in the KGB document, then it must be possible to match the code designations D through L with the names in Figure 4.11 so that the edges in $\mathcal{A}$ correspond exactly to the edges in $\mathcal{C}$. In general, we say that a graph $\mathcal{G}_1$ is **isomorphic** to a graph $\mathcal{G}_2$ when there is a one-to-one correspondence f between the vertices of $\mathcal{G}_1$ and $\mathcal{G}_2$ such that the vertices U and W are adjacent in $\mathcal{G}_1$ if and only if the vertices $f(U)$ and $f(W)$ are adjacent in $\mathcal{G}_2$. The function f is called an **isomorphism** of $\mathcal{G}_1$ with $\mathcal{G}_2$. Because the relation "is isomorphic to" is a symmetric relation on any set of graphs (see Exercise 38), we usually just say that the graphs $\mathcal{G}_1$ and $\mathcal{G}_2$ are **isomorphic**. Thus isomorphic graphs are essentially the same in the sense that, except for notation, they have the same vertices, and the same pairs of vertices are adjacent.

An examination of the graphs in Figure 4.12 shows that they are isomorphic by using the correspondence indicated in the figure. The graphs in Figures 4.10 and 4.12 are not isomorphic, however, because they have different numbers of vertices. In Figure 4.13, vertex C in $\mathcal{G}_1$ is adjacent to the vertices A, B, D, and E. Therefore, under any isomorphism of $\mathcal{G}_1$ with $\mathcal{G}_2$, the image of C would also need to have four adjacent vertices. Since there are no vertices of degree 4 in $\mathcal{G}_2$, we see that $\mathcal{G}_1$ and $\mathcal{G}_2$ are not isomorphic.

A .... a
B .... d
C .... g
D c
E .... f
F .... b
G .... e

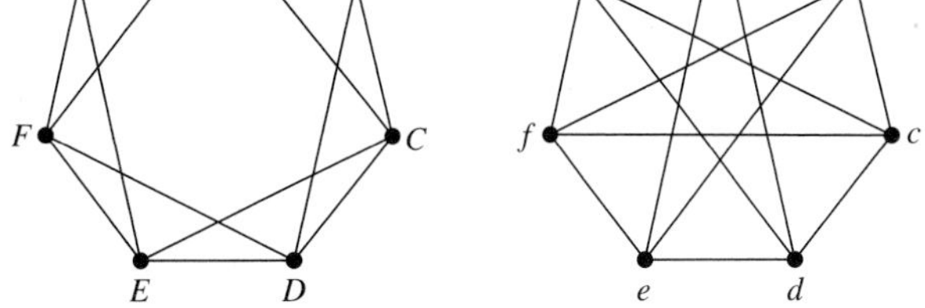

FIGURE 4.12

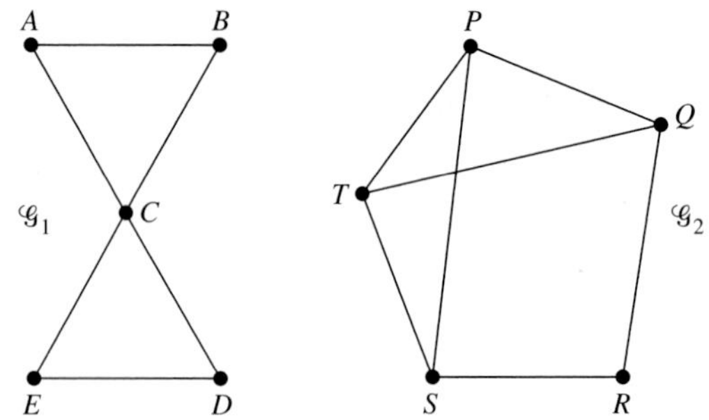

FIGURE 4.13

This last observation illustrates the following theorem:

Theorem 4.3 Let f be an isomorphism of graphs $\mathcal{G}_1$ and $\mathcal{G}_2$. For any vertex V in $\mathcal{G}_1$, the degrees of V and $f(V)$ are equal.

Proof. Suppose f is an isomorphism of $\mathcal{G}_1$ with $\mathcal{G}_2$ and that V is a vertex of degree k in $\mathcal{G}_1$. Then there are exactly k vertices $U_1, U_2, \ldots, U_k$ in $\mathcal{G}_1$ that are adjacent to V. Since f is an isomorphism, $f(U_1), f(U_2), \ldots, f(U_k)$ are adjacent to $f(V)$. Since there are no other vertices in $\mathcal{G}_1$ adjacent to V, there are no other vertices in $\mathcal{G}_2$ adjacent to $f(V)$. Thus $f(V)$ has degree k in $\mathcal{G}_2$. ■

As a consequence of this theorem, we see that the degrees of the vertices of isomorphic graphs must be exactly the same.

A property is said to be a graph isomorphism **invariant** if, whenever $\mathcal{G}_1$ and $\mathcal{G}_2$ are isomorphic graphs and $\mathcal{G}_1$ has this property, then so does $\mathcal{G}_2$. The properties "has n vertices," "has e edges," and "has a vertex of degree k" are all invariants. Thus one way to show two graphs are *not* isomorphic is to find an invariant property possessed by only one of the graphs. This is what was done in showing that the graphs in Figure 4.13 are not isomorphic.

Returning to the graphs in Figures 4.10 and 4.11, we will construct an explicit isomorphism of $\mathcal{A}$ with $\mathcal{C}$ by observing similarities between the graphs. Note that vertices L, D, and F form the only "triangle" in $\mathcal{A}$, that is, the only set of three vertices in which each pair is adjacent. Thus, to have an isomorphism, these vertices must correspond to Kutuzov, Lavrushka, and Dolokhov, in some order, since they form the only similar set in $\mathcal{C}$. In fact, since of these six vertices, only D and Kutuzov have degree 2, we must have $f(D) =$ Kutuzov. Moreover, since of L and F, only L is adjacent to another vertex of degree 2 (namely, G), and of Lavrushka and Dolokhov, only Dolokhov is joined to another vertex of degree 2 (namely Rostov), we must have $f(L) =$ Dolokhov, $f(F) =$ Lavrushka, and $f(G) =$ Rostov.

Continuing in this way, we find that, for f to be an isomorphism, we must have $f(J) =$ Telyanin, $f(I) =$ Balashev, $f(H) =$ Ippolit, $f(K) =$ Kuragin, and $f(E) =$ Willarski. It is easily checked that f is indeed an isomorphism.

Note that if there had been more than one isomorphism from $\mathcal{A}$ to $\mathcal{C}$, then a complete identification of the agents would not have been possible. For example, consider the graphs of agents $\mathcal{R}$ and contacts $\mathcal{S}$ in Figure 4.14.

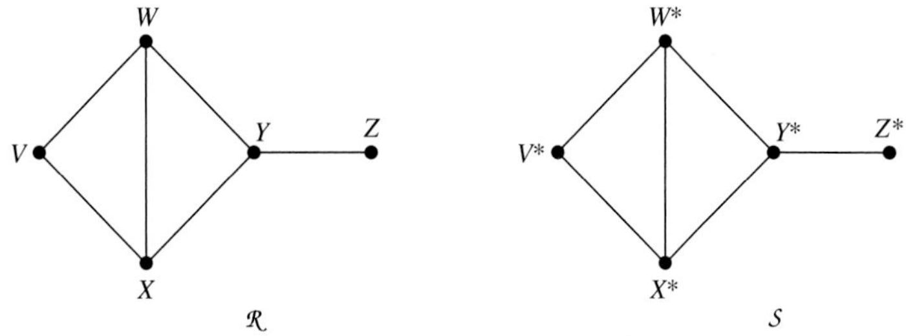

FIGURE 4.14

Notice that in addition to the obvious isomorphism sending V, W, X, Y, and Z into V^*, W^*, X^*, Y^*, and Z^*, respectively, there is also an isomorphism that is the same, except that W maps to X^* and X to W^*. Hence, in this case, it would be impossible to deduce whether W^* or X^* is the identity of agent W.

EXERCISES 4.1

In Exercises 1–4, list the set of edges and set of vertices for each graph.

1. 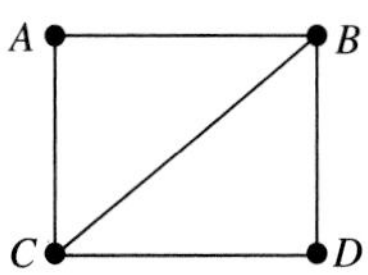

2. 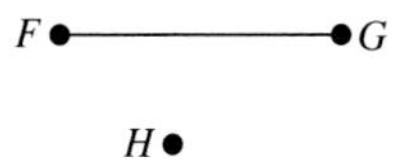

3. 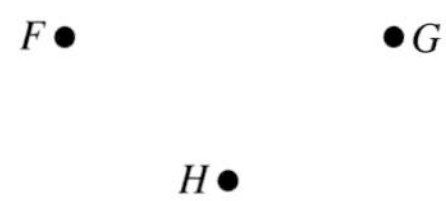

4. 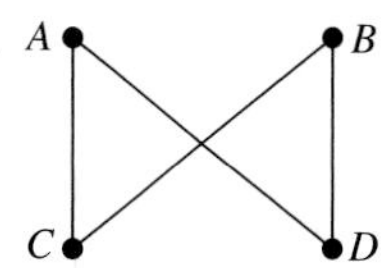

In Exercises 5–8, draw a diagram representing the graph with the set $\mathcal{V}$ of vertices and the set $\mathcal{E}$ of edges.

5. $\mathcal{V} = \{A, B, C, D\}$, $\mathcal{E} = \{\{B, C\}, \{C, A\}, \{B, D\}\}$

6. $\mathcal{V} = \{X, Y, Z, W\}$, $\mathcal{E} = \{\{X, Y\}, \{X, Z\}, \{Y, Z\}, \{Y, W\}\}$

7. $\mathcal{V} = \{G, H, J\}$, $\mathcal{E} = \emptyset$

8. $\mathcal{V} = \{A, X, B, Y\}$, $\mathcal{E} = \{\{A, X\}, \{X, B\}, \{B, Y\}, \{Y, A\}\}$

In Exercises 9–14, determine whether a graph is indicated.

9. 

10.

11. 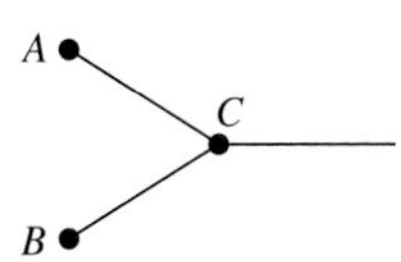

12. 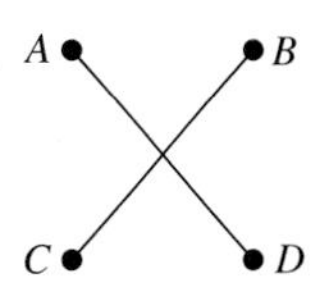

13. $\mathcal{V} = \{A, B, C, D\}$, $\mathcal{E} = \{\{A, B\}, \{A, A\}\}$

14. $\mathcal{V} = \{A, B\}$, $\mathcal{E} = \{\{A, B\}, \{B, C\}\}$

15. Construct the graph where the vertices are you, your parents, and your grandparents with a relationship of "are the same sex."

16. Construct the graph where the vertices are you, your parents, and your grandparents with a relationship of "born in the same country."

17. There is a group of 6 students, Alice, Bob, Carol, Dean, Santos, and Tom, where Alice and Carol are always feuding, likewise for Dean and Carol, and for Santos, Tom, and Alice. Draw the graph to represent this situation.

18. Draw the graph with $\mathcal{V} = \{1, 2, \ldots, 10\}$ as its set of vertices and

$$\mathcal{E} = \{\{x, y\}: x, y \text{ in } \mathcal{V},\ x \neq y, \text{ and } x \text{ divides } y \text{ or } y \text{ divides } x\}$$

as its set of edges.

In Exercises 19 and 20, list the vertices adjacent to A and give the degree of A. Repeat for the vertex B.

19.

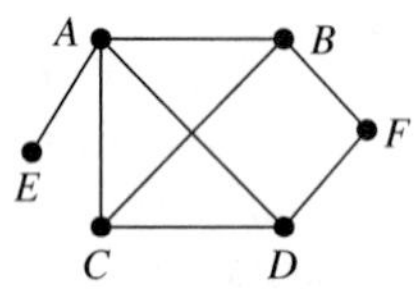

20.

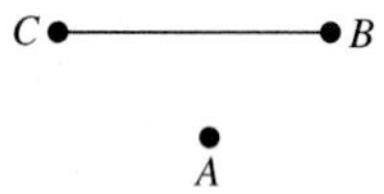

21. Draw graphs where

(a) there are 4 vertices, each with degree 1.

(b) there are 4 vertices, each with degree 2.

22. Show that there are an even number of vertices with odd degree in any graph.

23. How many edges does $\mathcal{K}_3$ have? $\mathcal{K}_4$? $\mathcal{K}_5$? $\mathcal{K}_n$ in general?

24. Can there be a graph with 8 vertices and 29 edges? Justify your answer.

25. How many vertices are there in a graph with 10 edges if each vertex has degree 2?

In Exercises 26–29, find the adjacency matrix and the adjacency list for each graph.

26. $\mathcal{K}_4$

27.

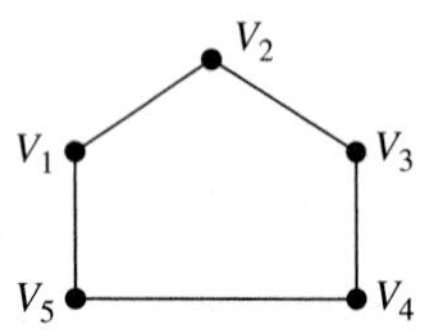

28.

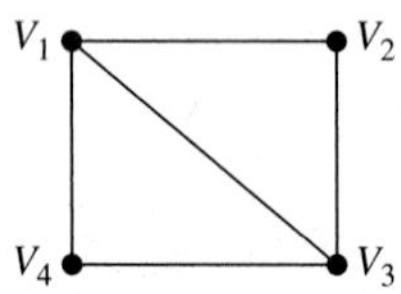

29.

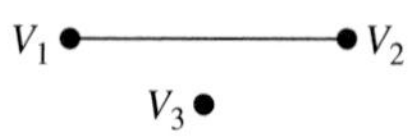

In Exercises 30 and 31, construct the graph for each adjacency matrix. Label the vertices $V_1, V_2, V_3, \ldots$.

30.
$$\begin{bmatrix} 0 & 1 & 1 & 0 & 1 \\ 1 & 0 & 1 & 1 & 1 \\ 1 & 1 & 0 & 1 & 0 \\ 0 & 1 & 1 & 0 & 0 \\ 1 & 1 & 0 & 0 & 0 \end{bmatrix}$$

31.
$$\begin{bmatrix} 0 & 1 & 1 & 1 \\ 1 & 0 & 1 & 1 \\ 1 & 1 & 0 & 1 \\ 1 & 1 & 1 & 0 \end{bmatrix}$$

In Exercises 32 and 33, construct the graph for each adjacency list.

32. V_1: V_2, V_4, V_5
V_2: V_1, V_3
V_3: V_2, V_5
V_4: V_1
V_5: V_1, V_3

33. V_1: V_2, V_3
V_2: V_1, V_4
V_3: V_1, V_4
V_4: V_2, V_3

34. What does it mean when the adjacency matrix of a graph contains only zeros?

In Exercises 35–37, can each matrix be an adjacency matrix?

35. $$\begin{bmatrix} 0 & 1 & 1 & 0 \\ 1 & 0 & 1 & 0 \\ 1 & 1 & 0 & 1 \\ 0 & 0 & 1 & 0 \end{bmatrix}$$

36. $$\begin{bmatrix} 0 & 1 & 0 & 1 & 0 \\ 1 & 0 & 0 & 1 & 1 \\ 0 & 0 & 0 & 0 & 1 \\ 1 & 1 & 0 & 0 & 0 \\ 0 & 0 & 0 & 0 & 0 \end{bmatrix}$$

37. $$\begin{bmatrix} 1 & 1 & 1 \\ 1 & 1 & 1 \\ 1 & 1 & 1 \end{bmatrix}$$

38. Show that the relation "is isomorphic to" is an equivalence relation on any set of graphs.

39. Will two graphs that have the same number of vertices always be isomorphic? Justify your answer.

40. Will two graphs that have the same number of edges always be isomorphic? Justify your answer.

41. Will two graphs having the same number of vertices of degree k for each nonnegative integer k always be isomorphic? Justify your answer.

42. Are the pairs of graphs in (a), (b), and (c) isomorphic? Justify your answer.

(a)
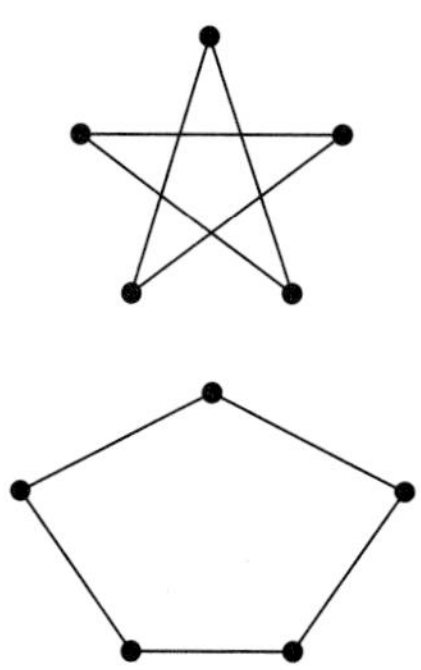

(b)
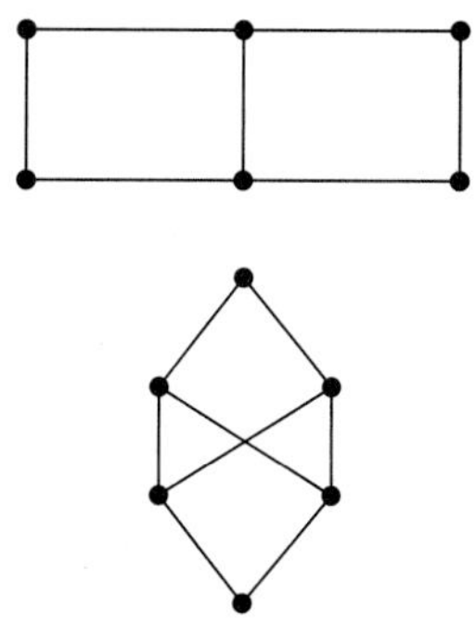

(c)
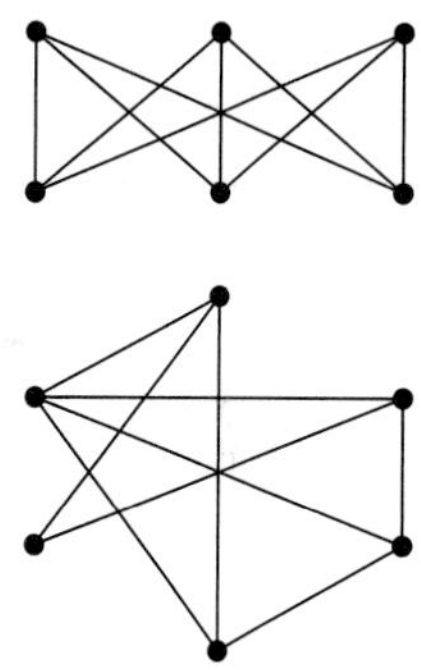

43. Are the pairs of graphs in (a), (b), and (c) isomorphic? Justify your answer.

(a)
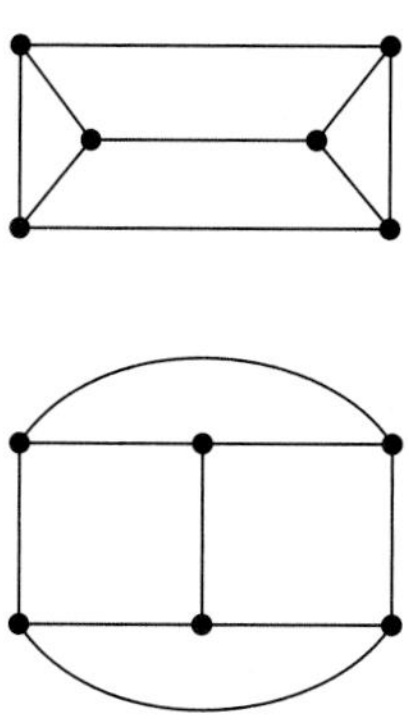

(b)
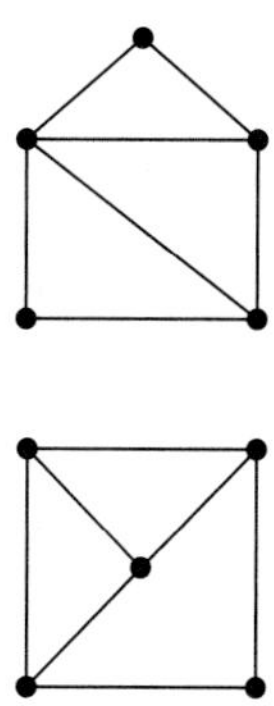

(c)
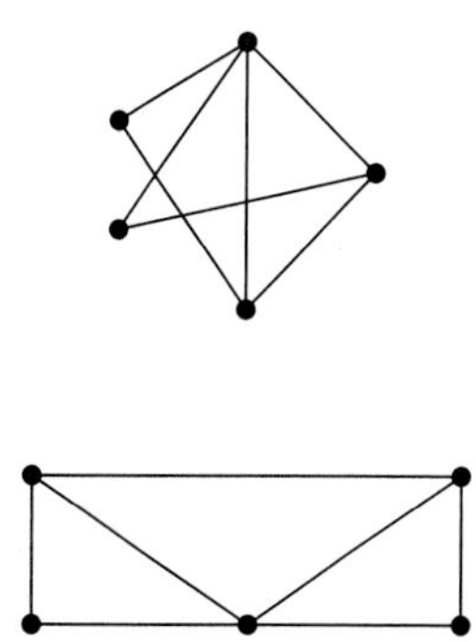

44. Draw all the nonisomorphic graphs with three vertices.

45. Draw all the nonisomorphic graphs with four vertices.

46. Draw all the nonisomorphic graphs with five vertices of degrees 1, 2, 2, 2, and 3.

47. Draw all the nonisomorphic graphs with six vertices of degrees 1, 1, 1, 2, 2, and 3.

48. Consider a graph whose vertices are the closed intervals $[m, n]$, where $m < n$ are distinct integers between 1 and 4, inclusive. Let two distinct vertices be adjacent if the corresponding intervals have at least one point in common. Draw the graph.

49. Suppose a graph has n vertices, each with degree at least 1. What is the smallest number of edges the graph can have? Justify your answer.

50. Suppose a graph has n vertices, each with degree at least 2. What is the smallest number of edges it can have? Justify your answer.

51. A graph has m edges with $m \geq 2$. What is the smallest number of vertices it can have? Justify your answer.

52. For $n > 3$, let $\mathcal{V} = \{1, 2, \ldots, n\}$ and

$$\mathcal{E} = \{\{x, y\}: x, y \text{ in } \mathcal{V},\ x \neq y, \text{ and } x \text{ divides } y \text{ or } y \text{ divides } x\}.$$

What vertices of this graph have degree 1?

53. Suppose Mr. and Mrs. Lewis attended a bridge party one evening. There were three other married couples in attendance, and several handshakes took place. No one shook hands with himself or herself, no spouses shook hands, and no two people shook hands more than once. When each other person told Mr. Lewis how many hands he or she shook, the answers were all different. How many handshakes did Mr. and Mrs. Lewis each make?

54. Prove that if a graph has at least two vertices, then there are two distinct vertices that have the same degree.

4.2 PATHS AND CIRCUITS

As we have seen, graphs can be used to describe a variety of situations. In many cases we want to know whether it is possible to go from one vertex to another by following edges. In other cases it may be necessary to perform a test that involves finding a route through all the vertices or over all the edges. While many situations can be described by graphs as we have defined them, there are others where it may be necessary to allow an edge from a vertex to itself or to allow more than one edge between vertices. For example, when a road system is being described, there can be two roads, an interstate highway and an older two-lane road, between the same two towns. There could even be a scenic route starting and ending at the same town. To describe these situations, we need to generalize the concept of a graph. A **multigraph** consists of a nonempty finite set of vertices and a set of edges, where we allow an edge to join a vertex to itself or to a different vertex, and where we allow several edges joining the same pair of vertices. An edge from a vertex to itself is called a **loop**. When there is more than one edge between two vertices, these edges are called **parallel edges**.

It is important to note that a graph is a special kind of multigraph. Thus all the definitions given for multigraphs apply to graphs, as well.

Example 4.6 The diagram in Figure 4.15 represents a multigraph, but not a graph, because there are two parallel edges k and m between the vertices Y and Z and a loop h at vertex X. ■

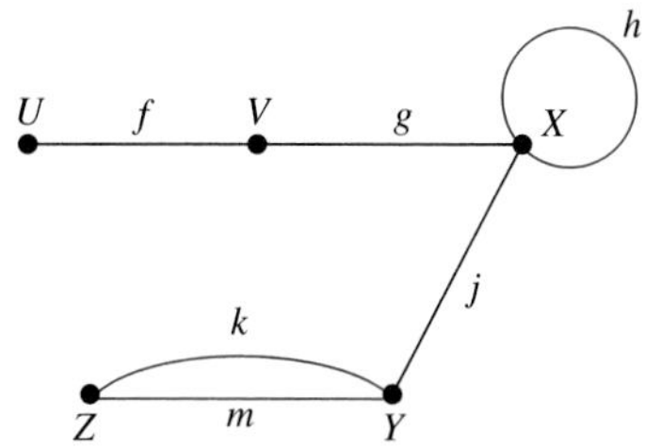

FIGURE 4.15

In a multigraph the number of edges incident with a vertex V is called the **degree** of V and is denoted as $\deg(V)$. A loop on a vertex V is counted twice in $\deg(V)$. Thus, in Figure 4.15, $\deg(Y) = 3$ and $\deg(X) = 4$.

Suppose $\mathcal{G}$ is a multigraph and U and V are vertices, not necessarily distinct. A ***U*-*V* path**, or a **path from *U* to *V***, is an alternating sequence

$$V_1, e_1, V_2, e_2, V_3, \ldots, V_n, e_n, V_{n+1}$$

of vertices and edges in which the first vertex V_1 is U, the last vertex V_{n+1} is V, and the edge e_i joins V_i and V_{i+1} for $i = 1, 2, \ldots, n$. The **length** of this path is n, the number of edges listed. We note that U is a path to itself of length 0.

In a path the vertices need not be distinct, and some of the edges can be the same. When there can be no chance of confusion, a path can be represented by the vertices $V_1, V_2, \ldots, V_{n+1}$ only or by the edges $e_1, e_2, \ldots, e_n$ only. Note that in a graph it is always sufficient to list only the vertices or the edges.

Example 4.7 In Figure 4.15, U, f, V, g, X is a path of length 2 from U to X. This path can also be written as f, g. Likewise, f, g, h is a path of length 3 from U to X, and U, f, V, f, U is a path of length 2 from U to U. The path Z, m, Y cannot be described by just listing the vertices Z, Y, since it would not be clear which edge between Z and Y, k or m, is part of the path. ■

A path provides a way of describing how to go from one vertex to another by following edges. A U-V path need not be an efficient route; that is, it may repeat vertices or edges. However, a ***U*-*V* simple path** is a path from U to V in which no vertex and, hence, no edge is repeated.

There are no simple paths of length 1 or more from a vertex to itself. Furthermore, a simple path does not have loops or pairs of parallel edges in it. In some sense, a simple path is an efficient route between vertices, whereas a path allows wandering back and forth, repeating vertices and edges.

Example 4.8 For the multigraph in Figure 4.16, the edges a, c, d, j form a simple path from U to Z, whereas a, c, m, d, j is a path from U to Z that is not a simple path, because the vertex W is repeated. Similarly, e, i is a simple path from X to Z, but f, i, j is a path from X to Z that is not a simple path. Note also that c, p, f, i, e, n is a path from V to U that is not simple, but deleting f, i, e produces a simple path c, p, n from V to U. This illustrates Theorem 4.4. ■

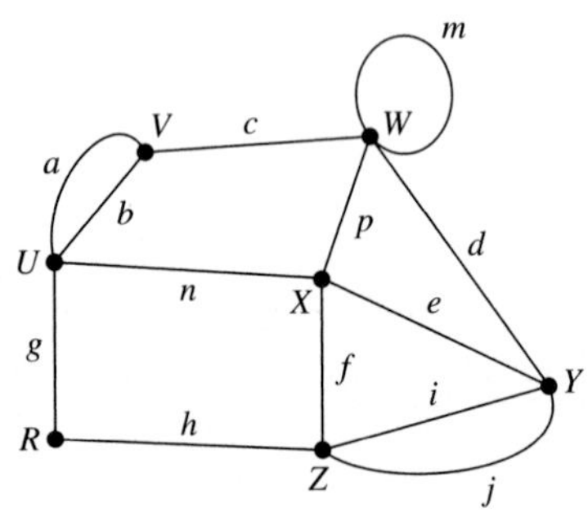

FIGURE 4.16

Theorem 4.4 Every U-V path contains a U-V simple path.

Proof. Let us suppose that $U = V_1, e_1, V_2, \ldots, e_n, V_{n+1} = V$ is a U-V path. In the special case that $U = V$ we can choose our U-V simple path to be just the vertex U. So suppose that $U \neq V$. If all of the vertices $V_1, V_2, \ldots, V_{n+1}$ are different initially, then our path is already a U-V simple path. Thus let us suppose that at least two of the vertices are the same, say, $V_i = V_j$, where $i < j$. See Figure 4.17 for an illustration of how to form a simple path from V_i to V_j.

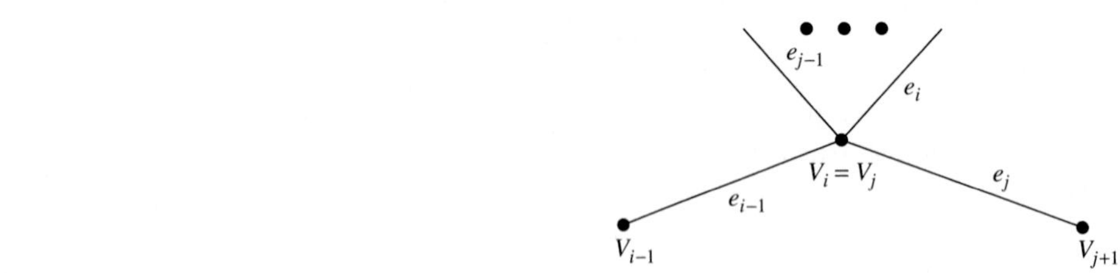

FIGURE 4.17

We delete e_i, $V_{i+1}, \ldots, e_{j-1}$, V_j from the original path. What has been deleted is the part that is between vertex V_i and edge e_j. This still leaves a path from U to V. If there are only distinct vertices left after this deletion, then we are done. If there are still repetitions among the remaining vertices, we delete vertices and edges as before. Because the number of vertices is finite, this process will eventually end and give a U-V simple path from U to V. ■

A multigraph is called **connected** if there is a path between every pair of vertices. Thus in a connected multigraph we can go from any one vertex to another by following some route along the edges.

Example 4.9 The multigraph in Figure 4.16 is connected, since a path can be found between any two vertices. However, the graph in Figure 4.18 is not connected, since there is no path from vertex U to vertex W. ■

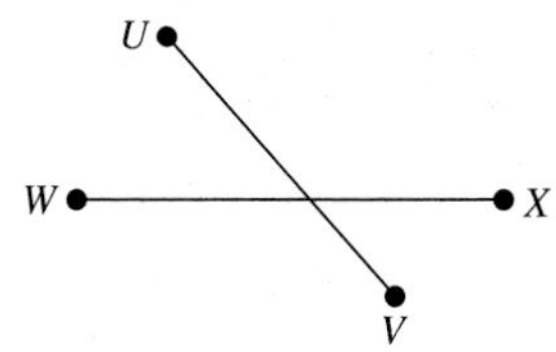

FIGURE 4.18

A **cycle** is a path $V_1, e_1, V_2, e_2, \ldots, V_n, e_n, V_{n+1}$, where $n > 0$, $V_1 = V_{n+1}$, and all the vertices $V_1, V_2, \ldots, V_n$ and all the edges $e_1, e_2, \ldots, e_n$ are distinct. Thus a cycle of length 3 or more cannot have loops or parallel edges as part of it.

Example 4.10 For the multigraph in Figure 4.16, the edges a, c, p, n form a cycle. Likewise, the edges g, b, c, p, f, h form a cycle. Furthermore, the edges f, p, d, e, n, g, h do not form a cycle, because the vertex X is used twice. ■

Euler Circuits and Paths

In testing a communication network, it is often necessary to examine each link (edge) in the system. In order to minimize the cost of such a test, it is desirable to devise a route that goes through each edge exactly once. Similarly, when devising a garbage pick-up route (where the garbage is picked up along both sides of the street with one pass), we will want to go over each street exactly once. Thus, when modeled by a multigraph (with corners as the vertices and streets as the edges), we want a path that includes every edge exactly once.

Because the mathematician Leonhard Euler was the first person known to consider this concept, a path in a multigraph $\mathcal{G}$ that includes exactly once all the edges of $\mathcal{G}$ and has different first and last vertices is called an **Euler path**. A path that includes exactly once all the edges of $\mathcal{G}$ and has the same initial and terminal vertices is called an **Euler circuit**.

Example 4.11 For the graph in Figure 4.19(a), the path a, b, c, d is an Euler circuit, since all the edges are included and each edge is included exactly once. However, the graph in Figure 4.19(b) has neither an Euler path nor circuit because, to include all three

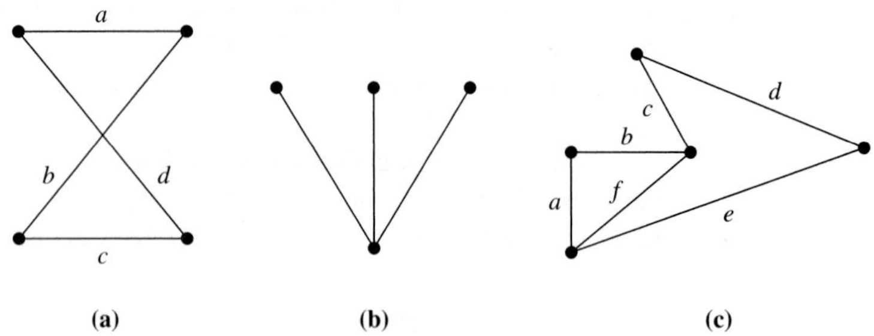

FIGURE 4.19

of the edges in a path, we would have to backtrack and use an edge twice. For the graph in Figure 4.19(c), there is an Euler path a, b, c, d, e, f, but not an Euler circuit. ■

As we proceed along an Euler circuit, each time a vertex is reached along some edge, there must be another edge for us to exit that vertex. This implies that the degrees of the vertices must all be even. In fact, as we will see shortly, the converse statement is also true: Whenever a multigraph is connected and the degree of each vertex is even, then the graph has an Euler circuit. The next example shows how an Euler circuit can be constructed in such a case.

Example 4.12 The multigraph shown in Figure 4.20(a) is connected, and the degree of each vertex is even. Therefore an Euler circuit can be constructed as follows: Select any vertex U, and construct a path $\mathcal{C}$ from U to U by randomly selecting unused edges for as long as possible. For example, if we start at G, we may construct the path

$$\mathcal{C}: G, h, E, d, C, e, F, g, E, j, H, k, G.$$

The edges in this path $\mathcal{C}$ are shown in color in Figure 4.20(b).

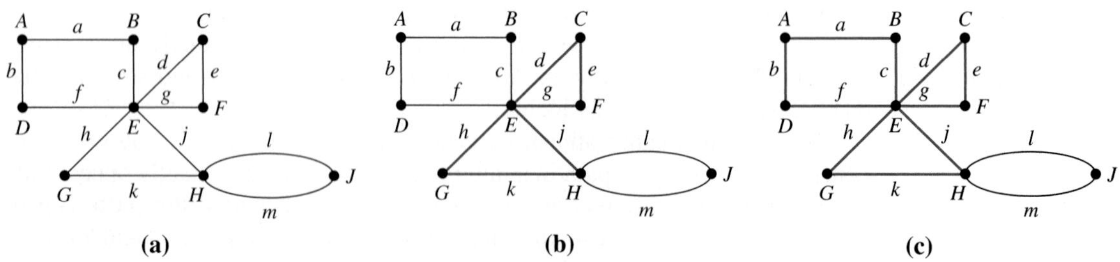

FIGURE 4.20

Note that such a path must return to the starting vertex because the degree of each vertex is even and the number of vertices is finite. In addition, every edge incident with the starting vertex must be included in this path. If, as is the case here, this path $\mathcal{C}$ is not an Euler circuit, then there must be edges not in $\mathcal{C}$. Moreover, since the multigraph is connected, there must be a vertex in $\mathcal{C}$ that is incident with an edge not in $\mathcal{C}$. In our case, the vertices E and H are vertices in $\mathcal{C}$ that are incident with an edge not in $\mathcal{C}$. Arbitrarily choose one of these, say, E, and construct a path $\mathcal{P}$ from E to E in the same manner that $\mathcal{C}$ was constructed. One possibility for $\mathcal{P}$ is

$$\mathcal{P}\colon\ E, c, B, a, A, b, D, f, E.$$

We now enlarge $\mathcal{C}$ to include the path $\mathcal{P}$ by replacing any one occurrence of E in $\mathcal{C}$ by $\mathcal{P}$. For example, if we replace the first occurrence of E in $\mathcal{C}$, we obtain

$$\mathcal{C}'\colon\ G, h, E, c, B, a, A, b, D, f, E, d, C, e, F, g, E, j, H, k, G.$$

The edges in the enlarged path $\mathcal{C}'$ are shown in color in Figure 4.20(c). Notice that $\mathcal{C}'$, although larger than $\mathcal{C}$, is still not an Euler circuit. We now repeat the procedure used in the preceding paragraph. In this case H is the only vertex in $\mathcal{C}'$ that is incident with an edge not in $\mathcal{C}'$. So we construct a path $\mathcal{P}'$ from H to H, say, $\mathcal{P}'$: H, m, J, l, H. Enlarging $\mathcal{C}'$ as before to include this path $\mathcal{P}'$, we obtain the Euler circuit

$$G, h, E, c, B, a, A, b, D, f, E, d, C, e, F, g, E, j, H, m, J, l, H, k, G. \quad \blacksquare$$

The process illustrated in Example 4.12 always produces an Euler circuit in a connected multigraph in which the degree of each vertex is even. A formal description of this procedure is given next.

Euler Circuit Algorithm

This algorithm constructs an Euler circuit for a connected multigraph $\mathcal{G}$ in which every vertex has even degree.

Step 1 (start path)
- (a) Set $\mathcal{E}$ to be the set of edges of $\mathcal{G}$.
- (b) Select a vertex U, and set $\mathcal{C}$ to be the path consisting of just U.

Step 2 (expand the path)
while $\mathcal{E}$ is nonempty
- *Step 2.1* (pick a starting point for expansion)
  - (a) Set V to be a vertex in $\mathcal{C}$ that is incident with some edge in $\mathcal{E}$.
  - (b) Set $\mathcal{P}$ to be the path consisting of just V.
- *Step 2.2* (expand $\mathcal{P}$ into a path from V to V)
  - (a) Set $W = V$.
  - (b) **while** there is an edge e on W in $\mathcal{E}$
    - (a) Remove e from $\mathcal{E}$.
    - (b) Replace W with the other vertex on e.

(c) Append edge e and vertex W to path $\mathcal{P}$.

endwhile

Step 2.3 (enlarge $\mathcal{C}$)

Replace any one occurrence of V in $\mathcal{C}$ with path $\mathcal{P}$.

endwhile

Step 3 (output)

The path $\mathcal{C}$ is an Euler circuit.

The following theorem gives necessary and sufficient conditions for a connected multigraph to have an Euler circuit or path and justifies the Euler circuit algorithm:

Theorem 4.5 Suppose a multigraph $\mathcal{G}$ is connected. Then $\mathcal{G}$ has an Euler circuit if and only if every vertex has even degree. Furthermore, $\mathcal{G}$ has an Euler path if and only if every vertex has even degree except for two distinct vertices, which have odd degree. When this is the case, the Euler path starts at one and ends at the other of these two vertices of odd degree.

Proof. We will give a proof only in the case that $\mathcal{G}$ contains no loops. An easy modification establishes the result when there are loops.

Suppose the multigraph $\mathcal{G}$ has an Euler circuit. Every time this Euler circuit passes through a vertex, it enters along an edge and leaves along a different edge. Since every edge is used in an Euler circuit, every edge through a vertex can be paired as one of two, either coming in or going out. Thus each vertex has even degree.

Conversely, suppose each vertex has even degree. The Euler circuit algorithm constructs a path starting at V. This path must return to V, since, when we enter a different vertex along one edge, another edge must leave the vertex because its degree is even. Thus a V-V path is constructed. The algorithm proceeds by starting along unused edges on vertices on this path. Since the multigraph is connected, there always exists a path from any unused edge to the path already constructed; so each edge is eventually included, and an Euler circuit is formed.

If an Euler path exists between distinct vertices U and V, then, clearly, the degrees of U and V must be odd, while all the other vertices have even degree. Conversely, if only U and V have odd degrees in a connected multigraph, we can add an edge e between U and V. The new multigraph will have all degrees even, and so an Euler circuit will exist for it by what we have already proved. Removing e produces an Euler path between U and V. ■

From the last paragraph of the proof of Theorem 4.5, we see that the Euler circuit algorithm may be used to find an Euler path in a connected multigraph with exactly two vertices of odd degree by applying it to the multigraph formed by adding an edge between these two vertices.

Example 4.13 For the multigraph in Figure 4.21(a), an edge e is added between the two vertices U and V of odd degree. This results in the multigraph in Figure 4.21(b) for which an Euler circuit, say, e, a, d, c, b, can be found by using the Euler circuit algorithm. Deleting the edge e from this circuit gives the Euler path a, d, c, b between U and V for the multigraph in Figure 4.21(a). ■

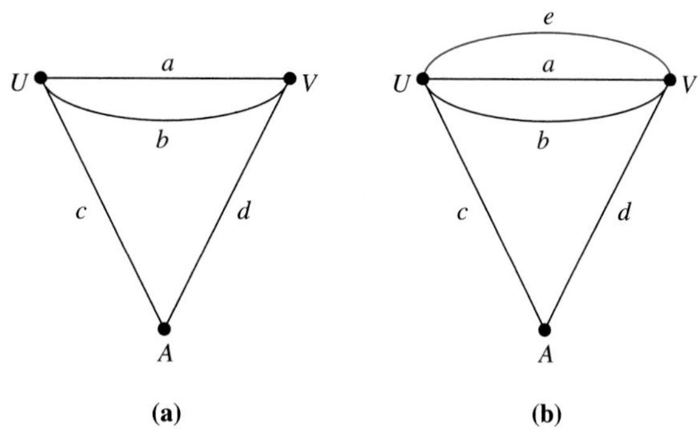

FIGURE 4.21

In analyzing the complexity of the Euler circuit algorithm, we will use picking an edge as an elementary operation. Since each of the e edges is used once, this algorithm is of order at most e. For a graph with n vertices,

$$e \leq \frac{1}{2}n(n-1) = \frac{1}{2}(n^2 - n),$$

because $C(n, 2) = \frac{1}{2}n(n-1)$ is the number of pairs of distinct vertices. (See Theorem 2.10.) Thus, for a graph with n vertices, this algorithm is of order at most n^2.

Hamiltonian Cycles and Paths

In Example 4.1 a graph was used to describe a system of nonstop flights. Suppose that a salesperson needs to visit each of the cities in this graph. In this situation time and money would be saved by visiting each city exactly once. What is needed for an efficient scheduling is a path that begins and ends at the same vertex and uses each vertex once and only once.

In the first part of this section, paths that used each edge once and only once were considered. Now we want to find a cycle that uses each *vertex* of a multigraph exactly once. But since we want to avoid repetition of vertices, loops and parallel edges will not be of any assistance. Consequently, we may assume that we are working with a graph. In a graph, a **Hamiltonian path** is a path that contains each vertex once and only once, and a **Hamiltonian cycle** is a cycle that includes each vertex. These are named after Sir William Rowan Hamilton, who developed a puzzle where the answer required the construction of this kind of cycle.

Example 4.14 Suppose the graph in Figure 4.22 describes a system of airline routes, where the vertices are towns and the edges represent the airline routes. The vertex U is the home base for a salesperson who must periodically visit all of the other cities. To be economical, the salesperson wants a path that starts at U, ends at U, and visits each of the other vertices exactly once. A brief examination of the graph shows that the edges a, b, d, g, f, e form a Hamiltonian cycle.

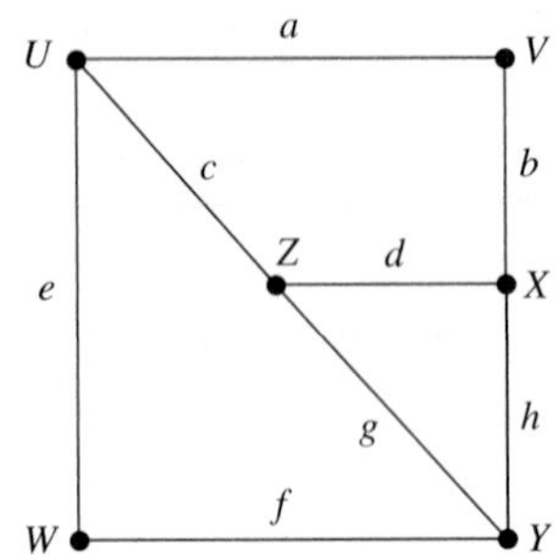

FIGURE 4.22

For the graph in Figure 4.1, there is no Hamiltonian cycle. The only way to reach New York or St. Louis is from Chicago, and once in New York or St. Louis, the only way to leave is to return to Chicago. ■

Relatively easy criteria exist to determine if there is an Euler circuit or an Euler path. All that has to be done is to check the degree of each vertex. Furthermore, there is a straightforward algorithm to use in constructing an Euler circuit or path. Unfortunately, the same situation does not hold for Hamiltonian cycles and paths. It is a major unsolved problem to determine necessary and sufficient conditions for a graph to have either of them. In general, it is very difficult to find a Hamiltonian cycle for a graph. There are, however, some conditions that guarantee the existence of a Hamiltonian cycle in a graph. We will provide an example of one of these theorems.

Theorem 4.6 Suppose $\mathcal{G}$ is a graph with n vertices, where $n > 2$. If, for each pair of nonadjacent vertices U and V, we have

$$\deg(U) + \deg(V) \geq n, \tag{4.1}$$

then $\mathcal{G}$ has a Hamiltonian cycle.

Proof. We will give a proof by contradiction. Suppose there exist graphs such that every pair of nonadjacent vertices satisfies (4.1), but which have no Hamiltonian cycle. Among all such graphs with n vertices, let $\mathcal{G}$ be one with a maximal number of edges. Then, if any edge is added to $\mathcal{G}$, the new graph has a Hamiltonian cycle. Because $\mathcal{G}$ has no Hamiltonian cycle, $\mathcal{G}$ is not a complete graph and so has nonadjacent vertices U and V. Let $\mathcal{G}'$ be the graph formed by adding the edge $\{U, V\}$ to $\mathcal{G}$.

By assumption, $\mathcal{G}'$ has a Hamiltonian cycle, and in fact every Hamiltonian cycle of $\mathcal{G}'$ contains the edge $\{U, V\}$. Removing this edge from such a cycle leaves a Hamiltonian path

$$U = U_1, U_2, \ldots, U_n = V$$

in $\mathcal{G}$.

We claim that, for $2 \leq j \leq n$, if the edge $\{U_1, U_j\}$ is in $\mathcal{G}$, then the edge $\{U_{j-1}, U_n\}$ is not. For if both these edges were in $\mathcal{G}$, then $\mathcal{G}$ would have the Hamiltonian cycle

$$U_1, U_j, U_{j+1}, \ldots, U_n, U_{j-1}, U_{j-2}, \ldots, U_1,$$

contrary to assumption. (See Figure 4.23.)

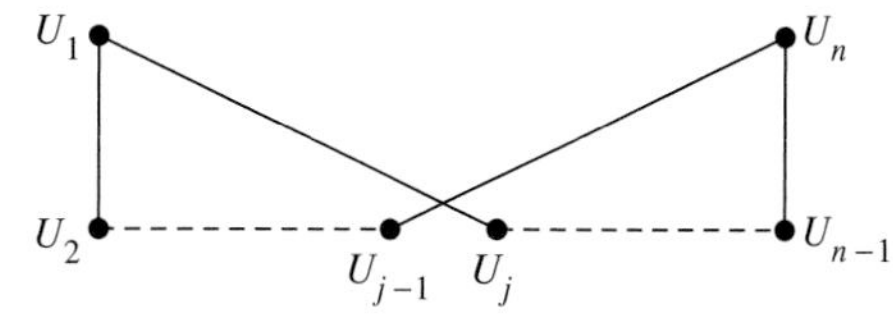

FIGURE 4.23

Now let d and d' be the respective degrees of U_1 and U_n in $\mathcal{G}$. Then there are d edges from U_1 to vertices U_j with $2 \leq j \leq n$. This gives d vertices U_{j-1}, $1 \leq j-1 \leq n-1$, not adjacent to U_n. Thus $d' \leq (n-1) - d$, and so $d + d' \leq n-1$, contradicting (4.1). ■

Example 4.15 It follows from Theorem 4.6 that if, in a graph with n vertices, the degree of each vertex is at least $\frac{n}{2}$, then the graph must have a Hamiltonian cycle. Thus the graph in Figure 4.24(a) has a Hamiltonian cycle because there are six vertices, each with degree 3. However, even though the theorem says there is a Hamiltonian cycle,

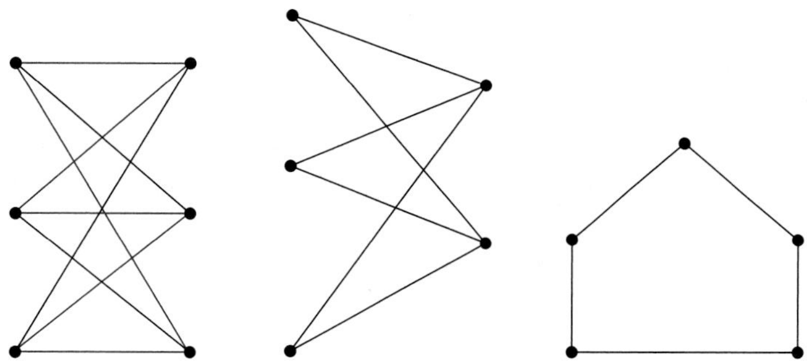

(a) (b) (c)

FIGURE 4.24

it does not tell us how to find one. Fortunately, in this case one can be found by a little bit of trial and error.

On the other hand, the graph with five vertices in Figure 4.24(b) does not have a Hamiltonian cycle, because, no matter where we start, we end up on the left side needing to go to a vertex on the left side, and there are no edges connecting the vertices on that side. Note that this graph does not satisfy the conditions of Theorem 4.6, because, among its five vertices, there are nonadjacent vertices of degree 2, so that (4.1) is not satisfied. On the other hand, the graph in Figure 4.24(c) also has five vertices, each with degree 2; yet it contains a Hamiltonian cycle. Thus, when (4.1) fails for some pair of nonadjacent vertices, it is not possible to conclude anything in general about the existence or nonexistence of a Hamiltonian cycle. ■

Example 4.16 There are several instances in which it is necessary to list all n-bit strings (a sequence of n symbols, each being a zero or one) in such a way that each n-bit string differs from the preceding string in exactly one position and the last n-bit string also differs from the first string in exactly one position. This kind of listing is called a **Gray code**. For $n = 2$, the listing 00, 01, 11, 10 is a Gray code, but 00, 01, 10, 11 is not, because string 2 and string 3 differ in more than one position, as do strings 1 and 4.

One way a Gray code is used is in determining the position of a circular disc after it stops rotating. In this situation a circular disc is divided into 2^n equal sectors, and an n-bit string is assigned to each sector. Figure 4.25 shows an assignment of the 3-bit strings to a disc divided into $2^3 = 8$ sectors.

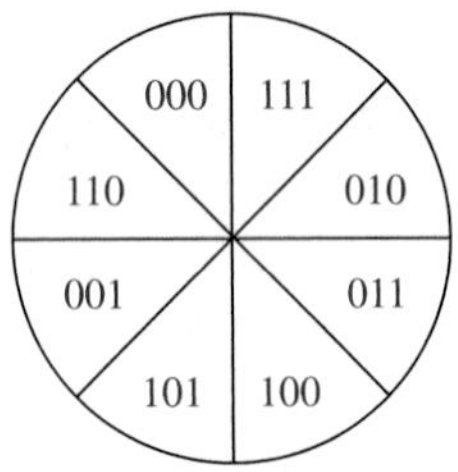

FIGURE 4.25

To determine which n-bit string is to be assigned, the circular disc is divided into n circular rings. Thus each sector is subdivided into n parts, each of which is treated in one of two ways: as, for example, opaque or translucent. Under the rotating disc are placed n electrical devices, such as photoelectric cells, that can determine what type of treated material is above it. Figure 4.26(a) illustrates how this can be done for the assignment in Figure 4.25, where the electrical device below a shaded region will send a 1, the device below an unshaded region will send a 0, and the regions are read from the outside to the center.

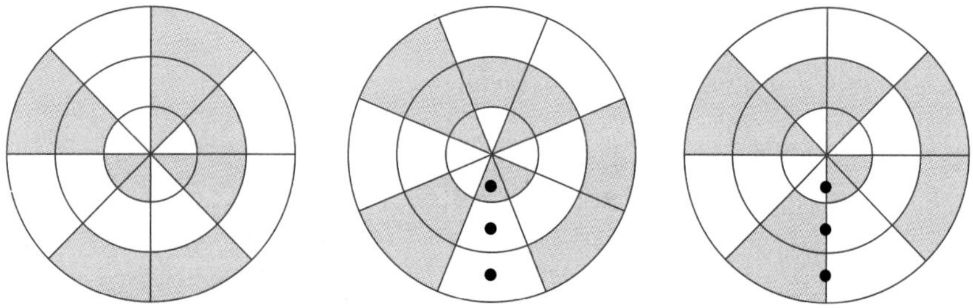

FIGURE 4.26

In Figure 4.26(b) the rotating disc has stopped so that the three electrical devices are completely contained within one sector. In this case the electrical devices will send 0 0 1, which indicates the sector in which the disc stopped. However, if the disc stopped as in Figure 4.26(c), there is a possibility for an incorrect reading. This time, the innermost device could send either 0 or 1, as could the other two devices. Because of this, any of the eight possible 3-bit strings could be sent when the location of the disc is as in Figure 4.26(c). To minimize this type of error, we want two adjacent sectors to be assigned n-bit strings that differ in exactly one position. Then, no matter whether 0 or 1 is sent from the parts of the adjacent sectors, there will be only two possible strings that could be sent to describe the location of the disc, and these two strings will identify the two adjacent sectors where the disc stopped.

To find a Gray code, we construct a graph with 2^n vertices representing the 2^n possible n-bit strings. Two vertices are connected with an edge if the corresponding n-bit strings differ in exactly one position. It can be proved that a graph constructed in this way always has a Hamiltonian cycle, and so this can be used to find a Gray code for any n. (See Figure 4.27 for the case $n = 3$, where a Hamiltonian cycle is indicated by the colored edges.) Thus, when $n = 3$, one Gray code is 000, 001, 011, 010, 110, 111, 101, 100. (The interested reader should consult suggested reading [11] for more details.) ■

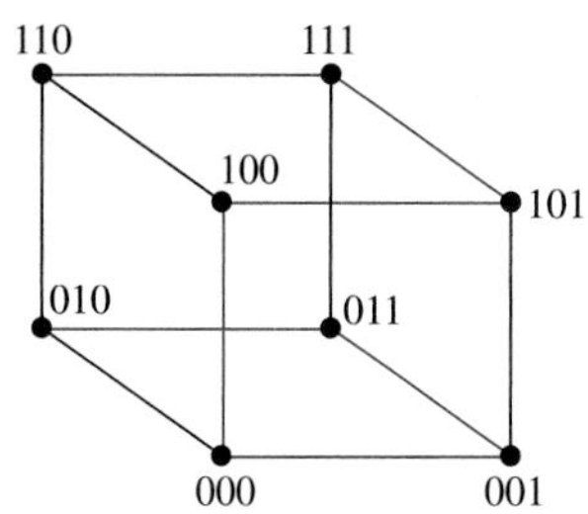

FIGURE 4.27

EXERCISES 4.2

In Exercises 1–4, determine whether the multigraph is a graph.

1. 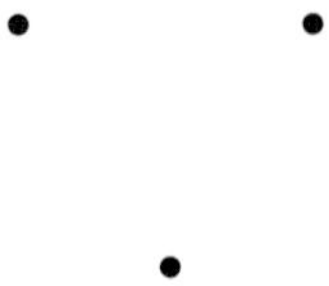

2. 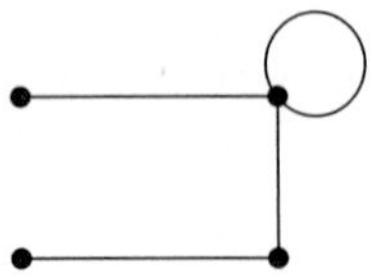

3. 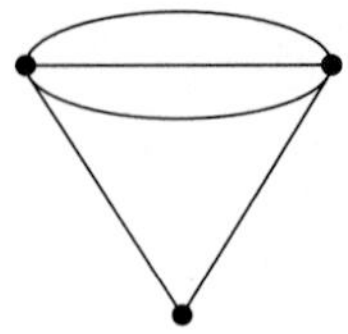

4. 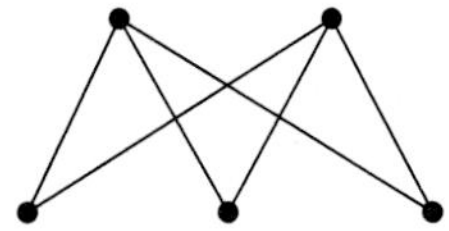

In Exercises 5–8, list the loops and parallel edges in the multigraph.

5. 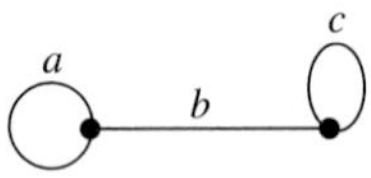

6. 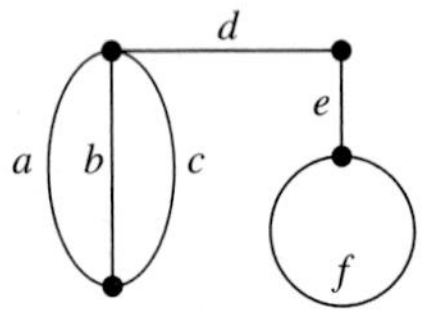

7. 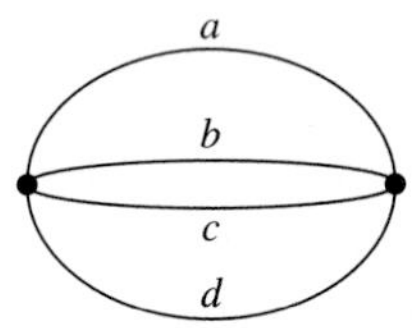

8. 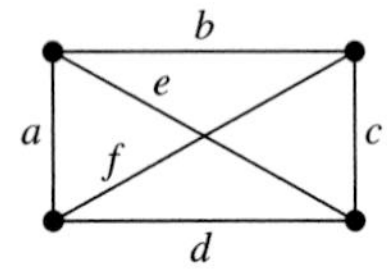

In Exercises 9 and 10, perform the following tasks:

(a) *List at least three different paths from A to D. Give the length of each.*

(b) *List the simple paths from A to D. Give the length of each.*

(c) *For each path you listed in (i), find a simple path from A to D contained in it.*

(d) *List the distinct cycles. (Two cycles are distinct if there is an edge in one that is not in the other.) Give the length of each.*

9. 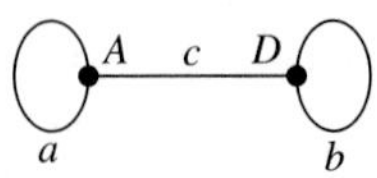

10. 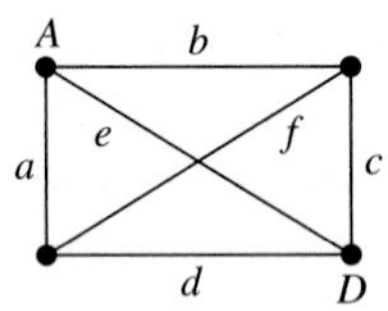

11. Give examples of multigraphs satisfying each of the following conditions:

(a) There are exactly two cycles.

(b) There is a cycle of length 1.

(c) There is a cycle of length 2.

In Exercises 12–17, determine whether the multigraph is connected.

12. 

13. 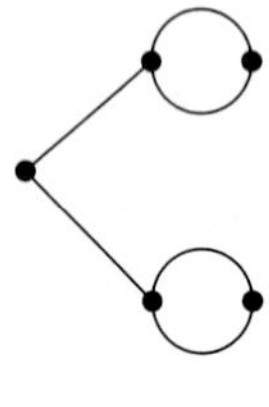

14. 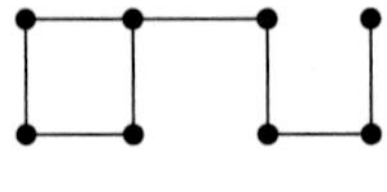

15. 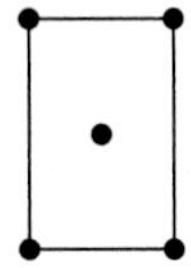

16.

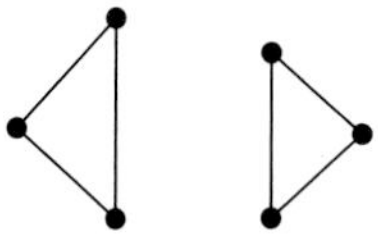

17.

In Exercises 18–23, determine whether the multigraph has an Euler path. If it does, construct one using the Euler circuit algorithm as in Example 4.13.

18.

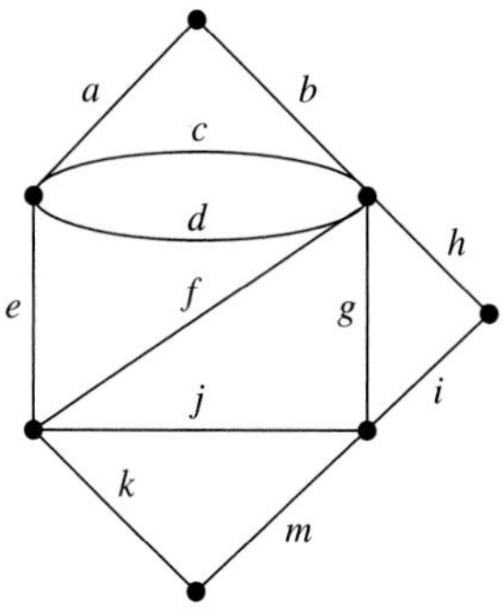

19.

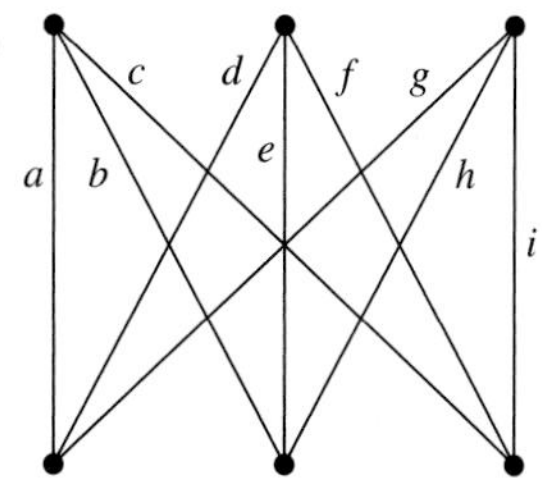

20.

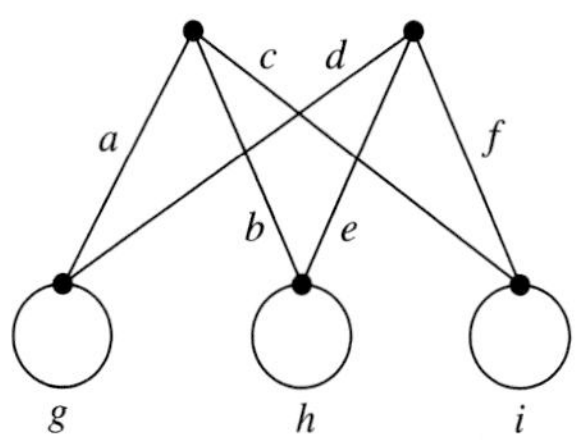

21.

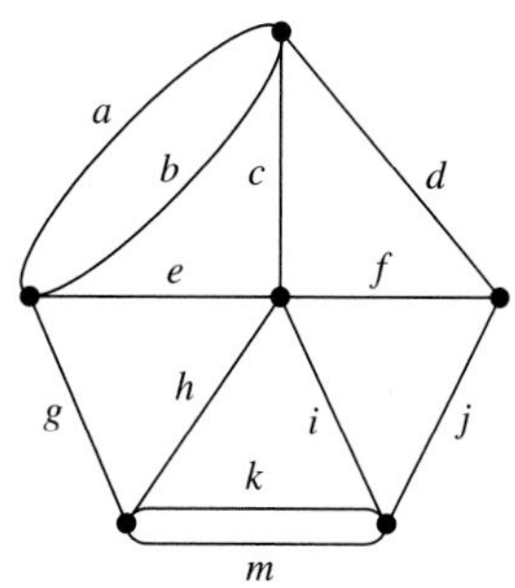

22.

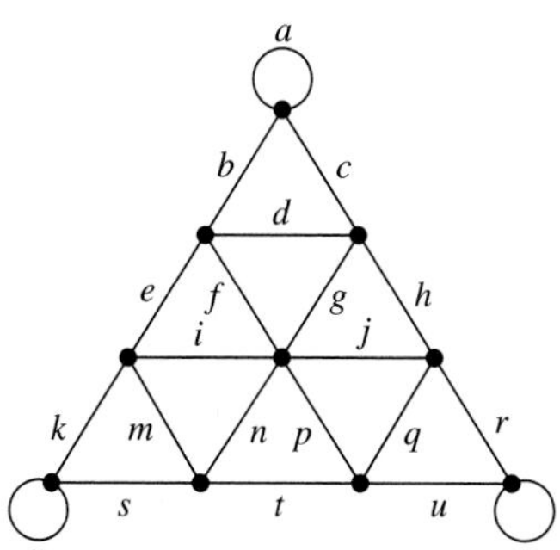

23.

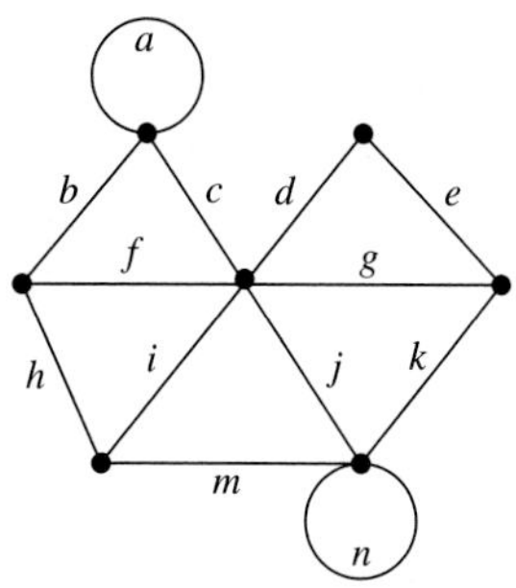

In Exercises 24–29, determine whether the multigraph in the indicated exercise has an Euler circuit. If it does, construct an Euler circuit by using the Euler circuit algorithm.

24. Exercise 18

25. Exercise 19

26. Exercise 20

27. Exercise 21

28. Exercise 22

29. Exercise 23

30. The city of Königsberg, located on the banks of the Pregel River, had seven bridges that connected islands in the river to the shores, as illustrated in the following diagram:

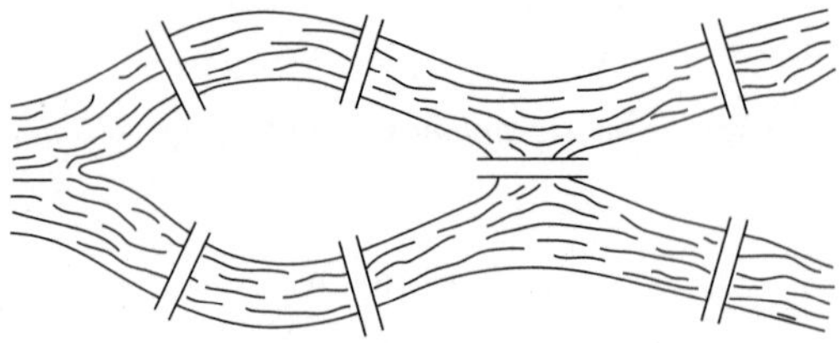

It was the custom of the townspeople to stroll on Sunday afternoons and, in particular, to cross over the bridges. The people of Königsberg wanted to know if it were possible to stroll in such a way that they could go over each bridge exactly once and return to the starting point. Is it? (*Hint:* Consider carefully what a vertex is to represent.) (This problem was presented to the famous mathematician, Leonhard Euler, and his solution is often credited with being the beginning of graph theory.)

31. Could the citizens of Königsberg find an acceptable route by building a new bridge? If so, how?

32. Could the citizens of Königsberg find an acceptable route by building two new bridges? If so, how?

33. Could the citizens of Königsberg find an acceptable route by tearing down a bridge? If so, how?

34. Could the citizens of Königsberg find an acceptable route by tearing down two bridges? If so, how?

An old childhood game asks children to trace a figure with a pencil without either lifting the pencil from the figure or tracing a line more than once. Determine whether this can be done for the figures in Exercises 35–38, assuming that you must begin and end at the same point.

35.

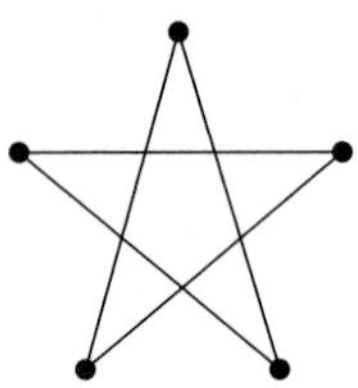

36.

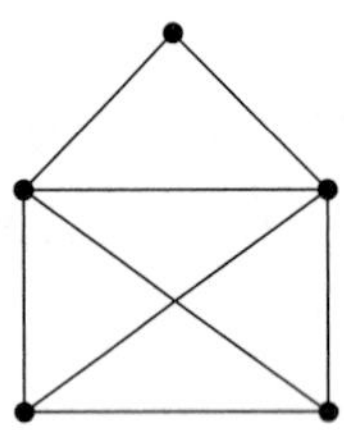

37.

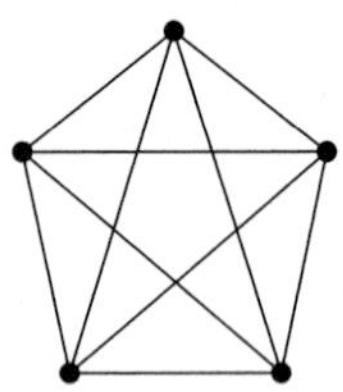

38.

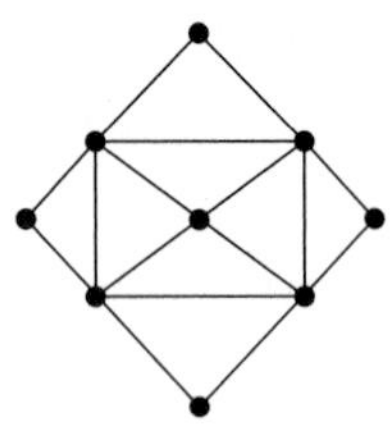

39. The accompanying graph has four vertices of odd degree; thus it has no Euler circuit or path. However, it is possible to find two distinct paths, one from A to B and the other from C to D, that use all the edges and have no edge in common. Find two such paths. (See Exercise 63.)

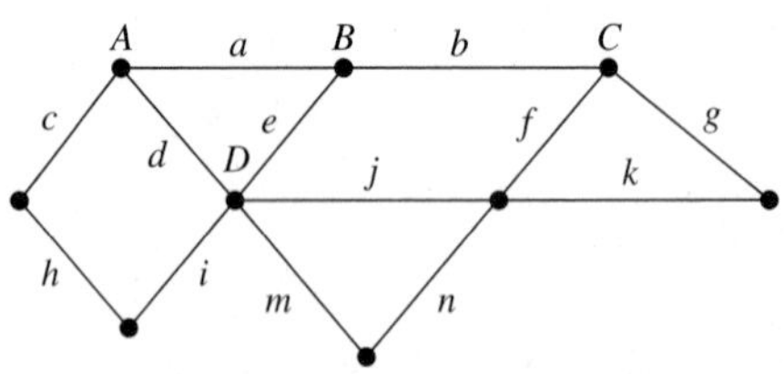

40. In 1859, Sir William Rowan Hamilton, a famous Irish mathematician, marketed a puzzle that consisted of a regular dodecahedron made of wood in which each corner represented a famous city. The puzzle was to find a route that traveled along the edges of the dodecahedron, visited each city exactly once, and returned to the original starting city. (To make the task somewhat easier, each corner had a nail in it, and one was to use string while tracing out a path.) A representation of this puzzle drawn in the plane is as follows:

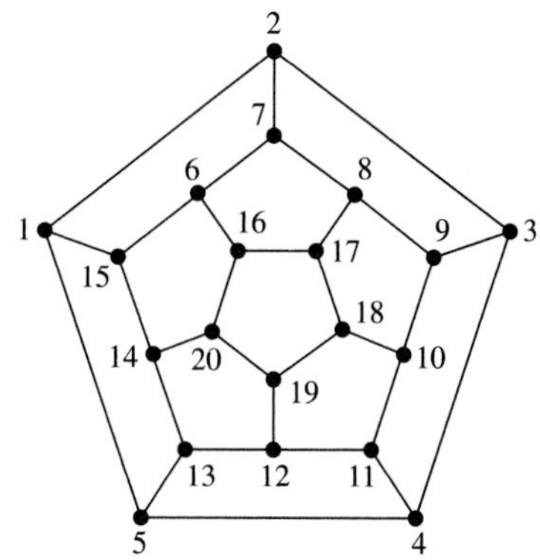

Can you find an answer to the puzzle?

41. Give examples of connected graphs satisfying each set of conditions.

(a) There is both an Euler circuit and a Hamiltonian cycle.

(b) There is neither an Euler circuit nor a Hamiltonian cycle.

(c) There is an Euler circuit, but not a Hamiltonian cycle.

(d) There is a Hamiltonian cycle, but not an Euler circuit.

(e) There is a Hamiltonian path, but not a Hamiltonian cycle.

42. Construct a Gray code for $n = 4$.

43. Draw all the nonisomorphic multigraphs having five vertices of degrees 1, 1, 2, 3, and 3.

44. Draw all the nonisomorphic multigraphs having four vertices of degrees 1, 2, 3, and 4.

45. Draw all the nonisomorphic multigraphs having six vertices of degrees 1, 1, 1, 2, 2, and 3.

46. Draw all the nonisomorphic multigraphs having five vertices of degrees 1, 2, 2, 2, and 3.

47. Is the property "has a cycle of length n" a graph isomorphism invariant? Justify your answer.

48. Is the property "has all vertices with even degrees" a graph isomorphism invariant? Justify your answer.

49. Is the property "has a Hamiltonian path" a graph isomorphism invariant? Justify your answer.

50. Are the following two graphs isomorphic? Justify your answer.

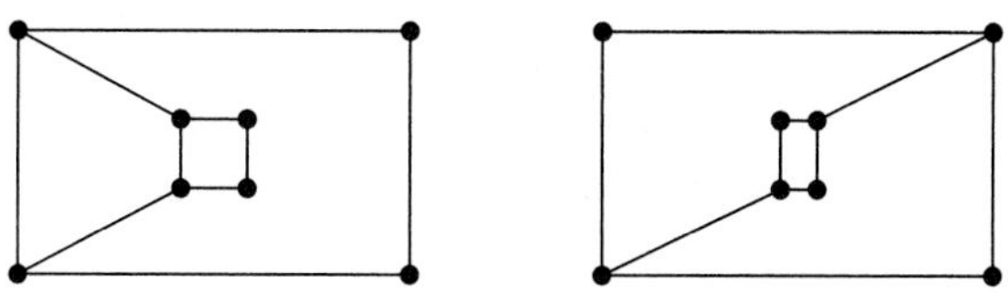

51. Are the following two graphs isomorphic? Justify your answer.

52. A **bipartite graph** is a graph in which the vertices can be divided into two disjoint nonempty sets A and B such that no two vertices in A are adjacent and no two vertices in B are adjacent. The **complete bipartite graph** $\mathcal{K}_{m,n}$ is a bipartite graph in which the sets A and B contain m and n vertices, respectively, and every vertex in A is adjacent to every vertex in B. For example, the graph $\mathcal{K}_{2,3}$ is as follows.

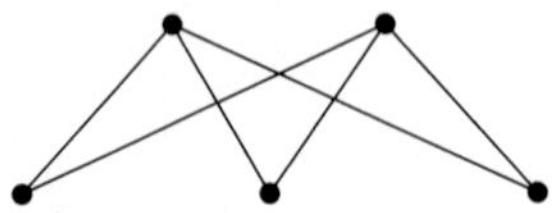

How many edges does $\mathcal{K}_{m,n}$ have?

53. For which m and n does $\mathcal{K}_{m,n}$ have an Euler circuit?

54. For which m and n does $\mathcal{K}_{m,n}$ have a Hamiltonian cycle?

55. Prove that $\mathcal{K}_n$ has a Hamiltonian cycle when $n > 2$.

56. In a multigraph with n vertices, what is the maximum length of a simple path?

57. Show that the relation "there is a path from vertex V to vertex U" is an equivalence relation on the set of vertices of a graph. The vertices in an equivalence class of this relation, along with the edges joining them, form a **component** of the graph.

58. Find the components of the following graphs (see Exercise 57):

(a)

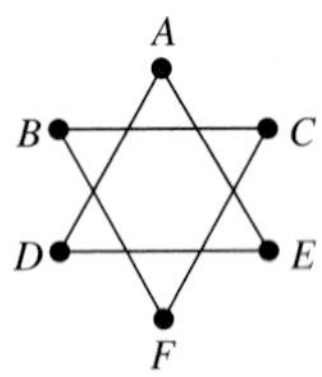

(b)

(c)

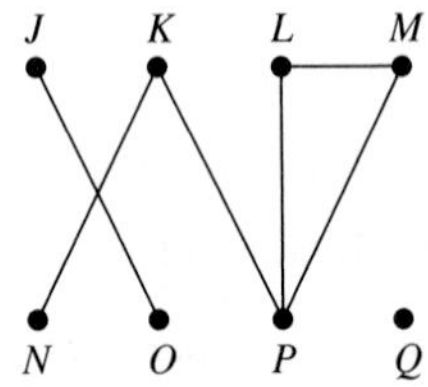

59. Find the components of the following graphs (see Exercise 57):

(a)

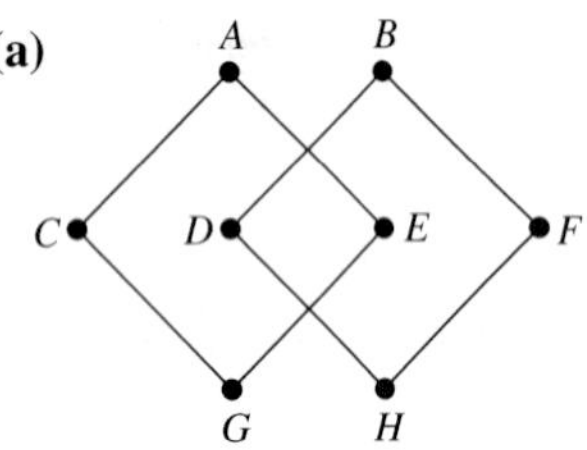

(b)

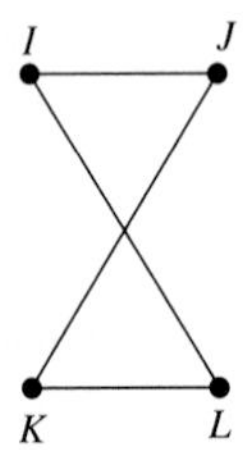

(c)

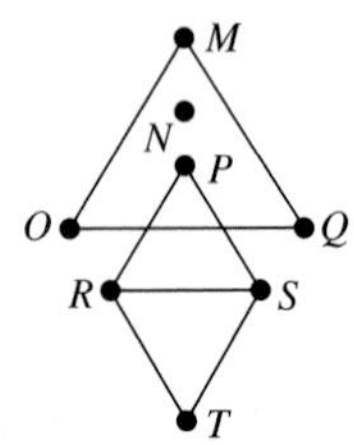

60. A dog show is being judged from pictures of the dogs. The judges would like to see pictures of the following pairs of dogs next to each other for their final decision: Arfie and Fido, Arfie and Edgar, Arfie and Bowser, Bowser and Champ, Bowser and Dawg, Bowser and Edgar, Champ and Dawg, Dawg and Edgar, Dawg and Fido, Edgar and Fido, Fido and Goofy, Goofy and Dawg.

(a) Draw a graph modeling this situation.

(b) Suppose that it is necessary to put pictures of the dogs in a row on the wall so that each desired pair of pictures appear together exactly once. (There are many copies of each picture.) What graph-theoretic object is being sought?

(c) Can the pictures be arranged on the wall in this manner? If so, how?

61. At a recent college party there were a number of young men and women present, some of whom had dated each other recently. This situation can be represented by a graph in which the vertices are the individuals in attendance, with adjacency being defined by having dated recently. If this graph has a Hamiltonian cycle, show that the number of men is the same as the number of women.
62. Prove Theorem 4.4 by mathematical induction.
63. Suppose a connected multigraph has the property that exactly four of its vertices have odd degree. Prove that there are two paths, one between two of these vertices and the other between the remaining two vertices, such that every edge is in exactly one of these two paths.
64. In a graph, prove that if there is a U-U path of odd length for some vertex U, then there is a cycle of odd length.
65. Prove that if a connected graph has n vertices, then it must have at least $n - 1$ edges.

4.3 SHORTEST PATHS AND DISTANCE

In this section we will consider ways to find a shortest path between vertices in a graph. The need to find such paths arises in many different situations.

We want to find a path of minimal length between two vertices S and T, that is, a path from S to T that has the fewest possible edges. This smallest possible number of edges in a path from S to T is called the **distance** from S to T. To find the distance from S to T, the general approach is to look first at S, then at the vertices adjacent to S, then at the vertices adjacent to these vertices, and so forth. By keeping a record of the way in which vertices are examined, we are able to construct a shortest path from S to T. To find the distance from S to every vertex T for which there is a path from S to T, we assign labels to some of the vertices in the graph. If a vertex V is assigned the label $3(U)$, then the distance from S to V is 3, and U is the predecessor of V on a shortest path from S to V (that is, a shortest path from S to V contains the edge $\{U, V\}$).

Example 4.17 For the graph in Figure 4.28, let us find the distance from S to each vertex for which there is a path from S. We begin by assigning S the label $0(-)$, which signifies that the distance from S to S is 0 and that there are no edges on this path. Next, we determine the vertices with distance 1 from S. These are A and B, which are both assigned the label $1(S)$, as shown in Figure 4.29.

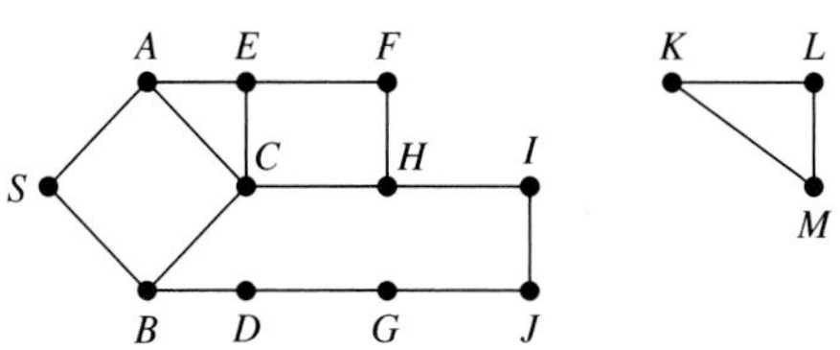

FIGURE 4.28

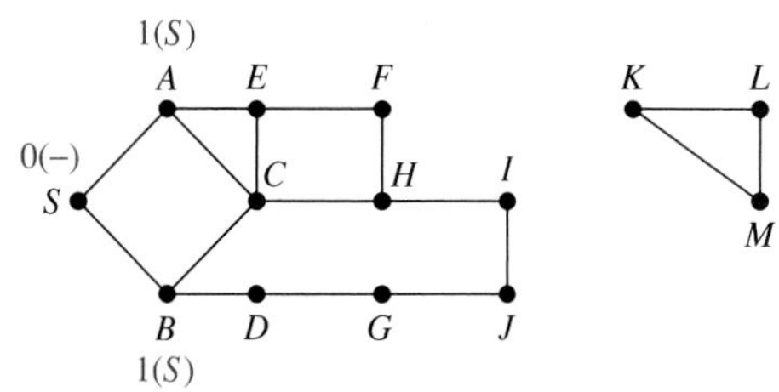

FIGURE 4.29

Having assigned labels to the vertices with distance 1 from S, we now determine the vertices with distance 2 from S. These are the vertices that are unlabeled and adjacent to a vertex whose distance from S is 1. For example, the unlabeled vertices C and E are adjacent to A, and so they are assigned the label $2(A)$. Likewise, the unlabeled vertex D is adjacent to B and so is given the label $2(B)$. The labels now appear as in Figure 4.30.

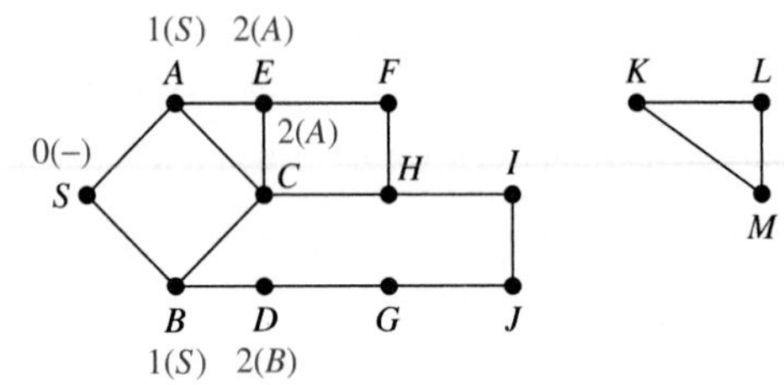

FIGURE 4.30

We continue in this manner until no labeled vertex is adjacent to an unlabeled vertex. If every vertex in the graph is labeled when this occurs, then the graph is connected. Otherwise, there is no path from S to any unlabeled vertex. For the graph in Figure 4.28, vertices A through J and S are eventually labeled as in Figure 4.31. At this point, we stop because no labeled vertex is adjacent to an unlabeled vertex. Note that there is no path from S to any of the unlabeled vertices (K, L, or M).

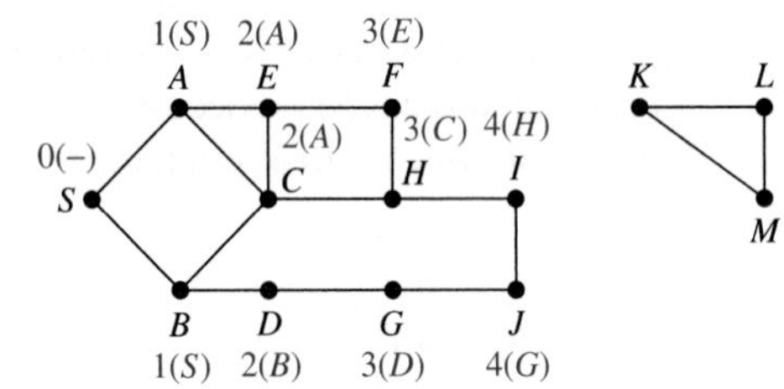

FIGURE 4.31

The label assigned to any labeled vertex gives its distance from S. For example, since the label assigned to I is $4(H)$, the distance from S to I is 4. Also, the predecessor of I is H, which means that a shortest path from S to I includes the edge $\{H, I\}$. Similarly, the predecessor of H is C, the predecessor of C is A, and the predecessor of A is S. Thus a shortest path from S to I includes the edges $\{H, I\}$, $\{C, H\}$, $\{A, C\}$, and $\{S, A\}$, and so a shortest path from S to I is S, A, C, H, I. In this graph, another shortest path from S to I exists, namely S, B, C, H, I. Which path is found depends on whether vertex C is labeled because it is adjacent to A or to B. ■

Here is a formal description of this process.

Breadth-First Search Algorithm

This algorithm determines the distance and a shortest path in a graph from vertex S to every other vertex for which there is a path from S. In the algorithm, $\mathcal{L}$ denotes the set of labeled vertices, and the *predecessor* of vertex A is a vertex in $\mathcal{L}$ that is used in labeling A.

Step 1 (label S)
  (a) Assign S the label 0, and let S have no predecessor.
  (b) Set $\mathcal{L} = \{S\}$ and $k = 0$.
Step 2 (label vertices)
  repeat
    Step 2.1 (increase the label)
      Replace k with $k + 1$.
    Step 2.2 (enlarge labeling)
      while $\mathcal{L}$ contains a vertex V with label $k - 1$ that
          is adjacent to a vertex W not in $\mathcal{L}$
            (a) Assign the label k to W.
            (b) Assign V to be the predecessor of W.
            (c) Include W in $\mathcal{L}$.
      endwhile
  until no vertex in $\mathcal{L}$ is adjacent to a vertex not in $\mathcal{L}$
Step 3 (construct a shortest path to a vertex)
  if a vertex T is in $\mathcal{L}$
    The label on T is its distance from S. A shortest path from S to T is formed by taking in reverse order T, the predecessor of T, the predecessor of the predecessor of T, and so forth, until S is reached.
  otherwise
    There is no path from S to T.
  endif

It can be shown that the label assigned to each vertex by the breadth-first search algorithm is its distance from S (see Exercise 18).

We will regard labeling a vertex and using an edge to find an adjacent vertex as the elementary operations in analyzing this algorithm. For a graph with n vertices and e edges, each vertex is labeled exactly once and each edge is used at most once to find an adjacent vertex. Hence there will be at most $n + e$ elementary operations. But since

$$n + e \le n + C(n, 2) = n + \frac{1}{2}n(n - 1),$$

we see that this algorithm is of order at most n^2.

Weighted Graphs

Frequently, when graphs are used to describe relationships between objects, a number is associated with each edge. For example, if a graph is being used to represent a highway system in the usual way, then a number can be assigned to each edge indicating the mileage between the two cities. This idea of assigning numbers to the edges is a very important one in applications.

A **weighted graph** is a graph in which a number called the **weight** is assigned to each edge. The **weight of a path** is the sum of weights of the edges in the path. When a weighted graph describes a highway system, with vertices representing cities and weights representing mileage between cities, the weight of a path is simply the total mileage between the cities representing the start and end of the path.

Example 4.18 The graph in Figure 4.32 is a weighted graph, since each edge has a number assigned to it. For example, the weight of the edge on A and B is 3 and the weight of the edge on D and F is 5. The weight of the path A, C, D, F is $4 + 2 + 5 = 11$, and the weight of the path F, D, B, E, D is $5 + 1 + 2 + 1 = 9$. ■

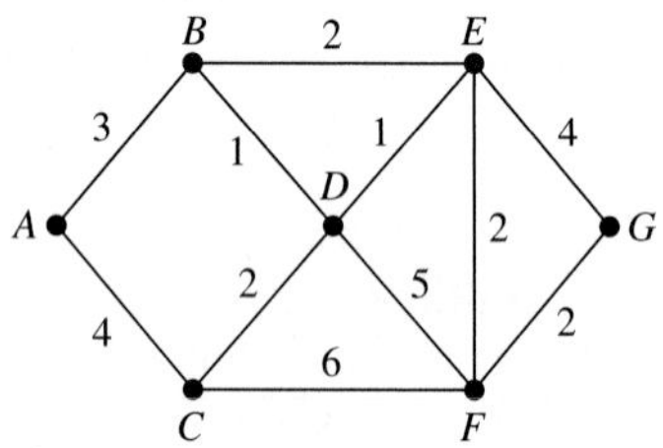

FIGURE 4.32

In many applications we need to find a path of smallest weight. However, there need not always be one. This kind of situation can occur if there is a cycle with negative weight.

Example 4.19 For the weighted graph in Figure 4.33, the path A, B, D, E has weight 2, and the path A, B, D, C, B, D, E has weight -2, which is a smaller weight than that of the first path. Note that as the cycle B, D, C is repeated, the weight of the path gets smaller and smaller. Thus there is no path of smallest weight between A and E. ■

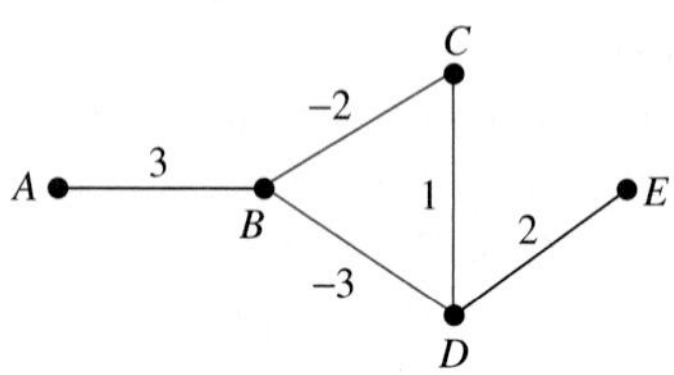

FIGURE 4.33

Consequently, we shall assume, unless explicitly stated otherwise, that weighted graphs do not have a cycle with negative weight. This assumption assures the existence of a path of smallest weight between two vertices if there is any path between them. Furthermore, a path of smallest weight between two vertices may be assumed to be simple, since any cycle of weight 0 could be removed, as in Theorem 4.4. A path of smallest weight is called a **shortest path** between those two vertices, and the weight of that path is called the **distance** between them.

When the weights assigned to edges are positive, as is the case with highway or airline mileage, there is an algorithm that finds the distance and a shortest path between two vertices S and T. In fact, it can be used to find the distance and a shortest path between S and all other vertices at the same time.

The idea of this algorithm is to find the vertex closest to S, then the second closest vertex to S, and so forth. In this way we can find the distance between S and all other vertices. In addition, if we keep a record of the vertices used in determining distances, it is possible to find a shortest path from S to any other vertex. This algorithm is due to E. Dijkstra, one of the pioneers in computer science.

Dijkstra's Algorithm

Let $\mathcal{G}$ be a weighted graph in which there is more than one vertex and all weights are positive. This algorithm determines the distance and a shortest path from vertex S to every other vertex in $\mathcal{G}$. In the algorithm, $\mathcal{P}$ denotes the set of vertices with permanent labels. The predecessor of vertex A is a vertex in $\mathcal{P}$ used to label A. The weight of the edge on vertices U and V is denoted by $W(U, V)$, and if there is no edge on U and V, we write $W(U, V) = \infty$.

Step 1 (label S)
 (a) Assign S the label 0, and let S have no predecessor.
 (b) Set $\mathcal{P} = \{S\}$.

Step 2 (label vertices)
 Assign to each vertex V not in $\mathcal{P}$ the (perhaps temporary) label $W(S, V)$, and let V have the (perhaps temporary) predecessor S.

Step 3 (enlarge $\mathcal{P}$ and revise labels)
 repeat
 Step 3.1 (make another label permanent)
 Include in $\mathcal{P}$ a vertex U having the smallest label of the vertices not in $\mathcal{P}$. (If there is more than one such vertex, arbitrarily choose any one of them.)
 Step 3.2 (revise temporary labels)
 For each vertex X not in $\mathcal{P}$ that is adjacent to U, replace the label on X with the smaller of the old label on X and the sum of the label on U and $W(U, X)$. If the label on X

was changed, let U be the new (perhaps temporary) predecessor of X.

until $\mathcal{P}$ contains every vertex of $\mathcal{G}$

Step 4 (find distances and shortest paths)
The label on a vertex Y is its distance from S. If the label on Y is ∞, then there is no path, and hence no shortest path, from S to Y. Otherwise, a shortest path from S to Y is formed by using in reverse order the vertices Y, the predecessor of Y, the predecessor of the predecessor of Y, and so forth, until S is reached.

The proof that this algorithm actually computes the distance between S and every other vertex can be found in Exercises 21–24.

In analyzing this algorithm for a graph with n vertices, we will consider assignments involving one vertex as being just one operation. So in step 1 there is just one operation, and in step 2 there are $n-1$ more. Step 3 is done $n-1$ times. Each time, at most $n-2$ comparisons are done on the labels to find the smallest one, and then at most one assignment occurs. Also, in revising the labels, we examine at most $n-1$ vertices, each of which requires an addition, a comparison, and two possible assignments, for a total of four operations. So, for step 3, there are at most $(n-1)[n-2+1+4(n-1)] = (n-1)(5n-5)$ operations. In step 4, looking up the distance and tracing back at most $n-1$ predecessors to find a shortest path takes at most n operations. From this, we see that there are at most

$$1 + (n-1) + (n-1)(5n-5) + n = 5n^2 - 8n + 5$$

operations, so the algorithm is of order at most n^2.

Example 4.20 For the weighted graph in Figure 4.34, we want to find a shortest path and the distance from S to every other vertex.

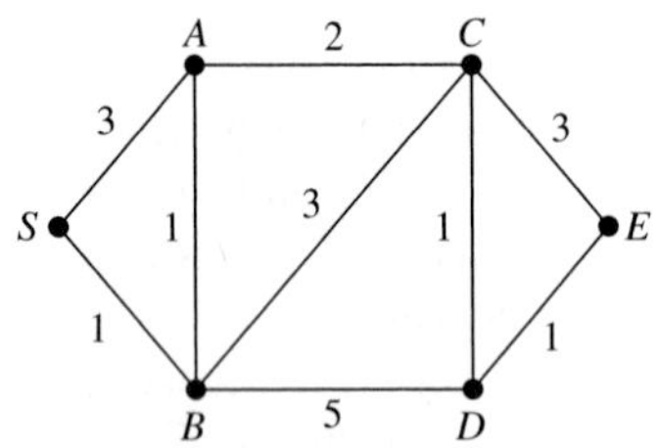

FIGURE 4.34

In step 1 we set $\mathcal{P} = \{S\}$ and assign S the label 0. We indicate this on the graph by writing the label and predecessor (in parentheses) beside S. We use an asterisk to denote that S is in $\mathcal{P}$. The graph now looks like that in Figure 4.35.

Next, in step 2, we assign the label $W(S, V)$ and the predecessor S to every other vertex V. Recall that $W(S, V) = \infty$ when there is no edge joining S and V. The graph now looks like that in Figure 4.36.

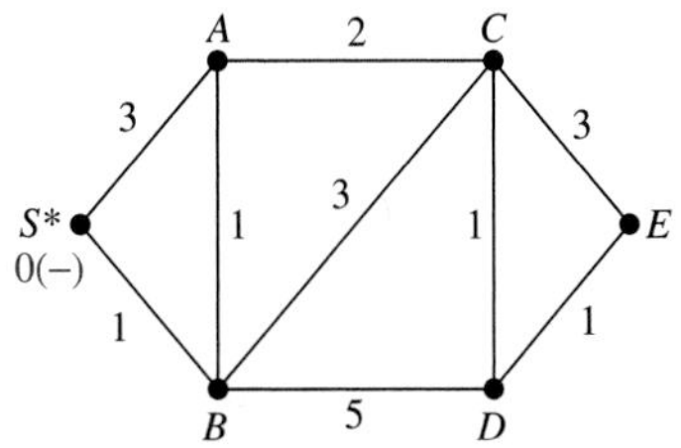

FIGURE 4.35

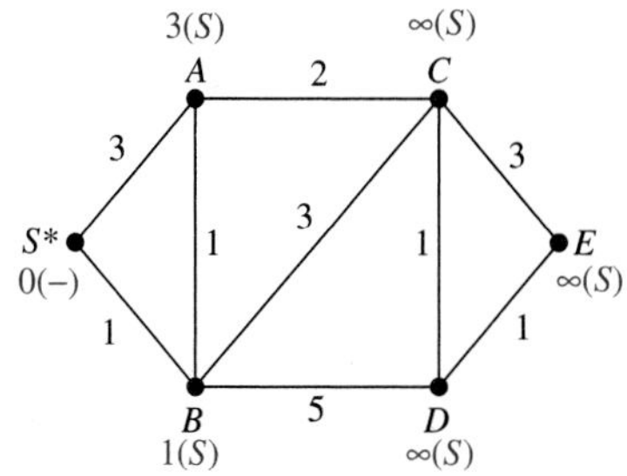

FIGURE 4.36

Now we perform step 3. The vertex not in $\mathcal{P}$ with the smallest label is B, and so we include B in $\mathcal{P}$. The vertices not in $\mathcal{P}$ and adjacent to B are A, C, and D; and we replace the label on each such vertex X by the minimum of the old label and the sum of the label on B and $W(S, B)$. These numbers are as follows:

Vertex X	*Old Label*	*(Label on B) + W(B, X)*	*Minimum*
A	3	$1 + 1 = 2$	2
C	∞	$1 + 3 = 4$	4
D	∞	$1 + 5 = 6$	6

Since each label is changed, we also replace the predecessor of each of these vertices by B, producing Figure 4.37.

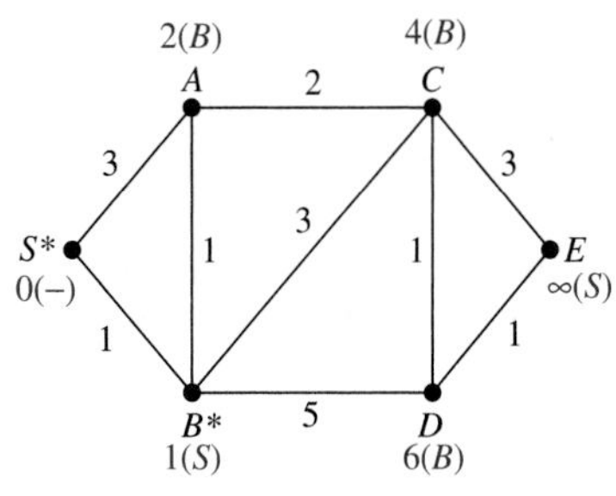

FIGURE 4.37

We continue in this way until $\mathcal{P}$ contains every vertex in the weighted graph. The following table shows the labels, predecessors, and vertices added to $\mathcal{P}$ at each stage (no entry in a column indicates no change from the previous stage):

Labels and Predecessors							
Vertex	S	A	B	C	D	E	*Vertex Added to* $\mathcal{P}$
	$0(-)$	$3(S)$	$1(S)$	$\infty(S)$	$\infty(S)$	$\infty(S)$	S
		$2(B)$		$4(B)$	$6(B)$		B
							A
					$5(C)$	$7(C)$	C
						$6(D)$	D
							E

The final graph is shown in Figure 4.38. In this figure the label on each vertex gives the distance between it and S, and a path of this length can be found by backtracking through the predecessors of the vertices. For example, the distance from S to E is 6, and the path S, B, C, D, E has this length. ■

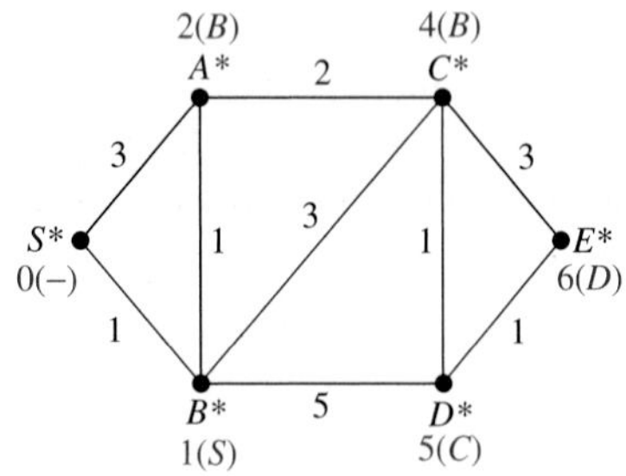

FIGURE 4.38

Number of Paths

We conclude this section by considering the number of paths between two vertices, or, alternatively, how many paths of length m there are between a pair of vertices. One answer to these questions involves powers of the adjacency matrix of the graph.

Theorem 4.7 For a graph $\mathcal{G}$ with vertices labeled $V_1, V_2, \ldots, V_n$ and adjacency matrix A, the number of paths of length m from V_i to V_j is the i, j entry of A^m.

Before we show how the theorem is proved for $m = 1, 2$, and 3, we present an example to illustrate the theorem.

Example 4.21 The graph in Figure 4.39(a) has the adjacency matrix A given in Figure 4.39(b).

To find the number of paths of length 2, we compute the product

$$A^2 = AA = \begin{bmatrix} 0 & 1 & 1 & 1 \\ 1 & 0 & 1 & 1 \\ 1 & 1 & 0 & 0 \\ 1 & 1 & 0 & 0 \end{bmatrix} \begin{bmatrix} 0 & 1 & 1 & 1 \\ 1 & 0 & 1 & 1 \\ 1 & 1 & 0 & 0 \\ 1 & 1 & 0 & 0 \end{bmatrix} = \begin{bmatrix} 3 & 2 & 1 & 1 \\ 2 & 3 & 1 & 1 \\ 1 & 1 & 2 & 2 \\ 1 & 1 & 2 & 2 \end{bmatrix}.$$

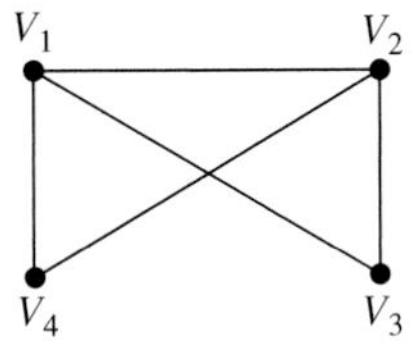

(a)

$$A = \begin{bmatrix} 0 & 1 & 1 & 1 \\ 1 & 0 & 1 & 1 \\ 1 & 1 & 0 & 0 \\ 1 & 1 & 0 & 0 \end{bmatrix}$$

(b)

FIGURE 4.39

That the 3, 4 entry is 2 means there are two paths of length 2 between V_3 and V_4, namely, V_3, V_1, V_4 and V_3, V_2, V_4. Likewise, the 1, 3 entry being 1 means that there is only one path of length 2 between V_1 and V_3, namely, V_1, V_2, V_3. The number of paths of length 3 is given by the product $A^2 \cdot A = A^3$, computed as follows:

$$A^3 = A^2 A = \begin{bmatrix} 3 & 2 & 1 & 1 \\ 2 & 3 & 1 & 1 \\ 1 & 1 & 2 & 2 \\ 1 & 1 & 2 & 2 \end{bmatrix} \begin{bmatrix} 0 & 1 & 1 & 1 \\ 1 & 0 & 1 & 1 \\ 1 & 1 & 0 & 0 \\ 1 & 1 & 0 & 0 \end{bmatrix} = \begin{bmatrix} 4 & 5 & 5 & 5 \\ 5 & 4 & 5 & 5 \\ 5 & 5 & 2 & 2 \\ 5 & 5 & 2 & 2 \end{bmatrix}$$

Since the 1, 2 entry of A^3 is 5, there are five paths of length 3 between V_1 and V_2, namely, V_1, V_2, V_1, V_2; V_1, V_2, V_4, V_2; V_1, V_2, V_3, V_2; V_1, V_3, V_1, V_2; and V_1, V_4, V_1, V_2. ■

We now give the proof of Theorem 4.7 for $m = 1, 2$, and 3. Let a_{ij} denote the i, j entry of A. The number of paths of length 1 between V_i and V_j is either 0 or 1, depending on whether there is an edge joining these vertices. But this is the same as a_{ij}, which is 1 when there is an edge joining V_i and V_j, and 0 otherwise. So the i, j entry of A gives the number of paths of length 1 from V_i to V_j.

For a path of length 2 between V_i and V_j, there needs to be a vertex V_k for which there is an edge joining V_i and V_k and an edge joining V_k and V_j. In terms of the adjacency matrix, this is the same as saying that there is an index k such that both a_{ik} and a_{kj} are 1, or equivalently, $a_{ik}a_{kj} = 1$. Thus the number of paths of length 2 between V_i and V_j is the number of k's where $a_{ik}a_{kj} = 1$. This number is the value of

$$a_{i1}a_{1j} + a_{i2}a_{2j} + \cdots + a_{in}a_{nj},$$

since each term in the sum is 1 or 0. But this sum is also the i, j entry in A^2, the product of A with A, and thus the i, j entry in A^2 is the number of paths of length 2 between V_i and V_j.

For a path of length 3 between V_i and V_j, there are vertices V_p and V_k with edges joining V_i and V_p, joining V_p and V_k, and joining V_k and V_j. But this means there is a path of length 2 between V_i and V_k and an edge on V_k and V_j. If b_{ik}

denotes the i, k entry of A^2, then the number of paths V_i, V_p, V_k, V_j of length 3 between V_i and V_j is $b_{ik}a_{kj}$. Thus the total number of paths of length 3 between V_i and V_j is the value of

$$b_{i1}a_{1j} + b_{i2}a_{2j} + \cdots + b_{in}a_{nj},$$

which is the same as the i, j entry in $A^2 \cdot A = A^3$. Hence the i, j entry in A^3 is the number of paths of length 3 between V_i and V_j.

The proof of the general case of Theorem 4.7 is left as an exercise; the last paragraph suggests what is to be done for the inductive step.

EXERCISES 4.3

In Exercises 1–4, use the breadth-first search algorithm to determine the distance and a shortest path from S to T in the graph. Use alphabetical order when there is a choice for a predecessor.

1. 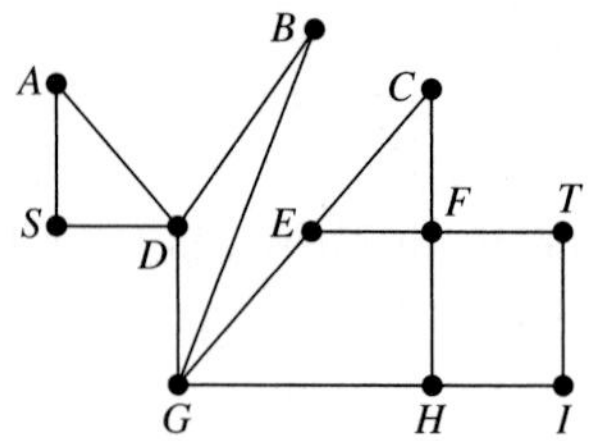

2. 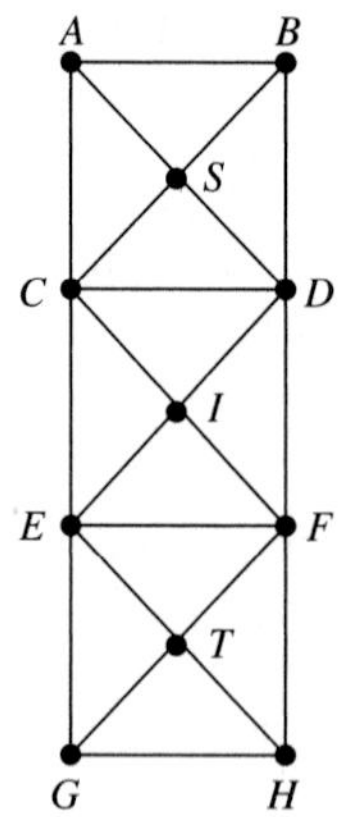

3. 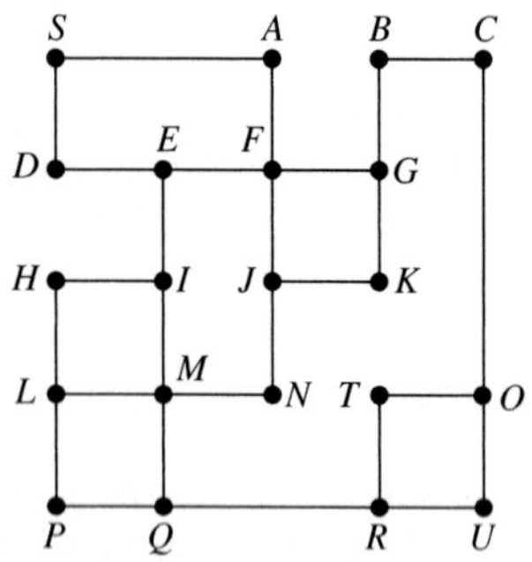

4. 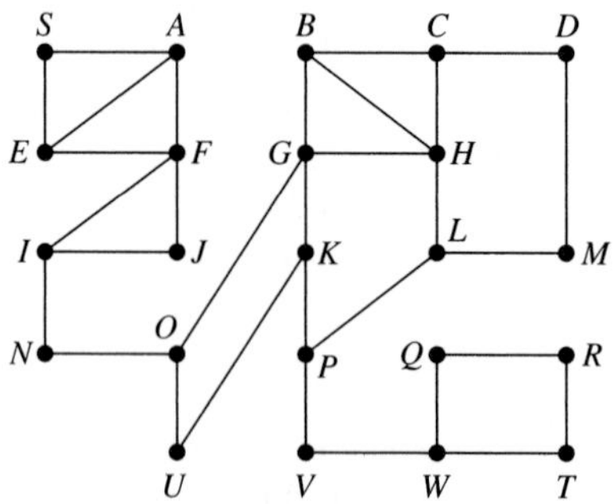

In Exercises 5–8, determine the distance from S to all the other vertices in the weighted graph. Find a shortest path from S to A.

5.

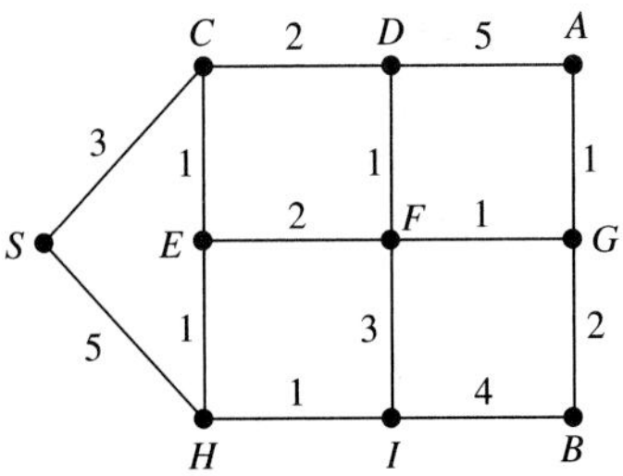

6.

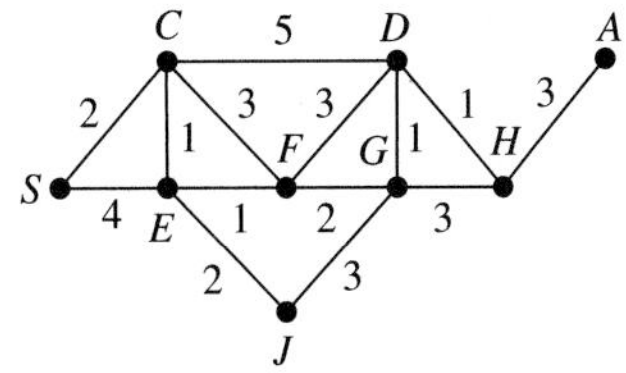

7.

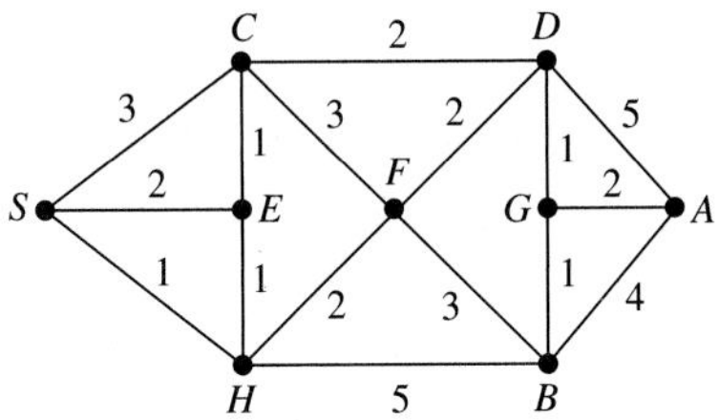

8.

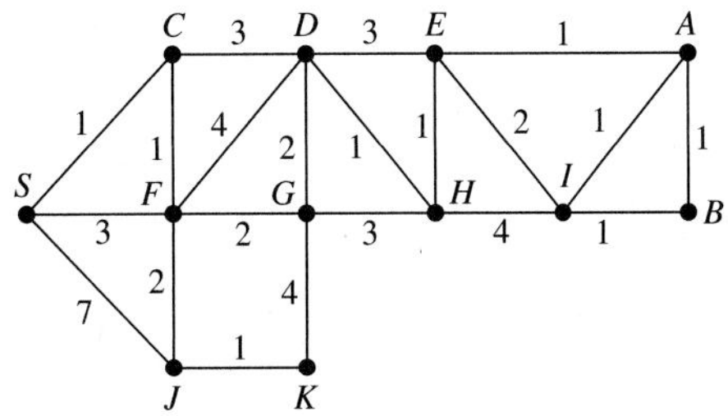

In Exercises 9–12, find a shortest path from S to T that goes through the vertex A in the weighted graph. Explain your procedure.

9.

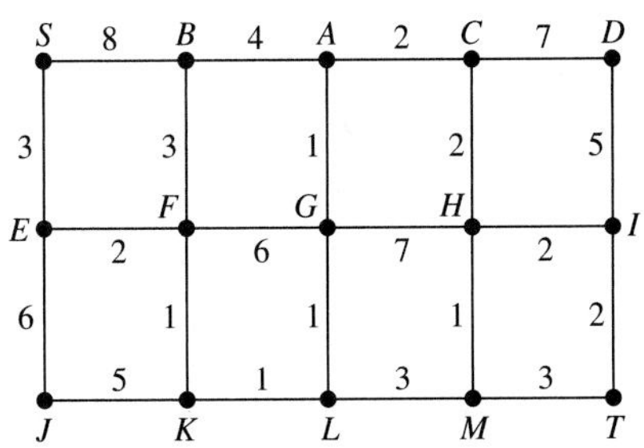

10.

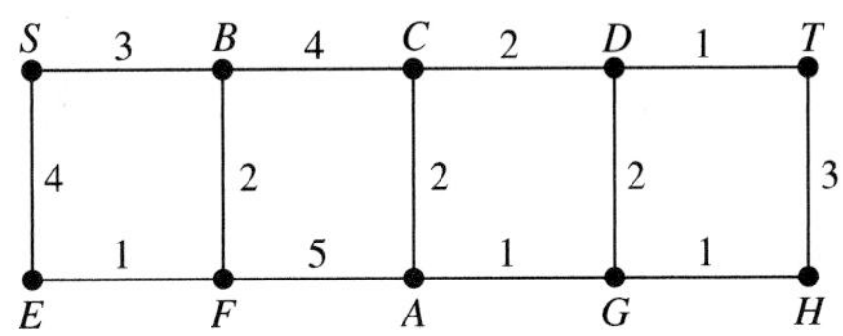

11.

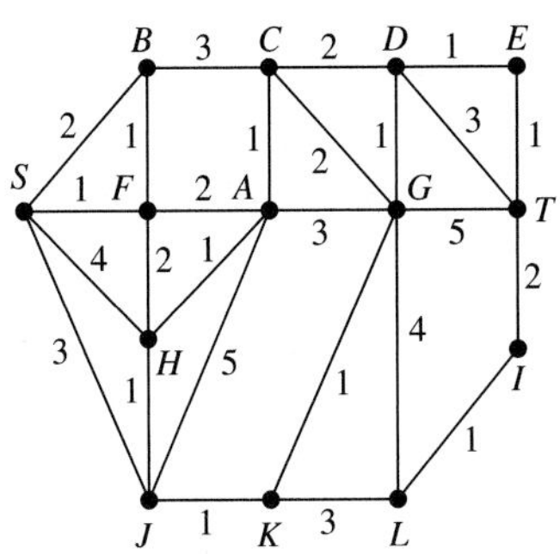

12.

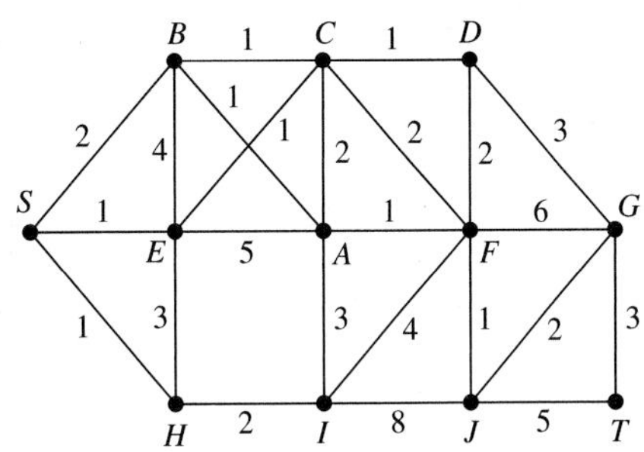

Use Theorem 4.7 to solve Exercises 13–16.

13. For the following graph, determine the number of paths of lengths 1, 2, 3, and 4 from V_1 to V_2, and from V_2 to V_3:

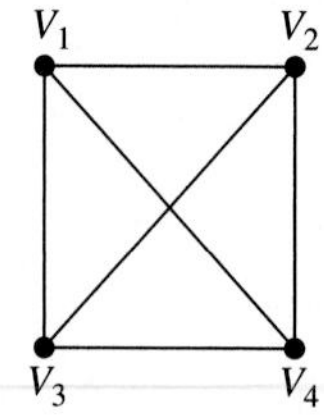

14. For the following graph, determine the number of paths of lengths 1, 2, 3, and 4 from V_1 to V_2, and from V_1 to V_3:

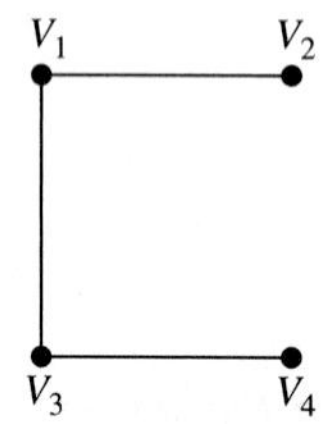

15. For the following graph, determine the number of paths of lengths 1, 2, 3, and 4 from V_1 to V_1, and from V_4 to V_3:

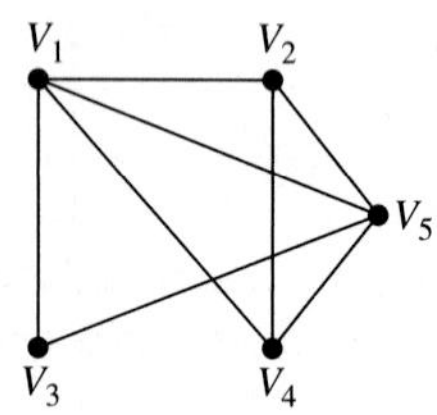

16. For the following graph, determine the number of paths of lengths 1, 2, 3, and 4 from V_1 to V_3, and from V_2 to V_4:

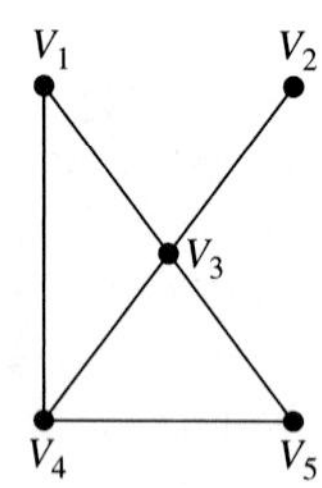

17. If A is the adjacency matrix of a labeled graph $\mathcal{G}$, what does the i, j element of $A + A^2 + A^3$ describe?

18. Prove that the label given to each vertex by the breadth-first search algorithm is its distance from S.

19. Prove Theorem 4.7 by mathematical induction.

20. Two weighted graphs are said to be **isomorphic** when there is an isomorphism between the underlying graphs such that the edges joining corresponding vertices have the same weight. Give an example of two nonisomorphic weighted graphs for which the underlying graphs are isomorphic.

Exercises 21–24 provide a proof of the validity of Dijkstra's algorithm. Assume in them that $\mathcal{G}$ is a weighted graph with all weights $W(U, V)$ positive, and that S is a vertex of $\mathcal{G}$.

21. Suppose each vertex V of $\mathcal{G}$ is assigned a label $L(V)$, which is either a number or ∞. Assume that $\mathcal{P}$ is a set of vertices of $\mathcal{G}$ containing S such that (i) if V is in $\mathcal{P}$, then $L(V)$ is the length of a shortest path from S to V and such a path exists with all its vertices in $\mathcal{P}$, and (ii) if V is not in $\mathcal{P}$, then $L(V)$ is the length of a shortest path from S to V subject to the restriction that V is the only vertex of the path not in $\mathcal{P}$. Let U be a vertex not in $\mathcal{P}$ with minimal label among such vertices. Show that a shortest path from S to U contains no element not in $\mathcal{P}$, except U.
22. Show that, under the assumptions of Exercise 21, the length of a shortest path from S to U is $L(U)$.
23. Assume that the hypotheses of Exercise 21 apply, and let $\mathcal{P}'$ be the set formed by U and the elements of $\mathcal{P}$. Show that $\mathcal{P}'$ satisfies property (i) of Exercise 21, and show that if V is not in $\mathcal{P}'$, then the length of a shortest path from S to V, all of whose vertices (except V) are in $\mathcal{P}'$, is the minimum of $L(V)$ and $L(U) + W(U, V)$.
24. Prove that Dijkstra's algorithm gives the length of a shortest path from S to each vertex of $\mathcal{G}$ by mathematical induction on the number of elements in $\mathcal{P}$. Let the induction hypothesis be that $\mathcal{P}$ is a set of vertices containing S and satisfying properties (i) and (ii) of Exercise 21.

4.4 COLORING A GRAPH

In Sections 4.1, 4.2, and 4.3 we discussed several situations described by graphs or multigraphs. Sometimes the situation in which a graph can be used is somewhat unexpected. Two such examples follow.

Example 4.22 Suppose that a chemical manufacturer needs to ship a variety of chemical products from a refinery to a processing plant. Shipping will be by rail, but according to EPA regulations, not all of these chemical products can be shipped together in one railroad car because of the possibility of their mixing together and creating a violent reaction should an accident occur. How can these products be shipped? In order to minimize expenses, the manufacturer wants to use the smallest possible number of railroad cars. What is this number? ■

Example 4.23 The state senate has a number of major standing committees, with every senator on one or more of these. Each committee meets every week for an hour. Each senator must be able to attend each meeting of a committee he or she is on, and so no two committees can meet at the same time if they have a member in common. The clerk of the senate is responsible for scheduling these meetings. How should the clerk schedule these committee meetings so that the senators can attend their major committee meetings and yet keep the number of meeting times as small as possible? ■

In these examples there are objects (chemical products or committees) and relationships (cannot travel in the same railroad car, or cannot meet at the same time) existing among them. Since this is the basic idea of a graph, it seems natural to describe each of these examples by a graph. In the first example the vertices are the chemical products, and an edge is drawn between two vertices whenever they represent chemical products that cannot be in the same railroad car. In the second example the vertices are the committees, and an edge is drawn between two vertices whenever some senator is on both of these committees.

To illustrate this idea further, let us assume in Example 4.22 that there are six chemical products P_1, P_2, P_3, P_4, P_5, and P_6 and that P_1 cannot ride in the same railroad car as P_2, P_3, or P_4; P_2 also cannot ride with P_3 or P_5; P_3 also cannot be with P_4; and P_5 cannot be with P_6. The graph that is described is found in Figure 4.40, where the vertices represent the six products and the edges join pairs of products that cannot ride together.

The question still remains: What is the smallest number of railroad cars needed? In the graph in Figure 4.40, products represented by adjacent vertices are to be in different cars. For example, product P_1 could be in car 1. Then, because P_1 and P_2 are adjacent, a different car is needed for P_2, say, car 2. Since P_3 is adjacent to both P_1 and P_2, another car is needed for P_3, say, car 3. But a new car is not needed for P_4; car 2 can be used again. Likewise, for P_5, a new car is not needed, as either car 1 or car 3 can be used. Let car 1 be chosen. Then, for P_6, car 2 or car 3 can be picked, say, car 2. The graph in Figure 4.41 shows how the vertices are labeled so that incompatible chemical products travel in different cars. Furthermore, because P_1, P_2, and P_3 are adjacent to each other, at least three different railroad cars must be used; so three is the smallest number of railroad cars that can be used.

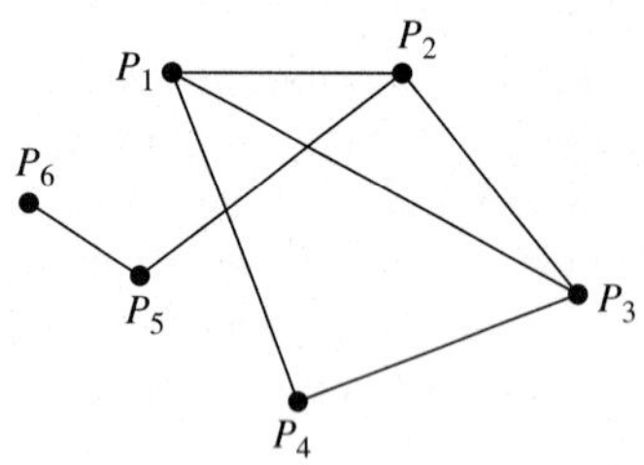

FIGURE 4.40

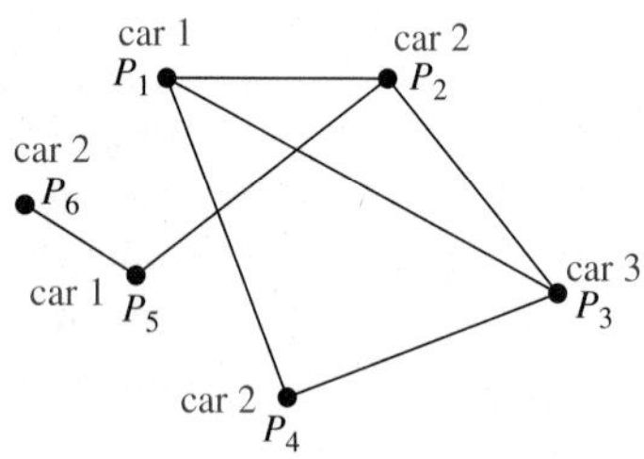

FIGURE 4.41

We have assigned labels to the vertices of a graph so that adjacent vertices have different labels. This idea occurs frequently in graph theory; for historical reasons, the labels are called **colors**. To **color a graph** means to assign a color to each vertex so that adjacent vertices have different colors. Asking what is the smallest number of railroad cars needed in Example 4.22 is the same as asking what is the smallest number of colors needed to color the graph in Figure 4.40, with a color corresponding to a railroad car.

When a graph can be colored with n colors, but not with a smaller number of colors, it is said to have **chromatic number** n. Thus the graph in Figure 4.40 has chromatic number 3.

Example 4.24 The graph in Figure 4.42(a) has chromatic number 2, since the vertices V_1, V_3, and V_5 can be colored with one color (say, red), and the other three vertices with a second color (blue), as shown in Figure 4.42(b). In general, if a cycle has an even number of vertices, then it can be colored with two colors. ■

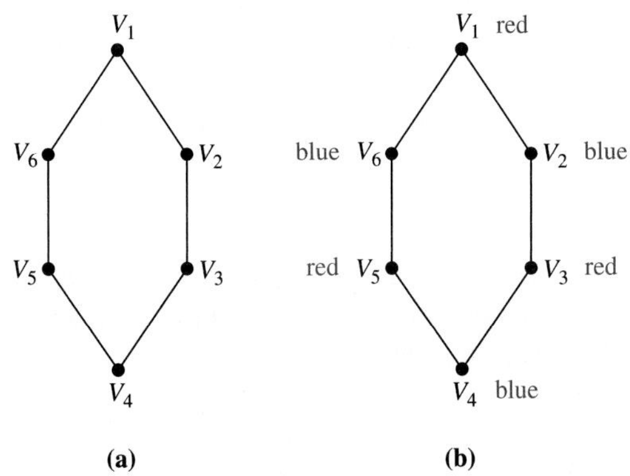

FIGURE 4.42

Example 4.25 When a cycle has an odd number of vertices, such as that in Figure 4.43(a), then three colors must be used. If we try to alternate colors, as was done in Figure 4.42, with the color red assigned to vertices V_1 and V_3 and the color blue assigned to vertices V_2 and V_4, then it is not possible to use either red or blue for V_5. Using three colors to color a cycle with an odd number of vertices is illustrated in Figure 4.43(b). ■

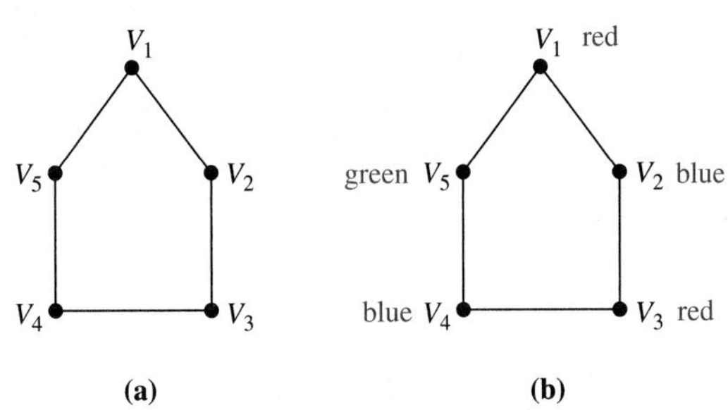

FIGURE 4.43

Example 4.26 The complete graph $\mathcal{K}_n$ with n vertices can be colored with n colors. But since every vertex is adjacent to every other vertex, a smaller number of colors will not work. Thus $\mathcal{K}_n$ has chromatic number n. ■

Example 4.27 The graph in Figure 4.44(a) can be colored with two colors, as indicated in Figure 4.44(b). ■

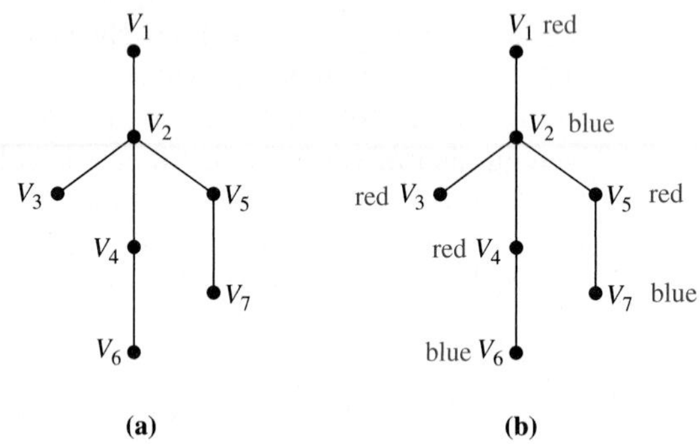

FIGURE 4.44

Example 4.28 The graph in Figure 4.45 has chromatic number 2, since the vertices on the left can be colored with one color and the vertices on the right can be colored with a second. ■

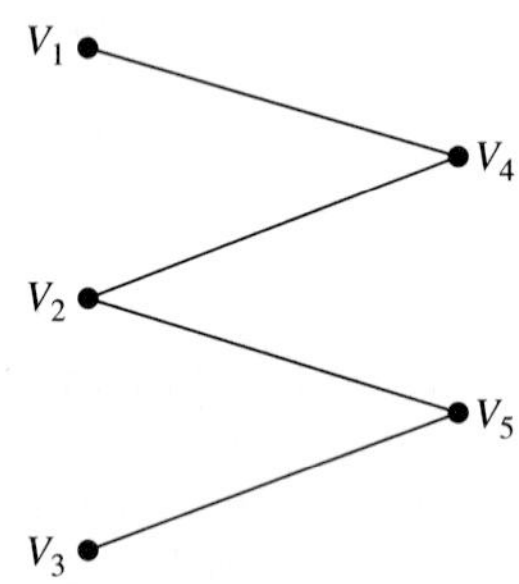

FIGURE 4.45

In general, it is very difficult to find the smallest number of colors needed to color a graph. One method is to list all the different ways to assign colors to the vertices of a graph, go through these ways one at a time to see which of them is a coloring, and then, finally, determine which colorings have the smallest number of colors. Unfortunately, even if the graph has a relatively small number of vertices, this becomes an extraordinarily time-consuming process, measured in centuries rather than minutes, even with the use of a supercomputer.

Nevertheless, there are a number of results that describe the chromatic number of a graph. For instance, as seen in Example 4.25, a cycle with an odd length has chromatic number 3. Thus any graph containing a cycle of this type needs at least three colors. The graph in Figure 4.41 is an example of this. When there are no cycles of odd length in a graph, then two colors are enough.

Theorem 4.8 A graph $\mathcal{G}$ can be colored with two colors if and only if it contains no cycle of odd length.

Proof. As noted before, when $\mathcal{G}$ has a cycle of odd length, then coloring $\mathcal{G}$ requires at least three colors. Hence if $\mathcal{G}$ can be colored with two colors, then it contains no cycle of odd length.

Conversely, suppose $\mathcal{G}$ has no cycle of odd length. We will show that $\mathcal{G}$ can be colored with two colors. Since any coloring of each component of $\mathcal{G}$ with two colors provides such a coloring of $\mathcal{G}$, we can assume that $\mathcal{G}$ is connected. (See Exercise 57 of Section 4.2 for the definition of a component.)

Choose an arbitrary vertex S of $\mathcal{G}$, and apply the breadth-first search algorithm to $\mathcal{G}$, starting with S. Since $\mathcal{G}$ is connected, every vertex gets labeled. Color each vertex red or blue, according as its label is even or odd.

We must show that no adjacent vertices, say, U and V, have the same color. By the way the breadth-first search algorithm works, labels on adjacent vertices cannot differ by more than 1. Since the labels on U and V are both even or both odd, they must be the same, say, m. Use the predecessors to trace shortest paths from U and V back to S. Let these paths first meet at the vertex W, with label k. (See Figure 4.46. We could have $W = S$ and $k = 0$.) Then the portions of these paths from W to U and W to V, along with the edge $\{U, V\}$, form a cycle of length $2(m - k) + 1$, which is odd, contrary to our assumption that $\mathcal{G}$ contains no odd cycle. ■

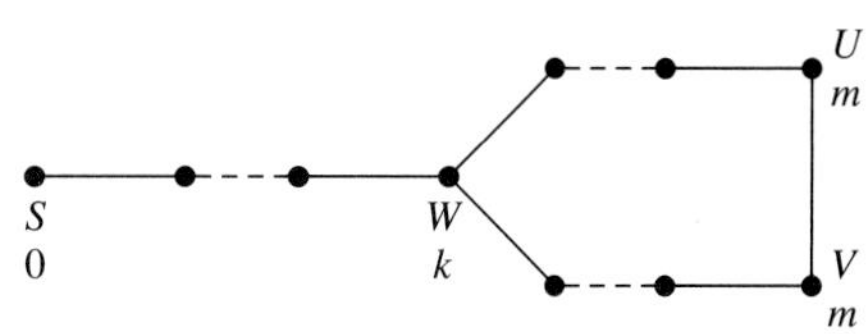

FIGURE 4.46

Example 4.29 The breadth-first search algorithm has been applied to the graph in Figure 4.47, starting with vertex V in the left component and X in the right. The resulting labels are shown. The coloring with two colors in Figure 4.48 is produced by coloring vertices red or blue, according as their labels are even or odd. ■

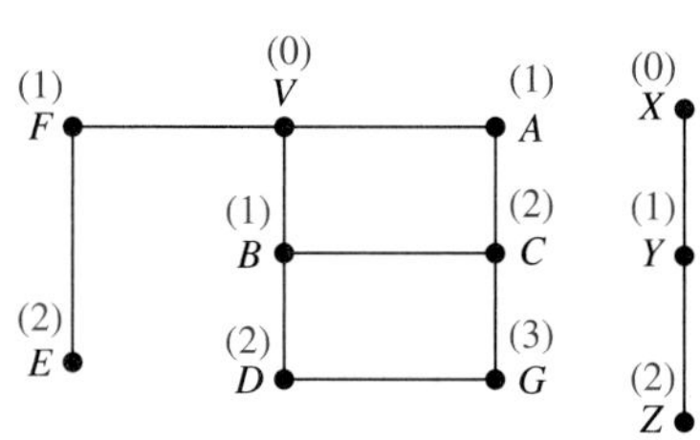

FIGURE 4.47

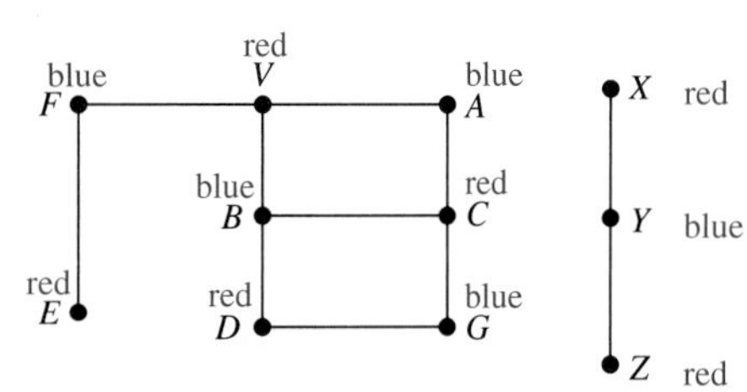

FIGURE 4.48

The next theorem gives an upper bound on the number of colors needed to color a graph.

Theorem 4.9 The chromatic number of a graph $\mathcal{G}$ cannot exceed one more than the maximum of the degrees of the vertices of $\mathcal{G}$.

Proof. Let k be the maximum of the degrees of the vertices of $\mathcal{G}$. We will show that $\mathcal{G}$ can be colored with $k+1$ colors. First, select a vertex V and assign a color to it. Next, pick some other vertex W. Since there are at most k vertices adjacent to W and there are at least $k+1$ colors available to choose from, there is at least one color (possibly, many) that has not been used on a vertex adjacent to W. Choose such a color. This process can be continued until all the vertices of $\mathcal{G}$ are colored. ■

Example 4.30 The procedure described in Theorem 4.9 may use more colors than are really necessary. The graph in Figure 4.49 has a vertex of degree 4, which is the maximum degree, and so, by Theorem 4.9, can be colored with $1+4=5$ colors. However, by using the procedure described in Theorem 4.8, it can be colored with two colors. ■

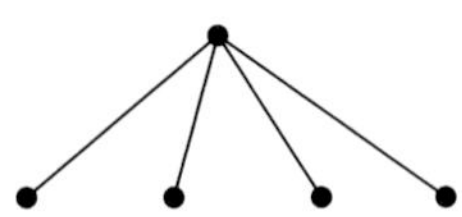

FIGURE 4.49

One of the most famous problems of the 19th century concerned the number of colors required to color a map. It is understood that, when coloring a map, countries with a common boundary other than a point are to be colored with different colors. The map is assumed to be drawn on a flat surface or globe, as opposed to a more complicated surface such as a doughnut. The usual approach to this problem is to let each country be a vertex of a graph and to join vertices representing countries with a common boundary other than a point. Then coloring the map is the same as coloring the vertices of this graph so that no two adjacent vertices have the same color. It was conjectured in 1852 that four colors would be enough to color any such map, but it was not until 1976 that Kenneth Appel and Wolfgang Haken, two mathematicians at the University of Illinois, verified this conjecture. Their verification required an exhaustive analysis of more than 1900 cases that took more than 1200 hours on a high-speed computer.

Example 4.31 Figure 4.50(a) is a portion of a map of the United States. The associated graph, obtained as described previously, is shown in Figure 4.50(b). This graph can be colored with three colors, as illustrated in Figure 4.50(c). ■

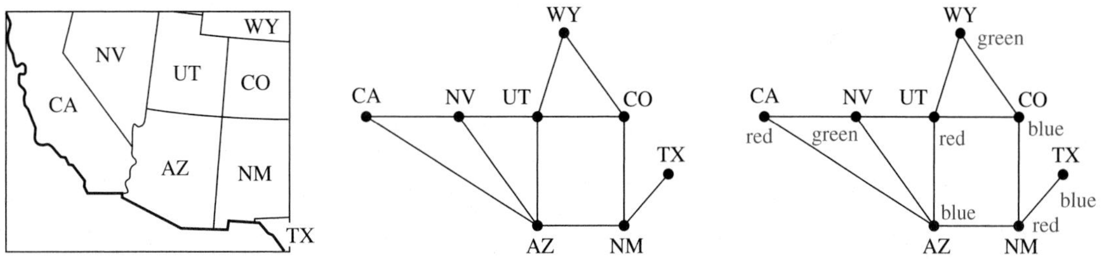

(a) (b) (c)

FIGURE 4.50

EXERCISES 4.4

In Exercises 1–8, find the chromatic number of the graph.

1. 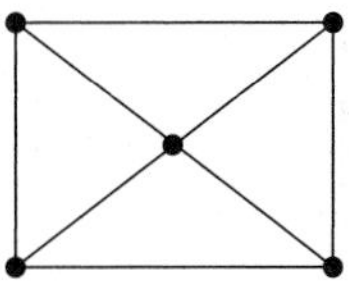

2. 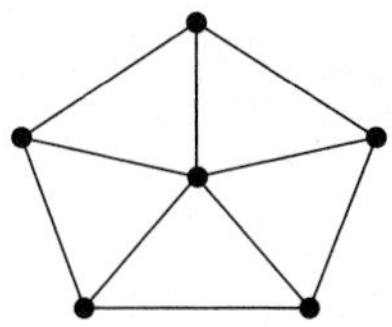

3. 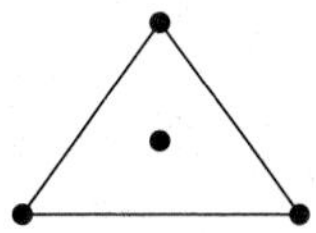

4. 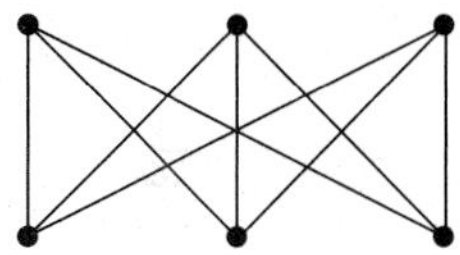

5. 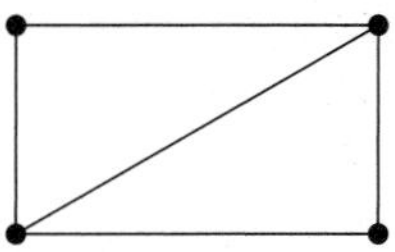

6. 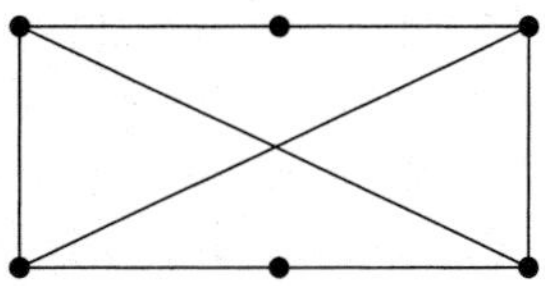

7. 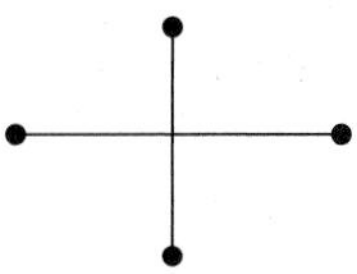

8. 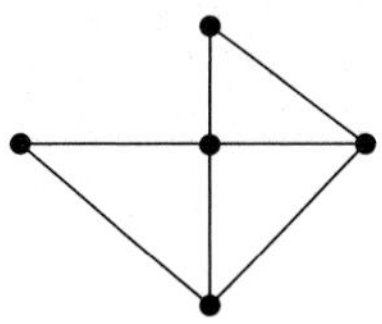

9. What does it mean for a graph to have chromatic number 1?
10. What is the chromatic number of $\mathcal{K}_{2,3}$? of $\mathcal{K}_{7,4}$? of $\mathcal{K}_{m,n}$? (See Exercise 52 of Section 4.2 for a definition of $\mathcal{K}_{m,n}$.)
11. Give examples of graphs for which
    (a) The chromatic number is one more than the maximum of the degrees of the vertices.
    (b) The chromatic number is not one more than the maximum of the degrees of the vertices.
12. It might be supposed that if a graph has a large number of vertices and each vertex has a large degree, then the chromatic number would have to be large. Show that this conjecture is incorrect by constructing a graph with at least 12 vertices, each of degree at least 3, that has chromatic number 2.
13. Using the process presented in the proof of Theorem 4.8, write a formal algorithm for coloring a graph with no cycles of odd length.

14. Show that, when the algorithm in Exercise 13 is applied to a graph with n vertices and e edges, the graph can be colored with at most $n + e$ elementary operations. (In analyzing the algorithm, consider the elementary operations to be coloring a vertex and using an edge.)

In Exercises 15–18, color the graph by using the algorithm in Exercise 13.

15.

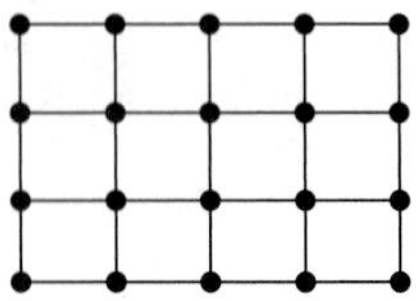

16.

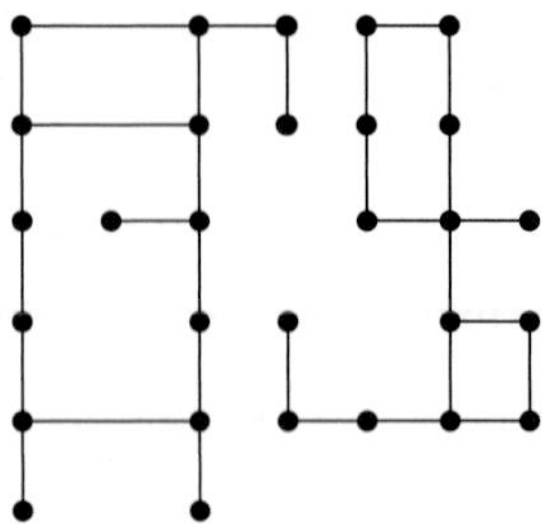

17.

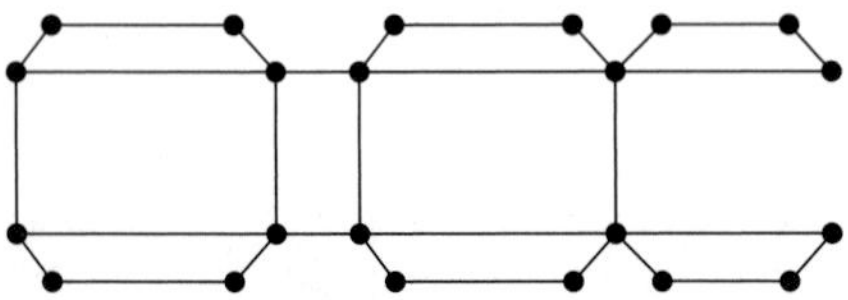

18.

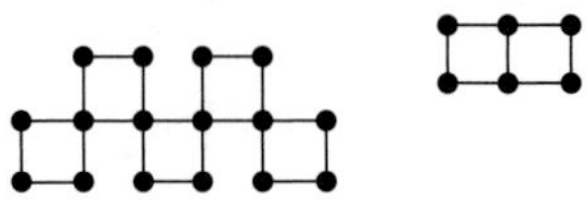

19. What is the chromatic number of the graph in Exercise 48 of Section 4.1?

20. Suppose $\mathcal{G}$ is a graph with three vertices. How many ways are there to assign three colors to the vertices (this need not be a coloring of the graph)? What if the graph has four vertices and four colors are available?

21. Generalize Exercise 20 to the case of a graph with n vertices and n colors.

22. Suppose $\mathcal{G}$ is a graph with n vertices and there are n available colors to assign to the vertices. If one operation consists of assigning colors to the vertices and checking if a coloring has been made, how long would it take a computer that can perform one billion operations per second to check all possible color assignments for a graph with 20 vertices? Would this be a good way to find a coloring using the least number of colors?

23. Color the following map with only three colors:

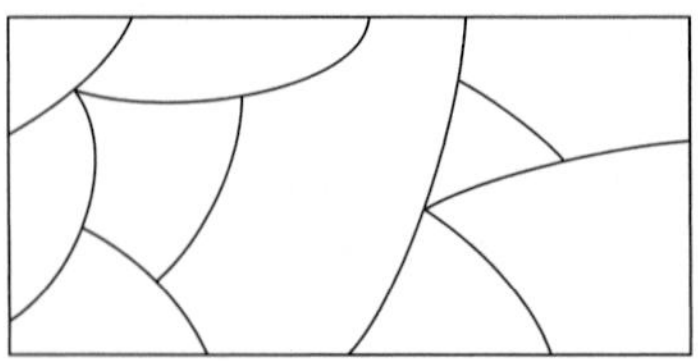

24. Color the following map with only three colors:

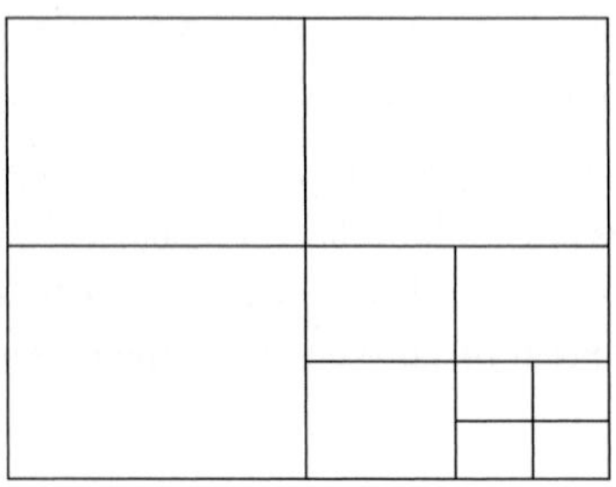

25. Color the following map with only four colors:

26. Color the following map with only four colors:

27. Solve Example 4.23 if there are five major committees: finance, budget, education, labor, and agriculture. The clerk of the senate needs only to consider state senators Brown, Chen, Donskvy, Geraldo, Smith, and Wang. The finance committee has members Chen, Smith, and Wang; the budget committee has members Chen, Donskvy, and Wang; the education committee has members Brown, Chen, Geraldo, and Smith; the labor committee has only Geraldo; and the agriculture committee has Donskvy and Geraldo.

28. By representing the following figure by a graph, determine the minimum number of colors needed to color each circle so that touching circles have different colors:

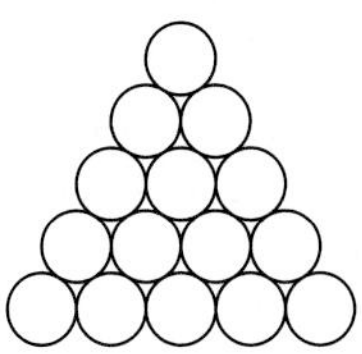

29. The zookeeper of a major zoo wants to redo the zoo in such a way that the animals live together in their natural habitat. Unfortunately, it is not possible to put all the animals together in one location, because some are predators of others. The letters in the following chart represent animals, and the dots show which are predators or prey of others:

	a	b	c	d	e	f	g	h	i	j
a		●			●					●
b	●			●			●			
c								●		●
d		●				●				
e	●								●	
f				●						●
g		●								
h			●						●	
i					●			●		●
j	●		●			●			●	

What is the minimum number of locations the zookeeper needs?

30. Is "can be colored with three colors" a graph isomorphism invariant?

31. Show that "has chromatic number 3" is a graph isomorphism invariant.

32. There are seven tour bus companies in the Los Angeles area. During a particular day, each visits at most three different locations from among Hollywood, Beverly Hills, Disneyland, and Universal Studios. The same location cannot be visited by more than one tour company on the same day. The first tour company visits only Hollywood, the second only Hollywood and Disneyland, the third only Universal Studios, the fourth only Disneyland and Universal Studios, the fifth Hollywood and Beverly Hills, the sixth Beverly Hills and Universal Studios, and the seventh Disneyland and Beverly Hills. Can these tours be scheduled only on Monday, Wednesday, and Friday?

33. Prove that if a graph with n vertices has chromatic number n, then the graph has $\frac{1}{2}n(n-1)$ edges.

34. Show that it is possible to assign one of the colors red and blue to each *edge* of $\mathcal{K}_5$ in such a way that no cycle of length 3 has all its edges the same color.

35. Show that the statement of Exercise 34 is incorrect if $\mathcal{K}_5$ is replaced by $\mathcal{K}_6$.

36. Prove Theorem 4.9 by mathematical induction on the number of vertices.

37. Suppose that each vertex of a graph $\mathcal{G}$ is such that if the vertex and the edges incident with it are removed from $\mathcal{G}$, then the resulting graph has a smaller chromatic number. Prove that if the chromatic number of $\mathcal{G}$ is k, then the degree of each vertex of $\mathcal{G}$ is at least $k-1$.

4.5 DIRECTED GRAPHS AND MULTIGRAPHS

In previous applications of graphs, an edge was used to represent a two-way or symmetric relationship between two vertices. However, there are situations where relationships hold in only one direction. In these cases the use of a line segment is not descriptive enough, and a directed line segment is needed.

Example 4.32 In many urban downtown areas the city streets are one-way. In such a case, it is necessary to use a directed line segment to indicate the legal flow of traffic. In Figure 4.51 major downtown locations are represented by dots, and two dots

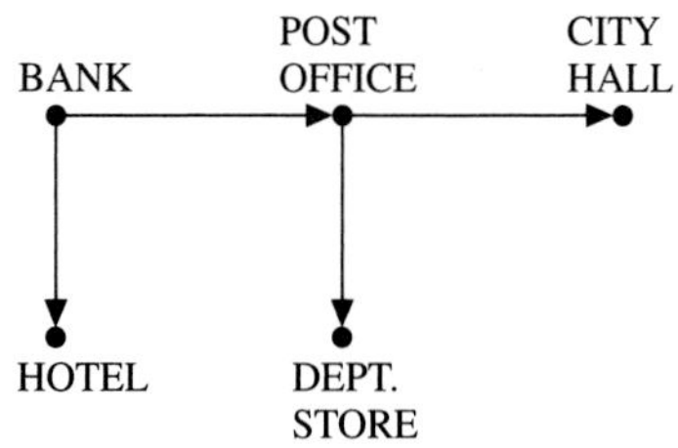

FIGURE 4.51

are connected by an arrow when it is possible to go from the first location to the second by means of a one-way street. For example, the arrow from BANK to HOTEL denotes that there is a one-way street from BANK to HOTEL. ■

Example 4.33 Although in a communication network there are routes where information can flow either way, there are also some where the flow is in just one direction. Within a microcomputer system, data usually can travel in either direction between CPU and the Memory, but only from the Input to the Memory and from the Memory to the Output. This type of situation can be represented by the diagram in Figure 4.52, where the arrows indicate how the data can flow. ■

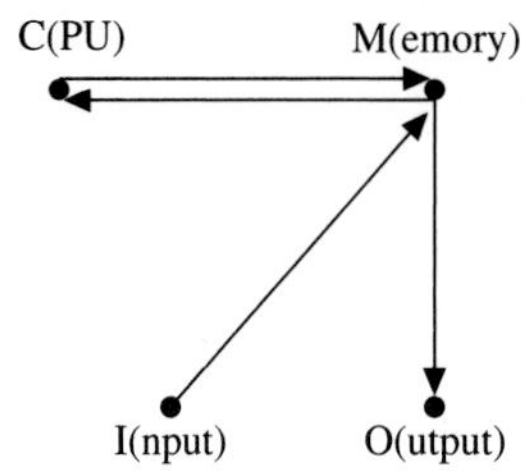

FIGURE 4.52

A **directed graph** is a finite nonempty set $\mathcal{V}$ and a set $\mathcal{E}$ of ordered pairs of distinct elements of $\mathcal{V}$. The elements of $\mathcal{V}$ are called **vertices** and the elements of $\mathcal{E}$ are called **directed edges**.

Figure 4.52 depicts a directed graph with vertices C, I, M, and O and directed edges (C, M), (M, C), (I, M), and (M, O). As is true for graphs, a directed graph can be described either by the use of sets or by the use of a diagram, where arrows between the vertices in $\mathcal{V}$ describe which ordered pairs of vertices are being included.

If there is a directed edge $e = (A, B)$, it is said that e is a **directed edge from *A* to *B***. In Figure 4.52 there is a directed edge from M to O, but no directed edge from O to M. Similarly, there is a directed edge from M to C and one from C to M.

Just as for graphs, two directed edges crossing in a diagram do not create a new vertex. Likewise, in this book the set of vertices is to be a finite set (although not all authors require this). Finally, a directed edge cannot go from a vertex to itself, nor can there be two or more directed edges from one vertex to another.

In a directed graph, the number of directed edges *from* vertex A is called the **outdegree** of A and is denoted as outdeg(A). Similarly, the number of directed edges *to* vertex A is called the **indegree** of A and is denoted by indeg(A). In Figure 4.52 we see that outdeg$(M) = 2$, indeg$(C) = 1$, and outdeg$(O) = 0$. Theorem 4.1 states that in a graph the sum of the degrees is equal to twice the number of edges. Because each directed edge leaves one vertex and enters a second vertex, we have a similar theorem for directed graphs.

Theorem 4.10 In a directed graph the following three numbers are equal: the sum of the indegrees of the vertices, the sum of the outdegrees of the vertices, and the number of directed edges.

Representations of Directed Graphs

As for graphs, a directed graph can be represented by a matrix. Suppose we have a directed graph $\mathcal{D}$ with n vertices labeled $V_1, V_2, \ldots, V_n$. Such a directed graph is called **labeled**. Form an $n \times n$ matrix in which the i, j entry is 1 if there is a directed edge from the vertex V_i to the vertex V_j and 0 if there is not. This matrix is called the **adjacency matrix** of $\mathcal{D}$ (with respect to the labeling) and is denoted by $A(\mathcal{D})$.

Example 4.34 Figure 4.53 contains a directed graph and its adjacency matrix. The 1, 4 entry is 0 because there is no directed edge from V_1 to V_4, but the 4, 1 entry is 1 because there is a directed edge from V_4 to V_1. Row 3 contains all zeros because there are no directed edges from the vertex V_3. Since there are no directed edges to the vertex V_4, column 4 also contains all zeros. ■

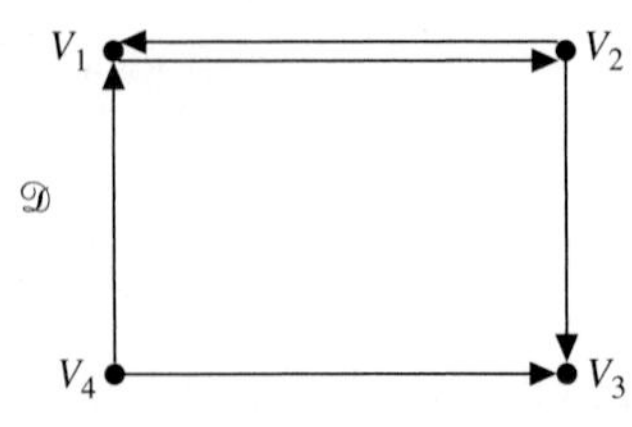

$$A(\mathcal{D}) = \begin{bmatrix} 0 & 1 & 0 & 0 \\ 1 & 0 & 1 & 0 \\ 0 & 0 & 0 & 0 \\ 1 & 0 & 1 & 0 \end{bmatrix}$$

FIGURE 4.53

The last two observations in the previous example suggest the following version of Theorem 4.2 for directed graphs (the proof follows from the definitions):

Theorem 4.11 The sum of the entries in row i of the adjacency matrix of a directed graph equals the outdegree of the vertex V_i, and the sum of the entries in column j equals the indegree of the vertex V_j.

Directed graphs can also be represented by **adjacency lists**. To form an adjacency list, we begin by labeling the vertices of the directed graph. Then we list the vertices in a column, and after each vertex we list the vertices to which there is a directed edge from the given vertex.

Example 4.35 For the directed graph in Figure 4.53, the adjacency list is as follows:

$$\begin{aligned} &V_1\colon V_2 \\ &V_2\colon V_1, V_3 \\ &V_3\colon \text{(none)} \\ &V_4\colon V_1, V_3 \end{aligned}$$

Since V_2 is the only vertex to which there is a directed edge from V_1, V_2 is the only vertex listed after V_1. Similarly, because the only directed edges from V_4 are to V_1 and V_3, these are the two vertices listed after V_4. ■

Directed Multigraphs

In Sections 4.2 and 4.3 we introduced the concepts of multigraph, weighted graph, path, simple path, and cycle. These concepts have analogs using directed edges. We will leave the concept of directed weighted graphs to the exercises. To illustrate the other definitions, we will consider the diagram in Figure 4.54.

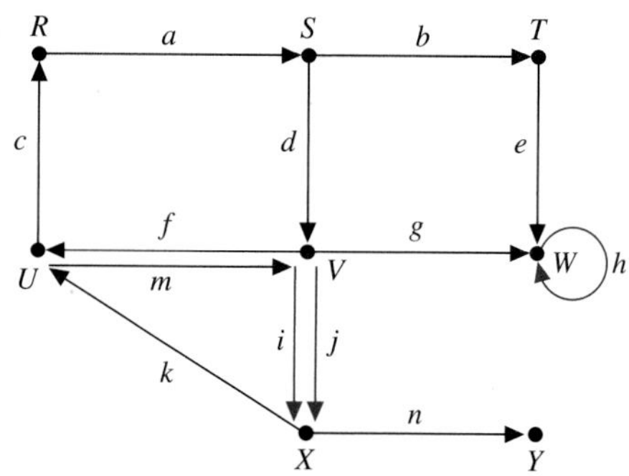

FIGURE 4.54

The diagram in Figure 4.54 describes a **directed multigraph**. Here, there is a **directed loop** h at the vertex W, and there are **parallel directed edges** i and j from V to X. Because a directed graph is also a special kind of directed multigraph, definitions developed for a directed multigraph also apply to a directed graph. Note that directed edges such as f and m are *not* parallel directed edges, whereas i and j are parallel directed edges.

An alternating sequence of vertices and directed edges

$$V_1, e_1, V_2, e_2, \ldots, V_n, e_n, V_{n+1}$$

is called a **directed path** from V_1 to V_{n+1} if $e_i = (V_i, V_{i+1})$ for each $i = 1, 2, \ldots n$. The **length** of this directed path is n, the number of directed edges. Thus R, a, S, b, T, e, W is a directed path from R to W of length 3, which can also be written

R, S, T, W or a, b, e. The directed path V, i X, k, U, m, V, j, X cannot be described by just using the vertices, because there are two directed edges from V to X. However, this directed path can be described by just listing the directed edges i, k, m, j. Also, T, b, S, d, V is not a directed path, since the directed edge b is not from T to S, but from S to T. Likewise, there is no directed path of positive length starting from Y. As before, a vertex is a directed path of length 0.

The directed path a, d, g is a **simple directed path** from R to W, that is, a directed path with no vertex repeated. The directed path a, d, i, k, m, g is not a simple directed path, because the vertex V is repeated. It is easily seen that a directed path contains a simple directed path between the same two vertices. This proof is similar to the proof of Theorem 4.4 for multigraphs and is omitted here.

Theorem 4.12 Every U-V directed path contains a U-V simple directed path.

The directed path a, d, f, c in Figure 4.54 is a **directed cycle** because it is a directed path of positive length from R to R in which no other vertex is visited twice. But b, e, g, d is not a directed cycle, because the directed edges g and d go in the wrong direction. Both h and f, m are considered to be directed cycles. The directed path k, m, f, c, a, d, j is not a directed cycle, because the vertices U and V appear twice.

A directed multigraph $\mathcal{D}$ is called **strongly connected** if, for every pair A and B of vertices in $\mathcal{D}$, there is a directed path from A to B. Thus, in a strongly connected directed multigraph, we can go from any vertex to any other by following some route along the directed edges.

Example 4.36 The directed multigraph in Figure 4.54 is not strongly connected, since there is no directed path from Y to any other vertex. The directed graph in Figure 4.55(a) is strongly connected, however, since a directed path can be found from any vertex to any other. On the other hand, the directed graph in Figure 4.55(b) is not strongly connected, because there is no directed path from A to C. ■

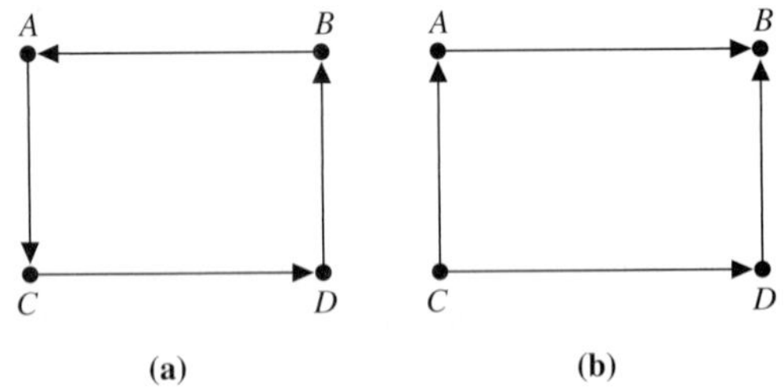

FIGURE 4.55

Example 4.37 Suppose the city council of a middle-sized city is concerned about traffic congestion in the downtown area. It instructs the city traffic engineer to turn each two-way street into a one-way street in the downtown area in such a way that there still will be a route from each downtown location to any other.

If a graph is used to represent the current downtown street system (where intersections are vertices and streets are edges), then the city traffic engineer is to assign a direction to each edge, transforming the graph into a directed graph. Since there is to be a route from any place to any other, we want this new directed graph to be strongly connected. For example, if the graph in Figure 4.56 represents the downtown streets, then assigning directions as in Figure 4.55(a) produces a directed graph that is strongly connected. Thus this assignment of directions satisfies the city council's requirement. On the other hand, the assignment of directions as in Figure 4.55(b) yields a directed graph that is not strongly connected, and so it does not satisfy the city council's requirement. ■

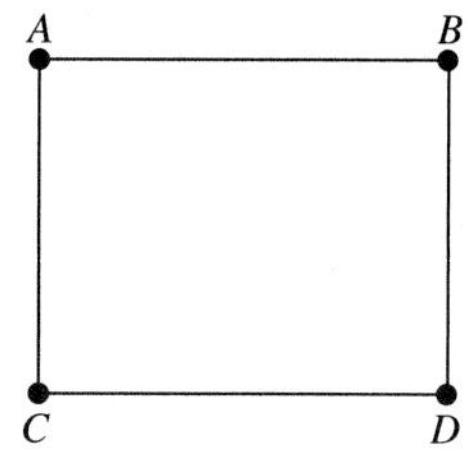

FIGURE 4.56

An important question is, When can directions be assigned to the edges of a graph to yield a directed graph that is strongly connected? In Figure 4.57 there is an example of a graph that cannot be transformed into a strongly connected directed graph. The source of difficulty is the edge joining A and B. For if we direct this edge from A to B, then we cannot find a route from a place on the right side to any place on the left side. A similar problem occurs if we direct this edge from B to A. This edge joining A and B possesses an interesting property: If it is removed from the graph, the graph is no longer connected.

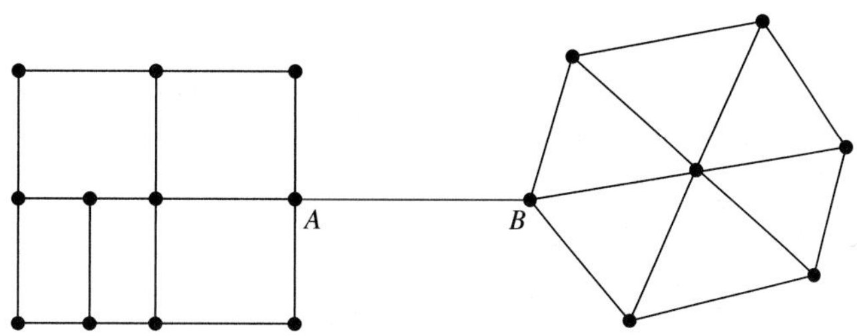

FIGURE 4.57

It can be proved that the absence of an edge whose removal disconnects a connected graph is equivalent to the existence of an assignment of directions to the edges to create a strongly connected directed graph. A proof can be found in suggested reading [9]. Thus, for the city traffic engineer in Example 4.37 to decide whether there is an acceptable pattern of one-way streets, it suffices for the engineer to find whether there is an edge whose removal will disconnect the graph.

The material in Section 4.2 can be modified in a natural way to work for directed graphs. To illustrate this statement, we will consider directed Euler paths and circuits and directed Hamiltonian paths and cycles.

Directed Euler Circuits and Paths

The ideas of a directed Euler path and a directed Euler circuit in a directed multigraph are similar to the corresponding ones in a multigraph. A directed path in a directed multigraph $\mathcal{D}$ that includes exactly once all the directed edges of $\mathcal{D}$ and has different initial and terminal vertices is called a **directed Euler path**. A directed path that includes exactly once all the directed edges of $\mathcal{D}$ and has the same first and last vertex is called a **directed Euler circuit**.

Recall that, in the proof of Theorem 4.5, constructing an Euler circuit required that each time we entered a vertex along an edge, there was another edge for us to leave on. This translated into the requirement that each vertex be of even degree. In constructing a directed Euler circuit, we require similarly that, for each directed edge going into a vertex, there must be another directed edge leaving that vertex. This implies that the indegree of each vertex is the same as the outdegree. These observations are summarized in the next theorem.

Theorem 4.13 Suppose the directed multigraph $\mathcal{D}$ has the property that, whenever the directions are ignored on the directed edges, the resulting multigraph is connected. Then $\mathcal{D}$ has a directed Euler circuit if and only if, for each vertex of $\mathcal{D}$, the indegree is the same as the outdegree. Furthermore, $\mathcal{D}$ has a directed Euler path if and only if every vertex of $\mathcal{D}$ has its indegree equal to its outdegree, except for two distinct vertices B and C, where the outdegree of B exceeds its indegree by 1 and where the indegree of C exceeds its outdegree by 1. When this is the case, the directed Euler path begins at B and ends at C.

The algorithm for constructing an Euler circuit in a multigraph may be modified in a natural way (by choosing an unused *directed* edge leaving the vertex) to construct directed Euler circuits and paths in directed multigraphs that satisfy the hypotheses of Theorem 4.13.

Example 4.38 In telecommunications, there is an interesting application of directed Euler circuits. (See [7] in the suggested readings.) Suppose there is a rotating drum with eight different sectors, where each sector contains either a 0 or a 1. Three detectors are placed so that they can read the contents of three adjacent sectors. (See Figure 4.58.)

The task is to assign 1s and 0s to the sectors so that a reading of the detectors describes the exact position of the rotating drum. Suppose the sectors are assigned

1s and 0s as in Figure 4.59. Then a reading of the detectors gives 010. If the drum is moved 1 sector clockwise, the reading becomes 101. However, if the drum is moved still another sector clockwise, the reading becomes 010 again. Thus two different positions of the rotating drum give the same reading. We want an assignment of 1s and 0s where this will not happen; that is, we want to arrange eight 1s and 0s in a circle so that every sequence of three consecutive entries is different.

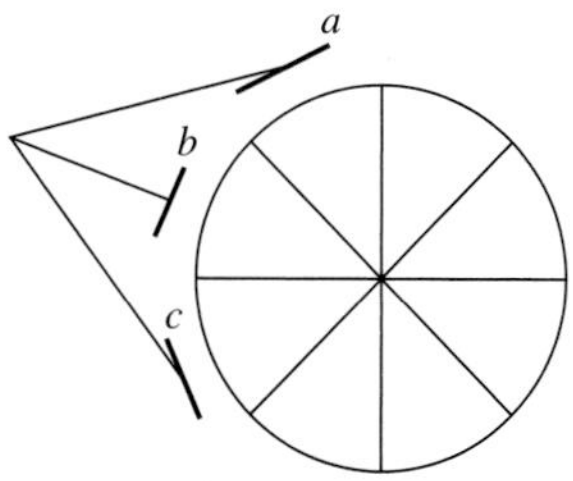

FIGURE 4.58

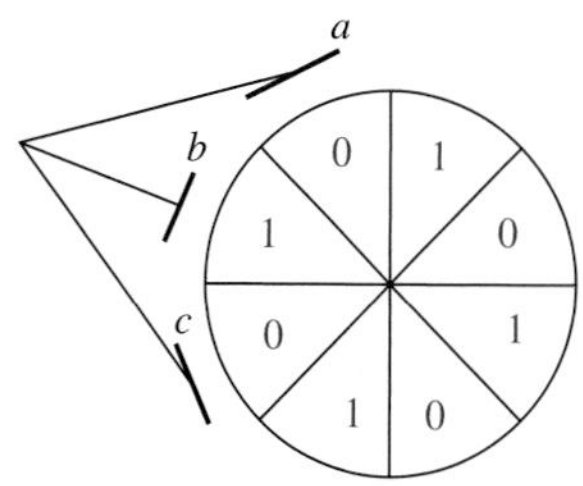

FIGURE 4.59

We will create a directed multigraph with 00, 01, 11, and 10 as vertices. From each vertex, construct two directed edges in the following manner: For the vertex *ab*, consider the two vertices *b*0 and *b*1 (obtained from *ab* by dropping *a* and appending 0 and 1 at the end). Construct a directed edge from vertex *ab* to the vertex *bc* (where *c* is either 0 or 1), and assign this directed edge the label *abc*. For example, there is a directed edge from 01 to 10 with label 010 and from 01 to 11 with label 011. This directed multigraph is shown in Figure 4.60. Note that the labels assigned to the directed edges are all different and would be an acceptable set of readings for the detectors.

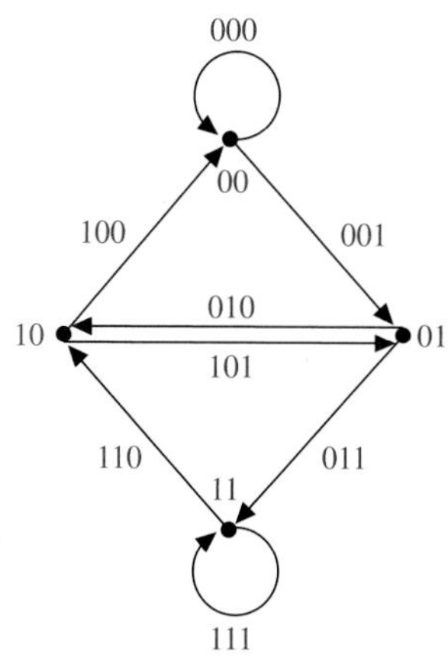

FIGURE 4.60

This is a directed multigraph such that when the directions on the directed edges are ignored, the resulting multigraph is connected. Also, the indegree equals the outdegree of each vertex, and so a directed Euler circuit exists. Using the modification of the Euler circuit algorithm indicated in the paragraph preceding this example, we start with vertex 01 and construct the directed Euler circuit 011, 111, 110, 101, 010, 100, 000, 001. For the directed edges in this directed Euler circuit, the last two digits in each label are the first two digits in the label of the next directed edge. Thus, if we select the first digit of the label of each directed edge in the directed Euler circuit, we get a sequence of eight numbers such that every sequence of three consecutive entries is different (because the labels are all different). For this example, this process gives the sequence 01110100. When this sequence is placed in the sectors of the rotating drum, the eight positions of the drum will give eight different readings. ■

Directed Hamiltonian Cycles and Paths

A **directed Hamiltonian cycle (path)** is a directed cycle (path) that includes each vertex exactly once. Because directed loops and parallel directed edges are not needed for a directed Hamiltonian cycle or path, we will assume that we are working with directed graphs rather than directed multigraphs. As with graphs, it is very difficult to decide whether there is a directed Hamiltonian cycle and, if so, to find one.

These concepts arise in connection with round-robin athletic competitions. In a round-robin contest, each team plays every other team exactly once, and a tie between two teams is not permitted. Such a competition can be described by a directed graph in which the teams are represented by vertices, and there is a directed edge from one vertex to another if the first team beats the second team. A directed graph of this kind is called a **tournament directed graph** or, more simply, a **tournament**. An alternative way of thinking of a tournament is that it is the result of taking the complete graph $\mathcal{K}_n$ and assigning a direction to each edge.

Example 4.39 Suppose there are three teams A, B, and C, where team A beat teams B and C, and team B beat team C. This is described in Figure 4.61(a). If, instead, team C beat team A, the tournament would be as in Figure 4.61(b). ■

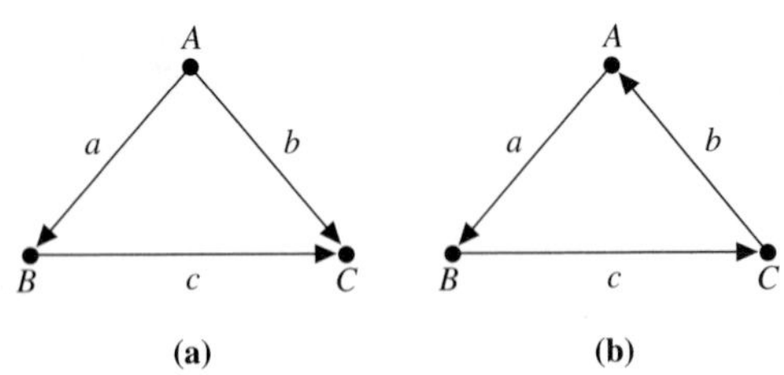

FIGURE 4.61

It may be desirable to find a ranking of the teams such that the first team beat the second team, the second team beat the third team, and so forth. Finding a ranking of the teams is the same as finding a directed Hamiltonian path for the tournament. It can be shown that every tournament has a directed Hamiltonian path. In Figure 4.61(a), the directed path a, c is Hamiltonian and thus provides a ranking of the teams. However, an examination of Figure 4.61(b) shows that there can be more than one directed Hamiltonian path. In fact, there are three: a, c; b, a; and c, b. This means that three separate rankings can be found. But also, note that in Figure 4.61(b) there is a directed cycle, whereas in Figure 4.61(a) there is not. In general, if a tournament has no directed cycles, then there is only one directed Hamiltonian path, which provides a unique ranking of the teams.

Further illustration of these points is found in Figure 4.62. For the tournament in Figure 4.62(a), there are no directed cycles and only one directed Hamiltonian path, namely, a, f, d, which gives the ranking A, B, C, D of the teams. In Figure 4.62(b) the tournament has a directed cycle, for example, a, f, d, e, and so there are several directed Hamiltonian paths, such as a, f, d and e, a, f.

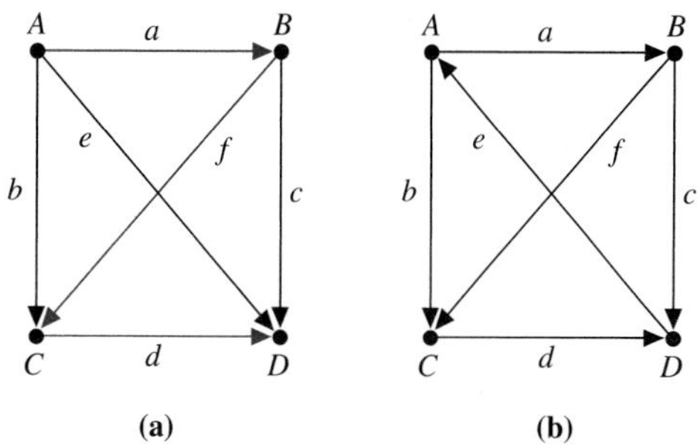

FIGURE 4.62

In Chapter 2 the concept of a relation on a set was introduced. When the set is finite, it is possible to depict the relation by a directed multigraph, which we will call the **directed multigraph of the relation**. In this multigraph the vertices correspond to the elements of the set, and there is a directed edge from x to y whenever x is related to y.

Consider, for instance, the round-robin competition between the three teams A, B, and C described in Example 4.39. This situation can be described as a relation R on set $S = \{A, B, C\}$, where $X\,R\,Y$ means that team X beat team Y. In this case the directed multigraph of relation R is precisely the tournament shown in Figure 4.61(a).

Another example of the directed multigraph of a relation follows.

Example 4.40 Let R be the relation on set

$$S = \{2, 3, 4, 5, 6\}$$

defined by $x\ R\ y$ whenever x divides y. Then R can be expressed as the following subset of $S \times S$:

$$R = \{(2, 2), (2, 4), (2, 6), (3, 3), (3, 6), (4, 4), (5, 5), (6, 6)\}.$$

Thus the directed multigraph of this relation has five vertices and eight directed edges. It is shown in Figure 4.63. ■

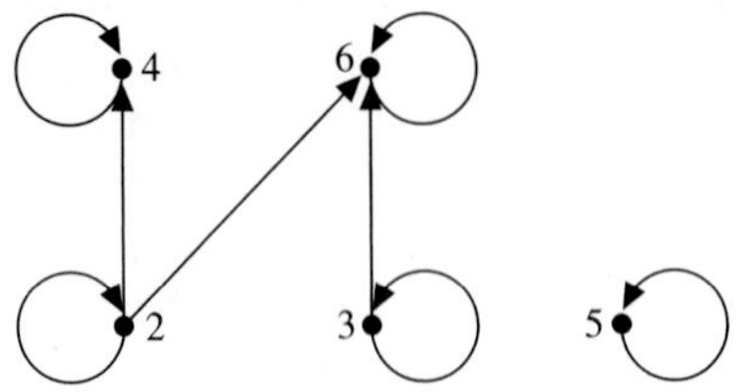

FIGURE 4.63

Other material in Sections 4.1–4.3 can also be modified in an appropriate way for directed graphs. Some examples are found in the exercises that follow.

EXERCISES 4.5

In Exercises 1–4, list the vertices and directed edges for the directed graph.

1.

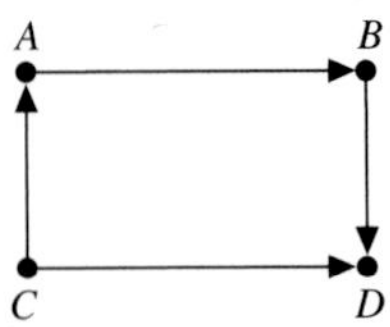

2.

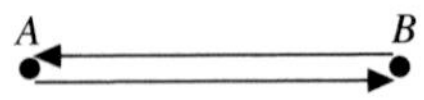

3.

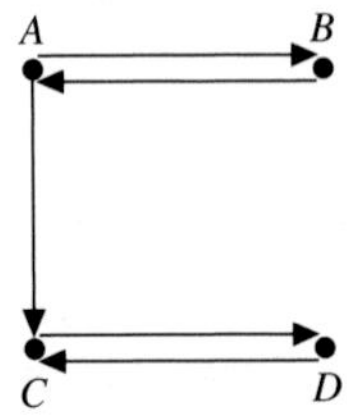

4.

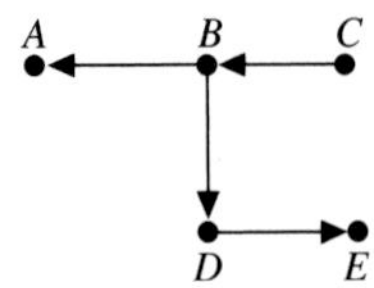

In Exercises 5–8, draw a diagram representing the directed graph with the set of vertices $\mathcal{V}$ and the set $\mathcal{E}$ of directed edges.

5. $\mathcal{V} = \{X, Y, Z, W, U\}$, $\mathcal{E} = \{(X, Y), (Z, U), (Y, X), (U, Z), (W, X), (Z, X)\}$

6. $\mathcal{V} = \{A, B, C, D\}$, $\mathcal{E} = \emptyset$

7. $\mathcal{V} = \{A, B, C\}$, $\mathcal{E} = \{(A, B), (B, C), (C, A), (B, A), (C, B)\}$

8. $\mathcal{V} = \{A, B, C, D\}$, $\mathcal{E} = \{(A, D), (D, B), (D, A)\}$

In Exercises 9–12, construct the labeled directed graph for the adjacency matrix.

9. $\begin{bmatrix} 0 & 1 & 0 & 1 \\ 1 & 0 & 1 & 0 \\ 0 & 0 & 0 & 0 \\ 0 & 1 & 1 & 0 \end{bmatrix}$

10. $\begin{bmatrix} 0 & 0 & 0 \\ 0 & 0 & 0 \\ 0 & 0 & 0 \end{bmatrix}$

11. $\begin{bmatrix} 0 & 1 & 1 & 1 \\ 1 & 0 & 1 & 1 \\ 1 & 1 & 0 & 1 \\ 0 & 1 & 0 & 0 \end{bmatrix}$

12. $\begin{bmatrix} 0 & 1 & 0 & 1 & 1 \\ 1 & 0 & 0 & 0 & 0 \\ 1 & 0 & 0 & 1 & 1 \\ 1 & 1 & 0 & 0 & 0 \\ 1 & 0 & 0 & 1 & 0 \end{bmatrix}$

In Exercises 13–16, list for the directed graph the other vertices on the directed edges to A, the other vertices on the directed edges from A, the indegree of A, and the outdegree of A.

13. 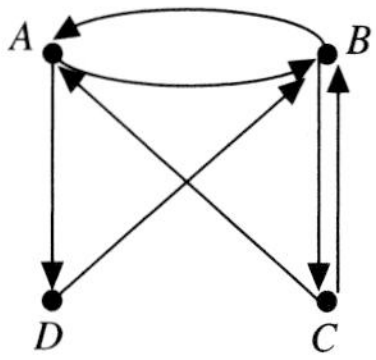

14. 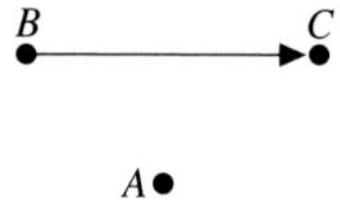

15. 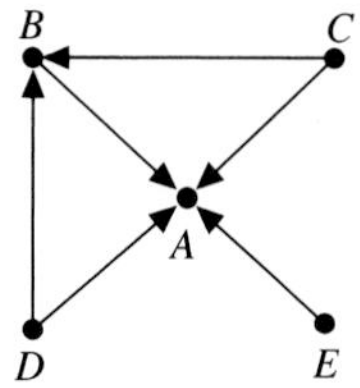

16. 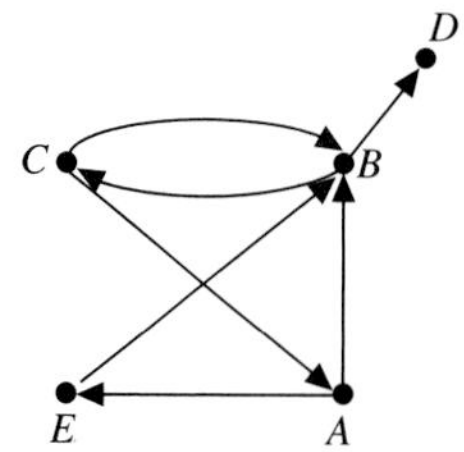

For the directed multigraphs in Exercises 17 and 18, perform the following tasks:

(a) *List the simple directed paths from A to B. Give the length of each.*

(b) *List the distinct directed cycles. (Two directed cycles are distinct if there is a directed edge in one, but not in the other.) Give the length of each.*

17. 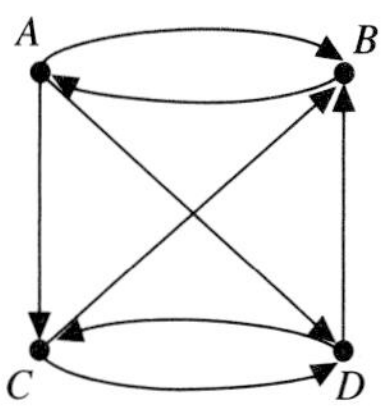

18. 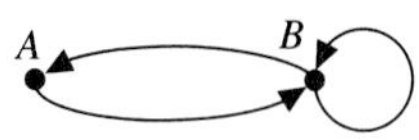

In Exercises 19–22, find the adjacency matrix and adjacency list for the directed graph in the indicated exercises. Order the vertices according to alphabetical order.

19. Exercise 13 **20.** Exercise 14 **21.** Exercise 15 **22.** Exercise 16

23. Let $\mathcal{V} = \{1, 2, \ldots, 10\}$ and $\mathcal{E} = \{(x, y)\colon x, y$ are in $V, x \neq y$, and x divides $y\}$. Draw the directed graph with vertices $\mathcal{V}$ and directed edges $\mathcal{E}$.

24. What does it mean if a row in an adjacency matrix for a directed graph contains only zeros? What if a column contains only zeros?

25. Draw all nonisomorphic directed graphs with two vertices.

26. Let $S = \{1, 2, 4, 8\}$ and $R = \{(1, 8), (2, 4), (8, 2), (4, 1), (2, 2), (8, 1)\}$ be the relation defined on S. Draw the directed multigraph of this relation.

27. Let $S = \{3, 5, 8, 10, 15, 24\}$ and R be the relation defined on S by $x\ R\ y$ whenever x divides y. Draw the directed multigraph of this relation.

28. Let S be the collection of all subsets of $\{1, 2, 3\}$ and R be the relation defined on S by $A\ R\ B$ whenever A is a subset of B or B is a subset of A. Draw the directed multigraph of this relation.

29. Describe the directed multigraph of a relation that is reflexive.

30. Describe the directed multigraph of a relation that is symmetric.

31. Describe the directed multigraph of a relation that is antisymmetric.

32. Construct the directed graph where the vertices are you, your parents, and your grandparents, using the relationship "is a child of."

33. Construct the directed graph, using "is a parent of" in place of "is a child of" in Exercise 32. How do the two directed graphs in Exercises 32 and 33 compare?

34. Susan has a fondness for chocolate desserts, in particular, pudding, pie, ice cream, eclairs, and cookies. Her preference is for pie over ice cream and cookies, eclairs over pie and cookies, cookies over pudding and ice cream, and pudding over eclairs, with no other preferences. Draw a directed graph to represent this situation.

In Exercises 35 and 36, determine whether the directed multigraphs in the indicated exercises are strongly connected.

35. Exercise 17

36. Exercise 18

37. Give an example of a directed graph with four vertices where every directed path of positive length has length 1.

38. In a directed multigraph with n vertices, what is the maximum length of a simple directed path?

In Exercises 39–42, determine whether a direction can be assigned to each edge of the graph resulting in a directed graph that is strongly connected. If so, give such an assignment.

39.

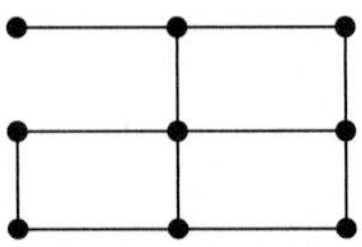

40.

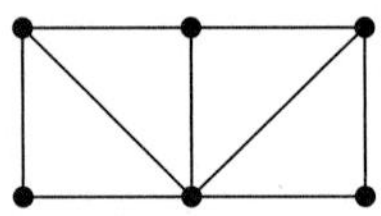

41.

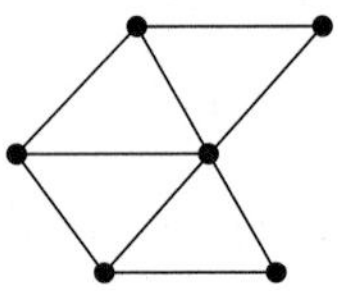

42.

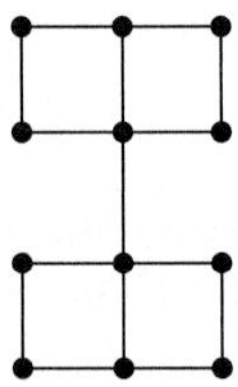

43. If a directed graph has a directed Hamiltonian cycle, why is it strongly connected?

In Exercises 44–49, determine whether the directed multigraph has a directed Euler path or circuit. If there is one, construct it by using the appropriate algorithm, as discussed in this section.

44.

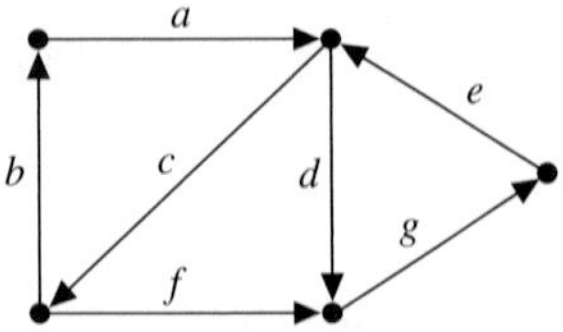

45.

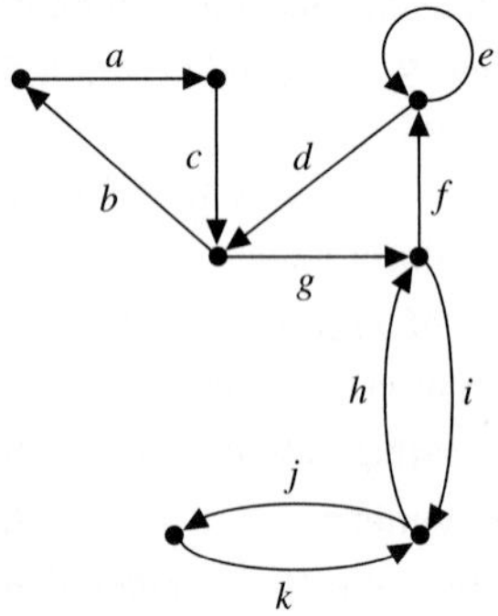

46.

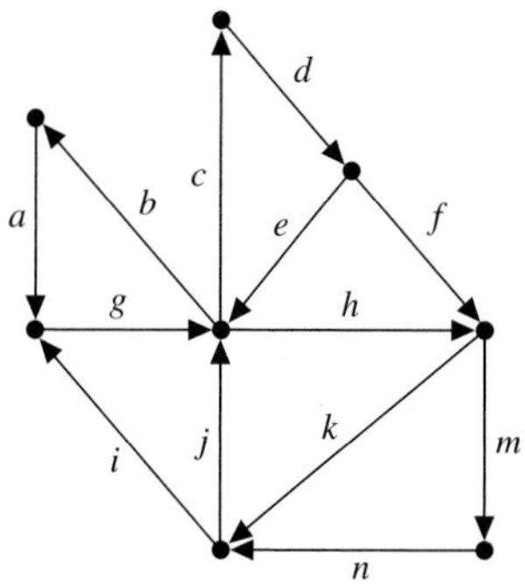

47.

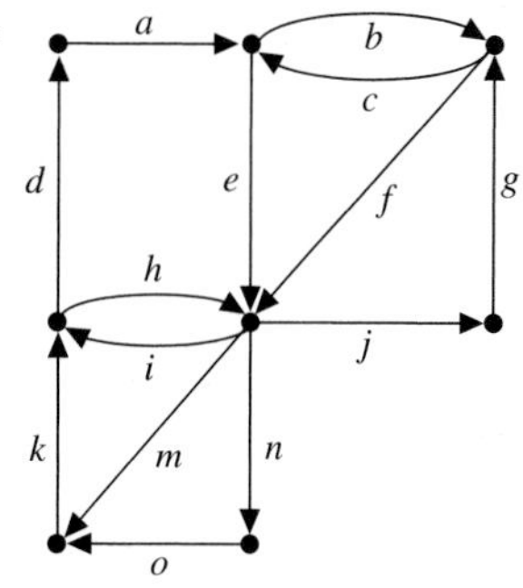

48.

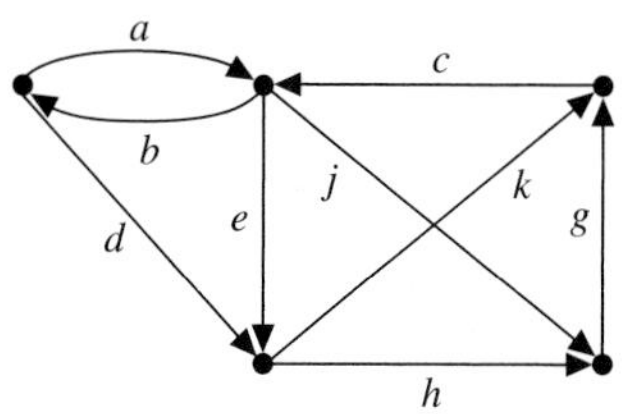

49.

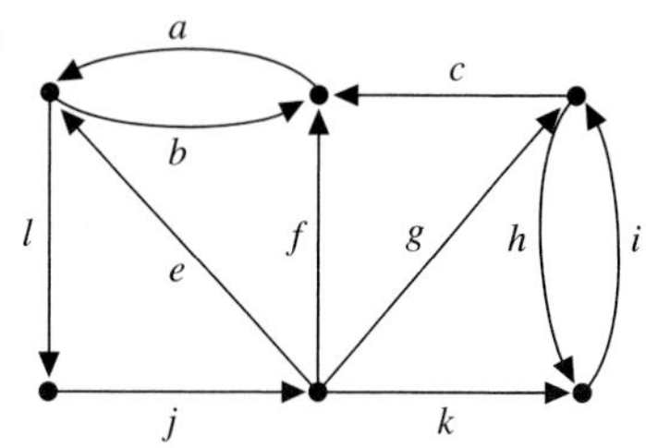

50. Suppose that in Example 4.38 the rotating drum has only four different sectors on the drum and two detectors. Using the procedure described in that example, find a sequence of four 0s and 1s to be used on the rotating drum so that every sequence of two consecutive entries is different.

51. Show that in a tournament with n vertices the sum of the outdegrees is $\frac{1}{2}n(n-1)$.

52. Show that in a round-robin contest with seven players there cannot be 23 winners.

In Exercises 53 and 54, find all the directed Hamiltonian paths in the tournament.

53.

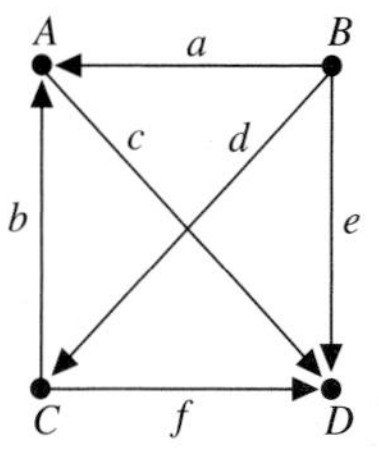

54.

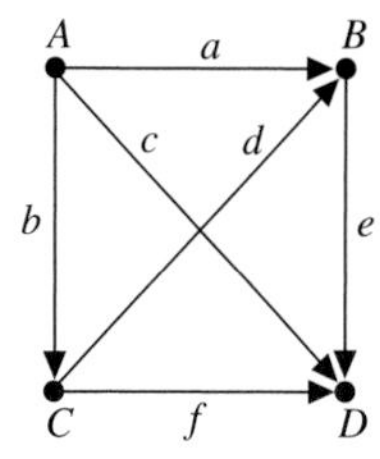

55. Suppose that Susan has established a preference between any two chocolate desserts (pie, pudding, ice cream, cookies, and eclairs). She prefers cookies over all of the others, ice cream over all but cookies, pie over pudding, and eclairs over pie and pudding. Is there a ranking to her preferences? How many?

56. In a tournament, the outdegree of a vertex is called the **score** (the number of wins for that team). In the following tournament, find a vertex with a maximum score and show that there is a directed path of length 1 or 2 from that vertex to any other:

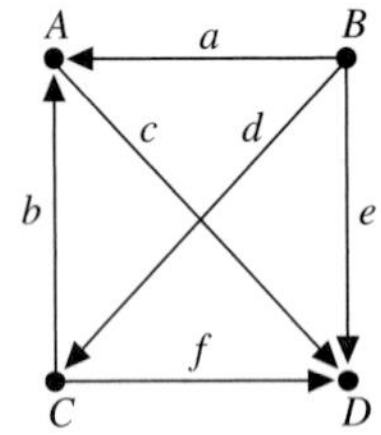

57. Repeat Exercise 56 for the following tournament:

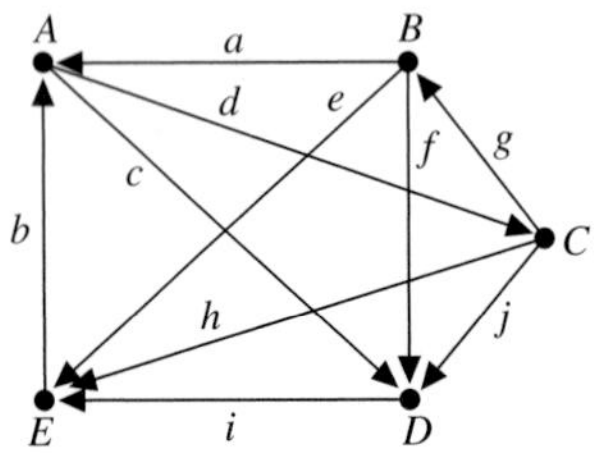

58. Can a tournament have two teams that lose every time?

59. Suppose that in 1978 the teams in the NFL National Conference North Division (the Bears, Lions, Packers, and Vikings) played a round-robin contest in which each team played every other team exactly once. If the Bears beat every other team, the Lions lost to every other team, and the Vikings beat everyone but the Bears, is there a ranking of the teams? Is it unique?

60. Write an algorithm for finding a directed Euler circuit.

61. Write a breadth-first search algorithm for directed graphs.

In Exercises 62–65, use the breadth-first search algorithm for directed graphs (see Exercise 61) to determine the distance and a shortest directed path from S to T in the directed graph.

62.

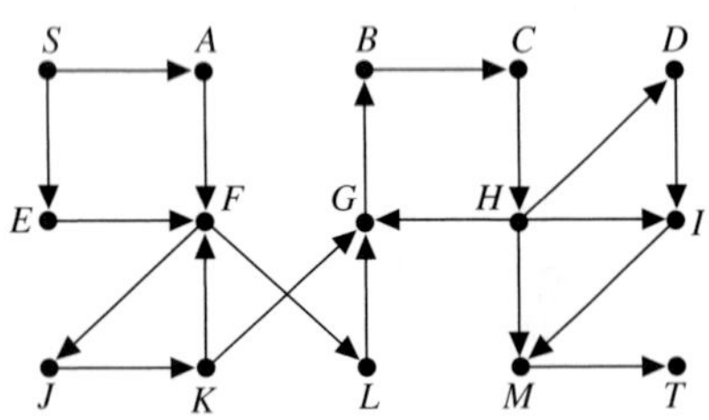

63.

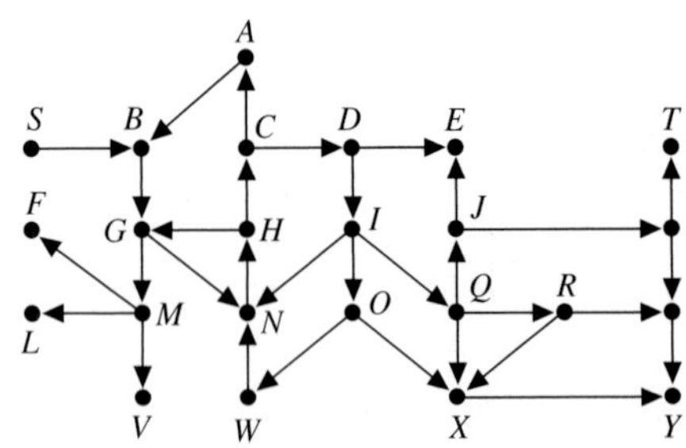

64.

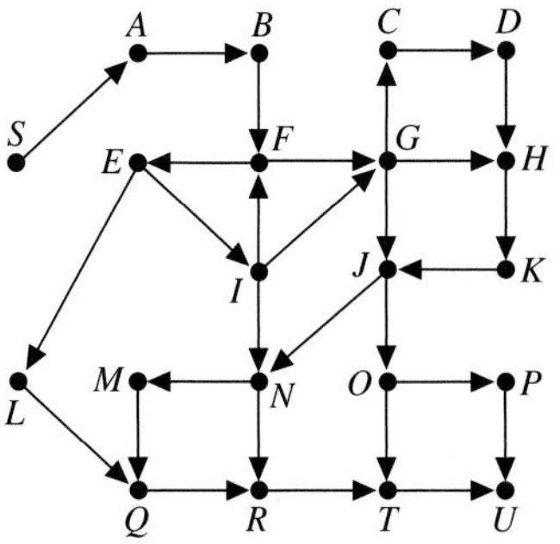

65.

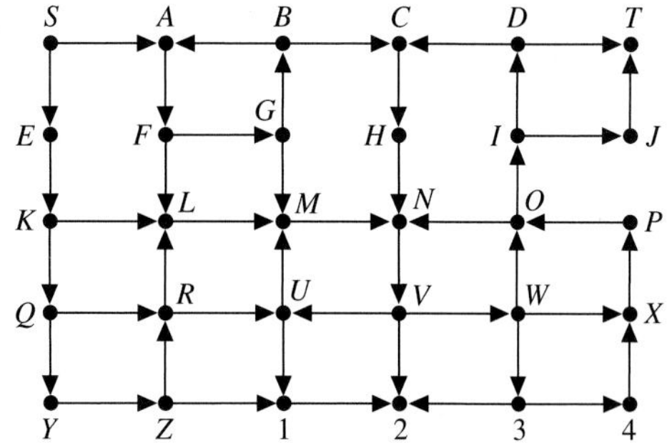

66. Write an algorithm for directed weighted graphs that finds the distance and a shortest directed path from a vertex S to every other vertex.

In Exercises 67–70, determine the distance from S to all the other vertices in the directed weighted graph. Find a shortest directed path from S to A.

67.

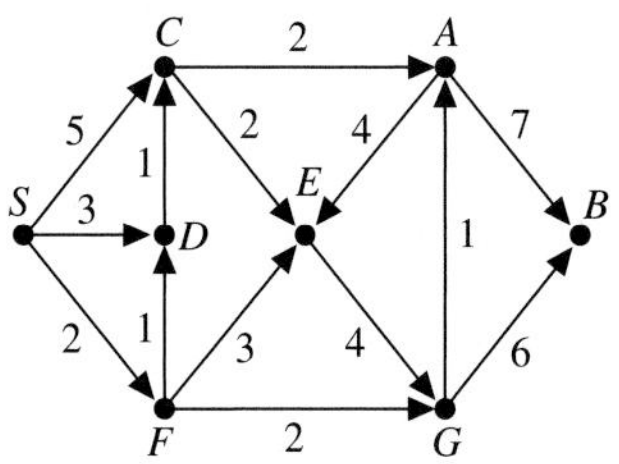

68.

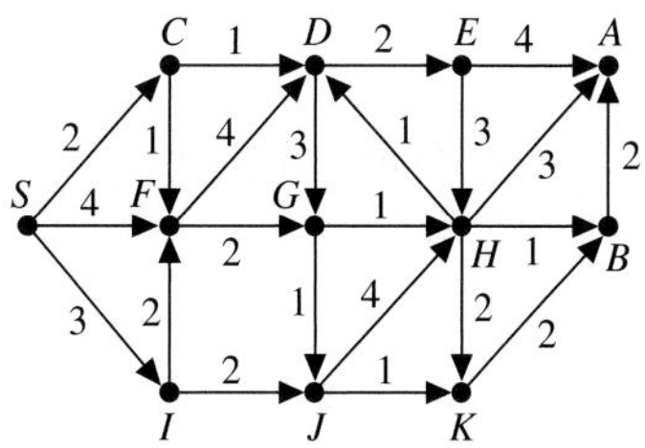

69.

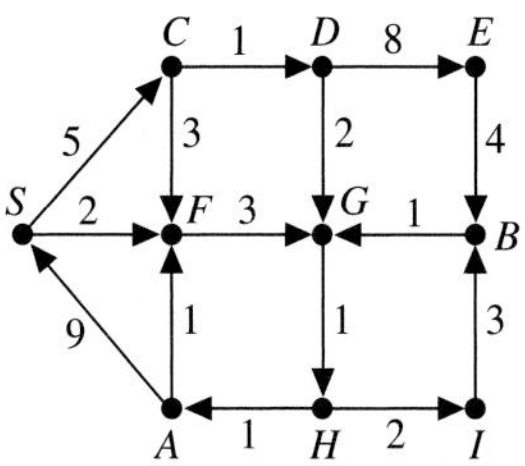

70.

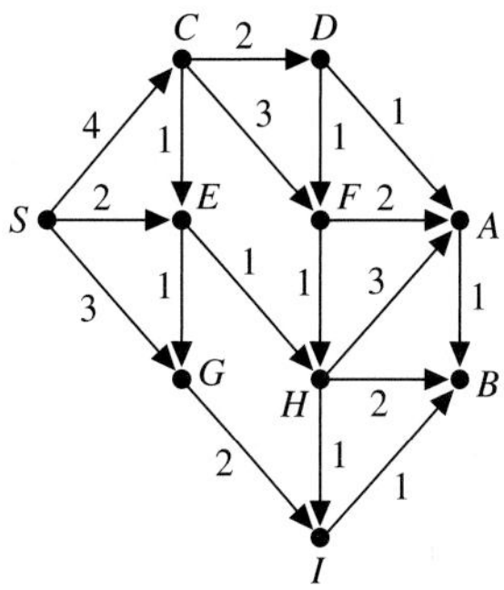

71. Show that, for a directed graph $\mathcal{D}$ with vertices $V_1, V_2, \ldots, V_n$ and adjacency matrix A, the number of directed paths of length m from V_i to V_j is the i, j entry of A^m.

72. For the following directed graph, determine the number of directed paths of lengths 1, 2, 3, and 4 from V_1 to V_3, and from V_2 to V_4:

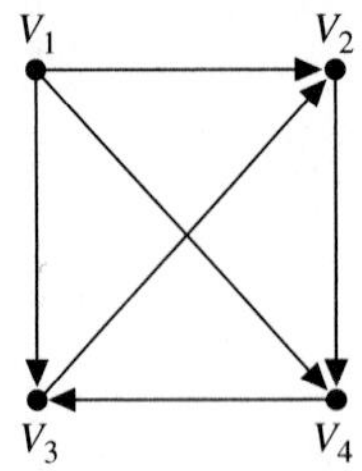

73. For the following directed graph, determine the number of directed paths of lengths 1, 2, 3, and 4 from V_1 to V_4, and from V_4 to V_1:

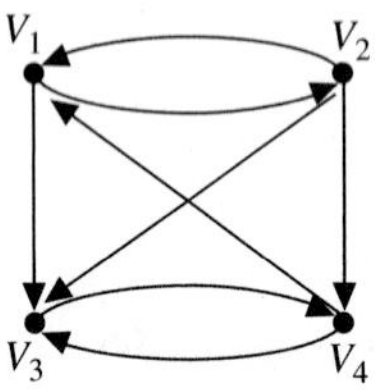

74. Write a definition for an isomorphism between directed graphs.

75. Give two examples of properties of directed graphs that are invariant under isomorphism. Justify your answer.

76. Determine whether the following pairs of directed graphs are isomorphic:

(a) 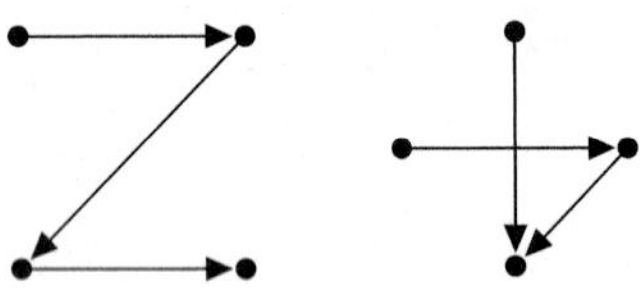

(b) 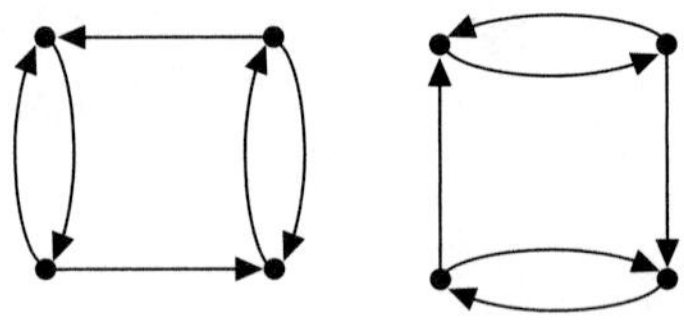

77. Determine whether the following pairs of directed graphs are isomorphic:

(a) 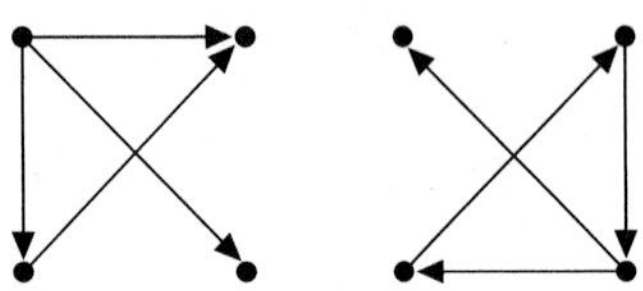

(b) 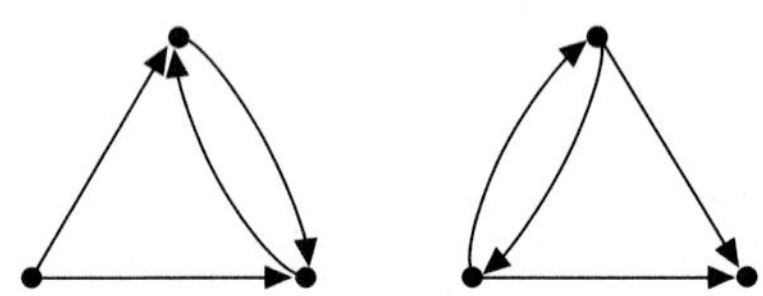

78. Prove Theorem 4.10.

79. Prove Theorem 4.11.

80. Prove Theorem 4.12.

81. Prove Theorem 4.13.

82. Prove that every tournament has a directed Hamiltonian path.

83. Prove that if A is a vertex of maximum score (see Exercise 56) in a tournament, then there is a directed path of length 1 or 2 from A to any other vertex.

84. Give a definition of isomorphic directed weighted graphs. (See Exercise 20 of Section 4.3 for the definition of isomorphic weighted graphs.) Give an example of two nonisomorphic directed weighted graphs, where the underlying graphs are isomorphic.

HISTORICAL NOTES

Leonhard Euler

The origin of graph theory is linked to Leonhard Euler's consideration of the bridges of Königsberg problem. (See Exercise 30 in Section 4.2.) This problem dealt with a long-standing puzzle in the eastern Prussian city of Königsberg. The center of the old city in Königsberg was situated on an island in the Pregel River just below the point where its two upper branches joined. Seven bridges connected the land between the branches, the island, and the parts of the city on the two banks, as shown in the picture on page 178. The puzzle for the citizens of Königsberg was to devise a walking route that crossed each of the bridges exactly once and ended at its starting place. Leonhard Euler (1707–1783) was the first to show this was impossible. In doing so, he characterized the situations where such paths and circuits were possible.

Euler's solution, while it did not use a graphical representation, used the type of combinatorial reasoning that distinguishes graph theoretic forms of mathematical reasoning. His work in describing how to create Eulerian paths and circuits for such situations used a method of "mentally removing" edges from graphs and considering the nature of the remaining structure. This representational process was central to the later proof, in 1752, of his famous formula for polyhedra and their related planar graphs, $f - e + v = 2$. (See Exercise 8 of the Supplementary Exercises for Chapter 4.) In his proof, Euler sliced off tetrahedral pieces of polyhedra associated with the planar graphs and noted that the

$$\text{(number of faces)} - \text{(number of edges)} + \text{(number of vertices)}$$

remains unchanged, eventually arriving at a tetrahedron. While there were some gaps in Euler's approach, they were filled in 1813 by Augustin-Louis Cauchy (1789–1857) [72].

William Rowan Hamilton

In 1859, Sir William Rowan Hamilton (1805–1865) marketed a puzzle that required that one find specified paths and circuits on a planar graph consisting of the edges and vertices related to a regular dodecahedron. (See Exercise 40 in Section 4.2.) The first problem was to find a cycle passing through each vertex once and only once. The puzzle was later offered in the form of a solid regular dodecahedron with pegs at the vertices and a string to mark out the edges in such a cycle. When the basic underlying graph is a weighted graph, the challenge of finding a Hamiltonian cycle of minimum weight is known as the traveling salesperson problem. To date, mathematicians have been unable to find necessary and sufficient conditions that characterize those graphs that have a Hamiltonian path or cycle.

The most famous problem in graph theory is the four-color problem. This problem deals with the minimum number of colors needed to color a map in such a way that neighboring countries are differently colored. The four-color problem was first examined by Francis Guthrie (1831–1899) in 1850. Augustus De Morgan learned of the problem through Guthrie's brother in 1852, and Arthur Cayley (1821–1895) posed it to the London Mathematical Society in 1878. While several incorrect proofs were offered over the years, resolution of the problem

Edsger Dijkstra

Dénes König

had to wait until the 1976 proof by Kenneth Appel and Wolfgang Haken of the University of Illinois. Their methods, implemented by computer, called for the checking of nearly 2000 cases through an involved algorithm.

These are but a few of the many different results that mark the first 250 years of work in graph theory. Much of this work involved creating algorithms for solving practical problems concerning graphs. Several individuals, such as Edsger Dijkstra (1930–2002), have contributed both to graph theory and computer science through their development and refinement of algorithms related to graph theory. The applications of graph theory have become increasingly important in business and industry. In 1936, the first book on graph theory, written by the Hungarian mathematician Dénes König (1884–1944), appeared [72]. Today, a multitude of such texts exist, along with several journals devoted to graph theory.

SUPPLEMENTARY EXERCISES

1. For a graph $\mathcal{G}$, the **complement** of $\mathcal{G}$ is the graph where the vertices are the same as the vertices of $\mathcal{G}$ and there is an edge between vertices A and B if and only if $\mathcal{G}$ does *not* have an edge between A and B. Find the complement of the following graph:

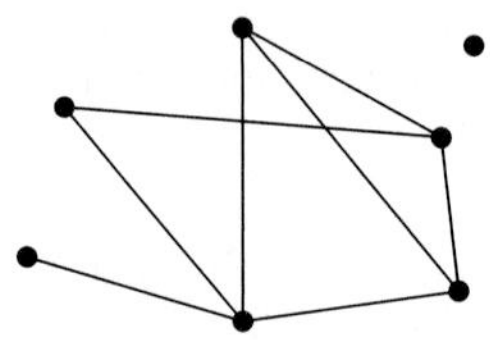

2. Is there a graph with five vertices, each with degree 1? Each with degree 2? Each with degree 3? Justify your answers.

3. Draw the graph with vertices X, Y, Z, W, R, and S, where X and R are adjacent, W, R, and S are adjacent to each other, and Y and Z are adjacent.

4. A few years ago the National Football League had two conferences, each with 13 teams. It was decided by the league office that each team would play a total of 14 games, 11 of which were to be with teams in their own conference, and the other 3 games with teams outside their own conference. Show that this is not possible.

5. Are the following graphs isomorphic? Justify your answer.

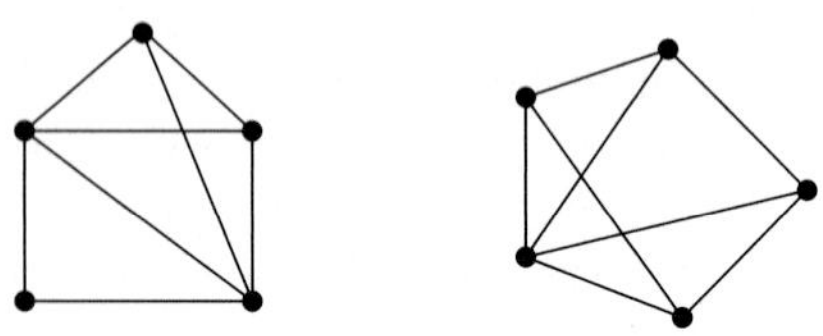

6. Consider the following graph:

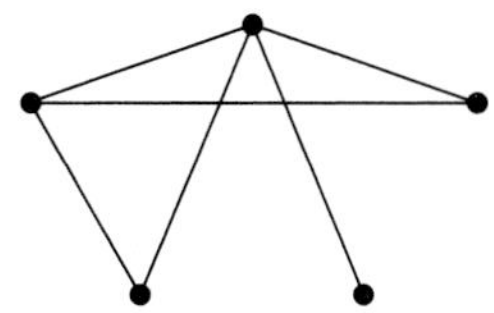

Label the vertices and construct the adjacency matrix. Then label the vertices in a different manner and construct the new adjacency matrix. Compare the two adjacency matrices and describe how they are related.

7. Construct the graph for the following adjacency list:

$$\begin{aligned} &V_1\colon V_2, V_3, V_5\\ &V_2\colon V_1, V_4, V_5\\ &V_3\colon V_1, V_4, V_5\\ &V_4\colon V_2, V_3, V_5\\ &V_5\colon V_1, V_2, V_3, V_4 \end{aligned}$$

Suppose a graph $\mathcal{G}$ is drawn in the plane so that the edges of $\mathcal{G}$ intersect only at the vertices of $\mathcal{G}$. Then $\mathcal{G}$ partitions the plane into a finite number of parts, called ***regions***. *In the following illustration, the regions are labeled A, B, C, D, E, F, H, and I.*

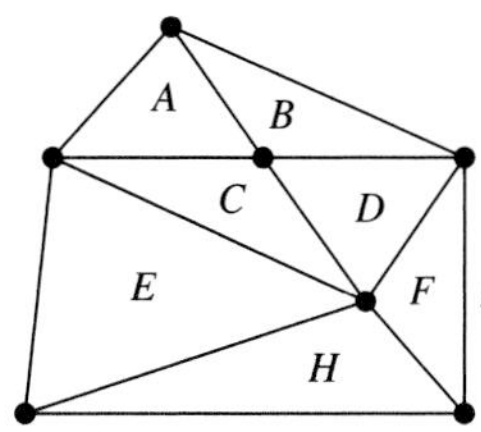

8. If $\mathcal{G}$ is such a connected graph in which e is the number of edges, v is the number of vertices, and f is the number of regions, prove that $f - e + v = 2$. (This result is called *Euler's formula*.)

9. The floor plan of the bottom level of a new home is as follows:

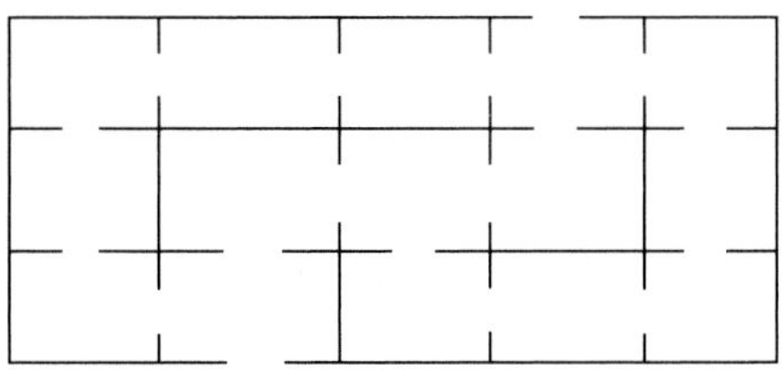

Is it possible to enter the house at the front, exit at the rear, and travel through the house going through each doorway exactly once?

10. Find a Hamiltonian cycle for $\mathcal{K}_4$ and for $\mathcal{K}_5$.

11. Could the citizens of Königsberg find an acceptable route by tearing down one bridge and building one new bridge? (See Exercise 30 in Section 4.2.)

12. Determine whether the multigraph that follows has an Euler path. If it does, construct one by using the Euler circuit algorithm.

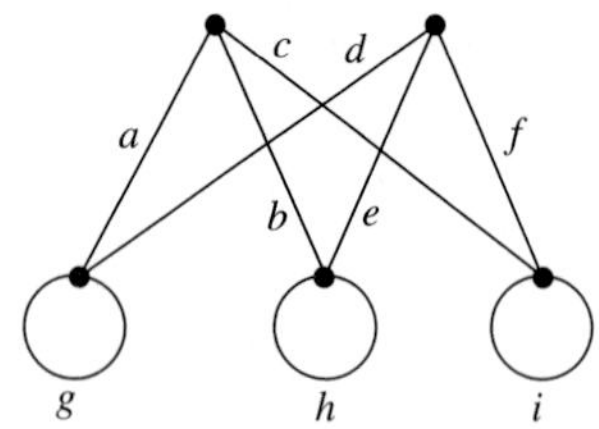

13. Determine whether the multigraph that follows has an Euler circuit. If it does, construct one by using the Euler circuit algorithm.

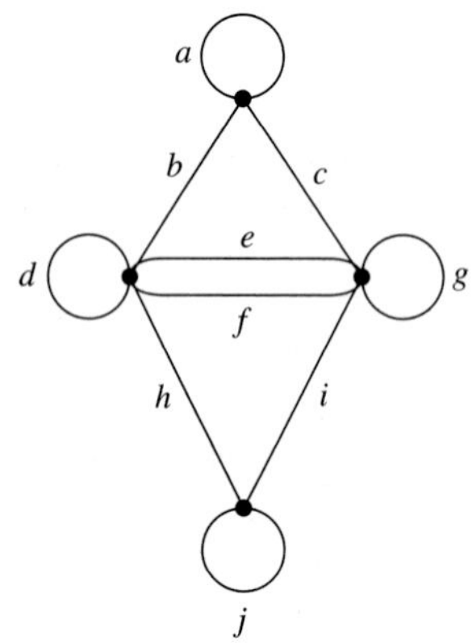

14. A street inspector wants to examine the streets in her region for potholes. If the map of her region is as follows, is it possible for her to devise a route to examine each street once and return to her office?

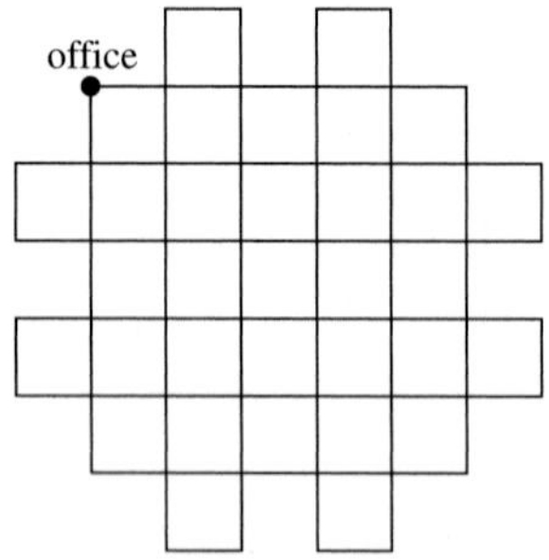

15. Is the property "is connected" a graph isomorphism invariant?

16. Is the property "has an Euler circuit" a graph isomorphism invariant?

17. Use the breadth-first search algorithm to determine the distance and a shortest path from S to T in the following graph:

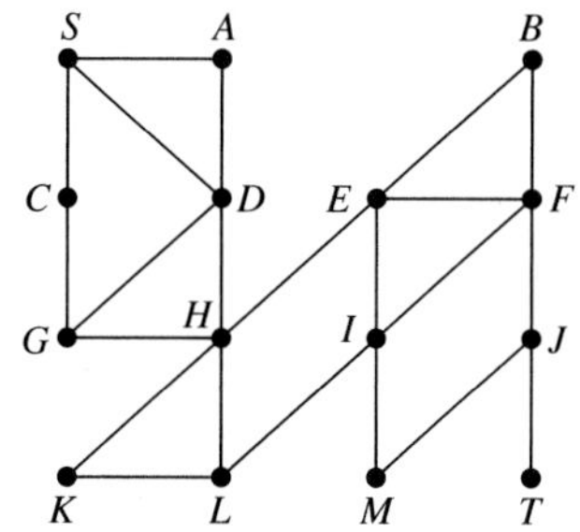

In Exercises 18 and 19, determine the distance from S to all the other vertices in the weighted graph. Find a shortest path from S to A and a shortest path from S to B.

18.

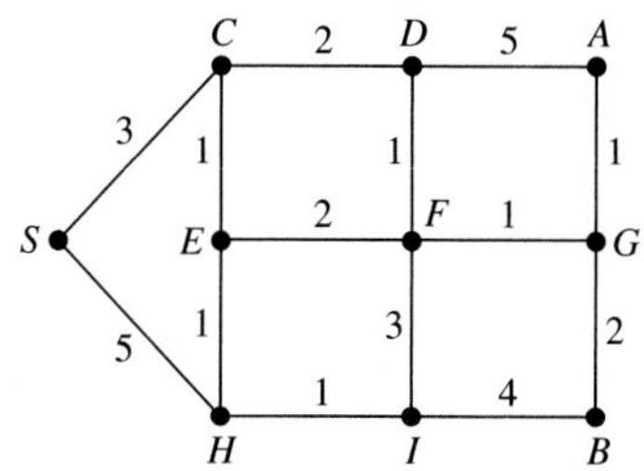

19.

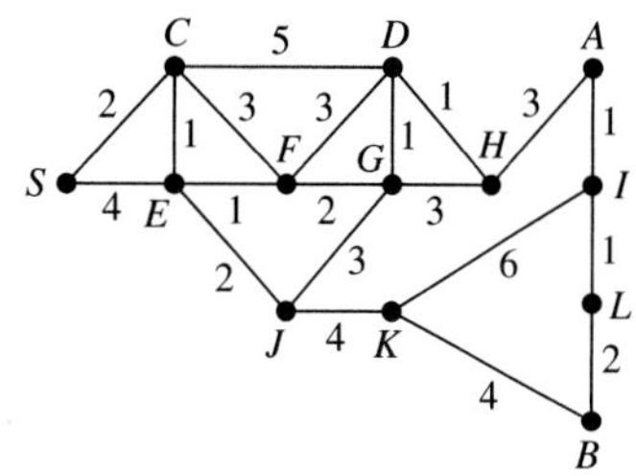

20. In the following weighted graph, find a shortest path from S to T that goes through the vertex A:

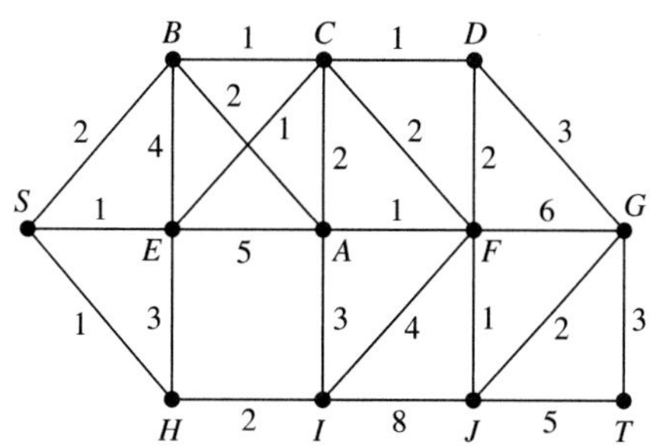

21. For the following graph, determine the number of paths of lengths 1, 2, 3, and 4 from V_1 to V_2 and from V_1 to V_4:

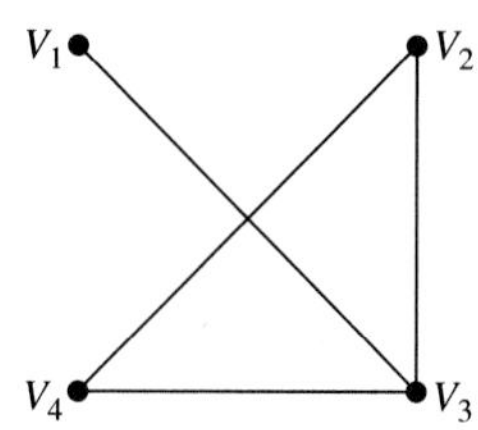

22. Find the chromatic number of each of the following graphs:

(a)

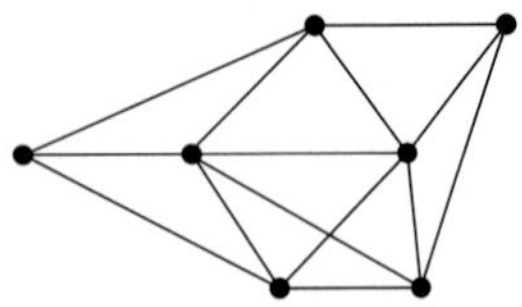

(b)

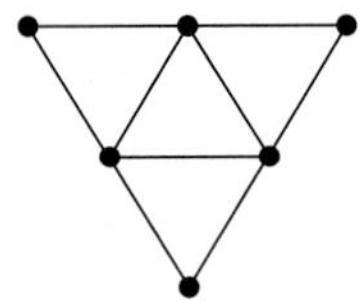

23. Color the following map:

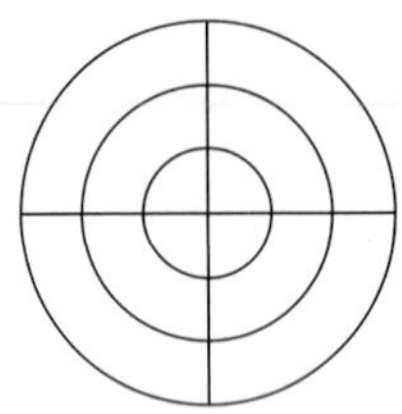

24. Is "has two vertices with the same color" a graph isomorphism invariant?

25. Show that a map formed by crossing a square with line segments can be colored with two colors.

26. Show that "has chromatic number n" is a graph isomorphism invariant.

27. There are locations in a computer memory where stacks are stored during the execution of a computer program. Furthermore, a location can store only one stack at a time. Suppose stacks $S_1, S_2, \ldots, S_{10}$ are to be constructed during the execution of a computer program and S_i and S_j will be in use at the same time if $i \equiv j \pmod 3$ or $i \equiv j \pmod 4$. What is the minimum number of locations for the stacks that will be needed during the execution of this computer program?

28. Suppose that the Illinois Electric Company has a power plant from which electrical power is sent along transmission lines to the surrounding communities. However, there is a continuing problem with power loss in these lines because of their deteriorated condition. The following table describes the loss along a transmission line from one community to another. (A dash in the table means there is no transmission line.)

To	*Plant*	*Normal*	*Hudson*	*Ospur*	*Kenney*	*Lane*	*Maroa*
Plant	—	3	2	6	1	4	5
Normal	—	—	4	4	—	3	—
Hudson	9	2	—	3	5	6	3
Ospur	3	—	—	—	6	9	4
Kenney	2	3	1	1	—	7	2
Lane	1	—	2	2	7	—	6
Maroa	6	2	3	4	2	2	—

What is the best route (one with least power loss) from the plant to each surrounding community?

29. Construct the directed graph for the following adjacency matrix:

$$\begin{bmatrix} 0 & 0 & 1 & 1 \\ 0 & 0 & 0 & 0 \\ 1 & 0 & 0 & 1 \\ 1 & 0 & 1 & 0 \end{bmatrix}.$$

30. In a large corporation, the chief executive officer communicates with her vice presidents, and they can communicate with her. Furthermore, the vice presidents can communicate with the directors, field managers, and division heads, but only the directors can communicate back. Also, the field managers and division heads can communicate with salespersons, but they can communicate back only with the field managers. Draw a directed graph to represent the communication lines among these positions.

31. Construct, if possible, a directed graph with six vertices where the outdegrees of the vertices are 2, 3, 4, 1, 0, and 5 and the indegrees of the vertices are 2, 4, 1, 1, 5, and 2.

32. Give an example of a directed graph with six vertices where every directed path is a simple directed path.

33. Work Example 4.38 if there are 16 different sectors and four detectors.

A tournament is ***transitive*** *if, whenever* (A, B) *and* (B, C) *are directed edges in the tournament, then so is* (A, C).

34. Prove that a tournament is transitive if and only if there are no directed cycles.

35. Prove that the scores in a transitive tournament with n vertices are $0, 1, 2, 3, \ldots, n-1$ (see Exercise 56 of Section 4.5).

36. Find all the directed Hamiltonian paths in the following tournament:

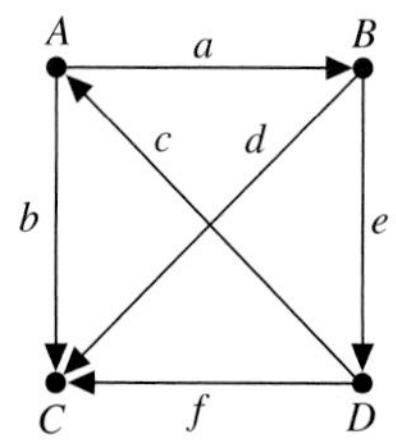

37. For the following directed graph, determine the number of directed paths of length 1, 2, 3, and 4 from V_1 to V_4, and from V_2 to V_5:

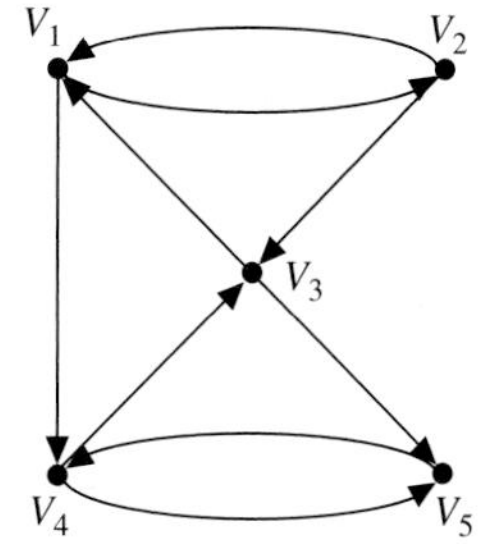

38. Show that a connected graph can be made into a strongly connected directed graph if and only if each edge is an edge of some cycle.

39. The directed edges of a directed graph can be considered as a relation on the set of vertices. (See Section 2.2.) When will this relation be reflexive? Symmetric? Transitive?

40. For the directed weighted graph that follows, determine the distance from S to all the other vertices in the directed graph. Find a shortest directed path from S to A. Find a shortest directed path from S to B.

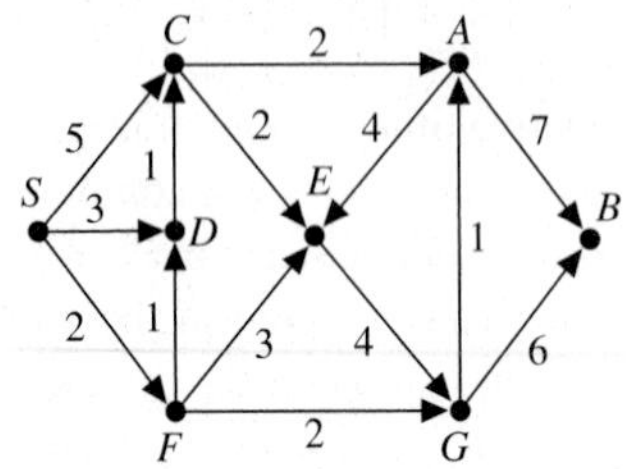

COMPUTER PROJECTS

Write a computer program having the specified input and output.

1. Given the adjacency matrix of a graph, find the degree of each vertex.

2. Given the adjacency matrix of a directed graph, find the indegree and outdegree of each vertex.

3. Given the adjacency matrix of a graph, find its adjacency list.

4. Given the adjacency list of a graph, find its adjacency matrix.

5. Given the adjacency matrix of a graph with vertices $V_1, V_2, \ldots, V_n$, find the number of paths of length m from V_i to V_j.

6. Determine whether a given graph is a complete graph.

7. Given a graph and U-V path, find a U-V simple path.

8. Given a multigraph, determine whether there are any loops or parallel edges.

9. Determine whether a given graph is a bipartite graph.

10. Use the breadth-first search algorithm to find the components of a given graph. (See Exercise 57 in Section 4.2.)

11. Given a graph, find a coloring of its vertices in which the number of colors does not exceed one more than the maximum of the degrees of the vertices.

12. Given a graph, use the breadth-first search algorithm to label its vertices.

13. Given a weighted graph $\mathcal{G}$ with positive weights and a vertex S, use Dijkstra's algorithm to determine the distance and a shortest path from S to every other vertex in $\mathcal{G}$.

14. Given a connected graph in which every vertex has even degree, find an Euler circuit.

15. Given a connected graph in which every vertex has even degree except for two vertices A and B, find an Euler path between A and B.

16. Given a graph with n vertices in which each vertex has degree greater than $\frac{n}{2}$, find a Hamiltonian cycle.

17. Given a graph and a positive integer n, determine whether the graph has a cycle of length n.

SUGGESTED READINGS

1. Bogart, Kenneth P. *Introductory Combinatorics*, 3d ed. San Diego: Academic Press, 1999.
2. Bondy, J.A. and U.S.R. Murty. *Graph Theory with Applications*. New York: Elsevier Science, 1979.
3. Chachra, Vinod, Prabhakar M. Ghare, and James M. Moore. *Applications of Graph Theory Algorithms*. New York: Elsevier Science, 1979.
4. Chartrand, Gary. *Graphs as Mathematical Models*. Boston: Prindle, Weber & Schmidt, 1977.
5. Even, Shimon. *Graph Algorithms*. New York: Freeman, 1984.
6. Harary, Frank. *Graph Theory*. Reading, MA: Addison-Wesley, 1994.
7. Liu, C.L. *Introduction to Combinatorial Mathematics*. New York: McGraw-Hill, 1968.
8. Ore, Oystein. *Graphs and Their Uses*, Washington, DC: Mathematical Association of America, 1963.
9. Polimeni, Albert D. and Joseph H. Straight. *Foundations of Discrete Mathematics*, 2d ed. Pacific Grove, CA: Brooks/Cole, 1990.
10. Roberts, Fred S. *Graph Theory and Its Applications to Problems of Society*. Philadelphia: SIAM, 1978.
11. Wilf, Herbert S. *Combinatorial Algorithms: An Update*. Philadelphia: SIAM, 1989.

CHAPTER

5

Trees

In Chapter 4 we studied several different types of graphs and their applications. A special class of graphs—trees—has been found to be very useful in computer science. Trees were first used in 1847 by Gustav Kirchhoff in his work on electrical networks. Later they were used by Arthur Cayley in the study of chemistry. Now trees are widely used in computer science as a way to organize and manipulate data.

5.1 PROPERTIES OF TREES

We begin this section by looking at some examples.

Example 5.1 In 1857, Arthur Cayley studied hydrocarbons, chemical compounds formed from hydrogen and carbon atoms. In particular, he investigated saturated hydrocarbons, which have k carbon atoms and $2k + 2$ hydrogen atoms. He knew that a hydrogen atom was bonded (chemically kept together) with one other atom, and each carbon atom was bonded with four other atoms. These compounds are usually represented pictorially as in Figure 5.1, where a line segment between two atoms indicates a bonding.

These chemical diagrams can be redrawn as graphs, as illustrated in Figure 5.2. Note that in these graphs we have followed the customary practice of using the same chemical symbol on different vertices representing the same element. However, it is not really necessary to label the vertices with C and H, since a vertex of degree 4 represents carbon and a vertex of degree 1 represents hydrogen. It was through the mathematical analysis of these graphs that Cayley predicted the existence of new saturated hydrocarbons. Later discoveries proved his predictions correct. ■

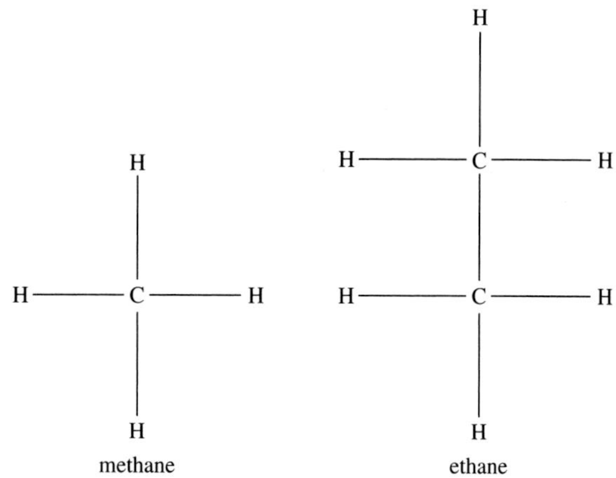

FIGURE 5.1

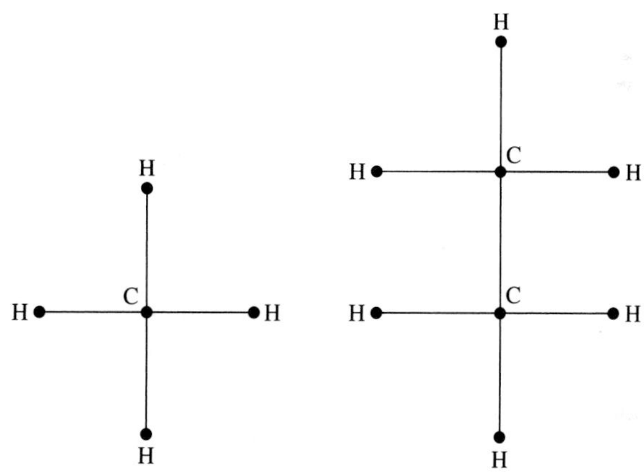

FIGURE 5.2

Example 5.2 Suppose we are planning the telephone network for an underdeveloped area, where the goal is to link together five isolated towns. We can build a telephone line between any two towns, but time and cost limitations restrict us to building as few lines as possible. It is important that each town be able to communicate with each other town, but it is not necessary that there be a direct line between any pair of towns, since it is possible to route calls through other towns. If we represent the towns by the vertices of a graph and the possible telephone lines by edges between the vertices, then the graph in Figure 5.3 represents all the possibilities we can have for the telephone lines. (This is merely the complete graph on five vertices.)

We need to select a set of edges that will give us a path between any two vertices and that has no more edges than necessary. One such set of edges is $\{a, b, c, d\}$, as illustrated in Figure 5.4. This choice of edges allows communication between any two towns. For example, to communicate between Y and

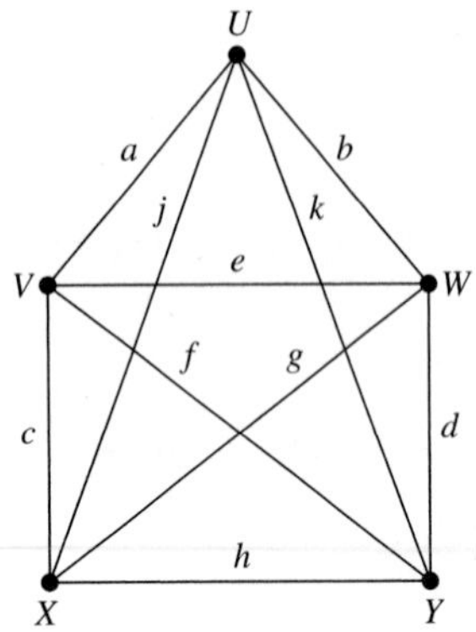

FIGURE 5.3

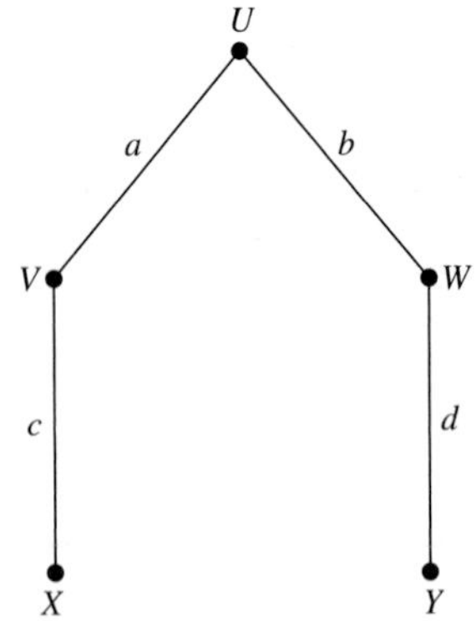

FIGURE 5.4

X, we can use edges d, b, a, c in that order. Notice that if any edge is deleted from this set, then it is not possible to communicate between some pair of towns. For example, if we use only edges a, b, and c, towns U and Y cannot communicate. Another set of acceptable edges is $\{e, g, h, k\}$. The sets $\{g, h, j, k\}$ and $\{a, b, e, h\}$ are not acceptable, because not every pair of towns can communicate. Also, the set $\{a, b, g, j, k\}$ is bigger than necessary because the edge g can be left out without disrupting communication between any two towns. ■

For the graphs in Figures 5.2 and 5.4, we note two common characteristics; namely, these graphs are connected (there is a path between any two vertices) and have no cycles. Any graph that is connected and has no cycles is called a **tree**. Additional examples of trees follow.

Example 5.3 Since each of the graphs in Figure 5.5 is connected and has no cycles, each is a tree. ■

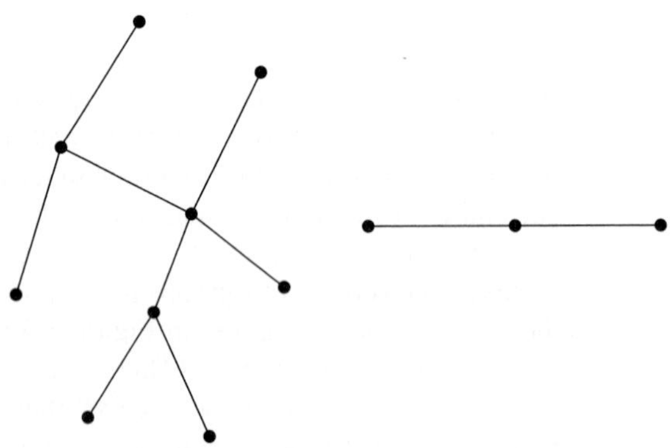

(a) (b)

FIGURE 5.5

Example 5.4 Neither of the graphs in Figure 5.6 is a tree. The graph in Figure 5.6(a) is not connected, and the graph in Figure 5.6(b) has a cycle. ■

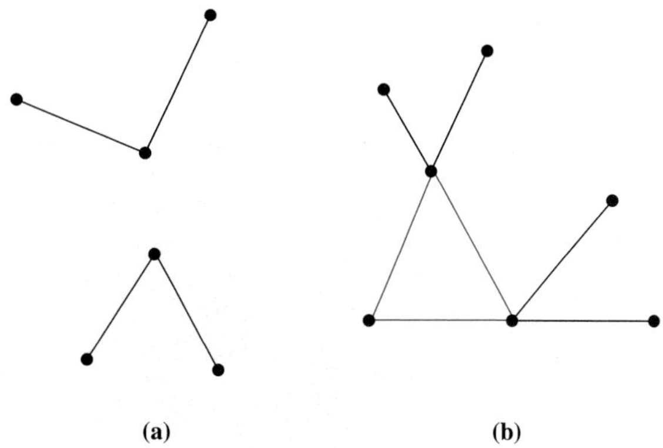

FIGURE 5.6

Theorem 5.1 Let U and V be vertices in a tree. Then there is exactly one simple path from U to V.

Proof. Since a tree is a connected graph, there is at least one path from U to V. Thus, by Theorem 4.4, there is a simple path from U to V.

We will now show that there cannot be two distinct simple paths from U to V. To do so, we will assume that there are distinct simple paths $\mathcal{P}_1$ and $\mathcal{P}_2$ from U to V and show that this leads to a contradiction. Since $\mathcal{P}_1$ and $\mathcal{P}_2$ are different, there must be a vertex A (possibly $A = U$) lying on both $\mathcal{P}_1$ and $\mathcal{P}_2$ such that the vertex B following A on $\mathcal{P}_1$ does not follow A on $\mathcal{P}_2$. In other words, $\mathcal{P}_1$ and $\mathcal{P}_2$ separate at A. (See Figure 5.7.) Now follow path $\mathcal{P}_1$ until we come to the first vertex C that is again on both paths. (The paths must rejoin because they meet again at V.) Consider the part of the simple path $\mathcal{P}_1$ from A to C and the part of the simple path $\mathcal{P}_2$ from C to A. These parts form a cycle. But trees contain no cycles, so we have a contradiction. It follows that there cannot be two distinct simple paths between any pair of vertices. ■

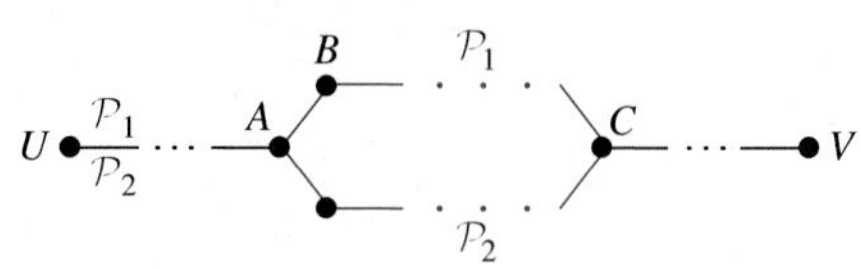

FIGURE 5.7

Looking at each previous example of a tree reinforces the idea that there is a *unique* simple path from any vertex to another. Notice also that in all of these examples every tree has at least two vertices of degree 1.

Theorem 5.2 In a tree $\mathcal{T}$ with more than one vertex, there are at least two vertices of degree 1.

Proof. Since $\mathcal{T}$ is a connected graph with at least two vertices, there is a simple path with at least two distinct vertices. Thus $\mathcal{T}$ contains a simple path with a maximal number of edges, say, from U to V, where U and V are distinct. If U had degree more than 1, then, since $\mathcal{T}$ has no cycles, a longer simple path would exist, and likewise for V. Thus U and V have degree 1. ■

For the tree in Figure 5.4, there are 5 vertices and 4 edges, and for the tree in Figure 5.5(a), there are 9 vertices and 8 edges. In fact, in each previous example of a tree, the number of vertices is one more than the number of edges. The next theorem establishes that this is always the case.

Theorem 5.3 A tree with n vertices has exactly $n - 1$ edges.

Proof. The proof will be by induction on n, the number of vertices. Because a tree is a graph, there are no loops in a tree. Hence there are no edges in a tree with only one vertex, and the theorem holds when $n = 1$.

Now assume that the theorem holds for all trees that have k vertices. We will prove that the theorem holds for a tree $\mathcal{T}$ with $k + 1$ vertices. By Theorem 5.2, there is a vertex V with degree 1. Remove the vertex V and the edge on V from the graph $\mathcal{T}$ to obtain a new graph $\mathcal{T}'$. (See Figure 5.8.) This graph $\mathcal{T}'$ has k vertices and is still a tree. (Why?) Thus, by the induction assumption, $\mathcal{T}'$ has $k - 1$ edges. But then, $\mathcal{T}$ has k edges.

By mathematical induction, the theorem holds for all positive integers n. ■

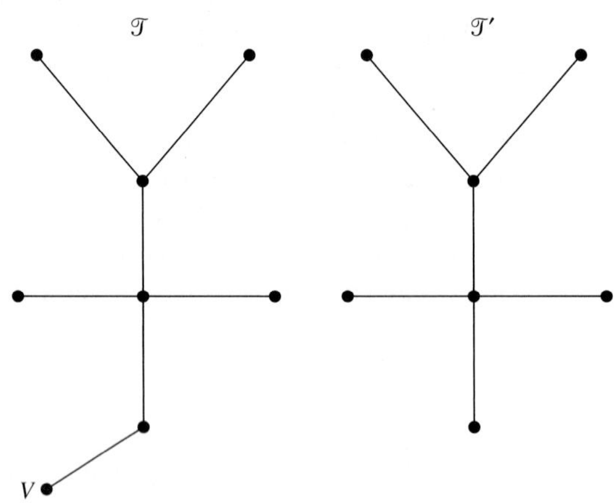

FIGURE 5.8

Example 5.5 An intelligence agency has established a network of 10 spies engaging in industrial espionage. It is important that each spy be able to communicate with any other, either directly or indirectly through a chain of others. Establishing secret locations to exchange messages is difficult, and the agency wants to keep the number of these meeting places as small as possible.

Yet, for reasons of secrecy, no more than two spies should know about any particular meeting place. This communication network can be represented by a graph in which the vertices correspond to spies and an edge joins 2 vertices when the corresponding spies know about the same meeting place. In fact, this graph is a tree with 10 vertices, so there will need to be nine meeting places in all. ■

Theorem 5.4 (a) When an edge is removed from a tree (leaving all the vertices), the resulting graph is not connected and hence is not a tree.

(b) When an edge is added to a tree (without adding additional vertices), the resulting graph has a cycle and hence is not a tree.

Proof. If an edge is added or removed from a tree, the resulting new graph can no longer be a tree, by Theorem 5.3. Since removing an edge cannot create a cycle, nor can adding an edge disconnect the graph, both parts of the theorem follow. ■

Theorem 5.4 shows that a tree has just the right number of edges to be connected and not have any cycles. By looking at the tree in Figure 5.5(a), we can see how the deletion of any edge produces a disconnected graph by breaking the tree into two parts. In addition, we can see how the addition of an edge between two existing vertices creates a cycle in the new graph.

The next theorem gives some other ways of characterizing a tree. Its proof will be left to the exercises.

Theorem 5.5 The following statements are equivalent for a graph $\mathcal{T}$:

(a) $\mathcal{T}$ is a tree.

(b) $\mathcal{T}$ is connected, and the number of vertices is one more than the number of edges.

(c) $\mathcal{T}$ has no cycles, and the number of vertices is one more than the number of edges.

(d) There is exactly one simple path between each pair of vertices in $\mathcal{T}$.

(e) $\mathcal{T}$ is connected, and the removal of any edge of $\mathcal{T}$ results in a graph that is not connected.

(f) $\mathcal{T}$ has no cycles, and the addition of any edge between two nonadjacent vertices results in a graph with a cycle.

It is the equivalence of parts (a) and (b) in Theorem 5.5 that helps in the mathematical analysis of saturated hydrocarbons of the type C_kH_{2k+2}. (See Example 5.1.) We know that there will be k carbon atoms and $2k+2$ hydrogen atoms represented in the graph. Furthermore, since the atoms form a compound, the graph will be connected. Since each vertex represents an atom, there will be

$k + (2k + 2) = 3k + 2$ vertices. Also, since a carbon atom has degree 4 and a hydrogen atom has degree 1, the sum of the degrees is $4k + (2k + 2) = 6k + 2$. By Theorem 4.1, the number of edges is $\frac{1}{2}(6k + 2) = 3k + 1$, which is one less than the number of vertices. Hence, by Theorem 5.5(b), the graph representing the chemical compound is always a tree. Knowing this, Cayley used information about trees to predict the existence of new saturated hydrocarbons. The interested reader should consult suggested reading [2] for more details.

EXERCISES 5.1

In Exercises 1–8, determine whether each graph is a tree.

1.

2.

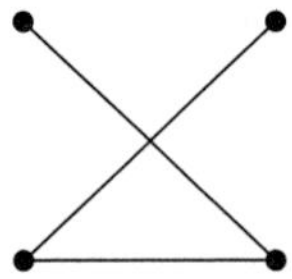

3.

4.

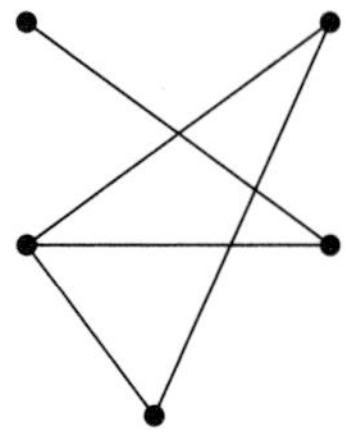

5.

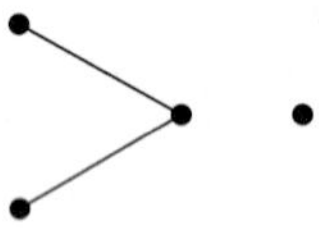

6.

7.

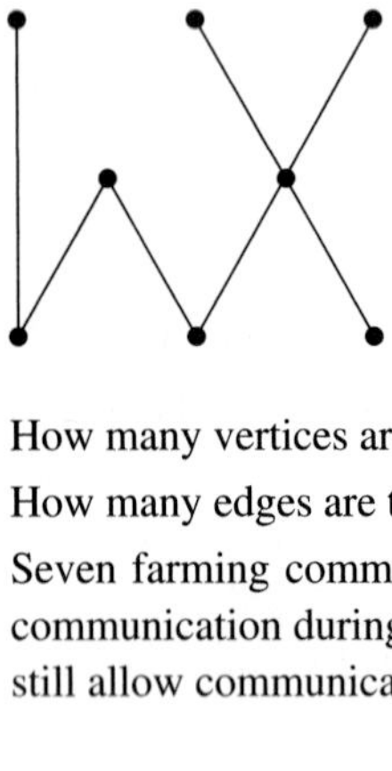

8.

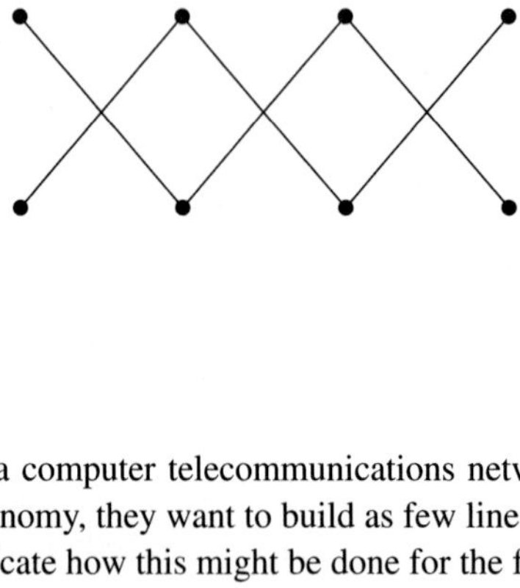

9. How many vertices are there in a tree with 15 edges?

10. How many edges are there in a tree with 21 vertices?

11. Seven farming communities in Iowa want to develop a computer telecommunications network to facilitate communication during a farm crisis. For reasons of economy, they want to build as few lines as possible, but still allow communication between any two towns. Indicate how this might be done for the following map:

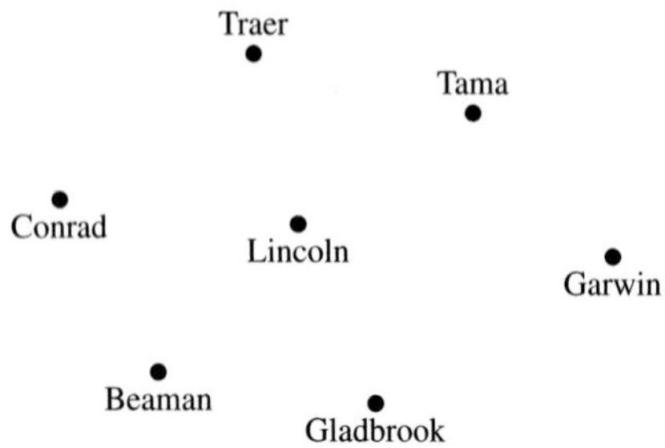

12. As few trails as possible are to be built between houses in a primitive community so that it is possible for a resident to go from any house to any other. If there are 34 houses, how many trails need to be built? Since it is considered bad luck to live at the end of a trail, can the trails be constructed so that no house is so situated?

13. A farmer needs to irrigate the fields in which his crops are growing. In the following map, the fields are the enclosed areas and edges represent earthen walls between the fields:

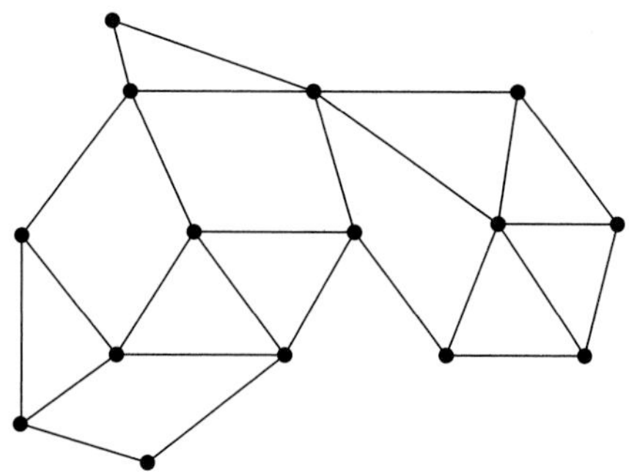

Because he lacks modern equipment, his method of irrigation is to break holes in the walls and let water from the outside cover the entire field. He wants to irrigate each field and to break as few walls as possible. In how many walls should he break holes?

14. Draw a graph that is not a tree for which the number of vertices is one more than the number of edges.

15. Draw a tree with at least 6 vertices that has exactly 2 vertices of degree 1.

16. What is the smallest number of edges in a connected graph with n vertices?

17. What is the largest number of vertices in a connected graph with n edges?

18. How many simple paths of nonzero length are there in a tree with n vertices, where $n \geq 2$? (Regard two simple paths as the same if they have the same edges.)

19. Prove that the graph $\mathcal{T}'$ in the proof of Theorem 5.3 is a tree.

20. Prove that if an edge is deleted from a cycle in a connected graph, the graph remains connected.

21. For which n is $\mathcal{K}_n$ a tree? (The graph $\mathcal{K}_n$ is defined in Section 4.1.)

22. Prove by mathematical induction on the number of vertices that any tree can be drawn on a sheet of paper so that its edges do not intersect, except at vertices.

23. There are two saturated hydrocarbons of the type C_4H_{10}: butane and isobutane. Draw trees representing the chemical structure of each.

24. Draw a graph representing a saturated hydrocarbon with 5 carbon atoms.

25. Can a tree with 13 vertices have 4 vertices of degree 3, 3 vertices of degree 4, and 6 vertices of degree 1?

26. How many vertices of degree 1 are there in a tree with 3 vertices of degree 4, 1 vertex of degree 3, 2 vertices of degree 2, and no vertices of degree more than 4?

As trees on the vertices labeled A, B, and C, the two trees in figures (a) and (b) that follow are the same, since they both have the same set of edges, namely, $\{A, B\}$ and $\{B, C\}$. The trees in figures (a) and (c) are distinct, since they do not have the same set of edges. For example, $\{A, C\}$ is an edge of the tree in figure (c), but not of the tree in figure (a).

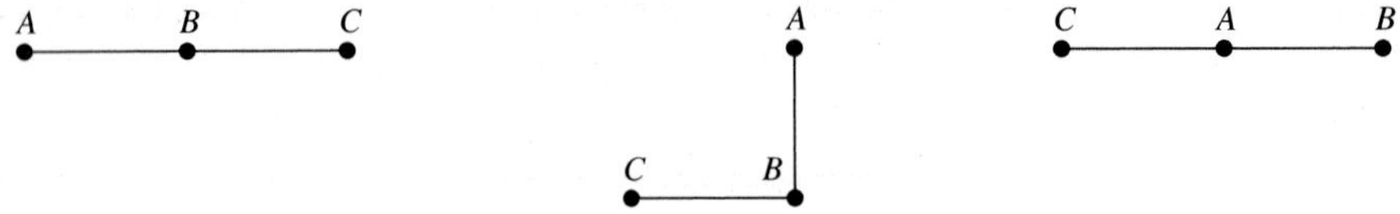

(a) **(b)** **(c)**

27. Draw the three distinct trees with 3 labeled vertices. (Use 1, 2, 3 as the labels.)

28. Draw the 16 distinct trees with 4 labeled vertices. (Use 1, 2, 3, 4 as the labels.)

In order to count the number of distinct trees with vertices labeled $1, 2, \ldots, n$ *we establish a one-to-one correspondence between each such tree and a list* $a_1, a_2, \ldots, a_{n-2}$, *where* $1 \leq a_i \leq n$ *for* $i = 1, 2, \ldots, n-2$. *The following algorithm shows how to get such a list from a labeled tree* $\mathcal{T}$:

Prufer's Algorithm

This algorithm constructs a list $a_1, a_2, \ldots, a_{n-2}$ of numbers for a tree with n labeled vertices, where $n \geq 3$ and the labels are $1, 2, \ldots, n$.

Step 1 (initialization)
- (a) Set $\mathcal{T}$ to be the given tree.
- (b) Set $k = 1$.

Step 2 (choose a_k)

while $\mathcal{T}$ has more than two vertices
- *Step 2.1* (find a vertex of degree 1)
  - Select from $\mathcal{T}$ the vertex X of degree 1 that has the smallest label.
- *Step 2.2* (make a new tree)
  - (a) Find the edge e on X, and let W denote the other vertex on e.
  - (b) Set a_k to be the label on W.
  - (c) Delete the edge e and vertex X from $\mathcal{T}$ to form a new tree $\mathcal{T}'$.
- *Step 2.3* (change $\mathcal{T}$ and k)
  - (a) Replace $\mathcal{T}$ with $\mathcal{T}'$.
  - (b) Replace k with $k + 1$.

endwhile

For example, the list for the following tree is 6, 5, 1, 5, 6:

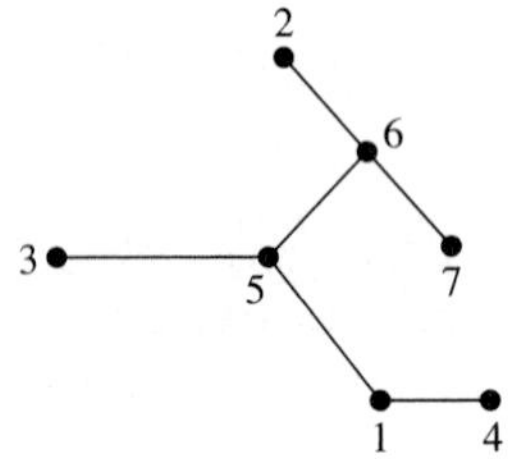

In Exercises 29–32, use Prufer's algorithm to find the list for each tree in the indicated exercise or graph.

29. Exercise 27

30. Exercise 28

31.

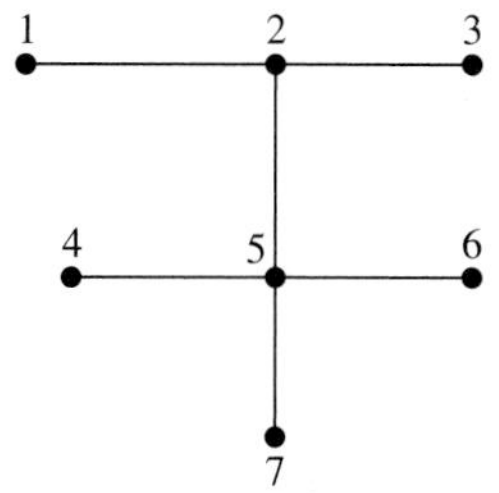

32.

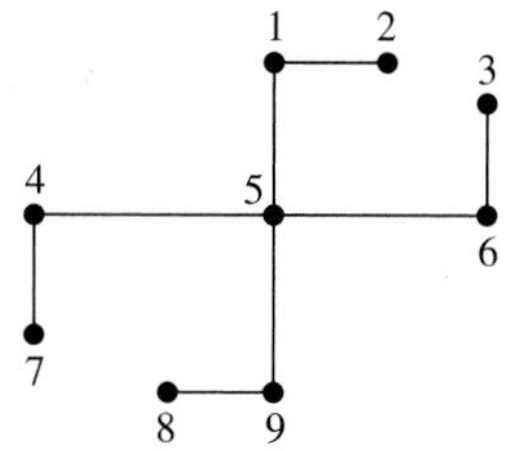

33. We can construct a tree from a list L of $n - 2$ numbers taken from $N = \{1, 2, \ldots, n\}$ as follows (here we assume that the vertices of the tree are labeled $1, 2, \ldots, n$): Pick the smallest number k from N that is not in the list L and construct an edge on that number and the first number in the list L. Then delete the first number in L, delete k from N, and repeat the process. When L is exhausted, construct an edge joining the two numbers remaining in N. For example, the tree generated by the list 6, 5, 1, 5, 6 is pictured before Exercise 29. Construct the tree for the list 2, 2, 2, 2.

In Exercises 34–37, repeat Exercise 33 for each list.

34. 1, 2, 3, 4

35. 1, 2, 3, 2, 1

36. 4, 3, 2, 1

37. 3, 5, 7, 3, 5, 7

38. Assuming that Prufer's algorithm establishes a one-to-one correspondence between trees with vertices labeled $1, 2, \ldots, n$ and lists as described in Exercise 33, prove that the number of distinct trees with vertices $1, 2, \ldots, n$ is n^{n-2} for $n > 1$.

Exercises 39–44 establish in a cyclical fashion a proof of Theorem 5.5.

39. Prove that part (a) implies part (b) in Theorem 5.5.

40. Prove that part (b) implies part (c) in Theorem 5.5.

41. Prove that part (c) implies part (d) in Theorem 5.5.

42. Prove that part (d) implies part (e) in Theorem 5.5.

43. Prove that part (e) implies part (f) in Theorem 5.5.

44. Prove that part (f) implies part (a) in Theorem 5.5.

45. Give an inductive proof of Theorem 5.3 that does not use Theorem 5.2. (*Hint:* Use mathematical induction on the number of edges.)

46. Use Theorem 5.3 to give an alternative proof of Theorem 5.2.

47. Show that Prufer's algorithm establishes a one-to-one correspondence between trees with vertices labeled $1, 2, \ldots, n$ and lists as described in Exercise 33.

5.2 SPANNING TREES

In Example 5.2 in Section 5.1, we found a tree that contained all the vertices of the original graph. This is an idea that appears in many applications, including those that involve power lines, pipeline networks, and road construction.

Example 5.6 Suppose an oil company wants to build a series of pipelines between six storage facilities in order to be able to move oil from one storage facility to any of the other five. Because the construction of a pipeline is very expensive, the company wants to construct as few pipelines as possible. Thus the company does not mind if oil has to be routed through one or more intermediate facilities. For environmental reasons, it is not possible to build a pipeline between some pairs of storage facilities. The graph in Figure 5.9(a) shows the pipelines that can be built.

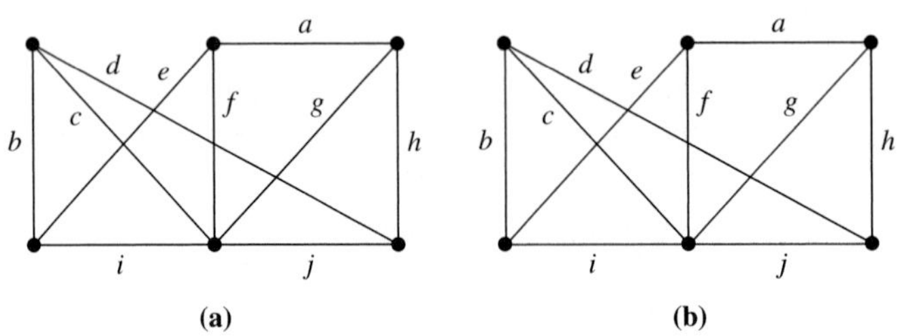

FIGURE 5.9

The task is to find a set of edges which, together with the incident vertices, form a connected graph containing all the vertices and having no cycles. This will allow oil to go from any storage facility to any other without unnecessary duplication of routes and, hence, unnecessary building costs. Thus a tree containing all the vertices of a graph is again being sought. One selection of edges is b, e, g, i, and j, as illustrated by the colored edges in Figure 5.9(b). ■

A **spanning tree** of a graph $\mathcal{G}$ is a tree (formed by using edges and vertices of $\mathcal{G}$) containing all the vertices of $\mathcal{G}$. Thus, in Figure 5.9, the edges b, e, g, i, and j and their incident vertices form a spanning tree for the graph. We shall follow the customary practice of describing a tree by listing only its edges, with the understanding that its vertices are those incident with the edges. Thus, in Figure 5.9, we would say that the edges b, e, g, i, and j form a spanning tree for the graph.

If a graph is a tree, then its only spanning tree is itself. But in general, a graph may have more than one spanning tree. For example, the edges a, b, c, d, and e also form a spanning tree for the graph in Figure 5.9(a).

There are several ways to find a spanning tree for a graph. One is to get rid of cycles by removing edges. This process is illustrated in the following example:

Example 5.7 The graph in Figure 5.10(a) is not a tree, because it contains cycles such as a, b, e, d. In order to obtain a tree, our procedure will be to delete an edge in each cycle. Deleting b from the cycle a, b, e, d gives the graph in Figure 5.10(b), which is still not a tree, because of the cycle c, e, d. So we delete an edge in this cycle, say, e. The resulting graph in Figure 5.10(c) is now a tree. This, then, is a spanning tree for the original graph. ■

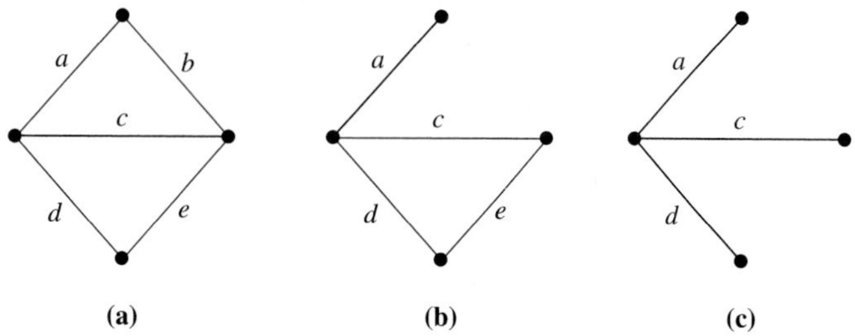

FIGURE 5.10

If a connected graph has n vertices and e edges, with $e \geq n$, we must delete $e - n + 1$ edges in order to obtain a spanning tree. By performing these deletions, we change the number of edges from e to $e - (e - n + 1) = n - 1$, which is the number of edges in a tree with n vertices.

The Breadth-First Search Algorithm

The method previously described is not the only way to find a spanning tree. There are many others, and some of these are easier to program on a computer because they do not require that cycles be found. One of these methods uses the breadth-first search algorithm, which was discussed in Section 4.3.

Recall that in the breadth-first search algorithm we start with a vertex S. Then we find the vertices adjacent to S and assign them the label $1(S)$. (The label given to a vertex by the breadth-first search algorithm indicates its distance from S and its predecessor on a shortest path from S.) Next, we look at each unlabeled vertex that is adjacent to a vertex V with label 1; these vertices are then given the label $2(V)$. We continue in this manner until there are no more unlabeled vertices adjacent to labeled vertices.

Let $\mathcal{T}$ denote the set of edges that join each labeled vertex to its predecessor. The labeling process in step 2.2 of the breadth-first search algorithm guarantees that the edges in $\mathcal{T}$ form a connected graph. Furthermore, each edge in $\mathcal{T}$ joins two vertices labeled with consecutive integers, and no vertex in $\mathcal{L}$ is joined by an edge in $\mathcal{T}$ to more than one vertex with a smaller label. Therefore no collection of edges in $\mathcal{T}$ forms a cycle. Because in a connected graph every vertex is eventually labeled, the edges in $\mathcal{T}$ form a tree that includes every vertex in the graph, and so $\mathcal{T}$ is a spanning tree for the graph. (As before, we are referring to $\mathcal{T}$ as a tree, with the understanding that the vertices of the tree are those incident with the edges.)

Example 5.8 We shall apply the breadth-first search algorithm to find a spanning tree for the graph in Figure 5.11.

We may start the breadth-first search algorithm at any vertex, say, K, which is labeled 0(–). The vertices adjacent to K are A and B, and these are labeled $1(K)$. Next, we label the unlabeled vertices adjacent to A and B, which are D and E. These are labeled $2(A)$ and $2(B)$, respectively. We continue in this manner until all the vertices are labeled. One possible set of labels is shown in Figure 5.12. The edges that join each vertex to its predecessor (which is indicated in the label on the vertex) then form a spanning tree for the graph. These edges are shown in color in Figure 5.12. ■

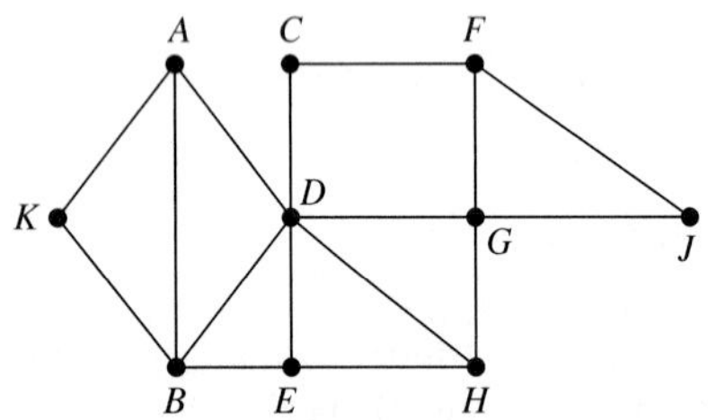

FIGURE 5.11

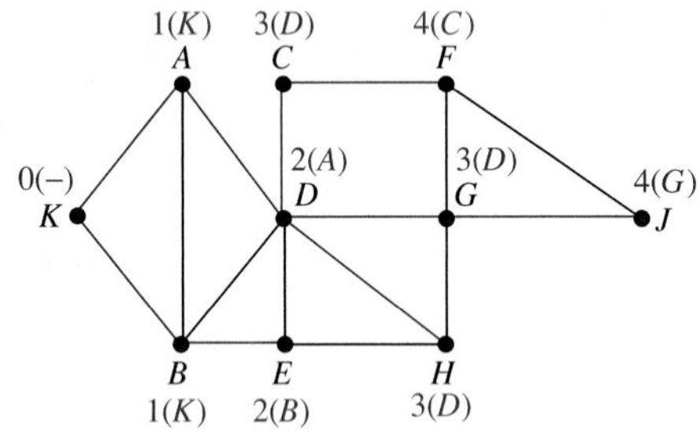

FIGURE 5.12

We should note that, when using the breadth-first search algorithm, there are places where predecessors are chosen arbitrarily. Different choices lead to different spanning trees. In Example 5.8, for instance, instead of choosing the edges $\{D, H\}$ and $\{C, F\}$, we could have chosen the edges $\{E, H\}$ and $\{F, G\}$. This would give the spanning tree shown in color in Figure 5.13.

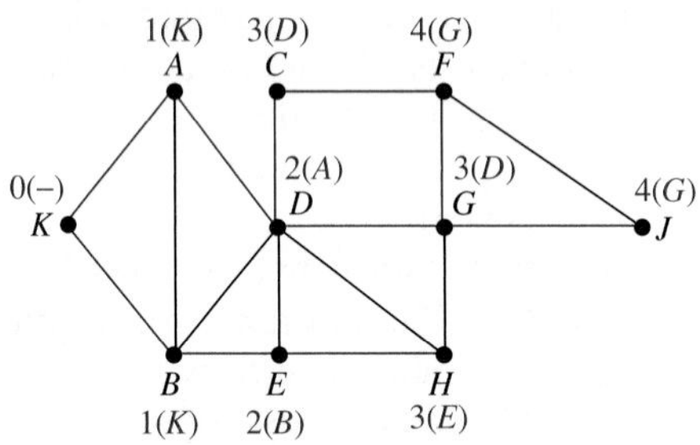

FIGURE 5.13

A simple path from the starting vertex S to any other vertex that uses only edges in a spanning tree is a shortest path in the original graph between these vertices. (Recall from Section 4.3 that the label given to each vertex by the breadth-first search algorithm is its distance from S.) For this reason, a spanning tree constructed by means of the breadth-first search algorithm is sometimes called a **shortest path tree**.

In the examples so far, the graphs have had spanning trees. However, this is not always the case, as the next example shows.

Example 5.9 The graph shown in Figure 5.14 does not have a spanning tree, because it is not possible to choose edges that connect all the vertices. In particular, we cannot find edges that can be used to make a path from A to E. ■

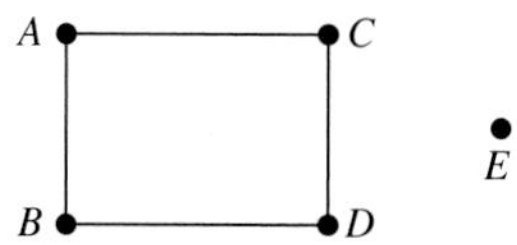

FIGURE 5.14

In previous examples we have seen that the existence of a spanning tree is related to the connectedness of the graph. This relationship is made explicit in the next theorem.

Theorem 5.6 A graph is connected if and only if it has a spanning tree.

Proof. Suppose that the graph $\mathcal{G}$ has a spanning tree $\mathcal{T}$. Since $\mathcal{T}$ is a connected graph containing all the vertices in $\mathcal{G}$, there is a path between any two vertices U and V in $\mathcal{G}$ using edges from $\mathcal{T}$. But since the edges of $\mathcal{T}$ are also edges of $\mathcal{G}$, we have a path between U and V using edges in $\mathcal{G}$. Hence $\mathcal{G}$ is connected.

Conversely, suppose $\mathcal{G}$ is connected. Applying the breadth-first search algorithm to $\mathcal{G}$ yields a set $\mathcal{L}$ of vertices with labels and a set $\mathcal{T}$ of edges connecting the vertices in $\mathcal{L}$. Moreover, $\mathcal{T}$ is a tree. Since $\mathcal{G}$ is connected, each vertex of $\mathcal{G}$ is labeled. Thus $\mathcal{L}$ contains all the vertices of $\mathcal{G}$, and $\mathcal{T}$ is a spanning tree for $\mathcal{G}$. ■

We will now discuss two types of spanning trees that occur frequently in applications.

Minimal and Maximal Spanning Trees

When pipelines are to be constructed between oil storage facilities, it is likely that the cost of building each pipeline is not the same. Because of terrain, distance, and other factors, it may cost more to build one pipeline than another. We can describe this problem by a weighted graph (discussed in Section 4.3), in which the weight of each edge is the cost of building the corresponding pipeline. Figure 5.15 depicts such a weighted graph. The problem is to build the cheapest set of pipelines. In other words, we want to find a spanning tree in which the sum of the costs of all the edges is as small as possible.

In a weighted graph, the **weight of a tree** is the sum of the weights of the edges in the tree. A **minimal spanning tree** in a weighted graph is a spanning tree for which the weight of the tree is as small as possible. In other words, a

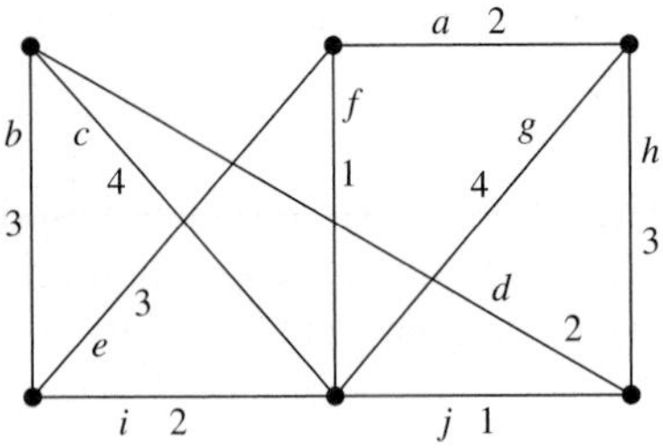

FIGURE 5.15

minimal spanning tree is a spanning tree such that no other spanning tree has a smaller weight.

Example 5.10 For the weighted graph in Figure 5.15, the edges b, c, e, g, and h form a spanning tree with weight $3+4+3+4+3=17$. The edges a, b, c, d, and e form another spanning tree with weight $2+3+4+2+3=14$. The edges a, d, f, i, and j form yet another spanning tree, which has weight 8. Since this spanning tree uses the five edges with the smallest weights, there can be no spanning tree with smaller weight. Thus, the edges a, d, f, i, and j form a minimal spanning tree for this weighted graph. ■

In Example 5.10, we were able to find a minimal spanning tree by trial and error. However, for a weighted graph with a large number of vertices and edges, this is not a very practical approach. One systematic approach would be to find all the spanning trees of a connected weighted graph, compute their weights, and then select a spanning tree with the smallest weight. Although this approach will always find a minimal spanning tree for a connected weighted graph, checking out all the possibilities can be a very time-consuming task, even for a supercomputer. A natural way to try to construct a minimal spanning tree is to build a spanning tree by using edges of smallest weights. This approach is illustrated in Example 5.11.

Example 5.11 For the weighted graph in Figure 5.16(a), we begin with any vertex, say, A, and select the edge of smallest weight on it, which is b. To continue building a tree, we look at the edges a, c, e, and f touching edge b and select the one with the smallest weight, which is f. The next edges to look at are a, c, e, and g, the ones touching b and f, the edges already selected. There are two with the smallest weight, e and g, and we select one arbitrarily, say, e. The next edges we consider are a, c, and d. (The edge g is not considered any longer, for its inclusion will form a cycle with e and f.) The edge with the smallest weight is a, and so it is added to the tree. These four edges a, b, e, f form a spanning tree (see Figure 5.16(b)), which also turns out to be a minimal spanning tree. ■

The method in Example 5.11 is due to Prim and will always produce a minimal spanning tree. Prim's algorithm builds a tree by selecting any vertex and then

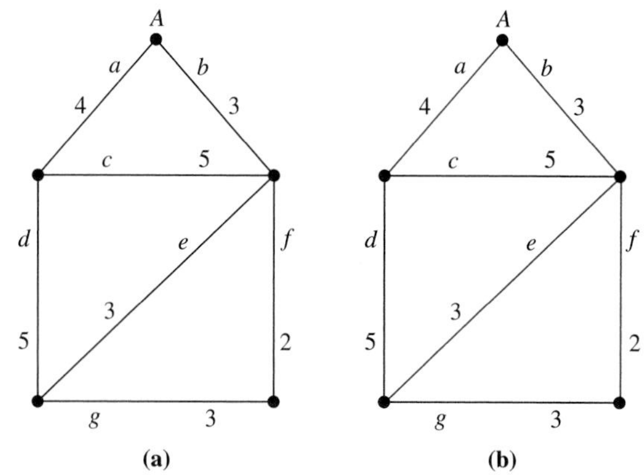

FIGURE 5.16

an edge of smallest weight on that vertex. We then extend the tree by choosing an edge of smallest weight that forms a tree with the previously chosen edge. We further extend this tree by choosing an edge of smallest weight that forms a tree with the two previously chosen edges. This process is continued until a spanning tree is obtained, which turns out to be a minimal spanning tree. The process can be formalized as follows:

Prim's Algorithm

This algorithm finds a minimal spanning tree, if one exists, for a weighted graph with n vertices. In the algorithm, $\mathcal{T}$ is a set of edges that form a tree and $\mathcal{L}$ is the set of vertices incident with the edges in $\mathcal{T}$.

Step 1 (select starting vertex)
Select a vertex U, and set $\mathcal{L} = \{U\}$ and $\mathcal{T} = \emptyset$.
Step 2 (enlarge $\mathcal{T}$)
while some edge has one vertex in $\mathcal{L}$ and one not in $\mathcal{L}$
(a) Choose an edge of smallest weight from among those with one vertex in $\mathcal{L}$ and the other not in $\mathcal{L}$. (Ties can be broken arbitrarily.)
(b) Place the edge in $\mathcal{T}$.
(c) Place its vertices in $\mathcal{L}$ (one of these is already in $\mathcal{L}$).
endwhile
Step 3 (is there a minimal spanning tree?)
if $|\mathcal{L}| < n$
The graph is not connected, and so it has no minimal spanning tree.
otherwise

The edges in $\mathcal{T}$ and their incident vertices form a minimal spanning tree.
endif

In step 2 of Prim's algorithm, the selection of an edge with one vertex in $\mathcal{L}$ and the other not in $\mathcal{L}$ guarantees that there are no cycles formed by any collection of edges in $\mathcal{T}$. Thus, at the end of each iteration of the loop in step 2, the edges in $\mathcal{T}$ and the vertices in $\mathcal{L}$ form a tree. Furthermore, when $\mathcal{L}$ contains all the vertices of $\mathcal{G}$, a spanning tree is formed. As usual, we will denote this tree by $\mathcal{T}$. The proof that Prim's algorithm yields a minimal spanning tree is found at the end of this section.

Example 5.12 Prim's algorithm will be applied to the weighted graph in Figure 5.17. We start with vertex F, and set $\mathcal{L} = \{F\}$ and $\mathcal{T} = \emptyset$. Since there are edges that have one vertex in $\mathcal{L}$ and the other not in $\mathcal{L}$, we perform (a), (b), and (c) of step 2. The edges on F that do not have their other vertex in $\mathcal{L}$ are a, b, f, and g (see Figure 5.18), and, of these, a is the one of smallest weight. Therefore a is included in $\mathcal{T}$ and its vertices are included in $\mathcal{L}$. Thus $\mathcal{L} = \{F, C\}$ and $\mathcal{T} = \{a\}$. Since there are edges that have one vertex in $\mathcal{L}$ and the other not in $\mathcal{L}$, we continue step 2. The edges having exactly one vertex in $\mathcal{L}$ are b, d, e, f, and g. (See Figure 5.19.) Of these, e has the smallest weight and is therefore included in $\mathcal{T}$, and the vertex E is included in $\mathcal{L}$. Now $\mathcal{L} = \{F, C, E\}$ and $\mathcal{T} = \{a, e\}$. Again, there are edges with exactly one vertex in $\mathcal{L}$, and so we continue step 2. This time, the edges to consider are b, d, g, and j. (See Figure 5.20.) Notice that edge f is not considered, for it has both of its vertices in $\mathcal{L}$. Of the edges b, d, g, and j, there are two with the smallest weight, namely, b and d. Let us arbitrarily choose b and include it in $\mathcal{T}$ and B in $\mathcal{L}$. Thus $\mathcal{L} = \{F, C, E, B\}$ and $\mathcal{T} = \{a, e, b\}$.

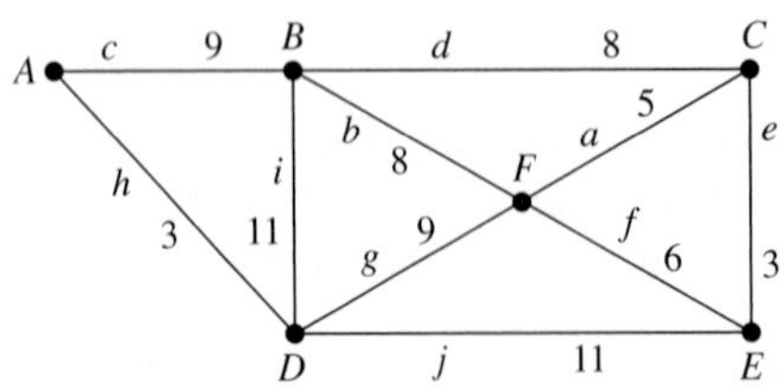

FIGURE 5.17

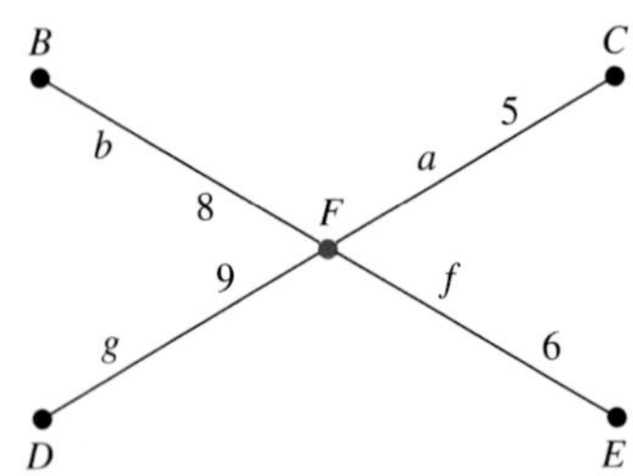

FIGURE 5.18

Again, there are edges having one vertex in $\mathcal{L}$ and the other not in $\mathcal{L}$, and so step 2 continues. The edges with exactly one vertex in $\mathcal{L}$ are c, g, i, and j. Of these, both c and g have the smallest weight. Suppose that we choose c. Then

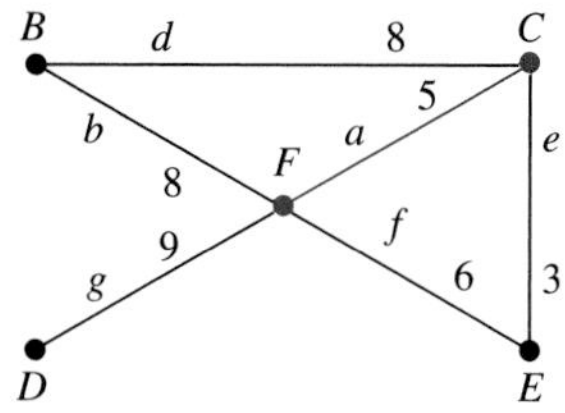

FIGURE 5.19

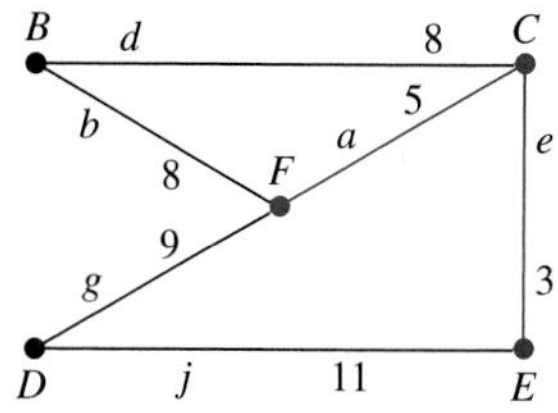

FIGURE 5.20

we include c in $\mathcal{T}$ and A in $\mathcal{L}$, making

$$\mathcal{L} = \{F, C, E, B, A\} \qquad \text{and} \qquad \mathcal{T} = \{a, e, b, c\}.$$

As step 2 continues, the edges to consider are g, h, i, and j. The one with the smallest weight is h, and so it is inserted in $\mathcal{T}$, and D is inserted in $\mathcal{L}$. Now

$$\mathcal{L} = \{F, C, E, B, A, D\} \qquad \text{and} \qquad \mathcal{T} = \{a, e, b, c, h\}.$$

Since there is no longer an edge with exactly one vertex in $\mathcal{L}$ (because $\mathcal{L}$ contains all the vertices of the weighted graph), we proceed to step 3. It tells us that the edges in $\mathcal{T}$ and their incident vertices form a minimal spanning tree, as illustrated in Figure 5.21. The weight of this spanning tree is 28. ■

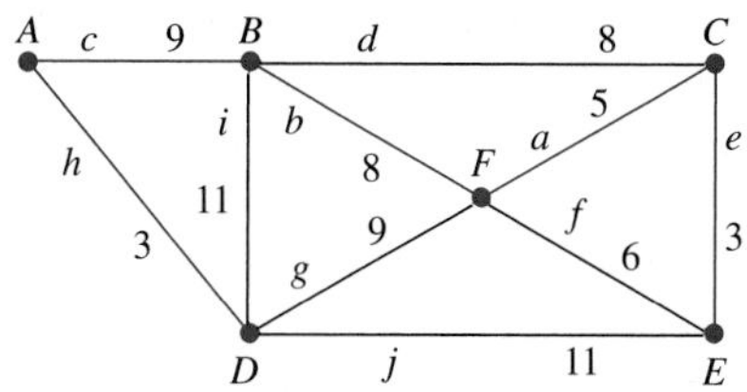

FIGURE 5.21

There are two places in the preceding example where we have a choice of edges with the same least weight. Step 2(a) of the algorithm indicates that any edge of least weight could be chosen in such cases. If other choices are made, different minimal spanning trees would be constructed. For example, if in Example 5.12 we choose edge d instead of b, followed by the choices of g and h, the minimal spanning tree in Figure 5.22 results. Thus we see that minimal spanning trees need not be unique.

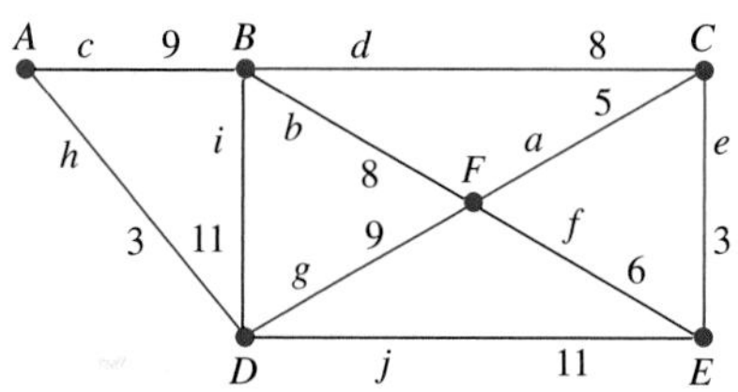

FIGURE 5.22

Prim's algorithm is an example of what is called a *greedy algorithm*, since at each iteration we do the thing that seems best at that step (extending a tree by including an available edge of smallest weight). In Prim's algorithm this approach does lead to a minimal spanning tree, although in general a greedy algorithm need not produce the best possible result. (See Exercises 36 and 37.)

In analyzing the complexity of Prim's algorithm for a weighted graph with n vertices and e edges, we will consider comparing the weights of two edges as the basic operation. At each iteration of the loop in step 2, there will be at most $e - 1$ comparisons made in order to find an edge of smallest weight having one vertex in $\mathcal{L}$ and one vertex not in $\mathcal{L}$. Step 2 is done at most n times, and so there are at most $n(e - 1)$ operations. Since

$$e \leq C(n, 2) = \frac{1}{2}n(n - 1),$$

our implementation of Prim's algorithm is of order at most n^3.

Another algorithm that can be used to find a minimal spanning tree is due to Kruskal. It is found in Exercises 5.2.

Let us return to Figure 5.15. Suppose now that the weights of the edges measure the profit that results when oil is pumped through the corresponding pipelines. Our problem is to find a spanning tree of pipelines that generates the most profit. Thus we want a spanning tree for which the sum of the weights of the edges is not as small as possible, but as large as possible.

A **maximal spanning tree** in a weighted graph is a spanning tree such that the weight of the tree is as large as possible. In other words, there is no spanning tree with larger weight. Fortunately, finding a maximal spanning tree is very similar to finding a minimal spanning tree. All that is needed is to replace the phrase "an edge of smallest weight" by the phrase "an edge of largest weight" in step 2(a) of Prim's algorithm.

Example 5.13 We will begin to construct a maximal spanning tree for the weighted graph in Figure 5.17 by picking the vertex F. Then $\mathcal{T} = \emptyset$ and $\mathcal{L} = \{F\}$. Examining the edges on F (see Figure 5.23), we pick one with the largest weight. This is the edge g, and so $\mathcal{T} = \{g\}$ and $\mathcal{L} = \{F, D\}$. The edges with one vertex in $\mathcal{L}$ and one vertex not in $\mathcal{L}$ are a, b, f, h, i, and j (see Figure 5.24). Of these, there are two with the largest weight, i and j. We choose one arbitrarily, say, i. So now, $\mathcal{T} = \{g, i\}$ and $\mathcal{L} = \{F, D, B\}$. Again, the process is repeated (see Figure 5.25) by choosing edge j (the edge of largest weight having one vertex in $\mathcal{L}$ and one not in $\mathcal{L}$). Now $\mathcal{T} = \{g, i, j\}$ and $\mathcal{L} = \{F, D, B, E\}$. Again, we look at the edges with a vertex in $\mathcal{L}$ and one not in $\mathcal{L}$. Of these, c is the edge with the largest weight, and so $\mathcal{T} = \{g, i, j, c\}$ and $\mathcal{L} = \{F, D, B, E, A\}$. One more iteration yields the choice of the edge d; therefore, $\mathcal{T} = \{g, i, j, c, d\}$ and $\mathcal{L} = \{F, D, B, E, A, C\}$, which is the set of all vertices. Hence $\mathcal{T}$ is a maximal spanning tree, as illustrated in Figure 5.26. The reader should check that the weight of this tree is 48. ■

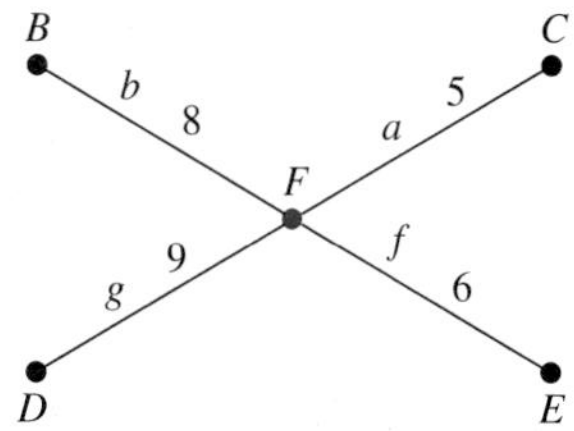

FIGURE 5.23

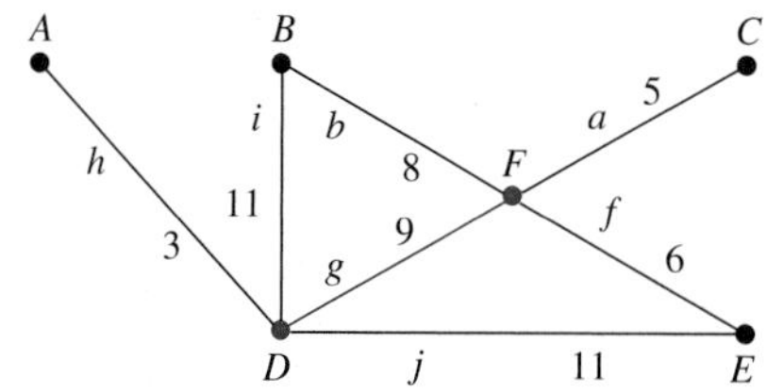

FIGURE 5.24

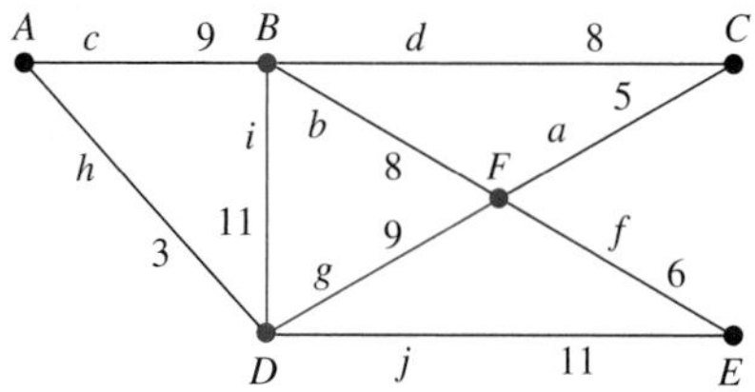

FIGURE 5.25

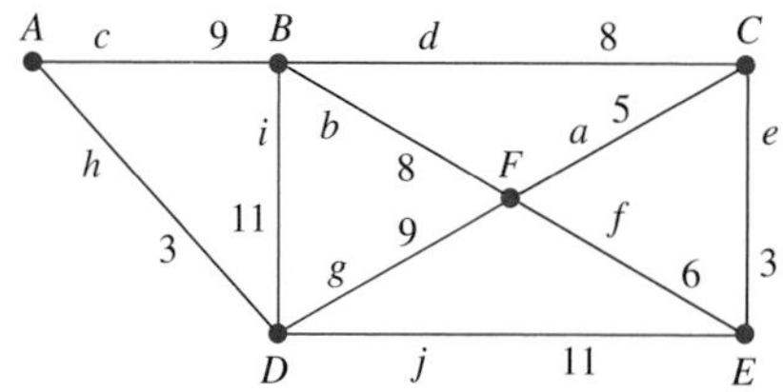

FIGURE 5.26

A Justification of Prim's Algorithm

Now we will prove that if Prim's algorithm is applied to a connected weighted graph with n vertices, it actually produces a minimal spanning tree. Let $\mathcal{T}$ be as in the algorithm—that is, $\mathcal{T}$ is a set to which we add edges one at a time until we get a spanning tree. We will prove that this spanning tree is minimal by induction on m, the number of edges in $\mathcal{T}$.

The induction hypothesis will be that when $\mathcal{T}$ has m edges, then $\mathcal{T}$ is contained in some minimal spanning tree. If $m = 0$, then $\mathcal{T}$ is the empty set. Since some minimal spanning tree exists, the hypothesis holds for $m = 0$.

Now suppose $\mathcal{T}$ is a set with k edges and that $\mathcal{T}$ is contained in a minimal spanning tree $\mathcal{T}'$. Let $\mathcal{L}$ be the corresponding set of vertices given by Prim's algorithm, and suppose $\{U, V\}$ is the next edge that it will put in $\mathcal{T}$, where U is in $\mathcal{L}$ but V is not in $\mathcal{L}$.

If $\{U, V\}$ is in $\mathcal{T}'$, then our induction hypothesis holds for $k + 1$, and we are done. Thus we suppose that $\{U, V\} \notin \mathcal{T}'$. Since $\mathcal{T}'$ is a spanning tree, there must be a path from U to V in $\mathcal{T}'$. Since $U \in \mathcal{L}$ and $V \notin \mathcal{L}$, this path must contain an edge e having one vertex in $\mathcal{L}$ and the other not.

Because Prim's algorithm chooses $\{U, V\}$ instead of e, the weight of $\{U, V\}$ must be less than or equal to the weight of e. Thus, if we form $\mathcal{T}''$ by adding $\{U, V\}$ to $\mathcal{T}'$ and taking out e, we do not increase its weight. Since $\mathcal{T}''$ is connected and has $n - 1$ edges, it is also a minimal spanning tree. But $\mathcal{T}''$ contains the $k + 1$ edges of $\mathcal{T} \cup \{\{U, V\}\}$. This proves the induction hypothesis for $k + 1$.

By the principle of mathematical induction, we see that the tree $\mathcal{T}$ produced by Prim's algorithm is always contained in a minimal spanning tree. But when the algorithm ends and $\mathcal{T}$ has $n - 1$ edges, this spanning tree can only be $\mathcal{T}$ itself. Thus Prim's algorithm produces a minimal spanning tree.

EXERCISES 5.2

In Exercises 1–6, use the breadth-first search algorithm to find a spanning tree for each connected graph. (Start with A, and use alphabetical order when there is a choice for a predecessor.)

1.

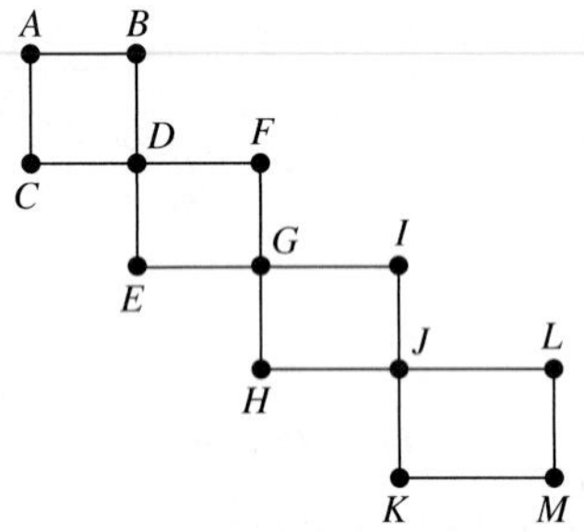

2.

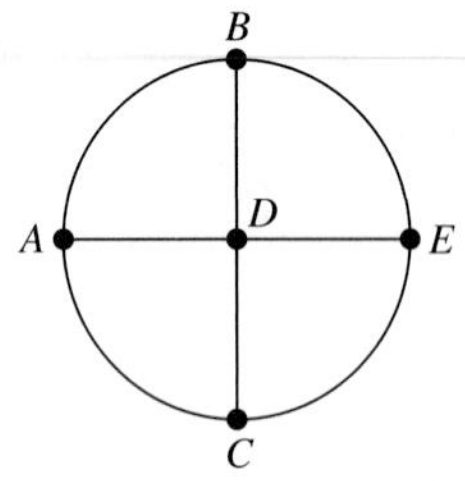

3.

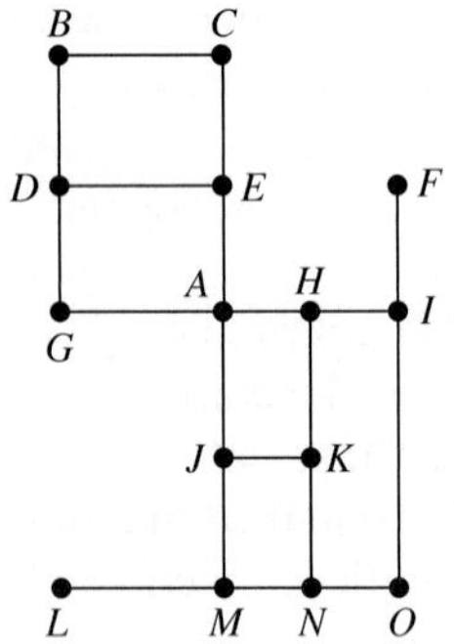

4.

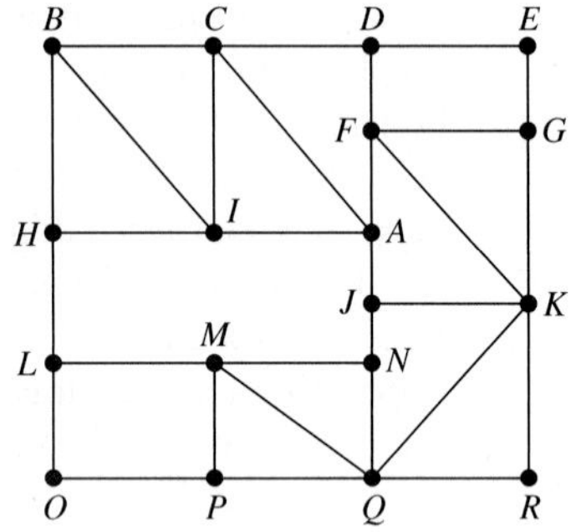

5.

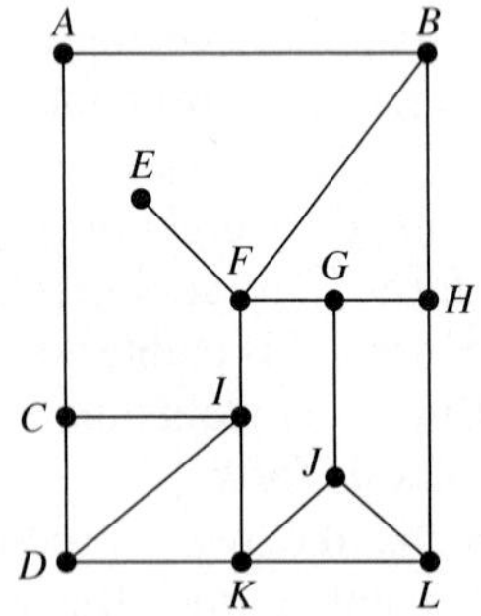

6.

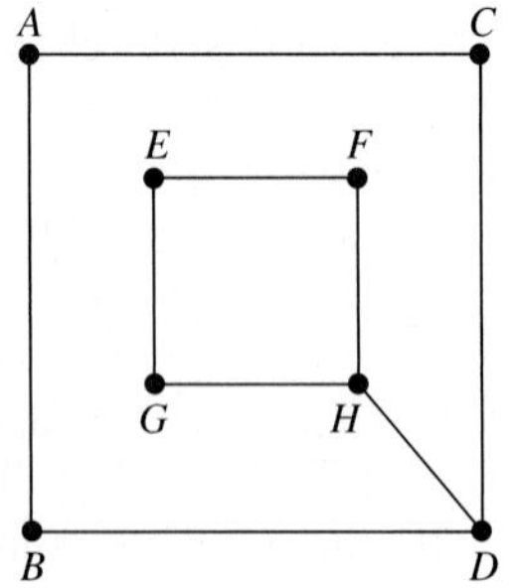

7. At its refinery, an oil company has seven major buildings that are connected by underground tunnels, as illustrated in the following diagram:

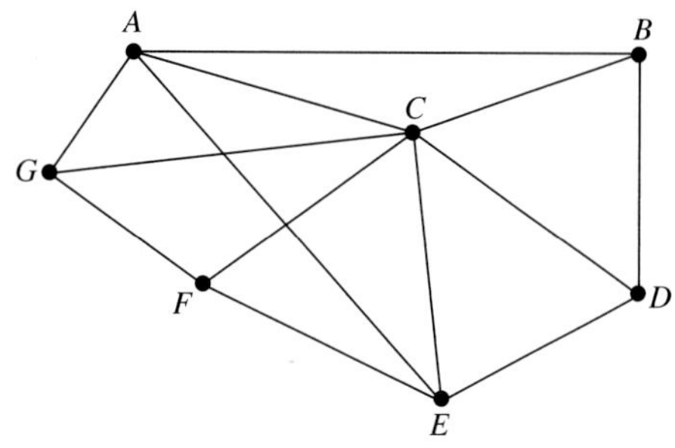

Because of the possibility of a major explosion, there is a need to reinforce some of these tunnels to avoid a potential cave-in. The company wants to be able to go from any building to any other in case of a major fire aboveground, but it wants to avoid reinforcing more tunnels than necessary. How can this be done?

8. One of the primary responsibilities of the National Security Agency is to assist other governmental agencies in providing secure computer communications. The Department of Agriculture does not ordinarily need to be concerned about this, but when estimates of future crop productions arrive, it is important that these be kept secret until the time of the public announcement. The map of computer links between reporting agencies for the Department of Agriculture follows.

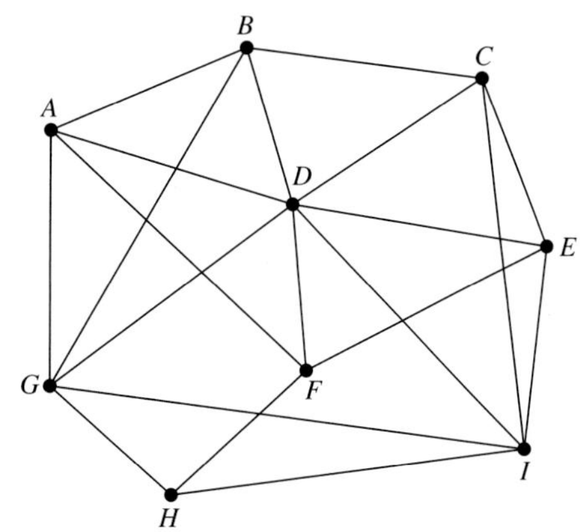

Realizing that there is a need for complete security only at certain times, the National Security Agency will make secure only the minimum number of lines. How can this be done?

9. Will any two spanning trees for a connected graph always have an edge in common? If so, give a proof, and if not, give a counterexample.

10. How many different spanning trees are there for a cycle with n vertices, where $n \geq 3$?

11. Prove that any edge whose removal disconnects a connected graph is part of every spanning tree.

12. Draw a spanning tree formed by applying the breadth-first search algorithm to $\mathcal{K}_n$.

13. Draw a spanning tree formed by applying the breadth-first search algorithm to $\mathcal{K}_{m,n}$.

14. Can a spanning tree formed by using the breadth-first search algorithm be a simple path?

15. A graph with vertices labeled $1, 2, \ldots, 9$ is described by giving its adjacency list:

1: 2, 3, 5, 7, 9
2: 1, 3, 4, 5, 9
3: 1, 2, 4, 6, 8
4: 2, 3, 5, 6
5: 1, 2, 6, 7
6: 3, 4, 5, 7, 9
7: 1, 5, 6, 8, 9
8: 3, 7
9: 1, 2, 6, 7

Use the breadth-first search algorithm to determine whether this graph is connected.

16. Repeat Exercise 15 with the following adjacency list:

1: 2, 5
2: 1, 3
3: 2, 6
4: 5, 6
5: 1, 4
6: 3, 4
7: 8, 9
8: 7, 9
9: 7, 8

Throughout the remaining exercises, if there is a choice of edges to use in forming a minimal or maximal spanning tree, select edges according to alphabetical order.

In Exercises 17–20, use Prim's algorithm to find a minimal spanning tree for each weighted graph. (Start at A.) Give the weight of the minimal spanning tree found.

17.

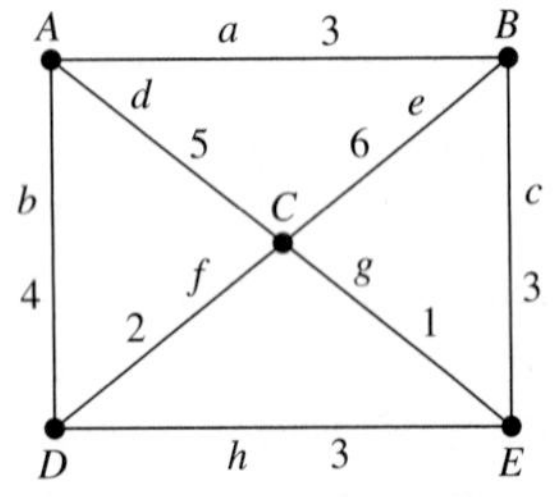

18.

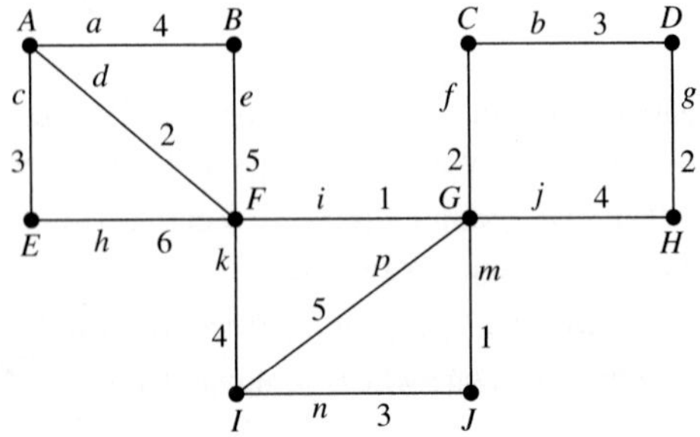

19.

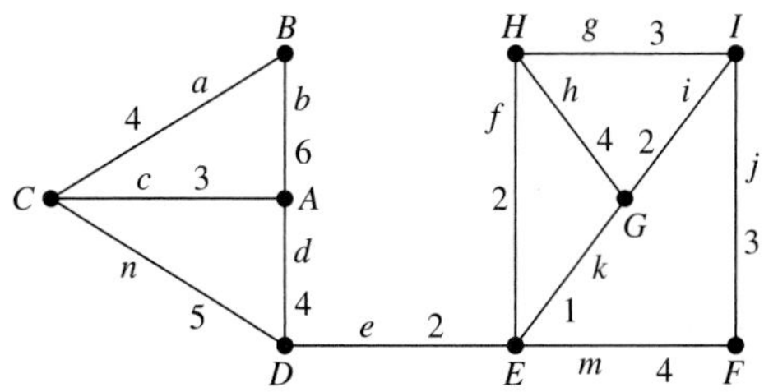

20.

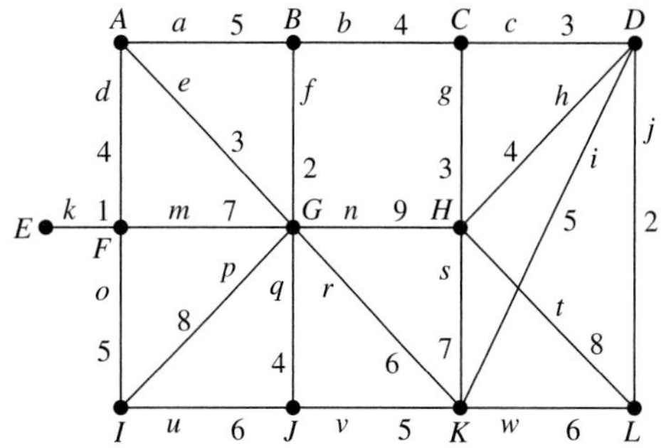

In Exercises 21–24, use Prim's algorithm to find a minimal spanning tree for the weighted graphs in the indicated exercises. Give the weight of the minimal spanning tree found.

21. Exercise 17 (Start at E.)

22. Exercise 18 (Start at H.)

23. Exercise 19 (Start at G.)

24. Exercise 20 (Start at H.)

In Exercises 25–28, use Prim's algorithm to find a maximal spanning tree for the weighted graphs in the indicated exercises. Give the weight of the maximal spanning tree found.

25. Exercise 17 (Start at A.)

26. Exercise 18 (Start at A.)

27. Exercise 19 (Start at F.)

28. Exercise 20 (Start at D.)

29. The Gladbrook Feed Company has seven bins of corn that must be connected by grain pipes so that grain can be moved from one bin to another. To minimize the cost of construction, it wants to build as few grain pipes as possible. The cost (in hundreds of thousands of dollars) of building a pipeline between two bins is given in the following table, where "–" indicates that no pipeline can be built:

Bin	*1*	*2*	*3*	*4*	*5*	*6*	*7*
1	—	4	—	6	2	—	3
2	4	—	5	2	—	3	1
3	—	5	—	7	—	2	2
4	6	2	7	—	4	1	—
5	2	—	—	4	—	1	—
6	—	3	2	1	1	—	2
7	3	1	2	—	—	2	—

How can the pipes be built at minimal cost?

30. FBI Special Agent Hwang is working with five informants who have infiltrated organized crime. She needs to make arrangements for the informants to communicate with each other, either directly or through others, but never in groups of more than two. For reasons of security, the number of meeting places must be kept as small as possible. Furthermore, each pair of informants has been assigned a danger rating that indicates the risk involved in their being seen together. The following table shows the informants and their danger ratings:

	Jones	*Brown*	*Hill*	*Ritt*	*Chen*
Jones	—	3	4	5	2
Brown	3	—	3	1	4
Hill	4	3	—	2	3
Ritt	5	1	2	—	4
Chen	2	4	3	4	—

How can Special Agent Hwang arrange communication so as to minimize the danger? Assume that the danger is proportional to the sum of the danger ratings of the individuals who meet directly.

31. Give an example of a connected weighted graph (not a tree) in which the same edge is part of every minimal spanning tree and every maximal spanning tree.

32. Modify Prim's algorithm to find a spanning tree that is minimal with respect to those containing a specified edge. Illustrate your modification with edge g in Exercise 17.

33. Repeat the second part of Exercise 32 with edge b in Exercise 19.

34. If the weights in a connected graph correspond to distances, does a minimal spanning tree give the shortest distance between any two vertices? If so, give a proof. If not, give a counterexample.

35. Can a minimal spanning tree in a connected weighted graph (not a tree) contain an edge of largest weight? If so, give an example. If not, give a proof.

36. In the knapsack problem of Section 1.3, explain why choosing the experiment with the highest rating does not give a good procedure.

37. Suppose that we want to mail a package and have stamps worth 1, 13, and 22 cents. If we want to make the necessary postage with the minimum number of stamps, the greedy algorithm approach is to use as many 22-cent stamps as possible, then as many 13-cent stamps as possible, and, finally, the necessary number of 1-cent stamps. Show that this approach need not result in the fewest stamps being used.

Kruskal's Algorithm

This algorithm finds a minimal spanning tree, if one exists, for a weighted graph $\mathcal{G}$ with n vertices, where $n \geq 2$. In the algorithm, $\mathcal{S}$ and $\mathcal{T}$ are sets of edges of $\mathcal{G}$.

Step 1 (initialization)
(a) Set $\mathcal{T} = \emptyset$.
(b) Let $\mathcal{S}$ be the set of all edges of $\mathcal{G}$.

Step 2 (enlarge $\mathcal{T}$)
while $|\mathcal{T}| < n - 1$ and $\mathcal{S}$ is not empty
(a) Choose an edge e of smallest weight from $\mathcal{S}$. (Ties can be broken arbitrarily.)

(b) If no cycle is formed by any of the edges in $\mathcal{T} \cup \{e\}$, then replace $\mathcal{T}$ with $\mathcal{T} \cup \{e\}$.

(c) Remove e from $\mathcal{S}$.

endwhile

Step 3 (is there a minimal spanning tree?)

if $|\mathcal{T}| < n - 1$

There is no minimal spanning tree for $\mathcal{G}$ because it is not connected.

otherwise

The edges in $\mathcal{T}$ and their incident vertices form a minimal spanning tree.

endif

In Exercises 38–41, use Kruskal's algorithm to find a minimal spanning tree for the weighted graphs in the indicated exercises.

38. Exercise 17 **39.** Exercise 18 **40.** Exercise 19 **41.** Exercise 20

42. Modify Kruskal's algorithm to find a spanning tree that is minimal with respect to all those containing a specified edge. Illustrate your modification with edge d in Exercise 17.

43. Repeat the second part of Exercise 42 with edge b in Exercise 19.

44. Prove that Kruskal's algorithm gives a minimal spanning tree for a connected weighted graph.

45. Prove that if the weights in a connected weighted graph are all different, then the weighted graph has exactly one minimal spanning tree.

5.3 DEPTH-FIRST SEARCH

In Section 5.2, we saw how breadth-first search can be used to find a spanning tree in a connected graph. This algorithm starts from one vertex and spreads out to all the adjacent vertices. From each of these, we spread out again to all the adjacent vertices that have not been reached and continue in this fashion until we can go no further. In this way, we obtain the distance from the initial vertex to each vertex and also a spanning tree.

Another algorithm for finding a spanning tree in a connected graph is the depth-first search algorithm. In this algorithm, we label the vertices with consecutive integers that indicate the sequence in which the vertices are encountered. The underlying idea of the algorithm is that, to find the vertex that should be labeled immediately after labeling vertex V, the first vertices to consider are the unlabeled ones adjacent to V. If there is an unlabeled vertex W adjacent to V, W is assigned the next label number, and the process of searching for the next vertex to label is begun with W. If V has no unlabeled adjacent vertices, we back up along the edge that we traveled to label V and continue backing up, if necessary, until we reach a vertex having an unlabeled adjacent vertex U. Vertex U is then assigned the next label number, and the process of searching for the next vertex to label is begun with U.

The key idea in the depth-first search algorithm is to back up when we have gone as far as we can. As an example of this process, consider the graph in Figure 5.27. We will assign to each vertex V a label that indicates both the sequence in which V is labeled and its predecessor (the vertex that we came from to reach V). We start at any vertex, say, A, and assign it the label 1(–) to indicate that it is the first vertex labeled and that it has no predecessor. Then, of the two adjacent vertices B and D, we arbitrarily choose one, say, B, and give it the label $2(A)$. Next, of the two unlabeled vertices adjacent to B, we arbitrarily choose C and give it the label $3(B)$. Since C is a vertex with no unlabeled adjacent vertices, we back up to B, the predecessor of C, and go next to D, giving it the label $4(B)$. When all the vertices are labeled, we can construct a spanning tree for the graph by selecting the edges (and their incident vertices) that join each vertex to its predecessor. These edges are shown in color in Figure 5.27.

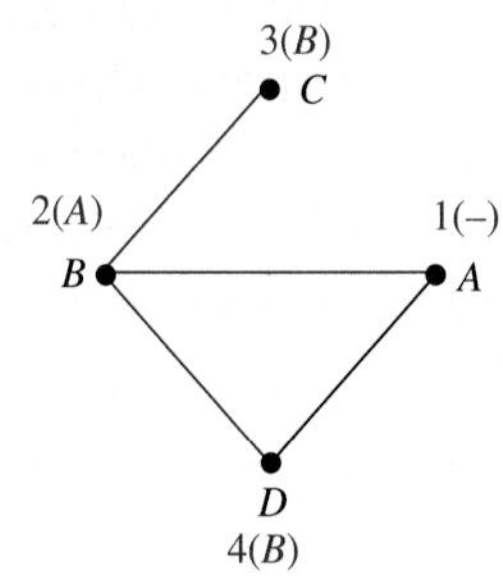

FIGURE 5.27

The following example demonstrates this technique on a more complicated graph:

Example 5.14 We will find a spanning tree for the graph in Figure 5.28 by using the depth-first search process. In this example, we will follow the convention that, when there is a choice of vertices, vertices will be chosen in alphabetical order. We begin by selecting a starting vertex, and using our convention, we choose A. Thus A is assigned the label 1(–), indicating that it is the first vertex labeled and that it has no predecessor. Now we select an unlabeled vertex adjacent to A. The possibilities are B and G, and, according to our convention, we choose B and assign it the label $2(A)$. (As illustrated in Figure 5.29, we display each vertex's label near the vertex, and show in color the edge joining a vertex to its predecessor.)

We now continue from B, selecting an adjacent unlabeled vertex from among F, J, and H. Here, we choose F and assign it the label $3(B)$. Continuing from

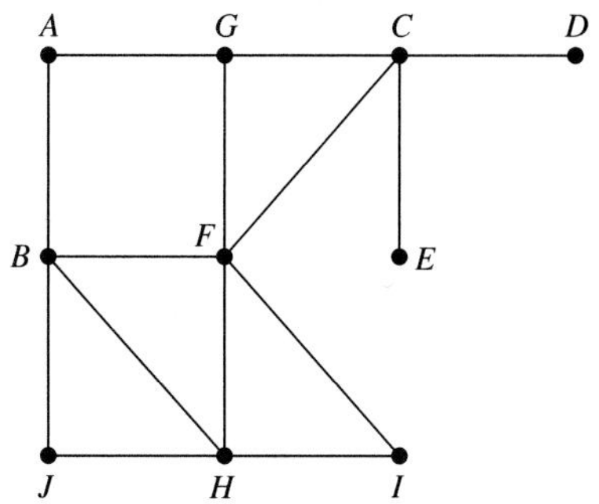

FIGURE 5.28

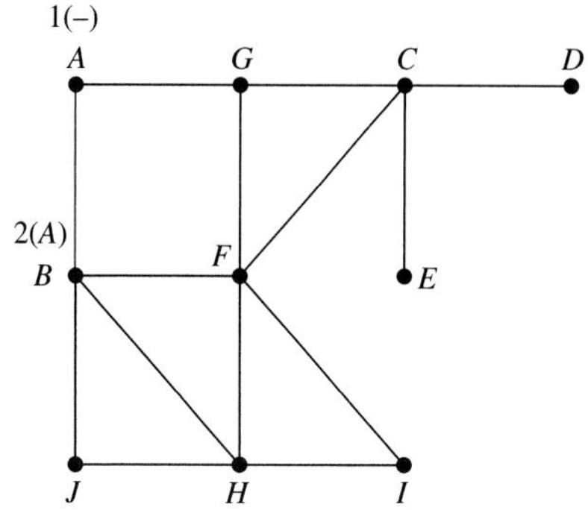

FIGURE 5.29

F, we select C and give it the label $4(F)$. Next, we select D and label it $5(C)$. The present situation is shown in Figure 5.30.

At this point, there are no unlabeled vertices adjacent to D, the last labeled vertex. Thus we must back up from D to its predecessor C. Since there are unlabeled vertices adjacent to C, we select one, namely, E, and label it $6(C)$. The current situation is shown in Figure 5.31.

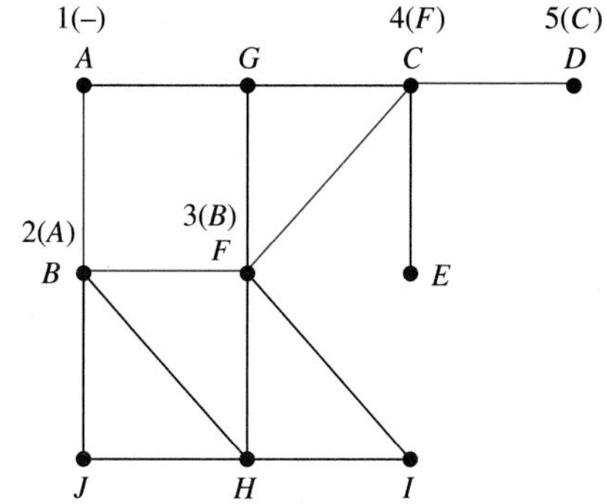

FIGURE 5.30

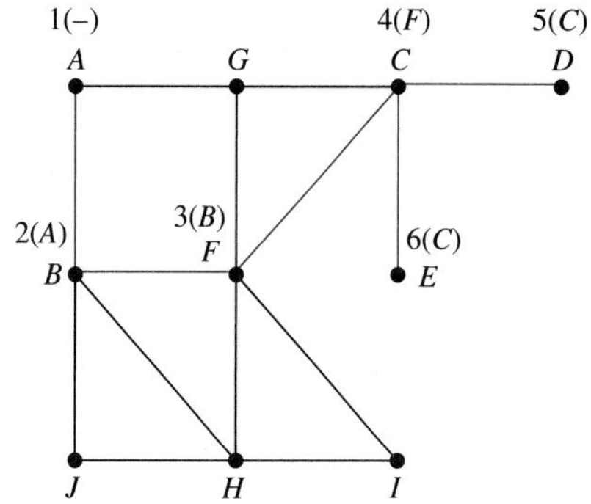

FIGURE 5.31

Because there are no unlabeled vertices adjacent to E, we back up to the predecessor C of E and select G next, giving it the label $7(C)$. See Figure 5.32.

Once again, we must back up, because there are no unlabeled vertices adjacent to G. Thus we return to C, the predecessor of G. This time, however, there are no unlabeled vertices adjacent to C, and so we are forced to continue backing up to F, the predecessor of C. Since there are unlabeled vertices adjacent to F, we continue labeling from F. We next select H and label it $8(F)$. Continuing from H, we choose I and give it the label $9(H)$. The present situation is shown in Figure 5.33.

Because there are no unlabeled vertices adjacent to I, we back up to H. Now we select J and assign it the label $10(H)$. At this point, every vertex is labeled

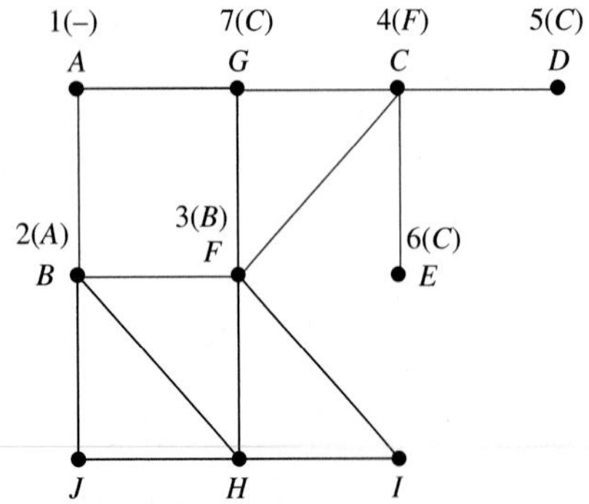

FIGURE 5.32

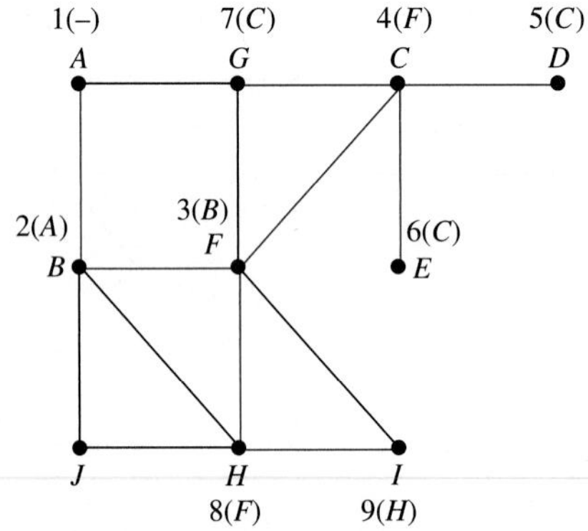

FIGURE 5.33

(see Figure 5.34), and so we stop. In Figure 5.34, the colored edges (and their incident vertices) form a spanning tree. ■

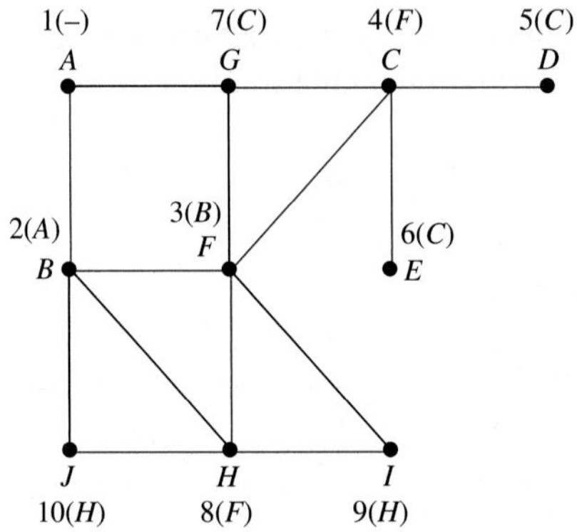

FIGURE 5.34

The procedure used in Example 5.14 is formalized next.

Depth-First Search Algorithm

This algorithm finds a spanning tree, if one exists, for a graph $\mathcal{G}$ with at least two vertices. In the algorithm, $\mathcal{L}$ is the set of vertices with labels, the *predecessor* of vertex Y is a vertex in $\mathcal{L}$ that is used in labeling Y, and $\mathcal{T}$ is the set of edges that join each vertex to its predecessor.

Step 1 (label the starting vertex)

(a) Select a vertex U, assign U the label 1, and let U have no predecessor.

(b) Set $\mathcal{L} = \{U\}$ and $\mathcal{T} = \emptyset$.

(c) Set $k = 2$ and $X = U$.

Step 2 (label other vertices)
repeat
Step 2.1 (label a vertex adjacent to X)
while there is a vertex Y not in $\mathcal{L}$ that is adjacent to X
(a) Place the edge $\{X, Y\}$ in $\mathcal{T}$.
(b) Assign X to be the predecessor of Y.
(c) Assign Y the label k.
(d) Include Y in $\mathcal{L}$.
(e) Replace k with $k + 1$.
(f) Now let X denote the vertex Y.
endwhile
Step 2.2 (back up)
Replace X with the predecessor of X.
until $X =$ null or every vertex of $\mathcal{G}$ is in $\mathcal{L}$

Step 3 (is there a spanning tree?)
if every vertex of $\mathcal{G}$ is in $\mathcal{L}$
The edges in $\mathcal{T}$ and their incident vertices form a spanning tree of $\mathcal{G}$.
otherwise
There is no spanning tree for $\mathcal{G}$ because $\mathcal{G}$ is not connected.
endif

There is a fundamental difference between breadth-first search and depth-first search. With breadth-first search, we fan out from each vertex to all the adjacent vertices, and this process is repeated at each vertex. Furthermore, at no time do we back up in order to continue the search. But, with depth-first search, we go out from a vertex as far as we can, and when unable to continue, we back up to the most recent vertex from which there was a choice; then we resume going out as far as we can.

An analogous situation can be found in two different ways to explore a cave with many tunnels. With the breadth-first search approach, a posse searches the cave, and whenever a tunnel branches off into several others, subgroups are formed to explore each of these simultaneously. With the depth-first search approach, one person explores the cave by leaving a phosphorous trail to mark where she has been. When there is a choice of tunnels, she chooses an unexplored one at random to be explored next. Upon reaching a dead end, she backtracks, using the marked trail to find the next unvisited tunnel.

Theorem 5.7 Let the depth-first search algorithm be applied to a graph $\mathcal{G}$.

(a) The edges in $\mathcal{T}$ and the vertices in $\mathcal{L}$ form a tree.

(b) Furthermore, if $\mathcal{G}$ is connected, this tree is a spanning tree.

Proof. (a) By the construction process of depth-first search, the edges of $\mathcal{T}$ and the vertices in $\mathcal{L}$ form a connected graph. In step 2, each time an edge is selected to be placed in $\mathcal{T}$, one vertex is in $\mathcal{L}$ and the other is not in $\mathcal{L}$. Thus this selection does not create any cycles using the other edges in $\mathcal{T}$. Consequently, at the end of the depth-first search algorithm, the graph formed by the edges in $\mathcal{T}$ and the vertices in $\mathcal{L}$ contains no cycles and is, therefore, a tree.

The proof of part (b) is left as an exercise. ■

We will follow our convention and refer to the tree in Theorem 5.7 formed by the edges in $\mathcal{T}$ and the vertices in $\mathcal{L}$ as simply $\mathcal{T}$. The tree $\mathcal{T}$ is called a **depth-first search tree**. The edges in $\mathcal{T}$ are called **tree edges**, and the other edges are called **back edges**. The labeling of the vertices is called a **depth-first search numbering**. Thus, in Figure 5.34, vertex F has depth-first search number 3, the edge on vertices F and C is a tree edge, and the edge on vertices F and G is a back edge. Of course, the designation of edges as tree and back edges, as well as the depth-first search numbering, depends upon the choices made during implementation of the algorithm.

In order to analyze the complexity of the depth-first search algorithm, we will regard labeling a vertex and using an edge as the elementary operations. For a graph with n vertices and e edges, each vertex is labeled at most once, and each edge is used at most twice, once in going from a labeled vertex to an unlabeled vertex and once in backing up to a previously labeled vertex. Hence, there will be at most

$$n + 2e \leq n + 2C(n, 2) = n + 2 \cdot \frac{1}{2}n(n - 1)$$

operations, and thus this algorithm is of order at most n^2.

Depth-first search can be used in many ways to solve problems involving graphs. Some of these ways will be presented in what follows.

Following Example 4.37 in Section 4.5, we investigated the problem of how to assign directions to the edges of a graph to create a strongly connected directed graph (a directed graph having a directed path between any two vertices). We stated that the absence of an edge (called a **bridge**) whose removal disconnects the graph is necessary and sufficient to guarantee that there is a way to assign directions to edges so as to produce a strongly connected directed graph. However, no procedure was given to determine whether a graph has a bridge. We will now describe how depth-first search can be used for this purpose.

We first apply depth-first search to a connected graph $\mathcal{G}$ to obtain a spanning tree $\mathcal{T}$. Observe that a bridge in $\mathcal{G}$ must be one of the edges in $\mathcal{T}$. Successively delete each edge in $\mathcal{T}$ from $\mathcal{G}$, and apply depth-first search to see if the resulting graph is connected. If not, the deleted edge must be a bridge.

Next, we will show how to assign directions to the edges of a connected graph $\mathcal{G}$ with no bridge so that $\mathcal{G}$ becomes a strongly connected directed graph. We begin by applying a depth-first search to $\mathcal{G}$ and assigning directions to the tree edges by going from the lower depth-first search number to the higher. Then we assign directions to back edges by going from the larger depth-first search number to the smaller. With these assignments, a strongly connected directed graph is formed.

Example 5.15 The depth-first search algorithm has been applied to a graph to yield the depth-first search numbering in Figure 5.35(a), where the tree edges are colored blue and the back edges black. Now we assign directions to all the edges, as previously described, producing the directed graph shown in Figure 5.35(b). Examination shows that it is strongly connected since there is a directed path from every vertex to every other vertex. ■

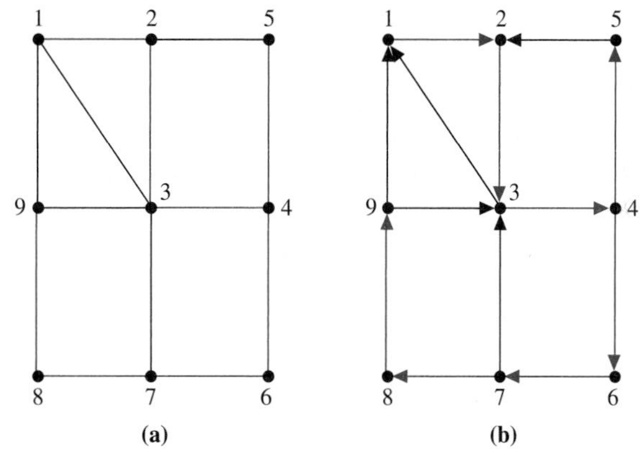

FIGURE 5.35

This example illustrates the next theorem.

Theorem 5.8 Suppose depth-first search is applied to a connected graph without a bridge. If directions are assigned to tree edges by going from the lower depth-first search number to the higher and to back edges by going from the higher number to the lower, then the resulting directed graph is strongly connected.

Backtracking

The process of going as far as possible before backing up is called *backtracking* when used as a general problem-solving strategy. It is often used as a systematic way to explore a large set of possibilities when looking for a solution or for all the solutions with particular characteristics. We will illustrate the idea of backtracking with two examples.

Although backtracking may not seem to be directly related to the depth-first search algorithm as applied to graphs, the set of possible solutions for a problem can often be represented as a graph. Depth-first search can then be used as a systematic way to search for solutions. This idea is illustrated in the next example.

Example 5.16 In the four-queens problem, we are asked to place four queens on a 4×4 chessboard so that no two queens can attack one another. This means that we must place four tokens on a 4×4 grid so that no two tokens are in the same row, column, or diagonal. We shall show how backtracking can be used to find a solution to this problem.

First we observe how the placement of the queens can be described by a graph. Each vertex will represent a placement of $n (n \geq 0)$ nonattacking queens placed in consecutive columns from left to right. An edge will connect two vertices when the two configurations differ by the placement of one additional queen. To help identify locations on the chessboard, we will think of it as a 4×4 matrix. We begin by placing a queen in the 1, 1 position. Then, in column 2, the only acceptable positions for a queen are positions 3, 2 or 4, 2 because position 1, 2 would result in two queens in the same row and position 2, 2 would result in two queens in the same diagonal. With the choice of the queen in position 3, 2, no further placements can be made, while the placement in position 4, 2 allows for the placement of another queen in position 2, 3. The rest of the graph can be completed in the same way, giving the graph in Figure 5.36. Now depth-first search can be used to search this graph for a solution to the problem—that is, a placement containing four nonattacking queens.

In practice, however, it is often desirable to search for solutions to a problem without first constructing a graph like that in Figure 5.36. We will illustrate this technique by reworking the four-queens problem just described. Our overall general search strategy will be to place queens on the chessboard in columns from left to right and in rows from top to bottom. If we are unable to place a queen in a column, we backtrack to the previous column and change the location of the queen there. If that does not work, we go back one more column. As before, to identify locations on the chessboard, we will think of it as a 4×4 matrix.

We begin by placing a queen in the 1, 1 position. (See Figure 5.37(a)). Then, going from top to bottom in column 2, we place a queen in position 3, 2 (Figure 5.37(b)) because position 1, 2 would result in two queens in the same row and position 2, 2 would result in two queens in the same diagonal. However, we now see that there is no spot in which to place a queen in column 3. So we backtrack to the queen in column 2 and move the queen to position 4, 2 (Figure 5.37(c)). Then in column 3 we are able to place a queen in position 2, 3 (Figure 5.37(d)). However, we are now unable to find a position in column 4 in which to place a queen. Since the location chosen in column 3 for the queen is the only one possible, it is necessary to backtrack to column 2. But again there is no place (other than position 3, 2, which was previously considered) in which

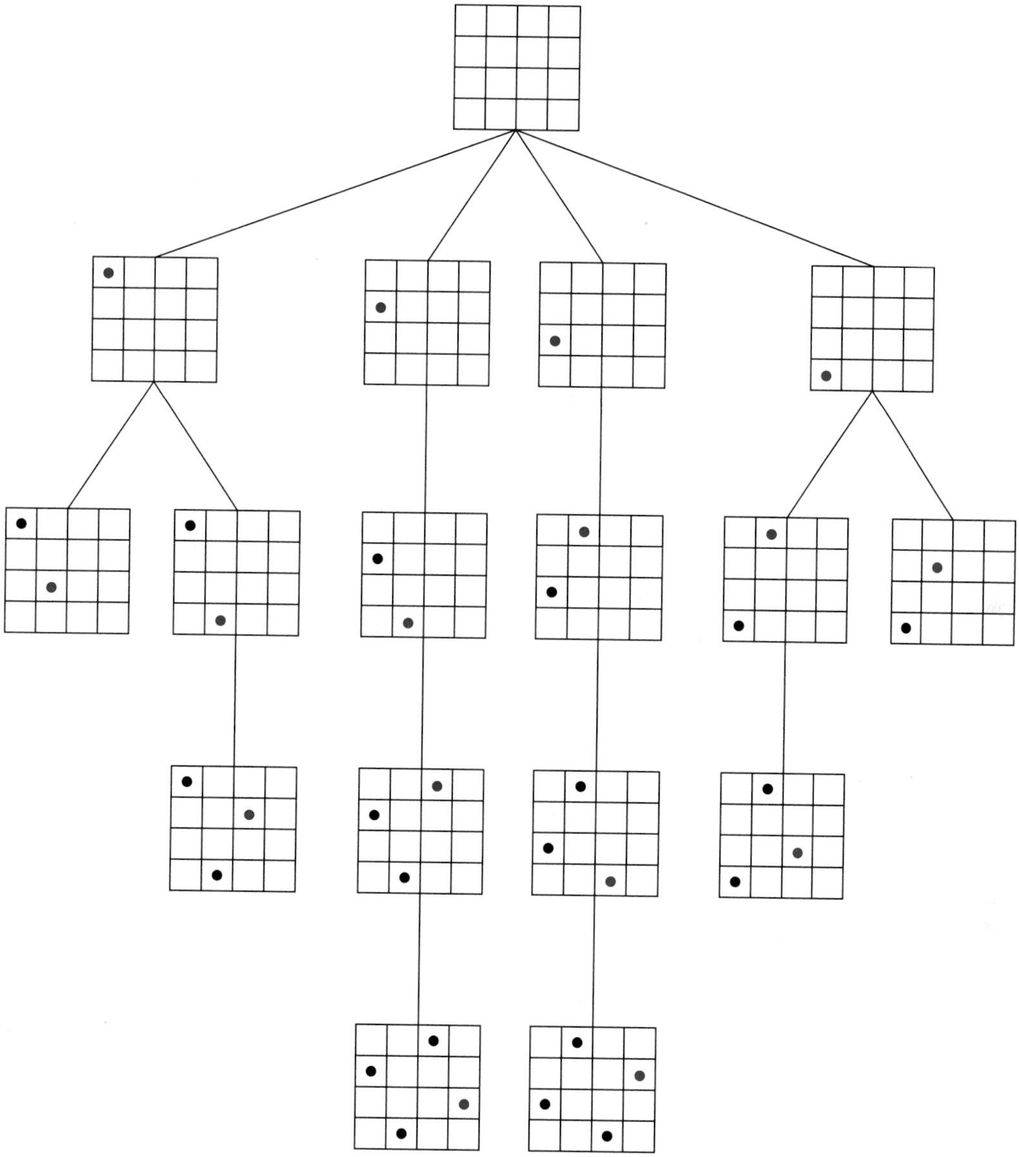

FIGURE 5.36

to move the queen in column 2. So we must backtrack all the way to column 1 and move the queen in column 1 to position 2, 1 (Figure 5.37(e)). Then the only location in column 2 for a queen is position 4, 2; the only location for a queen in column 3 is position 1, 3; and for column 4, the only location is position 3, 4 (Figure 5.37(f)). Thus a solution to the four-queens problem has been found. ■

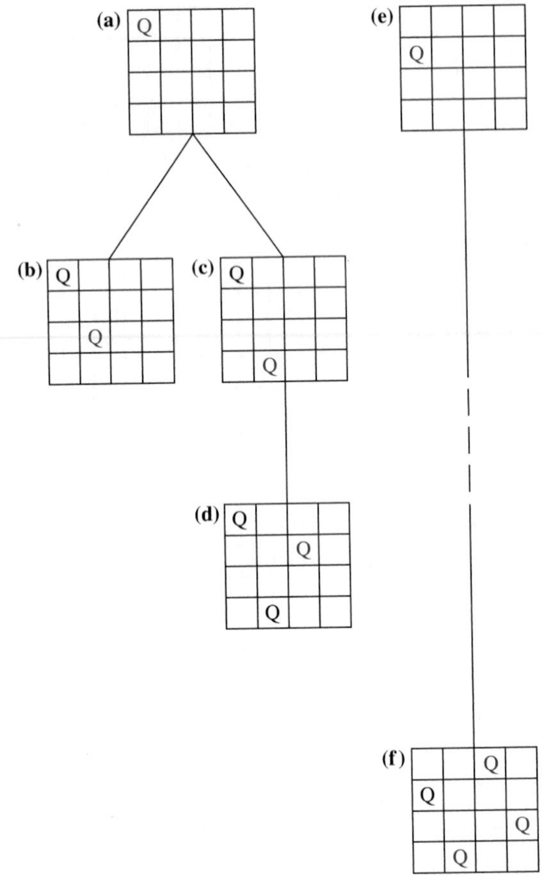

FIGURE 5.37

Example 5.17 Another use of backtracking is in providing a systematic way to go through a maze such as in Figure 5.38, where the colored lines indicate barriers. Since at each location we have four possible choices to make, our general strategy will be to go east, north, west, or south in that order without returning to a previously

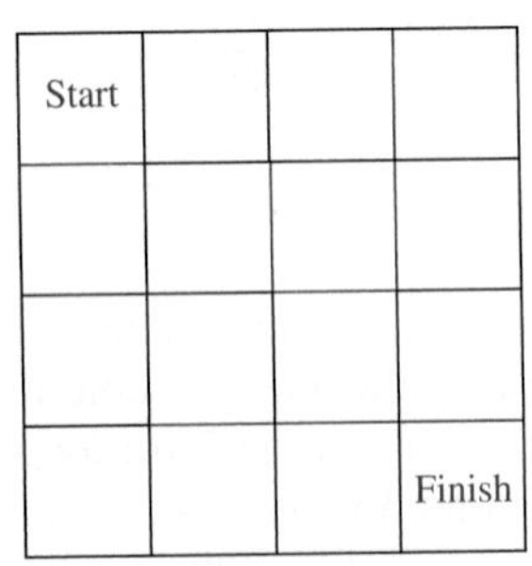

FIGURE 5.38

occupied location. When we can go no further, we backtrack to the last location where there was a choice and follow our priority list of directions. Again, we will use matrix notation to describe locations in the maze. From the starting position 1, 1, we go east, east, and then south, ending up in location 2, 3. We are unable to proceed any farther from this location, so we back up to location 1, 2. Then we go south, west, south, east, east, east, north, and north, ending in location 1, 4. Again unable to go any farther, we backtrack to location 3, 1. Then we go south, east, east, and east, ending up at position 4, 4. ■

For other applications of depth-first search, the interested reader should consult suggested reading [9] at the end of the chapter.

EXERCISES 5.3

Throughout these exercises, if there is a choice of vertices, choose the vertex that appears first in alphabetical order.

In Exercises 1–6, apply depth-first search to each graph to obtain a depth-first search numbering of the vertices.

1.

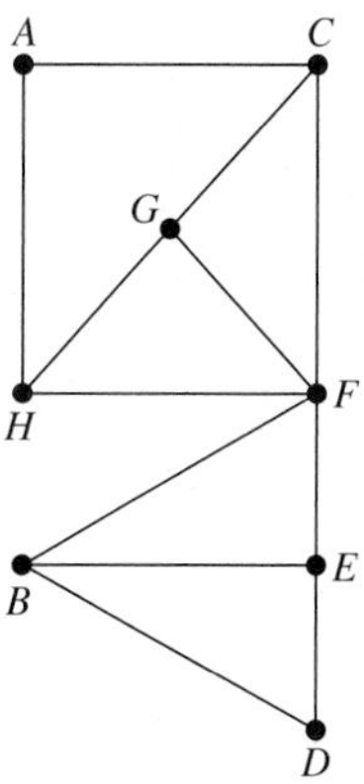

2.

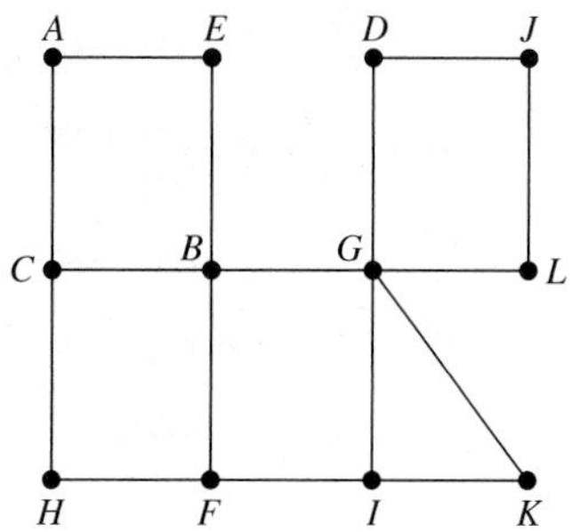

3.

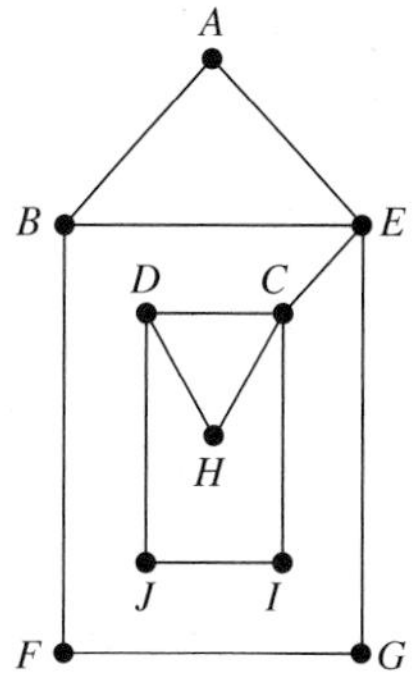

4.

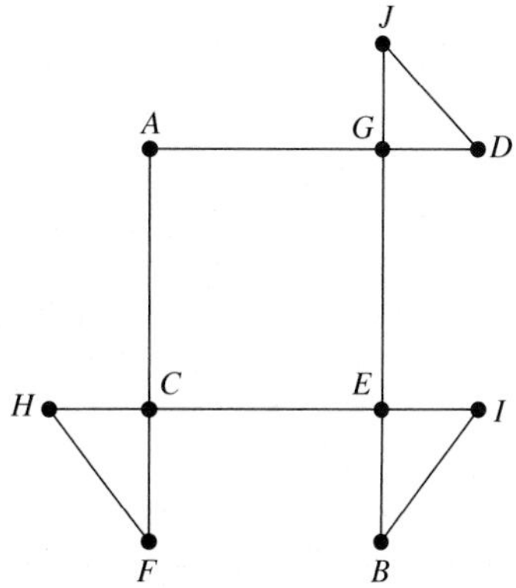

5.

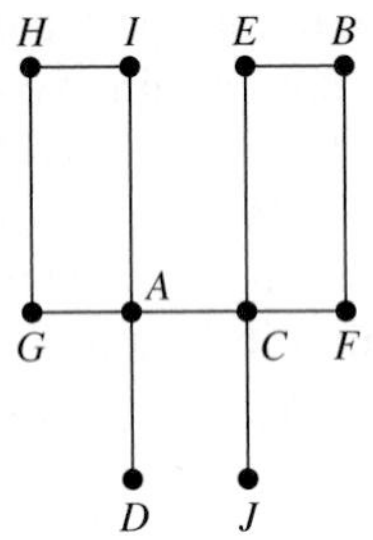

6.

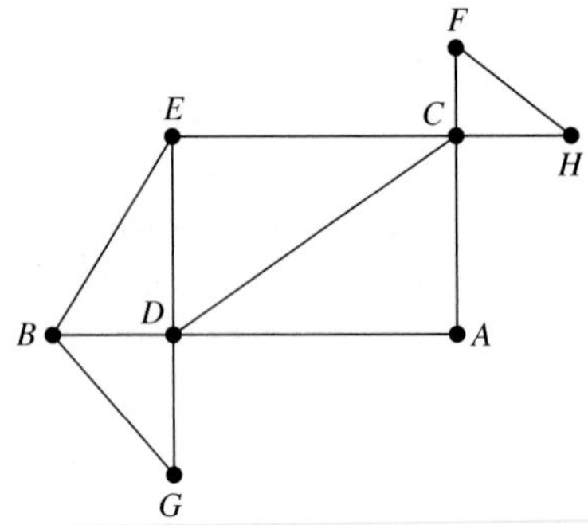

In Exercises 7–12, use the depth-first search numbering obtained for the graph in the indicated exercises to form the spanning tree described in Theorem 5.7.

7. Exercise 1 **8.** Exercise 2 **9.** Exercise 3
10. Exercise 4 **11.** Exercise 5 **12.** Exercise 6

In Exercises 13–18, use the depth-first search numbering obtained in the indicated exercises to list the back edges in the graphs.

13. Exercise 1 **14.** Exercise 2 **15.** Exercise 3
16. Exercise 4 **17.** Exercise 5 **18.** Exercise 6

In Exercises 19–22, determine whether there are any bridges, using the discussion preceding Example 5.15.

19.

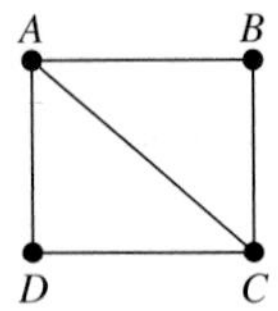

20.

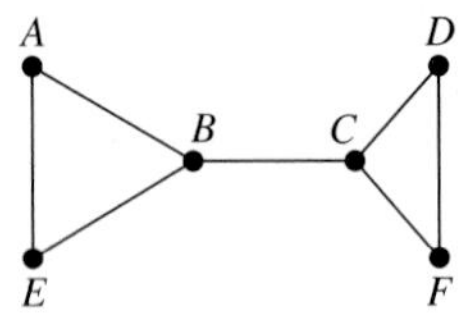

21.

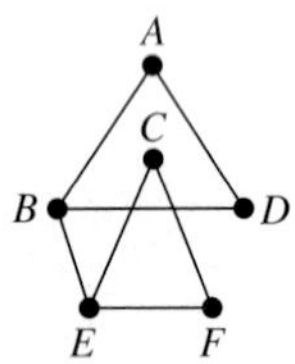

22.

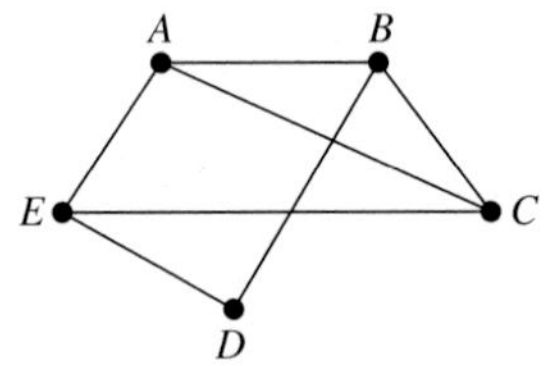

In Exercises 23–26, use Theorem 5.8 and the depth-first search numbering obtained in the indicated exercise to assign directions to the edges that will make each graph into a strongly connected directed graph.

23. Exercise 1 **24.** Exercise 2 **25.** Exercise 4 **26.** Exercise 6

27. The city manager of a community with a large university believes that something needs to be done to handle the heavy influx of automobile traffic on those days when students are checking into the dormitories. She instructs the chief of police to transform the current two-way street system into a system of one-way streets to handle the extra traffic, with the provision that students can still get from any place to any other. The campus area may be diagrammed as follows:

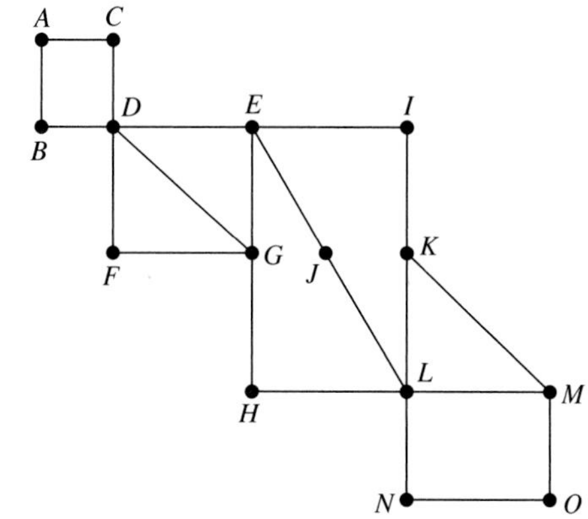

Is there a way for the chief of police to carry out these instructions? If so, how?

28. Can the spanning trees for a connected graph formed by breadth-first search and depth-first search be the same?

29. Prove part (b) of Theorem 5.7.

30. Show that every strongly connected directed graph with more than one vertex has at least one additional orientation of its edges under which it is strongly connected.

31. Label the vertices of $\mathcal{K}_3$ as 1, 2, 3. Apply depth-first search to $\mathcal{K}_3$, starting at 1. How many different depth-first search numberings are there?

32. Label the vertices of $\mathcal{K}_4$ as 1, 2, 3, 4. Apply depth-first search to $\mathcal{K}_4$, starting at 1. How many different depth-first search numberings are there?

33. Label the vertices of $\mathcal{K}_n$ as 1, 2, ..., n. Apply depth-first search to $\mathcal{K}_n$, starting at 1. How many different depth-first search numberings are there?

34. Suppose breadth-first search and depth-first search are applied to a connected graph starting at the same vertex. If b is the label assigned to a vertex by breadth-first search and d is the label assigned to the same vertex by depth-first search, what is the relationship between b and d? Why?

35. Let depth-first search be applied to a connected graph $\mathcal{G}$. Prove that every cycle of $\mathcal{G}$ contains a back edge, and every back edge is contained in a cycle of $\mathcal{G}$.

A vertex A is called an ***articulation point*** *of a connected graph $\mathcal{G}$ when the deletion of A and the edges incident on A creates a graph that is not connected. For example, the vertex A is an articulation point of the following graph:*

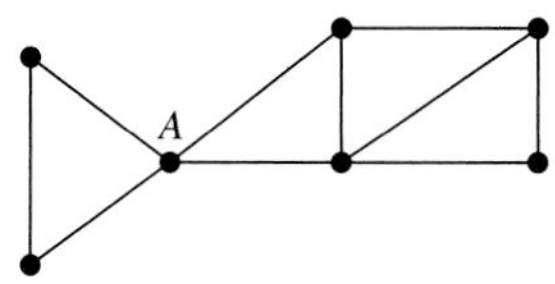

36. Prove that A is an articulation point of a connected graph $\mathcal{G}$ if and only if there exist vertices U and V such that U, V, and A are distinct and every path between U and V contains the vertex A.

37. Use backtracking to show that there is no solution to the two-queens problem.

38. Use backtracking to show that there is no solution to the three-queens problem.

39. Use backtracking to find a solution to the five-queens problem.

40. Use backtracking to show that it is not possible to fit seven dominoes (consisting of two unit squares) into a 4×4 chessboard that is missing opposite corners.

41. Use backtracking to construct a sequence of length 8 composed of the digits 1, 2, 3 with the property that nowhere in the sequence are there two adjacent subsequences that are identical.

5.4 ROOTED TREES

People have always been interested in learning about the descendants of historically important individuals. To assist in these investigations, a genealogical chart is often drawn. An example is given in Figure 5.39, where, for simplicity, only first names are used. It is understood that the downward lines represent the "is a parent of" relationship.

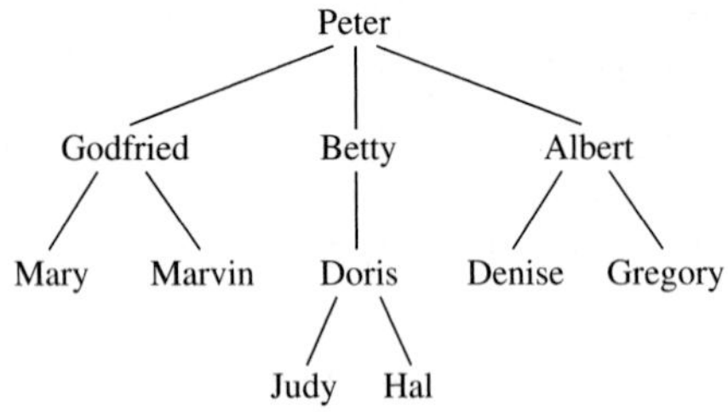

FIGURE 5.39

This chart can also be represented by a directed graph in which vertices represent individuals and directed edges begin at a parent and end at a child. Such a directed graph is shown in Figure 5.40.

Since all the arrows in Figure 5.40 point downward, it is not really necessary to draw the arrowheads on the edges, as long as the directions are understood to be downward. Figure 5.41 shows the corresponding directed tree without these arrowheads.

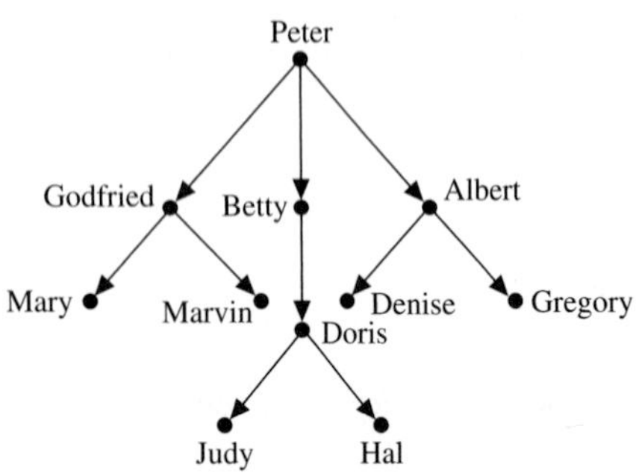

FIGURE 5.40

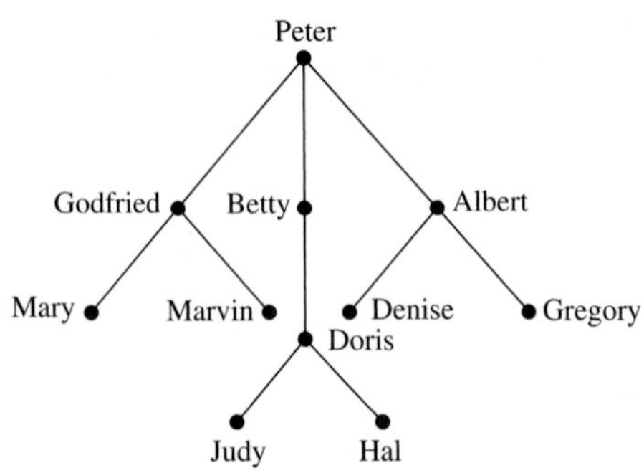

FIGURE 5.41

For the directed graph in Figure 5.40, there is one vertex with indegree 0, and all the other vertices have indegree 1. Furthermore, when the directions on the edges are ignored, we have a tree.

A **rooted tree** is a directed graph $\mathcal{T}$ satisfying two conditions: (1) When the directions of the edges in $\mathcal{T}$ are ignored, the resulting undirected graph is a tree; and (2) there is a unique vertex R such that the indegree of R is 0 and the indegree of any other vertex is 1. This vertex R is called the **root** of the rooted tree. The directed graph in Figure 5.40 is a rooted tree with Peter as its root. We will follow the customary practice of drawing rooted trees with the roots at the top and omitting arrowheads on the directed edges, with the understanding that edges are directed downward.

Example 5.18 The graph in Figure 5.42(a) is a rooted tree with root A, since (1) when the directions on the edges are ignored, the resulting graph is a tree; and (2) A has indegree 0, and all the other vertices have indegree 1. The usual way of drawing this tree is shown in Figure 5.42(b). ■

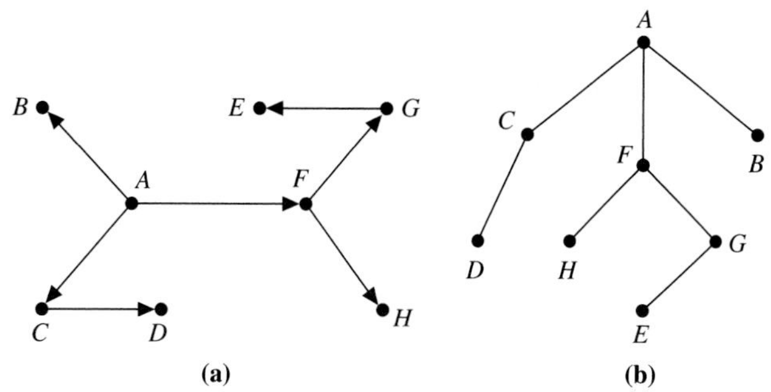

FIGURE 5.42

Rooted trees are often used to describe hierarchical structures. One such example occurs with the family tree of Peter. Another example is given next.

Example 5.19 A rooted tree can be used to describe the organization of a book by using "book" as the root and other vertices as subdivisions. In some books there are subsections of a section, and so another level of vertices could be added in this case. See Figure 5.43 for an illustration. ■

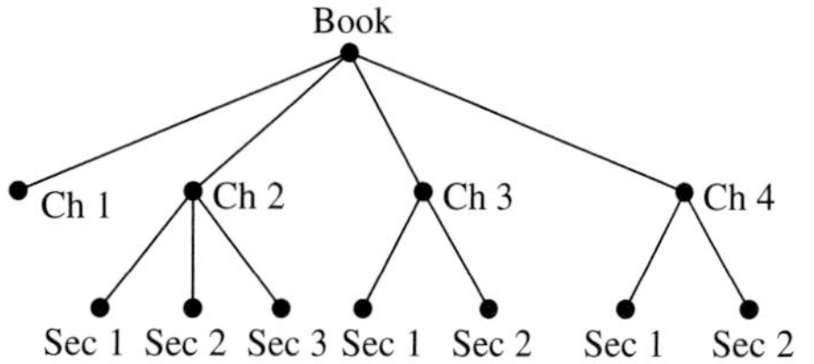

FIGURE 5.43

The next theorem gives some properties of rooted trees.

Theorem 5.9 In a rooted tree,

(a) The number of vertices is one more than the number of directed edges.

(b) There are no directed cycles.

(c) There is a unique simple directed path from the root to every other vertex.

Proof. Suppose $\mathcal{T}$ is a rooted tree with root R. The proofs of (a) and (b) follow immediately because $\mathcal{T}$ is a tree if the directions on the directed edges are ignored. Next we will show that there is a directed path (and, hence, there will be a simple directed path) from R to any other vertex $V \neq R$. Since the indegree of V is 1, there is a vertex $V_1 \neq V$ and a directed edge from V_1 to V. If $V_1 = R$, we are finished. If not, since the indegree of V_1 is 1, there is a vertex $V_2 \neq V_1$ and a directed edge from V_2 to V_1. Since there are no directed cycles, $V_2 \neq V$. If $V_2 = R$, then we are done. Otherwise, this process can be repeated, with each iteration generating a new vertex. Since the number of vertices is finite, we must eventually reach R. Thus we create a directed path from R to V. The uniqueness of a simple directed path from R to V follows immediately as in parts (a) and (b). ■

Family terms are used to describe the relationships among vertices in a rooted tree, just as they describe relationships in a genealogical chart. If, in a rooted tree, there is a directed edge from a vertex U to a vertex V, we say that U is a **parent** of V or V is a **child** of U. For a vertex V, the vertices other than V on the directed simple path from the root to V are called the **ancestors** of V, or, equivalently, we say that V is a **descendent** of these vertices. A **terminal vertex** is a vertex that has no children, and an **internal vertex** is one that has children. For the rooted tree in Figure 5.42, E is a child of G, and A, F, and G are ancestors of E. Also, F has H, G, and E as its descendants. Vertices B, D, E, and H are terminal vertices, and the others are internal vertices. Note that, in any rooted tree, the root has no ancestors, and every other vertex is a descendant of the root. A terminal vertex is a vertex with outdegree 0, and an internal vertex has nonzero outdegree.

We will now consider two examples where a rooted tree is used to obtain a solution to a problem.

Example 5.20 In Chapter 2 the concept of a partition of a set was introduced. To list all the partitions of $\{1, 2, \ldots, n\}$ requires a systematic approach, so as not to miss any possibility. The rooted tree in Figure 5.44 shows one such approach for $n = 4$. Here, the terminal vertices are the partitions. Do you recognize the pattern? The children of $\{1, 3\}, \{2\}$ are $\{1, 3, 4\}, \{2\}$; $\{1, 3\}, \{2, 4\}$; and $\{1, 3\}, \{2\}, \{4\}$. ■

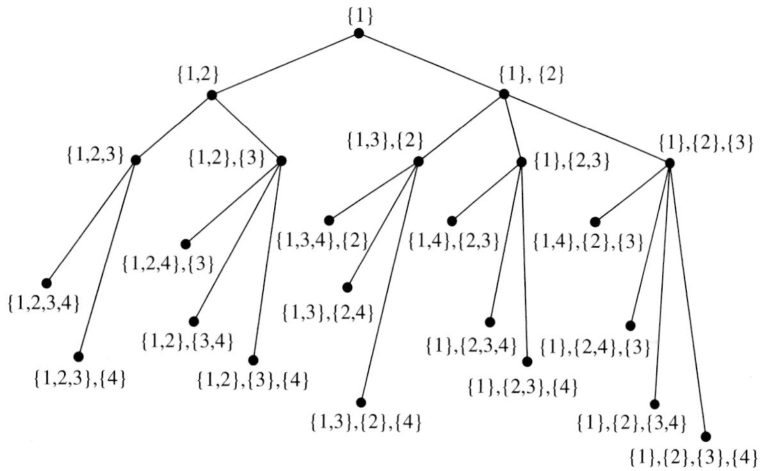

FIGURE 5.44

Example 5.21 Suppose we have seven identical coins and an eighth that looks the same, but is heavier. With the use of a balance scale, we want to identify the counterfeit coin in the smallest number of weighings. Let us label the coins $1, 2, \ldots, 8$. Note that when coins are placed on the two sides of the balance scale, either the left side will go down, the two sides will balance, or the right side will go down. We can construct a rooted tree as in Figure 5.45, giving a systematic approach for weighing the coins. The label beside each vertex indicates which coins are being weighed on each side of the balance scale. For example, $\{1, 2\}$–$\{3, 4\}$ means that coins 1 and 2 are weighed on the left side and coins 3 and 4 on the right. If the right side goes down, we proceed to the child on the right side for the next weighing, and similarly when the left side goes down. The terminal vertex indicates the heavy coin. For example, we begin by comparing the weight of coins 1, 2, 3, and 4 on the left with the weight of coins 5, 6, 7, and 8 on the right. If the balance tips to the left, we then compare coins 1 and 2 against coins 3 and 4. If in this weighing

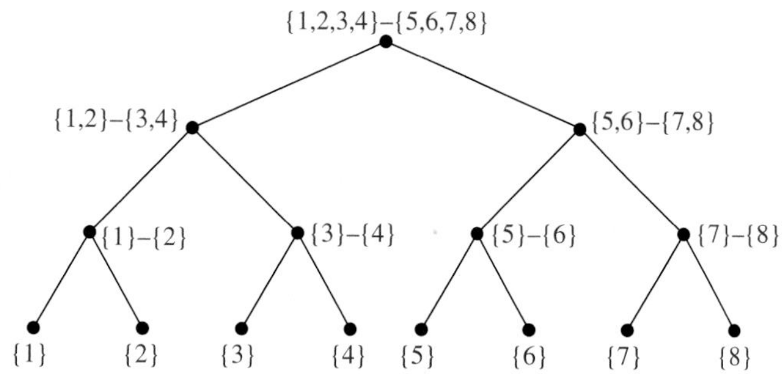

FIGURE 5.45

the right side goes down, we next compare coins 3 and 4. If this weighing shows the right side going down again, we reach the terminal vertex, indicating that 4 is the counterfeit coin. Since each terminal vertex is at the end of a simple directed path of length 3 from the root, we see that this scheme requires three weighings to find the counterfeit coin.

Could there be a different approach that will find the counterfeit coin with fewer weighings? Since a balance scale has three possible outcomes, we can build a rooted tree in which there are three children rather than just two, as we just did. Figure 5.46 gives one such possibility, where we proceed to the middle child when the two sides balance. Here, because each terminal vertex is at the end of a simple directed path of length two from the root, we can find the counterfeit coin with just two weighings. The trees in Figures 5.45 and 5.46 are called **decision trees** because of the way they structure a decision-making process. ■

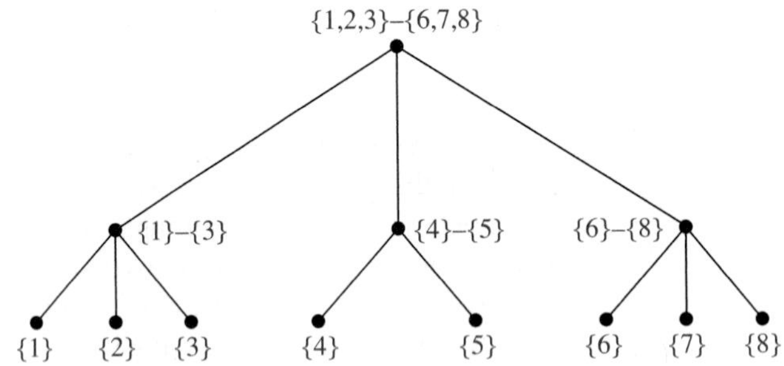

FIGURE 5.46

We will finish this section by describing the relationship between depth-first search and rooted trees. We will use the definitions from Section 5.3 and the notation of Theorem 5.7.

Theorem 5.10 If the depth-first search algorithm is applied to a graph, then the edges in $\mathcal{T}$, when oriented from the lower depth-first search number to the higher, form a rooted tree whose root is the vertex with depth-first search number 1.

Proof. Theorem 5.7 shows that $\mathcal{T}$ is a tree. Let R be the vertex with depth-first search number 1. Only during step 2.1 of depth-first search is a vertex assigned a depth-first search number and a tree edge going into it. This means that the root R has indegree 0, and each vertex in the tree other than R has indegree 1. ■

Example 5.22 In Figure 5.47(b), the vertices are labeled with a depth-first search numbering obtained by applying depth-first search to the graph in Figure 5.47(a). If we assign directions to the tree edges, as described in Theorem 5.10, and delete the back edges, we obtain the rooted tree in Figure 5.47(c). ■

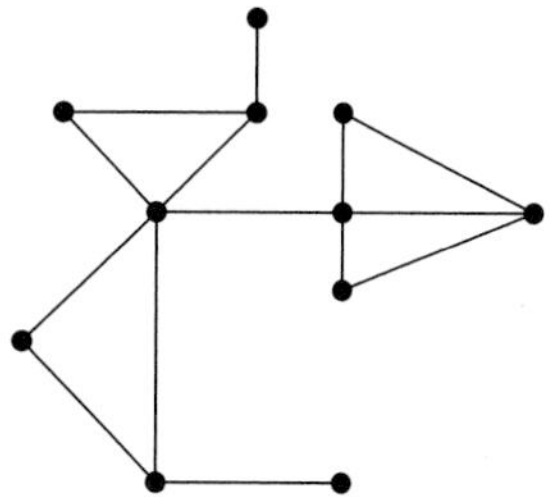

(a)

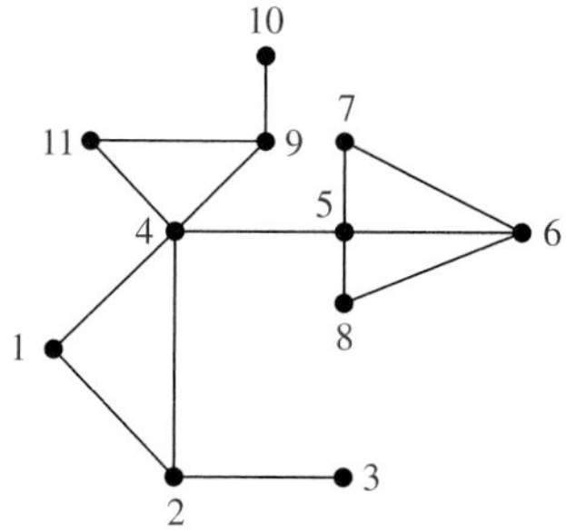

(b)

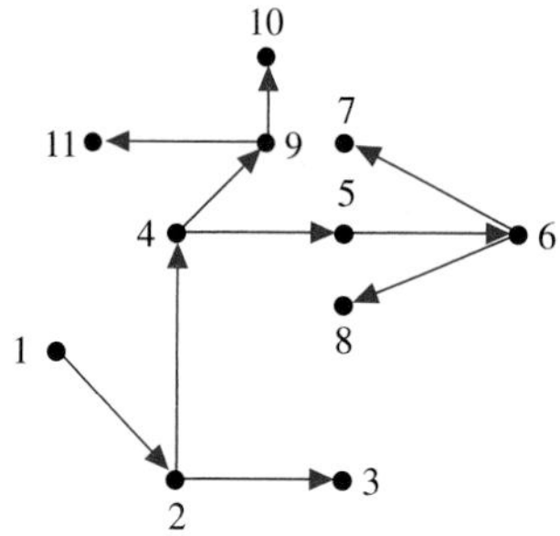

(c)

FIGURE 5.47

The ideas in Example 5.22 can also be used to find bridges in a graph. The interested reader should consult suggested reading [3] for more details.

EXERCISES 5.4

In Exercises 1–8, determine whether each directed graph is a rooted tree.

1. 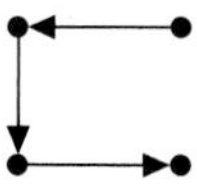

2. 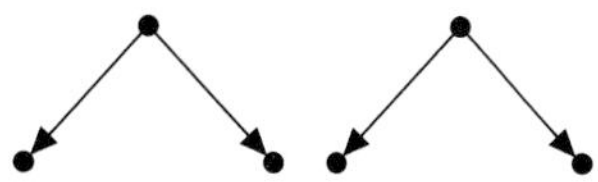

3. 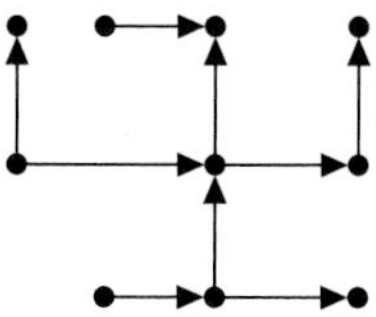

4. 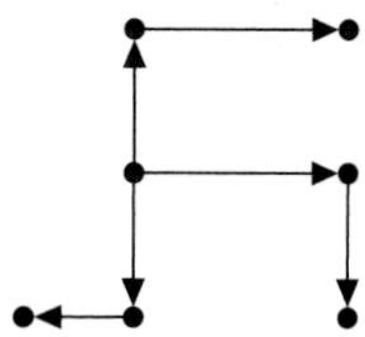

5. 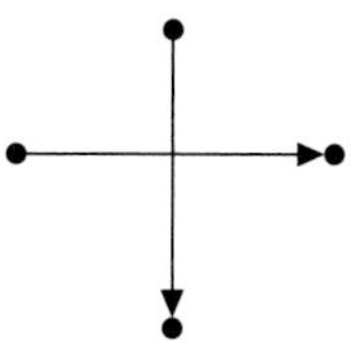

6. 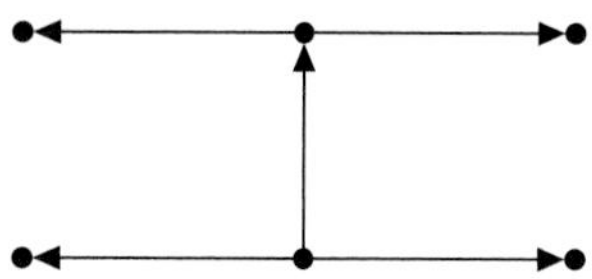

7. 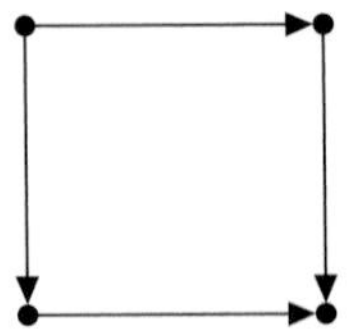

8. 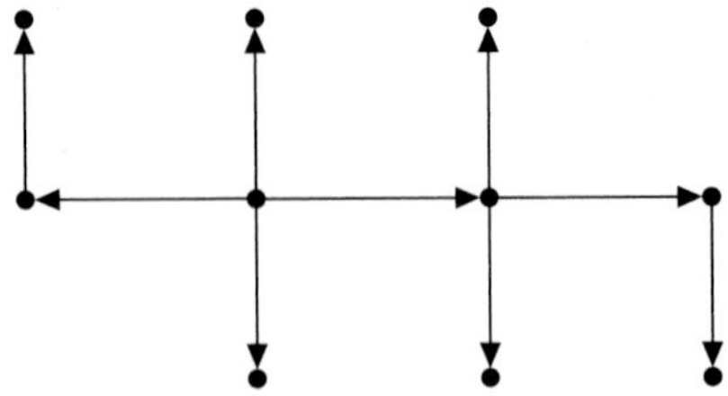

In Exercises 9–12, draw the rooted trees in the indicated exercises in the usual way, with the root at the top and without the arrowheads.

9. Exercise 1

10. Exercise 4

11. Exercise 6

12. Exercise 8

13. LISP is the primary programming language used in artificial intelligence. Over the past few years, one variation, COMMON LISP, has emerged as the principal version. There are eight basic data types and subtypes manipulated by COMMON LISP: S-expressions, atoms, lists, numbers, symbols, integers, ratios, and floating points. An S-expression can be an atom or a list, an atom can be a number or a symbol, and a number can be an integer, a ratio, or a floating point. Draw a rooted tree describing these relationships.

14. Draw a rooted tree for your mother and her descendants.

15. Tom and Sue are first cousins living in a state that allows first cousins to marry. If a child is born to this marriage, what effect would this have upon a genealogical chart in which the root is Tom and Sue's common grandfather?

16. It is known that a male bee has only a mother and that a female bee has both a mother and father. Draw a rooted tree that gives the ancestors of a male bee for four generations back, assuming that there was no mating between ancestors.

17. Write an algorithm describing how a tree with a vertex labeled R can be transformed into a rooted tree with root R. Illustrate your algorithm on the following tree:

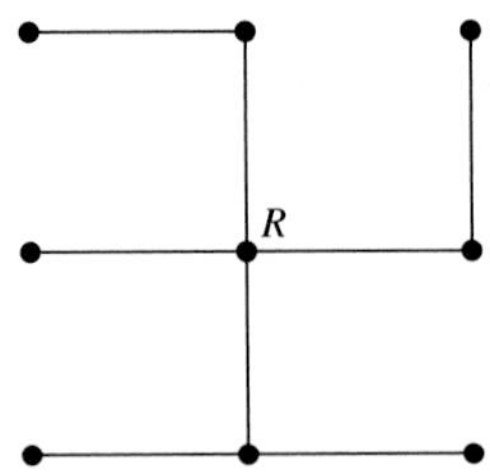

18. Repeat the second part of Exercise 17 for the following tree:

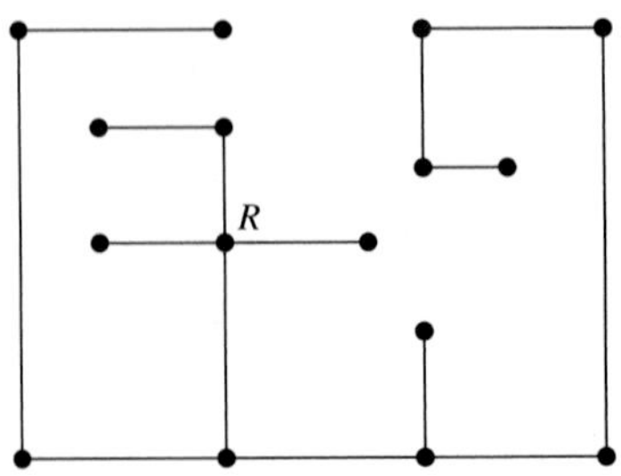

19. In how many ways can a tree with a vertex labeled R be transformed into a rooted tree with root R?

In Exercises 20–23, list the following for each rooted tree:

(a) *the root*

(b) *the internal vertices*

(c) *the terminal vertices*

(d) *the parent of G*

(e) *the children of B*

(f) *the descendents of D*

(g) *the ancestors of H*

20.

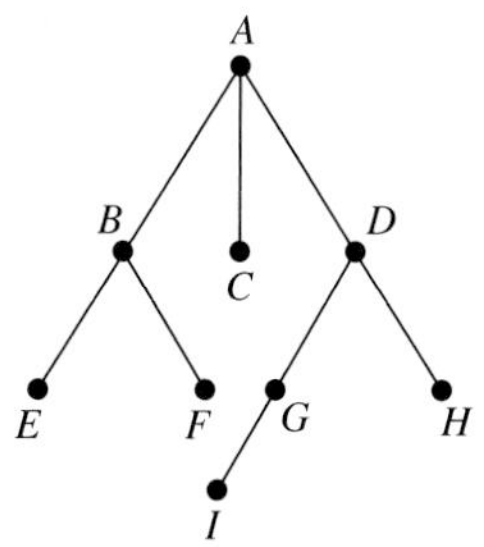

21.

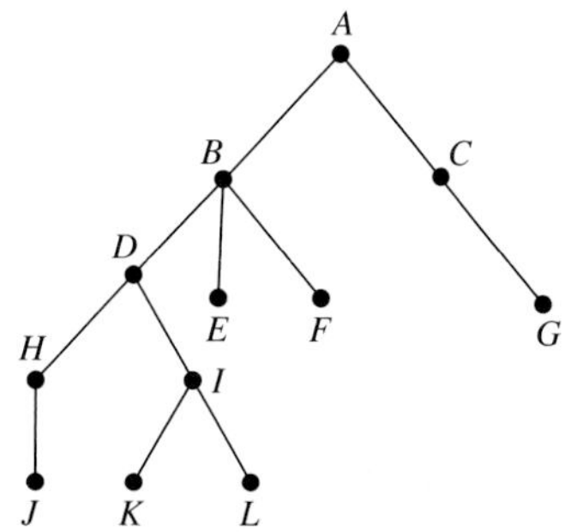

22.

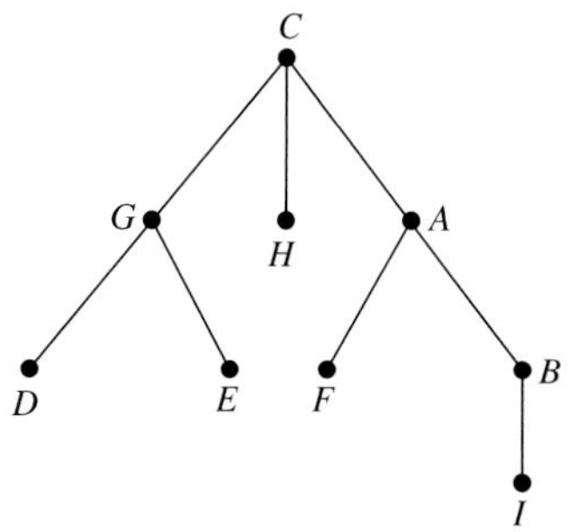

23.

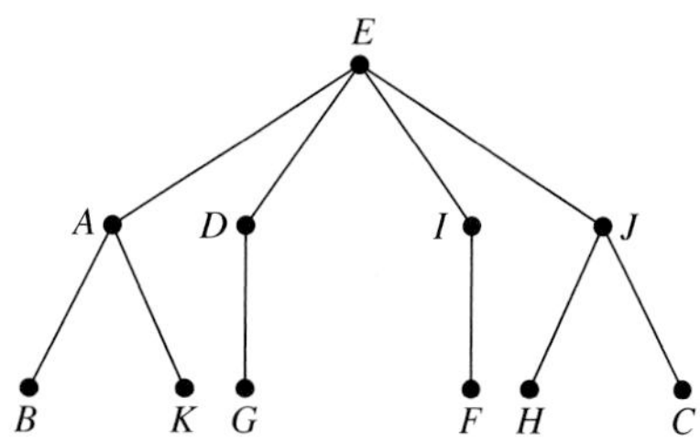

24. Draw a rooted tree with 7 vertices and having as many terminal vertices as possible.

25. Draw a rooted tree with 7 vertices and having as many internal vertices as possible.

26. Use Figure 5.44 to determine the number of partitions of $\{1, 2, 3, 4, 5\}$.

27. Draw a rooted tree describing all the possible outcomes for a two-game match between two chess players. (Remember that a chess game can end in a win, draw, or loss.)

28. Draw a rooted tree showing how to sort letters having a three-digit zip code in which the digits are 1 and 2.

29. Suppose we have three identical coins and a fourth that looks the same, but is lighter. Construct a decision tree that will find the counterfeit coin by using no more than two weighings on a balance scale.

30. Suppose we have 11 identical coins and a 12th that looks the same, but is lighter. Construct a decision tree that will find the counterfeit coin by using no more than three weighings on a balance scale.

31. Suppose we have three identical coins and a fourth that looks the same, but is heavier or lighter. Construct a decision tree that will find the counterfeit coin and determine whether it is heavier or lighter by using no more than three weighings on a balance scale.

32. Suppose we have seven identical coins and an eighth that looks the same, but is counterfeit (either heavier or lighter). Construct a decision tree that will find the counterfeit coin and determine whether it is heavier or lighter by using no more than three weighings on a balance scale.

33. In a rooted tree, the **level** of a vertex is defined to be the length of the simple directed path from the root to that vertex. What is the level of the root?

34. How many rooted trees are there with 2 vertices? with 3 vertices? with 4 vertices?

In Exercises 35–38, determine the level of each indicated vertex. (See Exercise 33 for the definition of "level.")

35. vertex F in the rooted tree of Exercise 20

36. vertex L in the rooted tree of Exercise 21

37. vertex H in the rooted tree of Exercise 22

38. vertex F in the rooted tree of Exercise 23

39. For the tree obtained by applying depth-first search to a connected graph, prove that the descendents (relative to the depth-first search tree) of any vertex V have larger depth-first search numbers than V.

5.5 BINARY TREES AND TRAVERSALS

Expression Trees

In previous examples and applications of rooted trees, it was not necessary to distinguish between the children of a parent. For instance, there was no need to designate a child as the first child or the second child. However, there are many situations where it is necessary to make such a distinction. For example, in an arithmetic expression such as $A - B$, the order of A and B is important. Thus, if we represent $A - B$ by a rooted tree in which the root represents the operation $(-)$ and the children represent the operands (A and B), then the order of the children is important.

A **binary tree** is a rooted tree in which each vertex has at most two children and each child is designated as being a **left child** or a **right child**. Thus, in a binary tree, each vertex may have 0, 1, or 2 children. When drawing a binary tree, we will follow customary practice and draw a left child to the left and below its parent and a right child to the right and below its parent. The **left subtree** of a vertex V in a binary tree is the graph formed by the left child L of V, the descendents of L, and the edges connecting these vertices. The **right subtree** of V is defined in an analogous manner.

Example 5.23 For the binary tree in Figure 5.48(a), A is the root. Vertex A has two children, a left child B and a right child C. Vertex B has one child, a left child D. Similarly,

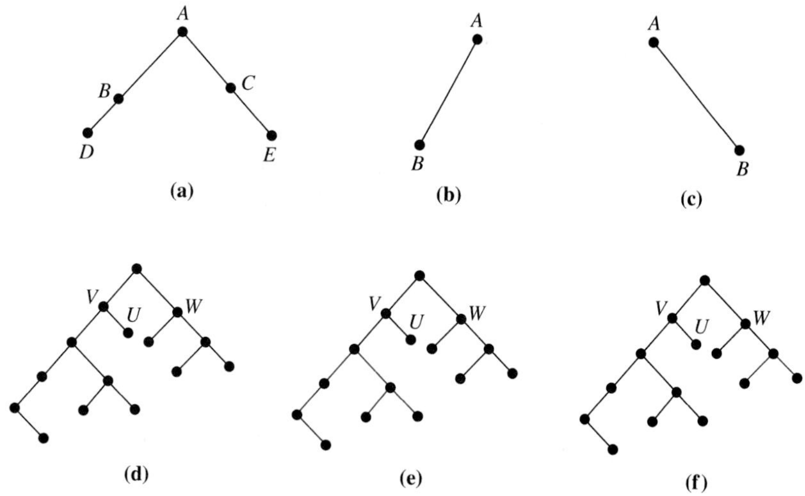

FIGURE 5.48

C has a right child E, but no left child. The binary tree in Figure 5.48(b), in which A has a left child B, is different from the one in Figure 5.48(c), in which B is a right child. For the binary tree in Figure 5.48(d), the left subtree of V is shown in color in Figure 5.48(e), and the right subtree of W is shown in color in Figure 5.48(f). The right subtree of V consists of the vertex U alone. ■

Binary trees are used extensively in computer science to organize data and describe algorithms. For example, during the execution of a computer program, it may be necessary to evaluate arithmetic expressions such as

$$(2 - 3 \cdot 4) + \left(4 + \frac{8}{2}\right).$$

Our knowledge of the conventions for the order of operations tells us how to proceed with this calculation: Scan from left to right, first doing multiplication and division and then addition and subtraction, with the understanding that parentheses have priority. However, when an expression needs to be evaluated frequently, this method cannot be used efficiently by a computer. An alternative approach is to represent an arithmetic expression by a binary tree and then process the data in some other way.

We will represent an arithmetic expression as a binary tree with the operations as internal vertices and the operands as terminal vertices. In this representation we let the root denote the final operation done in the expression, and we place the left operand as its left child and the right operand as its right child. If necessary, this process is repeated on these operands. The binary tree created by this process is called an **expression tree**.

Example 5.24 The expression $a * b$ (where $*$ denotes multiplication) is represented by the binary tree in Figure 5.49. Note that the operation $*$ is represented by an internal vertex, and the operands a and b are represented by terminal vertices. ■

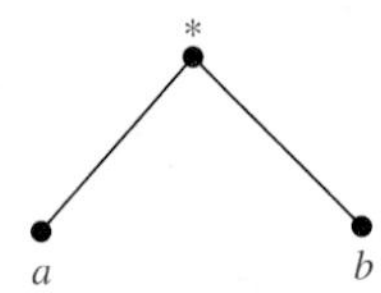

FIGURE 5.49

Example 5.25 The expression $a + b * c$ means $a + (b * c)$. The last operation to be performed is addition. Thus we first represent this expression by the binary tree in Figure 5.50(a). Repeating the process with the operand $b * c$ yields the expression tree in Figure 5.50(b). ■

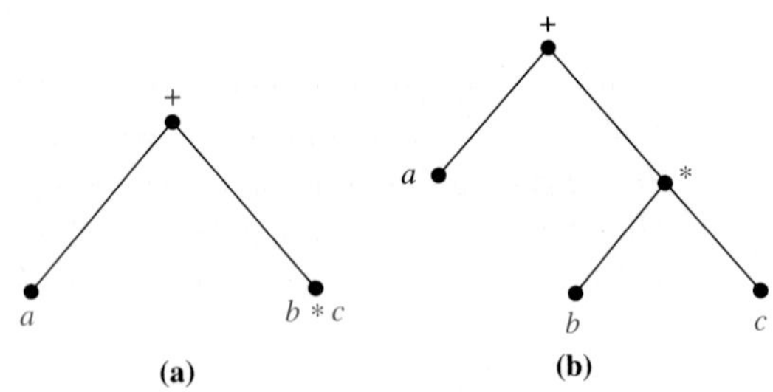

FIGURE 5.50

Example 5.26 The expression tree for $a + d * (b - c)$ is created by the sequence of binary trees in Figure 5.51. ■

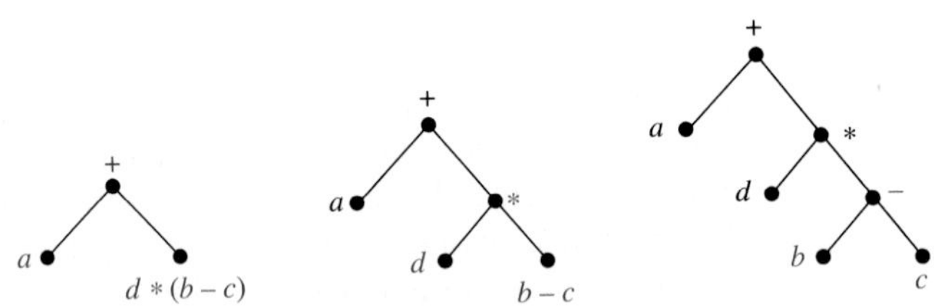

FIGURE 5.51

Example 5.27 The expression

$$(a + b * c) - \left(f - \frac{d}{e}\right)$$

is represented by the expression tree in Figure 5.52. ■

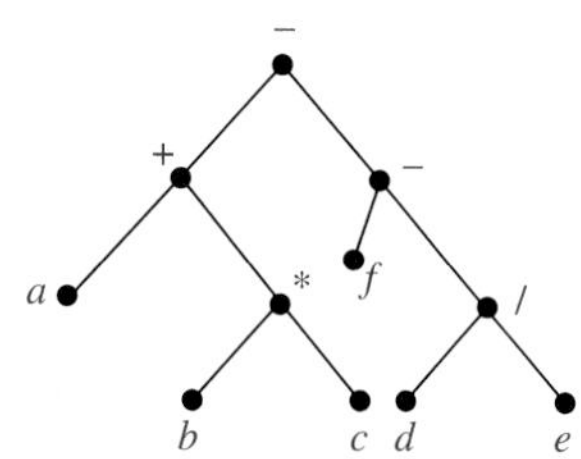

FIGURE 5.52

Preorder Traversal

We have seen that an arithmetic expression can be represented by an expression tree. Now we must process the expression tree in some way so as to obtain an evaluation of the original expression. We are looking for a systematic way to examine each vertex in the expression tree exactly once. Processing the data at a vertex is usually called **visiting a vertex**, and a search procedure that visits each vertex of a graph exactly once is called a **traversal** of the graph. For example, both breadth-first search and depth-first search are traversals of a connected graph because both are methods by which each vertex of the graph is visited (labeled) exactly once. Note that "visit" is used in a technical sense; merely considering a vertex in an algorithm does not necessarily constitute a visit.

We will consider a traversal of a binary tree characterized by visiting a parent before its children and a left child before a right child. (This holds for all the vertices in the binary tree.) Such a traversal is called a **preorder traversal**, and listing the vertices in the order they are visited is called a **preorder listing**.

Although it is possible to give a description of a preorder traversal by using the depth-first search algorithm,[1] we will state an algorithm for the preorder traversal that is consistent with the descriptions of the other traversals we will discuss. This is a **recursive** formulation of the preorder traversal, which means that in this description the algorithm refers to itself. This is analogous to the definition of $n!$ given in Section 2.5.

Preorder Traversal Algorithm

This algorithm gives a preorder listing of the vertices in a binary tree.

Step 1 (visit)
Visit the root.

Step 2 (go left)
Go to the left subtree, if one exists, and do a preorder traversal.

[1]Apply the depth-first search algorithm to a binary tree by starting at the root and always choosing a left child in preference to a right child. The order in which the vertices are labeled is the preorder listing.

Step 3 (go right)
Go to the right subtree, if one exists, and do a preorder traversal.

Example 5.28 For the binary tree in Figure 5.53(a), we start by visiting the root A. (We use the word "visit" to indicate when a vertex should be listed, and in the figures we show the order of visiting in parentheses near the vertex.) Then we go to the left subtree of A (see Figure 5.53(b)) and start the preorder traversal again. Now we visit the root B and go to the left subtree of B (see Figure 5.53(c)), where we start another preorder traversal. Next, we visit the root D. Since there is no left subtree of D, we go to the right subtree of D (which consists of just the vertex F) and again start a preorder traversal. Thus we visit the root F. Since there are no subtrees of F, we have completed the preorder traversal of the left subtree of B. Consequently, we next begin a preorder traversal of the right subtree of B (see Figure 5.53(d)). To do this, we visit the root E and then go to the left subtree of E (which consists of only the vertex G) to begin another preorder traversal. Thus we visit vertex G. Since G has no subtrees and E has no right subtree, both subtrees of B are traversed. This completes the traversal of the left subtree of A, and so we begin another preorder traversal on the right subtree of A. This consists only of visiting the root C and so completes the preorder traversal of the entire binary tree. The resulting preorder listing is A, B, D, F, E, G, C, with a labeling of the vertices as shown in Figure 5.53(e). ■

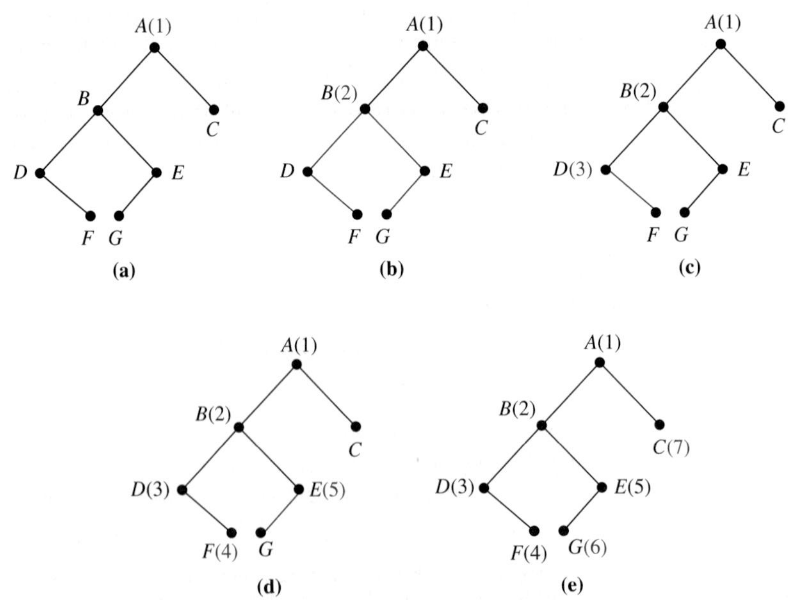

FIGURE 5.53

Example 5.29 Applying preorder traversal to the binary tree in Figure 5.54 yields the order of visiting shown in Figure 5.55. ■

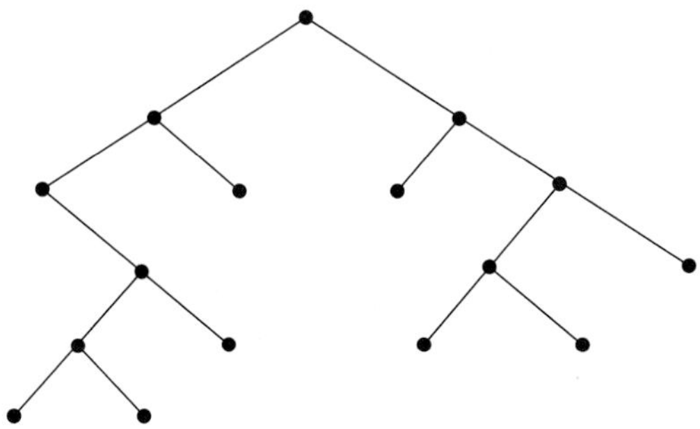

FIGURE 5.54

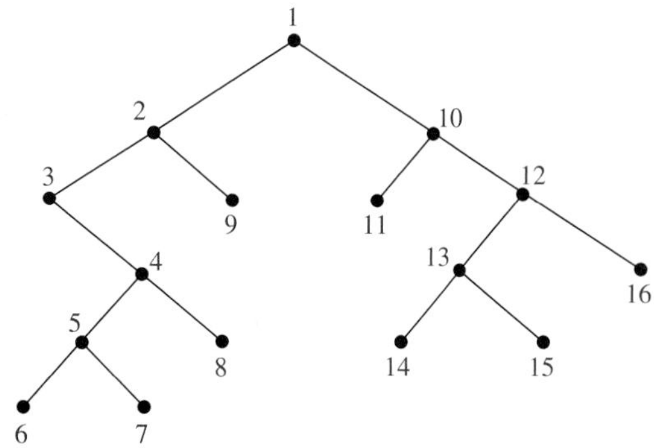

FIGURE 5.55

When a preorder traversal is performed on an expression tree, the resulting listing of operations and operands is called the **prefix form** or **Polish notation** for the expression. (The latter name is used in honor of the famous Polish logician Lukasiewicz.) For example, the four expressions in Examples 5.24, 5.25, 5.26, and 5.27 have as their Polish notations

$$* \, a \, b, \quad + a * b \, c, \quad + a * d - b \, c, \quad \text{and} \quad - + a * b \, c - f \, / \, d \, e,$$

respectively. An expression in Polish notation is evaluated according to the following rule: Scan from left to right until coming to an operation sign, say, T, that is followed by two successive numbers, say, a and b. Evaluate $T\ a\ b$ as $a\ T\ b$, and replace $T\ a\ b$ by this value in the expression. Repeat this process until the entire expression is evaluated. (Equivalently, an expression in Polish notation can be scanned from right to left until coming to two successive numbers followed immediately by an operation sign.)

Example 5.30 The expression $(2 - 3 * 4) + (4 + \frac{8}{2})$ is represented by the expression tree in Figure 5.56. The Polish notation for this expression (found by doing a preorder traversal on the expression tree) is $+ \ - \ 2 * 3\ 4 + 4 / 8\ 2$. The evaluation is performed as follows:

First, we evaluate $* \ 3\ 4$ and replace it by $3 * 4 = 12$. This substitution gives the new expression $+ \ - \ 2\ 12 + 4 / 8\ 2$.

Second, we evaluate $- \ 2\ 12$ and replace it by -10. This substitution yields the new expression $+ \ -10 + 4 / 8\ 2$, where we remember that the $-$ is part of -10 and is not a new operation.

Third, we evaluate $/ \ 8\ 2$ and replace these symbols with 4. Thus the current expression is $+ \ -10 + 4\ 4$.

Fourth, we evaluate $+ \ 4\ 4$ as 8. The expression now has the form $+ \ -10\ 8$.

Fifth, we evaluate $+ \ -10\ 8$ to obtain the final result, which is -2. ■

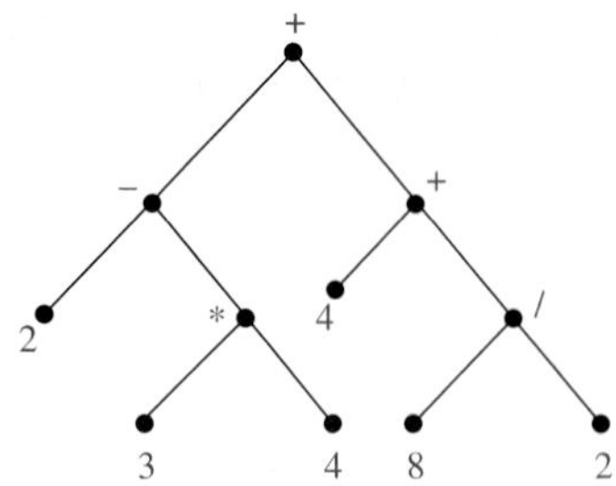

FIGURE 5.56

The Polish notation for an expression provides an unambiguous way to write it without the use of parentheses or conventions about the order of operations. Many computers are designed to rewrite expressions in this form.

Postorder Traversal

Readers who are familiar with hand calculators know that some require algebraic expressions to be entered in a form known as **reverse Polish notation** or **postfix form**, also introduced by Lukasiewicz. Unlike Polish notation, in which the operation sign precedes the operands, in reverse Polish notation the operation sign follows the operands. The reverse Polish notation for the expression in Example 5.30 is $2\ 3\ 4 * -\ 4\ 8\ 2\ /\ +\ +$. It is evaluated in a fashion similar to Polish notation, except that, as we scan from left to right, we look for two numbers immediately followed by an operation sign. (As with Polish notation, we could scan from right to left, looking for an operation sign followed immediately by two consecutive numbers.) The steps in evaluating the preceding expression are as follows:

$2\,3\,4\ *\ -\ 4\,8\,2\ /\ +\ +$	(First $3\ 4\ *$ is evaluated and replaced.)
$2\ 12\ -\ 4\,8\,2\ /\ +\ +$	(Next $2\ 12\ -$ is evaluated and replaced.)
$-10\ 4\,8\,2\ /\ +\ +$	(Then $8\ 2\ /$ is evaluated and replaced. Note that the first symbol is part of the number -10 and not an operation sign.)
$-10\ 4\,4\ +\ +$	(Next $4\ 4\ +$ is evaluated and replaced.)
$-10\ 8\ +$	(Finally we evaluate $-10\ 8\ +$ to obtain the final result.)
-2	

Again, we see that we can evaluate an expression without the need for parentheses and without worrying about the order of operations. Thus reverse Polish notation is an efficient method for use in hand calculators and computers. How can the reverse Polish notation for an expression be obtained from an expression tree?

By using a traversal called **postorder**, we can obtain the reverse Polish notation for an expression. The **postorder traversal** is characterized by visiting children before the parent and a left child before a right child. (This holds for all the vertices in the binary tree.) A systematic way to do this is described in the following recursive algorithm:

Postorder Traversal Algorithm

This algorithm gives a postorder listing of the vertices of a binary tree.

Step 1 (start)
Go to the root.

Step 2 (go left)
Go to the left subtree, if one exists, and do a postorder traversal.

Step 3 (go right)
Go to the right subtree, if one exists, and do a postorder traversal.

Step 4 (visit)
Visit the root.

Example 5.31 For the binary tree in Figure 5.57(a), we begin at the root A, go to the left subtree of A (see Figure 5.57(b)), and begin postorder traversal again. Thus we go to the left subtree of B (see Figure 5.57(c)) and start postorder traversal again. Since

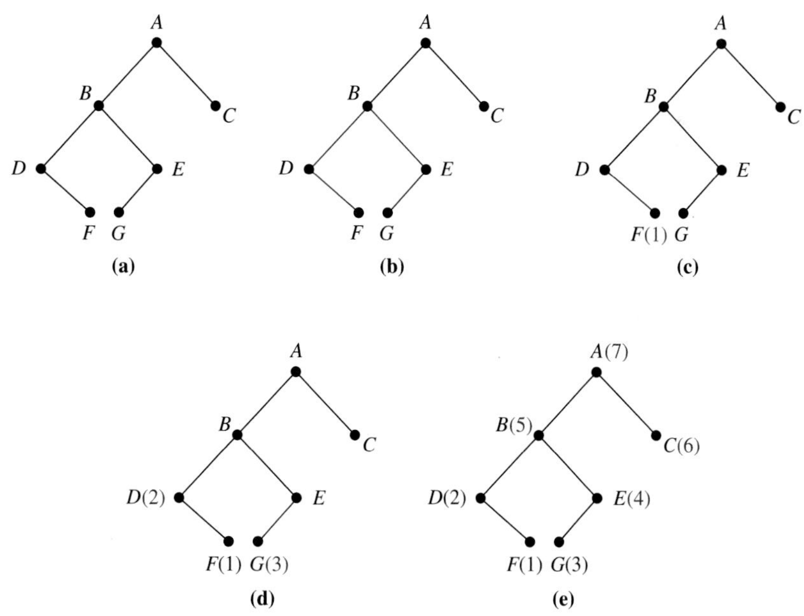

FIGURE 5.57

there is no left subtree of D, we go to the right subtree of D (which consists only of the vertex F) and begin postorder traversal again. Because F has no subtrees, we visit F. Since the right subtree of D has been traversed, we next visit D. (Again, we use the word "visit" to indicate when a vertex should be listed, and in the figures, we show the order of visiting in parentheses near the vertex.) Now the left subtree of B has been traversed, so we go to the right subtree of B (see Figure 5.57(d)) and start postorder traversal again. Next we go to the left subtree of E (which is just G) and start postorder traversal again. Since G has no subtrees, G is visited. Because the left subtree of E has been traversed and there is no right subtree of E, we visit E. Now both the left and right subtrees of B have been traversed; so we visit B. This completes the traversal of the left subtree of A, and so we go to the right subtree of A and start postorder traversal again. Because C has no subtrees, C is visited. Since both subtrees of A have been traversed, we visit A. This completes the postorder listing F, D, G, E, B, C, A, with a labeling of the vertices as shown in Figure 5.57(e). ■

Example 5.32 The postorder traversal applied to the binary tree in Figure 5.54 yields the order of visiting shown in Figure 5.58. ■

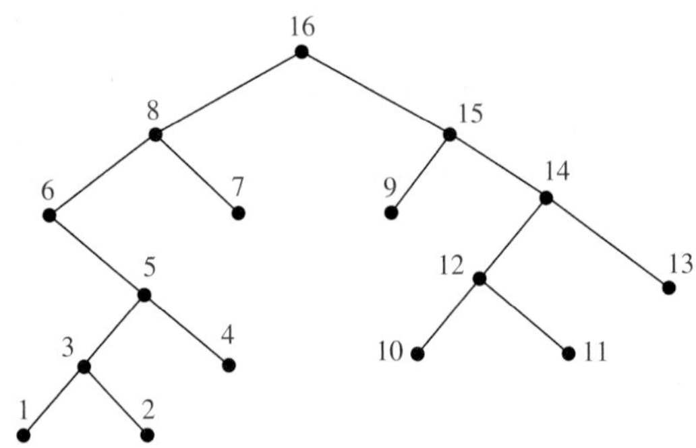

FIGURE 5.58

Inorder Traversal

We have seen how expression trees yield the Polish and reverse Polish notations for an expression. In these notations, the operation sign precedes or follows the operands, respectively. With the use of the inorder traversal, it is possible to obtain an expression with the operation sign between the operands. However, this traversal requires the careful insertion of parentheses in order to evaluate the expression properly.

The **inorder traversal** is characterized by visiting a left child before the parent and a right child after the parent. (This holds for all the vertices in the binary tree.) A systematic way to do this is described in the following recursive algorithm:

Inorder Traversal Algorithm

This algorithm gives an inorder listing of the vertices of a binary tree.

Step 1 (start)
Go to the root.

Step 2 (go left)
Go to the left subtree, if one exists, and do an inorder traversal.

Step 3 (visit)
Visit the root.

Step 4 (go right)
Go to the right subtree, if one exists, and do an inorder traversal.

Example 5.33 For the binary tree in Figure 5.59(a), we begin at the root A, go to the left subtree of A (see Figure 5.59(b)), and then start inorder traversal again. Next we go to the left subtree of B (see Figure 5.59(c)) and start inorder traversal again. Since there is no left subtree of D, we visit the root D. (Again, we use the word "visit" to indicate when a vertex should be listed, and in the figures, we show the order

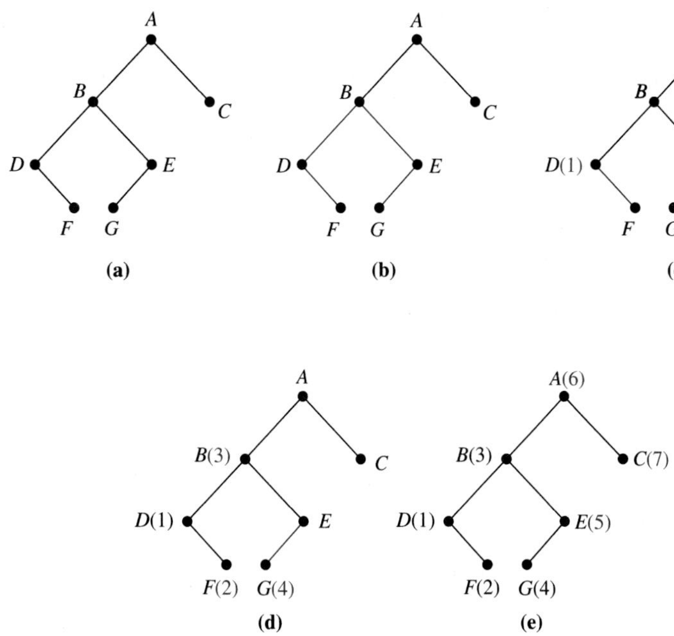

FIGURE 5.59

of visiting in parentheses near the vertex.) Then we go to the right subtree of D (which is just the vertex F) and start inorder traversal again. Since there is no left subtree of F, we visit the root F. Since F has no right subtree, we have traversed the left subtree of B. So we visit B, go to the right subtree of B (see Figure 5.59(d)), and do inorder traversal again. Thus we go to the left subtree of E (which is just the vertex G) and start inorder traversal again. Since G has no left subtree, we visit G. Because G has no right subtree, we have traversed the left subtree of E. Thus we visit E. Since E has no right subtree, we have traversed the right subtree of B. We have now traversed the left subtree of A. Hence we visit the root A and go to the right subtree of A (which consists only of the vertex C) and begin inorder traversal again. Since C has no left subtree, we visit C. This step completes the inorder traversal, giving the inorder listing D, F, B, G, E, A, C, with the labeling of vertices as shown in Figure 5.59(e). ■

Example 5.34 When inorder traversal is applied to the binary tree in Figure 5.54, the vertices are listed according to the numbering in Figure 5.60. ■

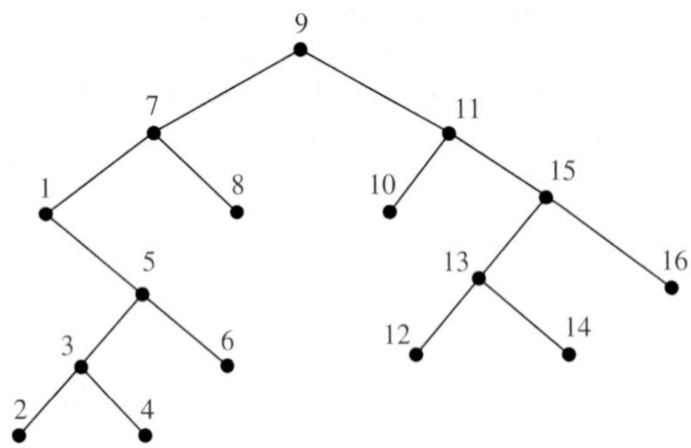

FIGURE 5.60

Example 5.35 Applying inorder traversal to the expression tree in Figure 5.56 yields the expression $2 - 3 * 4 + 4 + 8 / 2$. ■

Other uses of traversals can be found in suggested reading [7] at the end of the chapter.

EXERCISES 5.5

In Exercises 1–6, construct an expression tree for each expression.

1. $a * b + c$

2. $(4 + 2) * (6 - 8)$

3. $((a - b)/c) * (d + e/f)$

4. $(((6 - 3) * 2) + 7)/((5 - 1) * 4 + 8)$

5. $a * (b * (c * (d * e + f) - g) + h) + j$

6. $(((4 * 2)/3) - (6 - 7)) + (((8 - 9) * 8)/5)$

In Exercises 7–12, find the indicated subtrees.

7. the left subtree of vertex A

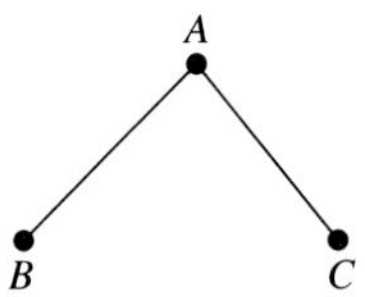

8. the right subtree of vertex A

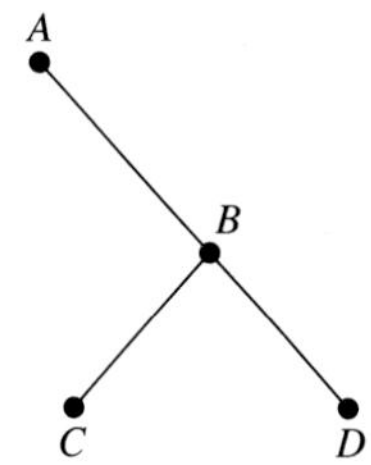

9. the left subtree of vertex C

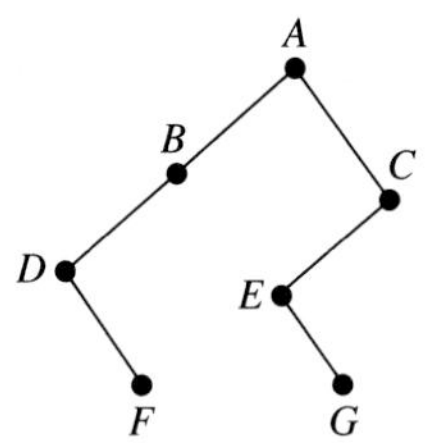

10. the right subtree of vertex E

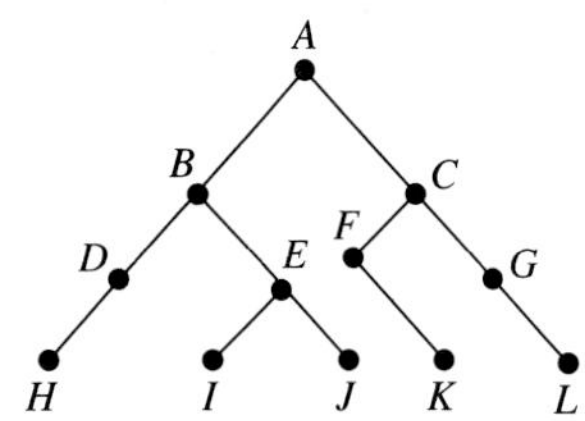

11. the left subtree of vertex E

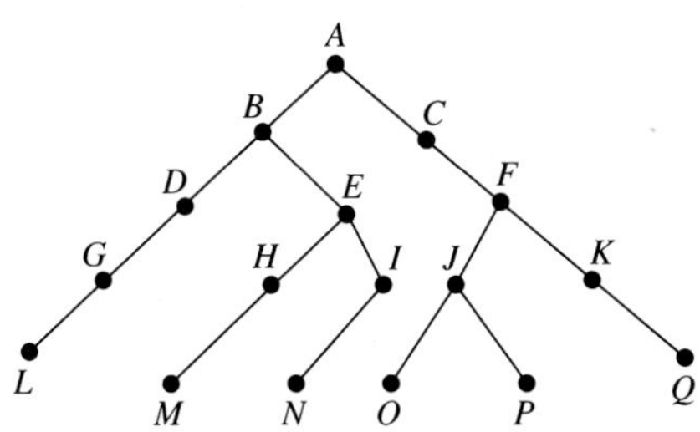

12. the right subtree of vertex D

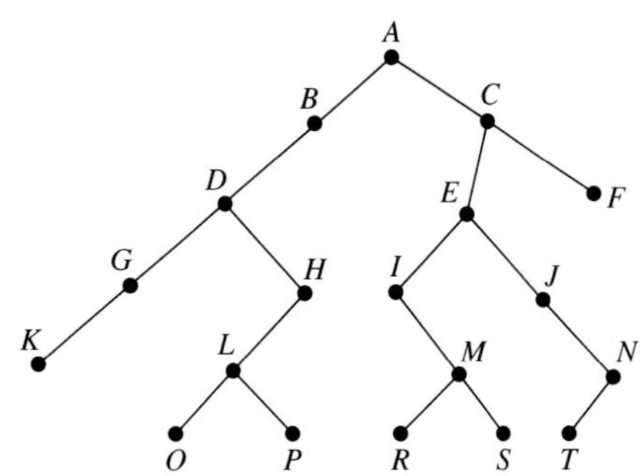

In Exercises 13–18, give the preorder listing of vertices for the binary trees in the indicated exercises.

13. Exercise 7
14. Exercise 8
15. Exercise 9
16. Exercise 10
17. Exercise 11
18. Exercise 12

In Exercises 19–24, give the postorder listing of vertices for the binary trees in the indicated exercises.

19. Exercise 7
20. Exercise 8
21. Exercise 9
22. Exercise 10
23. Exercise 11
24. Exercise 12

In Exercises 25–30, give the inorder listing of vertices for the binary trees in the indicated exercises.

25. Exercise 7
26. Exercise 8
27. Exercise 9
28. Exercise 10
29. Exercise 11
30. Exercise 12

In Exercises 31–36, find the Polish notation for the expressions in the indicated exercises.

31. Exercise 1
32. Exercise 2
33. Exercise 3
34. Exercise 4
35. Exercise 5
36. Exercise 6

In Exercises 37–42, find the reverse Polish notation for the expressions in the indicated exercises.

37. Exercise 1
38. Exercise 2
39. Exercise 3
40. Exercise 4
41. Exercise 5
42. Exercise 6

Evaluate the Polish notation expressions in Exercises 43–46.

43. $+ \; / \; 4 \; 2 + 5 \; 6$

44. $* + - + 4 \; 3 \; 6 \; 2 \; 8$

45. $+ * 4 \; / \; 6 \; 2 - + 4 \; 2 \; 5$

46. $+ * + 3 \; 4 - 1 \; 2 - 3 \; / \; 4 \; 2$

Evaluate the reverse Polish notation expressions in Exercises 47–50.

47. $4 \; 5 - 7 * 2 \; 3 + +$

48. $5 \; 6 \; 4 \; 2 \; 2 \; / + * -$

49. $2 \; 3 + 4 \; 6 - - 5 * 4 +$

50. $3 \; 4 + 1 \; 2 - * 4 \; 2 \; / \; 3 - +$

51. Construct an expression tree for the Polish notation expression

$$* + B \; D - A \; C.$$

52. Construct an expression tree for the Polish notation expression

$$* + B - D \; F + A \; C \; E.$$

53. Construct an expression tree for the reverse Polish notation expression

$$A \; C * B - D + .$$

54. Construct an expression tree for the reverse Polish notation expression

$$E \; D - A + B \; C - F * + .$$

55. Construct a binary tree for which the preorder listing of vertices is C, B, E, D, A and the inorder listing is B, E, C, A, D.

56. Construct a binary tree for which the preorder listing of vertices is E, C, A, D, B, F, G, H and the inorder listing is A, C, D, E, F, B, G, H.

57. Construct a binary tree for which the postorder listing of vertices is E, B, F, C, A, D and the inorder listing is E, B, D, F, A, C.

58. Construct a binary tree for which the postorder listing of vertices is D, H, F, B, G, C, A, E and the inorder listing is D, F, H, E, B, A, G, C.

59. Construct a binary tree with 7 vertices for which the preorder listing is the same as the inorder listing.

60. Construct a binary tree with 8 vertices for which the postorder listing is the same as the inorder listing.

61. Construct a binary tree for which the preorder listing is the same as the postorder listing.

62. Construct two distinct (nonisomorphic) binary trees that have 1, 2, 3 as their preorder listing of vertices.

63. Construct two distinct (nonisomorphic) binary trees that have 1, 2, 3 as their postorder listing of vertices.

64. Verify, for $n = 1, 2,$ and 3, that the number of binary trees with n vertices is

$$\frac{(2n)!}{n!\,(n+1)!}.$$

(Such numbers are called **Catalan numbers**.)

65. Prove that if vertex X is a descendant of vertex Y in a binary tree, then Y precedes X in the preorder listing of vertices and X precedes Y in a postorder listing.

66. Prove that if the preorder and the inorder listings of vertices of a binary tree are given, then it is possible to reconstruct the binary tree.

67. The **Fibonacci trees** are defined recursively as follows: Each of $\mathcal{T}_1$ and $\mathcal{T}_2$ is a single vertex, and for $n \geq 3$, $\mathcal{T}_n$ is a tree where the left subtree of the root is $\mathcal{T}_{n-1}$ and the right subtree is $\mathcal{T}_{n-2}$. Find and prove a formula for the number of vertices in $\mathcal{T}_n$.

5.6 OPTIMAL BINARY TREES AND BINARY SEARCH TREES

In this section we present two applications, both of which require the construction of a binary tree to solve a problem. These can be studied in either order.

Optimal Binary Trees

To represent symbols, computers use strings of 0s and 1s called **codewords**. For example, in the ASCII (American Standard Code for Information Interchange) code, the letter A is represented by the codeword 01000001, B by 01000010, and C by 01000011. In this system each symbol is represented by some string of eight bits, where a bit is either a 0 or a 1. To translate a long string of 0s and 1s into its ASCII symbols, we use the following procedure: Find the ASCII symbol represented by the first 8 bits, the ASCII symbol represented by the second 8 bits, etc. For example, 010000110100000101000010 is decoded as CAB.

For many purposes, this kind of representation works well. However, there are situations, as in large-volume storage, where this is not an efficient method. In a fixed length representation, such as ASCII, every symbol is represented by a codeword of the same length. A more efficient approach is to use codewords of variable lengths, where the symbols used most often have shorter codewords than the symbols used less frequently. For example, in normal English usage the letters E, T, O, and A are used much more frequently than the letters Q, J, X, and Z. Is there a way to assign the shortest codewords to the most frequently used symbols? If messages use only these eight letters, a natural assignment to try is

$$\begin{array}{llll} E: 0, & T: 1, & O: 01, & A: 11, \\ Q: 00, & J: 10, & X: 101, & Z: 011. \end{array}$$

Here, the shortest possible codewords are assigned to the most frequently used letters, and longer codewords are assigned to the other letters. This appears to be a more efficient approach than assigning all these letters a codeword of the same fixed length, which would have to be three or more. (Why?)

But how can we decode a string of 0s and 1s? For example, how should the string 0110110 be decoded? Should we start by looking at only the first digit, or the first two, or the first three? Depending upon the number of digits used, the first letter could be E, O, or Z. We see that, in order to use variable length codewords, we need to select representations that permit unambiguous decoding.

A way to do this is to construct codewords so that no codeword is the first part of any other codeword. Such a set of codewords is said to have the **prefix property**. This property is not enjoyed by the preceding choice of codewords, because the codeword for T is also the first part of the codeword for A. On the other hand, the set of codewords $S = \{000, 001, 01, 10, 11\}$ has the prefix property, since no codeword appears as the first part of another codeword. The method for decoding a string of 0s and 1s into codewords that have the prefix property is to read one digit at a time until this string of digits becomes a codeword, then repeat the process, starting with the next digit, and continue until the decoding is done. For example, using the set of codewords S above, we would decode the string 001100100011 as 001, 10, 01, 000, 11. Thus an efficient method of representation should use codewords such that (1) the codewords have the prefix property; and (2) the symbols used frequently have shorter codewords than those used less often.

Any binary tree can be used to construct a set of codewords with the prefix property by assigning 0 to each edge from a parent to its left child, and 1 to each edge from a parent to its right child. Following the unique directed path from the root to a terminal vertex will give a string of 0s and 1s. The set of all strings formed in this way will be a set of codewords with the prefix property, because we can find the unique directed path corresponding to any codeword by working down from the root of the binary tree, going left or right, according to whether a digit is 0 or 1. By definition, we finish at a terminal vertex, and so this codeword cannot be the first part of another codeword.

Example 5.36 For the binary tree in Figure 5.61(a), we assign 0s and 1s to its edges as shown in Figure 5.61(b). The directed paths from the root to all the terminal vertices then produce the codewords 000, 001, 01, 10, 11, as illustrated in Figure 5.62, where each codeword is written below the corresponding terminal vertex. ■

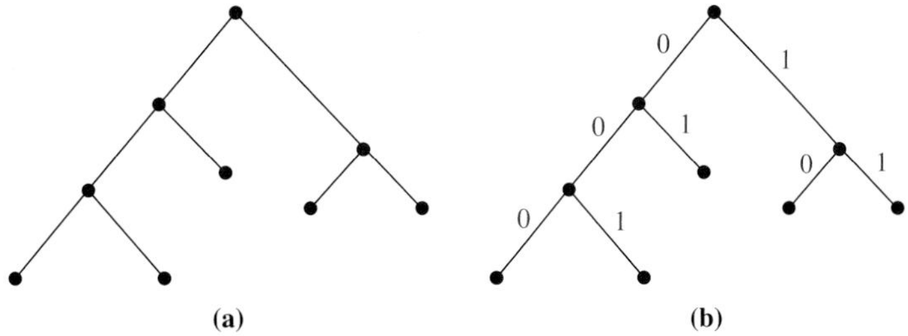

FIGURE 5.61

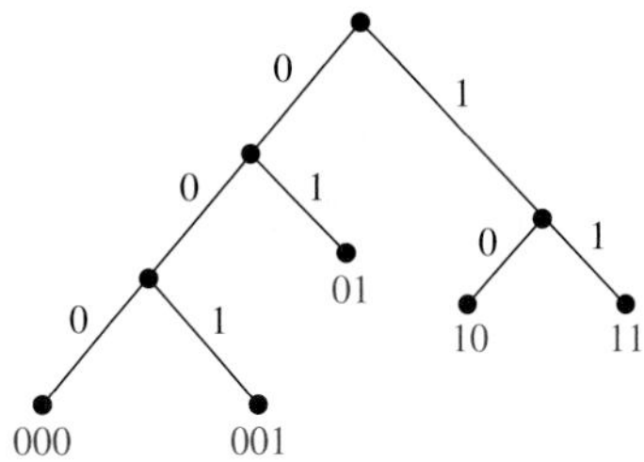

FIGURE 5.62

Thus, by using a binary tree, we have found a way to produce codewords that have the prefix property. It remains to find a method for assigning shorter codewords to the more frequently used symbols. If we have only the five symbols in Figure 5.62, then we want to use the codewords 01, 10, and 11 for the three most frequently used symbols. Notice that these codewords correspond to the terminal vertices that are closest to the root. Thus, to obtain an efficient method for representing symbols by variable length codewords, we can use a binary tree and assign the most frequently used symbols to the terminal vertices that are closest to the root.

We will restrict our discussion to those binary trees for which every internal vertex has exactly two children. Suppose $w_1, w_2, \ldots, w_k$ are nonnegative real numbers. A **binary tree for the weights** $w_1, w_2, \ldots, w_k$ is a binary tree with k terminal vertices labeled $w_1, w_2, \ldots, w_k$. A binary tree for the weights $w_1, w_2, \ldots, w_k$ has **weight** $d_1 w_1 + d_2 w_2 + \cdots + d_k w_k$, where d_i is the length of the directed path from the root to the vertex labeled w_i $(i = 1, 2, \ldots, k)$.

Example 5.37 The binary tree in Figure 5.63(a) is a binary tree for the weights 2, 4, 5, 6 and has weight $3 \cdot 6 + 3 \cdot 5 + 2 \cdot 4 + 1 \cdot 2 = 43$. In Figure 5.63(b) is another binary tree for the weights 2, 4, 5, 6, but its weight is $2(2 + 4 + 5 + 6) = 34$, since the distance from the root to each terminal vertex is 2. ■

For the coding problem, we want to find a binary tree of smallest possible weight in which the frequencies of the symbols to be encoded are the weights. A

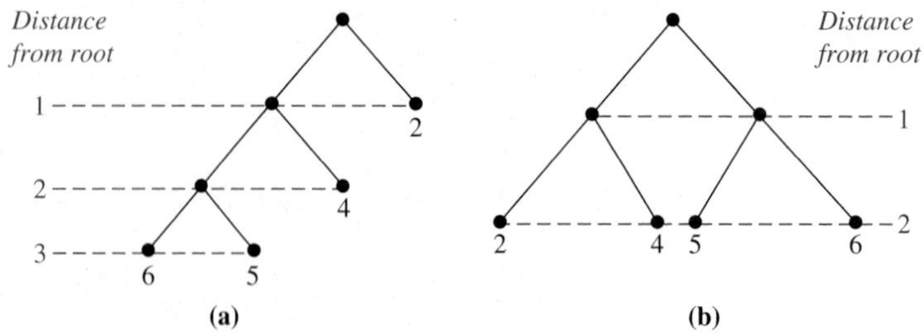

FIGURE 5.63

binary tree for the weights $w_1, w_2, \ldots, w_k$ is called an **optimal binary tree for the weights** $w_1, w_2, \ldots, w_k$ when its weight is as small as possible. Thus the binary tree in Figure 5.63(a) is not an optimal tree for the weights 2, 4, 5, 6, since there is another binary tree with smaller weight, namely, that in Figure 5.63(b).

The following algorithm due to David A. Huffman produces an optimal binary tree for the weights $w_1, w_2, \ldots, w_k$:

Huffman's Optimal Binary Tree Algorithm

For nonnegative real numbers $w_1, w_2, \ldots, w_k$, where $k \geq 2$, this algorithm constructs an optimal binary tree for the weights $w_1, w_2, \ldots, w_k$. In the algorithm, a vertex is referred to by its label.

Step 1 (create trees)
  (a) For $i = 1, 2, \ldots, k$ construct a tree consisting of one vertex that is labeled w_i.
  (b) Let $\mathcal{S}$ denote the set of trees constructed in this manner.
Step 2 (make a larger tree)
repeat
  Step 2.1 (select smallest weights)
    From $\mathcal{S}$, select two trees $\mathcal{T}_1$ and $\mathcal{T}_2$ with roots that have the smallest labels, say, V and W. (Ties can be broken arbitrarily.)
  Step 2.2 (combine the trees)
    (a) Construct the root of a binary tree, and assign the label $V + W$ to this root.
    (b) Make $\mathcal{T}_1$ the left subtree of this root.
    (c) Make $\mathcal{T}_2$ the right subtree of this root.
    (d) In $\mathcal{S}$, replace $\mathcal{T}_1$ and $\mathcal{T}_2$ with the tree having root labeled $V + W$.
until $|\mathcal{S}| = 1$

Example 5.38 We begin the construction of an optimal binary tree for the weights 2, 3, 4, 7, and 8 by constructing five binary trees, each having a single vertex that is labeled by one of the given weights. (See Figure 5.64.) The set consisting of these five trees is denoted $\mathcal{S}$. Now we select the two trees $\mathcal{T}_1$ and $\mathcal{T}_2$ with roots having the smallest labels, namely, 2 and 3. (For convenience, we will refer to a vertex by its label.)

FIGURE 5.64

We use $\mathcal{T}_1$ and $\mathcal{T}_2$ to form a new tree with root 5, and replace $\mathcal{T}_1$ and $\mathcal{T}_2$ in $\mathcal{S}$ by this new tree. The binary trees in $\mathcal{S}$ now have roots labeled 5, 4, 7, and 8, as shown in Figure 5.65. Next, we continue step 2 by combining the binary trees with roots labeled 5 and 4 to form a new binary tree with root 9. This tree replaces the two trees in $\mathcal{S}$ with labels 5 and 4. The binary trees now in $\mathcal{S}$ have roots labeled 9, 7, and 8. (See Figure 5.66.) As step 2 continues, we combine the binary trees with roots labeled 7 and 8 to form a binary tree with root 15. This new tree now replaces the trees in $\mathcal{S}$ with labels 7 and 8. At this point, $\mathcal{S}$ consists of two binary trees with roots labeled 9 and 15. (See Figure 5.67.) Combining these two binary trees gives the optimal binary tree for the weights 2, 3, 4, 7, and 8, shown in Figure 5.68. The weight of this tree is

$$2(4 + 7 + 8) + 3(2 + 3) = 53. \quad \blacksquare$$

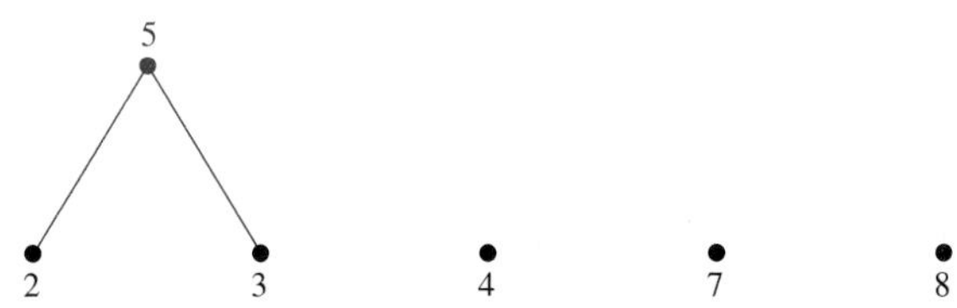

FIGURE 5.65

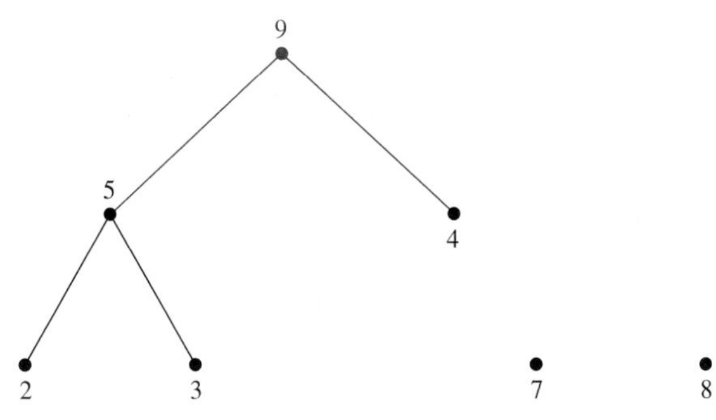

FIGURE 5.66

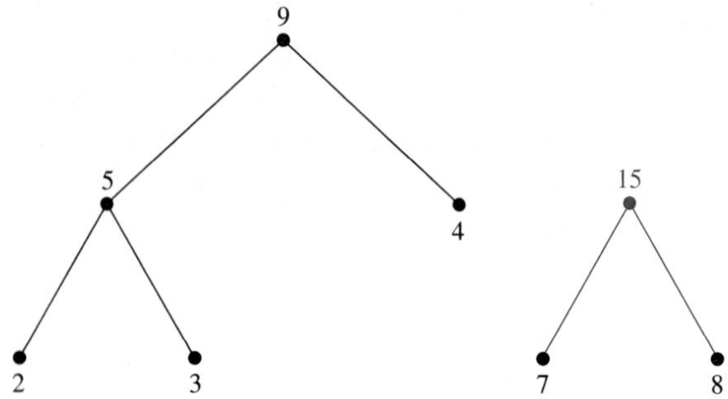

FIGURE 5.67

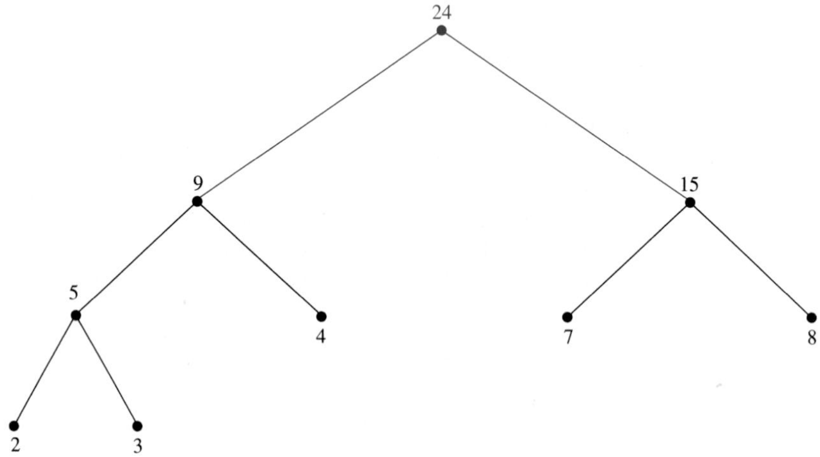

FIGURE 5.68

Example 5.39 Using Huffman's optimal binary tree algorithm, we can construct an optimal binary tree for the weights 2, 4, 5, and 6 in the steps shown in Figure 5.69. This tree has weight

$$1 \cdot 6 + 2 \cdot 5 + 3(2 + 4) = 34.$$

Note that in step 2.1 we could have chosen the binary tree with one vertex labeled 6, instead of the tree with three vertices labeled 2, 4, and 6. In this case we would have obtained a different optimal binary tree. ■

An analysis of this algorithm requires knowledge of sorting and inserting algorithms that we have not studied. Thus we state without proof (see suggested reading [4] at the end of the chapter) that Huffman's optimal binary tree algorithm is of order at most k^2, where k is the number of weights. A proof that the algorithm constructs an optimal binary tree is found in Exercises 45–47.

In order to find codewords with the prefix property such that the most frequently used symbols are assigned the shortest codewords, we construct an opti-

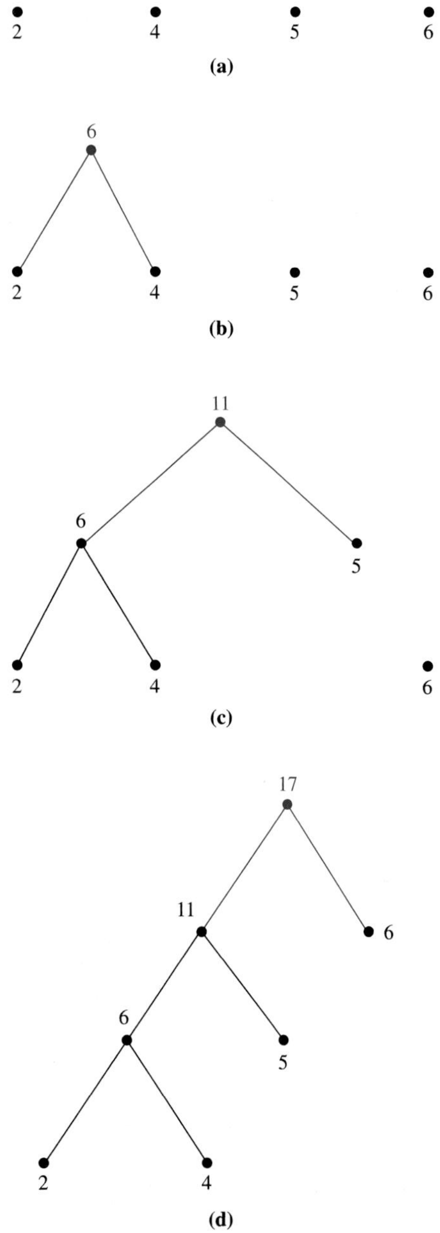

FIGURE 5.69

mal binary tree with the stated frequencies of the symbols as its weights. Then, by assigning 0s and 1s to the edges of this tree, as described in Example 5.36, codewords can be efficiently assigned to the various symbols.

Example 5.40 Suppose the characters E, T, A, Q, and Z have expected usage rates of 32, 28, 20, 4, and 1, respectively. In Figure 5.70 we see an optimal binary tree with weights 1, 4, 20, 28, 32, created by Huffman's optimal binary tree algorithm. Furthermore, each symbol has been placed in parentheses next to its usage rate. Then 0s and 1s are assigned to the edges of the tree so that codewords with the prefix property are formed at the terminal vertices. (See Figure 5.71.) Thus we see that E should be assigned the codeword 1, T should be assigned the codeword 01, A should be assigned 001, Q should be assigned 0001, and Z should be assigned 0000. ■

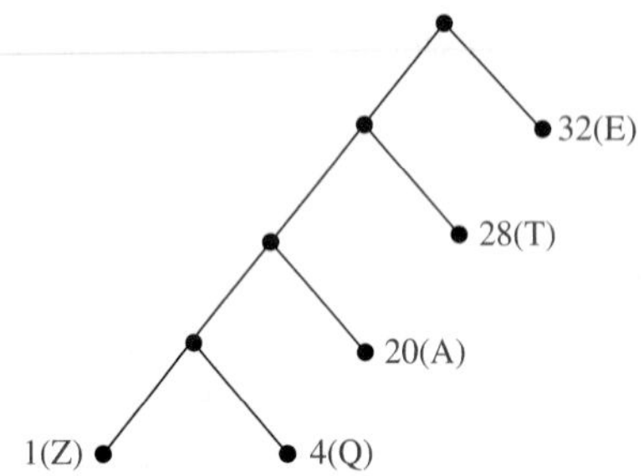

FIGURE 5.70

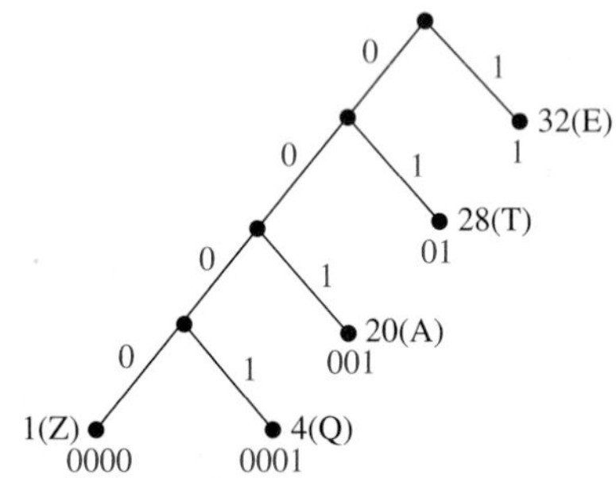

FIGURE 5.71

A binary tree describing the codewords can also be used to decode a string of 0s and 1s. To do so, we start at the root and follow the directed path from the root that is indicated by the 0s and 1s. When a terminal vertex is reached, the string is then decoded by the codeword at that vertex. Then this process is begun again at the root with the next digit. For example, to decode the string 00101 with the tree in Figure 5.71, we start at the root and go to the left child, then to the left child again, and then to the right child, which is a terminal vertex with the codeword 001 and symbol A. Then we go back to the root and decode the remaining bits 01, which correspond to the symbol T.

Another application of Huffman's optimal binary tree algorithm arises in regard to the merging of sorted lists. Suppose we have two sorted lists of numbers L_1 and L_2 that we want to merge together into one sorted list. Recall from Theorem 2.9 that if L_1 and L_2 have n_1 and n_2 numbers, respectively, then these two sorted lists can be merged into one sorted list with at most $n_1 + n_2 - 1$ comparisons. Now suppose we have three sorted lists L_1, L_2, and L_3 containing 150, 320, and 80 numbers, respectively. By merging only two lists at a time, how can we merge these three lists into one sorted list so that we minimize the number of comparisons needed? One way is to merge L_1 and L_2 into a third list with 470 numbers, which requires at most 469 comparisons. Then we merge this new list with L_3, which requires at most $470 + 80 - 1 = 549$ comparisons. Altogether, this merge pattern requires at most $469 + 549 = 1018$ comparisons. This process can be represented by the binary tree in Figure 5.72(a), where the labels represent the sizes of the lists.

A second merge pattern is to merge L_1 and L_3 first, followed by merging this new list with L_2. This merge pattern requires at most $229 + 549 = 778$

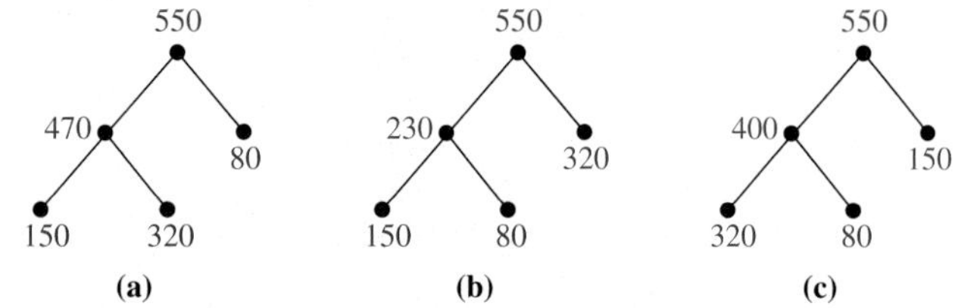

FIGURE 5.72

comparisons and is represented in Figure 5.72(b). Finally, we can merge L_2 and L_3 first and then merge the result with L_1. This requires at most $399 + 549 = 948$ comparisons and is represented in Figure 5.72(c). Thus we see that the second merge pattern requires the fewest comparisons. Furthermore, we observe that the optimal pattern of merging (the one requiring the fewest comparisons) occurs when the smallest lists are used first, since, in this way, fewer comparisons are made overall. In fact, the number of times an item is sorted is the distance of its list from the root in the binary trees shown in Figure 5.72. Thus the optimal merging pattern corresponds to the tree of minimum weight in Figure 5.72, where the weight of a vertex is the number of items in the corresponding list. Hence the optimal merge pattern can be found by following the construction in Huffman's optimal binary tree algorithm.

Example 5.41 In order to merge five sorted lists with 20, 30, 40, 60, and 80 numbers optimally, we begin by merging together the two sorted lists with the smallest number of items. These are the sorted lists with 20 and 30 numbers, which result in a new sorted list of 50 numbers and use at most 49 comparisons. Now we consider the four sorted lists with 50, 40, 60, and 80 numbers and combine the lists with 50 and 40 items; this merging yields a new sorted list of 90 items using at most 89 comparisons. Next, from the three sorted lists with 90, 60, and 80 numbers, we merge the lists with 60 and 80 numbers to obtain a new sorted list of 140 items using at most 139 comparisons. Finally, we combine the sorted lists with 90 and 140 numbers, giving one sorted list with 230 numbers and using at most 229 comparisons. This optimal merge pattern uses at most 506 comparisons. A binary tree representation of this optimal merge pattern is given in Figure 5.73. ■

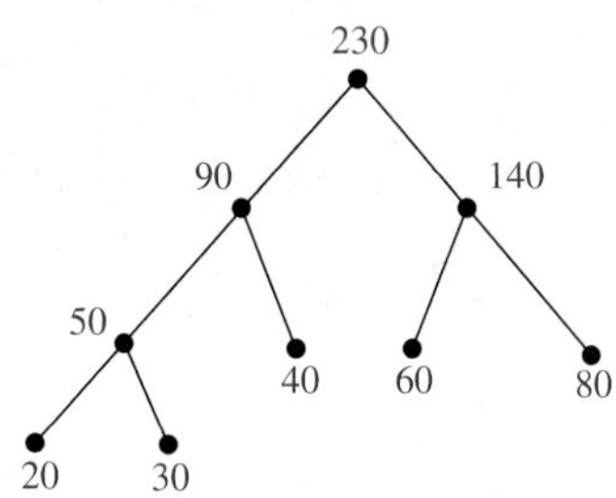

FIGURE 5.73

Binary Search Trees

Maintaining a large data set is a common problem for data processors. Such maintenance consists not only of updating the data set by adding and deleting, but also of searching the data for a particular piece of information. Suppose, for instance, that the Acme Manufacturing Company maintains a list of its customers. When an order is received, the company must search this list to determine whether the order is from an old or a new customer. On the one hand, if the order is from a new customer, then this customer's name must be added to the list. On the other hand, when a customer goes out of business, that customer's name must be removed from the list.

One way to maintain such lists is to keep the data in the order in which they are received. For example, if Acme Manufacturing has 10 customers named Romano, Cohen, Moore, Walters, Smith, Armstrong, Garcia, O'Brien, Young, and Tucker, the company can keep these names in an array in the given order. This method enables items to be added to the list easily; if Jones becomes a customer, this new name can be added to the end of the existing array. However, this method makes it very time-consuming to determine whether a particular name is in the list. Determining that Kennedy is not a customer, for instance, will require checking every name in the list. Of course, the amount of checking required is minimal when Acme has only 10 customers, but if Acme has a million customers, checking every name on the list is prohibitive.

Another approach is to keep the list in alphabetical order. For example, Acme Manufacturing's list of customer names can be stored as Armstrong, Cohen, Garcia, Moore, O'Brien, Romano, Smith, Tucker, Walters, and Young. With this method, it is easy to search the list for a particular name. (The procedure used in the proof of Theorem 2.8 can be adapted to give an efficient searching method.) However, adding or deleting from the list is more difficult because of the need to reposition the entries when an item is added or deleted. For example, if Acme gains a new customer named Baker, then we need to insert this name as the second entry in the list. This insertion requires repositioning every name in the original list except Armstrong's. Again, this process is prohibitive if the list is very long.

A third approach is to store data at the vertices of a binary tree. For example, the list of Acme Manufacturing's customer names can be stored as in Figure 5.74. This binary tree is arranged so that if a vertex U belongs to the left subtree of vertex V, then U precedes V in alphabetical order; and if a vertex W lies in the right subtree of V, then W follows V in alphabetical order. Adding a new name to this tree is simple because we need only to include one new vertex and edge in the tree, and searching the tree for a particular name requires no more than four comparisons if we search the tree properly.

In order to generalize this example, suppose that we have a list of distinct numbers or words. We will use the symbol $\leq$ to denote the usual numerical or alphabetical (dictionary) order. For example, $7 \leq 9$ and ABGT $\leq$ ACE. A **binary**

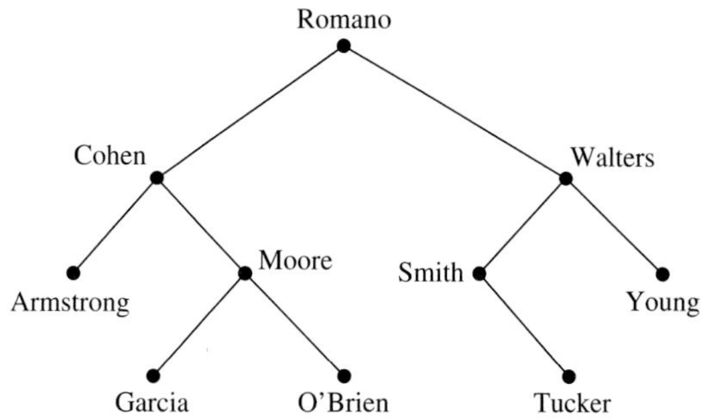

FIGURE 5.74

search tree for the list is a binary tree in which each vertex is labeled by an element of the list such that

(1) No two vertices have the same label.
(2) If vertex U belongs to the left subtree of vertex V, then $U \leq V$.
(3) If vertex W belongs to the right subtree of vertex V, then $V \leq W$.

Thus, for each vertex V, all descendents of V in the left subtree of V precede V, and all descendents of V in the right subtree of V follow V.

Example 5.42 One possible binary search tree for the list 1, 2, 4, 5, 6, 8, 9, 10 is given in Figure 5.75(a). Another possibility is shown in Figure 5.75(b). ■

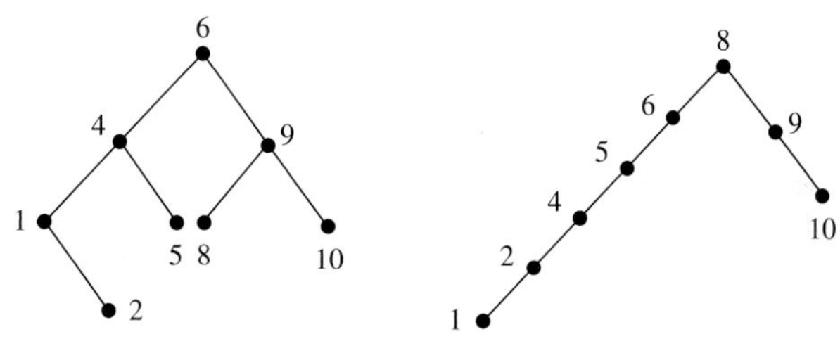

FIGURE 5.75

Example 5.43 A binary search tree for the list

if, this, for, break, else, while, throw

is illustrated in Figure 5.76. ■

There is a systematic way to construct a binary search tree for a list. The basic idea is to put smaller elements as left children, and larger elements as right children.

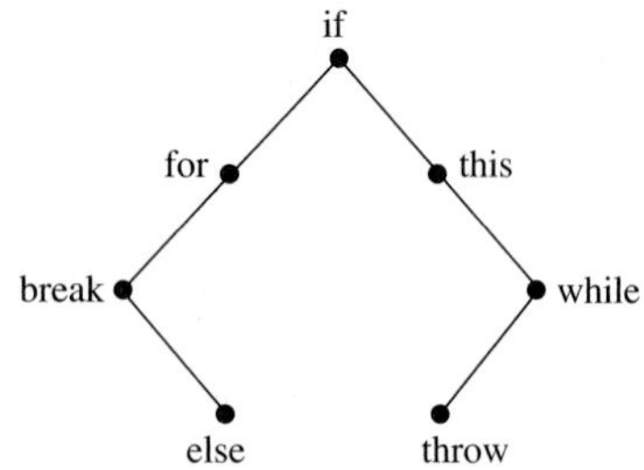

FIGURE 5.76

Binary Search Tree Construction Algorithm

This algorithm constructs a binary search tree in which the vertices are labeled $a_1, a_2, \ldots, a_n$, where $a_1, a_2, \ldots, a_n$ are distinct and $n \geq 2$. In the algorithm, a vertex is referred to by its label.

Step 1 (construct the root)
  (a) Construct the root of the binary tree, and label it a_1.
  (b) Set $k = 1$.
Step 2 (insert elements into the tree)
  while $k < n$
    Step 2.1 (find the insertion point)
      Step 2.1.1 (initialization for descent)
        (a) Let V denote the root of the tree.
        (b) Replace k with $k + 1$.
      Step 2.1.2 (descend the tree)
        repeat
          Perform exactly one of the following three steps:
          (a) (go left)
            if $a_k < V$ and V has a left child W
              Replace V with W.
            endif
          (b) (go right)
            if $a_k > V$ and V has a right child W
              Replace V with W.
            endif
          (c) (stay here)
            if neither (a) nor (b) is possible
              Do nothing.
            endif
        until either $a_k < V$ and V has no left child or $a_k > V$ and
          V has no right child

Step 2.2 (insert a_k)
if $a_k < V$
Construct a left child for V and label it a_k.
otherwise
Construct a right child for V and label it a_k.
endif
endwhile

The construction described in the binary search tree construction algorithm does produce a binary tree. Furthermore, labels for the left descendants (those on the left side) are smaller than the label for the parent, and labels for the right descendants are larger. Thus the algorithm yields a binary search tree.

Example 5.44 The result of using the binary search tree construction algorithm on the list 5, 9, 8, 1, 2, 4, 10, 6 is shown in Figure 5.77. ■

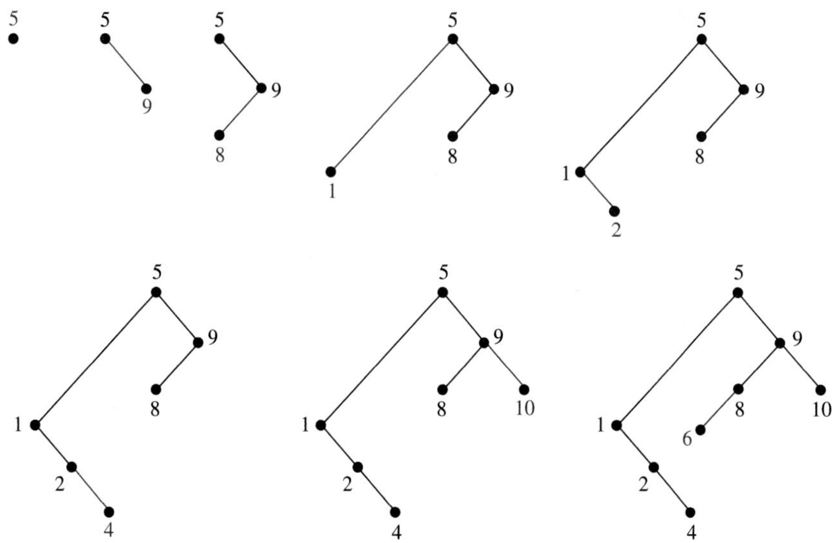

FIGURE 5.77

Example 5.45 For the list of words in the sentence DISCRETE MATH IS FUN BUT HARD, the algorithm yields the binary search tree in Figure 5.78. ■

If an additional item is to be added to the binary search tree, then we can simply use the binary search tree construction algorithm one more time with that item. For example, to add SOMETIMES to the end of the list of words in the sentence DISCRETE MATH IS FUN BUT HARD, we would repeat the algorithm, using the word SOMETIMES, with the binary search tree in Figure 5.78 to obtain the one in Figure 5.79. This procedure for adding an item to a binary search tree is an efficient one.

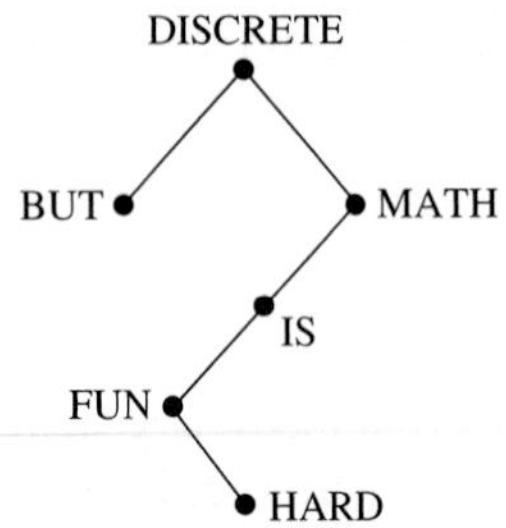

FIGURE 5.78

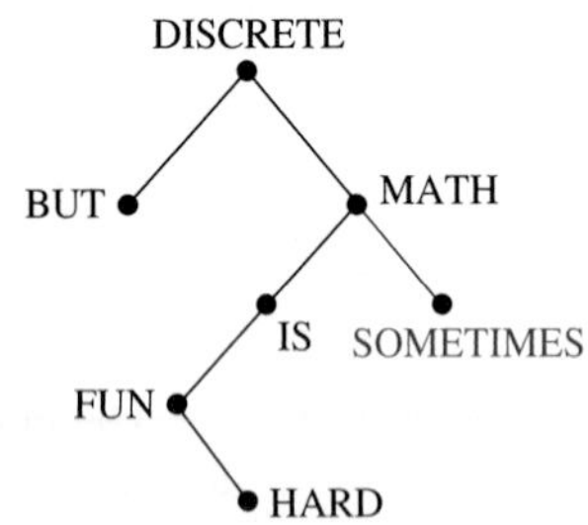

FIGURE 5.79

To determine whether an item is in a binary search tree (or equivalently, a list), we follow closely the process used to construct a binary search tree. Specifically, we compare the item with the root and go left if it is smaller and go right if it is larger. This process is repeated until we either match some item in the tree or find that the item is not in the tree. This procedure is formalized in the following algorithm:

Binary Search Tree Search Algorithm

This algorithm will examine a binary search tree to decide whether a given element a is in the tree.

Step 1 (initialization)
Let V be the root of the binary search tree.

Step 2 (descend the tree)
while ($a < V$ and V has a left child) or ($a > V$ and V has a right child)
if $a < V$ and V has a left child
Replace V with its left child.
otherwise
Replace V with its right child.
endif
endwhile

Step 3 (is a in the tree?)
if $a \neq V$
Element a is not in the tree.
otherwise
Element a is in the tree.
endif

Example 5.46 Let us use this algorithm to search for 7 in the last binary search tree of Figure 5.77. We begin by comparing 7 with the root of the tree. Since it is larger, we go to the right child 9 and make another comparison. This time, 7 is smaller, and so we go to the left child 8. Comparing 7 with 8, we go to the left child 6. Now comparing 7 with 6, we go to the right child of 6. Since there is none, 7 is not in the binary search tree. We also note that if we wanted to add 7 to this binary search tree at this time, it would become the right child of 6. ■

Example 5.47 To search for the word FUN in the tree of Figure 5.78, we begin with the root DISCRETE. The first comparison takes us to the right child MATH. From there, another comparison takes us to the left child IS. Again, we go to the left child FUN. This comparison results in a match. Hence we find that FUN is in the tree. ■

Applying the inorder traversal to the binary search tree in Figure 5.77 yields the inorder listing 1, 2, 4, 5, 6, 8, 9, 10, which is the usual numerical order for these numbers. Similarly, for the binary search tree in Figure 5.76, the inorder traversal gives the listing

break, else, for, if, this, throw, while,

which is the alphabetical order for these words. In general, when the inorder traversal is applied to a binary search tree, the resulting listing is the usual ordering of the elements. From this, the smallest and largest elements of the tree can be found. Thus, by going left as far as possible in a binary search tree, the last vertex reached is the smallest element in the tree. Similarly, by going right as much as possible, the last vertex reached is the largest element in the tree.

Deletions of items from a binary search tree can also be done efficiently. The details are left as exercises.

The construction of a binary search tree depends upon the order in which the items appear in the list. In other words, a different order for the items can produce a different binary search tree. For example, for the list 10, 9, 8, 6, 5, 4, 2, 1, which is the same set of items as in Example 5.44, the algorithm produces the binary search tree shown in Figure 5.80. It is easy to see that this tree offers no

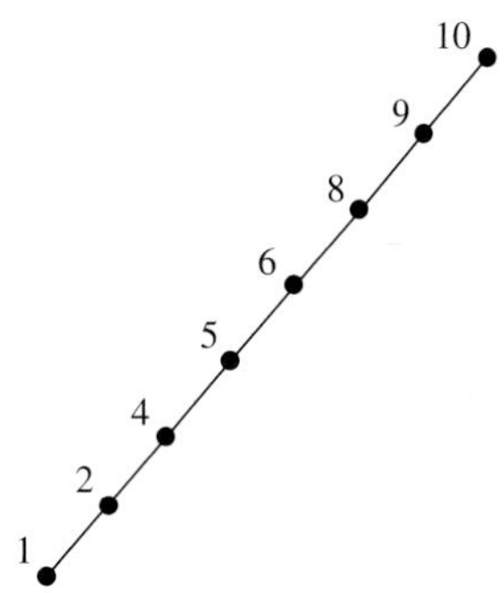

FIGURE 5.80

advantage over numerical order in storing the numbers, since a search may require a comparison with every item in the tree. However, there is an extensive literature on the construction of binary trees that make for efficient searching. For example, binary search trees can be constructed so that the more frequently accessed items are closer to the root than those items that are not as frequently accessed. Interested readers should consult suggested readings [7] and [9] at the end of the chapter.

EXERCISES 5.6

In Exercises 1–4, determine whether the given sets of codewords have the prefix property.

1. $\{0, 100, 101, 11, 1011\}$

2. $\{00, 11, 010, 100, 011\}$

3. $\{00, 101, 111, 10001, 1010\}$

4. $\{00, 110, 101, 01\}$

5. Can there be a set of 6 codewords with the prefix property that contains 0, 10, and 11?

6. Can there be a set of 6 codewords with the prefix property that contains 10, 00, and 110?

7. Determine values for a, b, and c so that $\{00, 01, 101, a10, bc1\}$ is a set of 5 codewords with the prefix property.

8. Determine values for a, b, c, and d so that $\{00, 0a0, 0bc, d0, 110, 111\}$ is a set of 6 codewords with the prefix property.

For the values of n given in Exercises 9–14, draw a binary tree in which each vertex has 0 or 2 children and that generates, as in Example 5.36, a set of n codewords with the prefix property. Label the vertices with the codewords.

9. $n = 2$

10. $n = 3$

11. $n = 4$

12. $n = 7$

13. $n = 8$

14. $n = 9$

In Exercises 15–18, draw a binary tree that generates, as in Example 5.36, the given codewords at the terminal vertices.

15. $1, 00, 011, 0100, 0101$

16. $101, 00, 11, 011, 100, 010$

17. $1111, 0, 1110, 110, 10$

18. $1100, 000, 1111, 1101, 0010, 10, 0011$

19. Decode the message 11101010000010110 with the assignment A: 010, B: 111, M: 000, N: 110, and T: 10.

20. Decode the message 110010111011110 with the assignment O: 0, B: 10, R: 110, I: 1110, N: 11110, and T: 11111.

21. Decode the message 10010111010101001011 with the assignment A: 111, E: 0, N: 1010, O: 1011, and T: 100.

22. Decode the message 00111100010000111 with the assignment B: 1100, D: 111, E: 1101, J: 0011, N: 0000, O: 01, S: 0010, and T: 0001.

In Exercises 23–26, decode the messages by using the given binary trees.

23. message: 01110111

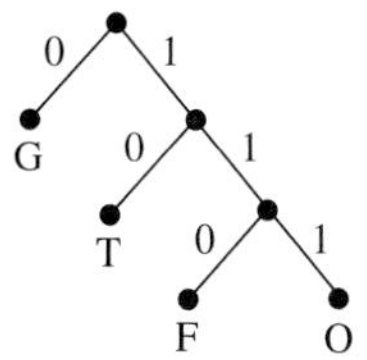

24. message: 11001001011000

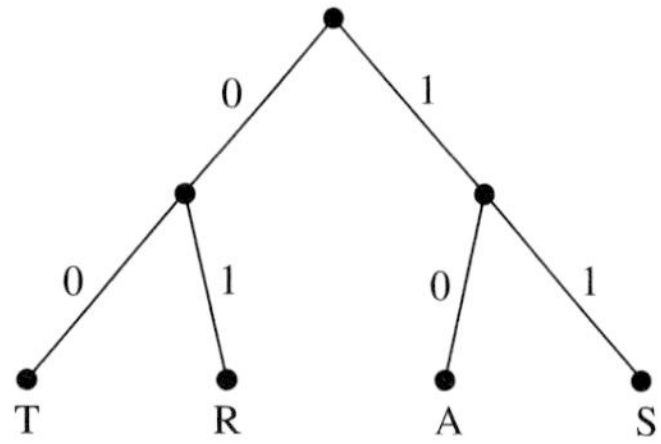

25. message: 01111111011101001100

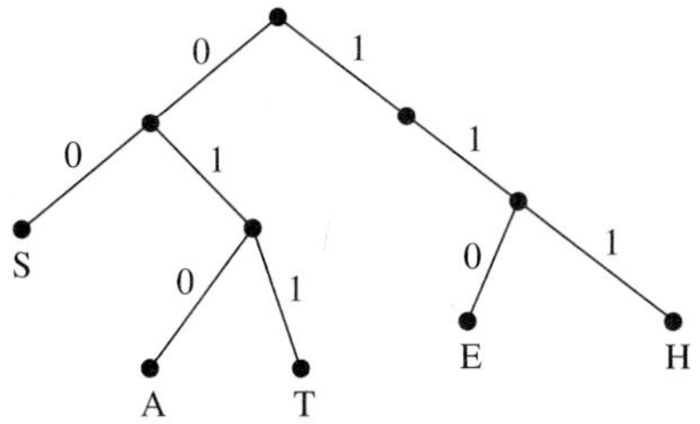

26. message: 000010010011000011 01001

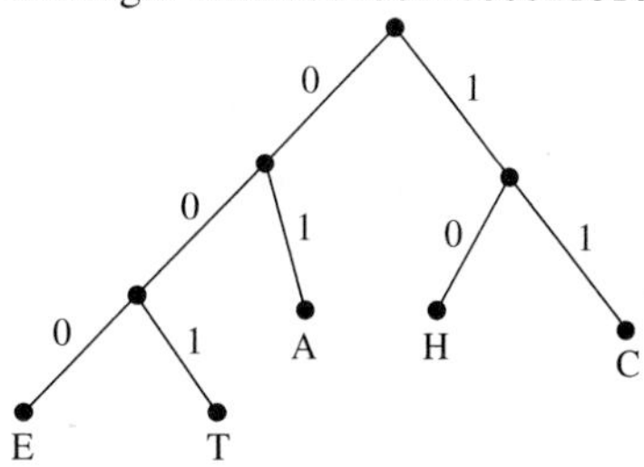

Locate a copy of the ASCII code in a computer programming book and use it to decode the messages given in Exercises 27–30.

27. 0100010001001111010000111

28. 01001000010011110100110101000101

29. 0101000101010101010010010100010101010100

30. 01001000010001010100110001010000

In Exercises 31–34, construct an optimal binary tree for the given weights. In the construction, when there is a choice of trees having roots with the same label, select the tree having the greater number of vertices.

31. 2, 4, 6, 8, 10

32. 4, 6, 8, 14, 15

33. 1, 4, 9, 16, 25, 36

34. 10, 12, 13, 16, 17, 17

In Exercises 35–38, determine the smallest maximum number of comparisons needed to merge sorted lists with the given numbers of items into one sorted list.

35. four lists with 20, 30, 40, and 50 items

36. five lists with 15, 25, 35, 40, and 50 items

37. six lists with 20, 40, 60, 70, 80, and 120 items

38. seven lists with 10, 30, 40, 50, 50, 50, and 70 items

In Exercises 39–42, in the construction of an optimal binary tree, when there is a choice of trees having roots with the same label, select the tree having the greater number of vertices.

39. The National Security Agency is helping American diplomats in foreign countries send coded messages back to the State Department in Washington, D.C. These messages are to be sent using the characters R, I, H, V, with an expected usage rate of 40, 35, 20, 5, respectively, per 100 characters. Find an assignment of codewords that minimizes the number of bits needed to send a message.

40. Tom and Susan are exchanging love letters during class. In order to prevent others from reading these sweet words of romance, the messages are coded using only the characters T, A, I, L, P, and J, with an expected usage rate of 34, 27, 21, 10, 6, 2, respectively, per 100 characters. Find an assignment of codewords that minimizes the amount of time (and hence, the number of bits) needed to send a message.

41. NASA is receiving information from one of its space probes. This information is in the form of numbers that represent pictures. (Each number corresponds to a shade of white, black, or gray.) The numbers used are 1, 2, 3, 4, 5, 6, 7, 8, 9, and 10, with expected usage rates of 125, 100, 75, 40, 60, 180, 20, 120, 150, and 130, respectively, per 1000 colored dots. Find an assignment of codewords for these numbers that minimizes the number of bits needed for the storage of this information.

42. The Gregory Computer Company has received a contract to store nursing data for all hospitals in the Bloomington, Illinois, area. Even though the storage will be on hard disks, the high volume of data makes it important that the data be stored efficiently. An analysis of sample data shows that only certain symbols are used: rn, @, c, s, po, os, od, tid, qod. Furthermore, the analysis shows a usage rate of 7, 12, 4, 9, 10, 8, 2, 18, 30, respectively, per 100 symbols. Find an assignment of codewords that minimizes the number of bits needed to store this data.

43. Prove that, for any positive integer n, there exists a binary tree with n terminal vertices in which each vertex has 0 or 2 children.

44. In a binary tree in which each vertex has 0 or 2 children, prove that the number of terminal vertices is one more than the number of internal vertices.

Exercises 45–47 provide a proof that Huffman's optimal binary tree algorithm creates an optimal binary tree. Suppose $w_1, w_2, \ldots, w_k$ are nonnegative real numbers and $w_1 \leq w_2 \leq \ldots \leq w_k$.

45. Prove that if T is an optimal binary tree for the weights $w_1, w_2, \ldots, w_k$, and if $w_i < w_j$, then the distance from the root to w_i is greater than or equal to the distance from the root to w_j.

46. Prove that there is an optimal binary tree for the weights $w_1, w_2, \ldots, w_k$, where w_1 and w_2 are children of the same parent.

47. Prove that if T is an optimal binary tree for the weights $w_1 + w_2, w_3, \ldots, w_k$, then the tree obtained by replacing the terminal vertex $w_1 + w_2$ by a binary tree with two children w_1 and w_2 is an optimal binary tree for the weights $w_1, w_2, \ldots, w_k$.

In Exercises 48–53, construct a binary search tree for the items in the order given.

48. The accounting department in the Busby Insurance Company has 8 divisions, with 11, 15, 8, 3, 6, 14, 19, and 10 staff members in them. Construct a binary search tree for the number of staff in these units.

49. Techniques and concepts in the programming language C++ are described by words such as

list, static, or, char, this, endl, heap, do, parameter, else.

Construct a binary search tree for these words.

50. Techniques and concepts in the programming language C++ are described by words such as

object, char, variable, string, pointer, function, virtual, Boolean,
parse, global, template, class, range, array.

Construct a binary search tree for these words.

51. The mathematics department has 13 faculty members, with 14, 17, 3, 6, 15, 1, 20, 2, 5, 10, 18, 7, and 16 years of teaching experience. Construct a binary search tree for the years of teaching experience by the faculty.

52. In a survey of 15 mathematics departments it was found that there were 18, 9, 27, 20, 30, 15, 4, 13, 25, 31, 2, 19, 7, 5, and 28 faculty members. Construct a binary search tree for the sizes of the faculty.

53. ASCII code is used to represent more than just the alphabet. It is also used to represent the symbols), :, %, −, #, <, @, ?, $, (, !, and &. The corresponding ASCII codewords can be interpreted as binary numbers (with decimal values 41, 58, 37, 45, 35, 60, 64, 63, 36, 40, 33, and 38, respectively) and, hence, can be used to provide an ordering of these symbols. Construct a binary search tree for these symbols.

54. Construct a binary search tree for the letters of the alphabet so that at most 5 comparisons are needed to locate any specified letter.

55. In the binary search tree of Exercise 49, draw the directed path required to show that the word "filenames" is not in the tree. Then indicate where it would be added to the tree.

56. In the binary search tree of Exercise 48, draw the directed path required to show that the numeral 16 is not in the tree. Then indicate where 16 would be added to the tree.

57. In the binary search tree of Exercise 51, draw the directed path required to show that the numeral 4 is not in the tree. Then indicate where 4 would be added to the tree.

58. In the binary search tree of Exercise 50, draw the directed path required to show that the word "path" is not in the tree. Then indicate where it would be added to the tree.

59. In the binary search tree of Exercise 53, draw the directed path required to show that > (with decimal number 62) is not in the tree. Then indicate where > would be added to the tree.

60. In the binary search tree of Exercise 52, draw the directed path required to show that the numeral 8 is not in the tree. Then indicate where 8 would be added to the tree.

61. Suppose that the vertices of a binary tree are assigned distinct elements from a list of either numbers or words with the following property: If L is the left child of a vertex V, then $L \leq V$, and if R is the right child of a vertex V, then $V \leq R$. Must the binary tree with this assignment be a binary search tree for the list?

Deletion of a terminal vertex V from a binary search tree is accomplished as follows: Delete the vertex V and the edge joining V to its parent.

62. Draw the binary search tree obtained by deleting the numeral 6 from the binary search tree in Exercise 48.

63. Draw the binary search tree obtained by deleting "parameter" from the binary search tree in Exercise 49.

When a binary search tree has a root R with only one child, deletion of the root is accomplished as follows: Delete R and the edge joining R to its child.

64. Draw the binary search tree obtained by deleting the root from the following binary search tree:

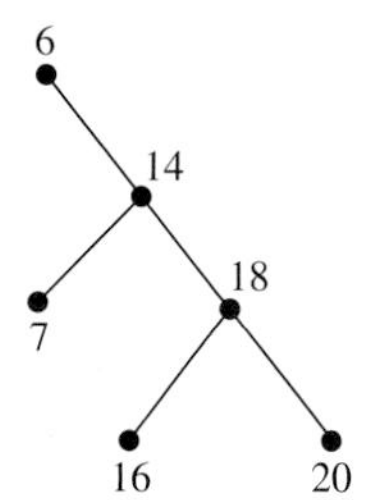

65. Repeat Exercise 64 for the following binary search tree:

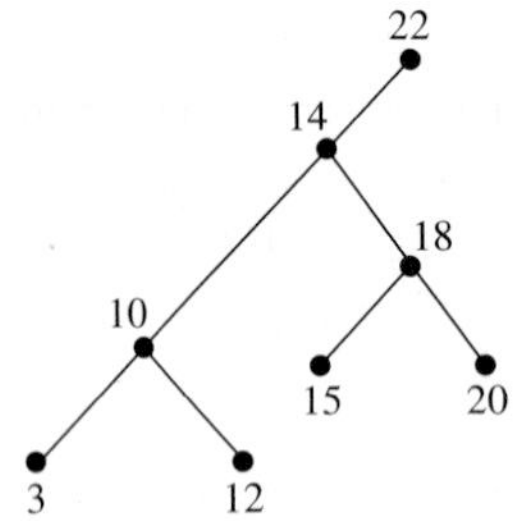

Suppose that V is a vertex in a binary search tree such that V is not the root and V has only one child C. Deletion of V from the binary search tree is accomplished as follows: Delete V and the edge joining V to C, and replace the tree formed by V and its descendants by the tree formed by C and its descendants.

66. Draw the binary search tree obtained by deleting < from the binary search tree in Exercise 53.

67. Repeat Exercise 66 for the vertex 3 in the binary search tree of Exercise 48.

In a binary search tree, deletion of a vertex V with 2 children is accomplished as follows: Find the largest item L in the left subtree of V. If L has no left child, delete L and the edge joining L to its parent, and replace V by L. If L has a left child C, delete L and the edge joining L to C, replace the tree formed by L and its descendants by the tree formed by C and its descendants, and then replace V by L.

68. Draw the tree obtained by deleting the numeral 9 from the binary search tree in Exercise 52.

69. Draw the tree obtained by deleting the numeral 3 from the binary search tree in Exercise 51.

70. Draw the tree obtained by deleting the word "object" from the binary search tree in Exercise 50.

71. Draw the tree obtained by deleting the word "variable" from the binary search tree in Exercise 50.

72. Draw the tree obtained by deleting the numeral 27 from the binary search tree in Exercise 52.

73. Draw the tree obtained by deleting the numeral 4 from the binary search tree in Exercise 52.

74. Prove that the inorder listing of the vertices in a binary search tree gives the natural order for the elements in the tree.

HISTORICAL NOTES

Unlike the real-world motivation of the Königsberg bridge problem, the study of trees got its start in considerations related to operators in differential calculus. The intuitive concept of a tree was first employed by the Germans G.K.C. Von Staudt

Arthur Cayley

(1798–1867) and Gustav Kirchhoff (1824–1887) in separate articles published in 1847. Kirchhoff's article dealt with an extension of Ohm's laws for electrical flow. However, the introduction of the term "tree" and the mathematical development of the concept came from the English mathematician Arthur Cayley (1821–1895) in 1857.

Cayley noted that the number of rooted trees with n edges could be found by observing the effect of removing the root vertex from the tree and examining the remaining collection of rooted trees. This observation, combined with some of the fundamental ideas concerning generating functions, led to a formula. Cayley's research in this area continued through the mid-1870s when he discovered a method for counting the number of unrooted trees.

Around the same time, the French mathematician Camille Jordan (1838–1922) began a systematic study of graphs. As part of this, he focused on the question of when two graphs are essentially the same, but have different representations—that is, when the graphs are isomorphic. In the special case when the vertices can be relabeled to result in an isomorphism, the mapping is called an *automorphism*. Jordan noted that the complete graph $\mathcal{K}_n$ has $n!$ automorphisms. He also noted that certain trees have one or more special vertices, called *bicentroids*, that are preserved by automorphisms. This work drew the notice of Cayley, who used Jordan's concepts in 1881 to develop a more elegant proof of his result on the number of rooted trees. One of Cayley's last major contributions to the development of trees was his proof in 1889 that the number of ways of joining n separate labeled vertices to form a tree is given by n^{n-2}. (See Exercise 38 in Section 5.1.) This result was proved independently in 1918 by the German Heinz Prüfer (1896–1934) [72].

James Joseph Sylvester

Other attempts to develop an algebra of graphs—trees, specifically—were made by James Joseph Sylvester (1814–1897) and William Kingdon Clifford (1845–1879). Their aim was to develop and enumerate the different compounds made possible by joining atoms of various substances. While the graphical representations resulting from these efforts has had an enormous impact on chemistry, the enumeration attempts eventually failed. One side product of these efforts was the use of the word "graph," which first appeared in an article in *Nature* authored by Sylvester in 1878.

While Kirchhoff's initial work focused on electrical networks, his ideas were not lost on the American mathematicians George David Birkhoff (1884–1944) and Oswald Veblen (1880–1960). In 1922, Veblen's analysis of Kirchhoff's work resulted in a theorem that every connected graph contains a tree, called a spanning tree, which includes every vertex of the graph. In 1956, Joseph B. Kruskal (1928–), and in 1957, Robert C. Prim (1921–), coworkers at the Bell Telephone Laboratories in Murray Hill, New Jersey, developed the algorithms that bear their names for finding a minimum spanning tree in a weighted graph. Their work opened new approaches to network designs for communication systems [72].

SUPPLEMENTARY EXERCISES

1. At the Illinois FBI office, Special Agent Jones is working with seven informants who have infiltrated a gambling ring. She needs to arrange for the informants to communicate with each other in groups of two in such a way that messages can be passed on to others. For secrecy, the number of meeting places must be kept as small as possible. How many meeting places must Agent Jones find?
2. What is the smallest number of colors needed to color a tree with n vertices, where $n \geq 2$?
3. If $m \geq 2$, for which n is $\mathcal{K}_{m,n}$ a tree? (The graph $\mathcal{K}_{m,n}$ is defined in Exercise 52 of Exercises 4.2.)
4. Ask a chemist about the chemical structure of benzene, and draw a graph describing it. Is it a tree?
5. Prove that if there is a vertex of degree k in a tree, then there are at least k vertices of degree 1.
6. Prove that if a tree has n vertices with degrees $d_1, d_2, \ldots, d_n$, then the sum of the degrees is $2n - 2$.
7. Prove that if $d_1, d_2, \ldots, d_n$ are positive integers with sum equal to $2n - 2$, then some $d_i = 1$; and if $n \geq 3$, then some $d_i > 1$.
8. For $n \geq 2$, suppose $d_1, d_2, \ldots, d_n$ are positive integers with sum $2n - 2$. Prove that there is a tree with n vertices having degrees $d_1, d_2, \ldots, d_n$. (*Hint:* Use mathematical induction on n.)
9. Find all the spanning trees of the following graph:

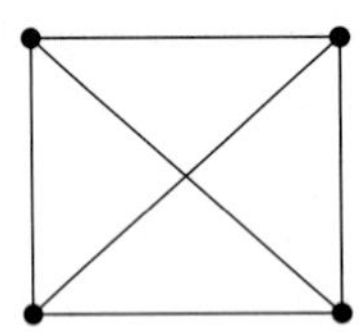

10. For a connected graph, show that the process of selecting edges so that an edge is not selected if it forms a cycle with edges already selected yields a spanning tree.
11. Suppose $\mathcal{G}$ is a connected graph with 10 vertices and 19 edges. What is the maximum number of edges that can be removed from $\mathcal{G}$ such that the remaining graph is still connected? Justify your answer.
12. For the accompanying graph, use the breadth-first search algorithm to find a spanning tree (start with B and use alphabetical order when there is a choice for a vertex):

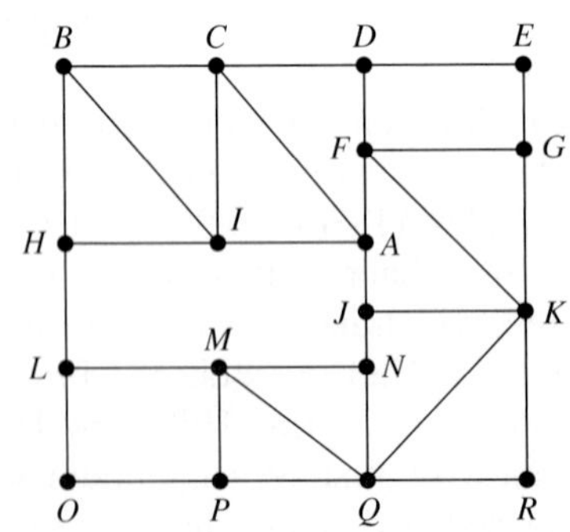

13. If $\mathcal{G}$ is a connected weighted graph and all the weights in $\mathcal{G}$ are distinct, must distinct spanning trees of $\mathcal{G}$ have distinct weights? Justify your answer.

14. Consider the following two weighted graphs:

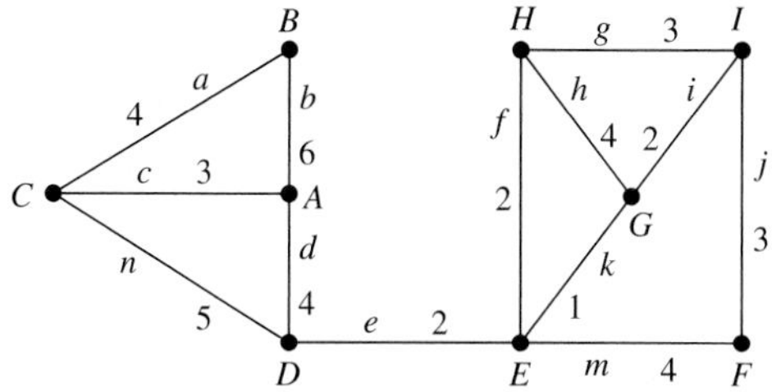

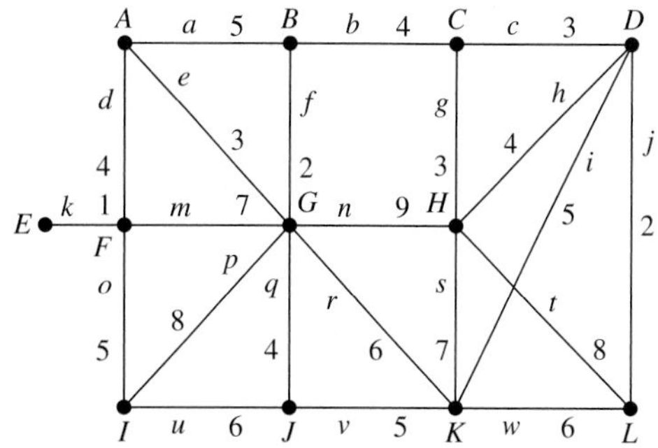

Use Prim's algorithm to find a minimal spanning tree for these graphs. (Start at G, and if there is a choice of edges to use in forming a minimal spanning tree, select edges according to alphabetical order.) Give the weight of the minimal spanning tree found.

15. Show that if $\mathcal{G}$ is a connected weighted graph and e is an edge of minimum weight incident on some vertex V, then there is a minimal spanning tree containing e.

16. Can a maximal spanning tree in a connected weighted graph (not a tree) contain an edge of smallest weight? If so, give an example. If not, give a proof.

17. Can Kruskal's algorithm be modified to find a spanning tree that is minimal with respect to all those containing two specified edges? Justify your answer. (See Exercise 42 in Section 5.2.)

18. Apply depth-first search to each of the following graphs to obtain a depth-first search numbering of the vertices (if there is a choice of vertices, choose the vertex that appears first in alphabetical order):

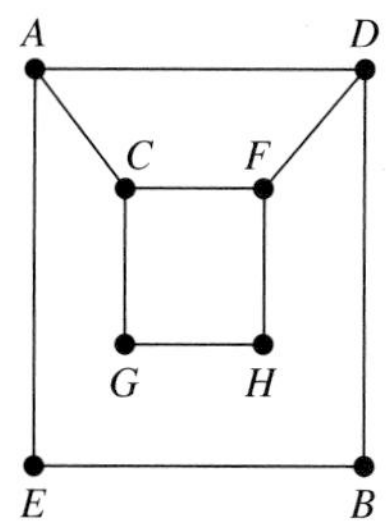

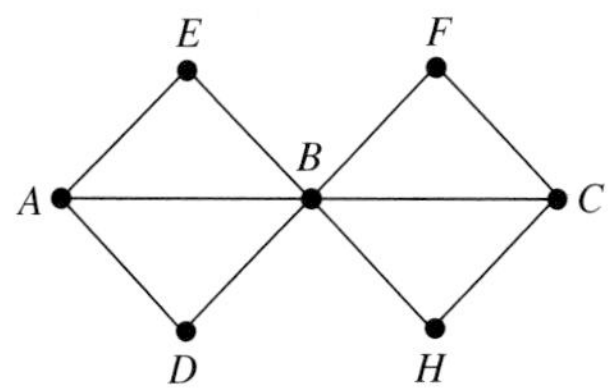

19. Use the depth-first search numbering obtained for each graph in Exercise 18 to form the spanning tree described in Theorem 5.7.

20. Use Theorem 5.8 and the depth-first search numbering obtained in Exercise 18 to assign directions to the edges that will make each graph in Exercise 18 into a strongly connected directed graph.

21. Show that an edge $\{A, B\}$ in a connected graph is a bridge if and only if every path from A to B includes $\{A, B\}$.

22. Show that an edge of a connected graph is a bridge if and only if it is not in any cycle.

23. Use backtracking to find a solution to the six-queens problem. (See Example 5.16.)

24. Use backtracking to construct a sequence of length 10 composed of the digits 1, 2, 3 with the property that nowhere in the sequence are there two adjacent subsequences that are identical.

25. For positive integers p, q, and n, where $p + q = n$, does there exist a rooted tree with n vertices and having p internal and q terminal vertices? Justify your answer.

26. A foreign language facility has 28 CD players that must be connected to a wall socket with four outlets. Extension cords having four outlets each are to be used to make the connections. What is the least number of cords needed to get these CD players connected so that they can be used?

27. How many vertices are there in a rooted tree with p internal vertices, each having exactly q children? Justify your answer.

28. How many terminal vertices are there in a rooted tree with p internal vertices, each having exactly q children? Justify your answer.

29. Consider 3^n or fewer coins that are identical, except for one that is lighter. Show by mathematical induction that they can be tested in at most n weighings on a balance scale to find out which one is lighter.

30. Suppose $\mathcal{T}$ is a rooted tree in which every vertex has at most $k \geq 2$ children and the length of the longest path from the root to a terminal vertex is h. Prove that

(a) $\mathcal{T}$ has at most $\dfrac{k^{h+1} - 1}{k - 1}$ vertices, and

(b) if some vertex has k children, then $\mathcal{T}$ has at least $h + k$ vertices.

31. For the tree obtained by applying depth-first search to a connected graph, show that if a vertex with depth-first search number k has m descendants (relative to the depth-first search tree) with $m \geq 1$, then their depth-first search numbers are $k + 1, k + 2, \ldots, k + m$.

32. Prove that when depth-first search is applied to a connected graph, one of the vertices on a back edge is an ancestor of the other (relative to the depth-first search tree).

33. Prove that when depth-first search is applied to a connected graph, one of the vertices on an edge is an ancestor of the other (relative to the depth-first search tree).

For Exercises 34 and 35, let depth-first search be applied to a connected graph $\mathcal{G}$ as in Theorem 5.10, and let A be a vertex in $\mathcal{G}$.

34. If the depth-first search starts at A, prove that A is an articulation point of $\mathcal{G}$ if and only if A has more than one child. (See Exercise 36 of Section 5.3.)

35. If depth-first search does not start at A, prove that A is an articulation point of $\mathcal{G}$ if and only if, for some child C of A, there is no back edge between C or any of its descendants and an ancestor of A.

*The **height** of a binary tree is the maximum of the levels of its terminal vertices.*

36. If $\mathcal{T}$ is a binary tree with n vertices and height h, prove that $n \leq 2^{h+1} - 1$.

37. Prove that a binary tree of height h has at most 2^h terminal vertices.

*A binary tree with height h is called **balanced** when the only vertices with no children are at level h or h − 1.*

38. Construct a balanced binary tree with 8 vertices.

39. Construct two distinct binary trees, each with more than one vertex, so that each has the same preorder listing and the same postorder listing as the other.

40. A complete binary tree has vertices $\mathcal{V} = \{1, 2, 3, 4, 5, 6, 7, 8, 9, 10\}$ with a postorder listing of 4, 5, 2, 8, 9, 6, 10, 7, 3, 1. Construct the tree if the height of the tree is 3.
41. Write the associative law for multiplication in Polish notation and in reverse Polish notation.
42. Write a distributive law of multiplication over addition in Polish notation and reverse Polish notation.
43. Does the set {00, 01, 100, 1010, 1011, 11} of codewords have the prefix property? Justify your answer.
44. Can there be a set of 7 codewords with the prefix property that contains 11, 101, 0101?
45. Draw a binary tree that generates the following codewords at the terminal vertices: 11, 00, 10, 010, 0110, 01111, 01110.
46. Construct an optimal binary tree for the weights 1, 3, 5, 7, 9, 11, 13, 15. In the construction, select a vertex with children in preference to a vertex without children, and use as the left child the vertex of smaller weight or the vertex with more children if the two vertices have equal weights.
47. Construct a binary search tree for the words in the order given in the sentence "Gladly would he learn so that others can be taught."

COMPUTER PROJECTS

Write a computer program having the specified input and output.

1. Given a graph, determine if it is a tree.
2. Given a rooted tree, find the internal vertices, the terminal vertices, and the root.
3. Given a graph, use the breadth-first search algorithm to find a spanning tree, if one exists.
4. Given a graph, use the breadth-first search algorithm to determine whether it is connected.
5. Given a graph, use the depth-first search algorithm to assign labels to its vertices.
6. Given a graph, use the depth-first search algorithm to determine whether it is connected.
7. Given a graph, use the depth-first search algorithm to find a spanning tree, if one exists.
8. Given a graph, use the discussion before Example 5.15 to determine whether it has a bridge.
9. Given a weighted graph, use Prim's algorithm to find a minimal spanning tree, if one exists.
10. Given a weighted graph, use Kruskal's algorithm to find a minimal spanning tree, if one exists.
11. Given a weighted graph, use Prim's algorithm to find a maximal spanning tree, if one exists.
12. Given a binary tree, use the preorder traversal algorithm to give a preorder listing of the vertices.
13. Given a binary tree, use the postorder traversal algorithm to give a postorder listing of the vertices.
14. Given a binary tree, use the inorder traversal algorithm to give an inorder listing of the vertices.
15. Given an arithmetic expression in Polish notation, evaluate it.
16. Given an arithmetic expression in reverse Polish notation, evaluate it.
17. Given nonnegative real numbers $w_1, w_2, \ldots, w_n$, use Huffman's optimal binary tree algorithm to construct an optimal binary tree for the weights $w_1, w_2, \ldots, w_n$.
18. Given words $a_1, a_2, \ldots, a_n$, use the binary search tree algorithm to construct a binary search tree with vertices labeled $a_1, a_2, \ldots, a_n$.

19. Given a binary search tree and an element a, use the binary search tree search algorithm to decide whether a is in the tree.

20. Given a tree $\mathcal{T}$ with n labeled vertices, use Prufer's algorithm to construct a list of numbers that uniquely describes $\mathcal{T}$.

21. Given a positive integer n, use backtracking to determine a solution to the n-queens problem.

22. Given a graph, use Theorem 5.8 to transform it into a strongly connected directed graph, if possible.

SUGGESTED READINGS

1. Aho, Alfred V., John E. Hopcroft, and Jeffrey D. Ullman. *Data Structures and Algorithms.* Reading, MA: Addison-Wesley, 1983.
2. Balaban, A.T. *Chemical Applications of Graph Theory.* New York: Academic Press, 1976.
3. Bogart, Kenneth P. *Introductory Combinatorics*, 3d ed. San Diego: Academic Press, 1999.
4. Horowitz, Ellis and Sartaj Sahni. *Fundamentals of Computer Algorithms.* New York: Freeman, 1984.
5. ________. *Fundamentals of Data Structures in Pascal.* 4th ed. New York: Freeman, 1992.
6. Hu, T.C., T.C. Shing, and Y.S. Kuo. *Combinatorial Algorithms.* Reading, MA: Addison-Wesley, 1982.
7. Knuth, Donald E. *The Art of Computer Programming*, vol. 1: *Fundamental Algorithms*, 2d ed. Reading, MA: Addison-Wesley, 1973.
8. Liu, C.L. *Elements of Discrete Mathematics*, 2d ed. New York: McGraw-Hill, 1985.
9. Reingold, Edward, Jurg Nievergelt, and Narsingh Deo. *Combinatorial Algorithms.* Englewood Cliffs, NJ: Prentice-Hall, 1977.
10. Stubbs, Daniel and Neil W. Webre. *Data Structures with Abstract Data Types and Pascal.* Monterey, CA: Brooks/Cole, 1985.
11. Tarjan, Robert Endre. *Data Structures and Network Algorithms.* Philadelphia: SIAM, 1983.

CHAPTER

6 Matching

Many combinatorial problems involve matching items, subject to certain restrictions. An example is the problem of assigning airline pilots to flights (introduced in Section 1.2). Another example is the assignment of pairs of participants at a conference to rooms so that roommates have the same smoking preference and sex. Sometimes an optimal matching may be desired. For example, a basketball coach must assign a player to guard each player on the opposing team in such a way as to minimize the opponents' total score. Such problems will be treated in this chapter.

6.1 SYSTEMS OF DISTINCT REPRESENTATIVES

The same matching problem may be viewed in various ways. As an example, let us consider the summer schedule of classes in the English department at a small college. There is a demand for 6 courses. To keep things simple, we will call these Course 1, Course 2, . . . , Course 6. Certain professors are available to teach each course, as given in the following table:

Course	*Professors*
1	Abel, Crittenden, Forcade
2	Crittenden, Donohue, Edge, Gilmore
3	Abel, Crittenden
4	Abel, Forcade
5	Banks, Edge, Gilmore
6	Crittenden, Forcade

For brevity, we will denote the professors by A, B, C, D, E, F, and G, according to their initials. In order to distribute the summer teaching jobs as fairly as possible, it is decided that no professor should teach more than one course. The question is whether all 6 courses can be taught, subject to this restriction. If not, what is the maximum number of courses that can be taught?

This is a problem of exactly the same sort as that of assigning airline pilots in Section 1.2. With only 6 courses and 7 professors, we could probably find the answer by considering all possible matchings. One systematic way of doing this is the following: Let P_1 denote the set of professors available to teach Course 1, P_2 the set of professors available to teach Course 2, etc. Thus

$$\begin{aligned} P_1 &= \{A, C, F\}, \\ P_2 &= \{C, D, E, G\}, \\ P_3 &= \{A, C\}, \\ P_4 &= \{A, F\}, \\ P_5 &= \{B, E, G\}, \\ P_6 &= \{C, F\}. \end{aligned}$$

If we forget for the moment the restriction that no professor may teach more than one course, then a possible assignment of a professor to each course consists of a 6-tuple $(x_1, x_2, x_3, x_4, x_5, x_6)$ where $x_1 \in P_1, x_2 \in P_2$, etc. This is an element of the Cartesian product

$$P_1 \times P_2 \times P_3 \times P_4 \times P_5 \times P_6,$$

which has $3 \cdot 4 \cdot 2 \cdot 2 \cdot 3 \cdot 2 = 288$ elements. We need to know whether any of these 288 6-tuples has all its entries distinct (so that no professor teaches more than one course). Checking this without the help of a computer would be possible, but extremely tedious. As in the case of the pilot assignment problem, however, such crude methods of searching for a solution quickly get beyond the capability of even a computer as the number of items to be matched gets larger. For example, if there were 30 courses and 3 professors available for each, then the Cartesian product would contain 3^{30} elements, and it would take a computer checking one million of these per second more than six years to go through them all.

There is a name for the sort of sequence of distinct elements, one from each of a given sequence of sets, that we are seeking in this example. Let $S_1, S_2, \ldots, S_n$ be a finite sequence of sets, not necessarily distinct. By a **system of distinct representatives** for $S_1, S_2, \ldots, S_n$, we mean a sequence $x_1, x_2, \ldots, x_n$ such that $x_i \in S_i$ for $i = 1, 2, \ldots, n$, and such that the elements x_i are all distinct.

Example 6.1 Find all systems of distinct representatives for the sets $S_1 = \{1, 2, 3\}$, $S_2 = \{1, 3\}$, $S_3 = \{1, 3\}$, $S_4 = \{3, 4, 5\}$.

Notice that the elements chosen from S_2 and S_3 must be 1 and 3 in some order. There are four systems of distinct representatives:

$$2, 1, 3, 4$$
$$2, 3, 1, 4$$
$$2, 1, 3, 5$$
$$2, 3, 1, 5. \quad \blacksquare$$

Example 6.2 Find all systems of distinct representatives for the sets $S_1 = \{2, 3\}$, $S_2 = \{2, 3, 4, 5\}$, $S_3 = \{2, 3\}$, $S_4 = \{3\}$.

There are none. For if x_1, x_2, x_3, x_4 were a system of distinct representatives, then x_1, x_3, and x_4 would be 3 distinct elements of $S_1 \cup S_3 \cup S_4 = \{2, 3\}$, which is impossible. ■

Example 6.3 How many systems of distinct representatives does the sequence S, S, S, S have, where $S = \{1, 2, 3, 4\}$?

In this case a system of distinct representatives is simply a permutation of the integers 1, 2, 3, 4. By Theorem 1.1, there are exactly $4! = 24$ of these. ■

Hall's Theorem

Now we return to our problem of assigning a professor to each summer English course. We are looking for a system of distinct representatives for the sequence

$$P_1 = \{A, C, F\},$$
$$P_2 = \{C, D, E, G\},$$
$$P_3 = \{A, C\},$$
$$P_4 = \{A, F\},$$
$$P_5 = \{B, E, G\},$$
$$P_6 = \{C, F\}.$$

The problem seems small enough that we might expect to find the solution, if there is one, by simply trying different combinations. Yet perhaps the best we can come up with is to cover 5 of the 6 courses. For example, we might assign the first 5 courses as in the following list:

Course 1 to Abel
Course 2 to Donohue
Course 3 to Crittenden
Course 4 to Forcade
Course 5 to Banks

We might suspect that it is not possible to do better than this, but it is difficult to be certain. We would like a way to convince ourselves that no assignment of all 6 courses is possible without going through all 288 possibilities.

There is a way, and the key to it is to be found in Example 6.2. If we could discover some collection of sets chosen from P_1 through P_6, the union of which contained fewer elements than the number of sets in the collection, then we would know that a system of distinct representatives was impossible. Since this is a somewhat abstract idea, we will exhibit such a collection to make the argument more concrete. How such a collection might be found will be covered in a later section of this chapter.

The collection we have in mind is P_1, P_3, P_4, and P_6. Notice that

$$P_1 \cup P_3 \cup P_4 \cup P_6 = \{A, C, F\},$$

and the argument is the same as in Example 6.2. Suppose we had a system of distinct representatives $x_1, x_2, \ldots, x_6$. Then x_1, x_3, x_4, and x_6 would be 4 distinct elements lying in the union of the sets P_1, P_3, P_4, and P_6. But that is impossible, because this union contains only 3 elements. There are only 3 professors (Abel, Crittenden, and Forcade) available to teach 4 of the courses, and so an assignment where no professor teaches more than 1 course cannot be made.

We have found a general principle, which could be stated as follows: Suppose $S_1, S_2, \ldots, S_n$ is a finite sequence of sets, and suppose I is a subset of $\{1, 2, \ldots, n\}$ such that the union of the sets S_i for $i \in I$ contains fewer elements than the set I does. Then $S_1, S_2, \ldots, S_n$ has no system of distinct representatives. In our example (taking $S_i = P_i$ for $i = 1$ to 6), the set I is $\{1, 3, 4, 6\}$.

Finding such a set I enables us to be sure that no system of distinct representatives exists. The person responsible for assigning summer courses in our example will have to assign the same professor to teach 2 courses if all 6 courses are to be given. Professors with no summer employment may object that this is unfair, but the scheduler can use the set I to demonstrate to them that there is no way to cover all the courses otherwise.

If a sequence of sets has no system of distinct representatives, is there always some set I, as in the preceding discussion, that can be used to demonstrate this fact in a compact way? The answer is yes, but the proof is somewhat complicated. This is the content of a famous theorem due to Phillip Hall.

Theorem 6.1 ***Hall's Theorem*** The sequence of finite sets $S_1, S_2, \ldots, S_n$ has a system of distinct representatives if and only if, whenever I is a subset of $\{1, 2, \ldots, n\}$, then the union of the sets S_i for $i \in I$ contains at least as many elements as the set I does.

The "only if" part of this theorem amounts to the principle we have already discovered. The "if" part will be proved in Section 6.4 in a different context; for a direct proof, see Exercise 31.

Example 6.4 We will use Hall's theorem to show that the sequence

$$S_1 = \{A, C, E\},$$
$$S_2 = \{A, B\},$$
$$S_3 = \{B, E\}$$

has a system of distinct representatives. The subsets I of $\{1, 2, 3\}$ and the corresponding unions of sets S_i are as follows:

I	*Union of Sets* S_i, $i \in I$
$\emptyset$	$\emptyset$
$\{1\}$	$\{A, C, E\}$
$\{2\}$	$\{A, B\}$
$\{3\}$	$\{B, E\}$
$\{1, 2\}$	$\{A, B, C, E\}$
$\{1, 3\}$	$\{A, B, C, E\}$
$\{2, 3\}$	$\{A, B, E\}$
$\{1, 2, 3\}$	$\{A, B, C, E\}$

Since every set on the right has at least as many elements as the corresponding set on the left, the sequence has a system of distinct representatives. Of course, it is easy in this case to find one by inspection, for example, A, B, E. ■

Applying Hall's theorem to our course scheduling example would involve examining the $2^6 = 64$ subsets of $\{1, 2, 3, 4, 5, 6\}$ and computing the corresponding union of sets P_i for each. (Of course, we would find that no system of distinct representatives exists.) Although this may seem better than our previous method, which entailed looking at 288 possible assignments, it is still not practical for finding whether a system of distinct representatives exists, since, if there are n sets S_i, then there are 2^n sets I, and 2^n increases very quickly with n. Also, although the theorem tells us when a system of distinct representatives exists, it does not tell how to find one. Efficient methods for finding optimal matchings will be developed later in this chapter.

Readers interested in extensions of Hall's theorem should consult [7] in the suggested readings at the end of this chapter.

EXERCISES 6.1

In Exercises 1–6, tell how many systems of distinct representatives the given sequence of sets has.

1. $\{1, 2\}, \{2, 3\}, \{1, 3\}$

2. $\{1, 4\}, \{2\}, \{2, 3\}, \{1, 2, 3\}$

3. $\{1, 2, 3\}, \{1, 2, 3\}, \{1, 2, 3\}$

4. $\{1, 2, 3, 4, 5\}, \{1, 2, 3, 4, 5\}$

5. $\{1, 2, 5\}, \{2, 1\}, \{3, 4\}, \{1, 5\}, \{1, 2, 5\}, \{2, 4, 5\}$

6. $\{1, 2, 3\}, \{4, 5\}, \{6, 7\}$

In Exercises 7–10, a sequence of sets $S_1, S_2, \ldots, S_n$ is given. For each subset I of $\{1, 2, \ldots, n\}$, compute the union of the corresponding sets S_i, and determine from these unions whether the sequence has a system of distinct representatives or not.

7. $\{1, 2, 4\}, \{2, 4\}, \{2, 3\}, \{1, 2, 3\}$

8. $\{1, 2, 5\}, \{5, 1\}, \{1, 2\}, \{2, 5\}$

9. $\{1\}, \{1, 2\}, \{1, 2, 3\}, \{1, 3\}$

10. $\{4.5, 5, 7\}, \left\{\frac{9}{2}, 6, 7\right\}, \{\emptyset\}$

In Exercises 11–16, a sequence of sets $S_1, S_2, \ldots, S_n$ is given. Find a subset I of $\{1, 2, \ldots, n\}$ such that the union of the corresponding sets S_i has fewer elements than I does.

11. $\{1, 2\}, \{2, 3\}, \emptyset$

12. $\{1\}, \{1, 2\}, \{2, 3\}, \{2\}$

13. $\{1, 2, 3\}, \{1, 2, 4\}, \{1, 3, 4\}, \{1, 2, 3, 4\}, \{2, 3, 4\}$

14. $\{1, 2\}, \{2, 3\}, \{5\}, \{1, 3\}, \{4, 5\}, \{4, 5\}$

15. $\{2, 5, 7\}, \{1, 3, 4, 5\}, \{5, 7\}, \{2, 7\}, \{1, 3, 6\}, \{2, 5\}$

16. $\{1, 2\}, \{2, 4, 5, 7\}, \{1, 2, 3, 5, 6\}, \{1, 4, 7\}, \{2, 5, 7\}, \{1, 4, 5, 7\}, \{2, 4, 7\}$

17. Let $S_i = \{1, 2, \ldots, n\}$ for $i = 1, 2, \ldots, n$. How many systems of distinct representatives does the sequence $S_1, S_2, \ldots, S_n$ have?

18. Let $S_i = \{1, 2, \ldots, k\}$ for $i = 1, 2, \ldots, n$, where $n \le k$. How many systems of distinct representatives does $S_1, S_2, \ldots, S_n$ have?

19. Let $S_i = \{1, 2, \ldots, k\}$ for $i = 1, 2, \ldots, n$, where $k < n$. How many systems of distinct representatives does $S_1, S_2, \ldots, S_n$ have?

20. Show that if the nonempty set S_i has k_i elements for $i = 1, 2, \ldots, n$, then the sequence $S_1, S_2, \ldots, S_n$ has exactly $k_1 k_2 \cdots k_n$ systems of distinct representatives if and only if the sets S_i are pairwise disjoint.

21. Mr. Jones brought home 6 differently flavored jellybeans for his 6 children. However, when he got home, he found out that each child likes only certain flavors. Amy will eat only chocolate, banana, or vanilla, while Burt likes only chocolate and banana. Chris will eat only banana, strawberry, and peach, and Dan will accept only banana and vanilla. Edsel likes only chocolate and vanilla, and Frank will eat only chocolate, peach, and mint. Show that not every child will get a jellybean he or she likes.

22. Five girls go into a library to get a book. Jennifer wants to read only *The Velvet Room* or *Daydreamer*. Lisa wants only *Summer of the Monkeys* or *The Velvet Room*. Beth and Kim each want only *Jelly Belly* or *Don't Hurt Laurie!*, while Kara wants either one of the latter two books or else *Daydreamer*. If the library has only one copy of each book, can each girl take out a book she wants?

23. Show that if the union of the sets $S_1, S_2, \ldots, S_n$ contains more than n elements, and if the sequence $S_1, S_2, \ldots, S_n$ has a system of distinct representatives, then it has more than one.

24. Let S be a set with m elements, and let $S_i = S$ for $i = 1, 2, \ldots, n$. Show that the number of systems of distinct representatives of $S_1, S_2, \ldots, S_n$ is the same as the number of one-to-one functions from $\{1, 2, \ldots, n\}$ into $\{1, 2, \ldots, m\}$.

25. In the example in Section 1.2, there are 7 cities and a set of pilots who want to fly to each city. Either find a system of distinct representatives for this sequence of sets or else prove that none exists.

26. Let $S_1, S_2, \ldots, S_m$ and $T_1, T_2, \ldots, T_n$ be sequences of sets such that S_i and T_j are disjoint for all i and j. Show that the sequence $S_1, S_2, \ldots, S_m, T_1, T_2, \ldots, T_n$ has a system of distinct representatives if and only if $S_1, S_2, \ldots, S_m$ and $T_1, T_2, \ldots, T_n$ do.

27. Let $S_1, S_2, \ldots, S_n$ be a sequence of sets such that $|S_i| \geq i$ for $i = 1, 2, \ldots, n$. Show that the sequence has a system of distinct representatives.

28. Let $S_i = \{1, 2, \ldots, i\}$ for $i = 1, 2, \ldots, n$. How many systems of distinct representatives does $S_1, S_2, \ldots, S_n$ have?

29. Let $S_i = \{0, 1, 2, \ldots, i\}$ for $i = 1, 2, \ldots, n$. How many systems of distinct representatives does $S_1, S_2, \ldots, S_n$ have?

30. Suppose that $S_i \subseteq S_{i+1}$ for $i = 1, 2, \ldots, n-1$, and that $|S_i| = k_i$ for $i = 1, 2, \ldots, n$. How many systems of distinct representatives does $S_1, S_2, \ldots, S_n$ have?

31. A sequence of finite sets $S_1, S_2, \ldots, S_n$ is said to satisfy **Hall's condition** if, whenever $I \subseteq \{1, 2, \ldots, n\}$, then the number of elements in the union of the sets S_i, $i \in I$, is at least $|I|$. The "if" part of Hall's theorem amounts to the statement that any sequence satisfying Hall's condition has a system of distinct representatives. Prove this by using the strong induction principle on n. To prove the inductive step, consider two cases: (a) Whenever I is a nonempty subset of $\{1, 2, \ldots, k+1\}$ with fewer than $k+1$ elements, the union of the sets S_i for $i \in I$ has at least one more element than I does; and (b) for some nonempty subset I of $\{1, 2, \ldots, k+1\}$ with fewer than $k+1$ elements, the union of the sets S_i for $i \in I$ has the same number of elements as I.

32. For $r \leq n$, an $r \times n$ **Latin rectangle** is an $r \times n$ matrix that has the numbers $1, 2, \ldots, n$ as its entries, with no number occurring more than once in any row or column. An $n \times n$ Latin rectangle is called a **Latin square**. Show that if $r < n$, then it is possible to append $n - r$ rows to an $r \times n$ Latin rectangle to form a Latin square. (*Hint:* Use Hall's theorem.)

6.2 MATCHINGS IN GRAPHS

There is a symmetry in matching problems that is hidden when they are formulated in terms of sets, as in Section 6.1. For example, when we were trying to match a professor with each English course, we associated with each of the six courses a set—the set of professors who could teach that course. But we could just as well have turned the problem around and considered for each professor the set of courses he or she can teach. This symmetry is displayed better if we draw a graph, as we did in Figure 1.10 for the airline pilot problem. We will let the courses and professors be the vertices of the graph, and we will put an edge between a course and a professor whenever the professor can teach the course. The result is shown in Figure 6.1.

The graph we get is of a special form, since no edge joins a course to a course, or a professor to a professor. We say a graph with vertex set $\mathcal{V}$ and edge set $\mathcal{E}$ is **bipartite** in case $\mathcal{V}$ can be written as the union of two disjoint sets $\mathcal{V}_1$ and $\mathcal{V}_2$ such that each edge joins an element of $\mathcal{V}_1$ with an element of $\mathcal{V}_2$. The graph of Figure 6.1 is bipartite, since we could take $\mathcal{V}_1$ to be the set of courses and $\mathcal{V}_2$ to be the set of professors.

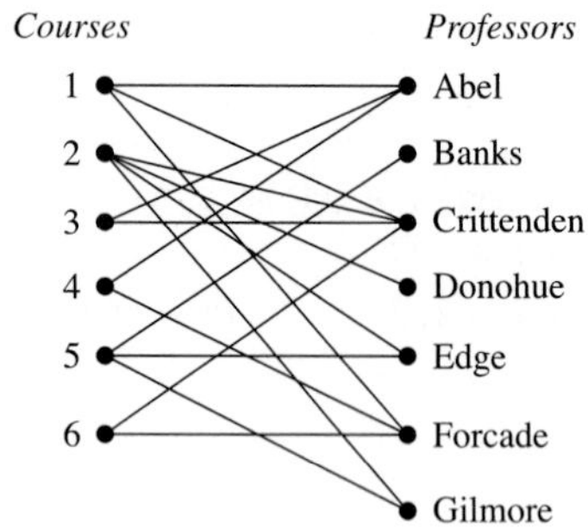

FIGURE 6.1

Example 6.5 The graph shown in Figure 6.2 is bipartite (even though it may not look it) because every edge goes between an odd-numbered vertex and an even-numbered one. Thus we could take $\mathcal{V}_1 = \{1, 3, 5, 7\}$ and $\mathcal{V}_2 = \{2, 4, 6, 8\}$. ■

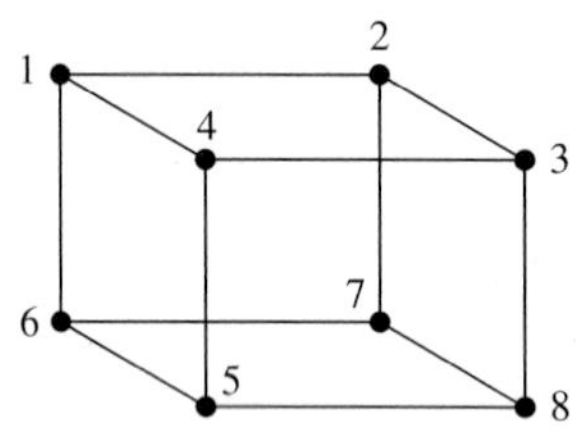

FIGURE 6.2

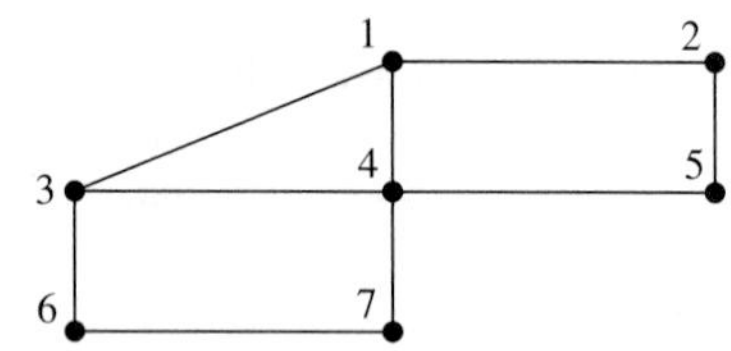

FIGURE 6.3

Example 6.6 The graph shown in Figure 6.3 is not bipartite, as we can see by considering the vertices 1, 3, and 4. If, for example, 1 is in $\mathcal{V}_1$, then 3 must be in $\mathcal{V}_2$. But then, 4 can be in neither of these sets. ■

In our course assignment problem, we wanted to pair up courses and professors. In terms of the graph representing the problem, this means that we want to choose a subset, say, $\mathcal{M}$, of the set of edges. No course can be taught by two professors, nor can a professor teach more than one course. This means that no vertex of the graph can be incident with more than one edge of $\mathcal{M}$. In this application we would like $\mathcal{M}$ to contain as many edges as possible. These considerations motivate the following definitions:

A **matching** of a graph is a set $\mathcal{M}$ of edges such that no vertex of the graph is incident with more than one edge of $\mathcal{M}$. A **maximum matching** is a matching such that no other matching contains more edges.

Example 6.7 The colored edges in Figure 6.4(a) form a matching of the bipartite graph pictured, since no two of them are incident with the same vertex. This matching of 3 edges is not a maximum matching, however, since Figure 6.4(b) shows another matching with 4 edges. Note that even though the first matching is not a maximum matching, no edge could be added to it and still have a matching. A maximum

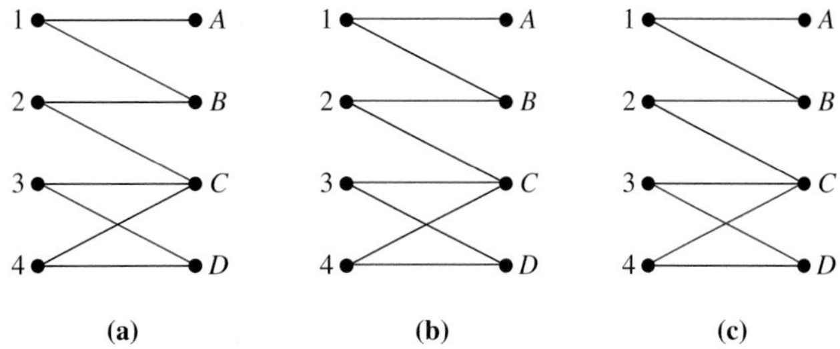

FIGURE 6.4

matching need not be unique. Figure 6.4(c) shows another maximum matching of the graph. ■

Our definition of a matching of a graph did not specify that the graph be bipartite. Finding a maximum matching is easier in the case of a bipartite graph, however, and many applications give rise to bipartite graphs. The following example gives a case when a maximum matching of a nonbipartite graph is desired:

Example 6.8 A group of United Nations peacekeeping soldiers is to be divided into 2-person teams. It is important that the 2 members of a team speak the same language. The following table shows the languages spoken by the 7 soldiers available:

Soldier	*Languages*
1	French, German, English
2	Spanish, French
3	German, Korean
4	Greek, German, Russian, Arabic
5	Spanish, Russian
6	Chinese, Korean, Japanese
7	Greek, Chinese

If we make a graph, putting an edge between 2 soldiers whenever they speak a language in common, the result is exactly the graph of Figure 6.3, which we saw was not bipartite. One matching is pictured in color in Figure 6.5. It is clearly a maximum matching, since only one soldier is unmatched. ■

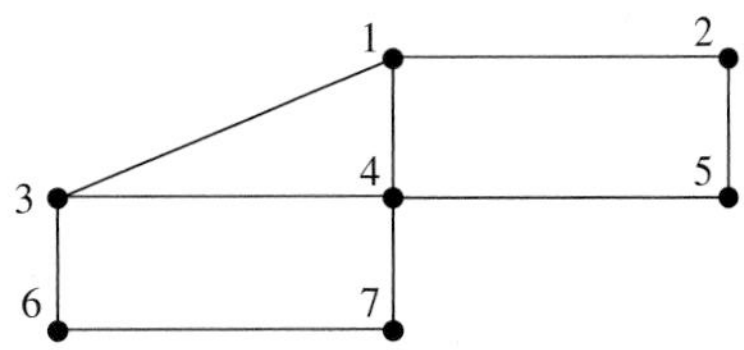

FIGURE 6.5

The Matrix of a Bipartite Graph

A convenient way to represent a bipartite graph where every edge joins a vertex of $\mathcal{V}_1$ to a vertex of $\mathcal{V}_2$ is by a matrix of 0s and 1s, with the rows corresponding to the elements of $\mathcal{V}_1$ and the columns to the elements of $\mathcal{V}_2$. We put a 1 in the matrix whenever the vertices corresponding to the row and column are joined by an edge and a 0 otherwise. For example, the matrix of the graph of Figure 6.4 is

$$\begin{array}{c} \\ 1 \\ 2 \\ 3 \\ 4 \end{array} \begin{array}{c} \begin{array}{cccc} A & B & C & D \end{array} \\ \begin{bmatrix} 1 & 1 & 0 & 0 \\ 0 & 1 & 1 & 0 \\ 0 & 0 & 1 & 1 \\ 0 & 0 & 1 & 1 \end{bmatrix} \end{array}$$

Of course, this matrix is uniquely determined only if we specify some order for the vertices in $\mathcal{V}_1$ and $\mathcal{V}_2$. Recall that a matching of a graph is a subset of its edges, and each edge corresponds to a 1 in the matrix. Two edges incident on the same vertex correspond to 1s in the same row or column of the matrix, depending on whether the vertex is in $\mathcal{V}_1$ or $\mathcal{V}_2$. Thus a matching of a bipartite graph corresponds to some set of 1s in the matrix of the graph, no two of which are in the same row or column. Matrices have their own terminology, however.

By a **line** of a matrix, we mean either a row or a column. Let A be a matrix. We say that a set of entries of A is **independent** if no two of them are in the same line. An independent set of 1s in A is a **maximum independent set** of 1s if no independent set of 1s in A contains more elements.

We will mark the 1s in a particular independent set with stars. The reader should check that the stars in the following three matrices mark independent sets corresponding to the three matchings shown in Figure 6.4:

$$\begin{bmatrix} 1 & 1^* & 0 & 0 \\ 0 & 1 & 1^* & 0 \\ 0 & 0 & 1 & 1^* \\ 0 & 0 & 1 & 1 \end{bmatrix} \qquad \begin{bmatrix} 1^* & 1 & 0 & 0 \\ 0 & 1^* & 1 & 0 \\ 0 & 0 & 1^* & 1 \\ 0 & 0 & 1 & 1^* \end{bmatrix} \qquad \begin{bmatrix} 1^* & 1 & 0 & 0 \\ 0 & 1^* & 1 & 0 \\ 0 & 0 & 1 & 1^* \\ 0 & 0 & 1^* & 1 \end{bmatrix}.$$

For example, since one of the edges in the matching shown in Figure 6.4(a) is $\{1, B\}$, a star is placed on the 1 in row 1 and column B of the first matrix.

Although the language is different, finding a maximum matching in a bipartite graph and finding a maximum independent set of 1s in a matrix of 0s and 1s are really the same problem, and we will use whichever form is more convenient. Graphs are sometimes more accessible to the intuition, while matrices may be better for computational purposes.

Coverings

Recall Example 6.8, where a group of soldiers was to be broken into 2-person teams speaking a common language. Suppose that before any teams are formed, some of the soldiers are to attend a meeting. It is desired that each possible team should have at least one member at the meeting.

Since each edge in the graph of Figure 6.5 represents a possible team, what we need is a set of vertices such that each edge of the graph is incident with at least one vertex in this set. We might want this set to be as small as possible so as to minimize the number of soldiers required to attend the meeting. Such considerations motivate the following definitions:

By a **covering** $\mathcal{C}$ of a graph, we mean a set of vertices such that every edge is incident with at least one vertex in $\mathcal{C}$. We say $\mathcal{C}$ is a **minimum covering** if no covering of the graph has fewer vertices. For example, the set $\{2, 3, 4, 5, 6\}$ may be seen to be a covering of the graph shown in Figure 6.5. This is not a minimum covering, however, since the covering $\{1, 3, 5, 7\}$ has fewer elements.

Example 6.9 Figure 6.6 represents the streets and intersections of the downtown area of a small city. A company wishes to place hot dog stands at certain intersections in such a way that no one in the downtown area will be more than one block from a stand. It would like to do this with as few stands as possible.

If we interpret Figure 6.6 as a graph with vertices at the intersections, then our problem is exactly one of finding a minimum covering. One covering is the set of vertices $\{1, 3, 6, 8, 9, 11\}$. We will see, as a consequence of the next theorem, that this is a minimum covering. ■

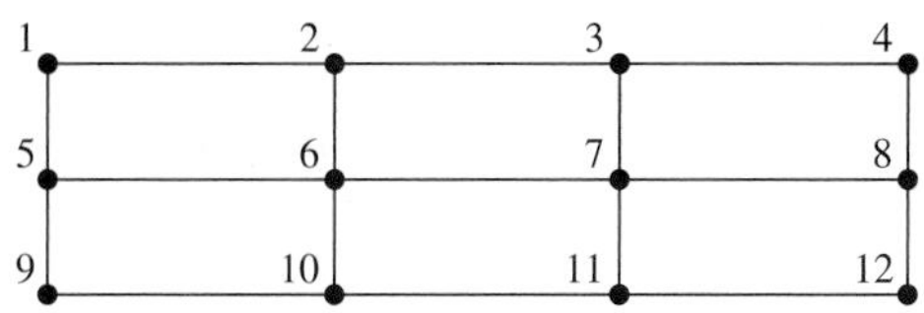

FIGURE 6.6

The next theorem gives a relation between the matchings and the coverings of a graph.

Theorem 6.2 Let a graph have a matching $\mathcal{M}$ and covering $\mathcal{C}$. Then $|\mathcal{M}| \le |\mathcal{C}|$. Moreover, if $|\mathcal{M}| = |\mathcal{C}|$, then $\mathcal{M}$ is a maximum matching and $\mathcal{C}$ is a minimum covering.

Proof. By the definition of a covering, every edge of the graph, and in particular every edge in $\mathcal{M}$, is incident with some vertex in $\mathcal{C}$. If the edge e is in $\mathcal{M}$, let $v(e)$ be a vertex in $\mathcal{C}$ incident with e. Notice that if e_1 and e_2 are distinct edges in $\mathcal{M}$, then $v(e_1)$ and $v(e_2)$ are also distinct, since, by definition, two edges in a matching cannot share a vertex. Thus there are at least as many vertices in $\mathcal{C}$ as edges in $\mathcal{M}$, and so $|\mathcal{M}| \le |\mathcal{C}|$.

Now suppose $|\mathcal{M}| = |\mathcal{C}|$. If $\mathcal{M}$ were not a maximum matching, there would be a matching $\mathcal{M}'$ with $|\mathcal{M}'| > |\mathcal{M}| = |\mathcal{C}|$, contradicting the first part of the theorem. Likewise, if $\mathcal{C}$ were not a minimum covering, there would be a covering with fewer than $|\mathcal{M}|$ vertices, leading to the same contradiction. ■

In light of the second part of this theorem, we can show that the covering given in Example 6.9 is a minimum covering by exhibiting a matching with the same number of elements, namely, six. One is indicated by the colored edges in Figure 6.7; of course, Theorem 6.2 also implies that it is a maximum matching.

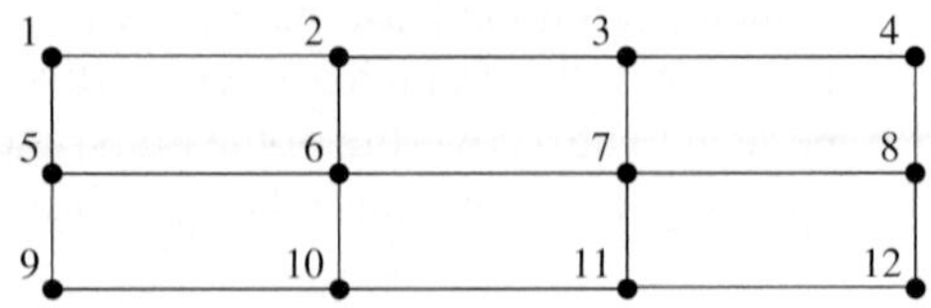

FIGURE 6.7

In the case of a bipartite graph, we can translate Theorem 6.2 into matrix language. The vertices of the graph correspond to the lines of its matrix, and an edge is incident with a vertex when the 1 corresponding to the edge is in the line corresponding to the vertex. Thus we define a **covering** of the 1s of a matrix of 0s and 1s to be a set of lines containing all the 1s of the matrix. It is a **minimum covering** if there is no covering with fewer lines. With these definitions, the next theorem is an immediate consequence of Theorem 6.2.

Theorem 6.3 If a matrix of 0s and 1s has an independent set of m 1s and a covering of c lines, then $m \leq c$. If $m = c$, then the independent set is a maximum independent set and the covering is a minimum covering.

Example 6.10 The Scientific Matchmaking Service has as clients 5 men—Bob, Bill, Ron, Sam, and Ed—and 5 women—Cara, Dolly, Liz, Tammy, and Nan. The company believes that 2 people are compatible exactly when their first names contain a common letter. On the basis of this rule, the company constructs the following matrix, in which a 1 means that the man and woman corresponding to the row and column are compatible:

	Cara	Dolly	Liz	Tammy	Nan
Bob	0	1	0	0	0
Bill	0	1	1	0	0
Ron	1	1	0	0	1
Sam	1	0	0	1	1
Ed	0	1	0	0	0

The company would like to match as many clients as possible; that is, it wants a maximum independent set of 1s. Since all the 1s lie in just 4 lines—namely, the

3rd and 4th rows and 2nd and 3rd columns—it is realized that no independent set of 1s can have more than 4 elements. An independent set with 4 elements does exist, however; one such set is

$$\begin{bmatrix} 0 & 1^* & 0 & 0 & 0 \\ 0 & 1 & 1^* & 0 & 0 \\ 1 & 1 & 0 & 0 & 1^* \\ 1^* & 0 & 0 & 1 & 1 \\ 0 & 1 & 0 & 0 & 0 \end{bmatrix}. \quad \blacksquare$$

EXERCISES 6.2

In Exercises 1–6, tell whether the graph is bipartite, and if so, give disjoint sets of vertices $\mathcal{V}_1$ and $\mathcal{V}_2$ so that every edge joins a vertex of $\mathcal{V}_1$ to a vertex of $\mathcal{V}_2$.

1.
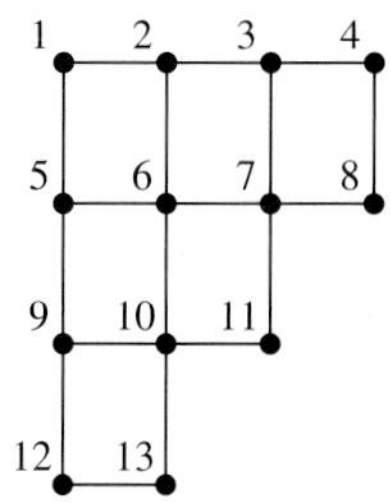

2.
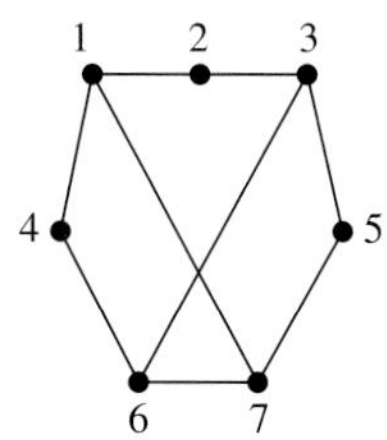

3.
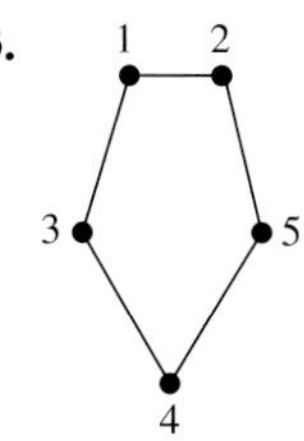

4.
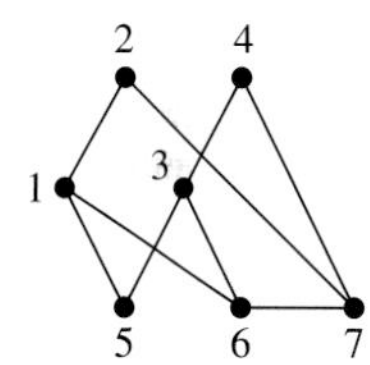

5.
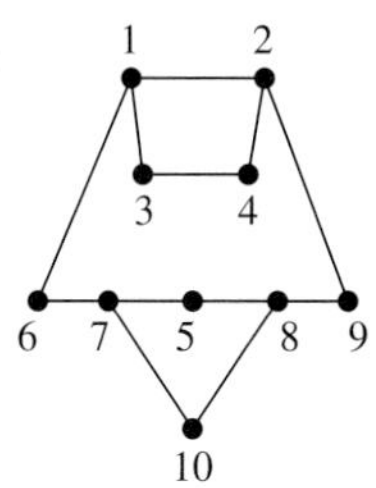

6.
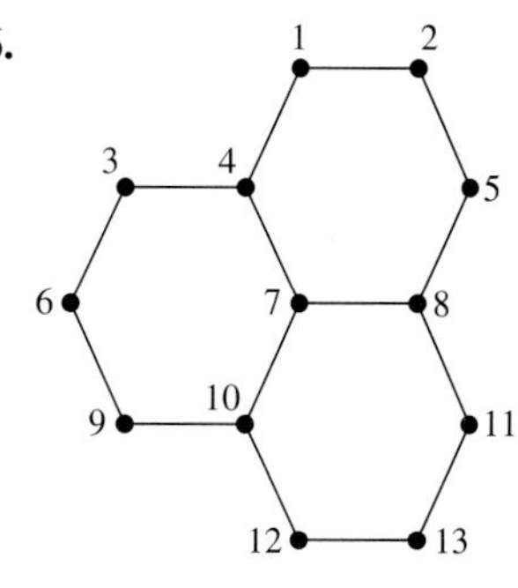

7. Give a maximum matching for each graph in Exercises 1, 2, and 3.

8. Give a maximum matching for each graph in Exercises 4, 5, and 6.

9. Give a minimum covering for each graph in Exercises 1, 2, and 3.

10. Give a minimum covering for each graph in Exercises 4, 5, and 6.

In Exercises 11–16, every edge of the graph joins a vertex of $\mathcal{V}_1 = \{1, 3, 5, \ldots\}$ to one of $\mathcal{V}_2 = \{2, 4, 6, \ldots\}$. Give the matrix of each graph. (Take the vertices in increasing order.)

11.

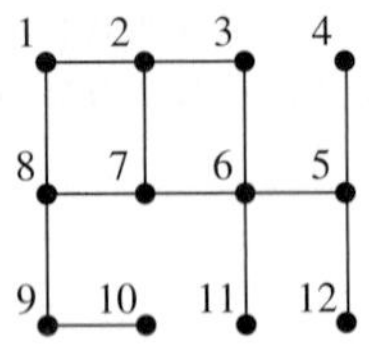

12.

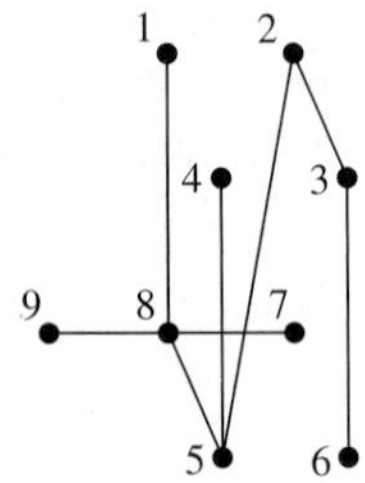

13.

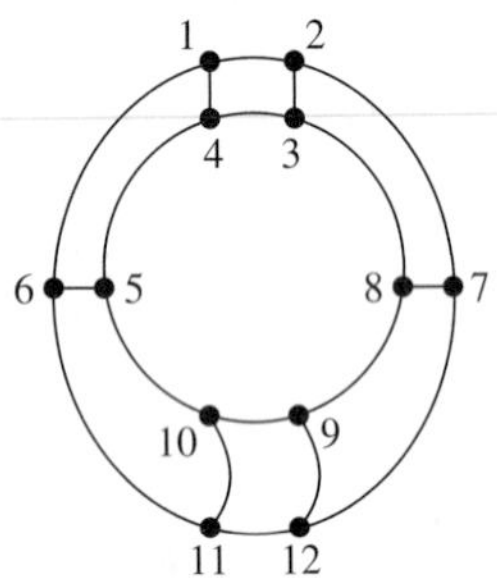

14.

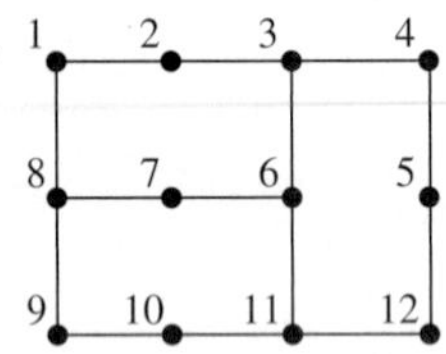

15.

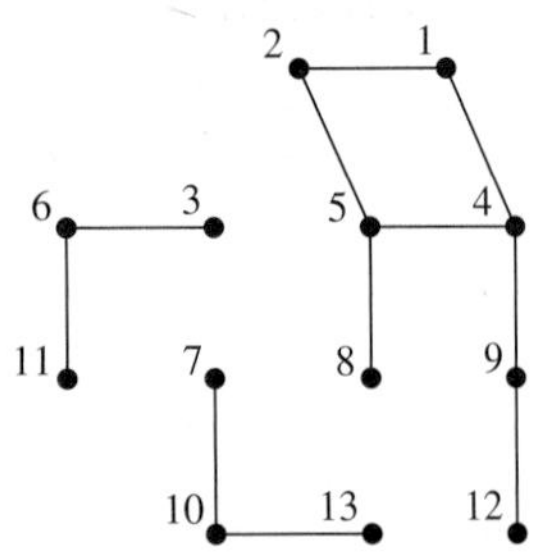

16.

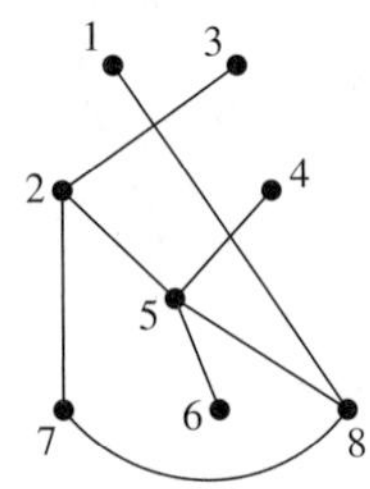

17. Find a maximum independent set of 1s for the matrices in Exercises 11, 12, and 13.

18. Find a maximum independent set of 1s for the matrices in Exercises 14, 15, and 16.

19. Find a minimum covering for the matrices of Exercises 11, 12, and 13.

20. Find a minimum covering for the matrices of Exercises 14, 15, and 16.

In Exercises 21 and 22, construct a bipartite graph and the corresponding matrix modeling the situation described. Indicate a maximum matching in the graph and the corresponding maximum independent set of 1*s in the matrix.*

21. Four airplane passengers want to read a magazine, but only five are available. Of these, Mr. Brown will only read *Time*, *Newsweek*, or *Fortune*; Ms. Garvey will only read *Newsweek* or *Organic Gardening*; Miss Rollo will only read *Organic Gardening* or *Time*; and Mrs. Onishi will only read *Fortune* or *Sunset*.

22. The Glumby family is going to Europe, and each member is to choose one country that he or she knows the language of to study beforehand. Mr. Glumby knows Russian and French; Mrs. Glumby knows only Russian; Sally knows French, German, and Spanish; and Tim knows only French.

In Exercises 23 and 24, construct a graph modeling the situation described, and find a maximum matching for it.

23. The church sewing circle wants to break into two-person groups to make altar cloths. The two people in a group should own the same brand of sewing machine. Ann has a Necchi; Beth has a Necchi and a Singer; Cora has a Necchi, a Singer, and a White; Debby has a Singer, a White, and a Brother; Ellie has a White and a Brother; and Felicia has a Brother.

24. The Weight Whittler Club wants to break up into two-person support groups. The weights of the two people in a group should differ by no more than 20 pounds. Andrew weighs 185, Bob 250, Carl 215, Dan 210, Edward 260, and Frank 205.

In Exercises 25 and 26, model the situation described with a bipartite graph, and construct the corresponding matrix. Find a minimum covering for the graph, and indicate the corresponding lines of the matrix.

25. The police department has a policy of putting an experienced officer together with a rookie in a squad car. The experienced officers are Anderson, Bates, Coony, and Dotson, and the rookies are Wilson, Xavier, Yood, and Zorn. Anderson always works with Wilson or Xavier; Bates with Xavier, Yood, or Zorn; Coony with Wilson; and Dotson with Wilson or Xavier. The captain, who does not know what the teams for the next month will be, would like to call as small a number of officers as possible to tell at least one member of each team its schedule.

26. In mixed doubles Roger always plays with Venus, Maria, or Serena; Andre plays with Lindsay; Lleyton always plays with Venus or Lindsay; and Andy always plays with Venus. The tournament director wants to tell each possible mixed-doubles pair its rating with as few phone calls as possible.

27. Show that the matrix of a bipartite graph is a submatrix of the adjacency matrix of that graph, as defined in Section 4.1. A **submatrix** of a matrix A is a matrix formed by removing some rows or columns (or both) from A.

28. Find a graph in which a maximum matching has fewer edges than a minimum covering has vertices.

29. Show that if a graph contains a cycle with an odd number of edges, then it is not bipartite.

30. Show that if a graph contains no cycle with an odd number of edges, then it is bipartite.

31. Show that a graph is bipartite if and only if it can be colored with two colors.

32. Consider $\mathcal{K}_{20}$, the complete graph on 20 vertices.

(a) How many edges are in a maximum matching for $\mathcal{K}_{20}$?

(b) How many vertices are in a minimum covering for $\mathcal{K}_{20}$?

33. Let a graph have a vertex set $\mathcal{V}$, edge set $\mathcal{E}$, and adjacency matrix $A = [a_{ij}]$. Show that a subset $\mathcal{C}$ of $\mathcal{V}$ is a covering if and only if $a_{ij} = 0$ whenever $i \notin \mathcal{C}$ and $j \notin \mathcal{C}$.

6.3 A MATCHING ALGORITHM

So far, our examples have been small enough that we could find a maximum matching by trial and error. For larger graphs, however, a better technique is needed; and, as was indicated in Section 6.1, simple exhaustion of all possibilities soon becomes impractical, even with a computer. There is an efficient algorithm for finding a maximum matching in a graph. For simplicity, we consider the algorithm only for the case of a bipartite graph. To make explaining the algorithm easier, we will present it as a method of finding a maximum independent set of 1s in a matrix of 0s and 1s. As we saw in the previous section, this is equivalent to the problem of finding a maximum matching in a bipartite graph.

We will give an example of the use of the algorithm before we state it in a more formal way later in this section. We start with some independent set of 1s. This set could be found by inspection and could even be the empty set! Starting with a larger independent set will speed up finding a maximum such set, however. The algorithm will either tell us that we have a maximum independent set of 1s, or else produce an independent set containing one more 1. We continue to apply the algorithm until a maximum independent set is reached.

For our example, we will use the matrix

$$
\begin{array}{cc}
 & \begin{array}{cccc} A & B & C & D \end{array} \\
\begin{array}{c} 1\\2\\3\\4 \end{array} & \begin{bmatrix} 1^* & 0 & 1 & 1 \\ 0 & 1^* & 0 & 0 \\ 1 & 1 & 0 & 0 \\ 0 & 1 & 0 & 0 \end{bmatrix}
\end{array}
$$

in which an independent set of 1s has been indicated. Notice that if any 1 is added to this set, it will no longer be independent. Our algorithm will involve performing two operations on some of the lines of this matrix, operations which we will call *labeling* and *scanning*. Once a line has been labeled, it will never be labeled again in one application of the algorithm, and the same is true for scanning. A line must be labeled before it can be scanned. We begin by labeling (with the symbol "#") all columns containing no starred 1s. (If there are no such columns, our set of starred 1s is already a maximum independent set.) In our example, this produces the following matrix:

$$
\begin{array}{cc}
 & \begin{array}{cccc} A & B & C & D \end{array} \\
\begin{array}{c} 1\\2\\3\\4 \end{array} & \begin{bmatrix} 1^* & 0 & 1 & 1 \\ 0 & 1^* & 0 & 0 \\ 1 & 1 & 0 & 0 \\ 0 & 1 & 0 & 0 \end{bmatrix} \\
 & \begin{array}{cccc} & & \# & \# \end{array}
\end{array}
$$

Now we scan each labeled column for *unstarred* 1s. In column C we find an unstarred 1 in the first row, so we label that row with a C to indicate that the unstarred 1 was found in column C. (In general, row labels are column names, and column labels are row names, except for the labels "#.") Then we put a check mark under column C to indicate that it has been scanned. The matrix now appears as follows:

$$\begin{array}{c|cccc|c} & A & B & C & D & \\ 1 & 1^* & 0 & 1 & 1 & C \\ 2 & 0 & 1^* & 0 & 0 & \\ 3 & 1 & 1 & 0 & 0 & \\ 4 & 0 & 1 & 0 & 0 & \\ & & & \#\surd & \# & \end{array}$$

When we scan column D, we also find an unstarred 1 in row 1. Since this row has already been labeled, we put a check mark under column D to indicate that it also has now been scanned.

Since all labeled columns have been scanned, we now turn our attention to the rows. Only row 1 has been labeled; so we scan it, now looking for *starred* 1s. There is one in column A; so we label this column with a 1 (the row scanned) and put a check mark after row 1 to show that it has been scanned:

$$\begin{array}{c|cccc|c} & A & B & C & D & \\ 1 & 1^* & 0 & 1 & 1 & C\surd \\ 2 & 0 & 1^* & 0 & 0 & \\ 3 & 1 & 1 & 0 & 0 & \\ 4 & 0 & 1 & 0 & 0 & \\ & 1 & & \#\surd & \#\surd & \end{array}$$

Since all labeled rows have been scanned, we go back to scanning columns. Column A is labeled but not scanned; so we scan it for unstarred 1s. There is one in row 3; we label that row with an A, since we found it when scanning column A:

$$\begin{array}{c|cccc|c} & A & B & C & D & \\ 1 & 1^* & 0 & 1 & 1 & C\surd \\ 2 & 0 & 1^* & 0 & 0 & \\ 3 & 1 & 1 & 0 & 0 & A \\ 4 & 0 & 1 & 0 & 0 & \\ & 1\surd & & \#\surd & \#\surd & \end{array}$$

We are now at a turning point in the algorithm. When we scan the labeled row 3, we find no starred 1, and so we mark this row with an exclamation point. This indicates that we will be able to improve on the independent set of 1s we started with. The labels on the lines of the matrix tell us exactly how to do this. Row 3 is labeled with an A, so we put a circle around the 1 in column A (and row 3). This column is labeled with a 1, so we put a circle around the starred 1 in row 1 (and column A). Row 1 is labeled with a C, so we put a circle around the 1 in column C (and row 1). Column C is labeled with the symbol "#," so we stop drawing circles at this point. Our matrix now appears as follows:

$$\begin{array}{c} \\ 1 \\ 2 \\ 3 \\ 4 \end{array}
\begin{array}{c} \begin{array}{cccc} A & B & C & D \end{array} \\ \left[\begin{array}{cccc} \textcircled{1*} & 0 & \textcircled{1} & 1 \\ 0 & 1^* & 0 & 0 \\ \textcircled{1} & 1 & 0 & 0 \\ 0 & 1 & 0 & 0 \end{array}\right] \end{array}
\begin{array}{l} \\ C\checkmark \\ \\ A! \\ \\ \end{array}$$

$$1\checkmark \quad \#\checkmark \#\checkmark$$

At this point, we find a larger independent set of 1s by reversing the stars on the circled 1s, that is, by adding a star to any circled 1 without a star, and removing the star from any circled 1 with a star. The result is an independent set of 1s with 3 elements instead of 2:

$$\begin{array}{c} \\ 1 \\ 2 \\ 3 \\ 4 \end{array}
\begin{array}{c} \begin{array}{cccc} A & B & C & D \end{array} \\ \left[\begin{array}{cccc} 1 & 0 & 1^* & 1 \\ 0 & 1^* & 0 & 0 \\ 1^* & 1 & 0 & 0 \\ 0 & 1 & 0 & 0 \end{array}\right] \end{array}$$

It is instructive to see what we have done in this example in terms of graphs. Figure 6.8(a) shows the bipartite graph corresponding to our matrix, with the matching of our original set of two 1s indicated in color.

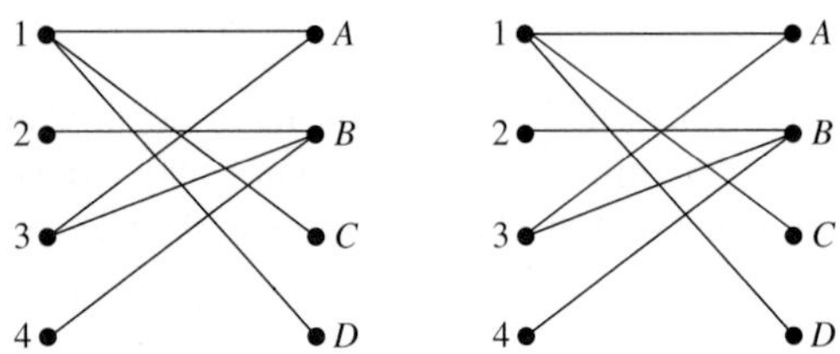

(a) (b)

FIGURE 6.8

The three positions circled in our matrix operations correspond to the edges $\{3, A\}$, $\{1, A\}$, and $\{1, C\}$ of the graph. These form a simple path from 3 to A to 1 to C. (See Figure 6.9(a).) Since the circled 1s of the matrix are alternately starred and unstarred, the edges of this path are alternately in and not in the original matching. Note that, when inserting the circles in our matrix, we start with an unstarred 1 (corresponding to a labeled row with no stars) and also end with an unstarred 1 (corresponding to a column labeled with the symbol "#" because it contained no stars). Thus the number of edges in the path must be odd; and reversing which of these edges are in the matching, as shown in Figure 6.9(b), increases the number of edges in our matching by one. The larger matching is shown in Figure 6.8(b).

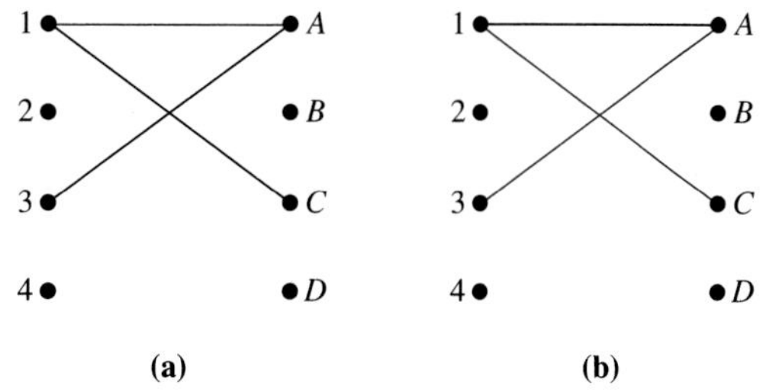

FIGURE 6.9

Applying the Algorithm to a Maximum Independent Set

Now we will apply the algorithm to the matrix with our new set of three starred 1s. Since only column D has no starred 1s, we start by labeling that column:

$$
\begin{array}{c|cccc|c}
 & A & B & C & D & \\
\hline
1 & 1 & 0 & 1^* & 1 & \\
2 & 0 & 1^* & 0 & 0 & \\
3 & 1^* & 1 & 0 & 0 & \\
4 & 0 & 1 & 0 & 0 & \\
\hline
 & & & & \# &
\end{array}
$$

Scanning this column for unstarred 1s leads us to label row 1:

$$
\begin{array}{c|cccc|c}
 & A & B & C & D & \\
\hline
1 & 1 & 0 & 1^* & 1 & D \\
2 & 0 & 1^* & 0 & 0 & \\
3 & 1^* & 1 & 0 & 0 & \\
4 & 0 & 1 & 0 & 0 & \\
\hline
 & & & & \#\surd &
\end{array}
$$

Scanning row 1 for starred 1s leads us to label column C:

$$
\begin{array}{c|cccc|c}
 & A & B & C & D & \\
\hline
1 & 1 & 0 & 1^* & 1 & D\surd \\
2 & 0 & 1^* & 0 & 0 & \\
3 & 1^* & 1 & 0 & 0 & \\
4 & 0 & 1 & 0 & 0 & \\
\hline
 & & & 1 & \#\surd &
\end{array}
$$

This is the most important point in the present application of the algorithm. When we scan column C, there is nothing to label, and we have the following matrix:

$$\begin{array}{c} \\ 1 \\ 2 \\ 3 \\ 4 \\ \\ \end{array}\begin{array}{c} \begin{array}{cccc} A & B & C & D \end{array} \\ \left[\begin{array}{cccc} 1 & 0 & 1^* & 1 \\ 0 & 1^* & 0 & 0 \\ 1^* & 1 & 0 & 0 \\ 0 & 1 & 0 & 0 \end{array}\right] \\ \begin{array}{cccc} & & 1\surd & \#\surd \end{array} \end{array}\begin{array}{l} \\ D\surd \\ \\ \\ \\ \\ \end{array}$$

All lines that are labeled have also been scanned, and there is nothing else we can do. This indicates that we started with a maximum independent set of 1s.

We now state our algorithm formally.

Independent Set Algorithm

Given an independent set of starred 1s in a matrix of 0s and 1s, this algorithm either indicates that this independent set is a maximum independent set, or else it finds a larger independent set.

Step 1 (start)
Label each column containing no starred 1.

Step 2 (scan and label)
repeat
Step 2.1 (scan columns)
For each column that is labeled but not scanned, look at every unstarred 1 in that column. If such a 1 is in an unlabeled row, then label that row with the name of the column being scanned. Mark the column to indicate that it has been scanned.
Step 2.2 (scan rows)
For each row that is labeled but not scanned, look for a starred 1 in that row. If there is a starred 1 in the row, then label the column containing the starred 1 with the name of the row being scanned. Mark the row to indicate that it has been scanned.
until either some labeled row contains no starred 1 or all the labeled rows and all the labeled columns have been scanned

Step 3 (enlarge the independent set if possible)
if some labeled row contains no starred 1
Step 3.1 (backtracking)
Find the first labeled row that contains no starred 1. Circle the 1 in this row and in the column that the row is labeled with. Circle the starred 1 in this column and the row that this column is labeled with. Then circle the unstarred 1 in this new row and in the column that this

row is labeled with. Continue in this manner until a 1 is circled in a column labeled in step 1.

Step 3.2 (larger independent set)

Reverse the stars on all the circled 1s. This gives an independent set of 1s with one more element than the original set.

otherwise

Step 3.3 (no improvement)

The present independent set is a maximum independent set.

end if

This algorithm is due to Ford and Fulkerson and can be found in suggested reading [3] at the end of this chapter. In the next section, we will prove that it does what it says it does. Of course, with a change of language the algorithm could just as well be applied to a graph, but there are complications if it is not bipartite. A modification of the algorithm that applies to arbitrary graphs can be found in suggested reading [2].

Let us examine the complexity of this algorithm. In our analysis, we will use the word "operation" in a somewhat vague way to indicate looking at some entry or row or column label of a matrix and perhaps taking some simple action such as applying or changing a symbol.

Suppose a matrix of 0s and 1s has m rows and n columns. Step 1 involves looking at all mn entries in the matrix, which we count as mn operations. After this, the algorithm alternates between steps 2.1 and 2.2, both of which involve scanning. In order to scan one of the n columns, we need to look at the m entries in that column, so all column scanning will take at most mn operations. Likewise, row scanning will take at most nm operations.

If we get to step 3.3, we are done, so we analyze steps 3.1 and 3.2. Backtracking will take at most $m + n$ operations, since each 1 we circle can be associated with a distinct row or column. Actually, we could combine step 3.2 into step 3.1 with no additional work, reversing the stars as we backtracked. Thus one application of the algorithm will take at most $3mn + m + n$ operations. To build up to a maximum independent set of 1s, the algorithm will have to be repeated at most $\min\{m, n\}$ times, even if we start with the empty set as our first independent set of 1s. Thus the complexity of the algorithm for finding a maximum independent set of ones in an m by n matrix is of order no more than $(3mn + m + n) \cdot \min\{m, n\}$. For the case $m = n = 30$, fewer than 90,000 operations would be necessary, and a fast computer could do the problem in less than one second.

Assigning Courses

As another example of the use of the algorithm, we will go back to the example of Section 6.1 that involved assigning English professors to courses. The matrix of the graph shown in Figure 6.1 is

	A	*B*	*C*	*D*	*E*	*F*	*G*
1	1*	0	1	0	0	1	0
2	0	0	1*	1	1	0	1
3	1	0	1	0	0	0	0
4	1	0	0	0	0	1*	0
5	0	1*	0	0	1	0	1
6	0	0	1	0	0	1	0

The independent set of 1s shown was chosen by taking the first available 1 in the first row, second row, etc., subject to the condition that we not choose two 1s in the same column. We will show what our matrix looks like after each step in the algorithm:

	A	*B*	*C*	*D*	*E*	*F*	*G*
1	1*	0	1	0	0	1	0
2	0	0	1*	1	1	0	1
3	1	0	1	0	0	0	0
4	1	0	0	0	0	1*	0
5	0	1*	0	0	1	0	1
6	0	0	1	0	0	1	0
				#	#		#

After step 1

	A	*B*	*C*	*D*	*E*	*F*	*G*	
1	1*	0	1	0	0	1	0	
2	0	0	1*	1	1	0	1	*D*
3	1	0	1	0	0	0	0	
4	1	0	0	0	0	1*	0	
5	0	1*	0	0	1	0	1	*E*
6	0	0	1	0	0	1	0	
				#√	#√		#√	

After step 2.1

	A	*B*	*C*	*D*	*E*	*F*	*G*	
1	1*	0	1	0	0	1	0	
2	0	0	1*	1	1	0	1	*D*√
3	1	0	1	0	0	0	0	
4	1	0	0	0	0	1*	0	
5	0	1*	0	0	1	0	1	*E*√
6	0	0	1	0	0	1	0	
		5	2	#√	#√		#√	

After step 2.2

	A	B	C	D	E	F	G	
1	1*	0	1	0	0	1	0	C
2	0	0	1*	1	1	0	1	D√
3	1	0	1	0	0	0	0	C
4	1	0	0	0	0	1*	0	
5	0	1*	0	0	1	0	1	E√
6	0	0	1	0	0	1	0	C
		5√	2√	#√	#√		#√	

After step 2.1

	A	B	C	D	E	F	G	
1	1*	0	1	0	0	1	0	C
2	0	0	(1*)	①	1	0	1	D√
3	1	0	①	0	0	0	0	C!
4	1	0	0	0	0	1*	0	
5	0	1*	0	0	1	0	1	E√
6	0	0	1	0	0	1	0	C
	1	5√	2√	#√	#√		#√	

After step 3.1

	A	B	C	D	E	F	G
1	1*	0	1	0	0	1	0
2	0	0	1	1*	1	0	1
3	1	0	1*	0	0	0	0
4	1	0	0	0	0	1*	0
5	0	1*	0	0	1	0	1
6	0	0	1	0	0	1	0

After step 3.2

If the algorithm is now applied with this new independent set of 1s, it tells us that we have a maximum independent set of 1s, and we end up with the following configuration:

	A	B	C	D	E	F	G	
1	1*	0	1	0	0	1	0	
2	0	0	1	1*	1	0	1	E√
3	1	0	1*	0	0	0	0	
4	1	0	0	0	0	1*	0	
5	0	1*	0	0	1	0	1	E√
6	0	0	1	0	0	1	0	
		5√		2√	#√		#√	

It should be noted that in steps 2.1 and 2.2 the order in which the labeled, but not scanned, lines are chosen may affect what larger independent set of 1s the algorithm produces. In computing answers for the examples in this section, we always chose the rows from top to bottom and the columns from left to right.

EXERCISES 6.3

Throughout these exercises, when applying the independent set algorithm, choose rows from top to bottom and columns from left to right. In Exercises 1 and 2, one stage in the application of the independent set algorithm is shown. Apply step 2.1 or 2.2, as appropriate.

1.
$$\begin{array}{c|cccc|c} & A & B & C & D & \\ \hline 1 & 0 & 1^* & 0 & 1 & D \\ 2 & 1^* & 0 & 0 & 1 & D \\ 3 & 1 & 1 & 0 & 0 & \\ 4 & 1 & 0 & 0 & 0 & \\ \hline & & & \#\checkmark & \#\checkmark & \end{array}$$

2.
$$\begin{array}{c|ccccc|c} & A & B & C & D & E & \\ \hline 1 & 1^* & 0 & 1 & 0 & 1 & C\checkmark \\ 2 & 0 & 1^* & 0 & 1 & 1 & E\checkmark \\ 3 & 1 & 1 & 0 & 0 & 0 & \\ 4 & 0 & 1 & 0 & 1^* & 0 & \\ \hline & 1 & 2 & \#\checkmark & & \#\checkmark & \end{array}$$

In Exercises 3 and 4, the matrix is ready for step 3.1. What entries should be circled?

3.
$$\begin{array}{c|cccc|c} & A & B & C & D & \\ \hline 1 & 0 & 1^* & 1 & 0 & C\checkmark \\ 2 & 1^* & 0 & 0 & 1 & D\checkmark \\ 3 & 1 & 1 & 0 & 0 & A! \\ 4 & 0 & 0 & 1^* & 1 & D\checkmark \\ \hline & 2\checkmark & 1 & 4\checkmark & \#\checkmark & \end{array}$$

4.
$$\begin{array}{c|ccccc|c} & A & B & C & D & E & \\ \hline 1 & 0 & 0 & 1^* & 0 & 1 & E\checkmark \\ 2 & 1^* & 0 & 1 & 0 & 0 & C\checkmark \\ 3 & 0 & 1^* & 0 & 1 & 0 & D\checkmark \\ 4 & 1 & 0 & 1 & 0 & 0 & C! \\ \hline & 2 & 3\checkmark & 1\checkmark & \#\checkmark & \#\checkmark & \end{array}$$

In Exercises 5–10, a matrix is given with an independent set of 1s. Use the independent set algorithm until it ends in step 3.3.

5. $\begin{bmatrix} 0 & 1^* & 0 & 1 \\ 1^* & 1 & 0 & 0 \\ 0 & 0 & 1^* & 1 \\ 1 & 1 & 1 & 0 \end{bmatrix}$

6. $\begin{bmatrix} 1^* & 0 & 0 & 1 \\ 1 & 0 & 1^* & 0 \\ 1 & 1^* & 0 & 0 \\ 0 & 1 & 1 & 0 \end{bmatrix}$

7. $\begin{bmatrix} 1^* & 1 & 0 & 1 & 1 \\ 1 & 0 & 0 & 0 & 1^* \\ 0 & 1^* & 0 & 1 & 0 \\ 1 & 1 & 0 & 0 & 1 \end{bmatrix}$

8. $\begin{bmatrix} 0 & 0 & 1^* & 1 & 0 \\ 0 & 1^* & 1 & 0 & 0 \\ 1^* & 1 & 0 & 0 & 0 \\ 0 & 1 & 1 & 0 & 0 \end{bmatrix}$

9. $\begin{bmatrix} 1^* & 1 & 1 & 1 & 1 \\ 1 & 0 & 0 & 0 & 0 \\ 0 & 1^* & 0 & 0 & 0 \\ 1 & 1 & 0 & 0 & 0 \\ 1 & 0 & 1^* & 0 & 1 \end{bmatrix}$

10. $\begin{bmatrix} 0 & 1^* & 0 & 1 & 0 \\ 0 & 0 & 1^* & 1 & 0 \\ 0 & 1 & 1 & 0 & 0 \\ 0 & 1 & 1 & 0 & 0 \\ 1^* & 0 & 1 & 1 & 1 \end{bmatrix}$

In Exercises 11–16, a bipartite graph is given with a matching. Convert it to a matrix and find a maximum matching by using the independent set algorithm, starting with the corresponding independent set of 1*s.*

11.

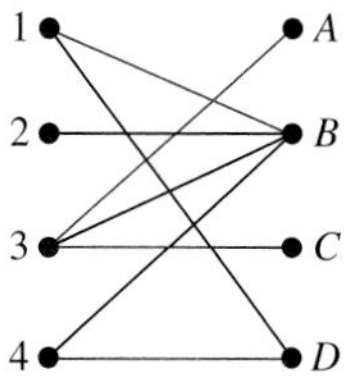

12.

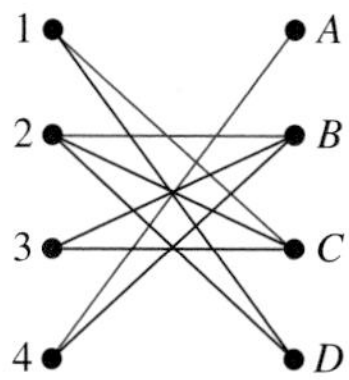

13.

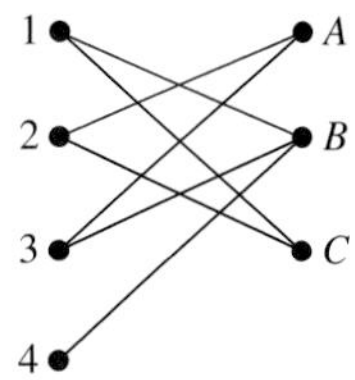

14.

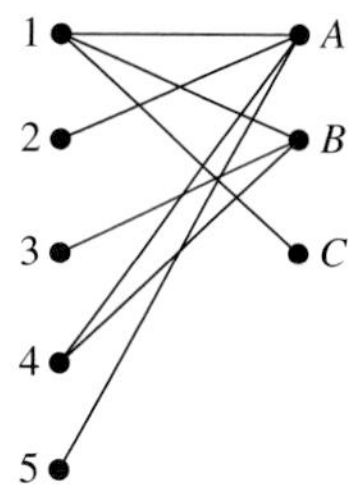

15.

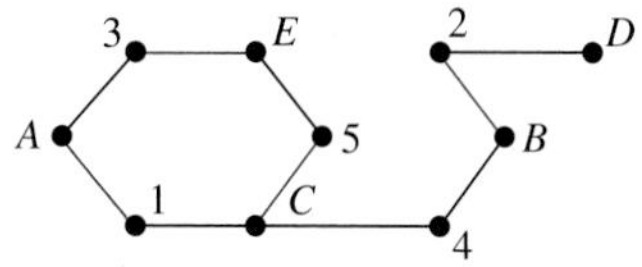

16.

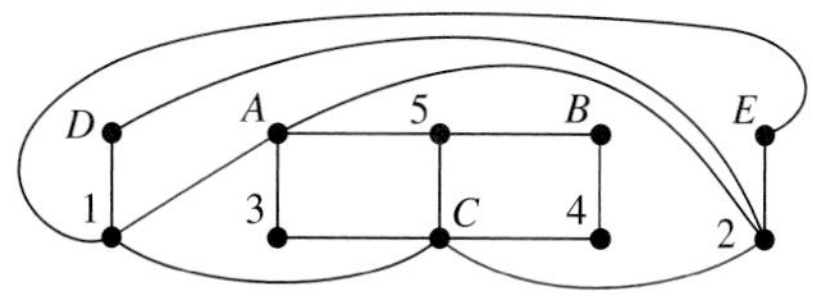

In Exercises 17–22, a sequence of sets is given with distinct elements in some of the sets starred. Convert it to a matrix and use the independent set algorithm to find a system of distinct representatives, if possible. Start with the corresponding independent set of 1*s.*

17. $\{B^*\}, \{A^*, B, C, D\}, \{A, B\}, \{B, D^*\}$

18. $\{C^*, D\}, \{A^*, B\}, \{A, D^*\}, \{A, C, D\}$

19. $\{W^*\}, \{Y^*, Z\}, \{W, Y\}, \{W, X^*, Y, Z\}$

20. $\{1^*, 3, 5\}, \{1, 4^*\}, \{2^*, 3, 5\}, \{1, 2, 4\}$

21. {carrot*, egg}, {apple*, banana, date, fennel}, {apple, carrot, egg*}, {apple, carrot, egg}

22. $\{5^*, 13\}, \{1^*, 6, 9\}, \{1, 5\}, \{1, 6^*, 13\}$

23. Five ships, the Arabella, Constantine, Drury, Egmont, and Fungo, arrive at five loading docks. For technical reasons, each dock can accept only certain ships. Dock 1 can accept only the Constantine or Drury. Likewise, Dock 2 can accept only the Egmont or Fungo; Dock 3 the Constantine, Egmont, or Fungo; Dock 4 the Arabella, Drury, or Fungo; and Dock 5 the Arabella, Constantine, or Egmont. The harbormaster sends the Constantine to Dock 1, the Egmont to Dock 2, the Fungo to Dock 3, and the Arabella to Dock 4. Use the independent set algorithm to improve on this, if possible.

24. A radio station wants to play an hour of rock music, followed by an hour each of classical, polka, and rap. Six disk jockeys are available, but each has his or her scruples. Only Barb, Cal, Deb, and Felicia are willing to play rock. Likewise, only Andy, Barb, Erika, and Felicia will play classical; only Barb, Deb, and Felicia will play polkas; and only Andy, Barb, and Deb will play rap. No disk jockey is allowed to work more than one hour per day. At present, the station manager plans to use Barb, Andy, and Deb for the first 3 hours, but has no one left for the rap hour. Use the independent set algorithm to find a better matching.

6.4 APPLICATIONS OF THE ALGORITHM

In this section we will prove that our independent set algorithm actually does what it claims to do. At the same time, we will find that the algorithm actually leads to more insight about the relations among independent sets, coverings, and systems of distinct representatives. We start with a sequence of short lemmas concerning the results of applying the algorithm.

Lemma 6.1 If the algorithm gets to step 3.1, then it produces an independent set of 1s with more elements than the original one.

Proof. What goes on in the backtracking process was indicated in the previous section. Schematically, a pattern such as that shown in Figure 6.10 emerges. According to the way the algorithm works, the circled symbols form an alternating sequence of unstarred and starred 1s, beginning and ending with an unstarred 1. Reversing the stars on these clearly increases the number of starred 1s by one. The new set is still independent, since, if a 1 given a star were in a line with any starred 1, the latter 1's star is removed in step 3.2. ■

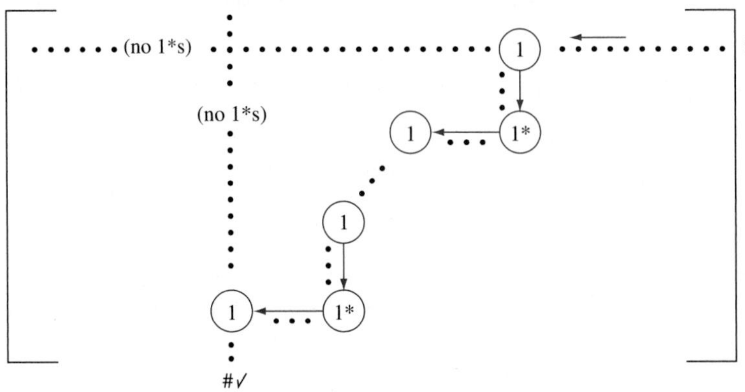

FIGURE 6.10

Now we prove some lemmas about what our matrix looks like if step 3.3 is reached and the algorithm indicates that we have a maximum set of starred 1s. Recall that in this case each line that has been labeled has also been scanned.

Lemma 6.2 If step 3.3 is reached, then the labeled rows and unlabeled columns form a covering.

Proof. If not, then some 1 is at the same time in an unlabeled row and labeled column. If this 1 is starred, then its column can only have been labeled when its row was scanned, contradicting the fact that the row is unlabeled. But if the 1 is unstarred, then when its column was scanned, its row would have been labeled—another contradiction. ■

Lemma 6.3 If step 3.3 is reached, then each labeled row and unlabeled column contains a starred 1.

Proof. On the one hand, each unlabeled column contains a starred 1, since columns that do not are labeled at step 1. On the other hand, if a labeled row contained no starred 1, we would go to step 3.1 instead of step 3.3. ■

Lemma 6.4 If step 3.3 is reached, then no starred 1 is in both a labeled row and unlabeled column.

Proof. If a starred 1 is in a labeled row, then its column is labeled when the row is scanned. ■

Theorem 6.4 The independent set algorithm increases the number of elements when applied to an independent set that is not a maximum independent set. When applied to an independent set that is a maximum independent set, it tells us so.

Proof. The flow of the algorithm is shown in Figure 6.11.

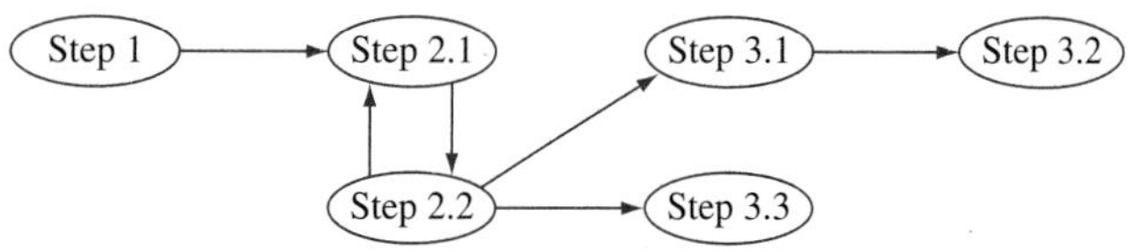

FIGURE 6.11

In steps 2.1 and 2.2, columns and rows are scanned. Since a matrix has only a finite number of lines, the algorithm eventually gets to step 3.2 or step 3.3. If it gets to step 3.2, then Lemma 6.1 tells us that it constructs an independent set with more elements than the one with which we started.

It remains to show that if the algorithm gets to step 3.3, then the independent set we started with is actually a maximum independent set. According to Lemma 6.2, the labeled rows and unlabeled columns form a covering. But Lemmas 6.3 and 6.4 say that the lines in this covering are in one-to-one correspondence with our independent set. Thus the covering and the independent set contain the same number of elements. Then, by Theorem 6.3, the covering is a minimum covering and the independent set is a maximum independent set, which is what we want to prove. ■

König's Theorem

The argument just given amounts to a proof of a famous theorem of graph theory, first stated in 1931 by D. König, who pioneered the area. We state it in both its matrix and bipartite graph forms.

Theorem 6.5 ***König's Theorem*** In a matrix of 0s and 1s, a maximum independent set of 1s contains the same number of elements as a minimum covering. Equivalently, in a bipartite graph, a maximum matching contains the same number of elements as a minimum covering.

When the independent set algorithm reaches step 3.3, it gives us a construction of a minimum covering, namely, the labeled rows and unlabeled columns.

Example 6.11 We will use the algorithm to find a minimum covering for the graph shown in Figure 6.12.

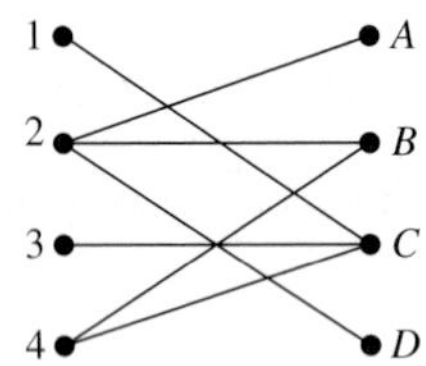

FIGURE 6.12

We convert the graph to the matrix that follows and, by inspection, find the independent set shown:

$$\begin{array}{c} \\ 1 \\ 2 \\ 3 \\ 4 \end{array} \begin{array}{c} \begin{array}{cccc} A & B & C & D \end{array} \\ \begin{bmatrix} 0 & 0 & 1^* & 0 \\ 1^* & 1 & 0 & 1 \\ 0 & 0 & 1 & 0 \\ 0 & 1^* & 1 & 0 \end{bmatrix} \end{array}$$

Applying the algorithm yields the following matrix:

$$\begin{array}{c} \\ 1 \\ 2 \\ 3 \\ 4 \\ \\ \end{array} \begin{array}{c} \begin{array}{cccc} A & B & C & D \end{array} \\ \begin{bmatrix} 0 & 0 & 1^* & 0 \\ 1^* & 1 & 0 & 1 \\ 0 & 0 & 1 & 0 \\ 0 & 1^* & 1 & 0 \end{bmatrix} \\ \begin{array}{cccc} 2\surd & & & \#\surd \end{array} \end{array} \begin{array}{l} \\ \\ D\surd \\ \\ \\ \\ \end{array}$$

We see that the matching we found by inspection was a maximum matching, since we have reached step 3.3. A minimum covering consists of the labeled rows and unlabeled columns, namely, row 2 and columns B and C. Thus vertices 2, B, and C form a minimum covering for the original graph. ■

Note that König's theorem applies only to bipartite graphs, even though matchings and coverings have been defined for arbitrary graphs. The reader should check that a maximum matching for the nonbipartite graph shown in Figure 6.13 contains 2 edges, while a minimum covering contains 3 vertices.

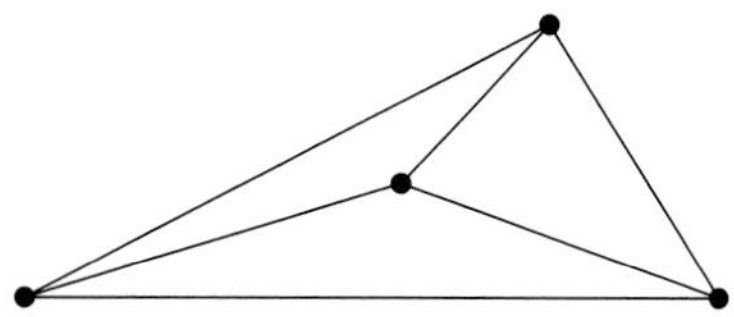

FIGURE 6.13

A Proof of Hall's Theorem

We can also use our conclusions about the algorithm to complete the proof of Hall's theorem, which was stated in Section 6.1. Recall that it remains to show that if $S_1, S_2, \ldots, S_n$ is a sequence of sets not having a system of distinct representatives, then there exists a subset I of $\{1, 2, \ldots, n\}$ such that the union of the sets S_i for $i \in I$ has fewer than $|I|$ elements.

Let the union of all the sets S_i for $i = 1, 2, \ldots, n$ be $\{t_1, t_2, \ldots, t_m\}$, where the t's are distinct. We construct a matrix of 0s and 1s, with rows corresponding to the sets S_i and columns corresponding to the elements t_j. Explicitly, the entry in row i and column j is to be 1 if $t_j \in S_i$ and 0 otherwise. (We have already constructed such matrices; for example, the sequence $P_1, P_2, \ldots, P_6$ of sets of professors who teach certain courses mentioned in Section 6.1 leads to the matrix of the second example in Section 6.3.)

We use our algorithm as many times as necessary to confirm that we have a maximum independent set of 1s in the matrix. Let r_L, r_U, c_L, and c_U denote the number of labeled rows, unlabeled rows, labeled columns, and unlabeled columns in this matrix after the last application of the algorithm, respectively. Certainly, $r_L + r_U = n$ (the number of sets), and $c_L + c_U = m$ (the number of elements). By Lemmas 6.2, 6.3, and 6.4, our maximum independent set has $r_L + c_U$ elements.

If our maximum independent set had n elements, it would correspond to a system of distinct representatives, so we may assume that

$$r_L + c_U < n = r_L + r_U.$$

Thus $c_U < r_U$.

We claim that the union of the r_U sets corresponding to the unlabeled rows contains fewer than r_U elements. The reason is that each 1 in an unlabeled row must be in an unlabeled column by Lemma 6.2. Thus the union of the corresponding sets contains at most c_U elements, and we know that $c_U < r_U$. Hence we can take I to be the numbers of the unlabeled rows. This completes the proof of Hall's theorem.

Notice that we have an actual construction of the set I with the use of the algorithm. For example, the problem of the courses and professors led to the following matrix:

	A	*B*	*C*	*D*	*E*	*F*	*G*	
1	1*	0	1	0	0	1	0	
2	0	0	1	1*	1	0	1	$E\surd$
3	1	0	1*	0	0	0	0	
4	1	0	0	0	0	1*	0	
5	0	1*	0	0	1	0	1	$E\surd$
6	0	0	1	0	0	1	0	
		$5\surd$		$2\surd$	$\#\surd$		$\#\surd$	

The unlabeled rows are the four rows 1, 3, 4, and 6. As we saw in Section 6.1, the union of the corresponding sets contains fewer than four elements. Note that the unlabeled columns correspond to the three professors able to teach any of these four courses.

Example 6.12 At a business meeting, 6 speakers are to be scheduled at 9, 10, 11, 1, 2, and 3 o'clock. Mr. Brown can only talk before noon. Ms. Krull can only speak at 9 or 2. Ms. Zeno cannot speak at 9, 11, or 2. Mr. Toomey cannot speak until 2. Mrs. Abernathy cannot speak between 10 and 3. Mr. Ng cannot speak from 10 until 2. The scheduler cannot seem to fit everyone in. Show that it is impossible to do so, and give a way the scheduler can convince the speakers of this fact.

We construct the following matrix, where the rows correspond to speakers and columns correspond to times:

	9	10	11	1	2	3
B	1*	1	1	0	0	0
K	1	0	0	0	1*	0
Z	0	1*	0	1	0	1
T	0	0	0	0	1	1*
A	1	0	0	0	0	1
N	1	0	0	0	1	1

The independent set indicated was found by inspection. Applying the algorithm produces first

$$\begin{array}{c} \\ B \\ K \\ Z \\ T \\ A \\ N \\ \\ \end{array}
\begin{array}{cccccc} 9 & 10 & 11 & 1 & 2 & 3 \\ \end{array}$$

$$\begin{array}{c|cccccc|c}
 & 9 & 10 & 11 & 1 & 2 & 3 & \\
\hline
B & \textcircled{1}^* & 1 & \textcircled{1} & 0 & 0 & 0 & 11\surd \\
K & 1 & 0 & 0 & 0 & 1^* & 0 & 9\surd \\
Z & 0 & 1^* & 0 & 1 & 0 & 1 & 1\surd \\
T & 0 & 0 & 0 & 0 & 1 & 1^* & \\
A & \textcircled{1} & 0 & 0 & 0 & 0 & 1 & 9! \\
N & 1 & 0 & 0 & 0 & 1 & 1 & 9 \\
\hline
 & B\surd & Z\surd & \#\surd & \#\surd & K & &
\end{array}$$

and then the matrix

$$\begin{array}{c|cccccc|c}
 & 9 & 10 & 11 & 1 & 2 & 3 & \\
\hline
B & 1 & 1 & 1^* & 0 & 0 & 0 & 10\surd \\
K & 1 & 0 & 0 & 0 & 1^* & 0 & \\
Z & 0 & 1^* & 0 & 1 & 0 & 1 & 1\surd \\
T & 0 & 0 & 0 & 0 & 1 & 1^* & \\
A & 1^* & 0 & 0 & 0 & 0 & 1 & \\
N & 1 & 0 & 0 & 0 & 1 & 1 & \\
\hline
 & & Z\surd & B\surd & \#\surd & & &
\end{array}$$

Notice that the rows corresponding to Krull, Toomey, Abernathy, and Ng are unlabeled. These four speakers all want to speak at 9, 2, or 3 (the unlabeled columns), which shows that all their restrictions cannot be accommodated. ■

The Bottleneck Problem

A foreman has 4 jobs that need to be done and 5 workers to whom he could assign them. The time in hours each worker would need to do each job is shown in the following table:

	Job 1	*Job 2*	*Job 3*	*Job 4*
Worker 1	3	7	5	8
Worker 2	6	3	2	3
Worker 3	3	5	8	6
Worker 4	5	8	6	4
Worker 5	6	5	7	3

He needs all 4 jobs finished as soon as possible and so is interested in making the maximum job time for the 4 workers chosen as small as possible.

Only one worker can do a job in 2 hours, so it is obviously impossible to get all 4 jobs done that fast. Three hours is more reasonable. Let us make a matrix of 0s and 1s, putting a 1 in each position corresponding to a job time of 3 hours or less:

$$\begin{bmatrix} 1 & 0 & 0 & 0 \\ 0 & 1 & 1 & 1 \\ 1 & 0 & 0 & 0 \\ 0 & 0 & 0 & 0 \\ 0 & 0 & 0 & 1 \end{bmatrix}.$$

We would like an independent set of 1s having 4 elements to correspond to the 4 jobs. Unfortunately, no such set exists. Since all the 1s lie in 3 lines (row 2 and columns 1 and 4), this fact is implied by König's theorem. Doing the jobs will take at least 4 hours, and so we add 1s to our matrix corresponding to the 4s in the original matrix:

$$\begin{bmatrix} 1 & 0 & 0 & 0 \\ 0 & 1 & 1 & 1 \\ 1 & 0 & 0 & 0 \\ 0 & 0 & 0 & 1 \\ 0 & 0 & 0 & 1 \end{bmatrix}.$$

The same reasoning shows that still no independent set of four 1s exists, so we add 1s corresponding to the 5s in the original matrix:

$$\begin{bmatrix} 1^* & 0 & 1 & 0 \\ 0 & 1^* & 1 & 1 \\ 1 & 1 & 0 & 0 \\ 1 & 0 & 0 & 1^* \\ 0 & 1 & 0 & 1 \end{bmatrix}.$$

The starred independent set was found by inspection. By applying the algorithm, we find the larger set that follows:

$$\begin{bmatrix} 1 & 0 & 1^* & 0 \\ 0 & 1^* & 1 & 1 \\ 1^* & 1 & 0 & 0 \\ 1 & 0 & 0 & 1^* \\ 0 & 1 & 0 & 1 \end{bmatrix}.$$

Thus the shortest time in which all the jobs can be completed is 5 hours.

Problems such as this are called **bottleneck** problems, since we are interested in making the job time of the slowest worker as small as possible. In other circumstances, we might be interested instead in minimizing the total time to do all the jobs. Such problems will be treated in the next section.

EXERCISES 6.4

In Exercises 1–4, a matrix of 0s and 1s is given with an independent set indicated. Use the independent set algorithm to find a minimum covering.

1. $\begin{bmatrix} 0 & 1^* & 0 & 1 \\ 1^* & 1 & 1 & 0 \\ 0 & 1 & 0 & 1^* \\ 0 & 0 & 0 & 1 \end{bmatrix}$

2. $\begin{bmatrix} 1^* & 0 & 0 & 0 \\ 0 & 1^* & 0 & 1 \\ 1 & 0 & 0 & 0 \\ 1 & 0 & 1^* & 1 \end{bmatrix}$

3. $\begin{bmatrix} 1^* & 0 & 1 & 0 & 0 \\ 0 & 0 & 1^* & 1 & 0 \\ 1 & 1^* & 0 & 1 & 1 \\ 1 & 0 & 1 & 1^* & 0 \\ 1 & 0 & 0 & 1 & 0 \end{bmatrix}$

4. $\begin{bmatrix} 0 & 1^* & 0 & 0 & 0 \\ 0 & 0 & 1^* & 0 & 0 \\ 0 & 1 & 1 & 0 & 0 \\ 0 & 0 & 1 & 0 & 0 \\ 1^* & 1 & 0 & 1 & 1 \end{bmatrix}$

In Exercises 5–8, a bipartite graph is given with a matching indicated. Use the independent set algorithm to find a minimum covering.

5.
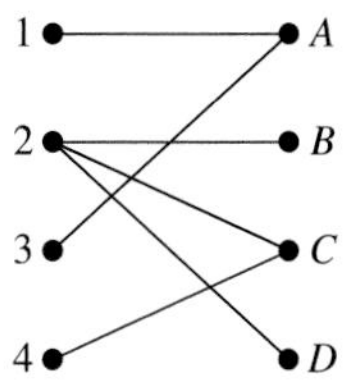

6.
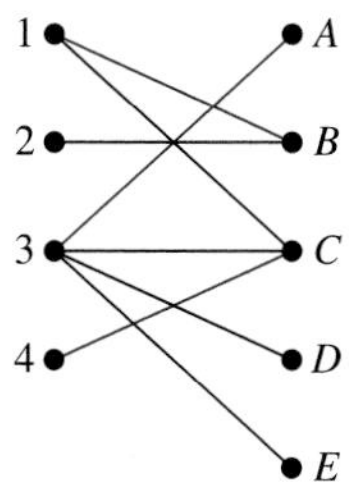

7.
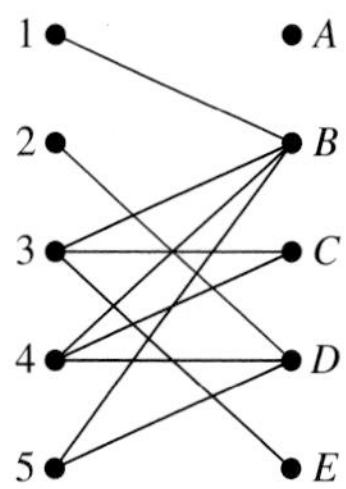

8.
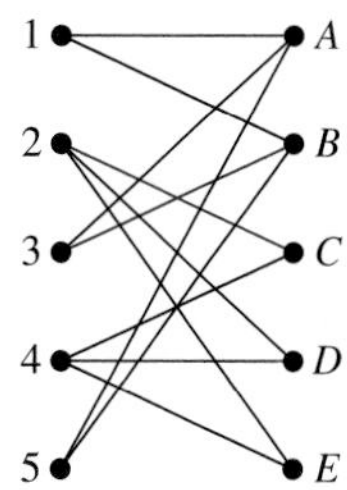

In Exercises 9–12, a sequence of sets $S_1, S_2, \ldots, S_n$ is given. Use the independent set algorithm to find, if possible, a subset I of $\{1, 2, \ldots n\}$ such that the union of the sets S_i for $i \in I$ has fewer elements than I.

9. $\{2, 4, 5\}, \{1, 3, 5\}, \{2, 3, 5\}, \{3, 4, 5\}, \{2, 3, 4\}$

10. $\{1, 2, 4\}, \{2, 3, 4, 5\}, \{2, 4, 6\}, \{1, 6\}, \{1, 4, 6\}, \{1, 2, 6\}$

11. $\{2, 7\}, \{1, 3, 6\}, \{5, 7\}, \{3, 4, 6\}, \{2, 5\}, \{2, 5, 7\}$

12. $\{1, 2\}, \{4, 6\}, \{0, 1, 3, 5, 6\}, \{1, 4, 7\}, \{2, 6\}, \{1, 4, 7\}, \{2, 6, 7\}$

13. A military commander must send a runner to each of four posts to notify them of a plan to attack. Because of differing terrain and skills, the time in hours for each runner to reach each post varies. Runner A takes 6 hours to get to Post 1, 5 hours to Post 2, 9 hours to Post 3, and 7 hours to Post 4. Runner B takes 4, 8, 7, and 8 hours to reach the four posts. Likewise, Runner C takes 5, 3, 9, and 8 hours; and runner D takes 7, 6, 3, and 5 hours. The attack cannot begin until all posts have gotten the message. What is the shortest time until it can begin?

14. One step of a manufacturing process takes 5 operations that can be done simultaneously. These take different times in minutes on the 5 machines available, as given in the following table:

	M1	*M2*	*M3*	*M4*	*M5*
Operation 1	6	7	3	6	2
Operation 2	6	3	4	3	3
Operation 3	2	5	3	7	4
Operation 4	3	4	2	6	3
Operation 5	4	7	2	7	6

How fast can the entire step be accomplished?

15. The following graph shows a city map:

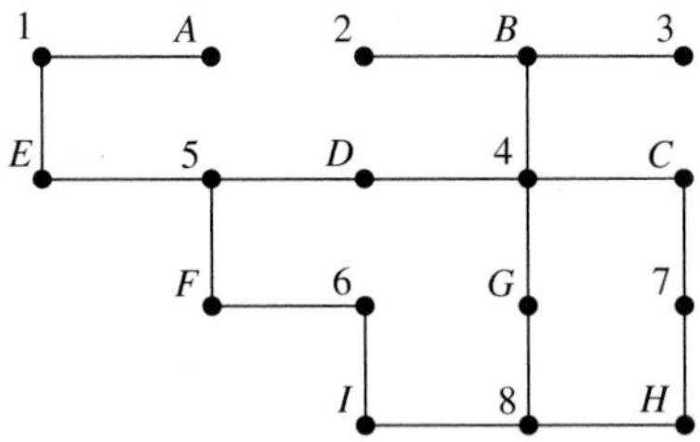

Adjacent vertices are one block apart. It is desired that a police officer be stationed at some of the vertices so that no one is more than one block from a police officer. Use the independent set algorithm to find the smallest number of police officers necessary to accomplish this, and tell where they should be positioned.

16. Show that if step 3.3 is reached in the independent set algorithm, then the number of labeled columns equals the number of labeled rows plus the number of columns containing no starred 1.

17. Show that if the independent set algorithm is applied to the matrix derived from a sequence of sets $S_1, S_2, \ldots, S_n$, as in the proof of Hall's theorem, and step 3.3 is reached, then the union of the sets corresponding to the unlabeled rows has exactly c_U elements, where c_U is the number of unlabeled columns.

18. Consider a bipartite graph where every edge is from a vertex in $\mathcal{V}_1$ to one in the disjoint set $\mathcal{V}_2$. If $\mathcal{S} \subseteq \mathcal{V}_1$, let $\mathcal{S}^*$ be the set of vertices of $\mathcal{V}_2$ adjacent to a vertex in $\mathcal{S}$. Show that the graph has a matching with $|\mathcal{V}_1|$ vertices if and only if $|\mathcal{S}^*| \geq |\mathcal{S}|$ whenever $\mathcal{S} \subseteq \mathcal{V}_1$.

6.5 THE HUNGARIAN METHOD

In the last section, we considered a problem of assigning 4 jobs to 5 workers in such a way that all 4 jobs got done as soon as possible. Although this might be our goal in special circumstances, a more common aim is to minimize the total time necessary to do the 4 jobs. If each worker were paid the same hourly rate, for example, this would minimize the labor cost for the project.

For simplicity, we will start with an example in which there are the same number of jobs and workers. The times in hours for each worker to do each job are given in the following table:

	Worker 1	*Worker 2*	*Worker 3*	*Worker 4*
Job 1	3	6	3	5
Job 2	7	3	5	8
Job 3	5	2	8	6
Job 4	8	3	6	4

An assignment of a worker to each job amounts to an independent set of four entries from the corresponding matrix, and we want the sum of the entries in that set to be as small as possible. For example, two possible assignments are as follows:

$$\begin{bmatrix} 3^* & 6 & 3 & 5 \\ 7 & 3^* & 5 & 8 \\ 5 & 2 & 8^* & 6 \\ 8 & 3 & 6 & 4^* \end{bmatrix} \qquad \begin{bmatrix} 3 & 6 & 3 & 5^* \\ 7 & 3^* & 5 & 8 \\ 5^* & 2 & 8 & 6 \\ 8 & 3 & 6^* & 4 \end{bmatrix}.$$

The first of these produces the sum $3 + 3 + 8 + 4 = 18$, and the second gives $5 + 3 + 5 + 6 = 19$; so the first independent set is better than the second for our purposes. Of course, other assignments might yield even smaller sums.

Suppose we subtract 3 from each entry in the first row of our matrix. Then two assignments for the new matrix are

$$\begin{bmatrix} 0^* & 3 & 0 & 2 \\ 7 & 3^* & 5 & 8 \\ 5 & 2 & 8^* & 6 \\ 8 & 3 & 6 & 4^* \end{bmatrix} \qquad \begin{bmatrix} 0 & 3 & 0 & 2^* \\ 7 & 3^* & 5 & 8 \\ 5^* & 2 & 8 & 6 \\ 8 & 3 & 6^* & 4 \end{bmatrix}.$$

Now the first set has the sum $0 + 3 + 8 + 4 = 15$, and the second has the sum $2 + 3 + 5 + 6 = 16$. The first assignment still has a sum 1 less than the second. The point is that, although subtracting the same number from each entry of the first row changes the problem, it does not change which positions give the answer. Since any independent set of four entries will have exactly one of them in the first row, the sum of the entries in any such set will be decreased by 3 by our operation.

Any assignment producing a minimum sum for the new matrix will also give a minimum sum for the original matrix. Furthermore, the same analysis applies to the other rows as well. In order to have entries as small as possible without introducing negative numbers, we will subtract from all entries in each row the smallest number in that row. This means subtracting 3 from the entries of the second row, 2 from those of the third row, and 3 from the entries of the fourth row. The resulting matrix is

$$\begin{bmatrix} 0 & 3 & 0 & 2 \\ 4 & 0 & 2 & 5 \\ 3 & 0 & 6 & 4 \\ 5 & 0 & 3 & 1 \end{bmatrix}.$$

The same argument applies to columns, so now we will subtract 1 from each entry of the fourth column, obtaining

$$\begin{bmatrix} 0 & 3 & 0 & 1 \\ 4 & 0 & 2 & 4 \\ 3 & 0 & 6 & 3 \\ 5 & 0 & 3 & 0 \end{bmatrix}.$$

Finding a four-entry independent set in this matrix will solve our original problem. Furthermore, now at least we might be able to recognize a solution. Suppose we could find an independent set of four 0s. This will clearly have minimum sum, since the matrix has no negative entries. Unfortunately, a maximum independent set of 0s has only three entries, as we can confirm with the independent set algorithm (modified to find an independent set of 0s instead of 1s):

$$\begin{array}{c} \\ 1 \\ 2 \\ 3 \\ 4 \\ \\ \end{array} \begin{array}{c} \begin{array}{cccc} A & B & C & D \end{array} \\ \begin{bmatrix} 0^* & 3 & 0 & 1 \\ 4 & 0^* & 2 & 4 \\ 3 & 0 & 6 & 3 \\ 5 & 0 & 3 & 0^* \end{bmatrix} \\ \begin{array}{cccc} 1\surd & & \#\surd & \end{array} \end{array} \begin{array}{c} \\ C\surd \\ \\ \\ \\ \\ \end{array}$$

We have reached step 3.3 of that algorithm, and so the independent set of three 0s indicated is a maximum independent set of 0s. Now we will show how to change the matrix so as to have a better chance of finding an independent set of four 0s. Later we will show why the solution to the minimum sum problem has not been changed.

Since a maximum independent set of 0s has fewer than four entries, there is a minimum covering consisting of fewer than four lines. In fact, by what we discovered in the last section, such a covering consists of the labeled rows and unlabeled columns of the preceding matrix. These lines are indicated as follows:

$$
\begin{array}{c}
\\ 1 \\ 2 \\ 3 \\ 4 \\ \\
\end{array}
\begin{array}{cccc}
A & B & C & D \\
\left[\begin{array}{cccc}
0^* & 3 & 0 & 1 \\
4 & 0^* & 2 & 4 \\
3 & 0 & 6 & 3 \\
5 & 0 & 3 & 0^*
\end{array}\right. & & & \left.\right] \\
1\surd & & \#\surd &
\end{array}
\quad C\surd
$$

Look at the entries not in any line of this covering. (By the definition of a covering, they are all positive.) The smallest of these is 2. Now we change our matrix as follows:

(1) Subtract 2 from each entry not in a line of the covering.
(2) Add 2 to each entry in both a row and column of the covering.
(3) Leave unchanged the entries in exactly one line of the covering.

The resulting matrix is

$$
\begin{bmatrix}
0 & 5 & 0 & 3 \\
2 & 0 & 0 & 4 \\
1 & 0 & 4 & 3 \\
3 & 0 & 1 & 0
\end{bmatrix}.
$$

Now we can find an independent set of four 0s and pick out the corresponding set in the original matrix. We obtain

$$
\begin{bmatrix}
0^* & 5 & 0 & 3 \\
2 & 0 & 0^* & 4 \\
1 & 0^* & 4 & 3 \\
3 & 0 & 1 & 0^*
\end{bmatrix}
\qquad
\begin{bmatrix}
3^* & 6 & 3 & 5 \\
7 & 3 & 5^* & 8 \\
5 & 2^* & 8 & 6 \\
8 & 3 & 6 & 4^*
\end{bmatrix}.
$$

The minimum sum for an independent set of four entries in the original matrix is $3 + 5 + 2 + 4 = 14$.

Of course, several questions need to be answered. One is whether the operation involving a minimum covering we just described is legitimate—that is, it does not change the solution to the minimum sum problem. Another is whether this operation even does any good for the purpose of producing an independent set of four 0s, since, although we subtract from some entries, we add to others. These questions will be answered after we state our method in a formal way.

Hungarian Algorithm

Starting with an $n \times n$ matrix with integer entries, this algorithm finds an independent set of n entries with minimum sum.

Step 1 (reduce the matrix)
(a) Subtract from each entry of each row the smallest entry in that row.
(b) Subtract from each entry of each column the smallest entry in that column.

Step 2 (determine a maximum independent set of 0s)
Find in the matrix a maximum independent set, S, of 0s.

Step 3 (enlarge the independent set if $|S| < n$)
while $|S| < n$
(a) Find a minimum covering for the 0s of the matrix.
(b) Let k be the smallest matrix entry not in any line of the covering.
(c) Subtract k from each entry not in a line of the covering.
(d) Add k to each entry in both a row and column of the covering.
(e) Replace S with a new maximum independent set of 0s.
endwhile

Step 4 (output)
The set S is an independent set of n entries with minimum sum.

Justification of the Hungarian Algorithm

First we will show why the loop in step 3 in the algorithm does not change which independent set is a solution. The reason is that this loop may be broken down into adding and subtracting numbers from rows and columns of the matrix, which we have already seen do not change which independent set is a solution. In particular, let k be the smallest (positive) entry not in any line of a covering. Let us subtract k from every entry of every row of the whole matrix, and then add k to every entry of every line of the covering, line by line. The net effect is exactly that of the loop of step 3. The number k is subtracted from each entry not in a line of the covering. If an entry is in a line of the covering exactly once, then it is not changed, since k is both subtracted from and added to it. Entries in both a row and column of the covering have k subtracted once but added twice, a net result of $+k$.

Now we address the question of whether step 3 does any good. It is conceivable that the algorithm could cycle through its parts forever without ever producing an independent set of n 0s. We will show that this cannot happen. After step 1, our matrix will contain only nonnegative integers as entries. We will show that the sum of all entries in the matrix will decrease whenever a loop of step 3 is performed. Obviously, if this sum were 0, then all matrix entries would be 0 and an independent set of n 0s would exist. Thus, if the algorithm went on forever, the sums of all matrix entries would give an infinite decreasing sequence of positive integers, which is impossible.

Step 3 continues only while no independent set of n 0s exists. Then a minimum covering will contain c rows and columns, where $c < n$. (This is a consequence of König's theorem.) Let us compute the effect of a loop of step 3 on the sum of all the entries of the matrix. As we have just seen, this amounts to subtracting k from each entry of the entire matrix and then adding k to each entry of each line of the covering. Since there are n^2 entries in the matrix, the subtraction decreases the sum of all entries by kn^2. Likewise, since there are c lines in the covering, each containing n entries, the addition increases the sum of all the entries by kcn. The net amount added to the sum of all the entries is

$$-kn^2 + kcn = kn(-n + c).$$

But this quantity is negative because $c < n$, and so the net effect is to decrease the sum of all entries, as claimed.

The reason this method is called "Hungarian" is to honor König, who was from Hungary, and upon whose theorem it is based. The algorithm is due to H.W. Kuhn.

Matrices That Are Not Square

Let us suppose that in our example a fifth worker becomes available, so that now our table becomes

	Worker 1	*Worker 2*	*Worker 3*	*Worker 4*	*Worker 5*
Job 1	3	6	3	5	3
Job 2	7	3	5	8	5
Job 3	5	2	8	6	2
Job 4	8	3	6	4	4

It is still reasonable to ask how to assign the four jobs in such a way as to make the sum of their times minimal; but our matrix is no longer square, and the algorithm applies only to square matrices. Of course, one worker is not going to get a job, and this simple idea provides a key to how to adapt the method. We introduce a fifth job, one requiring no time at all to do. This amounts to adding a row of 0s to the matrix, producing the following square matrix on the left:

$$\begin{bmatrix} 3 & 6 & 3 & 5 & 3 \\ 7 & 3 & 5 & 8 & 5 \\ 5 & 2 & 8 & 6 & 2 \\ 8 & 3 & 6 & 4 & 4 \\ 0 & 0 & 0 & 0 & 0 \end{bmatrix} \qquad \begin{bmatrix} 0 & 3 & 0 & 2 & 0 \\ 4 & 0 & 2 & 5 & 2 \\ 3 & 0 & 6 & 4 & 0 \\ 5 & 0 & 3 & 1 & 1 \\ 0 & 0 & 0 & 0 & 0 \end{bmatrix}.$$

The matrix on the right shows the result of applying step 1. Applying the independent set algorithm to this matrix yields the next matrix on the left. The matrix on the right shows the result of applying step 3 (with $k = 1$) to it.

$$\begin{array}{c} \\ 1 \\ 2 \\ 3 \\ 4 \\ 5 \\ \\ \end{array} \begin{array}{c} \begin{array}{ccccc} A & B & C & D & E \end{array} \\ \left[\begin{array}{ccccc} 0^* & 3 & 0 & 2 & 0 \\ 4 & 0^* & 2 & 5 & 2 \\ 3 & 0 & 6 & 4 & 0^* \\ 5 & 0 & 3 & 1 & 1 \\ 0 & 0 & 0^* & 0 & 0 \end{array}\right] \\ \begin{array}{ccccc} 1\surd & & 5\surd & \#\surd & \end{array} \end{array} \begin{array}{l} \\ C\surd \\ \\ \\ \\ D\surd \\ \\ \end{array} \qquad \begin{bmatrix} 0 & 4 & 0 & 2 & 1 \\ 3 & 0 & 1 & 4 & 2 \\ 2 & 0 & 5 & 3 & 0 \\ 4 & 0 & 2 & 0 & 1 \\ 0 & 1 & 0 & 0 & 1 \end{bmatrix}$$

An independent set of five 0s for this matrix, along with the corresponding set for the original matrix, is as follows:

$$\begin{bmatrix} 0^* & 4 & 0 & 2 & 1 \\ 3 & 0^* & 1 & 4 & 2 \\ 2 & 0 & 5 & 3 & 0^* \\ 4 & 0 & 2 & 0^* & 1 \\ 0 & 1 & 0^* & 0 & 1 \end{bmatrix} \qquad \begin{bmatrix} 3^* & 6 & 3 & 5 & 3 \\ 7 & 3^* & 5 & 8 & 5 \\ 5 & 2 & 8 & 6 & 2^* \\ 8 & 3 & 6 & 4^* & 4 \\ 0 & 0 & 0^* & 0 & 0 \end{bmatrix}.$$

By using the fifth worker, we can do all jobs in $3 + 3 + 4 + 2 = 12$ hours, instead of the previous minimum of 14 hours.

Independent Sets with Maximum Sum

A sweater factory has 4 workers and 4 machines on which sweaters can be made. The number of sweaters a worker can make in a day depends on the machine he or she uses, as indicated in the following table:

	Machine 1	*Machine 2*	*Machine 3*	*Machine 4*
Worker 1	3	6	7	4
Worker 2	4	5	5	6
Worker 3	6	3	4	4
Worker 4	5	4	3	5

In this case, we are looking for an independent set with 4 entries, the sum of which is a *maximum* instead of a minimum. We reduce this to a problem we already know how to solve, by multiplying the corresponding matrix by -1. The result is shown at the left:

$$\begin{bmatrix} -3 & -6 & -7 & -4 \\ -4 & -5 & -5 & -6 \\ -6 & -3 & -4 & -4 \\ -5 & -4 & -3 & -5 \end{bmatrix} \qquad \begin{bmatrix} 4 & 1 & 0 & 3 \\ 2 & 1 & 1 & 0 \\ 0 & 3 & 2 & 2 \\ 0 & 1 & 2 & 0 \end{bmatrix}.$$

Finding a maximum sum in the original matrix is equivalent to finding a minimum sum in this matrix. The negative entries cause no problems, since they disappear when we subtract the least entries of each row (here, −7, −6, −6, and −5). The result is the preceding matrix on the right. Thus a maximum sum problem may be solved by applying the Hungarian method to the negative of the original matrix. The reader should check that a maximum of 23 sweaters can be produced per day.

EXERCISES 6.5

In Exercises 1–8, find the smallest sum of an independent set of entries of the matrix with as many elements as the matrix has rows.

1. $\begin{bmatrix} 1 & 2 & 3 \\ 6 & 5 & 4 \\ 7 & 8 & 9 \end{bmatrix}$

2. $\begin{bmatrix} 1 & 4 & 3 & 8 \\ 2 & 7 & 9 & 3 \\ 8 & 2 & 5 & 5 \\ 6 & 6 & 4 & 7 \end{bmatrix}$

3. $\begin{bmatrix} 6 & 2 & 5 & 8 \\ 6 & 7 & 1 & 6 \\ 6 & 3 & 4 & 5 \\ 5 & 4 & 3 & 4 \end{bmatrix}$

4. $\begin{bmatrix} 2 & 3 & 5 & 1 & 2 \\ 4 & 3 & 5 & 4 & 2 \\ 3 & 6 & 3 & 1 & 4 \\ 3 & 6 & 4 & 5 & 4 \\ 4 & 2 & 4 & 5 & 4 \end{bmatrix}$

5. $\begin{bmatrix} 3 & 5 & 5 & 3 & 8 \\ 4 & 6 & 4 & 2 & 6 \\ 4 & 6 & 1 & 3 & 6 \\ 3 & 4 & 4 & 6 & 5 \\ 5 & 7 & 3 & 5 & 9 \end{bmatrix}$

6. $\begin{bmatrix} 0 & 1 & 0 & -1 & 1 \\ 3 & 0 & 4 & 4 & 5 \\ 1 & 3 & 7 & 4 & 7 \\ -1 & -2 & 2 & 3 & 3 \\ 2 & 4 & 7 & 5 & 9 \end{bmatrix}$

7. $\begin{bmatrix} 3 & 4 & 5 & 7 & 6 \\ 5 & 3 & 4 & 5 & 2 \\ 1 & 3 & 4 & 5 & 3 \\ 5 & 6 & 5 & 4 & 3 \end{bmatrix}$

8. $\begin{bmatrix} 5 & 6 & 2 & 3 & 4 & 3 \\ 6 & 4 & 4 & 2 & 0 & 3 \\ 5 & 4 & 5 & 2 & 6 & 6 \\ 5 & 6 & 1 & 4 & 7 & 6 \end{bmatrix}$

In Exercises 9–12, find the largest sum of an independent set of entries with as many elements as the matrix has rows.

9. $\begin{bmatrix} 5 & 4 & 2 & 3 \\ 3 & 1 & 4 & 3 \\ 1 & 1 & 1 & 3 \\ 5 & 3 & 6 & 3 \end{bmatrix}$

10. $\begin{bmatrix} 5 & 4 & 3 & 4 \\ 5 & 3 & 1 & 7 \\ 7 & 5 & 2 & 10 \\ 2 & 4 & 2 & 7 \end{bmatrix}$

11. $\begin{bmatrix} 6 & 5 & 3 & 1 & 4 \\ 2 & 5 & 3 & 7 & 8 \\ 8 & 3 & 7 & 5 & 4 \\ 7 & 1 & 5 & 3 & 8 \end{bmatrix}$

12. $\begin{bmatrix} 6 & 7 & 3 & 8 & 9 \\ 4 & 7 & 5 & 6 & 2 \\ 2 & 5 & 8 & 6 & 9 \end{bmatrix}$

13. A newspaper sports editor must send 4 of his reporters to 4 cities. From past experience, he knows what expenses to expect from each reporter in each city. He can expect Addams to spend \$700 in Los Angeles, \$500 in New York, \$200 in Las Vegas, and \$400 in Chicago. Hart can be expected to spend \$500, \$500, \$100, and \$600 in these cities; Young to spend \$500, \$300, \$400, and \$700; and Herriman to spend \$400, \$500, \$600, and \$500. How should the editor make the assignments to keep the total expenses to a minimum?

14. A supervisor has 5 salespeople who can be assigned to 5 different routes next month. Adam can be expected to sell \$9000 worth of goods on Route 1, \$8000 on Route 2, \$10,000 on Route 3, \$7000 on Route 4, and \$8000 on Route 5. Betty would sell \$6000, \$9000, \$5000, \$7000, and \$4000 on these routes; Charles would sell \$4000, \$5000, \$4000, \$8000, and \$2000; Denise would sell \$4000, \$7000, \$5000, \$4000, and \$2000; and Ed would sell \$5000, \$5000, \$7000, \$9000, and \$3000. What is the maximum total expected sales possible next month?

15. A foreman has 4 jobs and 5 workers he could assign them to. The time in hours each worker needs for each job is shown in the following table:

	Worker 1	*Worker 2*	*Worker 3*	*Worker 4*	*Worker 5*
Job 1	7	3	5	7	2
Job 2	6	1	4	2	6
Job 3	8	3	8	9	1
Job 4	7	2	1	5	6

After subtracting the minimum entries from the rows and columns of the corresponding matrix, we have the matrix

$$\begin{bmatrix} 0^* & 1 & 3 & 4 & 0 \\ 0 & 0^* & 3 & 0 & 5 \\ 2 & 2 & 7 & 7 & 0^* \\ 1 & 1 & 0^* & 3 & 5 \end{bmatrix},$$

in which the stars indicate a maximum independent set of 0s. The corresponding job assignment will require a total of $7+1+1+1=10$ hours. But, by assigning the jobs to workers 2, 4, 5, and 3, the total time could be reduced to $3+2+1+1=7$ hours. What is wrong?

HISTORICAL NOTES

Philip Hall

Philip Hall (1904–1982), who contributed Theorem 6.1, was a very gifted English mathematician. After receiving his doctorate in algebra, Hall worked on generalizations of the Sylow theorems in group theory and with correlation in statistics. In 1935, he published the paper that listed the necessary and sufficient conditions for the existence of a system of distinct representatives for a sequence of finite sets. During World War II, he worked with the famous British cryptography group at Bletchley Park.

The methods for linking systems of distinct representatives to matching are a combination of the work of Dénes König (1884–1944), a Hungarian mathematician, and the two American mathematicians Lester R. Ford Jr. (1927–) and Delbert R. Fulkerson (1924–1976). Ford and Fulkerson considered the question of whether or not there is a subset of edges in a bipartite graph with the property that every vertex meets just one of them. Their approach employed the independent set algorithm found in Section 6.3. When extended to apply to nonbipartite graphs, the result is equivalent to an algorithmic approach developed by König in 1931.

SUPPLEMENTARY EXERCISES

1. How many systems of distinct representatives does each of the following sequences of sets have?

   (a) $\{1, 2, 3, 4, 5\}, \{1, 2, 3, 4, 5\}, \{1, 2, 3, 4, 5\}$

   (b) $\{1, 2, 3, 4\}, \{1, 2, 3, 4\}, \{5, 6, 7\}$

   (c) $\{1, 2, 3\}, \{2, 3, 4\}, \{1, 2, 4\}, \{1, 3, 4\}, \{1, 2, 4\}$

2. Let $S_1 = \{1, 2, 5\}$, $S_2 = \{1, 5\}$, $S_3 = \{1, 2\}$, $S_4 = \{2, 3, 4\}$, and $S_3 = \{2, 5\}$. Give an argument to show that the sequence S_1, S_2, S_3, S_4, S_5 does not have a system of distinct representatives.

3. Tell whether each of the following graphs is bipartite or not, and if so, give disjoint sets $\mathcal{V}_1$ and $\mathcal{V}_2$ such that each edge joins a vertex in $\mathcal{V}_1$ to one in $\mathcal{V}_2$.

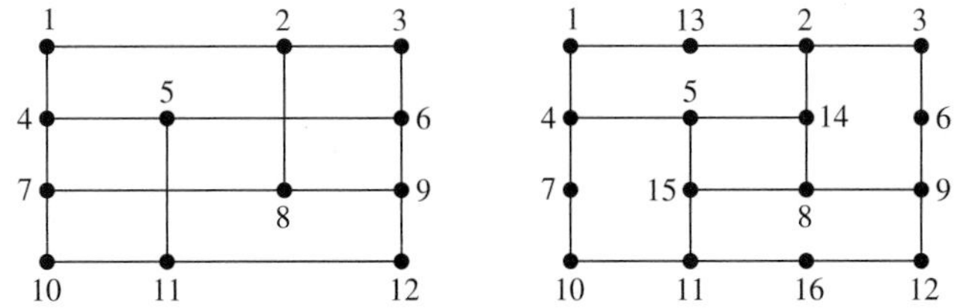

(a) (b)

4. Give a maximum matching for each graph in Exercise 3.

5. Give a minimum covering for each graph in Exercise 3.

6. The following graph is bipartite in that every edge joins a vertex in $\mathcal{V}_1 = \{1, 3, 6, 8\}$ to one in $\mathcal{V}_2 = \{2, 4, 5, 7\}$:

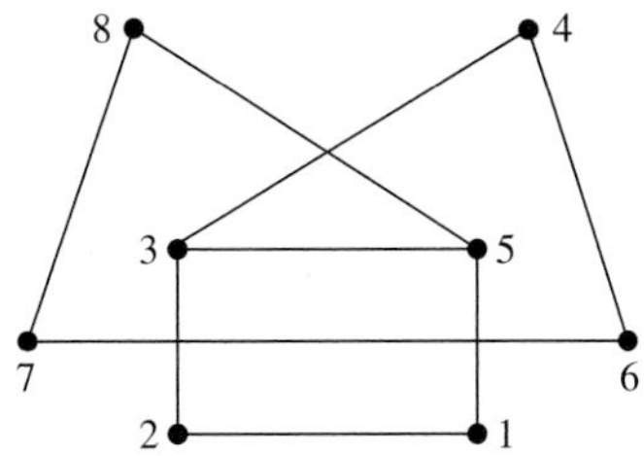

   Give the matrix of the graph.

7. Find a maximum independent set of 1s for the matrix of Exercise 6.

8. Find a minimum covering for the matrix of Exercise 6.

9. Use the independent set algorithm to find a maximum independent set of 1s for the following matrix, starting with the starred set of 1s:

$$\begin{bmatrix} 1^* & 0 & 0 & 0 & 1 \\ 0 & 0 & 1^* & 1 & 0 \\ 0 & 0 & 1 & 0 & 0 \\ 1 & 1^* & 0 & 1 & 1 \\ 0 & 0 & 1 & 0 & 0 \end{bmatrix}$$

   Then use the algorithm to find a minimum covering.

10. Convert the following bipartite graph into a matrix, and use the independent set algorithm to find a maximum matching and minimum covering for the graph, starting with the set of 1s corresponding to the given matching:

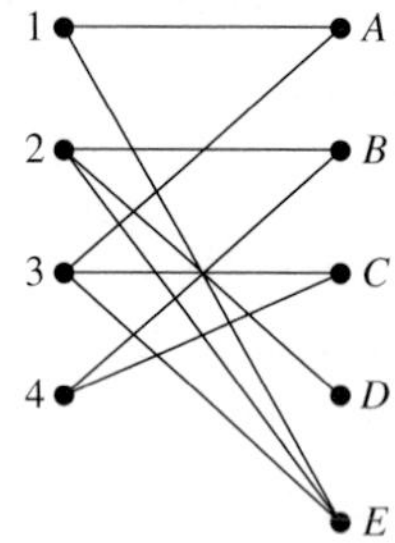

11. Convert the sequence of sets $\{w^*, y\}$, $\{x^*, z\}$, $\{v^*, z\}$, $\{w, x\}$, $\{v, y^*\}$ to a matrix, and use the independent set algorithm to find a system of distinct representatives if possible, starting with the set of 1s corresponding to the starred elements.

12. Dan, Ed, Fred, Gil, and Hal are at a dance with Ivy, June, Kim, Lil, and Mae. The only compatible dancing partners are Dan with Kim or Lil, Ed with Ivy or Mae, Fred with June or Mae, Gil with June or Lil, and Hal with Ivy or Kim. Those currently dancing are Dan with Kim, Ed with Ivy, Fred with June, and Gil with Lil, with Hal and Mae left out. Use the independent set algorithm to find everyone a partner.

13. The following table shows the number of hours it would take each of workers A, B, C, D, and E to do jobs 1, 2, 3, 4, and 5:

	A	B	C	D	E
1	7	5	8	6	4
2	4	3	5	4	6
3	5	8	6	7	3
4	6	7	3	4	5
5	4	3	6	5	3

What is the minimum time needed to do all five jobs?

14. Use the Hungarian algorithm to find an independent set of entries with minimum sum for each of the following matrices:

(a) $\begin{bmatrix} 1 & 3 & 5 & 7 \\ 2 & 8 & 6 & 4 \\ 5 & 4 & 1 & 2 \\ 6 & 4 & 3 & 5 \end{bmatrix}$ **(b)** $\begin{bmatrix} 3 & 4 & 7 & 8 \\ 7 & 5 & 6 & 5 \\ 5 & 3 & 4 & 4 \\ 8 & 6 & 5 & 7 \\ 4 & 2 & 8 & 9 \end{bmatrix}$

15. A car dealer has four salespeople, and each is assigned to sell a particular brand of car. Adam can sell 6 Hupmobiles, 8 Studebakers, 7 Packards, or 4 Hudsons per month. Beth can sell 7, 3, 2, or 5 of each brand; Cal 6, 7, 8, or 7; and Danielle 6, 4, 5, or 4. How should each be assigned a different brand to maximize the number of cars sold?

16. Prove that if a graph with v vertices has a matching $\mathcal{M}$, then $2|\mathcal{M}| \leq v$.

17. Suppose $\mathcal{M}$ is a matching of a graph such that there is a simple path $e_1, e_2, \ldots, e_n$ of odd length that begins and ends at vertices not incident with any edge in $\mathcal{M}$. Show that if $e_1, e_3, \ldots, e_n$ are not in $\mathcal{M}$, while $e_2, e_4, \ldots, e_{n-1}$ are in $\mathcal{M}$, then $\mathcal{M}$ is not a maximum matching.

18. Let a graph have a maximum matching with m edges and a minimum covering with c vertices. Show by mathematical induction on m that the greatest integer not exceeding $\frac{c+1}{2}$ is less than or equal to m.

COMPUTER PROJECTS

Write a computer program having the specified input and output.

1. Given a set S of m elements and n subsets $T_1, T_2, \ldots, T_n$ of S, generate all possible lists $x_1, x_2, \ldots, x_n$, where $x_i \in T_i$ for $i = 1, 2, \ldots, n$. For each list, check whether the elements x_i are all distinct. Apply the program to the example of professors and courses in Section 6.1 to confirm that the sets $P_1, P_2, \ldots, P_6$ there have no system of distinct representatives.

2. Given a graph with vertex set $\mathcal{V} = \{1, 2, \ldots, n\}$ and adjacency matrix $A = \{a_{ij}\}$, decide whether a given subset $\mathcal{W}$ of $\mathcal{V}$ is a covering or not. (See Exercise 33 of Section 6.2.)

3. Apply the independent set algorithm to an $m \times n$ matrix of 0s, 1s, and 2s, with no two 2s in a line. The 0s and 1s should be interpreted as in that algorithm, while the 2s correspond to starred 1s. Thus the program will either interchange some 1s and 2s to get a new matrix with a larger independent set of 2s or else determine that this is impossible. In the latter case, output the row and column numbers corresponding to a minimum covering.

4. Find a maximum independent set of 1s in a given matrix of 0s and 1s by repeatedly invoking the program of the previous exercise.

5. Solve the bottleneck problem, starting with an $m \times n$ matrix $A = [a_{ij}]$ of positive integers. (*Hint:* For $k = 1, 2, \ldots$ form a new matrix $B = [b_{ij}]$, where $b_{ij} = 0$ or 1 according as $a_{ij} > k$ or not. Apply the program of the previous exercise until k is sufficiently large so that B has an independent set with n elements.)

6. Given an $m \times m$ matrix, perform step 1 of the Hungarian algorithm to get a matrix with nonnegative entries and at least one 0 in each line.

7. Implement the Hungarian algorithm, given an $m \times m$ matrix. Use the program of Exercise 4. Note that an auxiliary matrix in which 0s and 1s correspond to positive and 0 entries will be needed to apply that program.

SUGGESTED READINGS

1. Berg, C. *Graphs and Hypergraphs*, 2d ed. New York: Elsevier Science, 1976.

2. Edmonds, J. "Paths, Trees, and Flowers." *Canad. J. Math.* 17, 1965, 449–467.

3. Ford, L.R. and D.R. Fulkerson. *Flows in Networks.* Princeton, NJ: Princeton University Press, 1962.

4. Hall, P. "On Representations of Subsets." *J. London Math. Soc.* 10, 1935, 26–30.

5. Kuhn, H.W. "The Hungarian Method for the Assignment Problem." *Naval Res. Logist. Quart.* 2, 1955, 83–97.

6. Lawler, E.L. *Combinatorial Optimization: Networks and Matroids.* New York: Holt, Rinehart and Winston, 1976.

7. Mirsky, L. *Transversal Theory.* New York: Academic Press, 1971.

8. Roberts, Fred S. *Applied Combinatorics.* Englewood Cliffs, NJ: Prentice-Hall, 1984.

CHAPTER

7

Network Flows

Many practical problems require the movement of some commodity from one location to another. For example, an oil company must move crude oil from the oil fields to its refinery, and a long-distance telephone company must move messages from one city to another. In both of these situations, there is a limitation to the amount of the commodity that can be moved at one time. The volume of crude oil that the oil company can move, for instance, is limited by the capacity of the pipeline through which the oil must flow. And the number of telephone calls that the phone company can handle is limited by the capacity of its cable and its switching equipment. This type of problem, in which some commodity must be moved from one location to another subject to the restriction that certain capacities not be exceeded, is called a **network flow** problem. In this chapter we will be primarily concerned with solving such problems.

7.1 FLOWS AND CUTS

When an oil company must ship crude oil from the oil fields to its refinery, there is one origin for the oil (the oil fields) and one destination (the refinery). However, there may be many different pipelines available through which the oil can be sent. Figure 7.1 shows this situation for an oil company with oil fields at Prudhoe Bay and a refinery in Seward, Alaska. (Here, the pipeline capacities are given in thousands of barrels per day.) This figure showing the possible routes from the oil fields to the refinery is a special type of weighted directed graph.

By a **transportation network**, or more simply a **network**, we mean a weighted directed graph satisfying the following three conditions:

(1) There is exactly one vertex having no incoming edges—that is, exactly one vertex with indegree 0. This vertex is called the **source**.

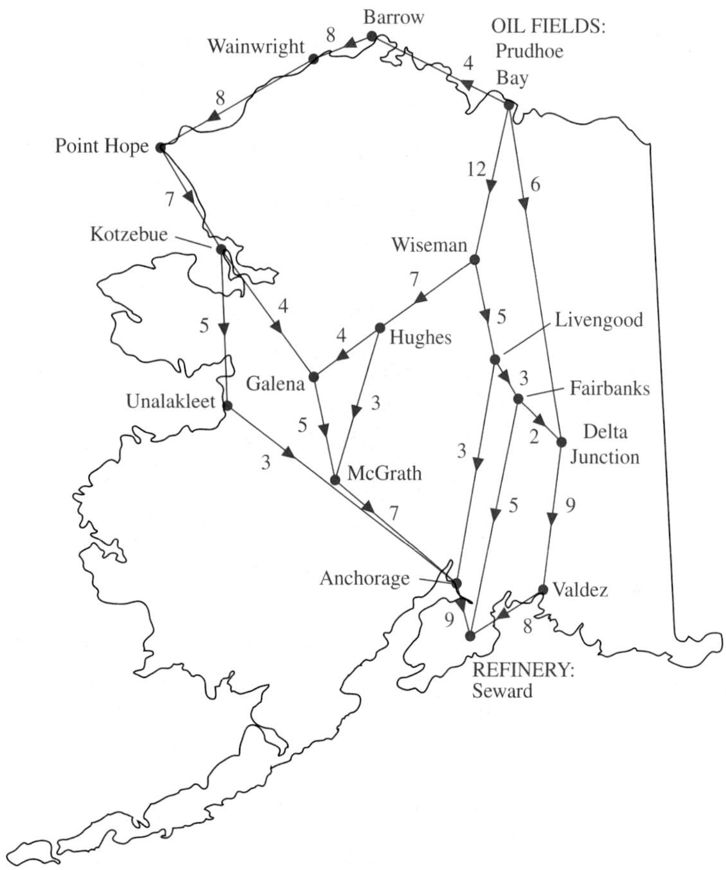

FIGURE 7.1

(2) There is exactly one vertex having no outgoing edges—that is, exactly one vertex with outdegree 0. This vertex is called the **sink**.
(3) The weight assigned to each edge is a nonnegative number.

In this context a directed edge of the network will be called an **arc**, and the weight of an arc will be called its **capacity**.

Example 7.1 Figure 7.2 shows a weighted directed graph with five vertices and seven arcs. The seven arcs are (A, B) with capacity 6, (A, C) with capacity 8, (A, D) with capacity 3, (B, C) with capacity 5, (B, D) with capacity 6, (C, E) with capacity 4, and (D, E) with capacity 9. Clearly, the capacity of each arc is a nonnegative number. Note that vertex A is the only vertex having no incoming arcs, and vertex E is the only vertex having no outgoing arcs. Thus the directed graph in Figure 7.2 is a transportation network with vertex A as its source and vertex E as its sink. ■

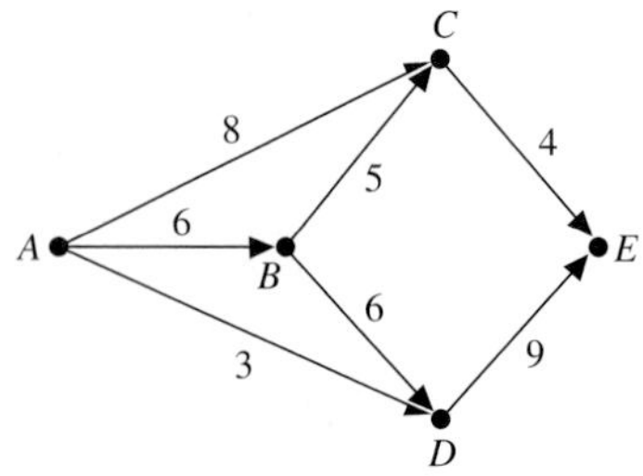

FIGURE 7.2

In a transportation network, we consider a commodity flowing along arcs from the source to the sink. The amount carried by each arc must not exceed the capacity of the arc, and none of the commodity can be lost along the way. Thus, at each vertex other than the source and the sink, the amount of the commodity that arrives must equal the amount of the commodity that leaves. We will formalize these ideas in the following definition:

Let $\mathcal{A}$ be the set of arcs in a transportation network $\mathcal{N}$, and, for each arc e in $\mathcal{A}$, let $c(e)$ denote the capacity of e. A **flow** in $\mathcal{N}$ is a function f that assigns to each arc e a number $f(e)$, called the **flow along arc** e, such that

(1) $0 \le f(e) \le c(e)$, and
(2) for each vertex V other than the source and sink, the total flow into V (the sum of the flows along all arcs ending at V) equals the total flow out of V (the sum of the flows along all arcs beginning at V).

Since the capacity of an arc is nonnegative, it is clear that the function f assigning the number 0 to each arc is always a flow in a transportation network. Consequently, every network has a flow.

Example 7.2 For the transportation network in Figure 7.2, the function f such that $f(A, B) = 6$, $f(A, C) = 0$, $f(A, D) = 3$, $f(B, C) = 4$, $f(B, D) = 2$, $f(C, E) = 4$, and $f(D, E) = 5$ is a flow. This flow is shown in Figure 7.3, where the first number on each arc is its capacity and the second number is the flow along that arc. Notice that each value of f is a nonnegative number that does not exceed the capacity of the corresponding arc. In addition, at vertices B, C, and D, the total flow into the vertex equals the total flow out of the vertex. For instance, the total flow into vertex B is 6 along arc (A, B); and the total flow out of vertex B is also 6: 4 along arc (B, C) and 2 along arc (B, D). Likewise, the total flow into vertex D is 5: 3 along arc (A, D) and 2 along arc (B, D); and the total flow out of vertex D is also 5 along arc (D, E). ■

In Figure 7.3 the total flow out of vertex A is 9: 6 along arc (A, B), 0 along arc (A, C), and 3 along arc (A, D). Notice that this number is the same as the total flow into vertex E, which is 4 along arc (C, E) and 5 along arc (D, E). This equality is a basic property of every flow.

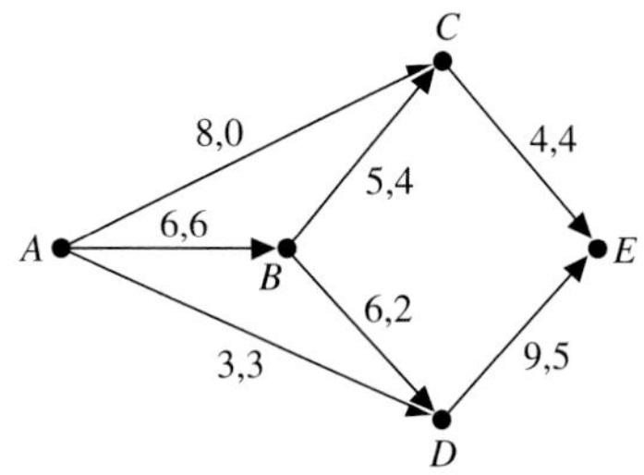

FIGURE 7.3

Theorem 7.1 For any flow in a transportation network, the total flow out of the source equals the total flow into the sink.

Proof. Let $V_1, V_2, \ldots, V_n$ denote the vertices of the network, with V_1 being the source and V_n being the sink. Let f be a flow in this network, and, for each k $(1 \le k \le n)$, define I_k to be the total flow into V_k and O_k to be the total flow out of V_k. Finally, let S denote the sum of the flows along every arc in the network.

For each arc $e = (V_j, V_k)$, $f(e)$ is included in the sum $I_1 + I_2 + \cdots + I_n$ exactly once (in the term I_k) and in the sum $O_1 + O_2 + \cdots + O_n$ exactly once (in the term O_j). Hence $I_1 + I_2 + \cdots + I_n = S$ and $O_1 + O_2 + \cdots + O_n = S$; so $O_1 + O_2 + \cdots + O_n = I_1 + I_2 + \cdots + I_n$. But for any vertex V_k other than the source and sink, $I_k = O_k$. Canceling these common terms in the preceding equation gives $O_1 + O_n = I_1 + I_n$. Now, $I_1 = 0$ because the source has no incoming arcs, and $O_n = 0$ because the sink has no outgoing arcs. Hence we see that $O_1 = I_n$—that is, the total flow out of the source equals the total flow into the sink. ∎

If f is a flow in a transportation network, the common value of the total flow out of the source and the total flow into the sink is called the **value** of the flow f.

In the network shown in Figure 7.1, in which crude oil is to be shipped through pipelines, the oil company would be interested in knowing how much oil can be sent per day from the oil fields to the refinery. Likewise, in any transportation network, it is important to know the amount of a commodity that can be shipped from the source to the sink without exceeding the capacities of the arcs. In other words, we would like to know the largest possible value of a flow in a transportation network. A flow having maximum value in a network is called a **maximal flow**.

In Section 7.2 we will present an algorithm for finding a maximal flow in a transportation network. In order to understand this algorithm better, we will first consider some of the ideas that are involved in finding a maximal flow. Suppose, for example, that we want to find a maximal flow in the transportation network shown in Figure 7.2. Because this network is so small, it will not be difficult to determine a maximal flow by a little experimentation. Our approach will be to find a sequence of flows with increasing values. We begin by taking the flow to be zero along every arc. Thus the current flow is as in Figure 7.4, where the numbers along each arc are the arc capacity and the current flow along the arc.

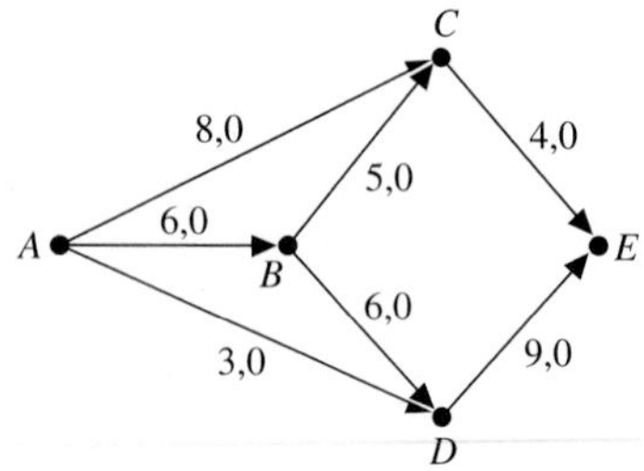

FIGURE 7.4

Now we will try to find a path from the source to the sink along which we can increase the present flow. Such a path is called a **flow-augmenting** path. In this case, since there is no arc along which the flow equals the capacity, any directed path from the source to the sink will suffice. Suppose that we choose the path A, C, E. By how much can we increase the flow along the arcs in this path? Because the capacities of the arcs (A, C) and (C, E) in this path are 8 and 4, respectively, it is clear that we can increase the flows along these two arcs by 4 without exceeding their capacities. Recall that we are only changing the flow along arcs in our chosen path A, C, E and that the flow out of vertex C must equal the flow into C. Consequently, if we tried to increase the flow along arc (A, C) by more than 4, then the flow along arc (C, E) would also need to be increased by more than 4. But a flow along arc (C, E) that is greater than 4 would exceed the capacity of this arc. Hence the largest amount by which we can increase the flow along the path A, C, E is 4. When we increase the flow in this manner, we obtain the flow shown in Figure 7.5.

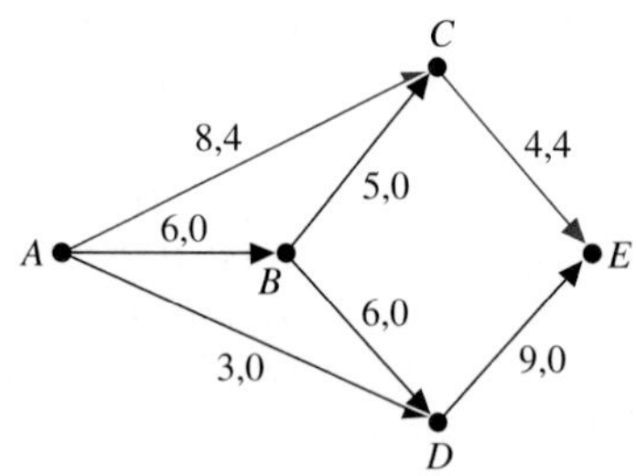

FIGURE 7.5

Now we will try to find another flow-augmenting path so that we can increase the present flow. Note that such a path cannot use arc (C, E), because the flow in this arc is already at its capacity. One acceptable path is A, D, E. For this path, we can increase the flow by as much as 3 without exceeding the capacity of any arc. (Why?) If we increase the flows in arcs (A, D) and (D, E) by 3, we obtain the new flow shown in Figure 7.6.

Again, we will try to find a flow-augmenting path. Path A, B, D, E is such a path. For this path, we can increase the flow by as much as 6 without exceeding the capacity of any arc. If we increase the flows in arcs (A, B), (B, D), and (D, E) by 6, we obtain the new flow shown in Figure 7.7.

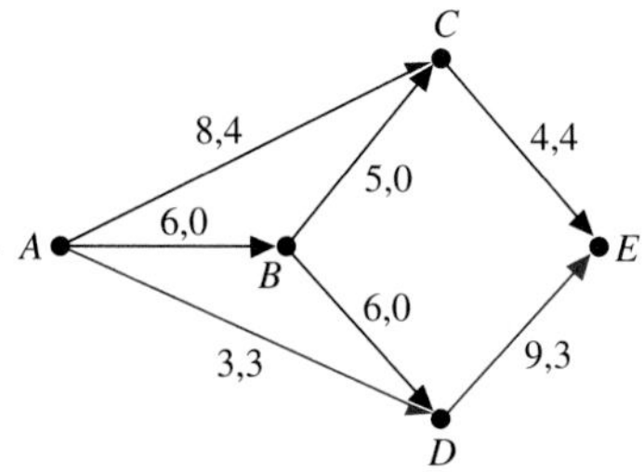

FIGURE 7.6

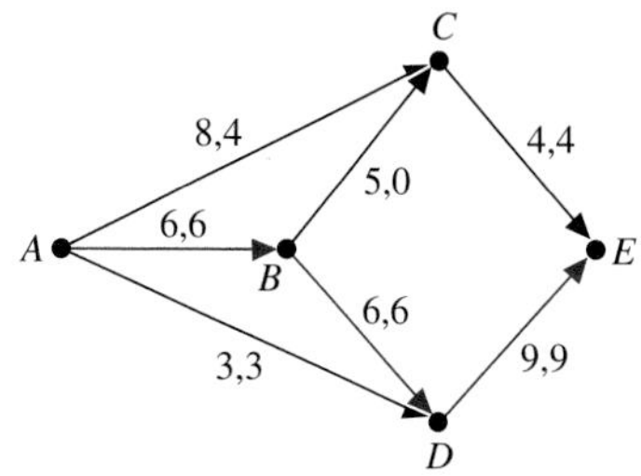

FIGURE 7.7

Is it possible to find another flow-augmenting path? Note that any path leading to the sink must use either arc (C, E) or arc (D, E) because these are the only arcs leading to the sink. But the flow along these arcs is already at the capacity of the arcs. Consequently, it is not possible to increase the flow in Figure 7.7 any further, and so this flow is a maximal flow. The value of this flow is 13, the common value of the flow out of the source and into the sink.

The argument used here to justify that there could be no flow having a value larger than 13 is an important one. As this argument suggests, the value of a maximal flow is limited by the capacities of certain sets of arcs. Recall once more the oil pipeline network in Figure 7.1. Suppose that, after analyzing this network, you have determined that the value of a maximal flow is 18, but your colleagues at the oil company are questioning your calculations. They point out that it is possible to ship 22 thousand barrels per day out of Prudhoe Bay and 22 thousand barrels per day into Seward; so they believe that there should be a flow having the value 22. How can you convince them that there can be no flow having a value greater than 18?

Suppose that the vertices of the network are partitioned into two sets $\mathcal{S}$ and $\mathcal{T}$ such that the source belongs to $\mathcal{S}$ and the sink belongs to $\mathcal{T}$. (Recall that this statement means that each vertex belongs to exactly one of the sets $\mathcal{S}$ or $\mathcal{T}$.) Since every path from the source to the sink begins at a vertex in $\mathcal{S}$ and ends at a vertex in $\mathcal{T}$, each such path must contain an arc that joins some vertex in $\mathcal{S}$ to some vertex in $\mathcal{T}$. So if we can partition the vertices of the network into sets $\mathcal{S}$ and $\mathcal{T}$ in such a way that the total capacity of the arcs going from a vertex in $\mathcal{S}$ to a vertex in $\mathcal{T}$ is 18, we will have proved that there can be no flow with a value greater than 18.

It can be seen in Figure 7.8 that such a partition is obtained by taking

$$\mathcal{T} = \{\text{Fairbanks, Delta Junction, Valdez, Seward}\}$$

and $\mathcal{S}$ to be the other cities in the figure. The heavy line in Figure 7.8 separates the cities in $\mathcal{S}$ (northwest of the line) from the cities in $\mathcal{T}$ (southeast of the

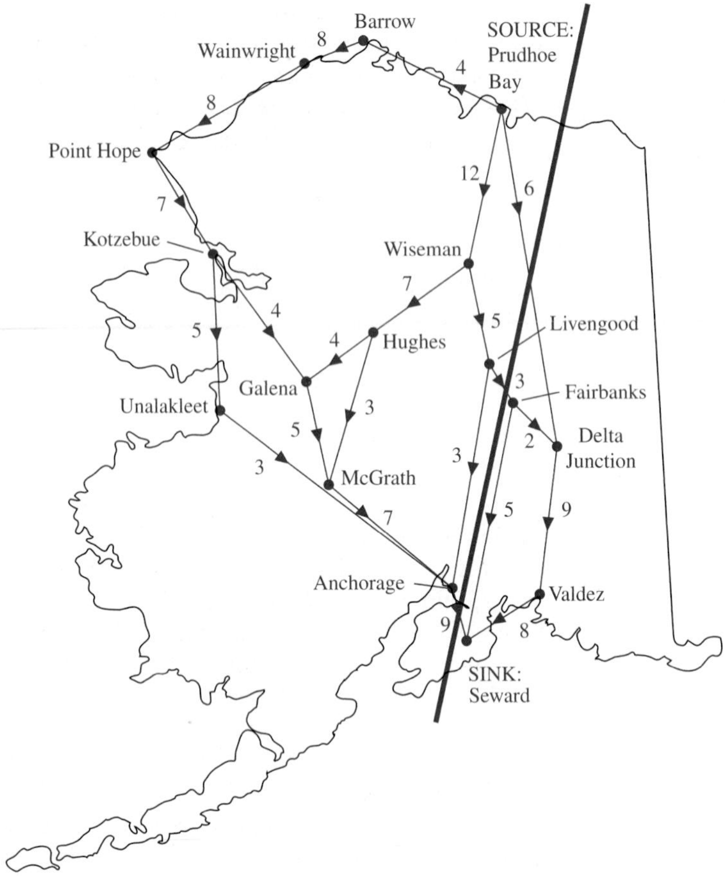

FIGURE 7.8

line). Notice that the only arcs joining a city in $\mathcal{S}$ to a city in $\mathcal{T}$ are those from Anchorage to Seward (with capacity 9), Livengood to Fairbanks (with capacity 3), and Prudhoe Bay to Delta Junction (with capacity 6). These arcs have a total capacity of $9 + 3 + 6 = 18$, and so no flow from a vertex in $\mathcal{S}$ to a vertex in $\mathcal{T}$ can exceed this number.

Generalizing from this example, we define a **cut** in a network to be a partition of its vertices into two sets $\mathcal{S}$ and $\mathcal{T}$ such that the source lies in $\mathcal{S}$ and the sink belongs to $\mathcal{T}$. The sum of the capacities of all arcs leading from a vertex in $\mathcal{S}$ to a vertex in $\mathcal{T}$ is called the **capacity** of the cut. Note that, in determining the capacity of the cut, we consider only the capacity of arcs leading from a vertex in $\mathcal{S}$ to a vertex in $\mathcal{T}$, and not those leading from a vertex in $\mathcal{T}$ to a vertex in $\mathcal{S}$.

Example 7.3 In Figure 7.8, let

$$\mathcal{S} = \{\text{Prudhoe Bay, Barrow, Wainwright, Point Hope, Kotzebue}\}$$

and $\mathcal{T}$ contain the cities not in $\mathcal{S}$. Then $\mathcal{S}$, $\mathcal{T}$ is a cut because Prudhoe Bay is in $\mathcal{S}$ and Seward is in $\mathcal{T}$. The arcs leading from a city in $\mathcal{S}$ to a city in $\mathcal{T}$ are Kotzebue to Unalakleet (with capacity 5), Kotzebue to Galena (with capacity 4), Prudhoe Bay to Wiseman (with capacity 12), and Prudhoe Bay to Delta Junction (with capacity 6). So the capacity of this cut is $5 + 4 + 12 + 6 = 27$. ■

Example 7.4 In Figure 7.9(a), $\mathcal{S} = \{A, B, C\}$ and $\mathcal{T} = \{D, E\}$ form a cut. The arcs leading from a vertex in $\mathcal{S}$ to a vertex in $\mathcal{T}$ are (A, D) with capacity 3, (B, D) with capacity 6, and (C, E) with capacity 4. Therefore the capacity of this cut is $3 + 6 + 4 = 13$.

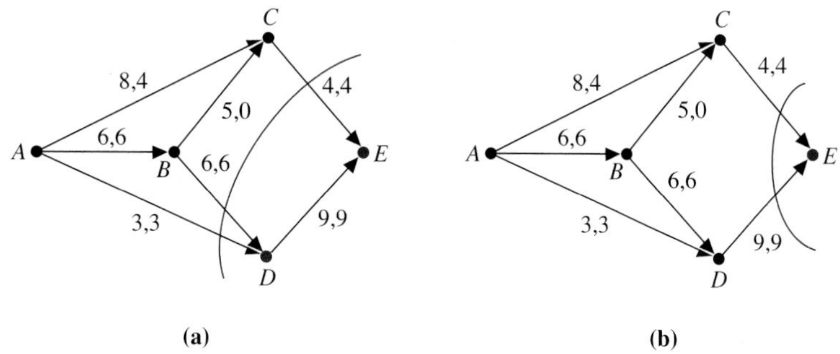

FIGURE 7.9

The sets $\mathcal{S}' = \{A, B, C, D\}$ and $\mathcal{T}' = \{E\}$ also form a cut. See Figure 7.9(b). In this case the arcs leading from a vertex in $\mathcal{S}'$ to a vertex in $\mathcal{T}'$ are (C, E) with capacity 4 and (D, E) with capacity 9. Thus this cut also has capacity 13. ■

In Figure 7.10, let $\mathcal{S}$, $\mathcal{T}$ be the cut with $\mathcal{S} = \{A, C, D\}$ and $\mathcal{T} = \{B, E\}$. Let us consider the total flow (not the capacities) along the arcs joining vertices in $\mathcal{S}$ and $\mathcal{T}$. Notice first that the total flow from $\mathcal{S}$ to $\mathcal{T}$ (that is, the total flow along arcs leading from a vertex in $\mathcal{S}$ to a vertex in $\mathcal{T}$) is $6 + 4 + 7 = 17$, the sum of the flows along the blue arcs (A, B), (C, E), and (D, E), respectively. Likewise, the total flow from $\mathcal{T}$ to $\mathcal{S}$ is $1 + 5 = 6$, the sum of the flows along the black arcs (B, C) and (B, D). The difference between the total flow from $\mathcal{S}$ to $\mathcal{T}$ and the

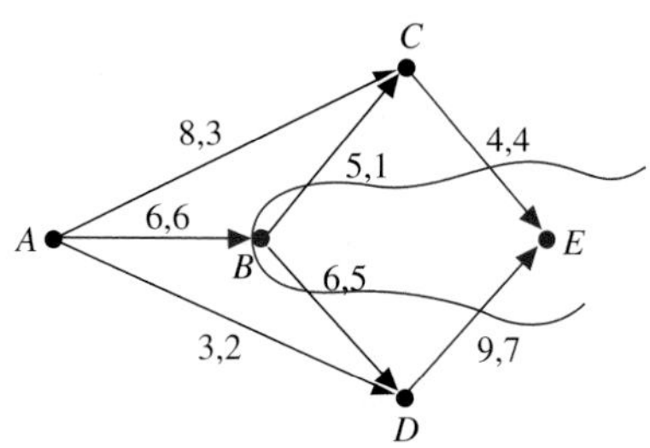

FIGURE 7.10

total flow from $\mathcal{T}$ to $\mathcal{S}$ is therefore $17 - 6 = 11$, which is the value of the flow shown in Figure 7.10. This equality is true in general, as the next theorem shows.

Theorem 7.2 If f is a flow in a transportation network and $\mathcal{S}, \mathcal{T}$ is a cut, then the value of f equals the total flow along arcs leading from a vertex in $\mathcal{S}$ to a vertex in $\mathcal{T}$ minus the total flow along arcs leading from a vertex in $\mathcal{T}$ to a vertex in $\mathcal{S}$.

Proof. If $\mathcal{U}$ and $\mathcal{V}$ are sets of vertices in the network, we will denote by $f(\mathcal{U}, \mathcal{V})$ the total flow along arcs leading from a vertex in $\mathcal{U}$ to a vertex in $\mathcal{V}$. Let a be the value of f. With this notation, the result to be proved can be written as $a = f(\mathcal{S}, \mathcal{T}) - f(\mathcal{T}, \mathcal{S})$. Note that if $\mathcal{V}_1 \cap \mathcal{V}_2 = \emptyset$, then

$$f(\mathcal{U}, \mathcal{V}_1 \cup \mathcal{V}_2) = f(\mathcal{U}, \mathcal{V}_1) + f(\mathcal{U}, \mathcal{V}_2);$$

likewise, if $\mathcal{U}_1 \cap \mathcal{U}_2 = \emptyset$, then

$$f(\mathcal{U}_1 \cup \mathcal{U}_2, \mathcal{V}) = f(\mathcal{U}_1, \mathcal{V}) + f(\mathcal{U}_2, \mathcal{V}).$$

By the definition of a flow, $f(\{V\}, \mathcal{S} \cup \mathcal{T}) - f(\mathcal{S} \cup \mathcal{T}, \{V\}) = 0$ if V is neither the source nor the sink, and $f(\{V\}, \mathcal{S} \cup \mathcal{T}) - f(\mathcal{S} \cup \mathcal{T}, \{V\}) = a$ if V is the source. Thus if we sum these equations for all V in $\mathcal{S}$, we obtain the equation $f(\mathcal{S}, \mathcal{S} \cup \mathcal{T}) - f(\mathcal{S} \cup \mathcal{T}, \mathcal{S}) = a$. Therefore

$$\begin{aligned} a &= f(\mathcal{S}, \mathcal{S} \cup \mathcal{T}) - f(\mathcal{S} \cup \mathcal{T}, \mathcal{S}) \\ &= [f(\mathcal{S}, \mathcal{S}) + f(\mathcal{S}, \mathcal{T})] - [f(\mathcal{S}, \mathcal{S}) + f(\mathcal{T}, \mathcal{S})] \\ &= f(\mathcal{S}, \mathcal{T}) - f(\mathcal{T}, \mathcal{S}). \quad ■ \end{aligned}$$

Corollary If f is a flow in a transportation network and $\mathcal{S}, \mathcal{T}$ is a cut, then the value of f cannot exceed the capacity of $\mathcal{S}, \mathcal{T}$.

Proof. Using the notation in the proof of Theorem 7.2, we have

$$a = f(\mathcal{S}, \mathcal{T}) - f(\mathcal{T}, \mathcal{S}) \le f(\mathcal{S}, \mathcal{T}),$$

since $f(\mathcal{T}, \mathcal{S}) \ge 0$. But the flow along any arc leading from a vertex in $\mathcal{S}$ to a vertex in $\mathcal{T}$ cannot exceed the capacity of that arc. Thus $f(\mathcal{S}, \mathcal{T})$ cannot exceed the capacity of the cut $\mathcal{S}, \mathcal{T}$. It follows that the value of f cannot exceed the capacity of the cut $\mathcal{S}, \mathcal{T}$. ■

The corollary to Theorem 7.2 is a useful result. It implies that the value of a maximal flow in a transportation network cannot exceed the capacity of *any* cut in the network. By using this fact, we can easily obtain an upper bound on the value of a maximal flow. In Section 7.3 we will be able to strengthen this result by showing that every transportation network contains at least one cut with capacity equal to the value of a maximal flow. (Notice, for instance, that Example 7.4 presents two cuts with capacity equal to the value of the maximal flow in the network shown in Figure 7.7.) This fact will enable us to prove that a particular flow is a maximal flow, as we did in analyzing the flow in Figure 7.7.

EXERCISES 7.1

In Exercises 1–6, tell whether the given weighted directed graph is a transportation network or not. If so, identify the source and sink. If not, tell why.

1.

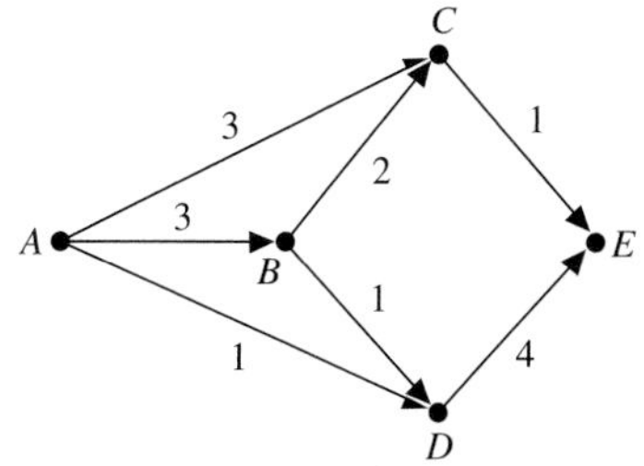

2.

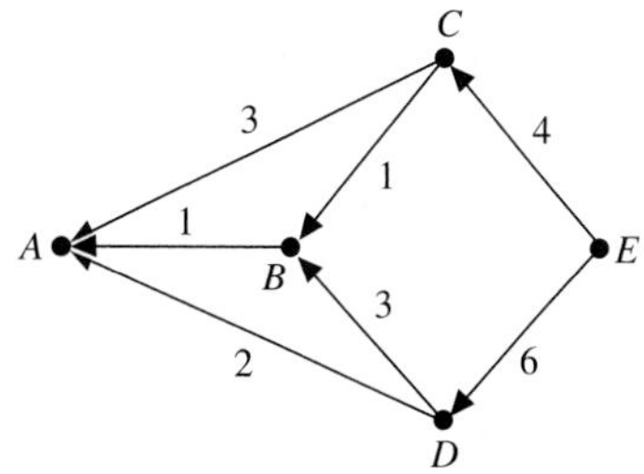

3.

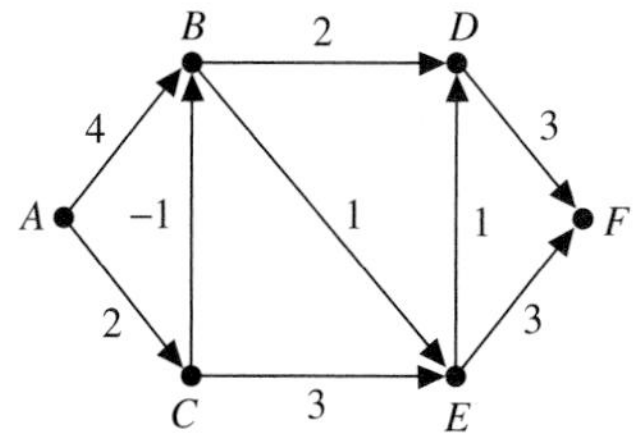

4.

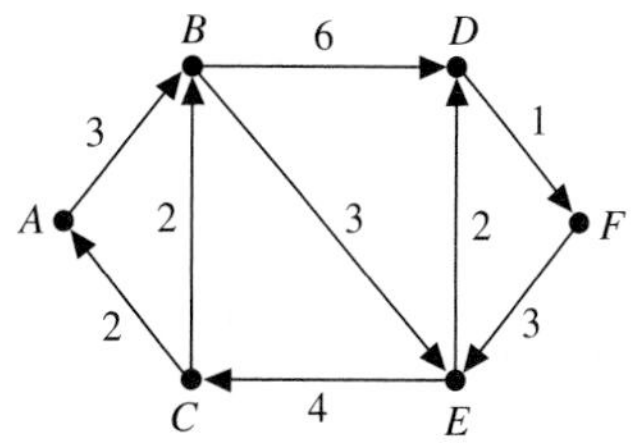

5.

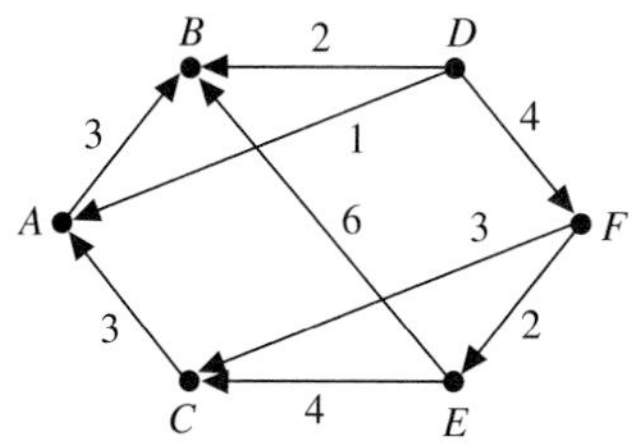

6.

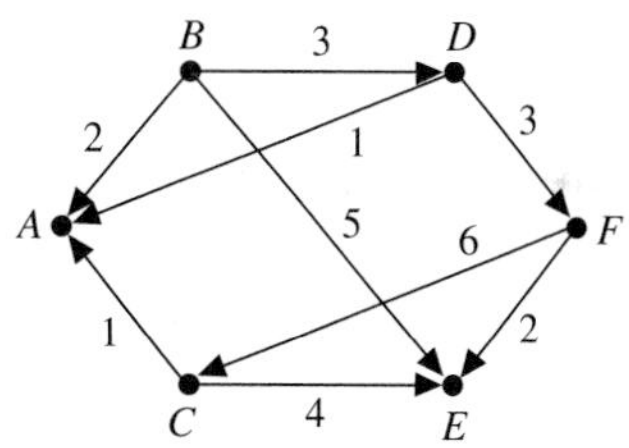

In Exercises 7–12, a transportation network is given. The first number along each arc gives the capacity of the arc. Tell whether the second set of numbers along the arcs is a flow for the network. If so, give the value of the flow. If not, tell why.

7.

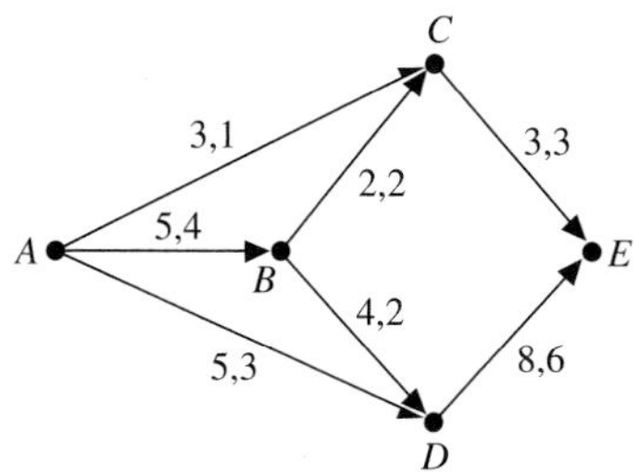

8.

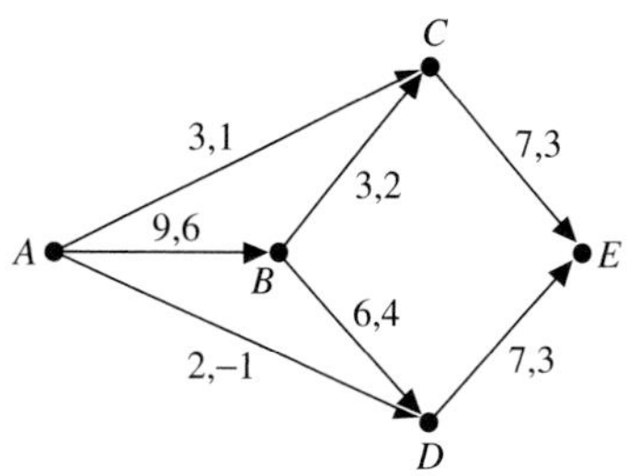

9.

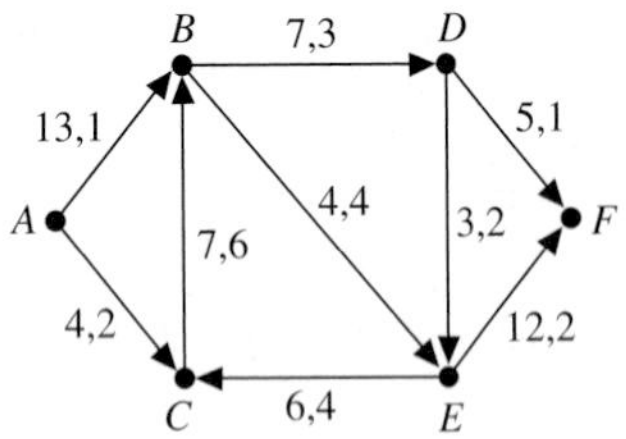

10.

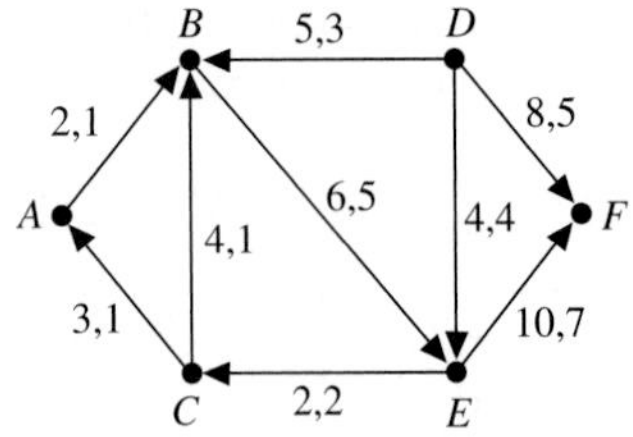

11.

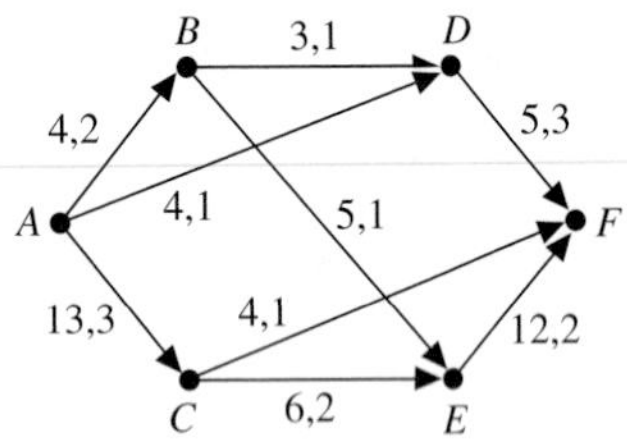

12.

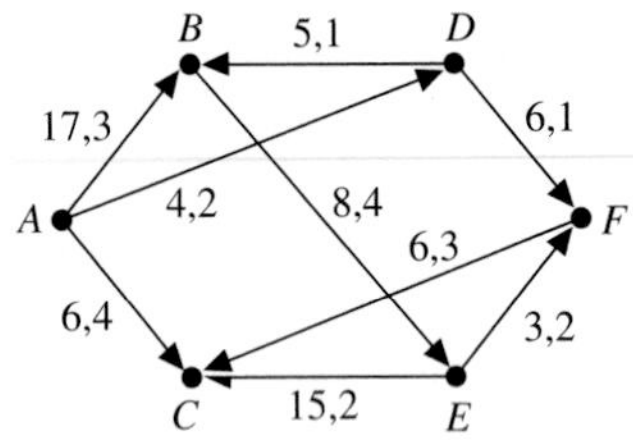

In Exercises 13–18, tell whether the given sets $\mathcal{S}$, $\mathcal{T}$ form a cut for the network indicated. If so, give the capacity of the cut. If not, tell why.

13. $\mathcal{S} = \{A, B\}$ and $\mathcal{T} = \{D, E\}$ for the network in Exercise 7

14. $\mathcal{S} = \{A, D\}$ and $\mathcal{T} = \{B, C, E\}$ for the network in Exercise 8

15. $\mathcal{S} = \{A, D, E\}$ and $\mathcal{T} = \{B, C, F\}$ for the network in Exercise 9

16. $\mathcal{S} = \{A, B, C, D\}$ and $\mathcal{T} = \{D, E, F\}$ for the network in Exercise 10

17. $\mathcal{S} = \{A, D, E\}$ and $\mathcal{T} = \{B, C, F\}$ for the network in Exercise 11

18. $\mathcal{S} = \{A, B, C\}$ and $\mathcal{T} = \{D, E, F\}$ for the network in Exercise 12

In Exercises 19–24, find by inspection a flow satisfying the given conditions.

19. A flow of value 11 for the network in Exercise 7

20. A flow of value 13 for the network in Exercise 8

21. A flow of value 11 for the network in Exercise 9

22. A flow of value 17 for the network in Exercise 10

23. A flow of value 18 for the network in Exercise 11

24. A flow of value 18 for the network in Exercise 12

In Exercises 25–30, find by inspection a cut satisfying the given conditions.

25. A cut of capacity 11 for the network in Exercise 7

26. A cut of capacity 13 for the network in Exercise 8

27. A cut of capacity 11 for the network in Exercise 9

28. A cut of capacity 17 for the network in Exercise 10

29. A cut of capacity 18 for the network in Exercise 11

30. A cut of capacity 18 for the network in Exercise 12

31. A telephone call can be routed from Chicago to Atlanta along various lines. The line from Chicago to Indianapolis can carry 40 calls at the same time. Other lines and their capacities are as follows: Chicago to St. Louis (30 calls), Chicago to Memphis (20 calls), Indianapolis to Memphis (15 calls), Indianapolis to Lexington (25 calls), St. Louis to Little Rock (20 calls), Little Rock to Memphis (15 calls), Little Rock to Atlanta (10 calls), Memphis to Atlanta (25 calls), and Lexington to Atlanta (15 calls). Draw a transportation network displaying this information.

32. A power generator at a dam can send 300 megawatts to substation 1, can send 200 megawatts to substation 2, and can send 250 megawatts to substation 3. In addition, substation 2 can send 100 megawatts to substation 1 and 70 megawatts to substation 3. Substation 1 can send at most 280 megawatts to the distribution center, and substation 3 can send at most 300 megawatts to the distribution center. Draw a transportation network displaying this information.

In Exercises 33–36, let $f(\mathcal{U}, \mathcal{V})$ be defined as in the proof of Theorem 7.2.

33. Find $f(\mathcal{U}, \mathcal{V})$ and $f(\mathcal{V}, \mathcal{U})$ if f is the flow in Exercise 10, $\mathcal{U} = \{B, C, D\}$, and $\mathcal{V} = \{A, E, F\}$.

34. Find $f(\mathcal{U}, \mathcal{V})$ and $f(\mathcal{V}, \mathcal{U})$ if f is the flow in Exercise 12, $\mathcal{U} = \{C, E, F\}$, and $\mathcal{V} = \{A, B, D\}$.

35. Give an example to show that if $\mathcal{V}_1$ and $\mathcal{V}_2$ are not disjoint, then it is possible that $f(\mathcal{U}, \mathcal{V}_1 \cup \mathcal{V}_2)$ may not equal $f(\mathcal{U}, \mathcal{V}_1) + f(\mathcal{U}, \mathcal{V}_2)$.

36. Prove that $f(\mathcal{U}, \mathcal{V} \cup \mathcal{W}) = f(\mathcal{U}, \mathcal{V}) + f(\mathcal{U}, \mathcal{W}) - f(\mathcal{U}, \mathcal{V} \cap \mathcal{W})$ for any sets of vertices $\mathcal{U}$, $\mathcal{V}$, and $\mathcal{W}$.

7.2 A FLOW AUGMENTATION ALGORITHM

In this section we present an algorithm for finding the maximal flow in a transportation network. This algorithm is based on a procedure formulated by Ford and Fulkerson and utilizes a modification suggested by Edmonds and Karp. (See suggested readings [5] and [3] at the end of this chapter.) The essence of the algorithm is described in Section 7.1:

(1) Begin with any flow, for example, the one having zero flow along every arc.
(2) Find a flow-augmenting path (a path from the source to the sink along which the present flow can be increased), and increase the flow along this path by as much as possible.
(3) Repeat step (2) until it is no longer possible to find a flow-augmenting path.

Some care must be taken in deciding whether there is a flow-augmenting path. Consider, for example, the transportation network in Figure 7.11, where the numbers along each arc are the capacity of the arc and the present flow along the arc, in that order. This flow was obtained by sending 4 units of flow along path A, B, C, E; 3 units along path A, D, E; and then 2 units along path A, B, D, E.

The value of the flow in Figure 7.11 is 9, and we know from the argument following Figure 7.7 that the value of a maximal flow in this network is 13.

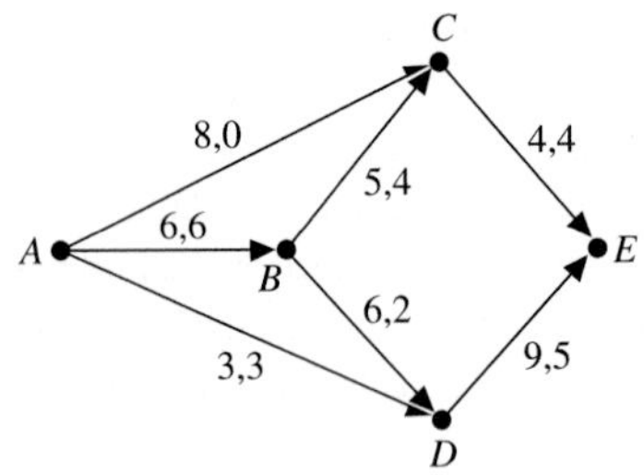

FIGURE 7.11

Consequently, we will look for a path from A to E along which the present flow can be increased. Clearly, the only way to increase the flow out of A is to use arc (A, C). But (C, E) is the only arc leading out of the vertex C, and the present flow along this arc equals its capacity. Thus we cannot increase the flow along arc (C, E). Therefore there is no *directed* path from A to E along which the flow can be increased. But if we allowed flow from vertex C to vertex B along the arc (B, C), we could send 4 units of flow along path A, C, B, D, E. This additional 4 units would give us a maximal flow from A to E.

How can we justify sending 4 units from C to B when the arc is directed from B to C? Since there are already 4 units of flow along arc (B, C), sending 4 units of flow from C to B has the effect of canceling the previous flow along (B, C). Thus, by sending 4 units of flow along path A, C, B, D, E, we obtain the maximal flow shown in Figure 7.7.

If we look at the network in Figure 7.11 more carefully, we can see that our first path A, B, C, E was not well chosen. By using the arc (C, E) as part of this path, we prevent the later use of arc (A, C). (Note that because there is no arc except (C, E) leaving vertex C, any flow sent into vertex C along arc (A, C) must leave along arc (C, E).) Therefore using arc (B, C) in the path A, C, B, D, E corrects the original poor choice of path A, B, C, E. Clearly, our algorithm will need some method to correct a poor choice of path from source to sink made earlier. In the version of the algorithm that follows, this correction occurs in step 2.2(b).

This algorithm, like the independent set algorithm of Section 6.3, is based on the labeling procedure devised by Ford and Fulkerson. Two operations, called *labeling* and *scanning*, are performed on vertices in the algorithm. As in the independent set algorithm, a vertex must be labeled before it can be scanned.

Flow Augmentation Algorithm

For a transportation network in which arc (X, Y) has capacity $c(X, Y)$, this algorithm either indicates that the current flow f is a maximal flow or else replaces f with a flow having a larger value.

Step 1 (label the source)
Label the source with the triple (source, $+$, ∞).

Step 2 (scan and label)
repeat
  Step 2.1 (select a vertex to scan)
  Among all the vertices that have been labeled, but not scanned, let V denote the one that was labeled first, and suppose that the label on V is $(U, \pm, a)$.
  Step 2.2 (scan vertex V)
  For each unlabeled vertex W, perform exactly one of the following three actions:
    (a) If (V, W) is an arc and $f(V, W) < c(V, W)$, assign to W the label $(V, +, b)$, where b is the smaller of a and $c(V, W) - f(V, W)$.
    (b) If (W, V) is an arc and $f(W, V) > 0$, assign to W the label $(V, -, b)$, where b is the smaller of a and $f(W, V)$.
    (c) If neither (a) nor (b) holds, do not label W.
  Step 2.3 (mark as scanned)
  Regard vertex V as having been scanned.
until either the sink is labeled or every labeled vertex has been scanned

Step 3 (increase the flow if possible)
if the sink is unlabeled
  The present flow is a maximal flow.
otherwise
  Step 3.1 (breakthrough)
  Now let V denote the sink, and suppose that the label on V is $(U, +, a)$.
  Step 3.2 (adjust the flow)
  repeat
    (a) **if** the label on V is $(U, +, b)$
        Replace $f(U, V)$ with $f(U, V) + a$.
      endif
    (b) **if** the label on V is $(U, -, b)$
        Replace $f(V, U)$ with $f(V, U) - a$.
      endif
    (c) Now let V denote the vertex U.
  until V is the source
endif

If the present flow is not a maximal flow, the algorithm uses breadth-first search to find a shortest flow-augmenting path (that is, one with the fewest arcs). Each vertex V along this path is labeled in one of two ways: $(U, +, a)$ or $(U, -, a)$. The first entry of the label, U, signifies that vertex U precedes V

on this path. The second entry of the label denotes that (U, V) is an arc or that (V, U) is an arc on this path, depending on whether the entry is $+$ or $-$, respectively. And the third entry of the label, a, is a positive number indicating how much the present flow can be increased (if the second entry of the label is $+$) or decreased (if the second entry of the label is $-$) without violating the restrictions in condition (1) of the definition of a flow for any arc along the path from the source to V.

We will illustrate the use of the flow augmentation algorithm by finding a maximal flow for the network discussed in Section 7.1. When we reach step 2.2 of the algorithm, we will examine the unlabeled vertices in alphabetical order. In order to begin the algorithm, we will take the flow to be 0 along every arc, as shown in Figure 7.12. (Again, the two numbers written beside each arc are the capacity and the present flow along that arc.)

In step 1 we assign the label (source, $+$, ∞) to the source, vertex A. In step 2 we note that only vertex A has been labeled, but not scanned. In step 2.2 we scan vertex A by examining the unlabeled vertices (B, C, D, and E) to see if any of them can be assigned labels. Note that vertex B is unlabeled, (A, B) is an arc, and the flow along this arc (0) is less than the capacity (6). Thus we can perform action (a) in step 2.2 on vertex B. Since 6 is the smaller of ∞ (the third entry in the label on A) and $6 - 0$, we label B with $(A, +, 6)$. Likewise, we can perform action (a) on the vertices C and D, which assigns them the labels $(A, +, 8)$ and $(A, +, 3)$, respectively. Because vertex E is not joined to vertex A, it cannot yet be given a label. This completes the scanning of vertex A. The current labels are shown in Figure 7.13.

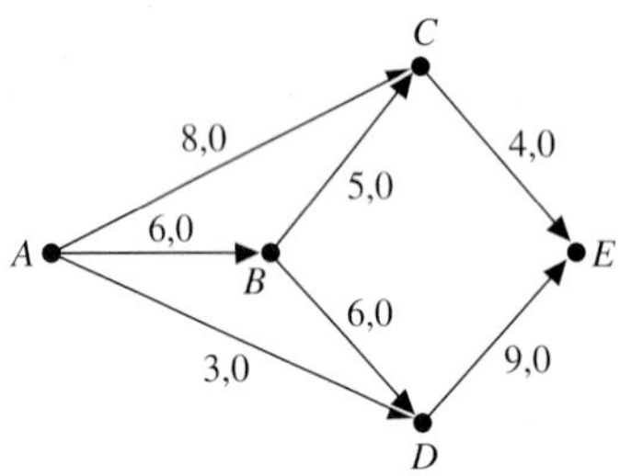

FIGURE 7.12

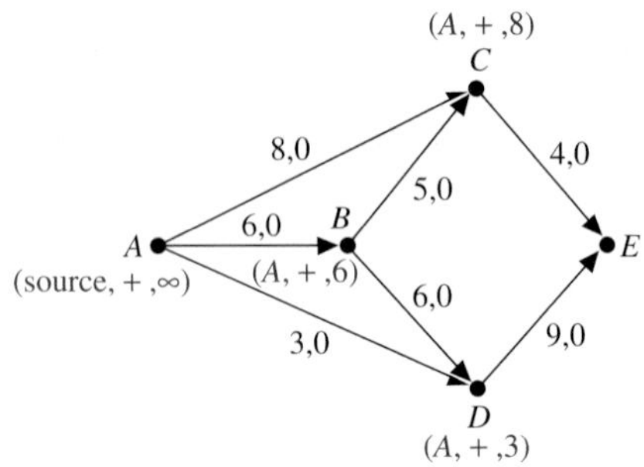

FIGURE 7.13

After scanning vertex A, we return to step 2.1. There are 3 vertices that have been labeled, but not scanned (namely, vertices B, C, and D). Of these, vertex B was the first one to be labeled, and so we scan vertex B. Because there are no unlabeled vertices joined by an arc to vertex B, no changes result from the scanning of vertex B. Therefore we return to step 2.1 once more. At this stage, there are two unlabeled vertices that have not been scanned (namely, vertices C and D), and, of these, C was the first to be labeled. Thus we scan vertex C in step 2.2. Since E is unlabeled, (C, E) is an arc, and the flow along this arc (0) is less than its capacity (4), we perform action (a). This action assigns to vertex E the label $(C, +, 4)$ because 4 is the smaller of 8 (the third entry in the label on C) and $4 - 0$ (the capacity minus the flow along arc (C, E)). Since there are no unlabeled vertices remaining, this completes the scanning of vertex C. The present labels are shown in Figure 7.14.

In the scanning of vertex C, the sink was labeled, so we proceed to step 3. The fact that the sink has been labeled $(C, +, 4)$ tells us that the current flow can be increased by 4 along a path through vertex C. The vertex that precedes C in this path is the first entry in the label on C, which is $(A, +, 8)$. Thus the path from the source to the sink along which the flow can be increased by 4 is A, C, E. When we increase the flow along the arcs in this path by 4, the resulting flow is as in Figure 7.15.

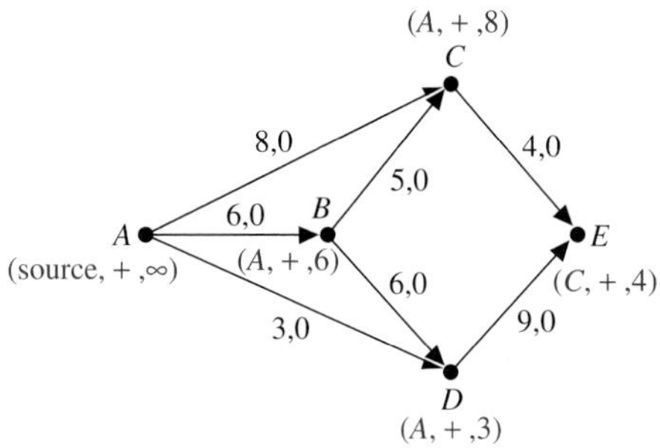

FIGURE 7.14

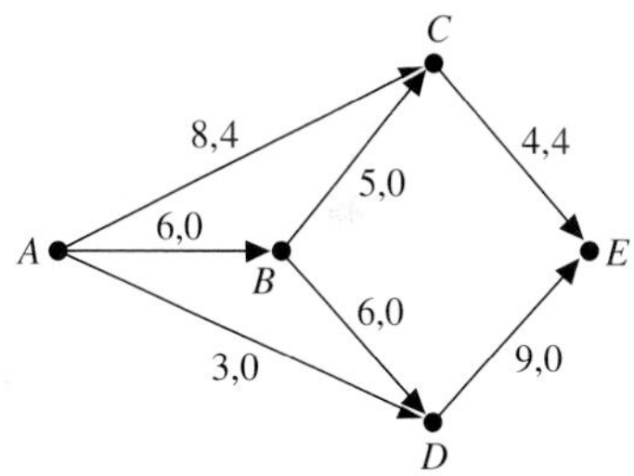

FIGURE 7.15

This finishes step 3, so the first iteration of the algorithm has been completed. We now remove all the labels and repeat the algorithm with the flow in Figure 7.15. As before, we assign to vertex A the label (source , $+$, ∞), and then in step 2.2 we scan vertex A. This results in vertices B, C, and D receiving the respective labels $(A, +, 6)$, $(A, +, 4)$, and $(A, +, 3)$. This completes the scanning of vertex A. Since B is the unscanned vertex that was labeled first, we now scan vertex B. As in the first iteration of the algorithm, no changes result from the scanning of vertex B. So we return to step 2.1. This time vertex C is the unscanned vertex that was labeled first. But unlike the first iteration, we cannot label vertex E, because the flow along arc (C, E) is not less than the capacity of the arc. Consequently, no changes result from scanning vertex C. Once more, we return to step 2.1. This time vertex D is the only labeled vertex that has not been scanned, so we scan vertex D. Since (D, E) is an arc along which the flow is less than the capacity, we perform action (a). As a result of this action, vertex E is labeled $(D, +, 3)$. This completes the scanning of vertex D. But now the sink has been labeled; so we proceed to step 3. (See Figure 7.16.)

Because the label on the sink is $(D, +, 3)$, we can increase the current flow by 3 along a path through vertex D. To find the vertex that precedes D in this path, we examine the label on D, which is $(A, +, 3)$. Since the first entry of this label is A, we see that the path along which the flow can be increased by 3 is A, D, E. When we increase the flow along the arcs in this path by 3, we obtain the flow shown in Figure 7.17. This completes the second iteration of the algorithm.

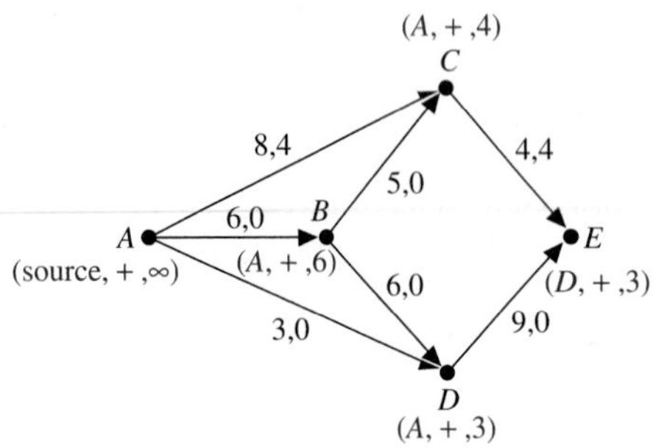

FIGURE 7.16

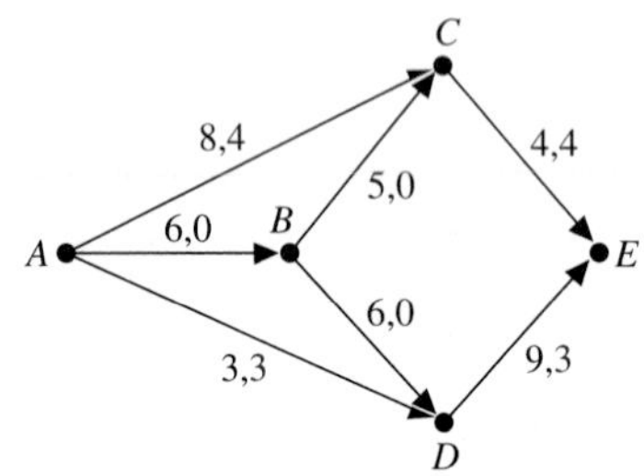

FIGURE 7.17

Again we discard all of the labels and perform another iteration of the algorithm. In this third iteration, we reach step 3 with the labels shown in Figure 7.18. From these labels, we see that the flow can be increased by 6 along the path A, B, D, E. By increasing the flows along the arcs in this path by 6, we obtain the flow in Figure 7.19.

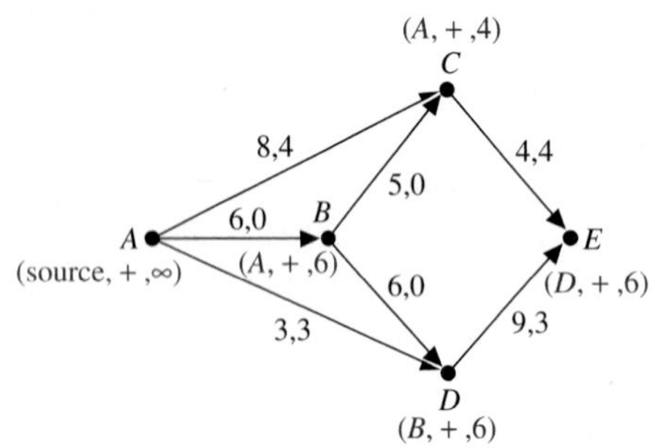

FIGURE 7.18

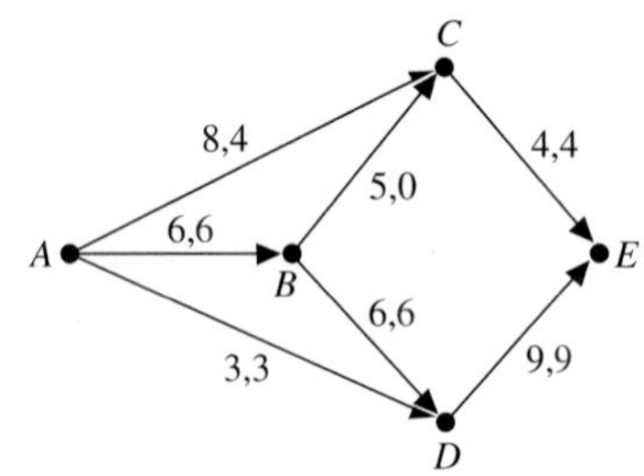

FIGURE 7.19

Again we discard all of the labels and perform another iteration of the algorithm. This time, however, we can label only vertex C when scanning vertex A. (See Figure 7.20.) Moreover, when vertex C is scanned, no changes occur. Consequently, all of the labeled vertices have been scanned. Thus step 3 assures us that the present flow (the one shown in Figure 7.19) is a maximal flow. When the algorithm ends, the set of labeled vertices $\mathcal{S} = \{A, C\}$ and the set of unlabeled vertices $\mathcal{T} = \{B, D, E\}$ form a cut. Notice that the capacity of this cut is $4 + 6 + 3 = 13$, which equals the value of the maximal flow in Figure 7.19. We will see in Section 7.3 that this is no coincidence: *When the flow augmentation algorithm ends with the sink unlabeled, the sets of labeled and unlabeled vertices always determine a cut with capacity equal to the value of a maximal flow.*

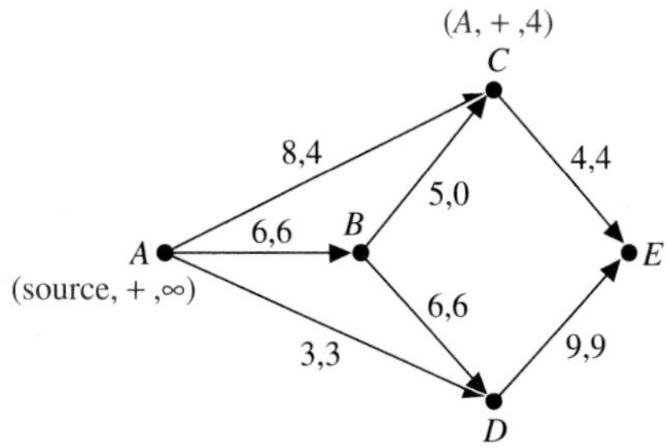

FIGURE 7.20

Example 7.5 We will use the flow augmentation algorithm to find a maximal flow for the network shown in Figure 7.21. When labeling vertices in step 2.2, we will consider them in alphabetical order.

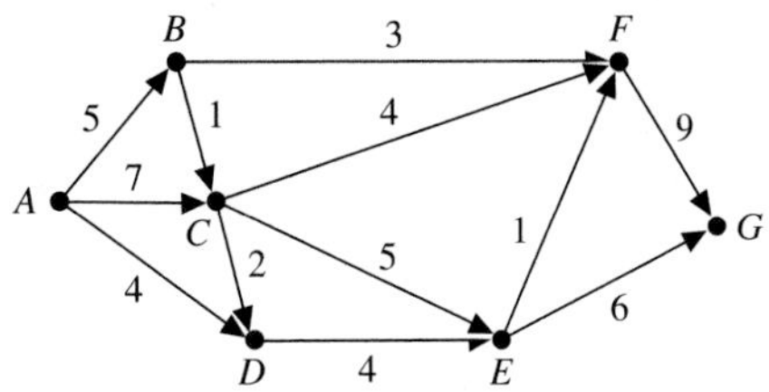

FIGURE 7.21

Iteration 1. The labels assigned in iteration 1 are shown in Figure 7.22. Thus we increase the flow by 3 along the path A, B, F, G.

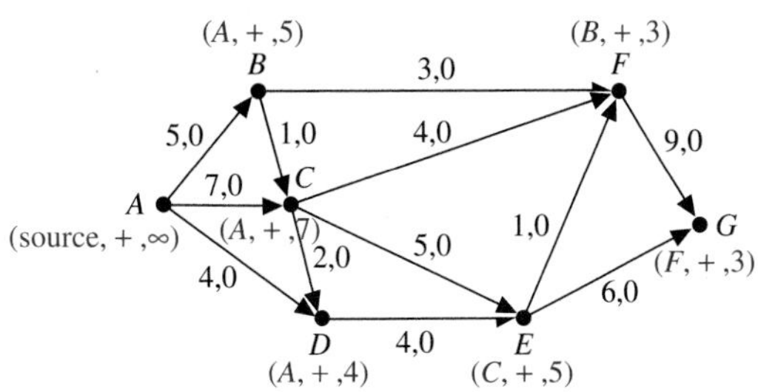

FIGURE 7.22

Iteration 2. The labels assigned in iteration 2 are shown in Figure 7.23. Thus we increase the flow by 5 along the path A, C, E, G.

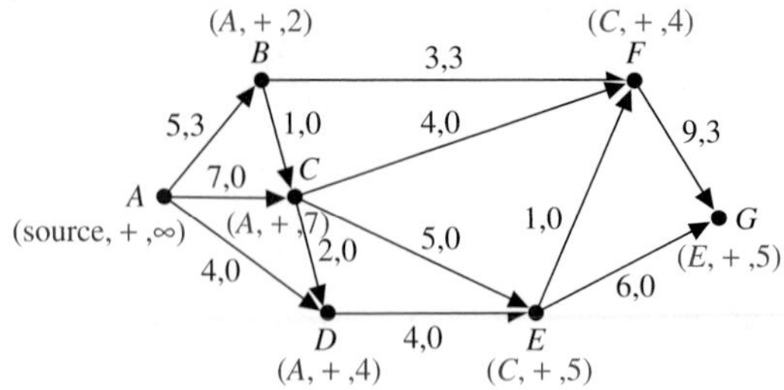

FIGURE 7.23

Iteration 3. The labels assigned in iteration 3 are shown in Figure 7.24. Thus we increase the flow by 2 along the path A, C, F, G.

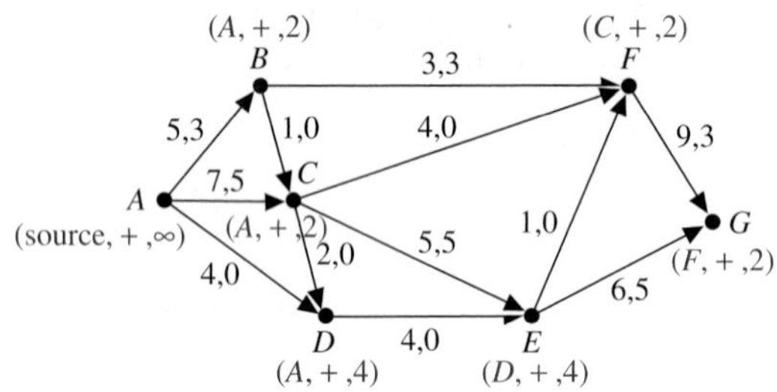

FIGURE 7.24

Iteration 4. The labels assigned in iteration 4 are shown in Figure 7.25. Thus we increase the flow by 1 along the path A, D, E, G.

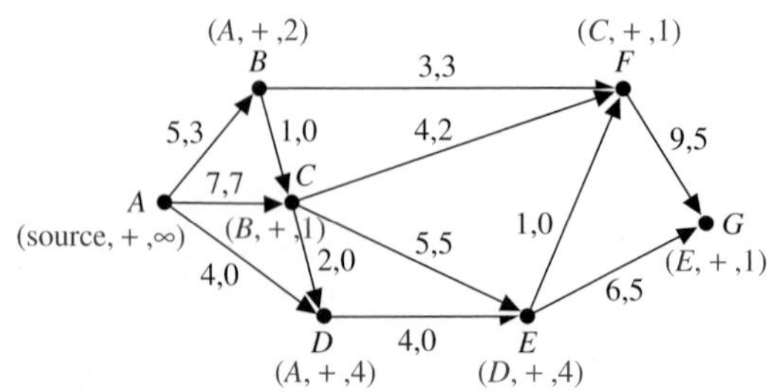

FIGURE 7.25

Iteration 5. The labels assigned in iteration 5 are shown in Figure 7.26. Thus we increase the flow by 1 along the path A, B, C, F, G.

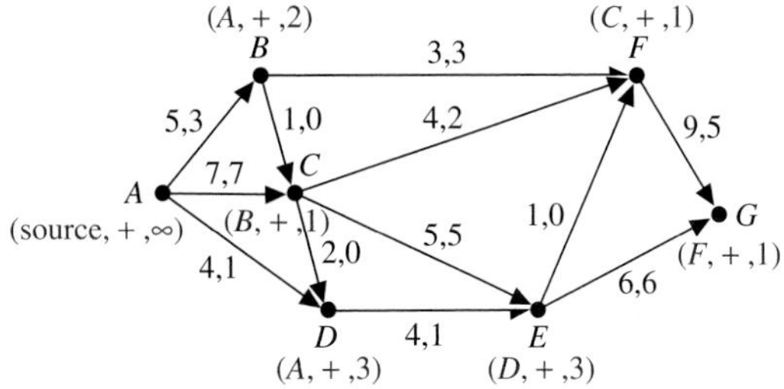

FIGURE 7.26

Iteration 6. The labels assigned in iteration 6 are shown in Figure 7.27. Thus we increase the flow by 1 along the path A, D, E, F, G.

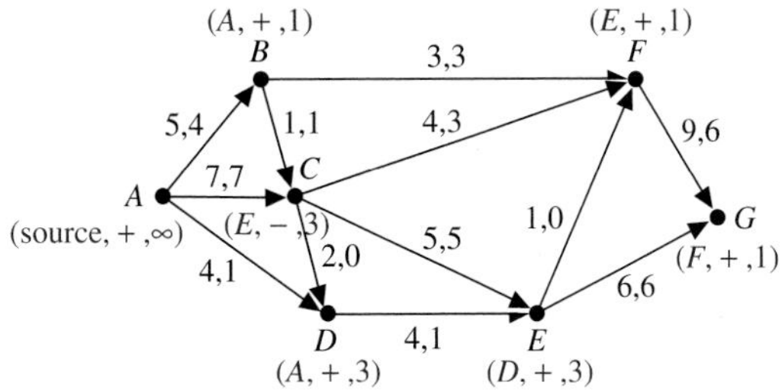

FIGURE 7.27

Iteration 7. The labels assigned in iteration 7 are shown in Figure 7.28. Thus we increase the flow by 1 along the path A, D, E, C, F, G. (Note that we are using arc (C, E) in the wrong direction to cancel 1 unit of the flow sent along this arc in iteration 2.)

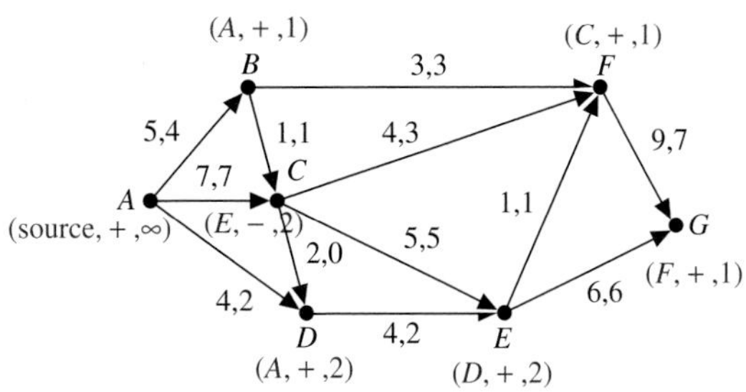

FIGURE 7.28

Iteration 8. The labels assigned in iteration 8 are shown in Figure 7.29.

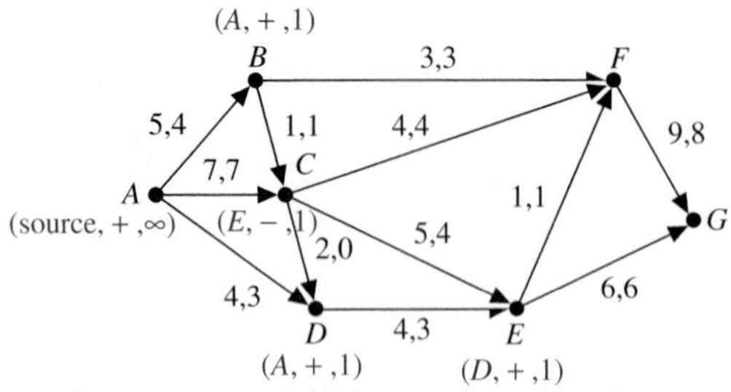

FIGURE 7.29

Since the sink is not labeled, the flow shown in Figure 7.29 is a maximal flow. The value of this maximal flow is 14. Note that the set $\mathcal{S} = \{A, B, C, D, E\}$ of labeled vertices and the set $\mathcal{T} = \{F, G\}$ of unlabeled vertices form a cut with capacity $3 + 4 + 1 + 6 = 14$. ■

Maximal flows need not be unique. For instance, the flow shown in Figure 7.30 is a maximal flow for the network in Example 7.5. This flow is different from the one shown in Figure 7.29.

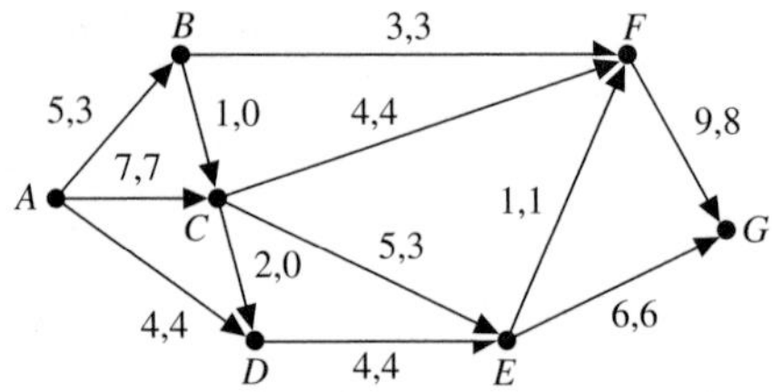

FIGURE 7.30

We conclude this section with a useful observation about the flow augmentation algorithm. Any flow in which the flow along each arc is an integer is called an **integral flow**. Suppose that all of the arc capacities in a network are integers and that we begin the flow augmentation algorithm with an integral flow. In this case the third entry of each label, which is assigned during step 2.2 of the algorithm, is a minimum of integers. Consequently, *if all the arc capacities are integers and we begin with zero flow along every arc, the maximal flow that results from repeated use of the flow augmentation algorithm is an integral flow.*

EXERCISES 7.2

Throughout these exercises, if there is a choice of vertices to label when using the flow augmentation algorithm, label the vertices in alphabetical order.

In Exercises 1–4, a network, a flow, and a flow-augmenting path are given. Determine the amount by which the flow can be increased along the given path.

1. Path: A, B, D, E

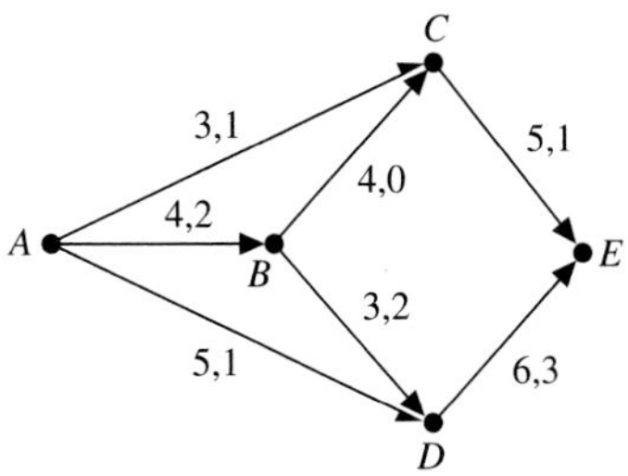

2. Path: A, B, C, E

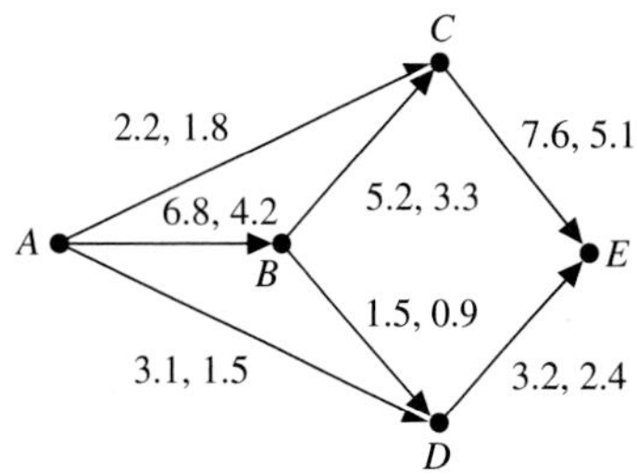

3. Path: A, B, E, D, F

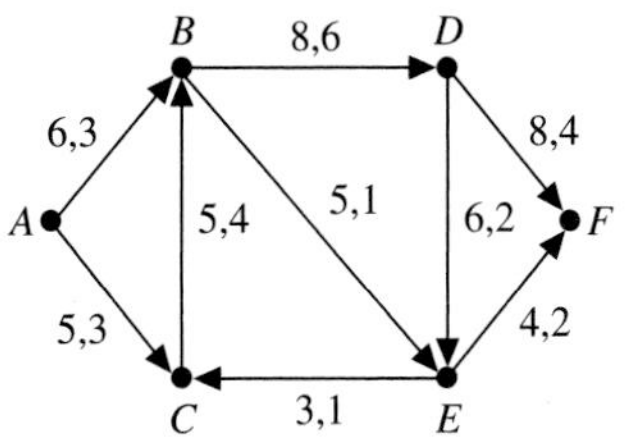

4. Path: D, B, C, E, F

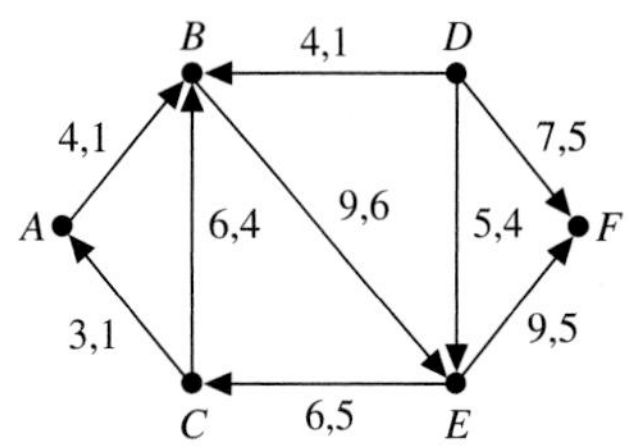

In Exercises 5–8, a network and flow are given. By performing the flow augmentation algorithm on this network and flow, we obtain the labels shown in each network. Determine a flow having a larger value than the given flow by performing steps 3.1 and 3.2 of the flow augmentation algorithm.

5.

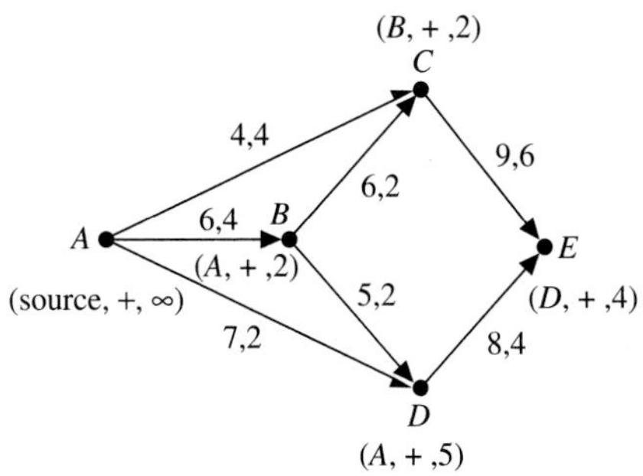

6.

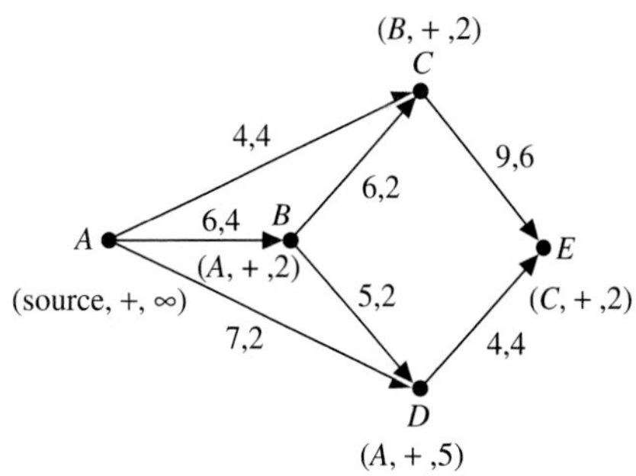

7.

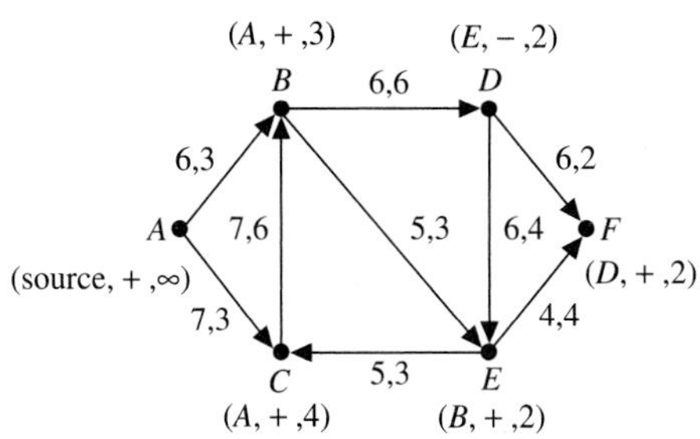

8.

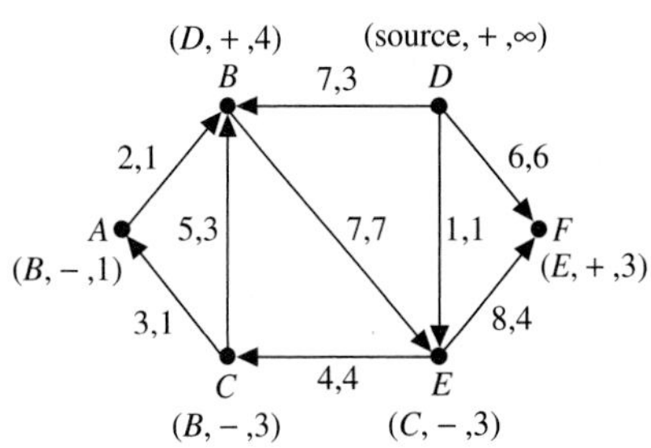

In Exercises 9–16, a network and flow are given. Use the flow augmentation algorithm to show that the given flow is maximal or else to find a flow with a larger value. If the given flow is not maximal, name the flow-augmenting path and the amount by which the flow can be increased.

9.

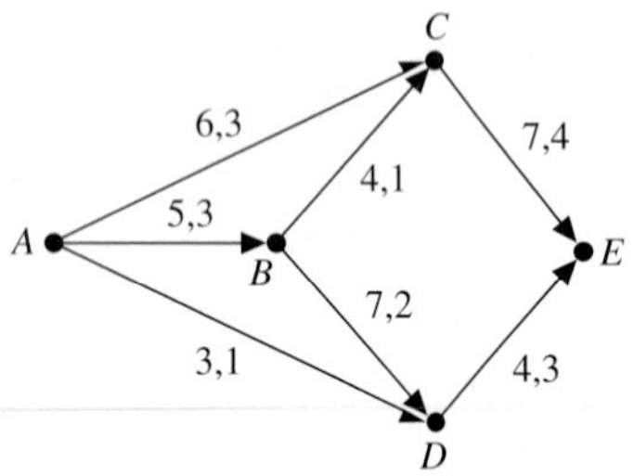

10.

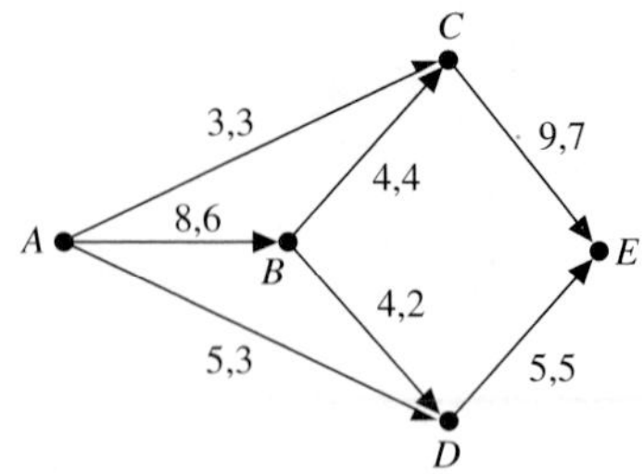

11.

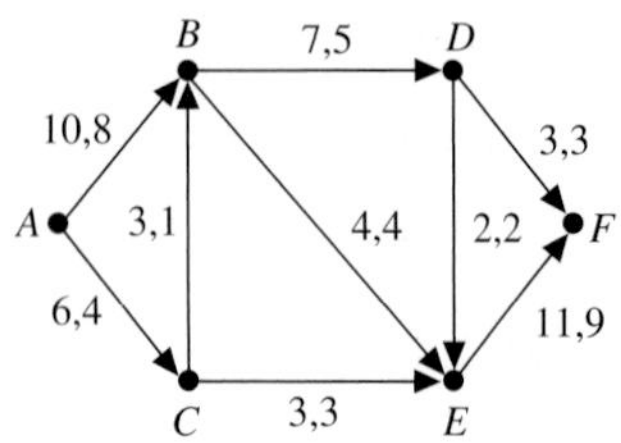

12.

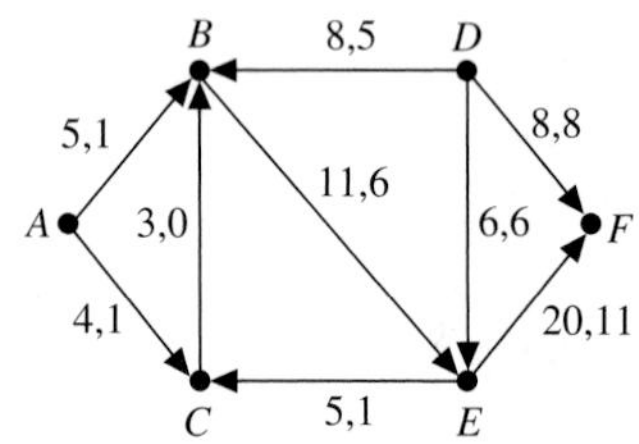

13.

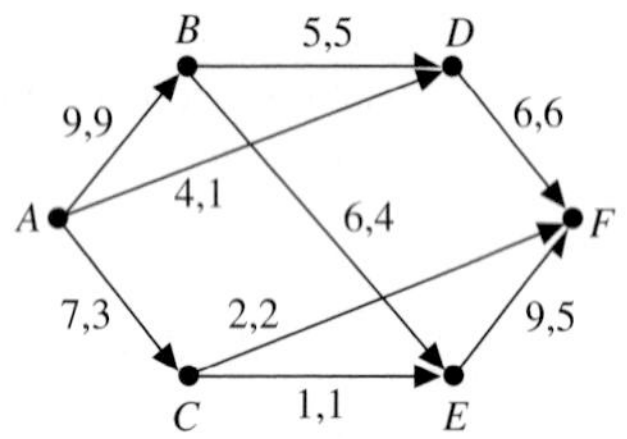

14.

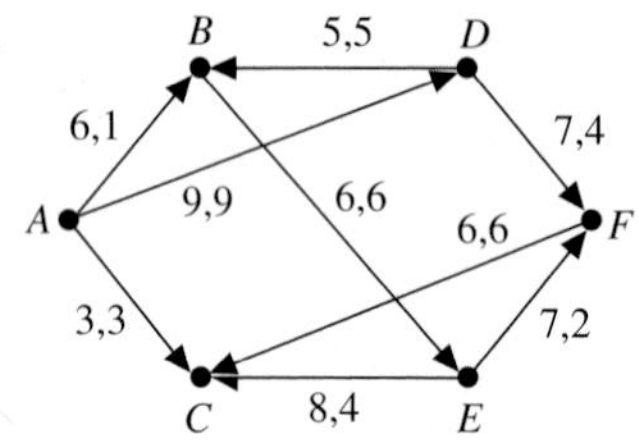

15.

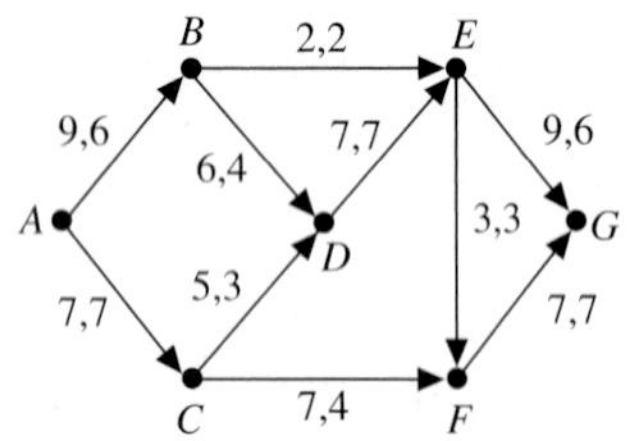

16.

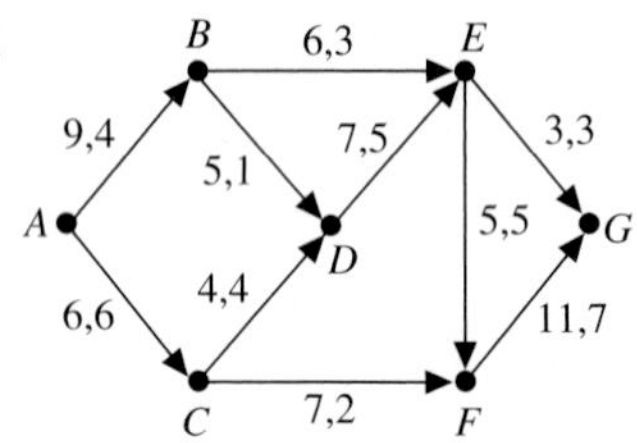

In Exercises 17–20, a transportation network and flow are given. Use the flow augmentation algorithm to find a maximal flow for each network.

17.

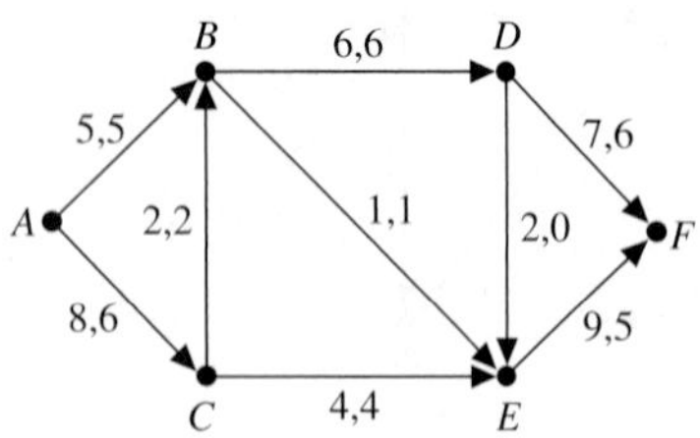

18.

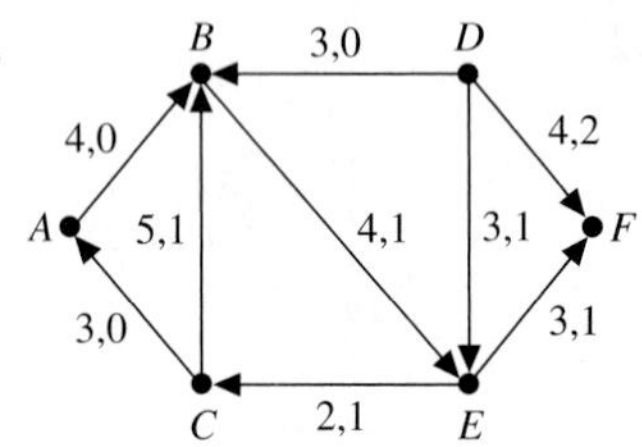

19.

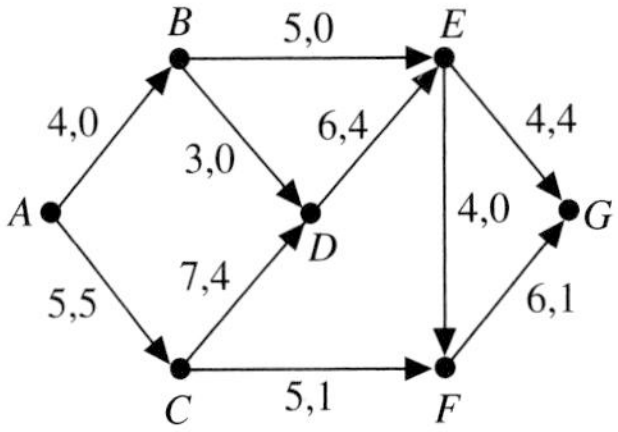

20.

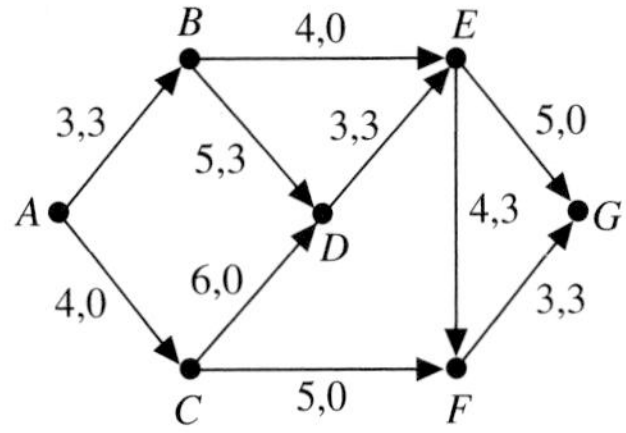

In Exercises 21–28, a transportation network is given. Find a maximal flow in each network by starting with the flow that is 0 along every arc and applying the flow augmentation algorithm.

21.

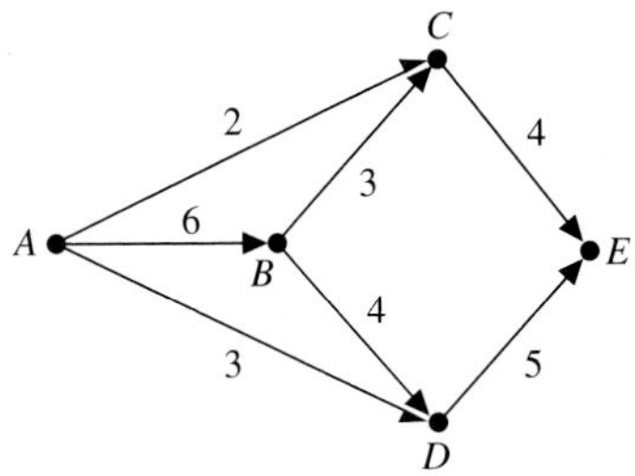

22.

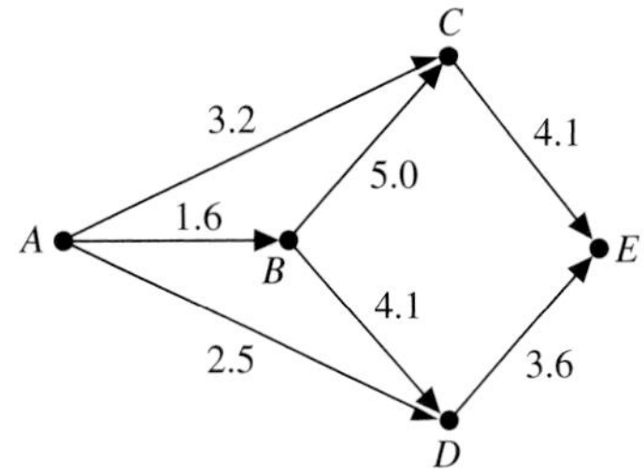

23.

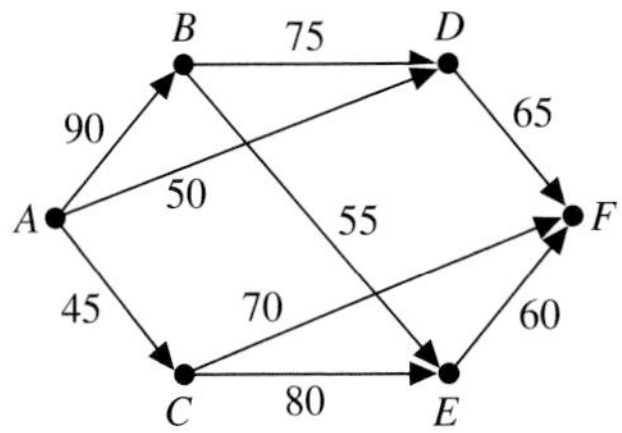

24.

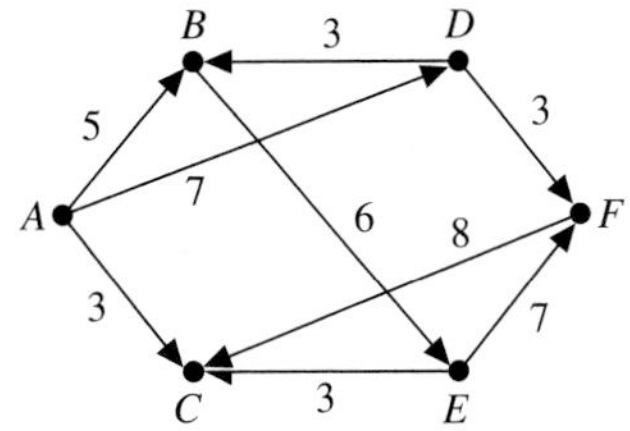

25.

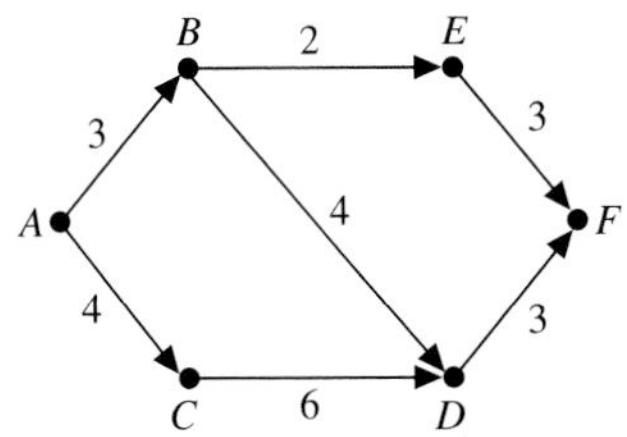

26.

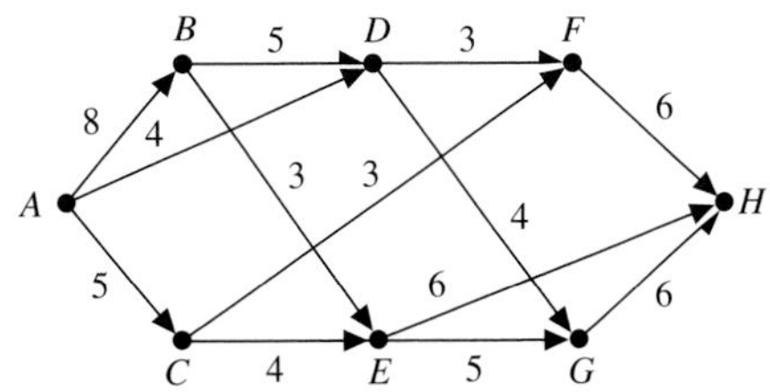

27.

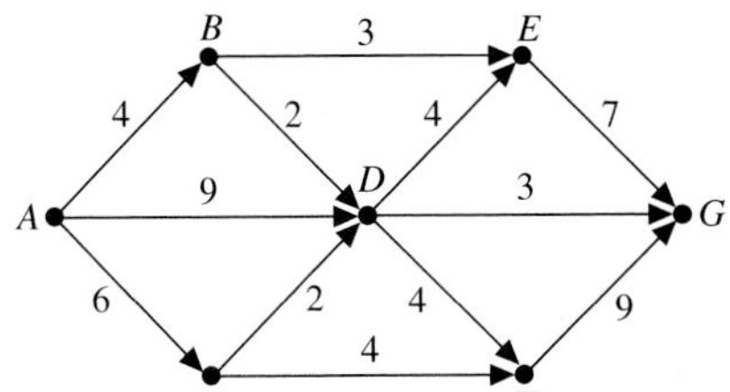

28.

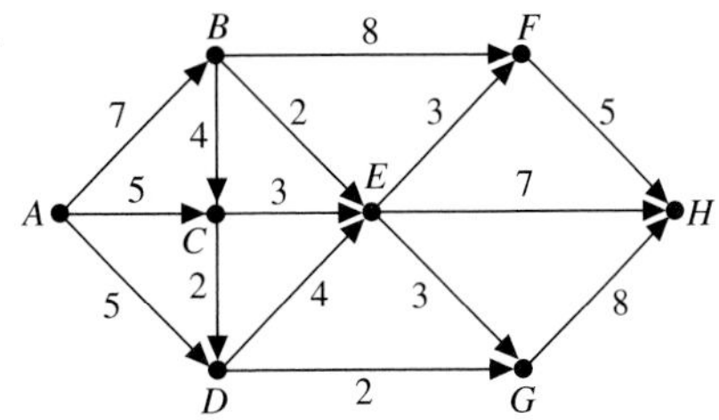

29. Give an example of a transportation network in which each arc has an integral capacity and there is a maximal flow such that the flow along some arcs is not an integer.

30. Consider a transportation network with source U and sink V in which each arc has capacity 1. Show that the value of a maximal flow is the maximum number of directed paths from U to V with no arcs in common.

7.3 THE MAX-FLOW MIN-CUT THEOREM

In this section we will show that the flow augmentation algorithm described in Section 7.2 does what we claim; that is, it either confirms that the current flow is a maximal flow or else finds a flow having a larger value. We will also verify the observation that when the algorithm ends, the sets of labeled and unlabeled vertices determine a cut with capacity equal to the value of a maximal flow. Such cuts are of special interest because they are cuts having the smallest possible capacity, as we will see in Theorem 7.4.

A cut in a transportation network is called a **minimal cut** if no other cut has a smaller capacity. The next theorem provides a method for detecting maximal flows and minimal cuts.

Theorem 7.3 In any transportation network, if f is a flow and $\mathcal{S}, \mathcal{T}$ is a cut such that the value of f equals the capacity of $\mathcal{S}, \mathcal{T}$, then f is a maximal flow and $\mathcal{S}, \mathcal{T}$ is a minimal cut.

Proof. Let f be a flow having value c, and let $\mathcal{S}, \mathcal{T}$ be a cut having capacity c. Let f' be any other flow in this network, and let the value of f' be v. Applying the corollary to Theorem 7.2 to f' and $\mathcal{S}, \mathcal{T}$, we see that $v \leq c$. Hence there is no flow in this network having a value greater than c, the value of f. It follows that f is a maximal flow.

Now let $\mathcal{S}', \mathcal{T}'$ be a cut in this network with capacity k. Applying the corollary to Theorem 7.2 to f and $\mathcal{S}', \mathcal{T}'$ gives $c \leq k$. Consequently, there is no cut in this network having a value less than c, the capacity of $\mathcal{S}, \mathcal{T}$. Therefore $\mathcal{S}, \mathcal{T}$ is a minimal cut. ■

In order to justify the validity of the flow augmentation algorithm, we need to show that

(1) When the algorithm ends with the sink unlabeled, the current flow is a maximal flow, and

(2) When the algorithm ends with the sink labeled, the original flow has been replaced by a flow with a larger value.

The proof of statement 2 will be left as an exercise (Exercise 24). Theorem 7.4 verifies statement 1 by showing that if an iteration of the flow augmentation algorithm ends with the sink unlabeled, then the present flow is a maximal flow.

Theorem 7.4 If, during an iteration of the flow augmentation algorithm, the sink is not labeled, then the present flow is maximal. Moreover, the sets of labeled and unlabeled vertices form a minimal cut having capacity equal to the value of the present flow.

Proof. Suppose that, during some iteration of the flow augmentation algorithm, the sink is not labeled. Let f denote the current flow, $c(X, Y)$ the capacity of arc (X, Y), $\mathcal{S}$ the set of labeled vertices, and $\mathcal{T}$ the set of unlabeled vertices. Then the source is in $\mathcal{S}$ and the sink is in $\mathcal{T}$; so $\mathcal{S}, \mathcal{T}$ is a cut.

Let (X, Y) be an arc leading from a vertex X in $\mathcal{S}$ to a vertex Y in $\mathcal{T}$. Since X is in $\mathcal{S}$, X has been labeled during this iteration of the flow augmentation algorithm. If $f(X, Y) < c(X, Y)$, then, when X was scanned, we would have labeled Y in step 2.2(a) of the algorithm. But Y is in $\mathcal{T}$ and hence is unlabeled; thus we must have $f(X, Y) = c(X, Y)$.

Now let (Y, X) be an arc leading from a vertex Y in $\mathcal{T}$ to a vertex X in $\mathcal{S}$. Since X is in $\mathcal{S}$, X has been labeled during this iteration of the algorithm. When X is scanned, we would have labeled Y in step 2.2(b) of the algorithm if $f(Y, X) > 0$. But Y is in $\mathcal{T}$ and so is unlabeled; thus we must have $f(Y, X) = 0$.

By Theorem 7.2, the value of f equals the total flow p along all arcs leading from a vertex in $\mathcal{S}$ to a vertex in $\mathcal{T}$ minus the total flow q along all arcs leading from a vertex in $\mathcal{T}$ to a vertex in $\mathcal{S}$. But the two preceding paragraphs show that p equals the capacity of the cut $\mathcal{S}, \mathcal{T}$ and $q = 0$. Therefore the value of f equals p, the capacity of the cut $\mathcal{S}, \mathcal{T}$. It then follows from Theorem 7.3 that f is a maximal flow and $\mathcal{S}, \mathcal{T}$ is a minimal cut. ■

Theorem 7.4 also proves our earlier assertion that when the flow augmentation algorithm ends with the sink unlabeled, the cut determined by the sets of labeled and unlabeled vertices is a minimal cut. Thus, for example, in Figure 7.20 we see that $\mathcal{S} = \{A, C\}$ and $\mathcal{T} = \{B, D, E\}$ form a minimal cut, and in Figure 7.29 we see that $\mathcal{S} = \{A, B, C, D, E\}$ and $\mathcal{T} = \{F, G\}$ form a minimal cut.

We have already seen that a network may have more than one maximal flow. Likewise, a network may have more than one minimal cut. In Figure 7.20, for instance, $\{A, B, C\}$ and $\{D, E\}$ is a different minimal cut from the one mentioned in the previous paragraph.

Example 7.6 A natural gas utility delivers gas to Little Rock from a source in Amarillo through the network of pipelines shown in Figure 7.31. In this diagram, the first number beside each pipeline is the capacity of the pipeline and the second is the present flow, both measured in hundreds of millions of cubic feet per day. The utility has proposed raising its rates to pay for additional pipelines. Although the Arkansas Regulatory Commission agrees that more than the present 14.7 hundred million cubic feet of gas are needed in Little Rock each day, it is not convinced that additional pipelines need to be built. It questions the need for more pipelines because most of the pipelines operated by the utility are not being used to capacity, and some are not being used at all. How should the utility argue for new pipelines?

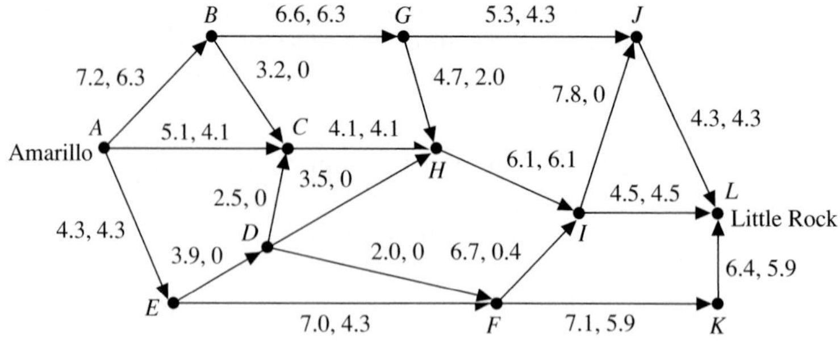

FIGURE 7.31

In order to justify its request for additional pipelines, the utility should apply the flow augmentation algorithm to the network and flow in Figure 7.31. By doing so, it will find that only the vertices A, B, C, G, H, and J are labeled. Consequently, the flow in Figure 7.31 is a maximal flow, and

$$\mathcal{S} = \{A, B, C, G, H, J\} \qquad \text{and} \qquad \mathcal{T} = \{D, E, F, I, K, L\}$$

form a minimal cut. The utility should, therefore, prepare a map, as in Figure 7.32, with A, B, C, G, H, and J in the northwestern region and D, E, F, I, K, and L in the southeastern region. This map indicates that only three pipelines (shown in color) carry gas from the northwestern region to the southeastern region, and each of these is being used to capacity. On this basis, the utility can argue the need for more pipelines from the northwestern region to the southeastern region. ■

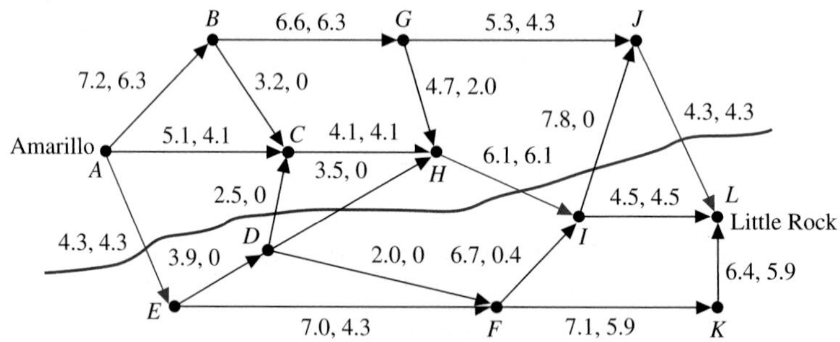

FIGURE 7.32

It is conceivable that the flow augmentation algorithm may never produce a maximal flow because no iteration occurs in which the sink cannot be labeled. Our next result, however, shows that this situation cannot occur if all the capacities in the network are rational numbers.

Theorem 7.5 If all the capacities in a transportation network are rational numbers and we start with zero flow along each arc, then repeated use of the flow augmentation algorithm produces a maximal flow in a finite number of iterations.

Proof. Suppose first that all the capacities in the network are integers. Let $\mathcal{S}$, $\mathcal{T}$ be the cut in which $\mathcal{S}$ consists of the source alone and $\mathcal{T}$ contains all the other vertices. Since all the arc capacities are integers, the capacity of the cut $\mathcal{S}$, $\mathcal{T}$ is an integer c.

Apply the flow augmentation algorithm, beginning with zero flow along every arc. Now consider any iteration of the algorithm in which the sink is labeled. The label on the sink must be of the form $(U, +, a)$ or $(U, -, a)$, where $a > 0$. Moreover, because all the capacities are integers, a is a minimum of integers and hence is an integer. Therefore $a \geq 1$, and each iteration of the algorithm in which the sink is labeled increases the value of the flow by at least 1. But by the corollary to Theorem 7.2, no flow in this network can have a value exceeding c. Hence, after at most $c + 1$ iterations, the flow augmentation algorithm must end with the sink unlabeled. But if the sink is unlabeled, then Theorem 7.4 guarantees that a maximal flow has been obtained.

Suppose now that all the capacities in the network are rational numbers. Find the least common denominator d of all the arc capacities, and consider the new network obtained by multiplying all of the original capacities by d. In this new network, all the capacities are integers. Applying the flow augmentation algorithm to the new network must therefore produce a maximal flow f in a finite number of steps, by the previous argument. But then this same sequence of steps will produce a maximal flow for the original network in which the flow along arc (X, Y) is $f(X, Y)/d$. (See Exercises 13–15.) ■

Theorem 7.5 can be proved without the requirement that the capacities be rational numbers. More generally, Edmonds and Karp (1972) have shown that the flow augmentation algorithm produces a maximal flow in no more than $\frac{1}{2}mn$ iterations, where m is the number of arcs and n is the number of vertices in the network. (See pages 117–119 of suggested reading [9] at the end of the chapter.) Note that, in each iteration of the algorithm, we consider an arc (V, W) at most twice, once in the proper direction from V to W and once in the opposite direction from W to V. Thus, if we count the number of times that an arc is considered before obtaining a maximal flow, the complexity of the flow augmentation algorithm is at most $\frac{2m(mn)}{2} = m^2n$. Since the number of arcs m cannot exceed $n(n-1)$, it follows that the complexity of the flow augmentation algorithm is at most $n^3(n-1)^2$.

We will end this section by proving a famous theorem discovered independently by Ford and Fulkerson and by Elias, Feinstein, and Shannon. (See suggested readings [6] and [4] at the end of the chapter.)

Theorem 7.6 *Max-Flow Min-Cut Theorem* In any transportation network, the value of a maximal flow equals the capacity of a minimal cut.

Proof. Let f be a maximal flow in a transportation network. Apply the flow augmentation algorithm to this network, with f as the current flow. Clearly, the sink will not be labeled, for otherwise we would obtain a flow having a greater value than f, which is a maximal flow. But if the sink is not labeled, then Theorem 7.4 shows that the sets of labeled and unlabeled vertices form a minimal cut having capacity equal to the value of f. ■

EXERCISES 7.3

In Exercises 1–4, give the capacity of the cut $\mathcal{S}, \mathcal{T}$ for the following network:

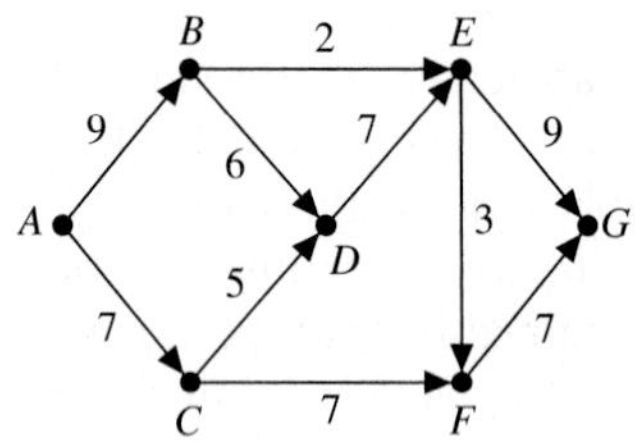

1. $\mathcal{S} = \{A, C, F\}$ and $\mathcal{T} = \{B, D, E, G\}$

2. $\mathcal{S} = \{A, B, E\}$ and $\mathcal{T} = \{C, D, F, G\}$

3. $\mathcal{S} = \{A, D, E\}$ and $\mathcal{T} = \{B, C, F, G\}$

4. $\mathcal{S} = \{A, E, F\}$ and $\mathcal{T} = \{B, C, D, G\}$

In Exercises 5–8, a network and a maximal flow are given. Find a minimal cut for the network by applying the flow augmentation algorithm to this network and flow.

5.

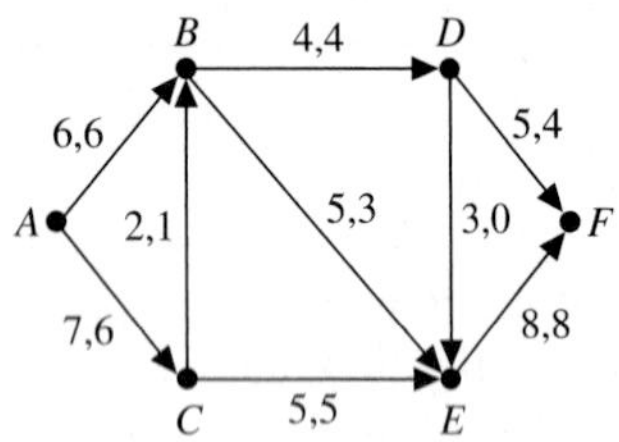

6.

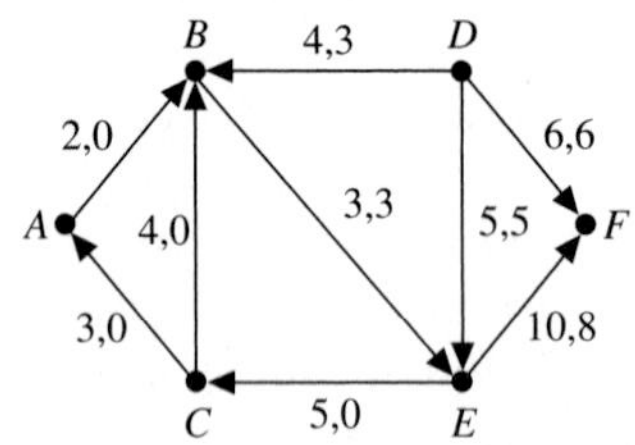

7.

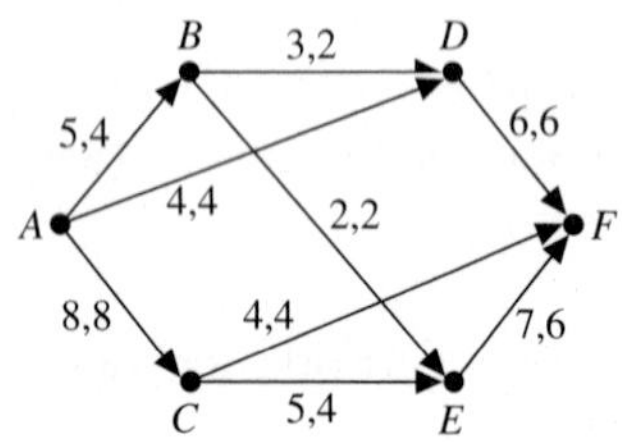

8.

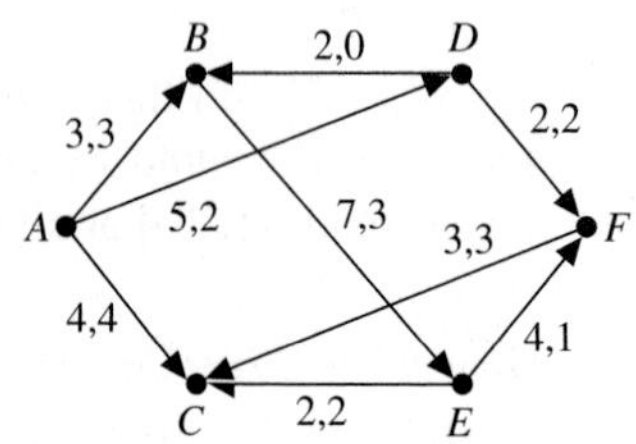

In Exercises 9–12, use the flow augmentation algorithm to find a minimal cut.

9.

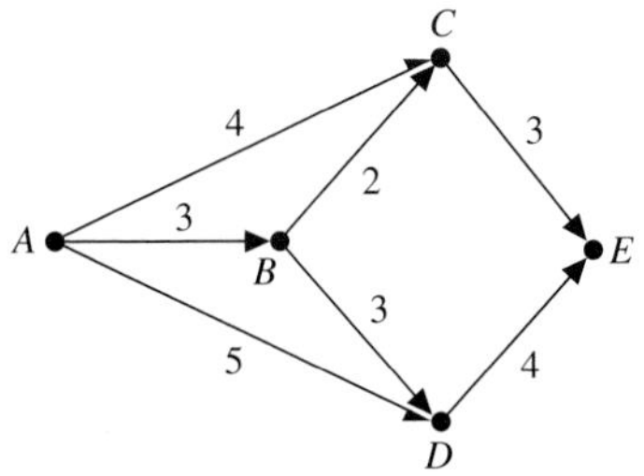

10.

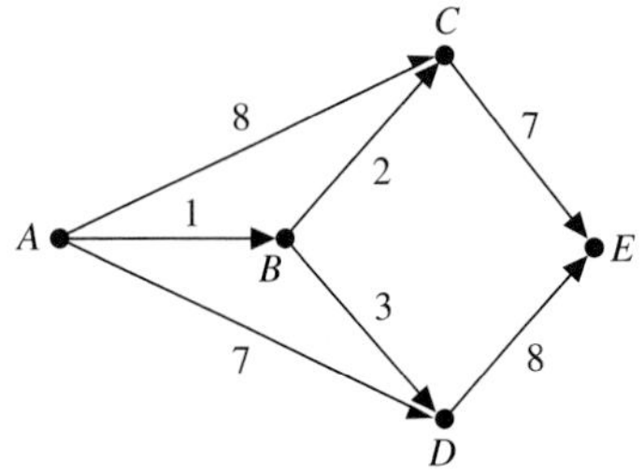

11.

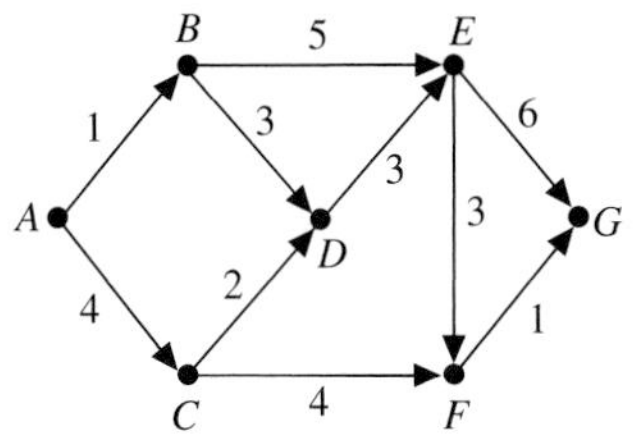

12.

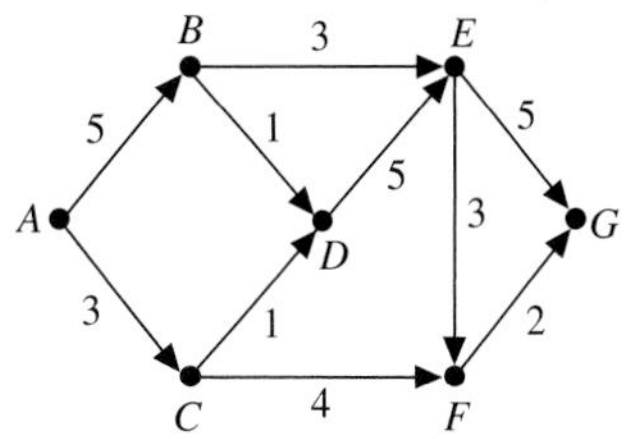

In Exercises 13–14, a network $\mathcal{N}$ with rational arc capacities is given. Let $\mathcal{N}'$ be the network obtained from $\mathcal{N}$ by multiplying all the capacities in $\mathcal{N}$ by d, the least common denominator of the capacities. Apply the flow augmentation algorithm to $\mathcal{N}'$, and use the result to determine a maximal flow for the original network $\mathcal{N}$.

13.

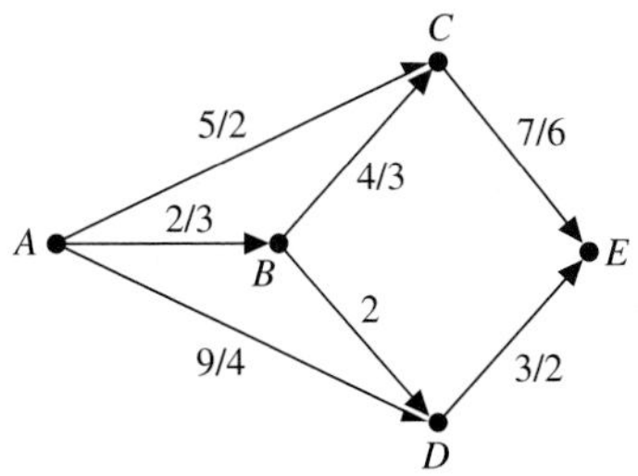

14.

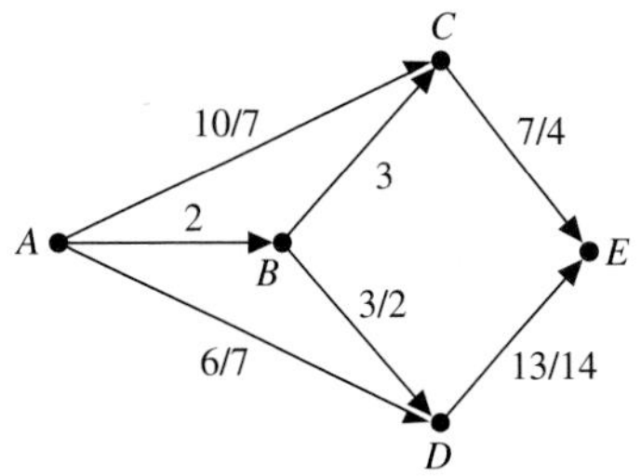

15. Let $\mathcal{N}$ be a transportation network and $d > 0$. Define $\mathcal{N}'$ as in Exercises 13 and 14 to be the network with the same directed graph as $\mathcal{N}$, but with all the arc capacities of $\mathcal{N}$ multiplied by d.

(a) Show that $\mathcal{S}, \mathcal{T}$ is a minimal cut for $\mathcal{N}'$ if and only if it is a minimal cut for $\mathcal{N}$.

(b) Prove that if v and v' are the values of maximal flows for $\mathcal{N}$ and $\mathcal{N}'$, respectively, then $v' = dv$.

(c) Show that f is a maximal flow for $\mathcal{N}$ if and only if f' is a maximal flow for $\mathcal{N}'$, where f' is defined by $f'(X, Y) = df(X, Y)$.

16. Suppose that $\mathcal{D}$ is a weighted directed graph having a nonnegative weight (capacity) on each directed edge. Show that if any two distinct vertices of $\mathcal{D}$ are designated as the source and the sink, then repeated use of the flow augmentation algorithm will produce a maximal flow from the source to the sink. (Thus the flow augmentation algorithm can be used even if conditions 1 and 2 in the definition of a transportation network are not satisfied.)

17. How many cuts are there in a transportation network with n vertices?

18. Let $\mathcal{D}$ be a directed graph, and let X and Y be distinct vertices in $\mathcal{D}$. Make $\mathcal{D}$ into a network with source X and sink Y by giving each directed edge a capacity of 1. Show that the value of a maximal flow in this network equals the minimum number n of directed edges that must be removed from $\mathcal{D}$ so that there is no directed path from X to Y. (*Hint:* Show that if $\mathcal{S}, \mathcal{T}$ is a minimal cut, then n equals the number of arcs from X to Y.)

In Exercises 19 and 20, use the result of Exercise 18 to find a minimal set of directed edges whose removal leaves no directed path from S to T.

19.

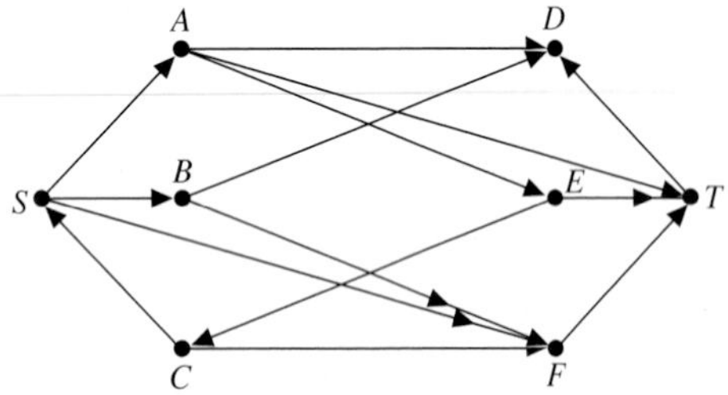

20.

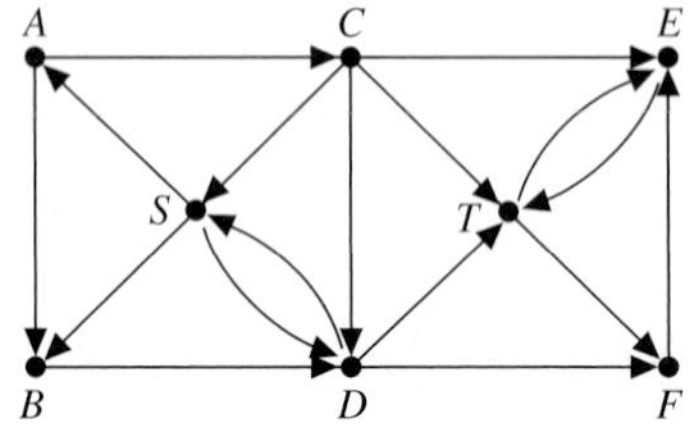

21. Consider an undirected graph $\mathcal{G}$ in which each edge $\{X, Y\}$ is assigned a nonnegative number $c(X, Y) = c(Y, X)$ representing its capacity to transmit the flow of some substance *in either direction.* Suppose that we want to find the maximum possible flow between distinct vertices S and T of $\mathcal{G}$, subject to the condition that, for any vertex X other than S and T, the total flow into X must equal the total flow out of X. Show that this problem can be solved with the flow augmentation algorithm by replacing each edge $\{X, Y\}$ of $\mathcal{G}$ by two directed edges (X, Y) and (Y, X), each having capacity $c(X, Y)$.

For the graphs in Exercises 22 and 23, use the method described in Exercise 21 to find the maximal possible flow from S to T if the numbers on the edges represent the capacity of flow along the edge in either direction.

22.

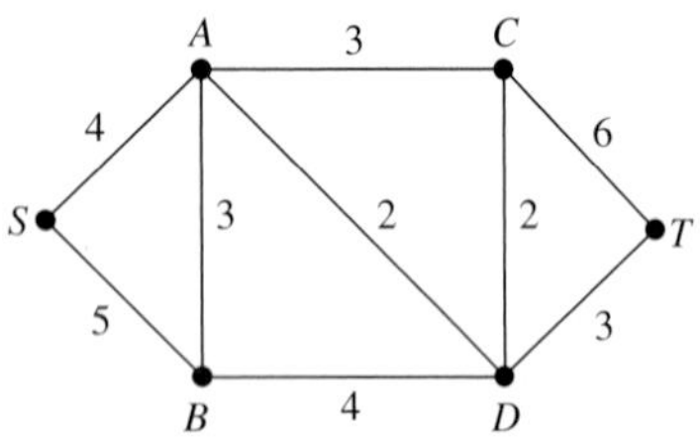

23.

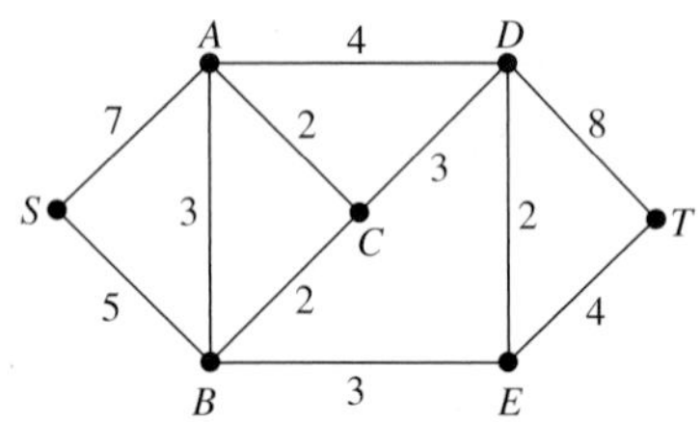

24. Prove that if the flow augmentation algorithm ends with the sink labeled, then the original flow has been replaced by a flow with a larger value.

25. Consider a transportation network with source S, sink T, and vertex set $\mathcal{V}$. Let $c(X, Y)$ denote the capacity of arc (X, Y). If $X, Y \in \mathcal{V}$, but (X, Y) is not an arc in the network, define $c(X, Y) = 0$. Show that the average capacity of a cut is

$$\frac{1}{4}\left(c(S, T) + \sum_{X \in \mathcal{V}} c(S, X) + \sum_{X \in \mathcal{V}} c(X, T) + \sum_{X, Y \in \mathcal{V}} c(X, Y)\right).$$

7.4 FLOWS AND MATCHINGS

In this section we will relate network flows to the matchings studied in Section 6.2. Recall from Sections 6.1 and 6.2 that a graph $\mathcal{G}$ is called **bipartite** if its vertex set $\mathcal{V}$ can be written as the union of two disjoint sets $\mathcal{V}_1$ and $\mathcal{V}_2$ in such a way that all the edges in $\mathcal{G}$ join a vertex in $\mathcal{V}_1$ to a vertex in $\mathcal{V}_2$. A **matching** of $\mathcal{G}$ is a subset $\mathcal{M}$ of the edges of $\mathcal{G}$ such that no vertex in $\mathcal{V}$ is incident with more than one edge in $\mathcal{M}$. In addition, a matching of $\mathcal{G}$ with the property that no matching of $\mathcal{G}$ contains more edges is called a **maximum matching** of $\mathcal{G}$.

From a bipartite graph $\mathcal{G}$, we can form a transportation network $\mathcal{N}$ as follows:

(1) The vertices of $\mathcal{N}$ are the vertices of $\mathcal{G}$ together with two additional vertices s and t. These vertices s and t are the source and sink for $\mathcal{N}$, respectively.
(2) The arcs in $\mathcal{N}$ are of three types:
  (a) There is an arc in $\mathcal{N}$ from s to every vertex in $\mathcal{V}_1$.
  (b) There is an arc in $\mathcal{N}$ from every vertex in $\mathcal{V}_2$ to t.
  (c) If X is in $\mathcal{V}_1$, Y is in $\mathcal{V}_2$, and $\{X, Y\}$ is an edge in $\mathcal{G}$, there is an arc from X to Y in $\mathcal{N}$.
(3) All arcs in $\mathcal{N}$ have capacity 1.

We call $\mathcal{N}$ the **network associated with** $\mathcal{G}$.

Example 7.7 In Figure 7.33, the vertex set $\mathcal{V} = \{A, B, C, W, X, Y, Z\}$ of the bipartite graph $\mathcal{G}$ is partitioned into the sets $\mathcal{V}_1 = \{A, B, C\}$ and $\mathcal{V}_2 = \{W, X, Y, Z\}$.

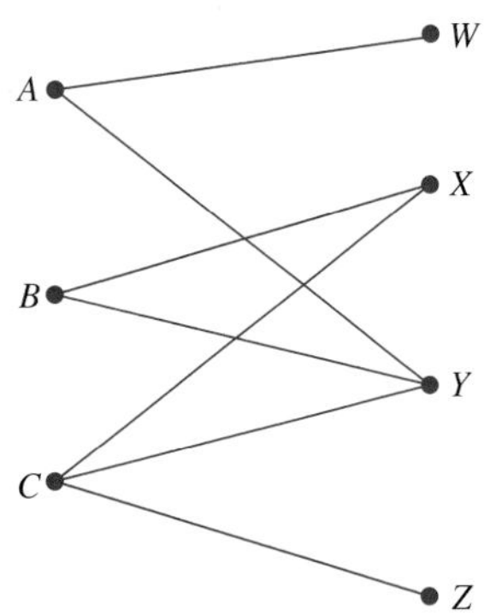

FIGURE 7.33

The network $\mathcal{N}$ associated with $\mathcal{G}$ is shown in Figure 7.34. Note that $\mathcal{N}$ contains a copy of $\mathcal{G}$ and two new vertices s and t, which are the source and the sink for $\mathcal{N}$, respectively. The edges of $\mathcal{G}$ that join vertices in $\mathcal{V}_1$ to vertices in $\mathcal{V}_2$ become arcs in $\mathcal{N}$ of capacity 1 directed from the vertices in $\mathcal{V}_1$ to the vertices in $\mathcal{V}_2$. The other arcs in $\mathcal{N}$ are directed from the source s to each vertex in $\mathcal{V}_1$ and from each vertex in $\mathcal{V}_2$ to the sink t; these arcs also have capacity 1. ■

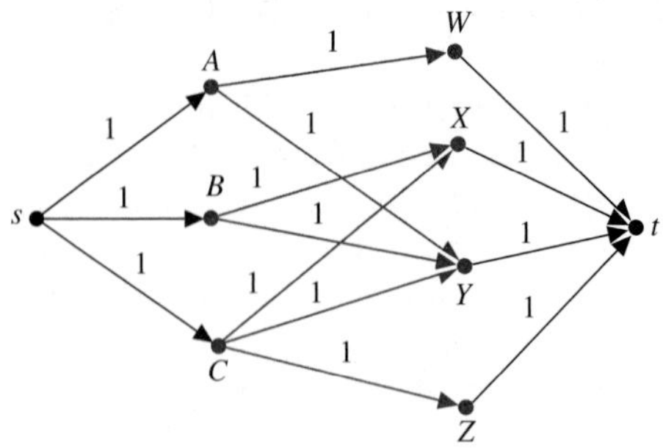

FIGURE 7.34

Consider the bipartite graph in Figure 7.35. The network associated with this graph is shown in Figure 7.36.

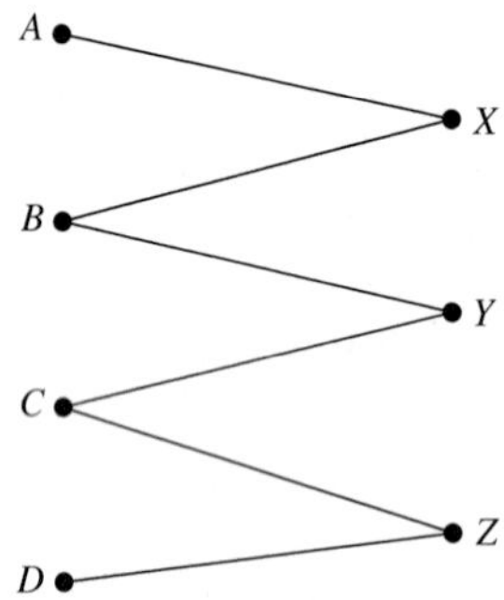

FIGURE 7.35

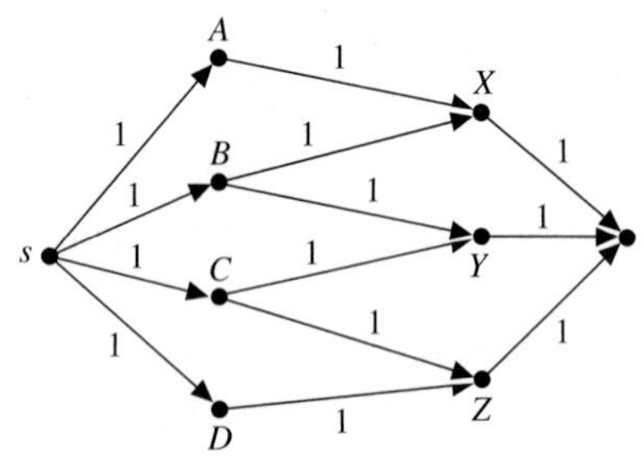

FIGURE 7.36

When the flow augmentation algorithm is applied to the network in Figure 7.36, the zero flow can be increased by 1 unit along the path s, A, X, t; by 1 unit along the path s, B, Y, t; and by 1 unit along the path s, C, Z, t. The resulting maximal flow is shown in Figure 7.37. Notice that this is an integral flow.

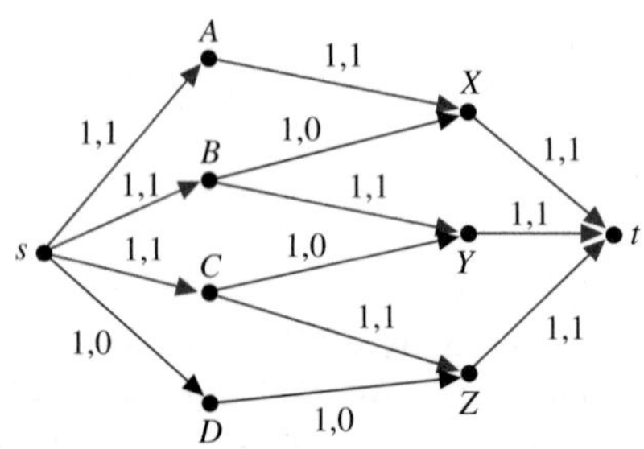

FIGURE 7.37

Thus we see that a maximal flow in the network shown in Figure 7.36 has value 3, and one maximal flow is obtained by sending

$$1 \text{ unit along } s, A, X, t; \qquad 1 \text{ unit along } s, B, Y, t;$$

and

$$1 \text{ unit along } s, C, Z, t.$$

If we disregard the source and sink in these three paths, we obtain the three arcs (A, X), (B, Y), and (C, Z). These arcs correspond to the edges $\{A, X\}$, $\{B, Y\}$, and $\{C, Z\}$ in Figure 7.35. Clearly, these edges are a maximum matching of the bipartite graph in Figure 7.35, because in this graph the set $\mathcal{V}_2$ contains only three vertices.

Thus we have obtained a maximum matching of a bipartite graph by using the flow augmentation algorithm on the network associated with the graph. Theorem 7.7 shows that this technique will always work.

Theorem 7.7 Let $\mathcal{G}$ be a bipartite graph, and let $\mathcal{N}$ be the network associated with $\mathcal{G}$.

(a) Every integral flow in $\mathcal{N}$ corresponds to a matching of $\mathcal{G}$, and every matching of $\mathcal{G}$ corresponds to an integral flow in $\mathcal{N}$. This correspondence is such that two vertices are matched in $\mathcal{G}$ if and only if there is 1 unit of flow along the corresponding arc in $\mathcal{N}$.

(b) A maximal flow in $\mathcal{N}$ corresponds to a maximum matching of $\mathcal{G}$.

Proof. Let the vertex set of $\mathcal{G}$ be written as the union of disjoint sets $\mathcal{V}_1$ and $\mathcal{V}_2$ such that every edge in $\mathcal{G}$ joins a vertex in $\mathcal{V}_1$ to a vertex in $\mathcal{V}_2$.

(a) Let f be an integral flow in $\mathcal{N}$, and let $\mathcal{M}$ be the set of edges $\{X, Y\}$ in $\mathcal{G}$ for which X is in $\mathcal{V}_1$, Y is in $\mathcal{V}_2$, and $f(X, Y) = 1$. To prove that $\mathcal{M}$ is a matching of $\mathcal{G}$, we must show that no vertex of $\mathcal{G}$ is incident with more than one edge in $\mathcal{M}$. Let U be any vertex in $\mathcal{G}$. Then U belongs to exactly one of the sets $\mathcal{V}_1$ or $\mathcal{V}_2$.

Assume without loss of generality that U belongs to $\mathcal{V}_1$ and that U is incident with the edge $\{U, V\}$ in $\mathcal{M}$. We will show that vertex U is not incident with any other edge in $\mathcal{M}$. Suppose that $\{U, W\}$ is another edge in $\mathcal{M}$. Then $f(U, V) = 1$ and $f(U, W) = 1$ by the definition of $\mathcal{M}$. Thus in $\mathcal{N}$ the total flow out of vertex U is at least 2. But in $\mathcal{N}$ the only arc entering U is (s, U), and this arc has capacity 1. So the total flow into vertex U does not equal the total flow out of vertex U, contradicting that f is a flow in $\mathcal{N}$. Hence U is incident with at most one edge in $\mathcal{M}$, and so $\mathcal{M}$ is a matching of $\mathcal{G}$. This proves that every integral flow in $\mathcal{N}$ corresponds to a matching of $\mathcal{G}$.

Now suppose that $\mathcal{M}$ is a matching in $\mathcal{G}$. Let $\mathcal{N}$ be the network associated with $\mathcal{G}$, and let s be the source in $\mathcal{N}$ and t be the sink. For each arc in $\mathcal{N}$, define a function f by

$$f(s, X) = 1 \text{ if } X \in \mathcal{V}_1 \text{ and there exists } Z \in \mathcal{V}_2 \text{ such that } \{X, Z\} \in \mathcal{M};$$
$$f(Y, t) = 1 \text{ if } Y \in \mathcal{V}_2 \text{ and there exists } W \in \mathcal{V}_1 \text{ such that } \{W, Y\} \in \mathcal{M};$$
$$f(X, Y) = 1 \text{ if } X \in \mathcal{V}_1,\ Y \in \mathcal{V}_2, \text{ and } \{X, Y\} \in \mathcal{M}; \text{ and}$$
$$f(U, V) = 0 \text{ otherwise.}$$

Since each arc e in $\mathcal{N}$ has capacity 1 and $0 \le f(e) \le 1$, f satisfies condition (1) in the definition of a flow.

Now consider any vertex X of $\mathcal{N}$ other than s and t. Such a vertex is a vertex of $\mathcal{G}$ and, hence, belongs to either $\mathcal{V}_1$ or $\mathcal{V}_2$. Assume without loss of generality that X belongs to $\mathcal{V}_1$. By the definition of f, either $f(s, X) = 0$ or $f(s, X) = 1$. If $f(s, X) = 0$, then there exists no $Z \in \mathcal{V}_2$ such that $\{X, Z\} \in \mathcal{M}$; so the total flow into X and the total flow out of X are both 0. On the other hand, if $f(s, X) = 1$, then there exists $Z \in \mathcal{V}_2$ such that $\{X, Z\} \in \mathcal{M}$. Because $\mathcal{M}$ is a matching of $\mathcal{G}$, Z is unique. Thus, in this case also, the total flow into X equals the total flow out of X, and so f satisfies condition (2) in the definition of a flow. It follows that f is a flow in $\mathcal{N}$. This proves that every matching of $\mathcal{G}$ corresponds to an integral flow in $\mathcal{N}$.

(b) Under the correspondence described in part (a), the total number of vertices in $\mathcal{V}_1$ that are matched with vertices in $\mathcal{V}_2$ is the value of the flow f. Thus $\mathcal{M}$ is a maximum matching of $\mathcal{G}$ if and only if f is a maximal flow in $\mathcal{N}$. ■

Example 7.8 Recall the example from Section 6.1 in which an English department wishes to assign courses to professors, one course per professor. The list of professors available to teach the courses is as follows:

Course	*Professors*
1	Abel, Crittenden, Forcade
2	Crittenden, Donohue, Edge, Gilmore
3	Abel, Crittenden
4	Abel, Forcade
5	Banks, Edge, Gilmore
6	Crittenden, Forcade

The English department would like to obtain a maximum matching so that it can offer the largest possible number of courses.

As in Section 6.2, we can represent this problem by a bipartite graph with the vertex set $\mathcal{V} = \{1, 2, 3, 4, 5, 6, A, B, C, D, E, F, G\}$, where we have denoted the professors by their initials. Here, the set $\mathcal{V}$ can be partitioned as the union of the disjoint sets of courses and professors, that is,

$$\mathcal{V}_1 = \{1, 2, 3, 4, 5, 6\} \qquad \text{and} \qquad \mathcal{V}_2 = \{A, B, C, D, E, F, G\}.$$

By drawing an edge between each professor and the courses he or she can teach, we obtain the graph shown in Figure 7.38. (This is the graph obtained previously in Figure 6.1.)

We will obtain a maximum matching for the graph in Figure 7.38 by using the flow augmentation algorithm. Let us begin by assigning professors A, C, F, and E to teach courses 1, 2, 4, and 5, respectively. This gives the matching with edges $\{1, A\}$, $\{2, C\}$, $\{4, F\}$, and $\{5, E\}$. The network associated with the graph in Figure 7.38 is shown in Figure 7.39. Here, all arcs are directed from the left to the right and have capacity 1. The matching $\{1, A\}$, $\{2, C\}$, $\{4, F\}$, $\{5, E\}$ obtained here corresponds to the flow shown in Figure 7.40, where arcs having a flow of zero are shown in black and those with a flow of 1 are shown in blue.

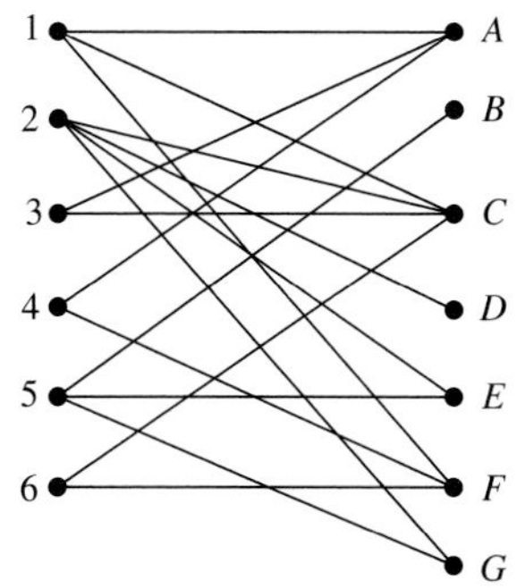

FIGURE 7.38

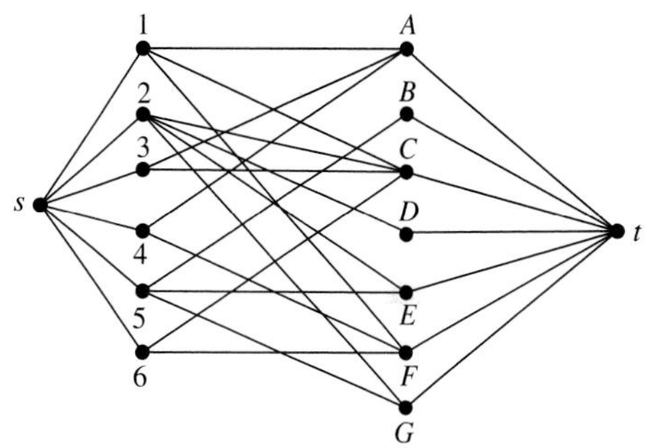

FIGURE 7.39

If we apply the flow augmentation algorithm to the network and flow in Figure 7.40, we find that $s, 3, C, 2, D, t$ is a flow-augmenting path. Increasing the flow by 1 along this path gives the flow in Figure 7.41.

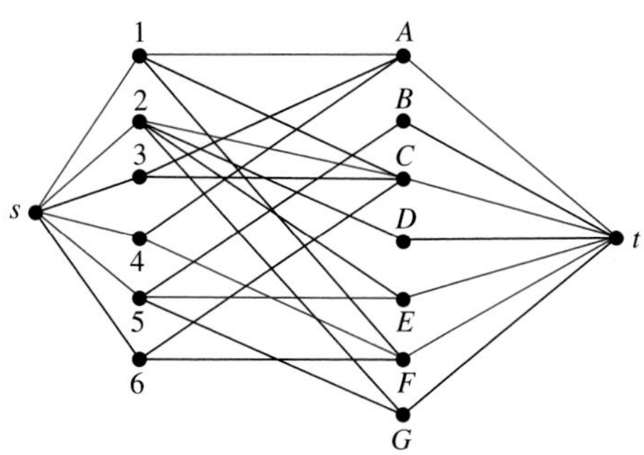

FIGURE 7.40

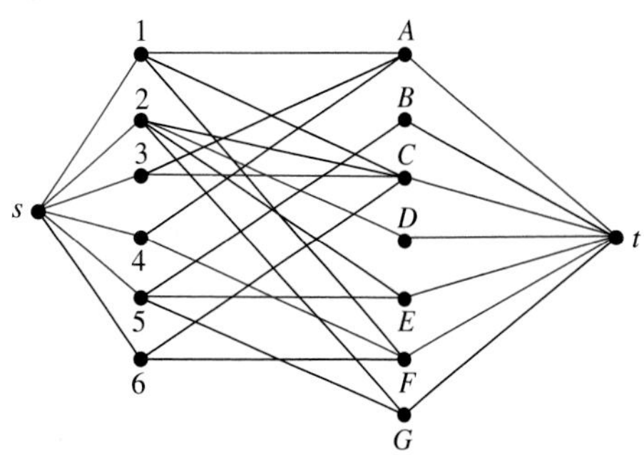

FIGURE 7.41

If another iteration of the flow augmentation algorithm is performed on the flow in Figure 7.41, only vertices s, 1, 3, 4, 6, A, C, and F will be labeled. Thus the flow shown in Figure 7.41 is a maximal flow. By Theorem 7.7, this means that the corresponding matching {1, A}, {2, D}{3, C}, {4, F}, and {5, E} is a maximum matching for the bipartite graph in Figure 7.38. Hence the English department can offer 5 of the 6 courses by assigning course 1 to Abel, course 2 to Donohue, course 3 to Crittenden, course 4 to Forcade, and course 5 to Edge. ■

EXERCISES 7.4

In Exercises 1–6, determine whether the given graph is bipartite or not. If it is, construct the network associated with the graph.

1.

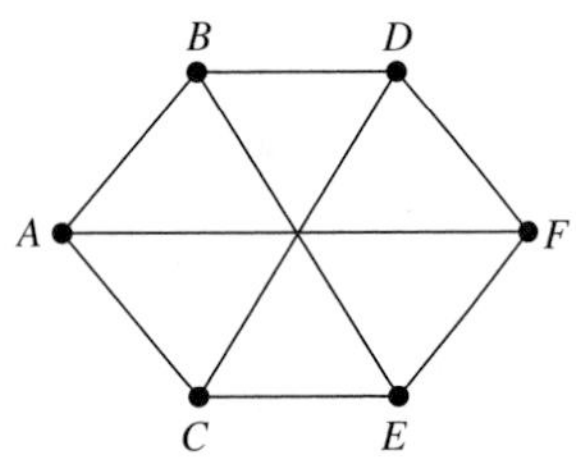

2.

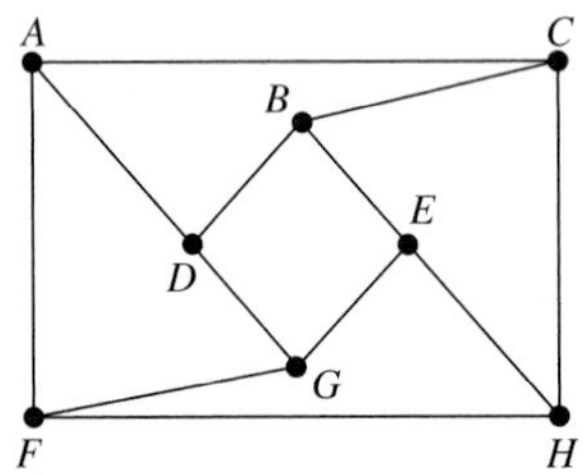

3.

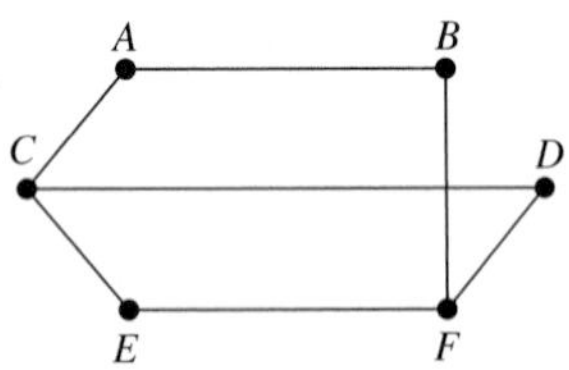

4.

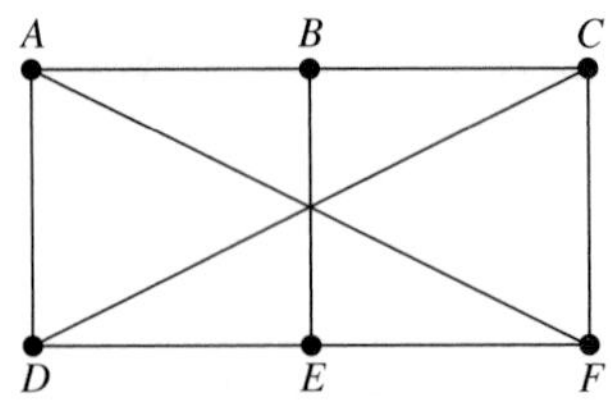

5.

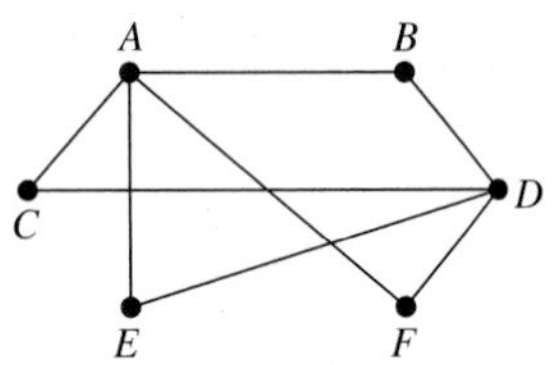

6.

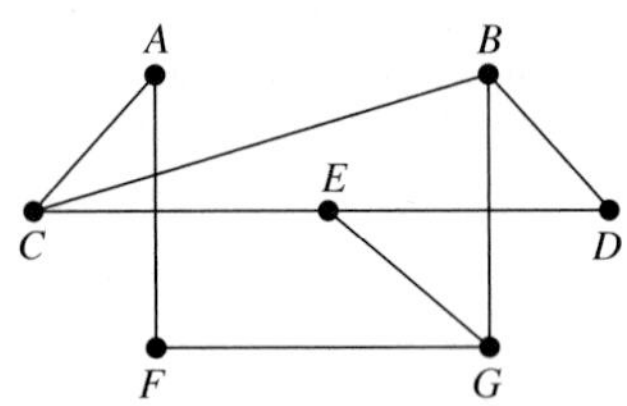

In Exercises 7–10, a bipartite graph is given with a matching indicated in color. Construct the network associated with the given graph, and use the flow augmentation algorithm to determine whether this is a maximum matching. If not, find a larger matching.

7.

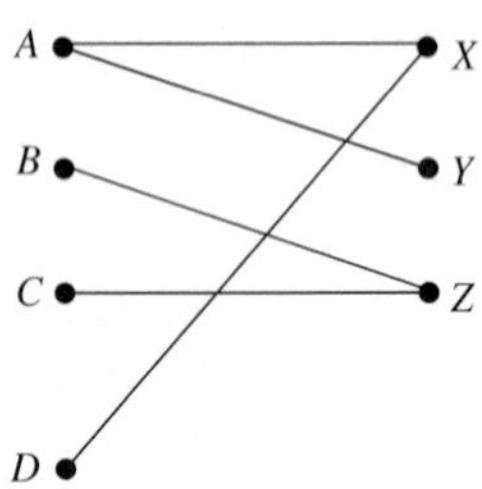

8.

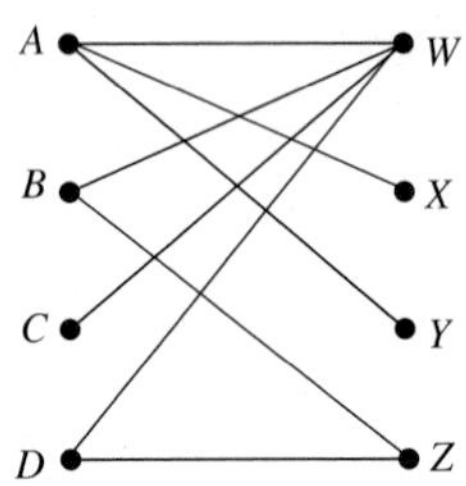

9. 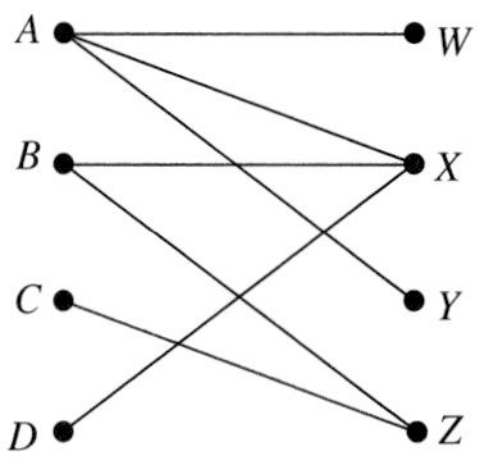

10. 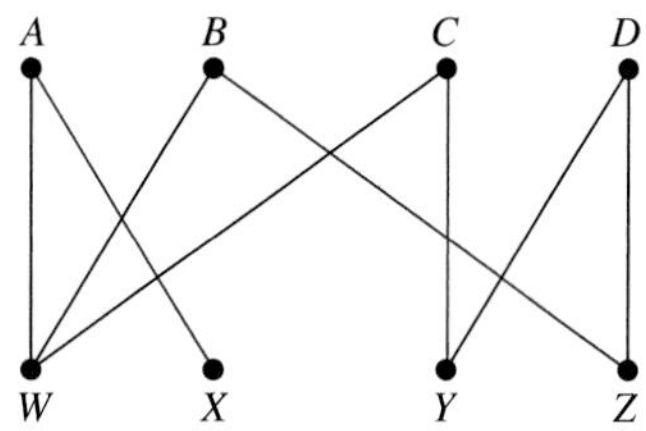

In Exercises 11–14, use the flow augmentation algorithm to find a maximum matching for the given bipartite graph.

11. 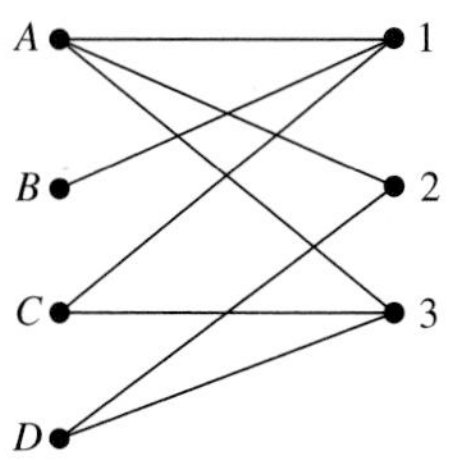

12. 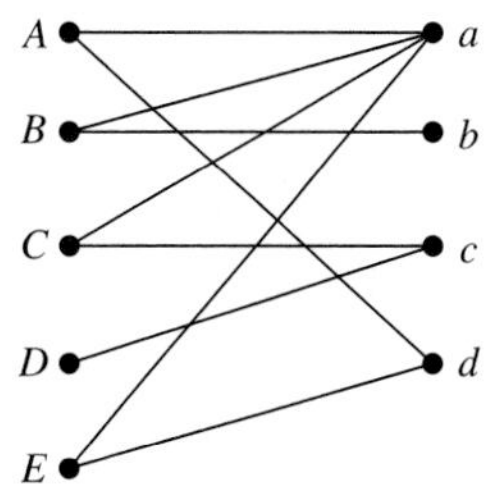

13. 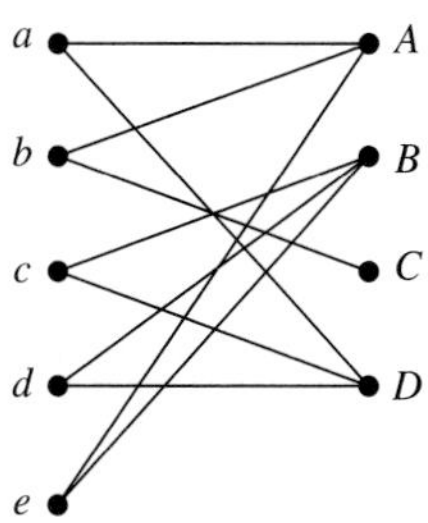

14. 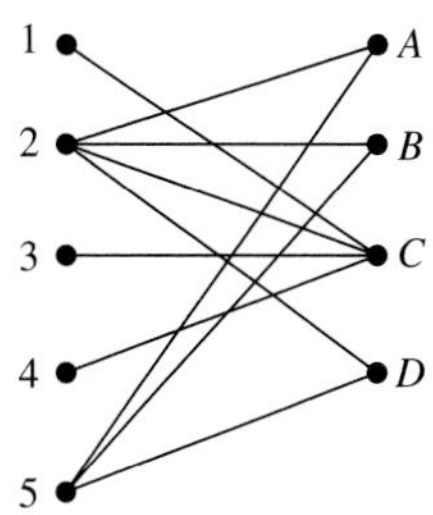

15. Four mixed couples are needed for a tennis team, and 5 men and 4 women are available. Andrew will not play with Flo or Hannah; Bob will not play with Iris; Flo, Greta, and Hannah will not play with Ed; Dan will not play with Hannah or Iris; and Cal will only play with Greta. Can a team be put together under these conditions? If so, how?

16. Five actresses are needed for parts in a play that require Chinese, Danish, English, French, and German accents. Sally does English and French accents; Tess does Chinese, Danish, and German; Ursula does English and French; Vickie does all accents except English; and Winona does all except Danish and German. Can the five roles be filled under these conditions? If so, how?

17. The five assistants in a mathematics department must decide which jobs each will do. Craig likes filing and collating, Dianne can distribute paychecks and help students, Gale types and collates, Marilyn enjoys typing and distributing the paychecks, and Sharon prefers to help students. Can the jobs be assigned so that every assistant is given one of his or her preferences? If so, how?

18. When the flow augmentation algorithm is applied to the network and flow in Figure 7.41, only the vertices, $s, 1, 3, 4, 6, A, C,$ and F will be labeled. What is the significance of courses 1, 3, 4, and 6 and professors A, C, and F in the context of Example 7.8?

19. Describe how to use the flow augmentation algorithm to determine whether a system of distinct representatives exists for a sequence of sets $S_1, S_2, \ldots, S_n$.

20. Apply the flow augmentation algorithm as described in Exercise 19 to find a system of distinct representatives, if possible, for the sequence of sets $\{3, 4\}, \{1, 5\}, \{2, 3\}, \{2, 5\}, \{1, 4\}$.

21. Apply the flow augmentation algorithm as described in Exercise 19 to find a system of distinct representatives, if possible, for the sequence of sets $\{2, 5\}, \{1, 6\}, \{3, 5\}, \{4, 6\}, \{2, 3\}, \{2, 3, 5\}$.

22. Five men and five women are attending a dance. Ann will dance only with Gregory or Harry, Betty will dance only with Frank or Ian, Carol will dance only with Harry or Jim, Diane will dance only with Frank or Gregory, and Ellen will dance only with Gregory or Ian. Is it possible for all 10 people to dance the last dance with an acceptable partner? If so, how?

23. The history department at a state college would like to offer six courses during the summer. Although there are seven professors available, only certain professors can teach each course. The list of courses and the professors capable of teaching them is as follows:

Course	*Professors*
American History	Getsi, Dammers, Kagle, Ericksen
British History	Duncan, Getsi, Harris
Latin American History	Duncan, Getsi
Oliver Cromwell	Duncan, Harris
Ancient History	White, Kagle, Harris
20th Century History	Getsi, Harris

Is there an assignment of professors to courses so that no professor teaches more than one course?

24. Suppose that the flow augmentation algorithm is applied to find a system of distinct representatives for a sequence of sets $S_1, S_2, \ldots, S_n$, as described in Exercise 19. If the algorithm is applied to a maximal flow with value less than n, prove that the number of labeled sets must exceed the number of elements in the union of the labeled sets.

In Exercises 25–28, let $\mathcal{G}$ be a bipartite graph in which the set of vertices is written as the union of two disjoint sets $\mathcal{V}_1$ and $\mathcal{V}_2$ such that all the edges in $\mathcal{G}$ join a vertex in $\mathcal{V}_1$ to a vertex in $\mathcal{V}_2$. For each subset $\mathcal{A}$ of $\mathcal{V}_1$, let $\mathcal{A}^*$ denote the set of vertices in $\mathcal{G}$ that are adjacent to some vertex in $\mathcal{A}$. The maximum value d of $|\mathcal{A}| - |\mathcal{A}^*|$ over all subsets $\mathcal{A}$ of $\mathcal{V}_1$ is called the ***deficiency*** *of $\mathcal{G}$.*

25. Prove that $d \geq 0$.

26. Prove that if $\mathcal{N}$ is the network associated with $\mathcal{G}$, then $\mathcal{N}$ has a flow with value $|\mathcal{V}_1| - d$.

27. Let $\mathcal{N}$ be the network associated with $\mathcal{G}$, and let s and t be the source and sink of $\mathcal{N}$, respectively. Let $\mathcal{A}$ be a subset of $\mathcal{V}_1$ such that $|\mathcal{A}| - |\mathcal{A}^*| = d$. Prove that the cut $\mathcal{S}, \mathcal{T}$ with

$$\mathcal{S} = \{s\} \cup \mathcal{A} \cup \mathcal{A}^* \qquad \text{and} \qquad \mathcal{T} = (\mathcal{V}_1 - \mathcal{A}) \cup (\mathcal{V}_2 - \mathcal{A}^*) \cup \{t\}$$

has capacity $|\mathcal{V}_1| - d$.

28. Deduce that a maximal flow in $\mathcal{N}$ has value $|\mathcal{V}_1| - d$ and hence that a maximum matching in $\mathcal{G}$ contains $|\mathcal{V}_1| - d$ edges.

In Exercises 29–31, let $\mathcal{G}$ be a bipartite graph in which the vertex set is written as the union of two disjoint sets $\mathcal{V}_1$ and $\mathcal{V}_2$ such that all the edges of $\mathcal{G}$ join a vertex in $\mathcal{V}_1$ to a vertex in $\mathcal{V}_2$. Suppose that the flow augmentation algorithm is performed on the network associated with $\mathcal{G}$ until the algorithm ends with the sink unlabeled.

29. Prove that if $X \in \mathcal{V}_1$, $Y \in \mathcal{V}_2$, and $f(X, Y) = 1$, then, in the last iteration of the flow augmentation algorithm, X is unlabeled or Y is labeled.

30. Prove that in Exercise 29 it is impossible that X is unlabeled *and* Y is labeled. Deduce that the number of unlabeled vertices in $\mathcal{V}_1$ plus the number of labeled vertices in $\mathcal{V}_2$ equals the value of the present flow f.

31. In the context of Exercise 30, show that there are no edges in $\mathcal{G}$ that join a labeled vertex $X \in \mathcal{V}_1$ to an unlabeled vertex $Y \in \mathcal{V}_2$. Deduce that the unlabeled vertices in $\mathcal{V}_1$ and the labeled vertices in $\mathcal{V}_2$ form a minimum covering of $\mathcal{G}$ in the sense of Section 6.2. (*Hint:* The label on X must be either $(s, +, 1)$ or $(Z, -, 1)$ for $Z \neq Y$.)

HISTORICAL NOTES

L.R. Ford, Jr.

The concept of flows in transportation networks is a recent mathematical discovery, with the majority of the work in this area appearing since 1960. The initial work in the field was provided by Lester Randolph Ford, Jr. (1927–) and Delbert Ray Fulkerson (1924–1976) in a series of papers, the first of which appeared in 1956 and 1957. Their seminal 1962 text *Flows in Networks* outlined the field. While others were later able to make some improvements to their algorithms, their basic approaches still define the way network flows are conceptualized.

SUPPLEMENTARY EXERCISES

In Exercises 1–8, find a maximal flow and a minimal cut in each transportation network by using the flow augmentation algorithm. Start with the flow that is 0 along every arc, and if there is a choice of vertices to label, label the vertices in alphabetical order.

1.

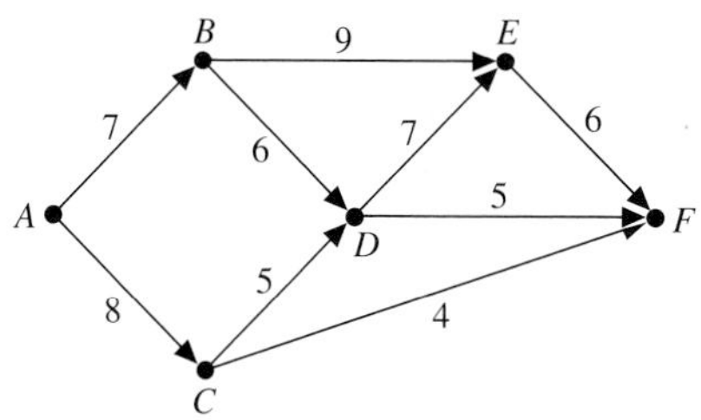

2.

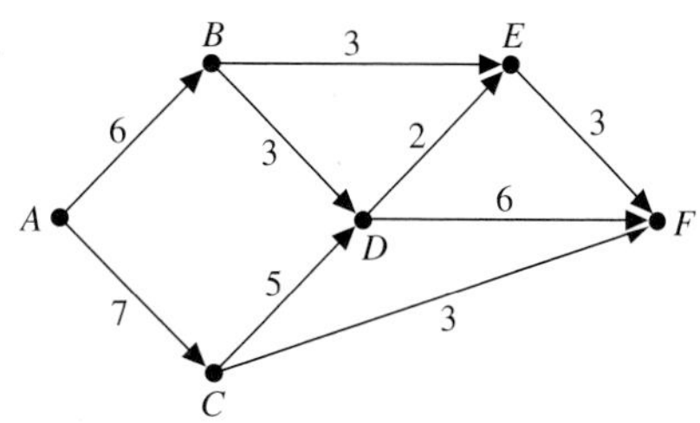

3.

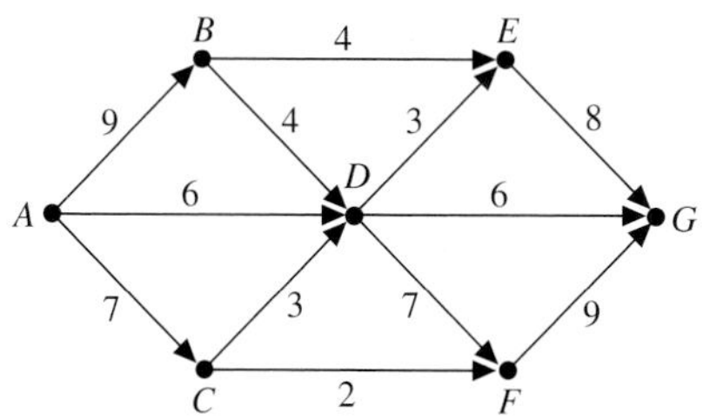

4.

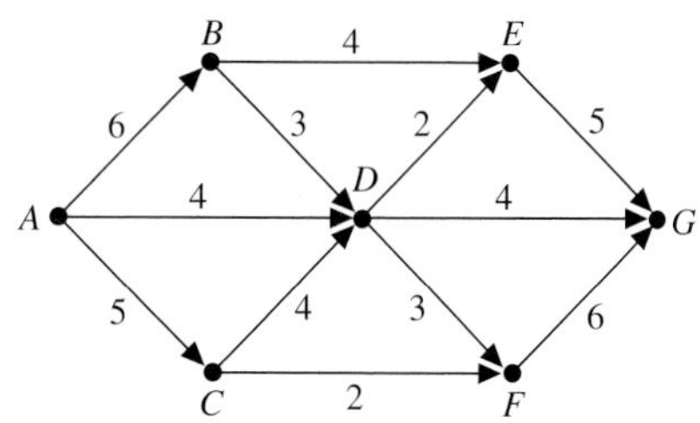

5.

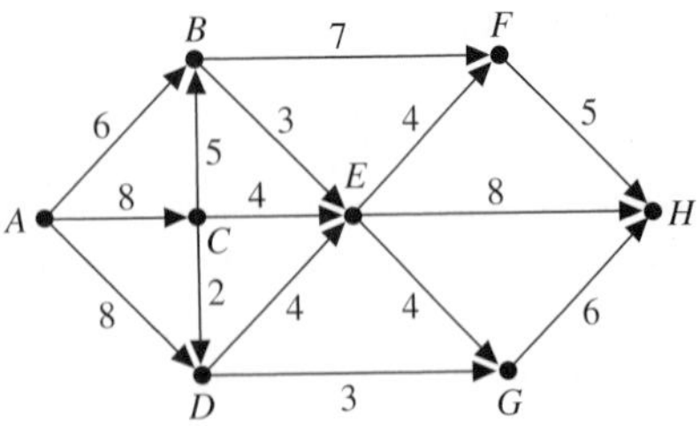

6.

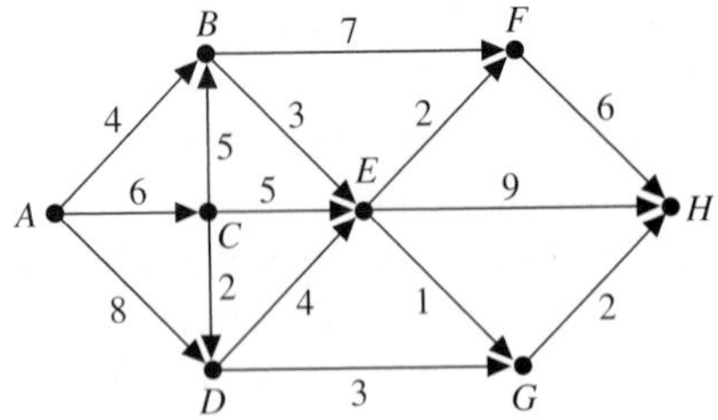

7.

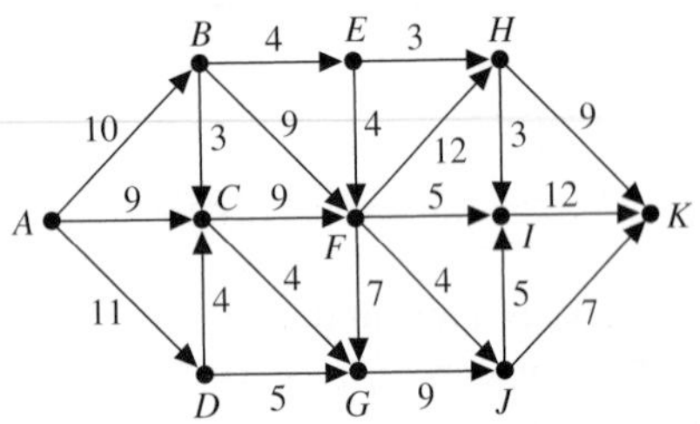

8.

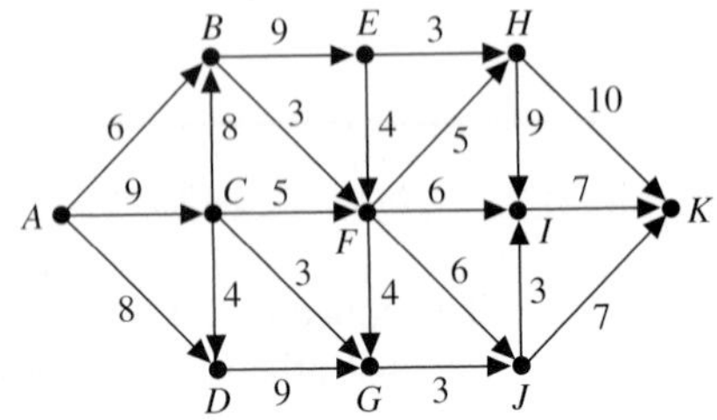

By a **multisource transportation network**, *we mean a weighted directed graph that satisfies the conditions in the definition of a transportation network, except that instead of its containing only one vertex of indegree* 0, *there is a nonempty finite set* $\mathcal{S}_0$ *of vertices with indegree* 0. *We say that* f *is a* **flow** *in such a network if*

(*i*) $0 \leq f(e) \leq c(e)$ *for each arc e, where* $c(e)$ *is the capacity of arc e, and*

(*ii*) *for each vertex* V *other than the sink and the elements of* $\mathcal{S}_0$, *the total flow into* V *equals the total flow out of* V.

The **value** *of such a flow is the total flow into the sink, and a flow is called a* **maximal flow** *if its value is as large as possible.*

9. Show that in a multisource transportation network the value of a flow equals the total flow out of all the vertices in $\mathcal{S}_0$.

10. Given a multisource transportation network $\mathcal{N}$, create from $\mathcal{N}$ a transportation network $\mathcal{N}'$ by introducing a new vertex u and edges with infinite capacity from u to each element in $\mathcal{S}_0$. Show that a maximal flow in $\mathcal{N}$ may be found by applying the flow augmentation algorithm to $\mathcal{N}'$.

By using Exercise 10, find a maximal flow for the multisource transportation networks given in Exercises 11 and 12.

11.

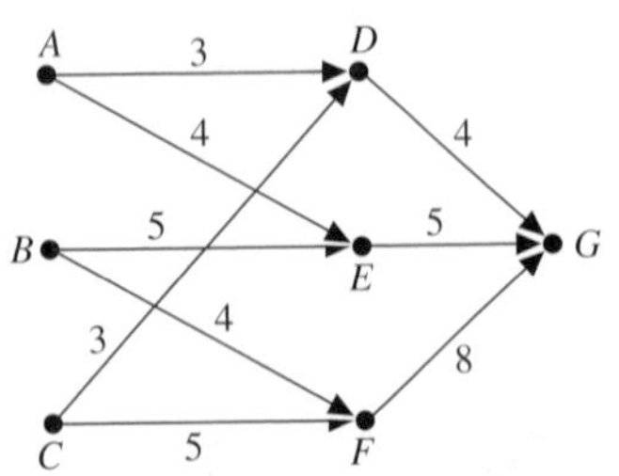

12.

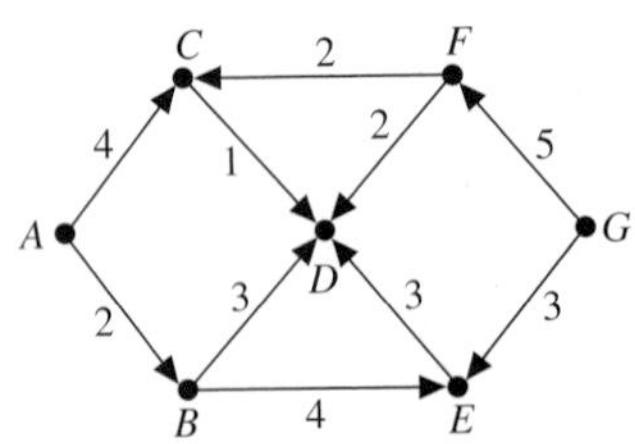

13. Generalize the concept of a transportation network to allow multiple sources and sinks; then define "flow," "value of a flow," and "maximal flow" for such a network. State and prove an analogue of Theorem 7.1 for such a network.

14. By a **network with vertex capacities**, we mean a transportation network $\mathcal{N}$, along with a function k from its set of vertices to the nonnegative real numbers. In such a network a flow must satisfy the additional restriction that, for each vertex V, neither the total flow into V nor the total flow out of V can exceed $k(V)$. (Of course, these totals are the same if V is a vertex other than the source or sink.) Show that the value of a maximal flow in such a network equals the value of a maximal flow in the ordinary transportation network $\mathcal{N}^*$, where $\mathcal{N}^*$ is formed as follows:

(i) For each vertex X in $\mathcal{N}$, include two vertices X' and X'' in $\mathcal{N}^*$.

(ii) For each vertex X in $\mathcal{N}$, include an arc (X', X'') in $\mathcal{N}^*$ with capacity $k(X)$.

(iii) For each arc (X, Y) in $\mathcal{N}$, include an arc (X'', Y') of the same capacity in $\mathcal{N}^*$.

(Note that if in $\mathcal{N}$ the source is s and the sink is t, then in $\mathcal{N}^*$ the source is s' and the sink is t''.)

For the networks with vertex capacities in Exercises 15–17, construct the network $\mathcal{N}^*$ described in Exercise 14.

15. $k(A) = 9, k(B) = 8, k(C) = 9, k(D) = 7, k(E) = 10$

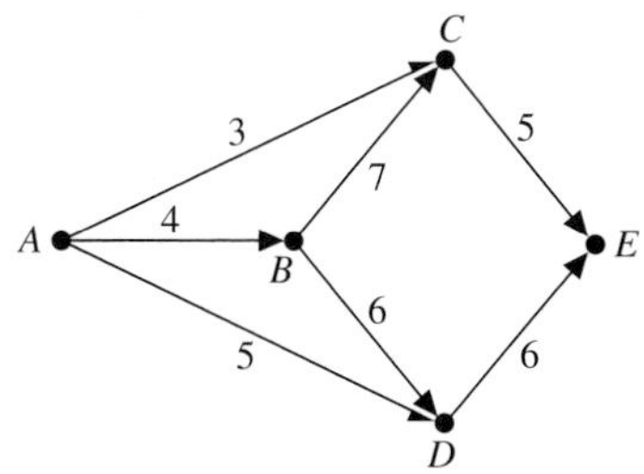

16. $k(A) = 8, k(B) = 4, k(C) = 7, k(D) = 7, k(E) = 6, k(F) = 9$

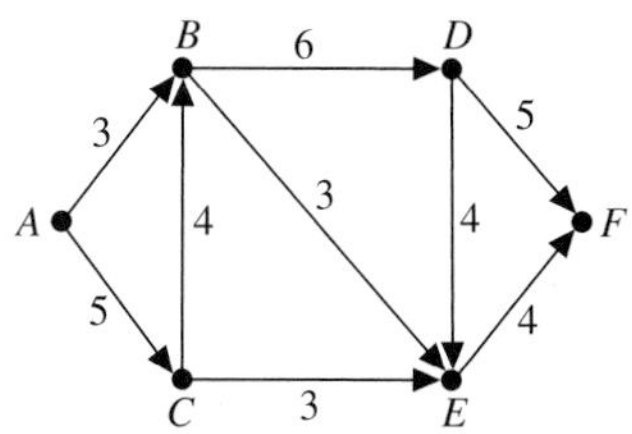

17. $k(A) = 16, k(B) = 9, k(C) = 6, k(D) = 5, k(E) = 8, k(F) = 7, k(G) = 15$

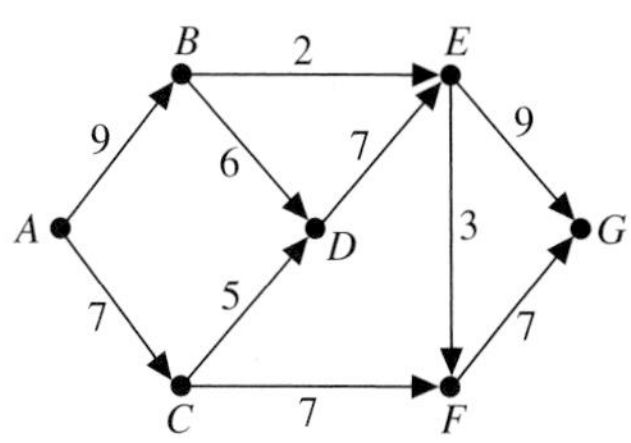

In Exercises 18 and 19, use the method of Exercise 14 to find a maximal flow for the network with vertex capacities in the indicated exercise.

18. Exercise 15

19. Exercise 16

20. Let s and t be vertices in a directed graph $\mathcal{D}$ having indegree 0 and outdegree 0, respectively. Make $\mathcal{D}$ into a transportation network $\mathcal{N}$ with vertex capacities by letting s and t have infinite capacity, letting the other vertices have capacity 1, and letting each arc have capacity 1. Use Supplementary Exercise 14 and Exercise 30 of Section 7.2 to show that the value of a maximal flow for $\mathcal{N}$ equals the number of directed paths from s to t that use no vertex other than s and t more than once.

21. Let m and n be positive integers. Make a transportation network with vertices $s, X_1, X_2, \ldots, X_m, Y_1, Y_2, \ldots, Y_n, t$ and an arc of infinite capacity from s to each X_i, an arc of infinite capacity from each Y_j to t, and an arc of capacity 1 from X_i to Y_j for every i and j. Prove that if f is an integral flow in this network, then there exists an $m \times n$ matrix of 0s and 1s having $f(s, X_i)$ 1s in row i and $f(Y_j, t)$ 1s in column j for every i and j.

22. Let A be an $m \times n$ matrix of 0s and 1s with u_i 1s in row i and v_j 1s in column j for every i and j. Prove that

$$u_1 + u_2 + \cdots + u_m = v_1 + v_2 + \cdots + v_n.$$

23. Let m and n be positive integers, and let u_i and v_j be nonnegative integers for $1 \le i \le m$ and $1 \le j \le n$. Suppose that

$$u_1 + u_2 + \cdots + u_m = v_1 + v_2 + \cdots + v_n.$$

Construct a network having the same arcs as in Exercise 21, but let the capacity of each arc (s, X_i) be u_i and the capacity of each arc (Y_j, t) be v_j. Show that if the value of a maximal flow in this network is $u_1 + u_2 + \cdots + u_m$, then there exists an $m \times n$ matrix of 0s and 1s with u_i 1s in row i and v_j 1s in column j for every i and j.

24. If in Exercise 23 the value of a maximal flow is not $u_1 + u_2 + \cdots + u_m$, prove that there does not exist an $m \times n$ matrix of 0s and 1s with u_i 1s in row i and v_j 1s in column j for every i and j.

25. Use Exercise 23 to construct a 4×6 matrix of 0s and 1s in which there are

(i) four 1s in rows 1, 2, and 4; and two 1s in row 3;

(ii) three 1s in columns 1, 3, and 6; two 1s in columns 2 and 4; and one 1 in column 5.

COMPUTER PROJECTS

Write a computer program having the specified input and output.

1. For a given transportation network, compute the capacity of every cut.
2. For a given transportation network, determine a maximal flow by repeated use of the flow augmentation algorithm, starting with the flow that is zero along every arc.
3. For a given bipartite graph, use the method described in Section 7.4 to produce a maximum matching.
4. For a given bipartite graph, use the method described in Section 7.4 to produce a minimum covering. (See Exercise 31 in Section 7.4.)
5. For a given bipartite graph, compute $|\mathcal{A}| - |\mathcal{A}^*|$ for each subset $\mathcal{A}$ of $\mathcal{V}_1$. (The notation is as in the instructions for Exercises 25–28 in Section 7.4.)

6. For a given network with multiple sources and sinks (see Supplementary Exercise 13), determine a maximal flow.
7. For a given network with vertex capacities (see Supplementary Exercise 14), determine a maximal flow.
8. Let m and n be positive integers, and let u_i for $1 \leq i \leq m$ and v_j for $1 \leq j \leq n$ be nonnegative integers such that

$$u_1 + u_2 + \cdots + u_m = v_1 + v_2 + \cdots + v_n.$$

Construct, if possible, an $m \times n$ matrix of 0s and 1s with u_i 1s in row i and v_j 1s in column j for every i and j. (See Supplementary Exercise 23.)

SUGGESTED READINGS

1. Bondy, J.A. and U.S.R. Murty. *Graph Theory with Applications.* New York: Elsevier Science, 1979.
2. Burr, Stefan A., ed. "The Mathematics of Networks." In *Proceedings of Symposia in Applied Mathematics*, vol. 26. Providence, RI: American Mathematical Society, 1982.
3. Edmonds, Jack and Richard M. Karp. "Theoretical Improvements in Algorithmic Efficiency for Network Flow Problems." *J. Assoc. Computing Machinery*, vol. 19, no. 2 (April 1972): 248–264.
4. Elias, P., A. Feinstein, and C.E. Shannon. "Note on Maximal Flow Through a Network." *IRE Transactions on Information Theory, IT-2* (1956): 117–119.
5. Ford, Jr., L.R. and D.R. Fulkerson. *Flows in Networks.* Princeton, NJ: Princeton University Press, 1962.
6. ________. "Maximal Flow Through a Network." *Canad. J. Math.* vol. 8, no. 3 (1956): 399–404.
7. Frank, Howard and Ivan T. Frisch. "Network Analysis." *Sci. Amer.* vol. 223, no. 1 (July 1970): 94–103.
8. Fulkerson, D.R., ed. *Studies in Graph Theory*, Part I. *MAA Studies in Mathematics*, vol. 11. The Mathematical Association of America, 1975.
9. Lawler, Eugene L. *Combinatorial Optimization: Networks and Matroids.* New York: Holt, Rinehart, and Winston, 1976.
10. Wilf, Herbert S. *Algorithms and Complexity.* Englewood Cliffs, NJ: Prentice-Hall, 1986.
11. Zadeh, Norman. "Theoretical Efficiency of the Edmonds-Karp Algorithm for Computing Maximal Flows." *J. Assoc. Computing Machinery*, vol. 19, no. 1 (January 1972): 184–192.

CHAPTER

8

Counting Techniques

As we saw in Sections 1.2 and 1.3, many combinatorial problems involve counting. Since the number of objects under consideration is often extremely large, it is desirable to be able to count a set of objects without having to list them all. In this chapter we will discuss several fundamental counting techniques that are frequently used in solving combinatorial problems. The reader should carefully review Sections 1.2 and 2.6, which contain several results we will refer to.

8.1 PASCAL'S TRIANGLE AND THE BINOMIAL THEOREM

One of the most basic problems that arises in combinatorial analysis is to count the number of subsets of an n-element set that contain r elements. Recall that we are denoting this number by $C(n, r)$. Theorem 2.10 shows that

$$C(n, r) = \frac{n!}{r!\,(n-r)!}. \tag{8.1}$$

Example 8.1 By (8.1), we see that the number of subsets containing two elements that can be formed from the set of vowels {a, e, i, o, u} is

$$C(5, 2) = \frac{5!}{2!\,3!} = \frac{5 \cdot 4 \cdot 3!}{2 \cdot 1 \cdot 3!} = \frac{5 \cdot 4}{2 \cdot 1} = 10.$$

The ten subsets in question are {a, e}, {a, i}, {a, o}, {a, u}, {e, i}, {e, o}, {e, u}, {i, o}, {i, u}, and {o, u}. ■

The proof of Theorem 2.10 shows that any subset R of $\{a_1, a_2, \ldots, a_n\}$ that contains r elements $(1 \le r \le n)$ is either

(1) a subset of $\{a_1, a_2, \ldots, a_{n-1}\}$ containing r elements (if $a_n \notin R$) or
(2) the union of $\{a_n\}$ and a subset of $\{a_1, a_2, \ldots, a_{n-1}\}$ containing $r - 1$ elements (if $a_n \in R$).

Thus the number of subsets of $\{a_1, a_2, \ldots, a_n\}$ containing r elements, $C(n, r)$, is the sum of the number of subsets of type (1), $C(n-1, r)$, and the number of subsets of type (2), $C(n-1, r-1)$. We have obtained Theorem 8.1.

Theorem 8.1 If r and n are integers such that $1 \le r < n$, then

$$C(n, r) = C(n-1, r-1) + C(n-1, r).$$

Example 8.2 It follows from Theorem 8.1 that $C(7, 3) = C(6, 2) + C(6, 3)$. We will verify this equation by evaluating $C(7, 3)$, $C(6, 2)$, and $C(6, 3)$, using (8.1):

$$C(6, 2) = \frac{6!}{2!\,4!} = \frac{6 \cdot 5 \cdot 4!}{2 \cdot 1 \cdot 4!} = \frac{6 \cdot 5}{2 \cdot 1} = 15,$$

$$C(6, 3) = \frac{6!}{3!\,3!} = \frac{6 \cdot 5 \cdot 4 \cdot 3!}{3 \cdot 2 \cdot 1 \cdot 3!} = \frac{6 \cdot 5 \cdot 4}{3 \cdot 2 \cdot 1} = 20, \text{ and}$$

$$C(7, 3) = \frac{7!}{3!\,4!} = \frac{7 \cdot 6 \cdot 5 \cdot 4!}{3 \cdot 2 \cdot 1 \cdot 4!} = \frac{7 \cdot 6 \cdot 5}{3 \cdot 2 \cdot 1} = 35.$$

So $C(6, 2) + C(6, 3) = 15 + 20 = 35 = C(7, 3)$. ■

The triangular array

$$\begin{array}{lccccccccc}
n = 0 & & & & & C(0,0) & & & & \\
n = 1 & & & & C(1,0) & & C(1,1) & & & \\
n = 2 & & & C(2,0) & & C(2,1) & & C(2,2) & & \\
n = 3 & & C(3,0) & & C(3,1) & & C(3,2) & & C(3,3) & \\
n = 4 & C(4,0) & & C(4,1) & & C(4,2) & & C(4,3) & & C(4,4) \\
 & & & & & \vdots & & & &
\end{array}$$

is called **Pascal's triangle**. Although this array was first known to the Chinese, its name comes from the French mathematician Blaise Pascal (1623–1662), whose paper *Traité du Triangle Arithmétique* developed many of the triangle's properties.

Note that the rows of the triangle are numbered beginning with row $n = 0$ and that the entries $C(n, r)$ for a fixed value of r lie along a diagonal extending from the upper right to the lower left.

Let us consider the entries of Pascal's triangle in more detail. Since every set has exactly one subset containing 0 elements (namely, the empty set), $C(n, 0) = 1$. Moreover, there is only one subset of n elements that can be formed from a set containing n elements (namely, the entire set); so $C(n, n) = 1$. Therefore the first and last numbers in every row of Pascal's triangle are 1s. In addition, Theorem 8.1 states that every entry that is not first or last in its row is the sum of the two nearest entries in the row above. For example, $C(4, 2) = C(3, 1) + C(3, 2)$ and $C(4, 3) = C(3, 2) + C(3, 3)$. By repeatedly using these properties, we can easily evaluate the entries in Pascal's triangle. The resulting numbers are arranged as follows:

$$\begin{array}{ccccccccc} & & & & 1 & & & & \\ & & & 1 & & 1 & & & \\ & & 1 & & 2 & & 1 & & \\ & 1 & & 3 & & 3 & & 1 & \\ 1 & & 4 & & 6 & & 4 & & 1 \\ & & & & \vdots & & & & \end{array}$$

Example 8.3 Continuing in the preceding triangle, we see that the numbers in the next row (the row $n = 5$) are

$$1, \quad 1+4=5, \quad 4+6=10, \quad 6+4=10, \quad 4+1=5, \quad \text{and} \quad 1. \quad \blacksquare$$

Pascal's triangle contains an important symmetry: Each row reads the same from left to right as it does from right to left. In terms of our notation, this statement means that $C(n, r) = C(n, n - r)$ for any r satisfying $0 \le r \le n$. Although this property is easily verified by computing $C(n, r)$ and $C(n, n - r)$ with the use of (8.1), we will prove this fact by a *combinatorial argument*, a proof based on the definition of these numbers. (This is the same type of argument used to establish Theorem 8.1.)

Theorem 8.2 If r and n are integers such that $0 \le r \le n$, then $C(n, r) = C(n, n - r)$.

Proof. Recall that $C(n, k)$ is the number of subsets containing k elements that can be formed from a set of n elements. Let S be a set of n elements. The function that assigns to an r-element subset $A \subseteq S$ the $(n - r)$-element subset $\bar{A}$, the complement of A with respect to S, is easily seen to be a one-to-one correspondence between the subsets of S that contain r elements and the subsets of S that contain $n - r$ elements. Hence the number of r-element subsets of S is the same as the number of subsets that contain exactly $n - r$ elements, that is, $C(n, r) = C(n, n - r)$. $\blacksquare$

The numbers $C(n, r)$ are called **binomial coefficients** because they appear in the algebraic expansion of the binomial $(x + y)^n$. More specifically, in this expansion, $C(n, r)$ is the coefficient of the term $x^{n-r}y^r$. Thus the coefficients that occur in the expansion of $(x + y)^n$ are the numbers in row n of Pascal's triangle. For example,

$$\begin{aligned}(x + y)^3 &= (x + y)(x + y)^2 = (x + y)(x^2 + 2xy + y^2)\\ &= x(x^2 + 2xy + y^2) + y(x^2 + 2xy + y^2)\\ &= (x^3 + 2x^2y + xy^2) + (x^2y + 2xy^2 + y^3)\\ &= x^3 + 3x^2y + 3xy^2 + y^3.\end{aligned}$$

Note that the coefficients (1, 3, 3, and 1) occurring in this expansion are the numbers in the $n = 3$ row of Pascal's triangle.

Theorem 8.3 ***The Binomial Theorem*** For every positive integer n,

$$(x + y)^n = C(n, 0)x^n + C(n, 1)x^{n-1}y + \cdots + C(n, n-1)xy^{n-1} + C(n, n)y^n.$$

Proof. In the expansion of

$$(x + y)^n = (x + y)(x + y)\cdots(x + y),$$

we choose either an x or a y from each of the n factors $x + y$. The term in the expansion involving $x^{n-r}y^r$ results from combining all the terms obtained by choosing x from $n - r$ factors and y from r factors. The number of such terms is, therefore, the number of ways to select a subset of r factors from which to choose y. (We will select x from each factor from which we do not choose y.) Hence the coefficient of $x^{n-r}y^r$ in the expansion of $(x + y)^n$ is $C(n, r)$. ▪

Example 8.4 Using the binomial theorem and the coefficients from the $n = 4$ row of Pascal's triangle, we see that

$$\begin{aligned}(x + y)^4 &= C(4, 0)x^4 + C(4, 1)x^3y + C(4, 2)x^2y^2 + C(4, 3)xy^3 + C(4, 4)y^4\\ &= x^4 + 4x^3y + 6x^2y^2 + 4xy^3 + y^4.\end{aligned}$$ ■

EXERCISES 8.1

Evaluate the numbers in Exercises 1–12.

1. $C(5, 3)$ **2.** $C(7, 2)$ **3.** $C(8, 5)$ **4.** $C(12, 7)$

5. the coefficient of x^2y^2 in the expansion of $(x + y)^4$

6. the coefficient of x^5y in the expansion of $(x + y)^6$

7. the coefficient of x^3y^9 in the expansion of $(x - y)^{12}$

8. the coefficient of x^5y^4 in the expansion of $(x - y)^9$

9. the coefficient of x^6y^4 in the expansion of $(x + 2y)^{10}$

10. the coefficient of x^4y^9 in the expansion of $(3x + y)^{13}$

11. the coefficient of x^3y^7 in the expansion of $(x - 3y)^{10}$

12. the coefficient of x^7y^2 in the expansion of $(2x - y)^9$

13. Write the numbers in the $n = 6$ row of Pascal's triangle.

14. Write the numbers in the $n = 7$ row of Pascal's triangle.

15. Evaluate $(x + y)^6$. **16.** Evaluate $(x + y)^7$. **17.** Evaluate $(3x - y)^4$. **18.** Evaluate $(x - 2y)^5$.

19. How many subsets containing four different numbers can be formed from the set $\{1, 2, 3, 4, 5, 6, 7\}$?

20. How many subsets containing eight different letters can be formed from the set {a, b, c, d, e, f, g, h, i, j, k, l}?

21. How many subsets of {b, c, d, f, g, h, j, k, l, m} contain five letters?

22. How many subsets of $\{2, 3, 5, 7, 11, 13, 17, 19, 23\}$ contain four numbers?

23. How many four-element subsets of $\{1, 2, 3, 4, 5, 6, 7, 8, 9, 10, 11, 12\}$ contain no odd numbers?

24. How many three-element subsets of {a, b, c, d, e, f, g, h, i, j, k} contain no vowels?

25. Use the binomial theorem to show that $C(n, 0) + C(n, 1) + \cdots + C(n, n) = 2^n$ for all nonnegative integers n.

26. Use (8.1) to verify that $C(n, r) = C(n, n - r)$ for $0 \le r \le n$.

27. Use the binomial theorem to show, for $n \ge 0$, that

$$C(n, 0) - C(n, 1) + C(n, 2) - C(n, 3) + \cdots + (-1)^n C(n, n) = 0.$$

28. Prove that $2^0C(n, 0) + 2^1C(n, 1) + \cdots + 2^nC(n, n) = 3^n$ for all positive integers n.

29. Prove that $rC(n, r) = nC(n - 1, r - 1)$ for $1 \le r \le n$.

30. Prove that $2C(n, 2) + n^2 = C(2n, 2)$ for $n \ge 2$.

31. Prove that, for any positive integer k and any nonnegative integer r,

$$C(k, 0) + C(k + 1, 1) + \cdots + C(k + r, r) = C(k + r + 1, r).$$

32. Prove that $C(r, r) + C(r + 1, r) + \cdots + C(n, r) = C(n + 1, r + 1)$ for $0 \le r \le n$. Why do you think the name "hockey stick formula" is used for this result?

33. Prove that $C(2n + 1, 0) + C(2n + 1, 1) + \cdots + C(2n + 1, n) = 2^{2n}$ for all nonnegative integers n.

34. Let p and r be integers such that p is prime and $1 \le r \le p - 1$. Prove that $C(p, r)$ is divisible by p.

35. Let k and n be nonnegative integers such that $k < \frac{n}{2}$. Prove that $C(n, k) \le C(n, k + 1)$.

36. Prove that $C(m, 2) + C(n, 2) \le C(m + n - 1, 2)$ for any integers $m, n \ge 2$.

37. Prove that the product of any n consecutive positive integers is divisible by $n!$.

38. Prove that $1 \cdot C(n, 0) + 2 \cdot C(n, 1) + \cdots + (n + 1) \cdot C(n, n) = 2^n + n2^{n-1}$ for all nonnegative integers n.

8.2 THREE FUNDAMENTAL PRINCIPLES

In this section we will introduce three basic principles that will find frequent use throughout this chapter. The first of these is a suprisingly simple existence statement that has many profound consequences.

Theorem 8.4 ***The Pigeonhole Principle*** If pigeons are placed into pigeonholes and there are more pigeons than pigeonholes, then some pigeonhole must contain at least two pigeons. More generally, if the number of pigeons is more than k times the number of pigeonholes, then some pigeonhole must contain at least $k + 1$ pigeons.

Proof. The first statement is a special case of the more general result, namely, the case in which $k = 1$. We will prove only the more general result.

Suppose that there are p pigeonholes and q pigeons. If no pigeonhole contains at least $k + 1$ pigeons, then each of the p pigeonholes contains at most k pigeons; so the total number of pigeons cannot exceed kp. Thus, if the number of pigeons is more than k times the number of pigeonholes (that is, if $q > kp$), then some pigeonhole must contain at least $k + 1$ pigeons. ■

Example 8.5 How many people must be selected from a collection of 15 married couples to ensure that at least two of the persons chosen are married to each other?

It is easy to see intuitively that if we choose any 16 persons from this collection of 15 couples, we must include at least one husband and wife pair. This conclusion is based on the pigeonhole principle. Let us place persons (the pigeons) into sets (the pigeonholes) in such a way that two persons are in the same set if and only if they are married to each other. Since there are only 15 possible sets, any distribution of 16 persons must place two people in the same set. Thus there must be at least one married couple included among the 16 persons. Note that if we choose fewer than 16 persons, a married couple may not be included (for instance, if we choose the 15 women). ■

Example 8.6 How many distinct integers must be chosen to assure that there are at least 10 having the same congruence class modulo 7?

This question involves placing integers (the pigeons) into congruence classes (the pigeonholes). Recall that there are 7 distinct congruence classes modulo 7, namely, [0], [1], [2], [3], [4], [5], and [6]. So if we want to guarantee that there are at least $10 = k + 1$ integers in the same congruence class, the generalized form of the pigeonhole principle states that we must choose more than $7k = 7 \cdot 9 = 63$ distinct integers. Hence at least 64 integers must be chosen. ■

Often, the pigeonhole principle is the key to the solution of a problem that requires producing a pair of elements with special properties. The next example is of this type.

Example 8.7 Choose any five points from the interior of an equilateral triangle having sides of length 1. Show that the distance between some pair of these points does not exceed $\frac{1}{2}$.

Subdivide the given triangle into 4 equilateral triangles with side $\frac{1}{2}$, as shown in Figure 8.1. Since there are 5 points and only 4 small triangles, some pair of points must lie in the same triangle. But it is easy to see that any 2 points lying in the same small triangle must be such that their distance apart does not exceed $\frac{1}{2}$. ■

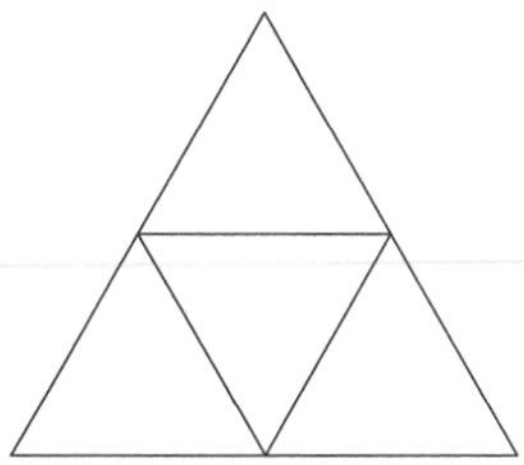

FIGURE 8.1

In contrast to the pigeonhole principle, which asserts that some pigeonhole contains a certain number of pigeons (an existence statement), the next two results tell us how to count the number of ways to perform certain procedures. The first theorem is a restatement of a result from Section 1.2 that enables us to count the number of ways of performing a procedure that consists of a sequence of operations.

Theorem 8.5 ***The Multiplication Principle*** Consider a procedure that is composed of a sequence of k steps. Suppose that the first step can be performed in n_1 ways; and for each of these, the second step can be performed in n_2 ways; and, in general, no matter how the preceding steps are performed, the ith step can be performed in n_i ways ($i = 2, 3, \ldots, k$). Then the number of different ways in which the entire procedure can be performed is $n_1 n_2 \cdots n_k$.

To illustrate the multiplication principle, suppose that a couple expecting a child has decided that if it is a girl, they will give it a first name of Jennifer, Karen, or Linda and a middle name of Ann or Marie. Since the process of naming the child can be divided into the two steps of selecting a first name and selecting a middle name, the multiplication principle tells us that there are $3 \cdot 2 = 6$ possible names that can be given. To see that this is the correct answer, we can enumerate the possibilities, as in Figure 8.2.

Example 8.8 A **bit** (or **binary digit**) is a zero or a one. An ***n*-bit string** is a sequence of n bits. Thus 01101110 is an 8-bit string. Information is stored and processed in computers in bit strings because a bit string can be regarded as a sequence of on or off settings for switches inside the computer.

Let us compute the number of 8-bit strings by using the multiplication principle. To do so, we will regard the 8-bit string as a sequence of 8 choices (choose

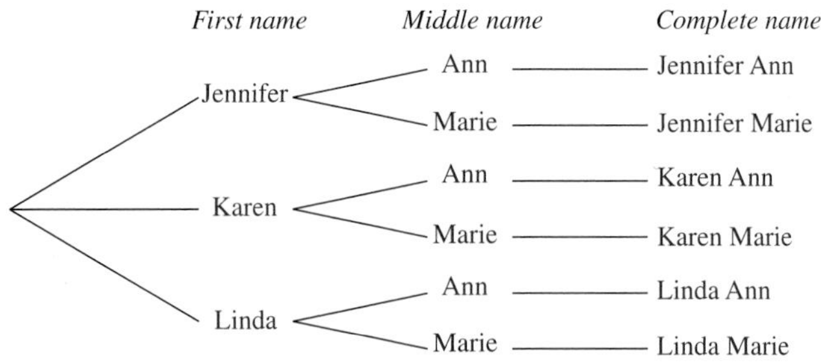

FIGURE 8.2

the first bit, then choose the second bit, and so forth). Because each bit can be chosen in two ways (namely, zero or one), the number of possible 8-bit strings is

$$2 \cdot 2 \cdot 2 \cdot 2 \cdot 2 \cdot 2 \cdot 2 \cdot 2 = 2^8 = 256.$$

Since an 8-bit string can be regarded as the binary representation of a nonnegative integer, this calculation shows that 256 nonnegative integers can be expressed with no more than 8 binary digits. More generally, a similar argument shows that the number of n-bit strings (and, hence, the number of nonnegative integers that can be expressed using no more than n binary digits) is 2^n. ■

Example 8.9 On January 20, 1996, telephones in the northern suburbs of Chicago were given a new area code (847). Previously, all the suburbs had the same area code (708). Other suburban and city regions are similarly divided. With the increased demand by businesses for telephone lines for computers, fax machines, and cellular phones, metropolitan areas are actually running out of phone numbers! How many different telephone numbers with the same area code are possible?

Within one area code, a local telephone number is a sequence of seven digits (0–9) with the restriction that the first and second digits cannot be 0 or 1. The number of possible local phone numbers can therefore be counted by using the multiplication principle. Each of the first two digits can be any of the numbers 2–9, and the remaining digits can be any value 0–9. Thus the number of possible local phone numbers, which equals the number of ways that the seven digits can be chosen, is

$$8 \cdot 8 \cdot 10 \cdot 10 \cdot 10 \cdot 10 \cdot 10 = 6{,}400{,}000. \quad ■$$

In using the multiplication principle, it is important to note that the number of ways to perform a step must not depend on the particular choice that is made at any previous step. That is, in the notation of Theorem 8.5, no matter how the first step is performed, there must be n_2 ways of performing the second step; no matter how the first two steps are performed, there must be n_3 ways of performing the third step, and so forth. Because of this restriction, some ingenuity may be required to obtain the correct solution to a problem, as in the next example.

Example 8.10 Suppose that we are to use the digits 1–8 without repetition to make five-digit numbers.

(a) How many different five-digit integers can be made?

(b) How many of the numbers in part (a) begin with 7?

(c) How many of the numbers in part (a) contain both 1 and 2?

(a) We can construct five-digit numbers by choosing a value for each of the five digits in the number. This amounts to filling each of the blanks

$$\_\_\_\_ \quad \_\_\_\_ \quad \_\_\_\_ \quad \_\_\_\_ \quad \_\_\_\_$$

with one of the digits 1–8. Clearly, there are 8 ways in which the first digit can be selected because any of the digits 1–8 can be used. There are only 7 ways to choose the second digit, however, because the first digit cannot be repeated. Similar reasoning shows that there are 6 ways to choose the third digit, 5 ways to choose the fourth digit, and 4 ways to choose the fifth digit. Hence the multiplication principle shows that the number of possible ways of making all five choices is

$$8 \cdot 7 \cdot 6 \cdot 5 \cdot 4 = 6720.$$

This is the number of five-digit numbers that can be formed from the digits 1–8 without repetition.

(b) To count the numbers that begin with 7, we can proceed as before, except that there is only one way to choose the first digit (since it must be 7). Therefore, of the 6720 integers in part (a),

$$1 \cdot 7 \cdot 6 \cdot 5 \cdot 4 = 840$$

begin with 7.

(c) The method used in the preceding parts cannot be used to count the number of five-digit numbers containing both 1 and 2. The reason this method fails is that the number of ways to select the fourth and fifth digits depends on earlier choices. For example, if the first three digits are 231, then there are 5 ways in which the fourth digit can be chosen (namely, 4, 5, 6, 7, or 8); but if the first three digits are 567, then there are only 2 ways in which the fourth digit can be chosen (namely, 1 or 2).

Consequently, we must look for another approach. Since the digits 1 and 2 must be used, we begin by deciding where to put them. We can then fill the three remaining positions with any of the digits 3–8. Thus we proceed as follows:

Choose a position for the 1 (in 5 possible ways).
Choose a position for the 2 (in 4 possible ways).
Choose a value to put in the first unfilled blank (using one of 6 possible digits).
Choose a value to put in the second unfilled blank (using one of 5 possible digits).
Choose a value to put in the third unfilled blank (using one of 4 possible digits).

Thus, by the multiplication principle, there are

$$5 \cdot 4 \cdot 6 \cdot 5 \cdot 4 = 2400$$

ways of making all five choices. Hence 2400 of the integers in part (a) contain both of the digits 1 and 2. ■

The second basic counting principle is concerned with the number of elements in the union of pairwise disjoint sets.

Theorem 8.6 ***The Addition Principle*** Suppose that there are k sets of elements with n_1 elements in the first set, n_2 elements in the second set, etc. If all of the elements are distinct (that is, if all pairs of the k sets are disjoint), then the number of elements in the union of the sets is $n_1 + n_2 + \cdots + n_k$.

To illustrate this result, let $A = \{1, 2, 3\}$ and $B = \{4, 5, 6, 7\}$. Since the elements in A and B are distinct, it follows from the addition principle that the number of elements in the union of sets A and B equals the number of elements in set A (which is 3) plus the number of elements in set B (which is 4). Therefore, in our example, the number of elements in $A \cup B$ is $3 + 4 = 7$. Clearly, this answer is correct because $A \cup B = \{1, 2, 3, 4, 5, 6, 7\}$. But note the necessity of distinct elements: If A had been the set $A = \{1, 2, 4\}$, then the answer would no longer have been 7, since in this case $A \cup B = \{1, 2, 4, 5, 6, 7\}$.

Example 8.11 Suppose that a couple expecting a child has decided to name it one of six names (Jennifer Ann, Jennifer Marie, Karen Ann, Karen Marie, Linda Ann, or Linda Marie) if it is a girl and one of four names (Michael Alan, Michael Louis, Robert Alan, or Robert Louis) if it is a boy. How many different names can the child receive?

Since the names to be given to a girl are different from those to be given to a boy, the addition principle states that the number of possible names is the sum of the number of the girls' names and the number of the boys' names. Thus the answer to the question posed above is $6 + 4 = 10$. ■

Example 8.12 How many integers between 1 and 100 (including 100) are even or end with 5?

Let A denote the set of even integers between 1 and 100, and let B denote the set of integers between 1 and 100 that end with 5. The number of integers between 1 and 100 that are even or end in 5 is then the number of elements in $A \cup B$. Now A contains 50 elements because every other number from 1 to 100 is even. And B contains 10 elements, since 5, 15, 25, 35, 45, 55, 65, 75, 85, and 95 are the only integers between 1 and 100 that end in 5. Moreover, the elements in A and B are distinct because a number ending in 5 cannot be even. Thus it follows from the addition principle that the number of integers between 1 and 100 that are even or end with 5 is $50 + 10 = 60$. ■

Often, the multiplication and addition principles are both needed to solve a problem, as in the following examples:

Example 8.13 In the Applesoft BASIC language, the name of a real variable consists of alphanumeric characters beginning with a letter. (An alphanumeric character is a letter A–Z or a digit 0–9.) Although variable names may be as long as 238 characters, they are distinguished by their first two characters only. (Thus RATE and RATIO are regarded as the same name.) In addition, there are seven reserved words (AT, FN, GR, IF, ON, OR, and TO) that are not legal variable names. We will use the multiplication and addition principles to determine the number of legal variable names that can be distinguished in Applesoft BASIC.

Clearly, there are 26 real variable names consisting of a single character, namely A–Z. Any other distinct name will consist of a letter followed by an alphanumeric character. It follows from the multiplication principle that the number of distinguishable names consisting of more than one character is $26 \cdot 36 = 936$, since there are 26 letters and 36 alphanumeric characters. Thus, by the addition principle, the number of one-character or two-character names is $26 + 936 = 962$. So there are 962 distinguishable real variable names in Applesoft BASIC, and hence $962 - 7 = 955$ legal names that can be distinguished. ■

Example 8.14 (a) How many 8-bit strings begin with 1011 or 01?

(b) How many 8-bit strings begin with 1011 or end with 01?

(a) An 8-bit string beginning with 1011 has the form

$$1\,0\,1\,1 - - - -,$$

where the dashes denote either zeros or ones. The number of 8-bit strings that begin with 1011 is equal to the number of ways that bits five through eight can be chosen. By reasoning as in Example 8.8, we find that this number is $2^4 = 16$. Likewise, the number of 8-bit strings that begin with 01 is $2^6 = 64$. Since the set of strings beginning with 1011 is disjoint from the set of strings beginning with 01, the addition principle shows that the number of strings beginning with 1011 or 01 is $16 + 64 = 80$.

(b) Although it is tempting to approach this problem as in part (a), the set of strings beginning with 1011 is not disjoint from the set of strings ending with 01; so the correct answer is not $16 + 64$ as before. Since the addition principle can be used only with disjoint sets, let us define sets of 8-bit strings A, B, and C as follows:

$$A = \{\text{strings beginning with 1011 and not ending with 01}\},$$
$$B = \{\text{strings ending with 01 and not beginning with 1011}\},$$
$$C = \{\text{strings beginning with 1011 and ending with 01}\}.$$

Clearly, all pairs of the sets A, B, and C are disjoint, and $A \cup B \cup C$ consists of the strings that begin with 1011 or end with 01. Therefore the addition principle states that the number of elements in $A \cup B \cup C$ is the sum of the sizes of A, B, and C. Note that strings in A begin with 1011 and end with 00, 10, or 11. Thus there is only one way to choose the first four bits of a string in A (namely, 1011), there are two ways to choose each of the fifth and sixth bits (0 or 1), and there are three ways to choose the last two bits (00, 10, or 11). Hence the number of strings in A is $1 \cdot 2 \cdot 2 \cdot 3 = 12$. Similar arguments show that the numbers of strings in B and C are $15 \cdot 2 \cdot 2 \cdot 1 = 60$ and $1 \cdot 2 \cdot 2 \cdot 1 = 4$, respectively. Therefore the number of strings in $A \cup B \cup C$ is $12 + 60 + 4 = 76$. ■

EXERCISES 8.2

1. How many people must there be in order to assure that at least two of their birthdays fall in the same month?

2. If a committee varies its meeting days, how many meetings must it schedule before we can guarantee that at least two meetings will be held on the same day of the week?

3. A drawer contains unsorted black, brown, blue, and gray socks. How many socks must be chosen in order to be certain of choosing two of the same color?

4. How many words must be chosen in order to assure that at least two begin with the same letter?

5. A conference room contains 8 tables and 105 chairs. What is the smallest possible number of chairs at the table having the most seats?

6. If there are 6 sections of Discrete Math with a total enrollment of 199 students, what is the smallest possible number of students in the section with the largest enrollment?

7. How many books must be chosen from among 24 mathematics books, 25 computer science books, 21 literature books, and 15 economic books in order to assure that there are at least 12 books on the same subject?

8. A sociologist intends to send a questionnaire to 32 whites, 19 blacks, 27 Hispanics, and 31 Native Americans. How many responses must she receive in order to guarantee that there will be at least 15 responses from the same ethnic group?

9. An automobile can be ordered with any combination of the following options: air conditioning, automatic transmission, bucket seats, rear window defogger, and CD player. In how many different ways can this car be equipped?

10. How many different pizzas can be ordered if a pizza can be selected with any combination of the following ingredients: anchovies, ham, mushrooms, olives, onion, pepperoni, and sausage?

11. Use the multiplication principle to determine the number of subsets of a set containing n elements.

12. How many different sequences of heads and tails can result if a coin is flipped 20 times?

13. A businessman must fly from Kansas City to Chicago on Monday and from Chicago to Boston on Thursday. If there are 8 daily flights from Kansas City to Chicago and 21 daily flights from Chicago to Boston, how many different routings are possible for this trip?

14. An interior decorator is creating layouts that consist of carpeting and draperies. If there are 4 choices of carpets and 6 choices of draperies, how many layouts must be made to show all of the possibilities?

15. Until recently, a telephone area code was a three-digit number that could not begin with 0 or 1 and must have 0 or 1 as its middle digit. How many such telephone area codes are possible?

16. How many different character strings of length three can be formed from the letters A, B, C, D, E, and F if

(a) letters can be repeated?

(b) letters cannot be repeated?

17. In how many different orders can 3 married couples be seated in a row of 6 chairs under the following conditions?

(a) Anyone may sit in any chair.

(b) Men must occupy the first and last chairs.

(c) Men must occupy the first three chairs and women the last three.

(d) Everyone must be seated beside his or her spouse.

18. On student recognition night, a high school will present awards to 4 seniors and 3 juniors. In how many different orders can the awards be presented under the following conditions?

(a) The awards can be presented in any order.

(b) Awards are presented to juniors before awards are presented to seniors.

(c) The first and last awards are presented to juniors.

(d) The first and last awards are presented to seniors.

19. In the Apple Pascal programming language, identifiers (that is, variable names, file names, and so forth) are subject to the following rules:

(i) The first character in an identifier must be a letter (capital or lower case).

(ii) Subsequent characters may be letters or digits (0, 1, . . . , 9).

If the reserved words IF, LN, ON, OR, and TO cannot be used as identifiers, how many different Apple Pascal identifiers contain exactly two characters?

20. In FORTRAN, unless an integer variable is explicitly declared, its name must begin with one of the letters I, J, K, L, M, or N. Subsequent characters can be any letter A, B, . . . , Z or digit 0, 1, . . . , 9. How many such integer variable names contain exactly four characters?

21. A men's clothing store has a sale on selected suits and blazers. If there are 30 suits and 40 blazers on sale, in how many ways may a customer select exactly one item that is on sale?

22. A restaurant offers a choice of 3 green vegetables or a potato prepared in one of 5 ways. How many different choices of vegetable can be made?

23. How many 8-bit strings begin with 1001 or 010?

24. How many 8-bit strings end with 1000 or 01011?

25. In the United States, radio station call letters consist of 3 or 4 letters beginning with either K or W. How many different sets of radio station call letters are possible?

26. Suppose that a license plate must contain a sequence of 2 letters followed by 4 digits, or 3 letters followed by 3 digits. How many different license plates can be made?

27. From among a group of 4 men and 6 women, 3 persons are to be appointed as a branch manager in different cities. How many different appointments can be made under the following circumstances?

(a) Any person is eligible for appointment.

(b) One man and two women are to be appointed.

(c) At least two men are to be appointed.

(d) At least one person of each sex is to be appointed.

28. Suppose that 3 freshmen, 5 sophomores, and 4 juniors have been nominated to receive scholarships of \$500, \$250, and \$100. How many different distributions of the three scholarships are possible under the following circumstances?

(a) Anyone may receive any scholarship.

(b) The \$500 scholarship is to be awarded to a freshman, the \$250 scholarship to a sophomore, and the \$100 scholarship to a junior.

(c) At least two scholarships are to be awarded to juniors.

(d) One scholarship is to be awarded to someone from each class.

29. How many 8-bit strings begin with 11 or end with 00?

30. How many 8-bit strings begin with 010 or end with 11?

31. The digits 1–6 are to be used to make four-digit numbers.

(a) How many such numbers can be made if repetition is allowed?

(b) How many such numbers can be made if repetition is not allowed?

(c) How many of the numbers in (b) begin with 3?

(d) How many of the numbers in (b) contain 2?

32. A university task force on remedial courses is to be formed from among 3 mathematics teachers, 4 English teachers, 2 science teachers, and 2 humanities teachers. The committee must contain at least 1 math teacher and at least 1 English teacher. How many different committees can be formed if

(a) two committees are considered the same when they contain precisely the same individuals?

(b) two committees are considered the same when they contain the same number of teachers from each discipline?

33. Show that if the 26 letters of the English alphabet are written in a circular array in any order whatsoever, there must be 5 consecutive consonants.

34. Prove that in any nonempty list of n integers (not necessarily distinct) there is some nonempty sublist having a sum that is divisible by n.

35. Let $S = \{a_1, a_2, \ldots, a_9\}$ be any set of 9 points in Euclidean space such that all three coordinates of each point are integers. Prove that, for some i and j $(i \neq j)$, the midpoint of the segment joining a_i and a_j has only integer coordinates.

36. Suppose that there are 15 identical copies of *The Great Gatsby* and 12 distinct biographies on a bookshelf.

(a) How many different selections of 12 books are possible?

(b) How many different selections of 10 books are possible?

8.3 PERMUTATIONS AND COMBINATIONS

Two types of counting problems occur so frequently that they deserve special attention. These problems are as follows:

(1) How many different arrangements (ordered lists) of r objects can be formed from a set of n distinct objects?
(2) How many different selections (unordered lists) of r objects can be made from a set of n distinct objects?

In this section we will consider these two questions in the case that repetition of the n distinct objects is not allowed. Section 8.4 will answer these same questions when repetition is permitted.

Recall from Section 1.2 that an arrangement or ordering of n distinct objects is called a **permutation** of the objects. If $r \leq n$, then the arrangement or ordering using r of the n distinct objects is called an ***r*-permutation**. Thus 3142 is a permutation of the digits 1, 2, 3, and 4, and 412 is a 3-permutation of these digits.

The number of different r-permutations of a set of n distinct elements is denoted $P(n, r)$. In Theorem 1.2, this number was found to be

$$P(n, r) = \frac{n!}{(n-r)!}. \tag{8.2}$$

Thus (8.2) gives us the answer to question (1).

Example 8.15 How many different three-digit numbers can be formed by using the digits 5, 6, 7, 8, and 9 without repetition?

This question asks for the number of 3-permutations from a set of 5 digits. This number is $P(5, 3)$. So, using (8.2), we see that the answer to the preceding question is

$$P(5, 3) = \frac{5!}{2!} = \frac{5 \cdot 4 \cdot 3 \cdot 2!}{2!} = 5 \cdot 4 \cdot 3 = 60. \quad \blacksquare$$

Example 8.16 In how many different orders can 4 persons be seated in a row of 4 chairs?

The answer to this question is the number of permutations of a set of 4 elements. Recalling that $0! = 1$, we see from (8.2) that this number is

$$P(4, 4) = \frac{4!}{0!} = \frac{4!}{1} = 4! = 24. \quad \blacksquare$$

Note that Example 8.16 could also have been answered by appealing to Theorem 1.1, which can be rewritten as $P(n, n) = n!$ in our present notation.

Let us now consider the second question. If $r \leq n$, then an unordered selection of r objects chosen from a set of n distinct objects is called an ***r*-combination** of the objects. Thus $\{1, 4\}$ and $\{2, 3\}$ are both 2-combinations of the digits 1, 2, 3, and 4. Note that, since combinations are unordered selections, the 2-combinations $\{1, 4\}$ and $\{4, 1\}$ are the same. In fact, an unordered selection of r elements from a set of n distinct elements is just a subset of the set that contains r elements. Thus the number of different r-combinations of a set of n distinct elements is $C(n, r)$. So, using (8.1), we see that the answer to question (2) is

$$C(n, r) = \frac{n!}{r!\,(n-r)!}.$$

Example 8.17 How many different 4-member committees can be formed from a delegation of 7 members?

Since a 4-member committee is just a selection of 4 members from the delegation of 7, the answer to this question is $C(7, 4)$. Using (8.1), we find that

$$C(7, 4) = \frac{7!}{4!\,3!} = \frac{7 \cdot 6 \cdot 5 \cdot 4!}{4! \cdot 3 \cdot 2 \cdot 1} = \frac{7 \cdot 6 \cdot 5}{3 \cdot 2 \cdot 1} = 35. \quad \blacksquare$$

Example 8.18 How many 8-bit strings contain exactly three 0s?

Note that an 8-bit string containing exactly three 0s is completely determined if we know the positions of the three 0s (since the other five positions must be filled with 1s). Thus the number of 8-bit strings containing exactly three 0s equals the number of different locations that the three 0s can occupy. But this number is the number of ways to choose three positions from among eight, which is $C(8, 3)$. So the number of 8-bit strings containing exactly three 0s is

$$C(8, 3) = \frac{8!}{3!\,5!} = \frac{8 \cdot 7 \cdot 6 \cdot 5!}{3 \cdot 2 \cdot 1 \cdot 5!} = \frac{8 \cdot 7 \cdot 6}{3 \cdot 2 \cdot 1} = 56. \quad \blacksquare$$

It is clear from (8.1) and (8.2) that

$$P(n, r) = r!\,C(n, r).$$

This equation can be interpreted combinatorially as follows: *The number of ways to arrange r objects from a set of n objects equals the number of ways to select r of the n objects and then arrange the selected objects in order.*

Many counting problems require distinguishing permutations from combinations. Since permutations are *ordered* lists, they arise in problems where the order of selection is significant, such as when the selected objects are to be treated differently. Combinations, on the other hand, are *unordered* lists and occur when the order of selection is irrelevant, such as when the selected objects are treated the same. Note the use of permutations and combinations in the next example.

Example 8.19 An investor is going to invest \$16,000 in 4 stocks chosen from a list of 12 prepared by her broker. How many different investments are possible if

(a) \$4000 is to be invested in each stock?

(b) \$6000 is to be invested in one stock, \$5000 in another, \$3000 in the third, and \$2000 in the fourth?

(a) Since each stock is to be treated the same, we need an *unordered* list of 4 stocks. Hence the number of investments in this case is

$$C(12, 4) = \frac{12!}{4!\,8!} = \frac{12 \cdot 11 \cdot 10 \cdot 9}{4 \cdot 3 \cdot 2 \cdot 1} = 495.$$

(b) Since each stock is to be treated differently, we need an *ordered* list of 4 stocks. Hence the number of investments is this case is

$$P(12, 4) = \frac{12!}{8!} = 12 \cdot 11 \cdot 10 \cdot 9 = 11{,}880. \quad \blacksquare$$

Counting problems often require that permutations or combinations be used together with the multiplication or addition principles. Examples 8.20–8.24 are of this type.

Example 8.20 Three men and three women are going to occupy a row of six seats. In how many different arrangements can they be seated so that men occupy the two end seats?

We can regard the assigning of seats as a two-step process: First fill the two end seats, and then fill the middle four seats. Since the end seats must be filled by two of the three men, there are $P(3, 2)$ different ways to occupy the end seats. The remaining four persons can fill the middle seats in any order, so there are $P(4, 4)$ different ways to fill the middle seats. Thus, by the multiplication principle, the number of ways to fill both the end seats and the middle seats is

$$P(3, 2) \cdot P(4, 4) = 6 \cdot 24 = 144. \quad \blacksquare$$

Example 8.21 An investor is going to purchase shares of 4 stocks chosen from a list of 12 prepared by her broker. How many different investments are possible if \$5000 is to be invested in each of two stocks and \$3000 in each of the others?

We can regard the choice of stocks as a two-step process by first choosing the stocks in which \$5000 is to be invested and then choosing the stocks in which \$3000 is to be invested. Clearly, the stocks in which \$5000 is to be invested can be chosen in $C(12, 2)$ ways. The stocks in which \$3000 is to be invested must be chosen from the remaining 10 stocks, and so this choice can be made in $C(10, 2)$ ways. The multiplication principle now gives the number of different investments to be

$$C(12, 2) \cdot C(10, 2) = 66 \cdot 45 = 2970. \quad \blacksquare$$

Example 8.22 From among a group of six men and nine women, how many three-member committees contain only men or only women?

The number of three-member committees containing only men is $C(6, 3)$, and the number of three-member committees containing only women is $C(9, 3)$. Since the set of committees containing only men is disjoint from the set of committees containing only women, the addition principle shows that the number of three-member committees containing only men or only women is

$$C(6, 3) + C(9, 3) = 20 + 84 = 104. \quad \blacksquare$$

Example 8.23 How many 8-bit strings contain six or more 1s?

If an 8-bit string contains six or more 1s, then the number of 1s that it contains must be six, seven, or eight. Reasoning as in Example 8.18, we see that the number of strings containing exactly six 1s is $C(8, 6)$, the number of strings containing exactly seven 1s is $C(8, 7)$, and the number of strings containing exactly eight 1s is $C(8, 8)$. So the addition principle shows that the number of 8-bit strings containing six or more 1s is

$$C(8, 6) + C(8, 7) + C(8, 8) = 28 + 8 + 1 = 37. \quad \blacksquare$$

Example 8.24 How many 8-bit strings with exactly two 1s are such that the 1s are not adjacent?

If an 8-bit string contains exactly two 1s, then it must also contain exactly six 0s. We will consider two cases, according to whether the last bit is 0 or 1. If the last bit is 0 and the two 1s are not adjacent, then each 1 is followed by at least one 0. Hence we can regard the bits to be arranged as two strings of 10 and four single 0s. The number of ways to arrange these six groups is the number of ways to choose positions for the four 0s from six locations, which is $C(6, 4)$. On the other hand, if the last digit is 1 and the two 1s are not adjacent, then we must arrange five 0s and one string of 10. (The other 1 is reserved for the last bit.) The number of such arrangements is the number of ways to choose positions for the five 0s from six locations, which is $C(6, 5)$. Hence, by the addition principle, the number of 8-bit strings with exactly two nonadjacent 1s is

$$C(6, 4) + C(6, 5) = 15 + 6 = 21. \quad \blacksquare$$

EXERCISES 8.3

Evaluate the numbers in Exercises 1–12.

1. $C(6, 3)$	**2.** $C(7, 4)$	**3.** $C(5, 2)$	**4.** $C(8, 4)$
5. $P(4, 2)$	**6.** $P(6, 3)$	**7.** $P(9, 5)$	**8.** $P(12, 3)$
9. $P(10, 4)$	**10.** $P(8, 3)$	**11.** $P(n, 1)$	**12.** $P(n, 2)$

13. How many different arrangements are there of the letters a, b, c, and d?

14. How many different arrangements are there of the letters in the word "number"?

15. How many different four-digit numbers can be formed by using the digits 1, 2, 3, 4, 5, and 6 without repetition?

16. How many different ways are there of selecting five persons from a group of seven persons and seating them in a row of five chairs?

17. How many different 3-member subcommittees can be formed from a committee with 13 members?

18. How many different 16-bit strings contain exactly four 1s?

19. How many different subsets of $\{1, 2, \ldots, 10\}$ contain exactly six elements?

20. How many different 4-person delegations can be formed from a group of 12 people?

21. Five speakers are scheduled to address a convention. In how many different orders can they appear?

22. Six persons are running for four seats on a town council. In how many different ways can these four seats be filled?

23. For marketing purposes, a manufacturer wants to test a new product in three areas. If there are nine geographic areas in which to test-market the product, in how many different ways can the test areas be selected?

24. An investor intends to buy shares of stock in 3 companies chosen from a list of 12 companies recommended by her broker. How many different investment options are there under the following circumstances?

    (a) Equal amounts will be invested in each company.

    (b) Amounts of \$5000, \$3000, and \$1000 will be invested in the chosen companies.

25. How many different committees consisting of three representatives of management and two representatives of labor can be formed from among six representatives of management and five representatives of labor?

26. In how many different sequences can we list 4 novels followed by 6 biographies if there are 8 novels and 10 biographies from which to choose?

27. An election will be held to fill three faculty seats and two student seats on a certain college committee. The faculty member receiving the most votes will receive a three-year term, the one receiving the second highest total will receive a two-year term, and the one receiving the third highest total will receive a one-year term. Both of the open student seats are for one-year terms. If there are nine faculty members and seven students on the ballot, how many different election results are possible, assuming that ties do not occur?

28. In how many different ways can 8 women be paired with 8 of 12 men at a dance?

29. Suppose that 3 freshmen, 4 sophomores, 2 juniors, and 3 seniors are candidates for four identical school service awards. In how many ways can the recipients be selected under the following conditions?

    (a) Any candidate may receive any award.

    (b) Only juniors and seniors receive awards.

    (c) One person from each class receives an award.

    (d) One freshman, two sophomores, and one senior receive awards.

30. Suppose that 3 freshmen, 5 sophomores, 4 juniors, and 2 seniors have been nominated to serve on a student advisory committee. How many different committees can be formed under the following circumstances?

    (a) The committee is to consist of any four persons.

    (b) The committee is to consist of one freshman, one sophomore, one junior, and one senior.

    (c) The committee is to consist of two persons: one freshman or sophomore and one junior or senior.

    (d) The committee is to consist of three persons from different classes.

31. Prove by a combinatorial argument that $2C(n, 2) + n^2 = C(2n, 2)$ for $n \geq 2$.

32. Prove by a combinatorial argument that $rC(n, r) = nC(n-1, r-1)$ for $1 \le r \le n$.

33. Prove that $C(n, m) \cdot C(m, k) = C(n, k) \cdot C(n-k, m-k)$ for $k \le m \le n$ by a combinatorial argument.

34. Prove that $C(n, 0)^2 + C(n, 1)^2 + \cdots + C(n, n)^2 = C(2n, n)$ for every positive integer n.

35. Prove that $C(1, 1) + C(2, 1) + \cdots + C(n, 1) = C(n+1, 2)$ for every positive integer n by a combinatorial argument.

36. State and prove a generalization of Exercise 35.

8.4 ARRANGEMENTS AND SELECTIONS WITH REPETITIONS

In this section we will learn how to count the number of arrangements of a collection that includes repeated objects and the number of selections from a set when elements can be chosen more than once. As we will see, both of these counting problems require the use of ideas from the two preceding sections.

Let us consider first the number of arrangements of a collection containing repeated indistinguishable objects. As a simple example of this type of problem, we will count the number of different arrangements of the letters in the word "egg." Since there are only three letters in "egg," it is not difficult to list all of the possible arrangements, namely,

egg   geg   gge.

Hence there are only 3 arrangements of the letters in "egg," compared with the $P(3, 3) = 6$ arrangements that we would expect if all the letters had been distinct. To see more clearly the effect of the repeated letters, let us capitalize the first "g" in "egg" and regard a capital letter as different from a lowercase letter. Then the six possible arrangements of the letters in "eGg" are as follows:

eGg   Geg   Gge
egG   geG   gGe.

Note that, because the two g's in the first list are identical, interchanging their positions does not produce different arrangements. But each arrangement in the first list gives rise to two arrangements in the second list, one with "G" preceding "g" and the other with "g" preceding "G." Thus the number of arrangements in the first list equals the number of arrangements in the second list divided by $P(2, 2) = 2$, the number of permutations of the two g's.

Another way to count the number of arrangements of the letters in "egg" is by thinking of an arrangement as having 3 positions and first choosing positions for the two g's and then choosing a position for the "e." Since the positions for the g's can be chosen in $C(3, 2)$ ways and the remaining position for the "e" can be chosen in only $C(1, 1)$ way, the multiplication principle then gives the number of possible arrangements as $C(3, 2) \cdot C(1, 1) = 3 \cdot 1 = 3$. This analysis and the one in the preceding paragraph lead to the same answer (see Exercise 35), which demonstrates the following result:

Theorem 8.7 Let S be a collection containing n objects of k different types. (Objects of the same type are indistinguishable, and objects of different types are distinguishable.) Suppose that each object is of exactly one type and that there are n_1 objects of type 1, n_2 objects of type 2, and, in general, n_i objects of type i. Then the number of different arrangements of the objects in S is

$$C(n, n_1)\cdot C(n-n_1, n_2)\cdot C(n-n_1-n_2, n_3)\cdots C(n-n_1 n_2-\ldots-n_{k-1}, n_k),$$

which equals

$$\frac{n!}{n_1!n_2!\cdots n_k!}.$$

The conclusion of this theorem states that the number of different arrangements of the objects in S equals the number of ways $C(n, n_1)$ to place the n_1 objects of type 1 in n possible locations, times the number of ways $C(n-n_1, n_2)$ to place the n_2 objects of type 2 in $n-n_1$ unused locations, times the number of ways $C(n-n_1-n_2, n_3)$ to place the n_3 objects of type 3 in $n-n_1-n_2$ unused locations, etc. By Exercise 35, this number can also be written in the form

$$\frac{n!}{n_1!n_2!\cdots n_k!}.$$

Note that $n = n_1 + n_2 + \cdots + n_k$ because we are assuming that each of the n elements in S belongs to exactly one of the k types.

Example 8.25 How many arrangements are there of the letters in the word "banana"?

Since "banana" is a six-letter word consisting of three types of letters (1 b, 3 a's, and 2 n's), the number of arrangements of its letters is

$$\frac{6!}{1!\,3!\,2!} = \frac{6\cdot 5\cdot 4\cdot 3!}{1\cdot 3!\cdot 2} = \frac{6\cdot 5\cdot 4}{2} = 60. \quad \blacksquare$$

Example 8.26 Each member of a 9-member committee must be assigned to exactly one of three subcommittees (the executive subcommittee, the finance subcommittee, or the rules committee). If these subcommittees are to contain 3, 4, and 2 members, respectively, how many different subcommittee appointments can be made?

Let us arrange the nine persons in alphabetical order and give each person a slip of paper containing the name of a subcommittee. Then the number of possible subcommittee appointments is the same as the number of arrangements of 9 slips of paper, 3 of which read "executive subcommittee," 4 of which read "finance subcommittee," and 2 of which read "rules subcommittee." By Theorem 8.7, this number is

$$\frac{9!}{3!\,4!\,2!} = 1260. \quad \blacksquare$$

Let us now consider the problem of counting the number of selections from a set when elements can be chosen more than once. As an example, suppose that seven persons in a hotel conference room call for refreshments from room service. If the choice of refreshments is limited to coffee, tea, or milk, how many different selections of seven refreshments are possible? Note that we are asking this question from the room service's point of view; that is, we are not interested in knowing who wants which beverage, but only in the total number of beverages of each type that are desired. For example, one such selection is for 4 coffees, 1 tea, and 2 milks. Thus we are selecting seven times from {coffee, tea, milk} with repetition allowed.

To answer this question, we will suppose that one of the seven persons in the room asks everyone which beverage he or she would like. In order to keep track of the answers, the responses are recorded on a tally sheet such as the following (note that we need only *two* vertical lines to divide our tally sheet into *three* columns):

Coffee	*Tea*	*Milk*

For example, an order for 4 coffees, 1 tea, and 2 milks would be recorded as follows:

Coffee	*Tea*	*Milk*
xxxx	x	xx

If we always list the beverages in this sequence, the beverage names can be omitted from the tally sheet because every order corresponds uniquely to some arrangement of seven x's and two |'s. For example, the order for 4 coffees, 1 tea, and 2 milks would be represented as xxxx|x|xx, and an order for 5 coffees, 2 teas, and 0 milks would appear as xxxxx|xx|. Hence the number of different refreshment orders is the same as the number of ways to arrange seven x's and two |'s, or equivalently, the number of ways to choose positions for seven x's from nine possible locations. Thus there are $C(9, 7) = 36$ different refreshment orders possible. (Since $C(9, 2) = C(9, 7) = 36$, we can also interpret the number of different refreshment orders as the number of ways to choose positions for two |'s from among nine positions.)

By using the same type of reasoning, we obtain the result in Theorem 8.8, in which s denotes the number of *s*elections and t denotes the number of *t*ypes of objects from which to choose. (In the beverage example, $s = 7$ and $t = 3$.)

Theorem 8.8 If repetition is allowed, the number of selections of s elements that can be made from a set containing t distinct elements is $C(s + t - 1, s)$.

Example 8.27 Suppose that we take five coins from a piggy bank containing many pennies, nickels, and quarters. How many different amounts of money might we get?

Note that, because we are selecting five coins, each possible choice of coins corresponds to a different amount of money. (This would not be the case if we selected six coins, for 1 quarter and 5 pennies have the same value as 6 nickels.) Thus, by Theorem 8.8, the answer to this question is

$$C(5+3-1,5) = C(7,5) = 21. \quad \blacksquare$$

Example 8.28 A bakery makes four different types of donuts.

(a) How many different assortments of one dozen donuts can be purchased?

(b) How many different assortments of one dozen donuts can be purchased that include at least one donut of each type?

(a) Since we are selecting 12 donuts from 4 types with repetition of the types allowed, we use Theorem 8.8 with $s = 12$ and $t = 4$. The number of possible choices is

$$C(s+t-1,s) = C(12+4-1,12) = C(15,12) = 455.$$

(b) Because at least one donut of each type must be included, let us begin by choosing one donut of each type. The number of possible assortments is then the number of different ways the remaining 8 donuts can be selected. As in (a), this number is

$$C(s+t-1,s) = C(8+4-1,8) = C(11,8) = 165. \quad \blacksquare$$

Example 8.29 How many 8-bit strings with exactly two 1s are such that the 1s are not adjacent?

The strings to be counted consist of two 1s and six 0s. Arrange the two 1s in a line. In order that the 1s not be adjacent, we insert a 0 between them. The present configuration is

$$1 \quad 0 \quad 1.$$

The string will be completely determined if we know the numbers of 0s before the first 1, between the two 1s, and after the second 1. Thus the number of different strings with the desired form is equal to the number of different ways to place the remaining five 0s into three positions. But the number of different ways to place the remaining five 0s into three positions equals the number of ways to choose 5 times with repetition from among 3 types of positions, which is

$$C(5+3-1,5) = C(7,5) = 21.$$

Compare this solution with that in Example 8.24. ■

Counting problems involving the distribution of objects can be interpreted as problems involving arrangements or selection with repetition. Usually, *problems involving the distribution of distinct objects correspond to arrangements with repetition, and problems involving the distribution of identical objects correspond to selections with repetition.* The following examples demonstrate the use of Theorems 8.7 and 8.8 in solving problems involving distributions:

Example 8.30 How many distributions of 10 different books are possible if Carlos is to receive 5 books, Doris is to receive 3 books, and Earl is to receive 2 books?

Distributing the 10 books is equivalent to lining them up in some order and inserting a piece of paper in each book marked with the recipient's name. Then the number of possible distributions is the same as the number of ways of arranging 5 slips of paper marked "Carlos," 3 slips marked "Doris," and 2 slips marked "Earl." Using Theorem 8.7, we see that this number is

$$\frac{10!}{5!\,3!\,2!} = 2520.$$

Note the similarity between this solution and that of Example 8.26. ■

Example 8.31 If 9 red balloons and 6 blue balloons are to be distributed to 4 children, how many distributions are possible if every child must receive a balloon of each color?

Let us distribute the red balloons first and the blue balloons second. Since every child must receive a red balloon, we give one to each child. Now we can distribute the remaining 5 red balloons in any way whatsoever. To decide who will receive each of these 5 balloons, we will think of selecting five times with repetition from a set containing the children's names. The number of possible selections is given by Theorem 8.8 to be $C(5+4-1, 5) = C(8, 5)$. Similar reasoning shows that the number of ways in which the blue balloons can be distributed so that every child receives at least one is $C(2+4-1, 2) = C(5, 2)$. Thus, by the multiplication principle, the number of possible distributions of the balloons in which every child receives a balloon of each color is

$$C(8, 5) \cdot C(5, 2) = 56 \cdot 10 = 560. \quad ■$$

In Section 8.3 we posed two basic counting problems:

(1) How many different arrangements (ordered lists) of r objects can be formed from a set of n distinct objects?

(2) How many different selections (unordered lists) of r objects can be formed from a set of n distinct objects?

Theorem 8.8 provides the answer to question (2) when repetition of items is permitted. The answer to question (1) in this case follows easily from the multiplication principle, for there are r objects to be chosen, and each object can be chosen in one of n ways. Hence the total number of arrangements of r objects that can be formed from a set of n distinct objects when repetition of items is permitted is

$$\underbrace{n \cdot n \cdots n}_{r \text{ factors}} = n^r.$$

The following chart summarizes the answers to the two previous questions:

Repetition of Items	*Number of Arrangements (ordered lists)*	*Number of Selections (unordered lists)*
Not permitted	$P(n, r)$	$C(n, r)$
Permitted	n^r	$C(n + r - 1, r)$

Note that in this context Theorem 8.7 gives the number of arrangements with repetition *when the number of items of each type is specified.*

EXERCISES 8.4

1. How many distinct arrangements of the letters in "redbird" are there?

2. How many distinct arrangements of the letters in "economic" are there?

3. How many different 7-digit numbers can be formed by using the digits in the number 5,363,565?

4. How many different 9-digit numbers can be formed by using the digits in the number 277,728,788?

5. How many different fruit baskets containing 8 pieces of fruit can be formed by using only apples, oranges, and pears?

6. How many different assortments of 6 boxes of cereal can be made by using packages of corn flakes, shredded wheat, and bran flakes?

7. How many different assortments of one dozen donuts can be purchased from a bakery that makes donuts with chocolate, vanilla, cinnamon, powdered sugar, and glazed icing?

8. How many different boxes containing 10 wedges of cheese can be made by using wedges of Cheddar, Edam, Gouda, and Swiss cheese?

9. A box contains 16 crayons, no 2 having the same color. In how many different ways can they be given to four children so that each child receives 4 crayons?

10. In how many different ways can 15 distinct books be distributed so that Carol receives 6, Don receives 4, and Ellen receives 5?

11. A committee's chairperson and secretary must telephone the other 7 members about a change in the committee's meeting time. In how many different ways can these telephone calls be made if the chairperson calls 3 people and the secretary calls 4?

12. Paula has bought 6 different CDs to give as Christmas gifts. In how many different ways can she distribute the CDs so that each of her three boyfriends receives 2 CDs?

13. In how many different ways can 8 identical pieces of construction paper be distributed to 4 children?

14. In how many different ways can 10 identical quarters be distributed to 5 people?

15. In how many different ways can 6 identical sticks of white chalk be distributed to 3 students so that each student receives at least one stick?

16. A father has 10 identical life insurance policies. He wants to name one of his 3 children as the beneficiary of each policy. In how many different ways can the beneficiaries be chosen if each child is to be named a beneficiary on at least 2 policies?

17. A concert pianist is preparing a recital that will consist of 1 Baroque piece, 3 classical pieces, and 3 romantic pieces. Assuming, for the sake of programming, that pieces of the same period are regarded as indistinguishable, how many different programs containing the 7 pieces can the pianist create?

18. In bridge a deal consists of distributing a 52-card deck into four 13-card hands. How many different deals are possible in bridge?

19. In how many different ways can 8 identical mathematics books and 10 identical computer science books be distributed among 6 students?

20. Twelve children are to be divided into groups of three to play different number games. In how many ways can the groups be chosen?

21. Ten diplomats are awaiting assignments to foreign embassies. If 3 of these diplomats are to be assigned to England, 4 to France, and 3 to Germany, in how many ways can the assignments be made?

22. In order to stagger the terms of service of 12 people elected to a new committee, 4 members are to be assigned a one-year term, 4 members are to be assigned a two-year term, and 4 members are to be assigned a three-year term. In how many different ways can these assignments be made?

23. How many 16-bit strings are there containing six 0s and ten 1s, with no consecutive 0s?

24. How many positive integer solutions are there to the equation $x + y + z = 17$?

25. In how many ways can 2 identical teddy bears and 7 distinct Cabbage Patch dolls be distributed to 3 children so that each child receives 3 gifts? (It is possible for the 3 gifts to include both teddy bears.)

26. How many numbers greater than 50,000,000 can be formed by rearranging the digits of the number 13,979,397?

27. How many positive integers less than 10,000 are such that the sum of their digits is 8?

28. How many distinct arrangements are there of two a's, one e, one i, one o, and seven x's in which no two vowels are adjacent?

29. How many positive integers less than 1,000,000 are such that the sum of their digits equals 12?

30. A domino contains two indistinguishable squares, each of which is marked with 0, 1, 2, 3, 4, 5, or 6 dots. How many different dominoes are possible?

31. In the following segment of a computer program, how many times is the PRINT statement executed?

```
FOR I: = 1 TO 10
    FOR J: = 1 TO I
        FOR K: = 1 TO J
            PRINT I, J, K
        NEXT K
    NEXT J
NEXT I
```

32. A pouch contains \$1 in pennies, \$1 in nickels, and \$1 in dimes. In how many different ways can 12 coins be selected from this pouch? (Assume that all coins of the same value are indistinguishable.)

33. A pinochle deck consists of 2 each of 24 different cards. How many different 12-card pinochle hands are possible?

34. If $m \geq n$, how many different ways are there to distribute m indistinguishable balls into n distinguishable urns with no urn left empty?

35. Prove that the two expressions in Theorem 8.7 are equal.

36. Use Exercise 31 of Section 8.1 to prove Theorem 8.8 by induction on t.

8.5 PROBABILITY

The subject of probability is generally accepted as having begun in 1654 with an exchange of letters between the great French mathematicians Blaise Pascal and Pierre de Fermat. During the next 200 years, probability was combined with statistics to form a unified theory of mathematical statistics, and it is in this context that any thorough discussion of probability must occur. Nevertheless, the history of probability is closely related to the history of combinatorics, the branch of mathematics concerned with counting. In this section we will discuss probability as an application of the combinatorial ideas presented in Sections 8.2, 8.3, and 8.4.

Intuitively, probability measures how likely something is to occur. In his important book *Theorie Analytique des Probabilités*, the French mathematician Pierre Simon Laplace (1749–1827) defined probability as follows: Probability is the ratio of the number of favorable cases to the total number of cases, assuming that all of the various cases are equally possible. Thus, according to Laplace's definition, probability measures the frequency with which a favorable case occurs. In this book we will study probability only in situations where this definition applies. Note that this definition requires that we know the number of favorable cases and the total number of cases and, therefore, requires the use of counting techniques.

By an **experiment**, we will mean any procedure that results in an observable outcome. Thus we may speak of the experiment of flipping a coin (and observing if it falls heads or tails) or the experiment of tossing a die (and noting the number of spots that show). A set consisting of all the possible outcomes of an experiment is called a **sample space** for the experiment. It is important to realize that there may be many possible sample spaces for an experiment. In the experiment of tossing an evenly balanced die, for instance, three possible sample spaces are

$$\{1, 2, 3, 4, 5, 6\}, \quad \{\text{even}, \text{odd}\}, \quad \text{and} \quad \{\text{perfect square}, \text{not a perfect square}\}.$$

Which of these sample spaces may be most useful depends on the particular type of outcomes that we wish to consider. But in order to use Laplace's definition of probability, we must be certain that the outcomes in the sample space are all equally likely to occur. This is the case for the outcomes in the first two sample spaces just discussed, but the outcomes in the third sample space are not equally likely, since there are only two perfect squares among the numbers 1 through 6 (namely, 1 and 4). Thus the sample space {perfect square, not a perfect square} will not prove useful for computing probabilities.

Any subset of a sample space is called an **event**. Thus, in the die-tossing experiment with sample space $\{1, 2, 3, 4, 5, 6\}$, the following sets are events:

$$A = \{1, 2, 4, 6\}, \qquad B = \{n\colon\ n \text{ is an integer and } 4 < n \le 6\}, \qquad \text{and}$$
$$C = \{n\colon\ n \text{ is an even positive integer less than } 7\}.$$

Recall that the number of elements in a finite set X is denoted $|X|$. For any event E in a finite sample space S consisting of equally likely outcomes, we define the **probability** of E, denoted $P(E)$, by

$$P(E) = \frac{|E|}{|S|}. \tag{8.3}$$

So, for the preceding events A, B, and C, we have

$$P(A) = \frac{4}{6} = \frac{2}{3}, \qquad P(B) = \frac{2}{6} = \frac{1}{3}, \qquad \text{and} \qquad P(C) = \frac{3}{6} = \frac{1}{2}.$$

Example 8.32 In the experiment of flipping a properly balanced coin three times, what is the probability of obtaining exactly two heads?

Since each flip of the coin has two possible results, heads (H) or tails (T), the multiplication principle shows that there are $2 \cdot 2 \cdot 2 = 8$ possible outcomes for three flips. The set

$$S = \{\text{HHH, HHT, HTH, HTT, THH, THT, TTH, TTT}\}$$

is a sample space for this experiment consisting of equally likely outcomes. The event of obtaining exactly two heads is the set $E = \{\text{HHT, HTH, THH}\}$. Thus the probability of obtaining exactly two heads is

$$P(E) = \frac{|E|}{|S|} = \frac{3}{8}. \quad \blacksquare$$

In Example 8.32, we obtained the desired probability by listing the outcomes in a sample space of equally likely outcomes. Usually, however, a sample space will be so large that we must use counting techniques to determine its size. Examples 8.33 through 8.37 are of this type. Notice that in these examples we begin by determining the sample space before counting the outcomes in the event of interest.

Example 8.33 Suppose that there are six applicants for a particular job, four men and two women, who are to be interviewed in a random order. What is the probability that the four men are interviewed before either woman?

To answer this question, we must decide on an appropriate sample space consisting of equally likely outcomes. Since the ordering of the interviews is important, the set S of all possible arrangements of the six interviews is the obvious choice. Let E denote the subset of S in which the men are interviewed before the women. Then the multiplication principle shows that the number of elements in E equals the number of arrangements of the men times the number of arrangements of the women. Thus, by (8.3), we have

$$P(E) = \frac{|E|}{|S|} = \frac{P(4,4) \cdot P(2,2)}{P(6,6)} = \frac{24 \cdot 2}{720} = \frac{1}{15}. \quad \blacksquare$$

Example 8.34 Suppose that there are 2 defective pens in a box of 12 pens. If we choose 3 pens at random, what is the probability that we do not select a defective pen?

In this problem, the set of all selections of 3 pens chosen from among the 12 in the box is the sample space S. The set of all selections of 3 pens chosen from among the 10 nondefective pens is the event E in which we are interested. Thus, by (8.3), we find that

$$P(E) = \frac{|E|}{|S|} = \frac{C(10,3)}{C(12,3)} = \frac{120}{220} = \frac{6}{11}. \quad \blacksquare$$

Example 8.35 What is the probability that a randomly chosen permutation of the letters in the word "computer" has no adjacent vowels?

Let the sample space be the set S of all permutations of the letters in the word "computer," and let E denote the subset of all such permutations in which no two vowels are adjacent.

To count the permutations in E, we first arrange the five consonants in one of $P(5,5) = 120$ ways, say,

$$\_\,\mathrm{p}\,\_\,\mathrm{t}\,\_\,\mathrm{c}\,\_\,\mathrm{r}\,\_\,\mathrm{m}\,\_.$$

Since no two vowels are adjacent, we must insert at most one vowel in each blank. The number of ways to choose positions for the three vowels is, therefore, $C(6,3) = 20$. Finally, we arrange the vowels in the chosen positions in $P(3,3) =$ 6 ways. Thus E contains 120(20)(6) = 14,400 permutations, so

$$P(E) = \frac{|E|}{|S|} = \frac{14{,}400}{P(8,8)} = \frac{14{,}400}{40{,}320} = \frac{5}{14}. \quad \blacksquare$$

Example 8.36 Suppose that we have 10 different novels, five by Hemingway and five by Faulkner, that we want to distribute so that Barbara receives 5, Cathy receives 2, and Danielle receives 3. If the individual novels are distributed at random, what is the probability that Barbara receives all five of the novels by Hemingway?

Here, the sample space S is the set of all distributions of 5 novels to Barbara, 2 to Cathy, and 3 to Danielle, and the event of interest is the set E of all such distributions in which Barbara receives all the Hemingway novels. Note that the distributions in which Barbara receives all the Hemingway novels are just those in which the Faulkner novels are distributed so that Cathy receives 2 and Danielle receives 3. So, by reasoning as in Example 8.30, we see that

$$P(E) = \frac{|E|}{|S|} = \frac{\left(\dfrac{5!}{2!\,3!}\right)}{\left(\dfrac{10!}{5!\,2!\,3!}\right)} = \frac{5!\,5!}{10!} = \frac{1}{252}. \quad \blacksquare$$

Example 8.37 We will compute the probability of being dealt each of the following hands if 5 cards are dealt from an ordinary 52-card deck:

(a) a flush (5 cards of the same suit), and

(b) a full house (3 cards of one denomination and 2 of another denomination).

In each case, the sample space S consists of all possible five-card hands. The number of these is

$$C(52, 5) = 2{,}598{,}960.$$

(a) We will count the number of different flushes. To obtain a flush, we must first choose a suit and then select 5 cards from that suit. Hence the multiplication principle shows that the number of different flushes is

$$C(4, 1) \cdot C(13, 5) = 4(1287) = 5148.$$

It follows that the probability of being dealt a flush is

$$\frac{5148}{2{,}598{,}960} \approx .00198.$$

(b) As in (a), we will count the number of possible full houses. To obtain a full house, we must choose a denomination, pick 3 cards of that denomination, select a different denomination, and pick 2 cards of that denomination. The number of possible full houses is

$$C(13, 1) \cdot C(4, 3) \cdot C(12, 1) \cdot C(4, 2) = 13 \cdot 4 \cdot 12 \cdot 6 = 3744.$$

Hence the probability of being dealt a full house is

$$\frac{3744}{2{,}598{,}960} \approx .00144.$$

Since the probability of obtaining a full house is less than the probability of obtaining a flush, a full house ranks higher than a flush in poker. ■

Because our definition of probability requires that the sample space consist of *equally likely* outcomes, care must be taken in using Theorem 8.8. For example, suppose that six identical cookies are to be distributed at random to three children. What is the probability that each child gets exactly two?

In this problem we must consider what "at random" means. Presumably, the first cookie is given to one of the children, with each child equally likely to get it, then the second cookie, etc. Thus the sample space S consists of all 6-element lists with entries chosen from the set $\{1, 2, 3\}$. For example, the list 2, 1, 3, 3, 3, 1 corresponds to giving the first cookie to child 2, the second cookie to child 1, etc. By the multiplication principle, $|S| = 3^6$.

The event E consists of all rearrangements of the list 1, 1, 2, 2, 3, 3; so

$$|E| = \frac{6!}{2!\,2!\,2!} = 90.$$

Thus the probability that each child receives two cookies is

$$P(E) = \frac{90}{3^6} = \frac{10}{81}.$$

Notice that the preceding analysis treats the six cookies individually (first cookie, second cookie, ...), despite their being identical. The wrong answer is obtained if the sample space S is taken to be all ways of distributing 6 indistinguishable objects into 3 sets C_1, C_2, and C_3. Then $|S| = C(6+3-1, 6) = 28$, by Theorem 8.8, and $|E| = 1$, so that

$$\frac{|E|}{|S|} = \frac{1}{28}.$$

The reason that this quotient is not $P(E)$ is that the elements of the sample space are not equally likely. For instance, a distribution in which each child receives two cookies is 90 times more likely than that the first child gets all six cookies.

EXERCISES 8.5

1. In the experiment of rolling a die, what is the probability of rolling a number greater than 1?
2. In the experiment of rolling a die, what is the probability of rolling a number divisible by 3?
3. If a coin is tossed five times, what is the probability that it will land heads each time?
4. If three dice are rolled, what is the probability that a 1 will appear on each die?
5. If a pair of dice is rolled, what is the probability that the sum of the spots that appear is 11?
6. If 4 coins are tossed, what is the probability that all of them land with the same side up?
7. If 5 coins are tossed, what is the probability that exactly 3 of them land tails?
8. If a coin is tossed 8 times, what is the probability that it will land heads exactly 4 times?
9. If 3 persons are chosen at random from a set of 5 men and 6 women, what is the probability that 3 women are chosen?
10. Suppose that a 4-digit number is created by using the digits 1, 2, 3, 4, and 5 as often as desired. What is the probability that it contains two 1s and two 4s?
11. In a 7-horse race, a bettor bet the trifecta, which requires that the first 3 horses be identified in order of their finish. What is the probability of winning the trifecta under these conditions by randomly guessing 3 horses?
12. If 4 persons are chosen at random from a class containing 8 freshmen and 12 sophomores, what is the probability that 4 freshmen are chosen?

13. What is the probability that a randomly chosen four-digit number contains no repeated digits?

14. What is the probability that a randomly chosen string of three letters contains no repeated letters?

15. If the letters of "sassafras" are randomly permuted, what is the probability that the four s's are adjacent and the three a's are adjacent?

16. In a consumer preferences test, 10 people were asked to name their favorite fruit from among apples, bananas, and oranges. If each person named a fruit at random, what would be the probability that no one named bananas?

17. If the personnel files of 5 employees are randomly selected, what is the probability that they are chosen in order of increasing salary? (Assume that no two employees have the same salary.)

18. In a particular group of people, 10 are right-handed and 4 are left-handed. If 5 of these people are chosen at random, what is the probability that exactly 1 left-handed person is selected?

19. What is the probability that a randomly chosen subset of $\{1, 2, 3, 4, 5, 6\}$ contains both 3 and 5?

20. A committee of 5 is to be formed from among 2 mathematics teachers, 2 English teachers, 2 science teachers, and 2 humanities teachers. If all such committees are equally likely, what is the probability that the committee contains at least 1 English teacher?

21. Thirteen sticks of chewing gum are to be given at random to 3 children. What is the probability that each child receives at least 4 sticks of gum?

22. Three \$10 bills, four \$5 bills, and six \$1 bills are randomly arranged in a stack. What is the probability that all of the \$5 bills are adjacent?

23. If a 5-member committee is selected at random from among 7 faculty and 6 students, what is the probability that it contains exactly 3 faculty and 2 students?

24. Suppose that we randomly distribute 5 distinct Cabbage Patch dolls and 3 identical teddy bears to 4 children. What is the probability that each child receives 2 gifts?

25. In a small garden, there is a row of 8 tomato plants, 3 of which are diseased. Assuming that the disease occurs at random in the plants, what is the probability that the 3 diseased plants are all adjacent?

26. If 10 quarters are distributed at random to 4 people, what is the probability that everyone receives at least 50 cents?

27. Each of 9 different books is to be given at random to Rebecca, Sheila, or Tom. What is the probability that Rebecca receives 2 books, Sheila receives 4, and Tom receives 3?

28. What is the probability that an odd number between 1000 and 9000 contains no repeated digits?

29. What is the probability that a randomly chosen permutation of the letters in the word "determine" has no adjacent e's?

30. Exactly 4 of 20 microcomputer diskettes are defective. If the diskettes are packaged in two boxes of 10, what is the probability that

(a) all the defective diskettes are packed in a particular box?

(b) 3 defective diskettes are packed in the same box?

(c) 2 defective diskettes are packed in each box?

In Exercises 31–34, compute the probability of being dealt each of the given hands if 5 *cards are dealt from an ordinary* 52*-card deck.*

31. a pair (2 cards of one denomination and 1 each of three other denominations)

32. two pairs (2 cards of one denomination, 2 of another denomination, and 1 of a third denomination)

33. three-of-a-kind (3 cards of one denomination and 1 each of two others)

34. a straight (5 cards of consecutive denominations, where an ace is the highest denomination)

35. A file contains 25 accounts numbered 1–25. If 5 of these accounts are selected at random for auditing, what is the probability that no 2 accounts with consecutive numbers are chosen?

36. In the Illinois State Lotto game, 6 of the integers $1, 2, \ldots, 54$ are picked to be the winning numbers. What is the probability that 3 consecutive numbers $n, n+1, n+2$ are picked and no pairs of consecutive numbers other than $n, n+1$ and $n+1, n+2$ are picked?

8.6* THE PRINCIPLE OF INCLUSION–EXCLUSION

The addition principle (Theorem 8.6) tells us how to find the number of elements in the union of pairwise disjoint sets in terms of the number of elements in the individual sets. In this section we present a similar result that enables us to count the number of elements in the union of any sets, whether pairwise disjoint or not.

The following simple example demonstrates the type of counting problem that we are considering. Suppose that a certain group of computer science students are all studying logic or mathematics. If 12 are studying logic, 26 are studying mathematics, and 5 are studying both logic and mathematics, how many students are in this group? If we let A denote the set of students studying logic and B denote the set of students studying mathematics, then the answer to this question is the number of elements in the set $A \cup B$. But since A and B are not disjoint, the addition principle cannot be used directly. It is not difficult, however, to see that the set B' of students studying mathematics, but not logic, contains $26 - 5$ elements. Now A and B' are disjoint and contain all of the students in the group. So the answer to our question is the number of elements in $A \cup B'$, which is $12 + (26 - 5)$ by the addition principle. (See Figure 8.3.)

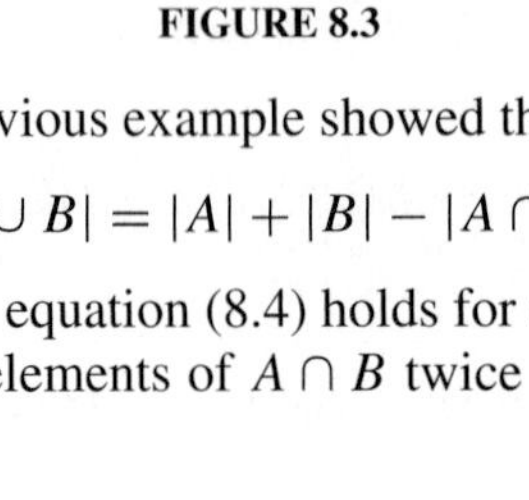

FIGURE 8.3

Our analysis in the previous example showed that

$$|A \cup B| = |A| + |B| - |A \cap B|. \tag{8.4}$$

It is not difficult to see that equation (8.4) holds for any finite sets A and B: The sum $|A| + |B|$ counts the elements of $A \cap B$ twice (once as members of A and once as members of B); so $|A| + |B| - |A \cap B|$ counts each element of $A \cup B$ exactly once.

Example 8.38 In Example 8.14 we used the addition principle to count the number of 8-bit strings that begin with 1011 or end with 01. Let us count them again by using (8.4).

Let A and B denote the sets of 8-bit strings that begin with 1011 and end with 01, respectively. Then $A \cap B$ is the set of strings that begin with 1011 and end with 01—that is, strings of the form 1011 – – 01. Since only the fifth and sixth bits are unspecified, the number of such strings is $2 \cdot 2 = 4$. But since $|A| = 2^4 = 16$ and $|B| = 2^6 = 64$ from Example 8.14, it follows from (8.4) that the number of 8-bit strings that begin with 1011 or end with 01 is

$$|A \cup B| = |A| + |B| - |A \cap B| = 16 + 64 - 4 = 76. \quad \blacksquare$$

Our objective in this section is to generalize (8.4) from two sets to r sets, $A_1, A_2, \ldots, A_r$. But let us first consider the case that $r = 3$. It is easy to see in Figure 8.4 that $(A_1 \cup A_2) \cap A_3 = (A_1 \cap A_3) \cup (A_2 \cap A_3)$. By using this fact and (8.4), we can obtain a formula for $|A_1 \cup A_2 \cup A_3|$ as follows:

$$\begin{aligned}
&|A_1 \cup A_2 \cup A_3| \\
&\quad = |(A_1 \cup A_2) \cup A_3| = |A_1 \cup A_2| + |A_3| - |(A_1 \cup A_2) \cap A_3| \\
&\quad = (|A_1| + |A_2| - |A_1 \cap A_2|) + |A_3| - |(A_1 \cap A_3) \cup (A_2 \cap A_3)| \\
&\quad = (|A_1| + |A_2| - |A_1 \cap A_2|) + |A_3| - (|A_1 \cap A_3| + |A_2 \cap A_3| - |A_1 \cap A_3 \cap A_2 \cap A_3|) \\
&\quad = |A_1| + |A_2| + |A_3| - |A_1 \cap A_2| - |A_1 \cap A_3| - |A_2 \cap A_3| + |A_1 \cap A_2 \cap A_3| \\
&\quad = (|A_1| + |A_2| + |A_3|) - (|A_1 \cap A_2| + |A_1 \cap A_3| + |A_2 \cap A_3|) + |A_1 \cap A_2 \cap A_3|.
\end{aligned}$$

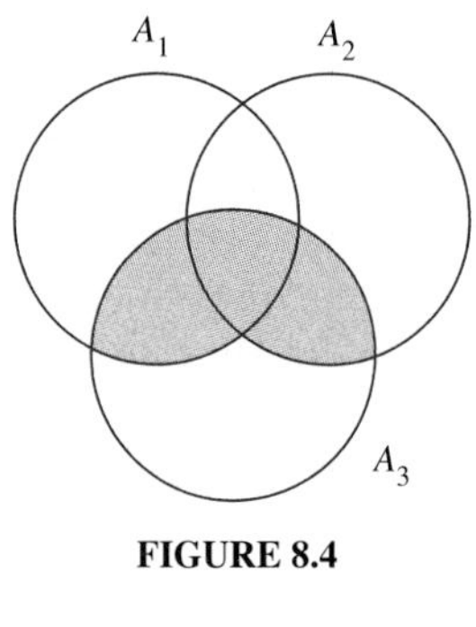

FIGURE 8.4

In order to generalize (8.4) to r sets, let us define n_s for $1 \le s \le r$ to be the sum of the sizes of all possible intersections of s sets chosen without repetition from among $A_1, A_2, \ldots, A_r$. (For $s = 1$, we define the "intersection" of a single set to be the set itself. Thus $n_1 = |A_1| + |A_2| + \cdots + |A_r|$.) Note that, since there are $C(r, s)$ ways to choose s sets from among $A_1, A_2, \ldots, A_r$, *each* n_s *is the sum of* $C(r, s)$ *terms.*

If $r = 3$, so that there are only three sets A_1, A_2, and A_3, we have

$$\begin{aligned} n_1 &= |A_1| + |A_2| + |A_3|, \\ n_2 &= |A_1 \cap A_2| + |A_1 \cap A_3| + |A_2 \cap A_3|, \text{ and} \\ n_3 &= |A_1 \cap A_2 \cap A_3|. \end{aligned}$$

With this notation, the formula just derived can be written

$$|A_1 \cup A_2 \cup A_3| = n_1 - n_2 + n_3.$$

Likewise, if $r = 4$ (there are four sets A_1, A_2, A_3, and A_4), we have

$$\begin{aligned} n_1 &= |A_1| + |A_2| + |A_3| + |A_4|, \\ n_2 &= |A_1 \cap A_2| + |A_1 \cap A_3| + |A_1 \cap A_4| + |A_2 \cap A_3| + |A_2 \cap A_4| + |A_3 \cap A_4|, \\ n_3 &= |A_1 \cap A_2 \cap A_3| + |A_1 \cap A_2 \cap A_4| + |A_1 \cap A_3 \cap A_4| + |A_2 \cap A_3 \cap A_4|, \text{ and} \\ n_4 &= |A_1 \cap A_2 \cap A_3 \cap A_4|. \end{aligned}$$

In this case, it can be shown that

$$|A_1 \cup A_2 \cup A_3 \cup A_4| = n_1 - n_2 + n_3 - n_4.$$

Then the desired generalization of (8.4) can be stated as follows:

Theorem 8.9 ***The Principle of Inclusion–Exclusion*** For any finite sets $A_1, A_2, \ldots, A_r$, define n_s for $1 \le s \le r$ to be the sum of the sizes of all possible intersections of s sets chosen without repetition from among $A_1, A_2, \ldots, A_r$. Then

$$|A_1 \cup A_2 \cup \cdots \cup A_r| = n_1 - n_2 + n_3 - n_4 + \cdots + (-1)^{r-1} n_r.$$

Proof. Let $m = |A_1 \cup A_2 \cup \cdots \cup A_r|$. We will show that

$$m - n_1 + n_2 - n_3 + \cdots + (-1)^r n_r = 0.$$

Let $a \in A_1 \cup A_2 \cup \cdots \cup A_r$, and suppose that a belongs to exactly k of the sets A_i. Then a is counted $C(k, 0) = 1$ time in m, $C(k, 1) = k$ times in n_1 (because a belongs to exactly k of the sets A_i), $C(k, 2)$ times in n_2 (because a belongs to exactly $C(k, 2)$ of the intersections $A_i \cap A_j$), $\ldots$, and $C(k, k) = 1$ time in n_k. Furthermore, if $s > k$, then a is not counted at all in n_s, because a does not belong to any intersection of more than k of the sets A_i. Hence the number of times that a is counted in $m - n_1 + n_2 - n_3 + \cdots + (-1)^r n_r$ is

$$C(k, 0) - C(k, 1) + C(k, 2) - C(k, 3) + \cdots + (-1)^k C(k, k).$$

But this value is $[1 + (-1)]^k = 0^k = 0$ by the binomial theorem. It therefore follows that

$$m = n_1 - n_2 + n_3 - n_4 + \cdots + (-1)^{r-1} n_r. \quad \blacksquare$$

Example 8.39 Among a group of programmers, 49 studied Pascal, 37 studied COBOL, and 21 studied FORTRAN. If 9 of these programmers studied Pascal and COBOL, 5 studied Pascal and FORTRAN, 4 studied COBOL and FORTRAN, and 3 studied Pascal, COBOL, and FORTRAN, how many programmers are in this group?

Let us denote the sets of programmers who studied Pascal, COBOL, and FORTRAN by P, C, and F, respectively (instead of A_1, A_2, and A_3). Then the number of programmers in the group is $|P \cup C \cup F|$. Now,

$$n_1 = |P| + |C| + |F| = 49 + 37 + 21 = 107,$$
$$n_2 = |P \cap C| + |P \cap F| + |C \cap F| = 9 + 5 + 4 = 18,$$

and

$$n_3 = |P \cap C \cap F| = 3.$$

So, by the principle of inclusion–exclusion, we have

$$|P \cup C \cup F| = n_1 - n_2 + n_3 = 107 - 18 + 3 = 92.$$

Hence there are 92 programmers in this group. ■

Example 8.40 How many positive integers less than 2101 are divisible by at least one of the primes 2, 3, 5, or 7?

Let A_1, A_2, A_3, and A_4 denote the sets of positive integers less than 2101 that are divisible by 2, 3, 5, and 7, respectively, and let n_s be defined as in the principle of inclusion–exclusion. Clearly,

$$|A_1| = \frac{2100}{2} = 1050, \qquad |A_2| = \frac{2100}{3} = 700,$$
$$|A_3| = \frac{2100}{5} = 420, \qquad \text{and} \qquad |A_4| = \frac{2100}{7} = 300.$$

Thus

$$\begin{aligned} n_1 &= |A_1| + |A_2| + |A_3| + |A_4| \\ &= 1050 + 700 + 420 + 300 \\ &= 2470. \end{aligned}$$

An element of $A_1 \cap A_2$ is divisible by both 2 and 3 and hence is divisible by 6. Therefore

$$|A_1 \cap A_2| = \frac{2100}{6} = 350,$$

and likewise,

$$|A_1 \cap A_3| = \frac{2100}{10} = 210, \qquad |A_1 \cap A_4| = \frac{2100}{14} = 150,$$
$$|A_2 \cap A_3| = \frac{2100}{15} = 140, \qquad |A_2 \cap A_4| = \frac{2100}{21} = 100,$$

and

$$|A_3 \cap A_4| = \frac{2100}{35} = 60.$$

Thus

$$\begin{aligned} n_2 &= |A_1 \cap A_2| + |A_1 \cap A_3| + |A_1 \cap A_4| + |A_2 \cap A_3| + |A_2 \cap A_4| + |A_3 \cap A_4| \\ &= 350 + 210 + 150 + 140 + 100 + 60 \\ &= 1010. \end{aligned}$$

Similar reasoning shows that

$$|A_1 \cap A_2 \cap A_3| = \frac{2100}{30} = 70, \quad |A_1 \cap A_2 \cap A_4| = \frac{2100}{42} = 50,$$

$$|A_1 \cap A_3 \cap A_4| = \frac{2100}{70} = 30, \quad \text{and} \quad |A_2 \cap A_3 \cap A_4| = \frac{2100}{105} = 20.$$

Hence

$$\begin{aligned} n_3 &= |A_1 \cap A_2 \cap A_3| + |A_1 \cap A_2 \cap A_4| + |A_1 \cap A_3 \cap A_4| + |A_2 \cap A_3 \cap A_4| \\ &= 70 + 50 + 30 + 20 \\ &= 170. \end{aligned}$$

Finally, we see that

$$n_4 = |A_1 \cap A_2 \cap A_3 \cap A_4| = \frac{2100}{210} = 10.$$

Thus, by the principle of inclusion–exclusion, the number of positive integers less than 2101 that are divisible by 2, 3, 5, or 7 is

$$\begin{aligned} |A_1 \cup A_2 \cup A_3 \cup A_4| &= n_1 - n_2 + n_3 - n_4 \\ &= 2470 - 1010 + 170 - 10 \\ &= 1620. \quad \blacksquare \end{aligned}$$

In many problems, there is a symmetry that makes the calculation of the numbers n_s easier than in Example 8.40. The following example is of this type:

Example 8.41 A bridge hand consists of 13 cards chosen from a standard 52-card deck. How many different bridge hands contain a void suit (that is, no cards in some suit)?

Let A_1, A_2, A_3, and A_4 denote the sets of bridge hands that contain no spades, no hearts, no diamonds, and no clubs, respectively. Then the number of bridge hands that contain a void suit is

$$|A_1 \cup A_2 \cup A_3 \cup A_4|.$$

Let n_s be defined as in the principle of inclusion–exclusion.

Since a bridge hand that contains no spades must consist of 13 cards chosen from among the 39 hearts, diamonds, and clubs, we see that

$$|A_1| = C(39, 13).$$

By the symmetry of the definition of the sets A_i, we see that

$$|A_1| = |A_2| = |A_3| = |A_4|,$$

so

$$\begin{aligned} n_1 &= |A_1| + |A_2| + |A_3| + |A_4| \\ &= C(4, 1) \cdot |A_1| \\ &= 4 \cdot C(39, 13). \end{aligned}$$

Likewise, a hand that is void in both spades and hearts must consist of 13 cards chosen from among the 26 diamonds and clubs; so $|A_1 \cap A_2| = C(26, 13)$. Again, by symmetry, all the sets $A_i \cap A_j$ have the same size. Thus

$$\begin{aligned} n_2 &= |A_1 \cap A_2| + |A_1 \cap A_3| + |A_1 \cap A_4| + |A_2 \cap A_3| + |A_2 \cap A_4| + |A_3 \cap A_4| \\ &= C(4, 2) \cdot |A_1 \cap A_2| \\ &= 6 \cdot C(26, 13). \end{aligned}$$

Similar reasoning shows that a hand that is void in three suits must consist of all the cards from the remaining suit; so

$$n_3 = C(4, 3) \cdot |A_1 \cap A_2 \cap A_3| = 4 \cdot C(13, 13).$$

Finally, no hand can be void in all four suits; so

$$A_1 \cap A_2 \cap A_3 \cap A_4 = \emptyset,$$

and hence $n_4 = 0$.

Therefore, by the principle of inclusion–exclusion, we have

$$\begin{aligned} |A_1 \cup A_2 \cup A_3 \cup A_4| &= n_1 - n_2 + n_3 - n_4 \\ &= 4 \cdot C(39, 13) - 6 \cdot C(26, 13) + 4 \cdot C(13, 13) - 0 \\ &= 4(8{,}122{,}425{,}444) - 6(10{,}400{,}600) + 4(1) - 0 \\ &= 32{,}427{,}298{,}180. \end{aligned}$$

Hence there are 32,427,298,180 different bridge hands containing a void suit. ■

In Examples 8.39–8.41, we were interested in determining the number of elements in

$$A_1 \cup A_2 \cup \cdots \cup A_r,$$

that is, the number of elements that belong to *at least one* of the sets A_i. When the sets A_i are subsets of a set U, we can also use the principle of inclusion–exclusion to find the number of elements in *none* of the sets A_i—that is, the number of elements in

$$\overline{(A_1 \cup A_2 \cup \cdots \cup A_r)} = \overline{A_1} \cap \overline{A_2} \cap \cdots \cap \overline{A_r}.$$

Suppose, for example, that we want to know the number of positive integers less than 2101 that are divisible by none of the numbers 2, 3, 5, or 7. In Example 8.40, we used the principle of inclusion–exclusion to calculate that there were 1620 positive integers less than 2101 that are divisible by at least one of the numbers 2, 3, 5, or 7. Hence the number of positive integers less than 2101 that are divisible by none of these is

$$2100 - 1620 = 480.$$

In the remaining examples of this section, we will illustrate this use of the principle of inclusion–exclusion.

Example 8.42 Among a group of 200 college students, 19 study French, 10 study German, and 28 study Spanish. If 3 study both French and German, 8 study both French and Spanish, 4 study both German and Spanish, and 1 studies French, German, and Spanish, how many of these students are not studying French, German, or Spanish?

Let U denote the set of all 200 students and F, G, and S denote the subsets of U consisting of the students who are studying French, German, and Spanish, respectively. Then the number of students in U who are not studying French, German, or Spanish is $|U| - |F \cup G \cup S|$. Now,

$$n_1 = |F| + |G| + |S| = 19 + 10 + 28 = 57,$$
$$n_2 = |F \cap G| + |F \cap S| + |G \cap S| = 3 + 8 + 4 = 15,$$

and

$$n_3 = |F \cap G \cap S| = 1.$$

Thus, by the principle of inclusion–exclusion,

$$|F \cup G \cup S| = n_1 - n_2 + n_3 = 57 - 15 + 1 = 43.$$

So $200 - 43 = 157$ students are not studying French, German, or Spanish. ■

Example 8.43 At McDonald's restaurants, a Happy Meal box contains one of four possible gifts. If you buy five Happy Meal boxes, what is the probability that you will receive every one of the four gifts?

Let U denote the set of all possible sequences in which five gifts can be obtained, and let A_i denote the subset of U consisting of all the sequences which do not include a gift of type i $(1 \le i \le 4)$. Then we must count the elements of U that are in none of the sets A_i.

Clearly, a sequence in A_1 must consist of only the second, third, and fourth gifts. Therefore $|A_1| = 3^5$, and by symmetry,

$$n_1 = C(4, 1) \cdot |A_1| = 4(3^5) = 4(243) = 972.$$

Likewise, a sequence in $A_1 \cap A_2$ must consist of only the third and fourth gifts. Therefore $|A_1 \cap A_2| = 2^5$, so

$$n_2 = C(4, 2) \cdot |A_1 \cap A_2| = 6(2^5) = 6(32) = 192.$$

Similar reasoning shows that

$$n_3 = C(4, 3) \cdot |A_1 \cap A_2 \cap A_3| = 4(1^5) = 4(1) = 4$$

and

$$n_4 = C(4, 4) \cdot |A_1 \cap A_2 \cap A_3 \cap A_4| = 1(0^5) = 1(0) = 0.$$

Hence the principle of inclusion–exclusion gives

$$\begin{aligned} |A_1 \cap A_2 \cap A_3 \cap A_4| &= n_1 - n_2 + n_3 - n_4 \\ &= 972 - 192 + 4 - 0 \\ &= 784. \end{aligned}$$

It follows that the number of elements of U that are in none of the sets A_i is

$$|U| - |A_1 \cup A_2 \cup A_3 \cup A_4| = 4^5 - 784 = 1024 - 784 = 240.$$

Thus the probability of collecting all four gifts if you buy five Happy Meal boxes is

$$\frac{240}{1024} \approx .234. \quad \blacksquare$$

A permutation of the integers $1, 2, \ldots, n$ such that no integer occupies its natural position is called a **derangement**. So 41532 is a derangement of the integers 1, 2, 3, 4, 5 because 1 is not the first digit, 2 is not the second digit, and so forth. Counting the number of derangements is a famous problem that can be solved by the use of the principle of inclusion–exclusion.

Example 8.44 How many derangements of the integers 1, 2, 3, 4 are there?

Let U denote the set of permutations of 1, 2, 3, 4; and let A_1 denote the set of members of U having a 1 as first digit, A_2 denote the set of members of U having a 2 as second digit, and so forth. Then a derangement of the integers 1, 2, 3, 4 is a member of U that is not in $A_1 \cup A_2 \cup A_3 \cup A_4$.

Note that any permutation in A_1 has the form 1 _ _ _, where the second, third, and fourth digits can be chosen arbitrarily. So the number of such permutations is $P(3, 3)$. Likewise, $|A_2| = |A_3| = |A_4| = P(3, 3)$.

Permutations in $A_1 \cap A_2$ have the form 1 2 _ _, and so there are $P(2, 2)$ of them. Likewise,

$$|A_1 \cap A_3| = |A_1 \cap A_4| = |A_2 \cap A_3| = |A_2 \cap A_4| = |A_3 \cap A_4| = P(2, 2).$$

Similar reasoning shows that

$$|A_1 \cap A_2 \cap A_3| = |A_1 \cap A_2 \cap A_4| = |A_1 \cap A_3 \cap A_4| = |A_2 \cap A_3 \cap A_4| = P(1, 1)$$

and

$$|A_1 \cap A_2 \cap A_3 \cap A_4| = 1.$$

Thus, by the principle of inclusion–exclusion, we have

$$\begin{aligned}|A_1 \cup A_2 \cup A_3 \cup A_4| &= 4 \cdot P(3, 3) - 6 \cdot P(2, 2) + 4 \cdot P(1, 1) - 1 \\ &= 4 \cdot 6 - 6 \cdot 2 + 4 \cdot 1 - 1 = 15.\end{aligned}$$

So the number of derangements of 1, 2, 3, 4 is

$$|U| - |A_1 \cup A_2 \cup A_3 \cup A_4| = P(4, 4) - 15 = 24 - 15 = 9. \quad \blacksquare$$

EXERCISES 8.6

1. In a survey of moviegoers it was found that 33 persons liked films by Bergman and 25 liked films by Fellini. If 18 of these persons liked both directors' films, how many liked films by Bergman or Fellini?
2. Among a group of children, 88 liked pizza and 27 liked Chinese food. If 13 of these children liked both pizza and Chinese food, how many liked pizza or Chinese food?
3. Among the 318 members of a local union, 127 liked their congressional representative and 84 liked their governor. If 53 of these members liked both their congressional representative and their governor, how many of these union members liked neither their congressional representative nor their governor?
4. In a particular dormitory, there are 350 college freshmen. Of these, 312 are taking an English course, and 108 are taking a mathematics course. If 95 of these freshmen are taking courses in both English and mathematics, how many are taking a course in neither English nor mathematics?
5. From a group of 650 residents of a city, the following information was obtained:

   310 were college-educated.
   356 were married.
   328 were homeowners.
   180 were college-educated and married.
   147 were college-educated and homeowners.
   166 were married and homeowners.
   94 were college-educated, married, and homeowners.

   How many of these residents were not college-educated, not married, and not homeowners?

6. In tabulating the 5681 responses to a questionnaire sent to her constituents, a congresswoman found that

3819 favored tax reform.
3307 favored a balanced budget.
2562 favored offshore drilling.
2163 favored tax reform and a balanced budget.
1985 favored tax reform and offshore drilling.
1137 favored a balanced budget and offshore drilling.
984 favored tax reform, a balanced budget, and offshore drilling.

How many of the respondents opposed tax reform, a balanced budget, and offshore drilling?

7. The following data were obtained from the fast-food restaurants in a certain city:

13 served hamburgers.
8 served roast beef sandwiches.
10 served pizza.
5 served hamburgers and roast beef sandwiches.
3 served hamburgers and pizza.
2 served roast beef sandwiches and pizza.
1 served hamburgers, roast beef sandwiches, and pizza.
5 served none of these three foods.

How many fast-food restaurants are there in this city?

8. The following information was found about the residents of a certain retirement community:

38 played golf.
21 played tennis.
56 played bridge.
8 played golf and tennis.
17 played golf and bridge.
13 played tennis and bridge.
5 played golf, tennis, and bridge.
72 did not play golf, tennis, or bridge.

How many residents are there in this retirement community?

9. Eight married couples came to a bridge party. Each woman was randomly paired with a man to be partners for the evening. What is the probability that exactly four husbands were paired with their wives?

10. List all the derangements of 1, 2, 3, 4.

11. While taking a 6-week summer math class, Alison frequently had dinner with seven friends from her hometown. She ate dinner with each friend (exactly) 15 times, every pair of friends 8 times, every set of three friends 6 times, every foursome 5 times, every set of five 4 times, and every set of six 3 times, but she never ate with all seven at once. On how many days did Alison have dinner with none of these friends?

12. How many sequences of five digits (0–9) contain at least one 4 and at least one 7?

13. For the following graph, determine the number of ways to assign one of k colors to the vertices so that no adjacent vertices receive the same color:

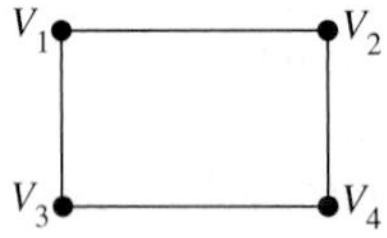

14. If three married couples are seated randomly in six chairs around a circular table, what is the probability that no couple is seated in adjacent seats?

15. How many positive integers less than 101 are square free—that is, divisible by no perfect square greater than 1?

16. How many sequences of six digits (0–9) contain at least one 3, at least one 5, and at least one 8?

17. The *sieve of Eratosthenes* is an ancient method for finding prime numbers in a list of integers $2, 3, \ldots, n$. First, cross from the list every multiple of 2 greater than 2. Then, cross from the list every multiple of the next prime (3) greater than that prime. Continue this process until no further crossing out is possible. The remaining integers are primes. Here is what the sieve looks like after crossing out multiples of 2 and 3 from the list $2, 3, \ldots, 20$:

	2	3	~~4~~	5	~~6~~	7	~~8~~	~~9~~	~~10~~
11	~~12~~	13	~~14~~	~~15~~	~~16~~	17	~~18~~	19	~~20~~

How many integers in the list $2, 3, \ldots, 1000$ are *not* crossed out after crossing out multiples of the primes 2, 3, 5, and 7?

18. At Brokaw Hospital, six babies were born to six different women on Monday through Thursday of a particular week. Assuming that each baby was equally likely to be born on any of the four days, what is the probability that there was at least one baby born on each day?

19. In how many ways can four married couples be seated in a row of eight chairs with no husband seated beside his wife?

20. How many arrangements of the numbers 1, 1, 2, 2, 3, 3, 4, 4 are there in which no adjacent numbers are equal?

21. How many five-card poker hands contain at least one card in each suit?

22. How many of the functions with domain $\{5, 6, 7, 8, 9, 10\}$ and codomain $\{1, 2, 3, 4\}$ are onto?

23. How many nonnegative integer solutions of $x_1 + x_2 + x_3 + x_4 = 12$ are there in which no x_i exceeds 4?

24. Suppose that five balls numbered 1, 2, 3, 4, and 5 are successively removed from an urn. A *rencontre* is said to occur if ball number k is the kth ball removed. What is the probability that no rencontres occur?

25. Let S be a set containing m elements, and let $n \geq m$ be a positive integer. Ordered lists of n items chosen from S are to be constructed in which each element of S appears at least once. Show that the number of such lists is

$$C(m,0)(m-0)^n - C(m,1)(m-1)^n + \cdots + (-1)^{m-1}C(m,m-1)(1)^n.$$

26. Two integers are called *relatively prime* if 1 is the only positive integer that divides both numbers. Show that if a positive integer n has $p_1, p_2, \ldots, p_k$ as its distinct prime divisors, then the number of positive integers that are less than n and relatively prime to n is

$$n - \frac{n}{p_1} - \frac{n}{p_2} - \cdots + \frac{n}{p_1 p_2} + \frac{n}{p_1 p_3} + \cdots + (-1)^k \frac{n}{p_1 p_2 \cdots p_k}.$$

27. Compute the number D_k of derangements of $1, 2, \ldots, k$.

28. For D_n as in Exercise 27, evaluate $D_{n+1} - (n+1)D_n$ when n is a positive integer.

For nonnegative integers n and m, define $S(n, m)$ to be the number of ways to distribute n distinguishable balls into m indistinguishable urns with no urns empty. These numbers are named ***Stirling numbers of the second kind*** *after the British mathematician James Stirling (1692–1770).*

29. Let n be a positive integer. Evaluate $S(n, 0)$, $S(n, 1)$, $S(n, 2)$, $S(n, n-2)$, $S(n, n-1)$, and $S(n, n)$.

30. Let X be a finite set containing n elements. How many partitions of X into k subsets are there? (See Section 2.2 for the definition of a partition.)

31. How many equivalence relations are possible on a set of n elements?

32. Prove that, for all integers $n > 0$ and $m > 1$,

$$S(n+1, m) = C(n, 0)S(0, m-1) + C(n, 1)S(1, m-1) + \cdots + C(n, n)S(n, m-1).$$

33. Prove that $S(n+1, m) = S(n, m-1) + m \cdot S(n, m)$ for all integers $n > 0$ and $m > 1$.

34. Use the result of Exercise 33 to describe a procedure for computing the numbers $S(n, m)$ that is similar to the manner in which the numbers $C(n, r)$ can be computed by using Pascal's triangle.

35. Prove that, for all positive integers n and m,

$$S(n, m) = \frac{1}{m!}[C(m, 0)(m-0)^n - C(m, 1)(m-1)^n + \cdots + (-1)^{m-1}C(m, m-1)(1)^n].$$

36. Let X and Y be finite sets containing n and m elements, respectively. How many functions with domain X and codomain Y are onto?

37. Let U be a finite set containing n_0 elements, and let $A_1, A_2, \ldots, A_r$ be subsets of U. Let n_s for $1 \le s \le r$ be as defined in the principle of inclusion–exclusion, and for $0 \le s \le r$ let p_s be the number of elements in U that belong to *precisely* s of the subsets $A_1, A_2, \ldots, A_r$. Prove that

$$\begin{aligned} p_s &= C(s, 0) \cdot n_s - C(s+1, 1) \cdot n_{s+1} + \cdots + (-1)^{r-s}C(r, r-s) \cdot n_r \\ &= \sum_{k=0}^{r-s} (-1)^k C(s+k, k) \cdot n_{s+k}. \end{aligned}$$

38. Use Exercise 37 to determine the number of different rearrangements of the letters in "correspondents" having exactly three pairs of identical letters in adjacent positions.

8.7* GENERATING PERMUTATIONS AND r-COMBINIATIONS

Unfortunately, there are many practical problems for which no efficient method of solution is known (such as the knapsack problem described in Section 1.3). In such cases, the only method of solution may be to perform an exhaustive search—that is, to systematically list and check all of the possibilities. Often, as in Section 1.2, listing all of the possibilities amounts to enumerating all the permutations or combinations of a set. In this section we will present procedures for listing all of the permutations and r-combinations of a set of n elements. For convenience, we will assume that the set in question is $\{1, 2, \ldots, n\}$.

The most natural order in which to list permutations is called **lexicographic order** (or **dictionary order**). To describe this order, let $p = (p_1, p_2, \ldots, p_n)$ and $q = (q_1, q_2, \ldots, q_n)$ be two different permutations of the integers $1, 2, \ldots, n$. Since p and q are different, they must differ in some entry. Let k denote the smallest index for which $p_k \neq q_k$. Then (reading from left to right) the first $k-1$ entries of p and q are the same, and the kth entries differ. In this case, we will say that p is **greater than** q in the lexicographic ordering if $p_k > q_k$. If p is greater than q in the lexicographic ordering, then we write $p > q$ or $q < p$. Thus, in the lexicographic order, we have

$$(2, 4, 1, 5, 3) > (2, 4, 1, 3, 5) \qquad \text{and} \qquad (3, 2, 4, 1, 5, 6) < (3, 2, 6, 5, 1, 4).$$

By using a tree diagram and choosing entries in numerical order, we can list all the permutations of $1, 2, \ldots, n$ in lexicographic order. Figure 8.5 depicts the case where $n = 3$. The permutations listed in the last column are in lexicographic order.

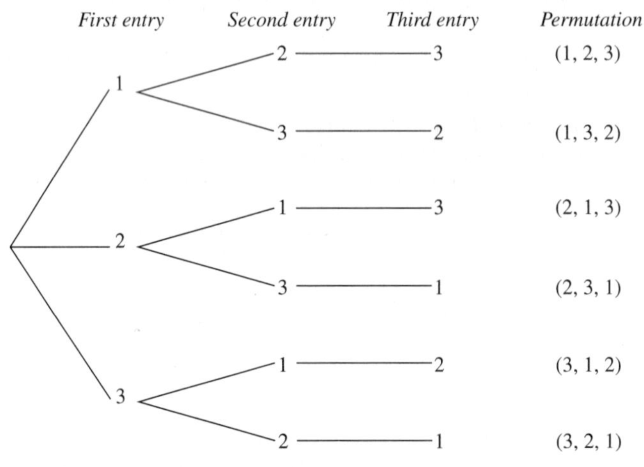

FIGURE 8.5

In order to have an efficient algorithm for listing permutations in lexicographic order, we must know how to find the successor of a permutation p in the lexicographic order—that is, the first permutation greater than p. Consider, for example, the permutation $p = (3, 6, 2, 5, 4, 1)$ of the integers 1 through 6. Let q denote the successor of p in the lexicographic ordering, and let r denote any permutation greater than q. Since $p < q < r$, q must agree with at least as much of p (from the left) as r does. Thus q must differ from p as far to the right in its list as possible. Clearly, we cannot rearrange the order of the last two entries of p (4 and 1) or the last three entries of p (5, 4, and 1) and obtain a greater permutation. But we can rearrange the last four entries of p (2, 5, 4, and 1) to get a greater permutation, and the least such rearrangement is 4, 1, 2, 5. Thus the successor of p in the lexicographic ordering is $q = (3, 6, 4, 1, 2, 5)$. Notice that the first two entries of q are the same as those of p and that the third entry of q is greater than that of p. Moreover, the third entry of q is the rightmost entry of p that exceeds the third entry of p. Finally, note that the entries of q to the right of the third entry are in increasing order.

More generally, consider a permutation $p = (p_1, p_2, \ldots, p_n)$ of the integers 1 through n. The successor of p in the lexicographic ordering is the permutation $q = (q_1, q_2, \ldots, q_n)$ such that

(1) The first $k - 1$ entries of q are the same as in p.
(2) The kth entry of q, q_k, is the rightmost entry of p that is greater than p_k.
(3) The entries of q that follow q_k are in increasing numerical order.

Therefore we can completely determine q from p if we know the value of k, the index of the entry of p to be changed. As we saw in our example, we want k to be chosen as large as possible. So, because of condition (2), we must choose k to be the largest possible index for which p_k is less than one of the entries that follow it. But then k is the largest index such that $p_k < p_{k+1}$. Thus, if we examine the entries of p from *right to left*, the entry of p to be changed is the first entry we reach that is less than the number to its right. In addition, since the entries of p to the right of the kth entry are in decreasing order, q_k equals the rightmost entry of p that exceeds p_k. If we now switch p_k with the rightmost entry of p that exceeds it, we obtain a new permutation in which the rightmost entries are the remaining entries of q in reverse order.

Example 8.45 Let us determine the permutation q of the integers 1 through 7 that is the successor of $p = (4, 1, 5, 3, 7, 6, 2)$. Scanning p from right to left, we see that the first entry we reach that is less than the number to its right is the fourth entry, which is 3. (So in the previous notation, $k = 4$.) Thus q has the form (4, 1, 5, ?, ?, ?, ?). Moreover, the fourth entry of q is the rightmost entry of p that exceeds the entry that is being changed (which is 3 in our case). Scanning p again from right to left, we see that the fourth entry of q is 6. Interchanging the positions of the 3 and 6 in p, we obtain (4, 1, 5, 6, 7, 3, 2). If we now reverse the order of the entries to the right of position k, we obtain (4, 1, 5, 6, 2, 3, 7), which is the successor of p. ■

The following algorithm uses the method described in the previous example to list all the permutations of $1, 2, \ldots, n$:

Algorithm for the Lexicographic Ordering of Permutations

This algorithm prints all the permutations of $1, 2, \ldots, n$ in lexicographic order. In the algorithm, $(p_1, p_2, \ldots, p_n)$ denotes the permutation currently being considered.

Step 1 (initialization)
    for $i = 0$ to n
        Set $p_i = i$.
    endfor
Step 2 (generate the permutations)
    repeat
        Step 2.1 (output)
            Print $(p_1, p_2, \ldots, p_n)$.
        Step 2.2 (find the index k of the leftmost entry to be changed)
            Find the largest index k for which $p_k < p_{k+1}$.
        Step 2.3 (is there something to change?)
            if $k > 0$
                Step 2.3.1 (determine the new value for p_k)
                    Find the largest index j for which $p_j > p_k$, and interchange the values of p_k and p_j.
                Step 2.3.2 (prepare to rearrange)
                    Set $r = k + 1$ and $s = n$.
                Step 2.3.3 (rearrange)
                    while $r > s$
                        (a) Interchange the values of p_r and p_s.
                        (b) Replace r with $r + 1$ and s with $s - 1$.
                    endwhile
            endif
    until $k = 0$

Although the lexicographic ordering is the most natural ordering for listing permutations, determining the successor of a given permutation in the lexicographic ordering requires several comparisons. For this reason, an algorithm that lists permutations in lexicographic order may be less efficient than one that lists the permutations in a different order. But since there are $n!$ permutations of the integers $1, 2, \ldots, n$, the complexity of any algorithm that lists these permutations will be at least $n!$. Readers who are interested in learning more efficient algorithms for listing permutations should consult suggested reading [3].

Lexicographic Enumeration of r-Combinations

In Section 1.4, we discussed an algorithm for generating all of the subsets of a set with n elements. Often, however, we need to consider only subsets of a specified size. We will now describe a procedure for generating all of the r-element subsets of $\{1, 2, \ldots, n\}$. As for permutations, we will list the subsets in lexicographic order. Because a subset is not an ordered array, we will understand this to mean that *the elements of a subset will be listed in increasing order as we read from left to right*. Thus we will write the subset $\{3, 6, 2, 4\}$ as $\{2, 3, 4, 6\}$.

In order to obtain an algorithm for listing subsets in lexicographic order, we need to determine the successor of any particular subset. Consider, for example, the 4-element subsets of $\{1, 2, 3, 4, 5, 6\}$. There are $C(6, 4) = 15$ such subsets, and they are listed as follows in lexicographic order from left to right:

$$\begin{array}{lllll}
\{1,2,3,4\}, & \{1,2,3,5\}, & \{1,2,3,6\}, & \{1,2,4,5\}, & \{1,2,4,6\}, \\
\{1,2,5,6\}, & \{1,3,4,5\}, & \{1,3,4,6\}, & \{1,3,5,6\}, & \{1,4,5,6\}, \\
\{2,3,4,5\}, & \{2,3,4,6\}, & \{2,3,5,6\}, & \{2,4,5,6\}, & \{3,4,5,6\}.
\end{array}$$

As was true for the lexicographic ordering of permutations, the successor of a subset S must differ from S as far to the right in its list of elements as possible. Thus, if the last element of a subset in the preceding list is not 6, the successor is obtained by adding 1 to the last element. For instance,

$$\begin{array}{llll}
\text{the successor of} & \{1,2,3,4\} & \text{is} & \{1,2,3,5\}, \\
\text{the successor of} & \{1,2,3,5\} & \text{is} & \{1,2,3,6\},
\end{array}$$

and

$$\text{the successor of} \quad \{1,3,4,5\} \quad \text{is} \quad \{1,3,4,6\}.$$

If the last element of a subset is 6, its successor will be obtained by a different procedure. Consider $\{1, 2, 5, 6\}$, for instance. Because the last element is 6, it cannot be increased. Likewise, the next-to-last element is 5, and so it cannot be increased. However, the third-from-last element can be increased from 2 to 3, and we finish the subset by listing consecutive integers beginning with 3. Thus

$$\text{the successor of} \quad \{1,2,5,6\} \quad \text{is} \quad \{1,3,4,5\},$$

and similarly,

$$\text{the successor of} \quad \{2,3,5,6\} \quad \text{is} \quad \{2,4,5,6\}.$$

More generally, consider an r-element subset $S = \{s_1, s_2, \ldots, s_r\}$ of $\{1, 2, \ldots, n\}$. The successor of S in the lexicographic ordering is a subset $T = \{t_1, t_2, \ldots, t_r\}$ such that

(1) The first $k - 1$ elements in T are the same as those in S.
(2) The kth element in T, t_k, is one more than s_k, the kth element in S.
(3) The elements $t_k, t_{k+1}, \ldots, t_r$ are consecutive integers.

So, as for permutations, we can completely determine T from S if we know s_k, the leftmost element in S to be changed. Our example illustrates that s_k is the rightmost element in S that can be increased—that is, the rightmost element in S that does not equal its maximum value. Note that the maximum value of

TABLE 8.1

Subset	*Index k*
$\{1, 2, 3, 4\}$	4
$\{1, 2, 3, 5\}$	4
$\{1, 2, 3, 6\}$	3
$\{1, 2, 4, 5\}$	4
$\{1, 2, 4, 6\}$	3
$\{1, 2, 5, 6\}$	2
$\{1, 3, 4, 5\}$	4
$\{1, 3, 4, 6\}$	3
$\{1, 3, 5, 6\}$	2
$\{1, 4, 5, 6\}$	1
$\{2, 3, 4, 5\}$	4
$\{2, 3, 4, 6\}$	3
$\{2, 3, 5, 6\}$	2
$\{2, 4, 5, 6\}$	1
$\{3, 4, 5, 6\}$	none

the last element in S is n, the maximum value of the next-to-last element in S is $n - 1$, and so forth. Hence s_k is the rightmost element in S that does not equal $n - r + k$. Then T is formed by replacing s_k by $s_k + 1$, s_{k+1} by $s_k + 2$, and so forth. For example, in the lexicographic ordering of the 5-element subsets of $\{1, 2, 3, 4, 5, 6, 7, 8\}$,

$$\text{the successor of} \quad S = \{1, 3, 4, 7, 8\} \quad \text{is} \quad T = \{1, 3, 5, 6, 7\}$$

because 4 (the third element in S) is the rightmost element that can be increased.

Although the procedure just described is easy for humans to implement, searching for the element s_k requires more comparisons than necessary. Consequently, we can make the procedure more efficient for a computer by determining k rather than s_k. Let us return to the list of all 4-element subsets of $\{1, 2, 3, 4, 5, 6\}$ considered earlier. Table 8.1 gives the value of k needed to determine the successor of each subset.

The observation that enables us to determine k more quickly is that *the value of k for the successor of a subset S is either r or one less than the value of k for S*. The following algorithm uses this method of determining k:

Algorithm for the Lexicographic Ordering of r-Combinations

This algorithm prints all the r-element subsets of $\{1, 2, \ldots, n\}$ in lexicographic order, where $1 \le r \le n$. In the algorithm, $\{s_1, s_2, \ldots, s_r\}$ denotes the subset currently being considered.

Step 1 (initialization)
  for $j = 1$ to r
    Set $s_j = j$.
  endfor
  if $r = n$
    Set $k = 1$.
  otherwise
    Set $k = r$.
  endif
Step 2 (create the subsets)
  repeat
    Step 2.1 (output)
      Print $\{s_1, s_2, \ldots, s_r\}$.
    Step 2.2 (find the index, k, of the first element to be changed)
      if $s_k \ne n - r + k$
        Set $k = r$.
      otherwise

Replace k with $k - 1$.
endif
Step 2.3 (determine the successor)
if $k \neq 0$
(a) Replace s_k with $s_k + 1$.
(b) **for** $i = k + 1$ to r
Replace s_i with $s_k + (i - k)$.
endfor
endif
until $k = 0$

It can be shown that this algorithm has order at most n^r. Therefore, for a fixed value of r, the algorithm for the lexicographic ordering of r-combinations is a "good" algorithm.

EXERCISES 8.7

For the permutations p and q in Exercises 1–6, determine whether $p < q$ or $p > q$ in the lexicographic ordering.

1. $p = (3, 2, 4, 1), q = (4, 1, 3, 2)$

2. $p = (2, 1, 3), q = (1, 2, 3)$

3. $p = (1, 2, 3), q = (1, 3, 2)$

4. $p = (2, 1, 3, 4), q = (2, 3, 1, 4)$

5. $p = (4, 2, 5, 3, 1), q = (4, 2, 3, 5, 1)$

6. $p = (2, 5, 3, 4, 1, 6), q = (2, 5, 3, 1, 6, 4)$

In Exercises 7–18, determine the successor of permutation p in the lexicographic ordering of the permutations of $1, 2, 3, 4, 5, 6$.

7. $p = (2, 1, 4, 3, 5, 6)$

8. $p = (3, 6, 4, 2, 1, 5)$

9. $p = (2, 1, 4, 6, 5, 3)$

10. $p = (3, 6, 5, 4, 2, 1)$

11. $p = (5, 6, 3, 4, 2, 1)$

12. $p = (5, 1, 6, 4, 3, 2)$

13. $p = (6, 5, 4, 3, 2, 1)$

14. $p = (1, 2, 3, 6, 5, 4)$

15. $p = (5, 2, 6, 4, 3, 1)$

16. $p = (4, 5, 6, 3, 2, 1)$

17. $p = (6, 3, 5, 4, 2, 1)$

18. $p = (2, 3, 1, 6, 5, 4)$

19. List the permutations of $1, 2, 3, 4$ in lexicographic order.

In Exercises 20–31, determine the successor of subset S in the lexicographic ordering of the 5-element subsets of $\{1, 2, 3, 4, 5, 6, 7, 8, 9\}$.

20. $S = \{1, 2, 4, 5, 6\}$

21. $S = \{1, 3, 5, 7, 9\}$

22. $S = \{1, 3, 6, 8, 9\}$

23. $S = \{2, 3, 5, 8, 9\}$

24. $S = \{2, 4, 6, 7, 9\}$

25. $S = \{3, 4, 5, 7, 8\}$

26. $S = \{3, 4, 7, 8, 9\}$

27. $S = \{3, 5, 7, 8, 9\}$

28. $S = \{4, 5, 6, 7, 8\}$

29. $S = \{4, 5, 7, 8, 9\}$

30. $S = \{4, 6, 7, 8, 9\}$

31. $S = \{5, 6, 7, 8, 9\}$

32. List all the 3-element subsets of $\{1, 2, 3, 4, 5, 6\}$ in lexicographic order.

HISTORICAL NOTES

Blaise Pascal

The roots of combinatorial enumerations reach back to at least the 79th problem of the Rhind papyrus (c. 1650 B.C.). Others lie in the work of Xenocrates of Chalcedon (396–314 B.C.) and his attempts to solve a problem involving permutations and combinations, and work by Oriental and Hindu mathematicians. Texts exist from the sixth century B.C. listing the values of combinations of tastes drawn from six basic qualities: sweet, acid, saline, pungent, bitter, and astringent. In his text *Lilavati*, the Hindu mathematician Bhaskara (ca. 1114–1185) wrote rules for the computation of combinations and the familiar $n!$ rule for permutations. There is also evidence that the Hindus were familiar with the binomial expansion of $(a + b)^n$ for small positive integers n [71, 73, 78, 86, 87].

Girolamo Cardano (1501–1576) stated the binomial theorem, and Blaise Pascal (1623–1662) presented the first known proof in his 1665 *Traité du Triangle Arithmétique*. Jacob Bernoulli (1654–1705) provided an alternative proof of the theorem in his *Ars Conjectandi* (1713) and is often incorrectly given credit for the first proof of the theorem. The arithmetic triangle, often referred to as Pascal's triangle, was known to the Chinese through Chu Shih-Chieh's *Ssu Yuan Yii Chien* in 1303 and is believed to have been known to others in the Orient before that date [73].

Abraham De Moivre (1667–1754) extended the binomial theorem in 1697 to the multinomial theorem, which governs the expansion of $(x_1 + x_2 + \cdots + x_r)^n$ for positive integers r and n. By 1730, De Moivre and the British mathematician James Stirling (1692–1770) had obtained the asymptotic result, now known as Stirling's formula, that

$$n! \approx \left(\frac{n}{e}\right)^n (2\pi n)^{1/2}$$

for large positive integers n.

At much the same time, the foundations of probability were forming. Early work in the area came from Cardano and Niccolo Tartaglia (1500–1557), whose work dealt with odds and gambling situations. Cardano published his *Liber de Ludo Alea*, a book on games of chance, in 1526. In it, he shows knowledge of independent events and the multiplication rule.

The Dutchman Christian Huygens (1629–1695) wrote *De Ludo Aleae* in 1657. In this book, he considered problems dealing with the probabilities associated with drawing colored balls from an urn.

Jacob Bernoulli

Bernoulli's *Ars Conjectandi*, published posthumously in 1713, included information on permutations and combinations, work on elementary probability, and the law of large numbers. With the work of Bernoulli, one sees the binomial theorem being used to compute binomial-based probabilities. This work was extended by Pierre Simon Laplace (1749–1827) in *Essai Philosophique sur les Probabilités* in 1814. Laplace's work gave special attention to the applications of probability to demography and other social science problems.

Abraham De Moivre is credited with the statement that the probability of a compound event is the product of the probabilities of its components. He also presented an analytical version of the principle of inclusion–exclusion in his 1718 work on probability, the *Doctrine of Chances*. However, the modern version of the principle of inclusion–exclusion is usually credited to the English/American mathematician James Joseph Sylvester (1814–1897) [86, 87].

SUPPLEMENTARY EXERCISES

Evaluate each of the expressions in Exercises 1–8.

1. $C(9, 7)$ **2.** $P(8, 5)$ **3.** $P(9, 4)$ **4.** $C(10, 6)$

5. $(x - 1)^6$ **6.** $(x + 2y)^7$ **7.** $(2x + 3y)^5$ **8.** $(5x - 2y)^4$

9. What is the successor of (8, 2, 3, 7, 6, 5, 4, 1) in the lexicographic ordering of the permutations of 1, 2, 3, 4, 5, 6, 7, 8?

10. What is the successor of $\{1, 3, 6, 7, 8\}$ in the lexicographic ordering of the 5-combinations of $\{1, 2, \ldots, 8\}$?

11. The first nine numbers in the $n = 17$ row of Pascal's triangle are 1, 17, 136, 680, 2380, 6188, 12376, 19448, and 24310. What are the remaining numbers in this row of Pascal's triangle?

12. Use your answer to Exercise 11 to determine the coefficient of x^{12} in the binomial expansion of $(x - 2)^{17}$.

13. Use your answer to Exercise 11 to evaluate $C(18, 12)$.

14. For a town's annual Easter egg hunt, 15 dozen eggs were hidden. There were 6 gold eggs, 30 pink eggs, 30 green eggs, 36 blue eggs, 36 yellow eggs, and 42 purple eggs. How many eggs must a child find in order to be assured of having at least 3 of the same color?

15. The snack bar at a movie theatre sells 5 different sizes of popcorn, 12 different candy bars, and 4 different beverages. In how many different ways can one snack be selected?

16. A woman has 6 different pairs of slacks, 8 different blouses, 5 different pairs of shoes, and 3 different purses. How many outfits consisting of one pair of slacks, one blouse, one pair of shoes, and one purse can she create?

17. A woman has 6 different pairs of slacks, 8 different blouses, 5 different pairs of shoes, and 3 different purses. Suppose that an outfit consists of one pair of slacks, one blouse, and one pair of shoes and may or may not include a purse. How many outfits can the woman create?

18. How many integers between 1500 and 8000 (inclusive) contain no repeated digits?

19. A pianist participating in a Chopin competition has decided to perform 5 of the 14 Chopin waltzes. How many different programs are possible consisting of 5 waltzes played in a certain order?

20. How many ways are there to select a subcommittee of 5 members from among a committee of 12?

21. If 2 distinct integers are chosen from among the numbers $1, 2, \ldots, 60$, what is the probability that their sum is even?

22. A committee of 4 is to be chosen at random from among 5 women and 6 men. What is the probability that the committee contains at least 3 women?

23. In a literature class of 12 graduate students, the instructor will choose 3 students to analyze *Howard's End*, 4 other students to analyze *Room with a View*, and the remaining 5 students to analyze *A Passage to India*. In how many different ways can the students be chosen?

24. How many nonnegative integer solutions are there to $x + y + z = 15$?

25. How many arrangements are there of all the letters in the word "rearrangement"?

26. If an arrangement of all the letters in the word "rearrangement" is chosen at random, what is the probability that all the r's are adjacent?

27. How many ways are there to select 4 novels from a list of 16 novels to be read for a literature class?

28. Nine athletes are entered in the conference high jump competition. In how many different ways can the gold, silver, and bronze medals be awarded?

29. A college student needs to choose one more course to complete next semester's schedule. She is considering 4 business courses, 7 physical education courses, and 3 economics courses. How many different courses can she select?

30. Suppose that 8 people raise their glasses in a toast. If every person clinks glasses exactly once with everyone else, how many clinks will there be?

31. Suppose that the digits 1–7 are to be used without repetition to make five-digit numbers.

- **(a)** How many different five-digit integers can be made?
- **(b)** What is the probability that if one of these numbers is chosen at random, the number begins with 6?
- **(c)** What is the probability that if one of these numbers is chosen at random, the number contains both the digits 1 and 2?

32. If a die is rolled six times, what is the probability of rolling two 3's, three 4's, and one 5?

33. A newly formed consumer action group has 30 members. In how many ways can the group elect

- **(a)** a president, vice president, secretary, and treasurer (all different)?
- **(b)** an executive committee consisting of 5 members?

34. A bakery sells 8 varieties of bagels. How many ways are there to select a dozen bagels if we must choose at least one of each type?

35. A university's alumni service award can be given to at most 5 persons per year. This year there are 6 nominees from which the recipients will be chosen. In how many different ways can the recipients be selected?

36. A candy company has an unlimited supply of cherry, lime, licorice, and orange gumdrops. Each box of gumdrops contains 15 gumdrops, with at least 3 of each flavor. How many different boxes are possible?

37. If 10 different numbers are chosen from among the integers $1, 2, \ldots, 40$, what is the probability that no two of the numbers are consecutive?

38. How many different assortments of two dozen cupcakes can be purchased if there are six different types of cupcakes from which to choose?

39. How many different assortments of two dozen cupcakes can be purchased if there are six different types of cupcakes from which to choose and we must choose at least two cupcakes of each type?

40. In Bogart's restaurant, the entrees include prime rib, filet mignon, ribeye steak, scallops, and a fish-of-the-day. Each dinner is served with salad and a vegetable. Customers may choose from 4 salad dressings and 5 vegetables, except that the seafood dishes are served with wild rice instead of the choice of a vegetable. In how many different ways can a dinner be ordered?

41. What is the probability that a randomly chosen integer between 10,000 and 99,999 (inclusive) contains a zero?

42. What is the probability that a randomly chosen list of 5 letters contains 3 different consonants and 2 different vowels? (Regard "y" as a consonant.)

43. In how many distinguishable ways can 4 identical algebra books, 6 identical geometry books, 3 identical calculus books, and 5 identical discrete math books be arranged on a shelf?

44. A shish kebab is to be made by placing on a skewer a piece of beef followed by 7 vegetables, each of which is either a mushroom, a green pepper, or an onion. How many different shish kebabs are possible if at least 2 vegetables of each type must be used?

45. Sixteen subjects are to be used in a test of 3 experimental drugs. Each experimental drug will be given to 4 subjects, and no subject will receive more than one drug. The 4 subjects who are not given an experimental drug will be given a placebo. In how many different ways can the drugs be assigned to the subjects?

46. The 12 guests of honor at an awards banquet are to be given corsages. Each guest of honor can choose the color: pink, red, yellow, or white. How many corsage orders to the florist are possible?

47. Fifteen geraniums are to be planted in a row. There are 4 red geraniums, 6 white geraniums, and 5 pink geraniums. Assuming that the plants are indistinguishable except for color, in how many distinguishable arrangements can the flowers be planted?

48. Let S be a 6-element subset of $\{1, 2, \ldots, 9\}$. Show that S must contain a pair of elements with sum 10.

49. Can the integers $1, 2, \ldots, 12$ be placed around a circle so that the sum of 5 consecutive numbers never exceeds 32?

50. How many nonnegative integer solutions are there to the equation

$$x_1 + x_2 + x_3 + x_4 = 28$$

with $x_1 \leq 8$, $x_2 \leq 6$, $x_3 \leq 12$, and $x_4 \leq 9$?

51. A pinochle deck consists of 48 cards, including two each of the aces of spades, hearts, diamonds, and clubs. What is the probability that a random 12-card pinochle hand contains at least one ace of each suit?

52. In how many ways can n married couples be seated in a row of $2n$ chairs with no husband seated beside his wife?

53. Give a combinatorial proof that $C(n, k) \leq 2^n$ for $0 \leq k \leq n$.

54. Prove that the largest entry in row n of Pascal's triangle exceeds $(1.5)^n$ for $n \geq 4$.

55. Evaluate $C(2, 2) + C(3, 2) + \cdots + C(n, 2)$, and verify your answer by mathematical induction.

56. If $(x_1 + x_2 + \cdots + x_k)^n$ is expanded and like terms are combined, how many terms will there be in the answer?

57. Prove the multinomial theorem: For any positive integers k and n,

$$(x_1 + x_2 + \cdots + x_k)^n = \sum \frac{n!}{n_1!n_2!\cdots n_k!} x_1^{n_1} x_2^{n_2} \cdots x_k^{n_k},$$

where the sum is taken over all nonnegative integer solutions of the equation $n_1 + n_2 + \cdots + n_k = n$.

58. Let $s_n = C(n, 0) + C(n-1, 1) + C(n-2, 2) + \cdots + C(n-r, r)$, where r denotes the greatest integer less than or equal to $\frac{n}{2}$.

(a) What pattern does s_n represent in Pascal's triangle?

(b) Formulate a conjecture about the value of s_n for all nonnegative integers n.

(c) Prove the conjecture made in part (b).

59. To each subset of $\{1, 2, \ldots, n\}$, assign one of n possible colors. Show that no matter how the colors are assigned, there are distinct sets A and B such that the four sets A, B, $A \cup B$, and $A \cap B$ are all assigned the same color. Show that the conclusion need not be true if there are $n+1$ colors available.

60. Complete the accompanying chart by giving the number of distributions of n balls into m *distinguishable* urns in the 8 cases indicated. In 6 of the 8 cases, whether the answer is zero or nonzero depends on $m > n$, $m = n$, or $m < n$. Note that *each of the n balls must be put into some urn.*

Are the Balls Distinguishable?	*More than One Ball per Urn Allowed?*	*May Urns Be Empty?*	*Number of Distributions of n Balls into m Distinguishable Urns*
Yes	Yes	Yes	
Yes	Yes	No	
Yes	No	Yes	
Yes	No	No	
No	Yes	Yes	
No	Yes	No	
No	No	Yes	
No	No	No	

COMPUTER PROJECTS

Write a computer program having the specified input and output.

1. Given integers n and r such that $0 \le r \le n$, compute the values of $P(n, r)$ and $C(n, r)$.
2. Given a nonnegative integer n, compute the numbers in rows $0, 1, \ldots, n$ of Pascal's triangle.
3. Given positive integers k and n, list all the nonnegative integer solutions to the equation $x_1 + x_2 + \cdots + x_k = n$.
4. Given a positive integer n, determine the probability that in a random selection of n people, no two have the same birthday. Assume that no one was born on February 29.
5. Given a positive integer n, list all the derangements of the integers $1, 2, \ldots, n$. (See Section 8.6 for the definition of a derangement.)
6. Given a positive integer n, use the sieve of Eratosthenes to determine all the primes less than or equal to n. (See Exercise 17 in Section 8.6.)
7. Given positive integers k and n, compute $S(n, k)$, the Stirling number of the second kind. (See Exercise 33 in Section 8.6.)

8. Given a positive integer n, list all the permutations of $1, 2, \ldots, n$ in lexicographic order.

9. Given positive integers r and n, list all the r-element subsets of $\{1, 2, \ldots, n\}$ in lexicographic order.

10. Given positive integers r and n, list all the r-permutations of $1, 2, \ldots, n$ in lexicographic order.

11. Given positive integers r and n, print all possible ordered lists of r items selected from the set $\{1, 2, \ldots, n\}$ if repetition of items is allowed.

12. Given positive integers r and n, print all possible unordered lists of r items selected from the set $\{1, 2, \ldots, n\}$ if repetition of items is allowed.

SUGGESTED READINGS

1. Beckenbach, E. *Applied Combinatorial Mathematics*. New York: Wiley, 1964.

2. Even, Shimon. *Algorithmic Combinatorics*. New York: Macmillan, 1973.

3. Nijenhuis, Albert and Herbert S. Wilf. *Combinatorial Algorithms*, 2d ed. New York: Academic Press, 1978.

4. Ryser, Herbert J. *Combinatorial Mathematics*, Carus Monograph Number 14. New York: Mathematical Association of America, 1963.

5. Whitworth, William Allen. *Choice and Chance*. Reprint of the 5th edition. New York: Hafner, 1965.

CHAPTER

9 Recurrence Relations and Generating Functions

In preceding chapters, we have seen several situations in which we wanted to associate with a set of objects a number such as the number of subsets of the set or the number of ways to arrange the objects in the set. Sometimes this number can be related to the corresponding number for a smaller set. In Section 2.6, for example, we saw that the number of subsets of a set with n elements is twice the number of subsets of a set with $n - 1$ elements. Often, such a relationship can be exploited to derive a formula for the number we are seeking. Techniques for doing so will be explored in this chapter.

9.1 RECURRENCE RELATIONS

An infinite ordered list is called a **sequence**. The individual items in the list are called **terms** of the sequence. For example,

$$0!, 1!, 2!, \ldots, n!, \ldots$$

is a sequence with first term $0!$, second term $1!$, and so forth. In this case, the nth term of the sequence is defined explicitly as a function of n, namely, $(n-1)!$.

In this chapter we will study sequences where a general term is defined as a function of preceding terms. An equation relating a general term to terms that precede it is called a **recurrence relation**. In Section 2.5 we saw that $n!$ could be defined recursively by specifying that

$$0! = 1 \qquad \text{and} \qquad n! = n(n-1)! \quad \text{for } n \geq 1.$$

In this definition, the equation

$$n! = n(n-1)! \qquad \text{for } n \geq 1$$

is a recurrence relation. It defines each term of the sequence of factorials as a function of the immediately preceding term.

In order to determine the values of the terms in a recursively defined sequence, we must know the values of a specific set of terms in the sequence, usually the beginning terms. The assignment of values for these terms gives a set of **initial conditions** for the sequence. In the case of the factorials, there is a single initial condition, which is that $0! = 1$. Knowing this value, we can then compute values for the other terms in the sequence from the recurrence relation. Thus

$$\begin{aligned} 1! &= 1(0!) = 1 \cdot 1 = 1, \\ 2! &= 2(1!) = 2 \cdot 1 = 2, \\ 3! &= 3(2!) = 3 \cdot 2 = 6, \\ 4! &= 4(3!) = 4 \cdot 6 = 24, \end{aligned}$$

and so on.

Another example of a sequence that is defined by a recurrence relation is the sequence of Fibonacci numbers. Recall from Section 2.5 that the Fibonacci numbers satisfy the recurrence relation

$$F_n = F_{n-1} + F_{n-2} \qquad \text{for } n \geq 3.$$

Because F_n is defined as a function of the two preceding terms, we must know two consecutive terms of the sequence in order to compute subsequent ones. For the Fibonacci numbers, the initial conditions are $F_1 = 1$ and $F_2 = 1$. Note that there are sequences other than the Fibonacci numbers that satisfy the same recurrence relation $s_n = s_{n-1} + s_{n-2}$, for example,

$$3, 4, 7, 11, 18, 29, 47, 76, \ldots.$$

Here, each term after the second is the sum of the two preceding terms, and so the sequence is completely determined by the initial conditions $s_1 = 3$ and $s_2 = 4$.

In this section we will examine other situations in which recurrence relations occur and illustrate how they can be used to solve problems involving counting.

Example 9.1 Let us consider from a recursive point of view the question of determining the number of edges e_n in the complete graph $\mathcal{K}_n$ with n vertices. We begin by considering how many new edges need to be drawn to obtain $\mathcal{K}_n$ from $\mathcal{K}_{n-1}$. The addition of one new vertex requires the addition of $n - 1$ new edges, one to each of the vertices in $\mathcal{K}_{n-1}$. (See Figure 9.1(a) for the case $n = 4$ and Figure 9.1(b) for the case $n = 5$.) Thus we see that the number of edges in $\mathcal{K}_n$ satisfies the recurrence relation

$$e_n = e_{n-1} + (n - 1) \qquad \text{for } n \geq 2.$$

In this equation, the value of e_n depends only on the preceding term e_{n-1}, and so we need only one value of e_n to use the recurrence relation. Since the complete graph with 1 vertex has no edges, we see that $e_1 = 0$. This is the initial condition for this sequence. ■

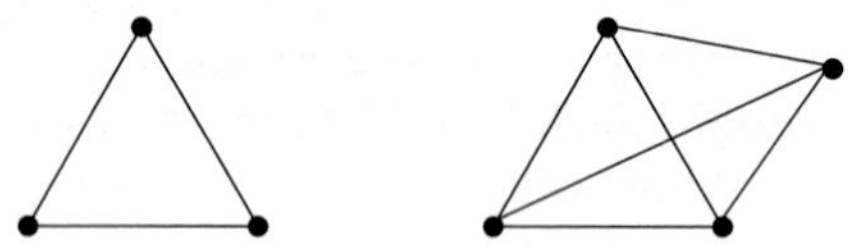

(a)

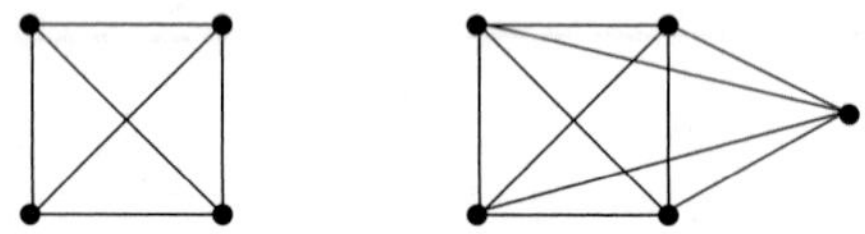

(b)

FIGURE 9.1

Example 9.2 The **Towers of Hanoi** game is played with a set of disks of graduated size with holes in their centers and a playing board having three spokes for holding the disks. (See Figure 9.2.) The object of the game is to transfer all the disks from spoke A to spoke C by moving one disk at a time, without ever placing a larger disk on top of a smaller one. What is the minimum number of moves required when there are n disks?

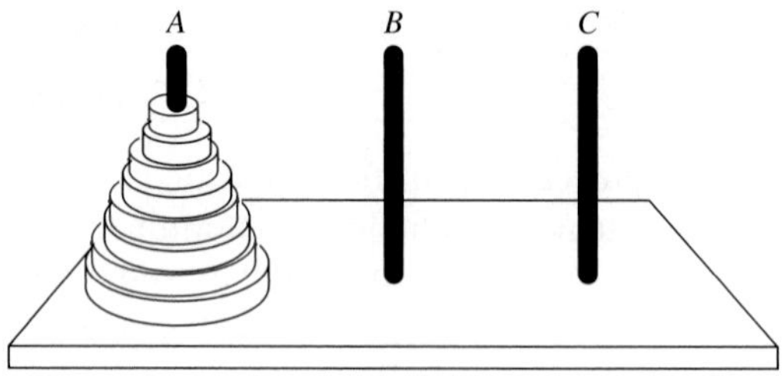

FIGURE 9.2

To answer the question, we will formulate a recurrence relation for m_n, the minimum number of moves to transfer n disks from one spoke to another. This will require expressing m_n in terms of previous terms m_i. It is easy to see that the most efficient procedure for winning the game with $n \geq 2$ disks is as follows (see Figure 9.3):

(1) Move the smallest $n - 1$ disks (in accordance with the rules) as efficiently as possible from spoke A to spoke B.
(2) Move the largest disk from spoke A to spoke C.
(3) Move the smallest $n - 1$ disks (in accordance with the rules) as efficiently as possible from spoke B to spoke C.

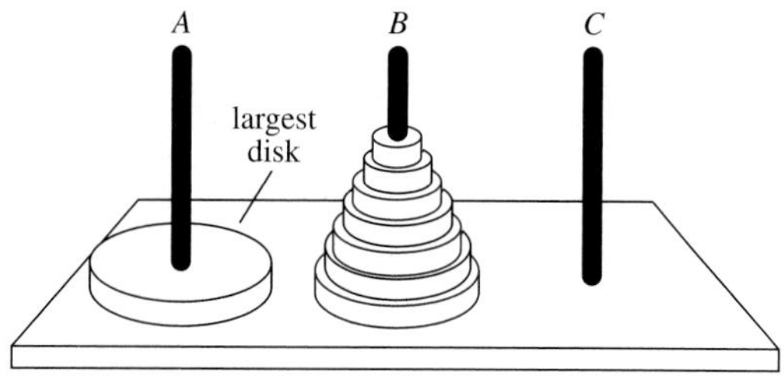

Step 1 Position after moving $n - 1$ disks from A to B

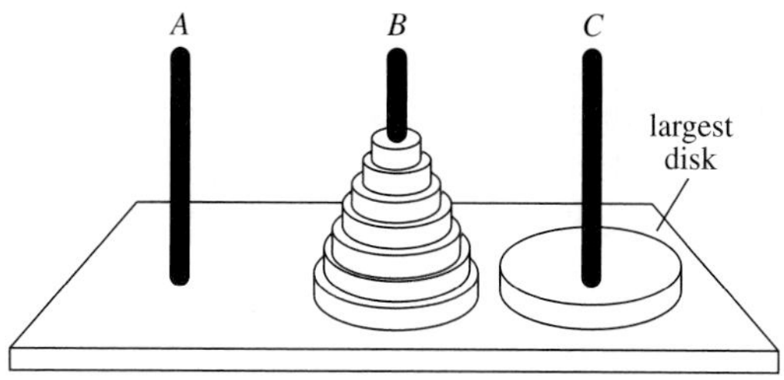

Step 2 Position after moving the bottom disk from A to C

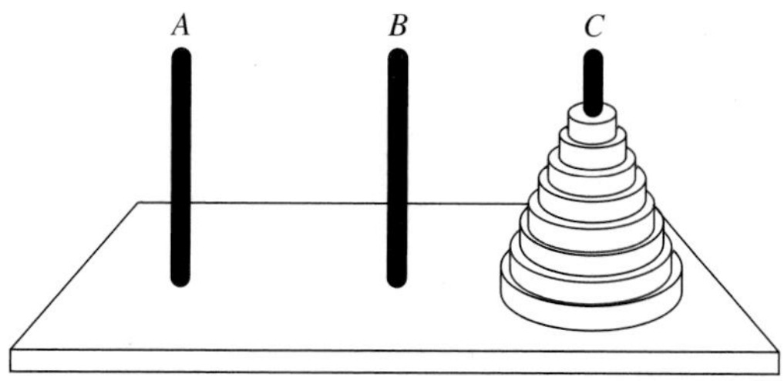

Step 3 Position after moving $n - 1$ disks from B to C

FIGURE 9.3

Since step 1 requires moving $n - 1$ disks from one spoke to another, the minimum number of moves required in step 1 is just m_{n-1}. It then takes one move to accomplish step 2 and another m_{n-1} moves to accomplish step 3. This analysis produces the recurrence relation

$$m_n = m_{n-1} + 1 + m_{n-1},$$

which simplifies to the form

$$m_n = 2m_{n-1} + 1 \qquad \text{for } n \geq 2.$$

Again, we need to know one value of m_n in order to use this recurrence relation. Because only one move is required to win a game with 1 disk, the initial condition for this sequence is $m_1 = 1$. By using the recurrence relation and the initial condition, we can determine the number of moves required for any desired number of disks. For example,

$$\begin{aligned} m_1 &= 1, \\ m_2 &= 2(1) + 1 = 3, \\ m_3 &= 2(3) + 1 = 7, \\ m_4 &= 2(7) + 1 = 15, \qquad \text{and} \\ m_5 &= 2(15) + 1 = 31. \end{aligned}$$

In Section 9.2 we will obtain an explicit formula that expresses m_n in terms of n. ■

Example 9.3 A carpenter needs to cover n consecutive 1-foot gaps between the centers of successive roof rafters with 1-foot and 2-foot boards, as shown in Figure 9.4. In how many ways can the carpenter complete his task?

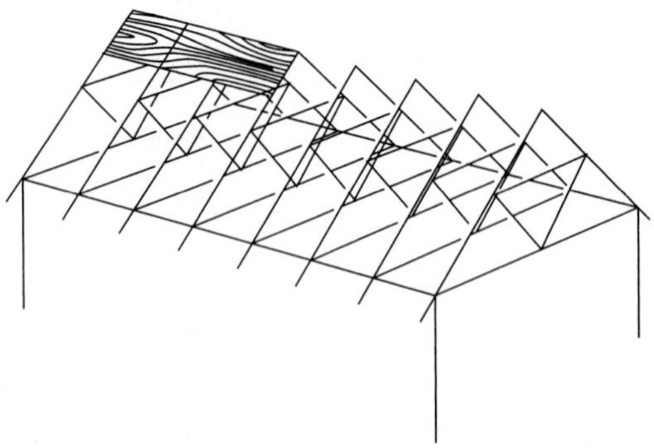

FIGURE 9.4

Our approach is to determine a recurrence relation and initial conditions for s_n, the number of ways n gaps can be covered. This requires expressing s_n as a function of previous terms s_i. Note that in order to cover n gaps, the carpenter must finish with either a 1-foot board or a 2-foot board. If the carpenter finishes with a 1-foot board, then he must have covered $n - 1$ gaps prior to using the last board. There are s_{n-1} ways to cover these gaps. On the other hand, if the carpenter finishes with a 2-foot board, then he must have covered $n - 2$ gaps prior to using the last board. There are s_{n-2} ways to cover these gaps. Since exactly one of these two cases must occur, the addition principle gives

$$s_n = s_{n-1} + s_{n-2} \qquad \text{for } n \geq 3.$$

To use this recurrence relation, we need to know two consecutive terms of the sequence. Clearly, the only way to cover a single 1-foot gap is with a single 1-foot board; so $s_1 = 1$. However, there are two ways to cover two 1-foot gaps: with a single 2-foot board or two 1-foot boards. Thus $s_2 = 2$. The number s_n of ways for the carpenter to complete his task is therefore given by the recurrence relation

$$s_n = s_{n-1} + s_{n-2} \qquad \text{for } n \geq 3,$$

subject to the initial conditions $s_1 = 1$ and $s_2 = 2$. Note the similarity between the numbers s_n and the Fibonacci numbers F_n. ■

Example 9.4 Recall from Section 8.6 that a permutation of the integers 1 through n in which no integer occupies its natural position is called a *derangement*. By enumerating all permutations of $1, 2, \ldots, n$, we find that there are no derangements of 1, there is one derangement of 1, 2 (namely, 2, 1), and there are two derangements of 1, 2, 3 (namely, 2, 3, 1 and 3, 1, 2). We have seen in Example 8.44 that the number D_n of derangements of the integers $1, 2, \ldots, n$ can be computed by using the principle of inclusion–exclusion. In this example we will use a recurrence relation to count derangements. The preceding comments show that $D_1 = 0$, $D_2 = 1$, and $D_3 = 2$.

To illustrate the general technique, we will list the derangements of the integers 1, 2, 3, 4 that begin with 2. These derangements are of two types. The first type is a derangement that has 1 in position 2. Here the situation is as follows:

Prohibited value:	1	2	3	4
Derangement:	2	1	?	?

It is easy to check that there is exactly one derangement of this type, namely, 2, 1, 4, 3. Notice that completing the derangement

2 1 ? ?

amounts to deranging the integers 3 and 4, and there is $D_2 = 1$ derangement of two integers.

The second type of derangement of 1, 2, 3, 4 that begins with 2 has 2 in position 1 and 1 *not* in position 2. Note that in this case we have two restrictions on the second position; neither 1 nor 2 can occur in the second position. But since 2 is in the first position, it cannot also occur in the second position. Thus the only restriction with which we must be concerned is that 1 not occur in the second position. Hence the situation can be represented by the following diagram:

Prohibited value:	1	1	3	4
Derangement:	2	?	?	?

In this case, it is easy to check that there are exactly two derangements of this type, namely, 2, 3, 4, 1 and 2, 4, 1, 3. Note that, because 1 cannot occur in the second position, 3 cannot occur in the third position, and 4 cannot occur in the fourth position, completing the derangement

2 ? ? ?

amounts to deranging the integers 1, 3, 4, and there are $D_3 = 2$ derangements of three integers. Thus there are, in all,

$$D_2 + D_3 = 1 + 2 = 3$$

derangements of 1, 2, 3, 4 beginning with 2.

In the general case, a derangement of $1, 2, \ldots, n$ must begin with k, where $k = 2, 3, \ldots, n$. For $n \geq 3$, there are two types, one in which integer 1 is moved to position k and one in which integer 1 is not moved to position k. If integer 1 is moved to position k, the situation is as follows:

Prohibited value:	1	2	...	$k-1$	k	$k+1$	...	n
Derangement:	k	?	...	?	1	?	...	?

Here, the remaining $n-2$ positions can be filled with the integers other than 1 and k to form a derangement in D_{n-2} ways. In the second type of derangement, we have the integer k in the first position and cannot have the integer 1 in position k. Since the integer k is in the first position, it cannot also be in the kth position, and so it is sufficient to require that the kth position not be filled by 1. We depict this situation as follows:

Prohibited value:	1	2	...	$k-1$	1	$k+1$	...	n
Derangement:	k	?	...	?	?	?	...	?

Thus we must place the $n-1$ integers other than k into $n-1$ positions, with no integer in a prohibited location. There are D_{n-1} ways to do this.

Thus, by the addition principle, there are

$$D_{n-2} + D_{n-1}$$

derangements of the integers 1 through n in which integer k is moved to position 1. But there are $n-1$ possible values of k (namely, $2, 3, \ldots, n$), so D_n must satisfy the recurrence relation

$$D_n = (n-1)(D_{n-2} + D_{n-1}) \qquad \text{for } n \geq 3.$$

To use this recurrence relation, we need two consecutive values of D_n. Since we have already seen that $D_1 = 0$ and $D_2 = 1$, these are the initial conditions. In the next section, we will obtain an explicit formula giving D_n as a function of n. ■

Example 9.5 A *stack* is an important data structure in computer science. It stores data, subject to the restriction that all insertions and deletions take place at one end of the stack (called the *top*). As a consequence of this restriction, the last item inserted into the stack must be the first item deleted, and so a stack is an example of a last-in-first-out structure.

We will insert all of the integers $1, 2, \ldots, n$ into a stack (in sequence) and count the possible sequences in which they can leave the stack. Note that each integer from 1 through n enters and leaves the stack exactly once. We will denote that integer k enters the stack by writing k and denote that integer k leaves the stack by writing $\overline{k}$. If $n = 1$, there is only one possible sequence, namely, $1, \overline{1}$. For $n = 2$, there are two possibilities:

	Order leaving the stack
(a) $1, 2, \overline{2}, \overline{1}$	2, 1
(b) $1, \overline{1}, 2, \overline{2}$	1, 2

Thus, if $n = 2$, there are two possible sequences in which the integers 1, 2 can leave a stack. Now consider the case $n = 3$. There are only five possibilities for inserting the integers 1, 2, 3 into a stack and deleting them from it:

	Order leaving the stack
(a) $1, 2, 3, \overline{3}, \overline{2}, \overline{1}$	3, 2, 1
(b) $1, 2, \overline{2}, 3, \overline{3}, \overline{1}$	2, 3, 1
(c) $1, 2, \overline{2}, \overline{1}, 3, \overline{3}$	2, 1, 3
(d) $1, \overline{1}, 2, 3, \overline{3}, \overline{2}$	1, 3, 2
(e) $1, \overline{1}, 2, \overline{2}, 3, \overline{3}$	1, 2, 3

Thus, of the six possible permutations of the integers 1, 2, 3, only five can result from the insertion and deletion of 1, 2, 3 by the use of a stack.

We will count the number c_n of permutations of $1, 2, \ldots, n$ that can result from the use of a stack in this manner. (Thus c_n is just the number of different ways that the integers 1 through n can leave a stack if they enter it in sequence.) The preceding paragraph shows that

$$c_1 = 1, \qquad c_2 = 2, \qquad c_3 = 5.$$

It is convenient also to define $c_0 = 1$. For an arbitrary positive integer n, we consider when the integer 1 is deleted from the stack. If it is the first integer deleted from the stack, then the sequence of operations begins

$$1, \overline{1}, 2, \ldots.$$

The number of permutations that can result from such a sequence of operations is just the number of possible ways that $2, 3, \ldots, n$ can leave a stack if they enter it in sequence. This number is $c_{n-1} = c_0 c_{n-1}$.

If 1 is the second integer deleted from the stack, then the first integer deleted from the stack must be 2. Thus the sequence of operations must begin

$$1, 2, \overline{2}, \overline{1}, 3, \ldots$$

The number of permutations that can result from such a sequence of operations is $c_1 c_{n-2}$.

If 1 is the third integer deleted from the stack, then the first two integers deleted from the stack must be 2 and 3. Thus 1 must enter the stack, 2 and 3 must enter and leave the stack in some sequence, then 1 must leave, and finally $4, 5, \ldots, n$ must enter and leave the stack in some sequence. The number of permutations that can result from such a sequence of operations is $c_2 c_{n-3}$.

In general, suppose that 1 is the kth integer deleted from the stack. Then the $k-1$ integers $2, 3, \ldots, k$ must enter and leave the stack in some sequence before integer 1 is deleted, and the $n-k$ integers $k+1, k+2, \ldots, n$ must enter and leave the stack in some sequence after integer 1 is deleted. The multiplication principle shows that the number of ways to perform these two operations is $c_{k-1}c_{n-k}$. Thus the addition principle gives

$$c_n = c_0 c_{n-1} + c_1 c_{n-2} + \cdots + c_{n-1} c_0 \qquad \text{for } n \geq 1.$$

Since we know that $c_0 = 1$, the preceding recurrence relation can be used to compute subsequent values of the sequence. For example,

$$\begin{aligned}
c_1 &= c_0 c_0 = 1 \cdot 1 = 1, \\
c_2 &= c_0 c_1 + c_1 c_0 = 1 \cdot 1 + 1 \cdot 1 = 2, \\
c_3 &= c_0 c_2 + c_1 c_1 + c_2 c_0 = 1 \cdot 2 + 1 \cdot 1 + 2 \cdot 1 = 5, \\
c_4 &= c_0 c_3 + c_1 c_2 + c_2 c_1 + c_3 c_0 = 1 \cdot 5 + 1 \cdot 2 + 2 \cdot 1 + 5 \cdot 1 = 14, \\
c_5 &= c_0 c_4 + c_1 c_3 + c_2 c_2 + c_3 c_1 + c_4 c_0 \\
&= 1 \cdot 14 + 1 \cdot 5 + 2 \cdot 2 + 5 \cdot 1 + 14 \cdot 1 = 42,
\end{aligned}$$

and so forth.

The numbers c_n are called **Catalan numbers**, after Eugene Charles Catalan (1814–1894), who showed that they represent the number of ways in which n pairs of parentheses can be inserted into the expression

$$x_1 x_2 \ldots x_{n+1}$$

to group the factors into n products of pairs of numbers. For example, the $c_3 = 5$ different groupings of $x_1x_2x_3x_4$ into three products of pairs of numbers are

$$((x_1x_2)x_3)x_4, \qquad x_1(x_2(x_3x_4)), \qquad (x_1(x_2x_3))x_4,$$
$$x_1((x_2x_3)x_4), \qquad \text{and} \qquad (x_1x_2)(x_3x_4).$$

The Catalan numbers occur in several basic problems of computer science. ■

The preceding examples have shown several situations in which recurrence relations arise in counting problems. Recurrence relations are also invaluable in examining change over time in discrete settings, as shown in the next example.

Example 9.6 A grain elevator company receives 200 tons of corn per week from farmers once harvest starts. The elevator operators plan to ship out 30 percent of the corn on hand each week once the harvest season begins. If the company has 600 tons of corn on hand at the beginning of harvest, what recurrence relation describes the amount of corn on hand at the end of each week throughout the harvest season?

If g_n represents the number of tons of corn on hand at the end of week n of the harvest season, we can express the situation described in the preceding paragraph by the recurrence relation

$$g_n = g_{n-1} - 0.30g_{n-1} + 200 \qquad \text{for } n \geq 1,$$

with the initial condition $g_0 = 600$—that is,

$$g_n = 0.70g_{n-1} + 200 \qquad \text{for } n \geq 1 \qquad \text{and} \qquad g_0 = 600.$$

The 0.70 coefficient of g_{n-1} reflects that 70 percent of the corn on hand is not shipped during the week, and the constant term 200 represents the amount of new corn brought to the elevator within the week. ■

Recurrence relations are also often used to study the current or projected status of financial accounts.

Example 9.7 The Thompsons are purchasing a new house costing \$200,000, with a down payment of \$25,000 and a 30-year mortgage. Interest on the unpaid balance of the mortgage is to be compounded at the monthly rate of 1 percent, and the monthly payments will be \$1800. How much will the Thompsons owe after n months of payments?

Let b_n denote the balance (in dollars) that will be owed on the mortgage after n months of payments. We will obtain a recurrence relation expressing b_n in terms of previous balances. Note that the balance owed after n months will equal the balance owed after $n - 1$ months plus the monthly interest minus one monthly payment. Symbolically, we have

$$b_n = b_{n-1} + 0.01b_{n-1} - 1800,$$

which simplifies to the form

$$b_n = 1.01b_{n-1} - 1800 \qquad \text{for } n \geq 1.$$

Since this equation expresses b_n in terms of b_{n-1} only, we need just one term to use this recurrence relation. Now the amount owed initially is the purchase price minus the down payment, and so the initial condition is $b_0 = 175{,}000$. ■

Recurrence relations, when applied to the study of change, as shown in Examples 9.6 and 9.7, are sometimes referred to as **discrete dynamical systems**. They are the discrete analogs of the differential equations used to study change in continuous settings.

The examples in this section have shown several situations in which recurrence relations arise. Other examples will be considered throughout this chapter. When a sequence is defined by a recurrence relation, it is sometimes possible to find an explicit formula that expresses the general term as a function of n. Sections 9.2 and 9.3 will be devoted primarily to this subject.

EXERCISES 9.1

In Exercises 1–12, determine s_5 if $s_0, s_1, s_2, \ldots$ is a sequence satisfying the given recurrence relation and initial conditions.

1. $s_n = 3s_{n-1} - 9$ for $n \geq 1$, $s_0 = 5$
2. $s_n = -s_{n-1} + n^2$ for $n \geq 1$, $s_0 = 3$
3. $s_n = 2s_{n-1} + 3n$ for $n \geq 1$, $s_0 = 5$
4. $s_n = 5s_{n-1} - 2^n$ for $n \geq 1$, $s_0 = 1$
5. $s_n = 2s_{n-1} + s_{n-2}$ for $n \geq 2$, $s_0 = 2$, $s_1 = -3$
6. $s_n = 5s_{n-1} - 3s_{n-2}$ for $n \geq 2$, $s_0 = -1$, $s_1 = -2$
7. $s_n = -s_{n-1} + ns_{n-2} - 1$ for $n \geq 2$, $s_0 = 3$, $s_1 = 4$
8. $s_n = 3s_{n-1} - 2ns_{n-2} + 2^n$ for $n \geq 2$, $s_0 = 2$, $s_1 = 4$
9. $s_n = 2s_{n-1} + s_{n-2} - s_{n-3}$ for $n \geq 3$, $s_0 = 2$, $s_1 = -1$, $s_2 = 4$
10. $s_n = s_{n-1} - 3s_{n-2} + 2s_{n-3}$ for $n \geq 3$, $s_0 = 2$, $s_1 = 3$, $s_2 = 4$
11. $s_n = -s_{n-1} + 2s_{n-2} + s_{n-3} + n$ for $n \geq 3$, $s_0 = 1$, $s_1 = 2$, $s_2 = 5$
12. $s_n = s_{n-1} - 4s_{n-2} + 3s_{n-3} + (-1)^n$ for $n \geq 3$, $s_0 = 3$, $s_1 = 2$, $s_2 = 4$

13. For the 1995–96 academic year, tuition at Stanford University was \$28,000 and had increased by at least 5.25% for each of the preceding 15 years. Assuming that the tuition at Stanford increases by 5.25% per year for the indefinite future, write a recurrence relation and initial conditions for t_n, the cost of tuition at Stanford n years after 1995.
14. Individual membership fees at the Evergreen Tennis Club were \$50 in 1970 and have increased by \$2 per year since then. Write a recurrence relation and initial conditions for m_n, the membership fee n years after 1970.
15. A restaurant chain had 24 franchises in 1975 and has opened 6 new franchises each year since then. Assuming that this trend continues indefinitely, write a recurrence relation and initial conditions for r_n, the number of restaurant franchises n years after 1975.
16. A bank pays 6% interest compounded annually on its passbook savings accounts. Suppose that you deposit \$800 in one of these accounts and make no further deposits or withdrawals. Write a recurrence relation and initial conditions for b_n, the balance of the account after n years.
17. A consumer purchased items costing \$280 with a department store credit card that charges 1.5% interest per month compounded monthly. Write a recurrence relation and initial conditions for b_n, the balance of the consumer's account after n months if no further charges occur and the minimum monthly payment of \$25 is made.
18. Tom, a new college graduate, has just been offered a job paying \$24,000 in the first year. Each year thereafter, the salary will increase by \$1000 plus a 5% cost of living adjustment. Write a recurrence relation and initial conditions for s_n, the amount of Tom's salary after n years of employment.
19. The process for cleaning up waste in a nuclear reactor core room eliminates 85% of the waste present in the area each week. If there is 1.7 kg of waste in the room at the beginning of the monitoring period and 2 kg of additional waste are generated each week, determine a recurrence relation and initial conditions describing the amount w_n of waste in the core room at the end of week n of the monitoring period.
20. The jabby bird is in danger of being placed on the endangered species list, as there are only 975 of the birds known to be in existence. A bird is placed on the list when the known population reaches 100. If 27% of the jabby bird population either dies or is taken by a poacher each year and only 5 new jabby birds are born, write a recurrence relation and initial conditions describing the number j_n of jabby birds at the end of n years.

21. Each day you buy exactly one of the following items: tape (costing \$1), a ruler (costing \$1), pens (costing \$2), pencils (costing \$2), paper (costing \$2), or a loose-leaf binder (costing \$3). Write a recurrence relation and initial conditions for the number s_n of different sequences in which you can spend exactly n dollars ($n \geq 1$).

22. Suppose that you have a large supply of 2-, 3-, and 5-cent stamps. Write a recurrence relation and initial conditions for the number s_n of different ways in which n cents worth of postage can be attached to an envelope if the order in which the stamps are attached matters. (Thus a 2-cent stamp followed by a 3-cent stamp is different from a 3-cent stamp followed by a 2-cent stamp.)

23. Write a recurrence relation and initial conditions for a_n, the number of arrangements of the integers $1, 2, \ldots, n$.

24. Write a recurrence relation and initial conditions for s_n, the number of subsets of a set with n elements.

25. Write a recurrence relation and initial conditions for s_n, the number of two-element subsets of a set with n elements.

26. Write a recurrence relation and initial conditions for the number s_n of n-bit strings having no two consecutive zeros. Compute s_6.

27. Write a recurrence relation and initial conditions for the number s_n of sequences of nickels, dimes, and quarters that can be inserted into a vending machine to purchase a soft drink costing $5n$ cents. How many sequences are there for a drink costing 50 cents?

28. For $n \geq 2$, a $6 \times n$ checkerboard can be covered by L-shaped pieces of the type shown in Figure 2.20. Write a recurrence relation and initial conditions for p_n, the number of L-shaped pieces needed to cover a $6 \times n$ checkerboard.

29. Write a recurrence relation and initial conditions for the number c_n of different ways to group $2n$ people into pairs to play n chess games.

30. Let p_n denote the number of permutations of $1, 2, \ldots, n$ in which each integer either occupies its natural position or is adjacent to its natural position. Write a recurrence relation and initial conditions for p_n.

31. For some positive integer n, draw n circles in the Euclidean plane such that every pair of circles intersects at exactly two points and no three circles have a point in common. Write a recurrence relation and initial conditions for r_n, the number of regions into which these circles divide the plane.

32. Suppose that you have an unlimited supply of red, white, blue, green, and gold poker chips, which are indistinguishable except for color. Write a recurrence relation and initial conditions for the number s_n of ways to stack n chips with no two consecutive red chips.

33. Write a recurrence relation and initial conditions for the number s_n of n-bit strings having no three consecutive zeros. Compute s_5.

34. Write a recurrence relation and initial conditions for s_n, the number of three-element subsets of a set with n elements.

35. Suppose that $2n$ points are marked on a circle and labeled $1, 2, \ldots, 2n$. Write a recurrence relation and initial conditions for the number c_n of ways to draw n nonintersecting chords joining two of these points.

36. Write a recurrence relation and initial conditions for s_n, the number of squares of any size that can be formed by using the blocks on an $n \times n$ checkerboard.

37. Write a recurrence relation and initial conditions for the number s_n of n-bit strings that do not contain the pattern 010. Then compute s_6.

38. Write a recurrence relation and initial conditions for the number s_n of n-bit strings that contain neither the pattern 1000 nor the pattern 0011.

9.2 THE METHOD OF ITERATION

In Example 9.2 we saw that the minimum number of moves required to shift n disks from one spoke to another in the Towers of Hanoi game satisfies the recurrence relation

$$m_n = 2m_{n-1} + 1 \qquad \text{for } n \geq 2$$

and the initial condition $m_1 = 1$. From this information, we can determine the value of m_n for any positive integer n. For example, the first few terms of the sequence defined by these conditions are

$$\begin{aligned} m_1 &= 1, \\ m_2 &= 2(1) + 1 = 2 + 1 = 3, \\ m_3 &= 2(3) + 1 = 6 + 1 = 7, \\ m_4 &= 2(7) + 1 = 14 + 1 = 15, \qquad \text{and} \\ m_5 &= 2(15) + 1 = 30 + 1 = 31. \end{aligned}$$

We can continue evaluating terms of the sequence in this manner, and so we can eventually determine the value of any particular term. This process can be quite tedious, however, if we need to evaluate m_n when n is large. In Example 9.7, for instance, we might need to know the unpaid balance of the mortgage after 20 years (240 months), which would require us to evaluate b_{240}. Although straightforward, this calculation would be quite time-consuming if we were evaluating the terms by hand in this manner.

We see, therefore, that it is often convenient to have a formula for computing the general term of a sequence defined by a recurrence relation without needing to calculate all of the preceding terms. A simple method that can be used to try to find such a formula is to start with the initial conditions and compute successive terms of the sequence, as just illustrated. If a pattern can be found, we can then guess an explicit formula for the general term and try to prove it by mathematical induction. This procedure is called the **method of iteration**.

We will use the method of iteration to find an explicit formula for the general term of the sequence satisfying the Towers of Hanoi recurrence

$$m_n = 2m_{n-1} + 1 \qquad \text{for } n \geq 2$$

with the initial condition $m_1 = 1$. We have already computed the first few terms of the sequence satisfying these conditions. Although it is possible to see a pattern developing from these computations, it is helpful to repeat these calculations *without* simplifying the results to a numerical value:

$$\begin{aligned}
m_1 &= 1 \\
m_2 &= 2(1) + 1 = 2 + 1 \\
m_3 &= 2(2+1) + 1 = 2^2 + 2 + 1 \\
m_4 &= 2(2^2 + 2 + 1) + 1 = 2^3 + 2^2 + 2 + 1 \\
m_5 &= 2(2^3 + 2^2 + 2 + 1) + 1 = 2^4 + 2^3 + 2^2 + 2 + 1 \\
&\vdots
\end{aligned}$$

From these calculations, we can guess an explicit formula for m_n:

$$m_n = 2^{n-1} + 2^{n-2} + \cdots + 2^2 + 2 + 1.$$

By using the familiar algebraic identity (see Example 2.51)

$$1 + x + x^2 + \cdots + x^n = \frac{x^{n+1} - 1}{x - 1},$$

this formula can be expressed in an even more compact manner:

$$m_n = \frac{2^n - 1}{2 - 1} = 2^n - 1.$$

At this point, the formula for m_n is nothing more than an educated guess. To verify that it does indeed give the correct values for m_n, we must prove by induction that the formula is correct. To do so, we must show that if a sequence $m_1, m_2, m_3, \ldots$ satisfies the recurrence relation

$$m_n = 2m_{n-1} + 1 \qquad \text{for } n \geq 2$$

and the initial condition $m_1 = 1$, then $m_n = 2^n - 1$ for all positive integers n. Clearly, the formula is correct for $n = 1$ because

$$2^1 - 1 = 2 - 1 = 1 = m_1.$$

Now we assume that the formula is correct for some nonnegative integer k; that is, we assume that

$$m_k = 2^k - 1.$$

It remains to show that the formula is correct for $k + 1$. From the recurrence relation, we know that

$$m_{k+1} = 2m_k + 1.$$

Hence

$$\begin{aligned}
m_{k+1} &= 2(2^k - 1) + 1 \\
&= 2^{k+1} - 2 + 1 \\
&= 2^{k+1} - 1,
\end{aligned}$$

which proves the formula is correct for $k+1$. It follows from the principle of mathematical induction that the formula

$$m_n = 2^n - 1$$

is correct for all positive integers n.

Certain formulas are very useful for simplifying the algebraic expressions that arise when using the method of iteration. One of these is the identity

$$1 + x + x^2 + \cdots + x^n = \frac{x^{n+1} - 1}{x - 1}$$

from Example 2.51. Another is the formula for the sum of the first n positive integers, namely,

$$1 + 2 + 3 + \cdots + n = \frac{n(n+1)}{2},$$

which was obtained in Exercise 11 of Section 2.5.

Example 9.8 In Example 9.1 we saw that the number e_n of edges in the complete graph $\mathcal{K}_n$ satisfies the recurrence relation

$$e_n = e_{n-1} + (n-1) \qquad \text{for } n \geq 2$$

and the initial condition $e_1 = 0$. We will use the method of iteration to obtain a formula for e_n. To begin, we use the recurrence relation to compute several terms of the sequence:

$$\begin{aligned} e_1 &= 0 \\ e_2 &= 0 + 1 \\ e_3 &= (0 + 1) + 2 \\ e_4 &= (0 + 1 + 2) + 3 \\ e_5 &= (0 + 1 + 2 + 3) + 4 \\ &\vdots \end{aligned}$$

From these calculations, we conjecture that

$$\begin{aligned} e_n &= 0 + 1 + 2 + \cdots + (n-1) \\ &= \frac{(n-1)n}{2} \\ &= \frac{n^2 - n}{2}. \end{aligned}$$

To verify that the formula is correct, we again need a proof by induction to show that if a sequence satisfies the recurrence relation

$$e_n = e_{n-1} + (n-1) \qquad \text{for } n \geq 2$$

and the initial condition $e_1 = 0$, then its terms are given by the formula

$$e_n = \frac{n^2 - n}{2}.$$

The formula is correct for $n = 1$ because

$$\frac{n^2 - n}{2} = \frac{1^2 - 1}{2} = 0 = e_1.$$

Assume that

$$e_k = \frac{k^2 - k}{2}$$

for some $k \geq 1$. Then

$$\begin{aligned} e_{k+1} &= e_k + [(k+1) - 1] \\ &= \frac{k^2 - k}{2} + k \\ &= \frac{k^2 - k}{2} + \frac{2k}{2} \\ &= \frac{(k^2 + 2k + 1) - (k+1)}{2} \\ &= \frac{(k+1)^2 - (k+1)}{2}. \end{aligned}$$

Thus the formula is correct for $k + 1$. It now follows from the principle of mathematical induction that the formula is correct for all positive integers n. ■

Example 9.9 Find a formula for p_n, the number of ways to group $2n$ people into pairs.

We begin by finding a recurrence relation and initial conditions for p_n. In order to group $2n$ people into pairs, we first select a person and find that person a partner. Since the partner can be taken to be any of the other $2n - 1$ persons in the original group, there are $2n - 1$ ways to form this first pair. We now are left with the problem of grouping the remaining $2n - 2$ persons into pairs, and the number of ways of doing this is p_{n-1}. Thus, by the multiplication principle, we have

$$p_n = (2n-1)p_{n-1} \qquad \text{for } n \geq 1.$$

Since two people can be paired in only one way, the initial condition is $p_1 = 1$.

Let us use the method of iteration to find an explicit formula for p_n. Because

$$\begin{aligned} p_1 &= 1 \\ p_2 &= 3(1) \\ p_3 &= 5(3)(1) \\ p_4 &= 7(5)(3)(1) \\ p_5 &= 9(7)(5)(3)(1), \end{aligned}$$

it appears that

$$p_n = (2n-1)(2n-3)\cdots(3)(1),$$

the product of the odd integers from 1 through $2n-1$. This expression can be written more compactly by using factorial notation. Since the even integers are missing from this product, we insert them into both the numerator and the denominator:

$$\begin{aligned} (2n-1)(2n-3)\cdots(3)(1) &= \frac{(2n)(2n-1)(2n-2)(2n-3)\cdots(3)(2)(1)}{(2n)(2n-2)\cdots(2)} \\ &= \frac{(2n)!}{(2)(n)(2)(n-1)\cdots(2)(1)} \\ &= \frac{(2n)!}{2^n n!}. \end{aligned}$$

Thus our conjecture is that

$$p_n = \frac{(2n)!}{2^n n!}.$$

We must prove that this formula is correct for all positive integers n by mathematical induction. For $n = 1$, the formula gives

$$\frac{(2n)!}{2^n n!} = \frac{2!}{2^1 \cdot 1!} = \frac{2}{2} = 1,$$

which is correct. Assume that

$$p_k = \frac{(2k)!}{2^k k!}$$

for some positive integer k. Then

$$\begin{aligned} p_{k+1} &= [2(k+1)-1]p_k \\ &= (2k+1)\frac{(2k)!}{2^k k!} \\ &= \frac{(2k+1)!}{2^k k!} \\ &= \frac{2k+2}{2(k+1)} \cdot \frac{(2k+1)!}{2^k k!} \\ &= \frac{(2k+2)!}{2^{k+1}(k+1)!}, \end{aligned}$$

proving the formula for $k+1$. Thus the formula is correct for all positive integers n by the principle of mathematical induction. ■

The process for finding a formula for the general term of the sequence of values associated with a recurrence relation is akin to finding the solution of a differential equation in a continuous setting. For this reason, the formula expressing the relation is sometimes called a **solution** to the recurrence relation.

In our examples so far, we verified formulas for recurrence relations that expressed s_n in terms of s_{n-1}, but no other s_i. When we want to verify a formula for a recurrence relation that expresses s_n in terms of s_i for $i \leq n-2$, then the *strong* principle of mathematical induction will be required.

Example 9.10 We will prove that if x_n satisfies $x_n = x_{n-1} + 2x_{n-2} + 2n - 9$ for $n \geq 2$ with the initial conditions $x_0 = 6$ and $x_1 = 0$, then

$$x_n = 3(-1)^n + 2^n + 2 - n \qquad \text{for } n \geq 0.$$

It is easily checked that the formula is correct for $n = 0$ and $n = 1$. Assume that the formula is correct for $n = 0, 1, \ldots, k$, where $k \geq 1$. Then

$$\begin{aligned} x_{k+1} &= x_k + 2x_{k-1} + 2(k+1) - 9 \\ &= [3(-1)^k + 2^k + 2 - k] + 2[3(-1)^{k-1} + 2^{k-1} + 2 - (k-1)] + 2k - 7 \\ &= -3(-1)^{k-1} + 2^k + 2 - k + 6(-1)^{k-1} + 2(2^{k-1}) + 4 - 2(k-1) + 2k - 7 \\ &= -3(-1)^{k-1} + 6(-1)^{k-1} + 2^k + 2^k + 2 - k + 4 + 2 - 7 \\ &= 3(-1)^{k-1} + 2(2^k) + 2 - k - 1 \\ &= 3(-1)^{k+1} + 2^{k+1} + 2 - (k+1), \end{aligned}$$

which verifies the formula for $k+1$. It follows from the strong principle of induction that the formula is correct for all nonnegative integers n. ■

There can be many formulas that agree with the beginning terms of a particular sequence. Here is a famous problem where it is easy to mistake the pattern of numbers. It can be shown that for any positive integer n, it is possible to draw n circles in the Euclidean plane such that every pair of circles intersects at exactly two points and no three circles have a point in common. Moreover, for any configuration of such circles, the number r_n of regions into which these circles divide the plane is the same. Let us determine a formula expressing r_n as a function of n.

Figure 9.5 shows that $r_0 = 1$, $r_1 = 2$, $r_2 = 4$, and $r_3 = 8$. From these numbers, it is natural to conjecture that $r_n = 2^n$. However, the formula $r_n = 2^n$ is *not* correct, because Figure 9.6 shows that $r_4 = 14$. To obtain a correct formula, we must discover a recurrence relation that relates the number of regions formed by n circles to the number of regions formed by $n - 1$ circles.

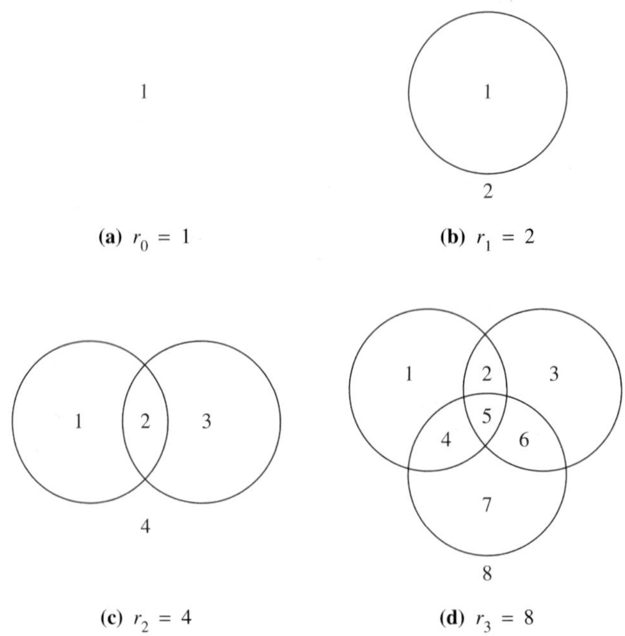

FIGURE 9.5

Suppose that we include a fourth circle with the three in Figure 9.5(d). This new circle (the inner circle in Figure 9.7) does not subdivide either region 5 or region 8 in Figure 9.5(d). We see, however, that this circle intersects each of the other three circles in two points each, and so the circle is subdivided into $2(3) = 6$ arcs, each of which subdivides a region into two new regions. The arc from A to B in Figure 9.7, for instance, subdivides region 3 in Figure 9.5(d) into the two regions marked 3 in Figure 9.7. This same situation occurs in general: If there are $n - 1 \geq 1$ circles satisfying the given conditions and another circle is drawn so that every pair of circles intersects at exactly two points and no three circles have a point in common, then the new circle forms $2(n-1)$ new regions. Hence we see that

$$r_n = r_{n-1} + 2(n-1) \qquad \text{for } n \geq 2.$$

Note that this recurrence relation is *not* valid for $n = 1$. Hence, in trying to find a formula for r_n, we cannot expect our formula to be valid for $n = 0$.

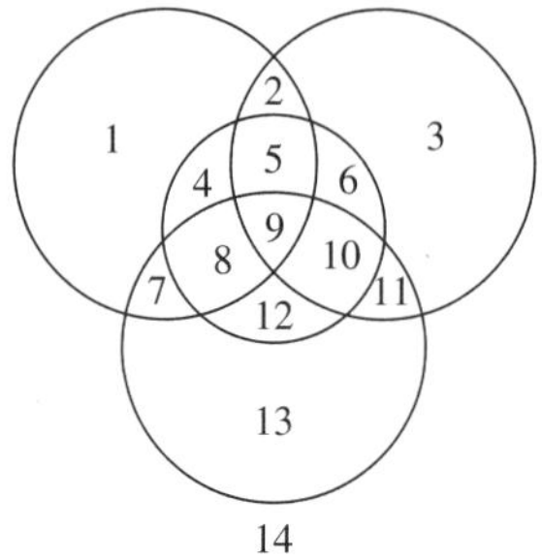

FIGURE 9.6

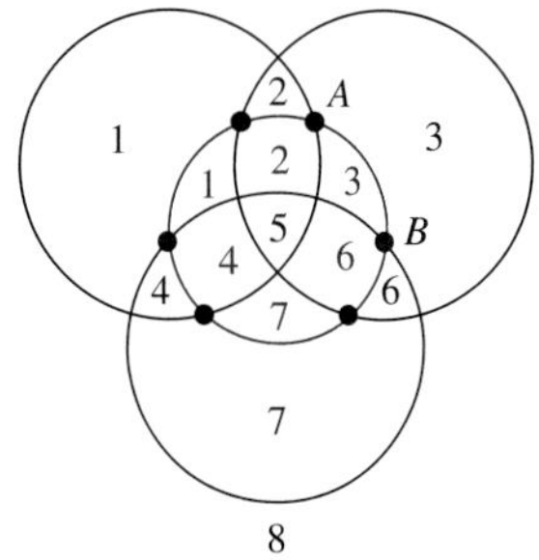

FIGURE 9.7

From this recurrence relation, we see that

$$\begin{aligned}
r_1 &= 2 \\
r_2 &= r_1 + 2(1) = 2 + 2(1) \\
r_3 &= r_2 + 2(2) = 2 + 2(1) + 2(2) \\
r_4 &= r_3 + 2(3) = 2 + 2(1) + 2(2) + 2(3) \\
r_5 &= r_4 + 2(4) = 2 + 2(1) + 2(2) + 2(3) + 2(4).
\end{aligned}$$

From these calculations, the pattern appears to be

$$r_n = 2 + 2[1 + 2 + \cdots + (n-1)].$$

Using the formula for the sum of the first k positive integers, namely,

$$1 + 2 + 3 + \cdots + k = \frac{k(k+1)}{2},$$

we can simplify the expression for r_n to the form

$$r_n = 2 + 2\frac{(n-1)n}{2} = 2 + (n-1)n = n^2 - n + 2.$$

Thus we conjecture that $r_n = n^2 - n + 2$ for $n \geq 1$. We will leave as an exercise the verification of this formula by induction. Note that, as expected, the formula

$$r_n = n^2 - n + 2$$

that we obtained is valid only for $n \geq 1$.

It is instructive to see what would happen if we tried to prove that the incorrect formula $r_n = 2^n$ satisfies the recurrence relation. Note that for $n = 1$, the formula is correct because $2^1 = 2 = r_1$. The difficulty arises in the inductive step. Assume that the formula is correct for some positive integer k; that is, assume that $r_k = 2^k$. Then

$$r_{k+1} = r_k + 2k = 2^k + 2k,$$

which is not equal to 2^{k+1} for every positive integer k. Since our induction proof breaks down, we must conclude that the general term of the recurrence relation

$$r_n = r_{n-1} + 2(n-1)$$

is not given by the formula $r_n = 2^n$.

We will conclude this section by obtaining a formula for the number of derangements of the integers 1 through n.

Example 9.11 In Example 9.4 we obtained the recurrence relation

$$D_n = (n-1)(D_{n-1} + D_{n-2}) \qquad \text{for } n \geq 3$$

and the initial conditions $D_1 = 0$ and $D_2 = 1$ for the number D_n of derangements of the integers $1, 2, \ldots, n$. It turns out that a sequence that satisfies this recurrence relation must also satisfy the relation

$$D_n = nD_{n-1} + (-1)^n.$$

To see why, note first that

$$\begin{aligned} D_n - nD_{n-1} &= D_n - (n-1)D_{n-1} - D_{n-1} \\ &= (n-1)D_{n-2} - D_{n-1} \\ &= -[D_{n-1} - (n-1)D_{n-2}]. \end{aligned}$$

It follows that

$$\begin{aligned} D_n - nD_{n-1} &= (-1)[D_{n-1} - (n-1)D_{n-2}] \\ &= (-1)^2[D_{n-2} - (n-2)D_{n-3}] \\ &= (-1)^3[D_{n-3} - (n-3)D_{n-4}] \\ &\;\;\vdots \\ &= (-1)^{n-2}[D_2 - 2D_1] \\ &= (-1)^{n-2}[1 - 2(0)] \\ &= (-1)^{n-2} \\ &= (-1)^n. \end{aligned}$$

The resulting recurrence relation

$$D_n = nD_{n-1} + (-1)^n$$

holds not only for $n \geq 3$, but also for $n = 2$. Hence it is valid for $n \geq 2$.

We will use the method of iteration to obtain a formula expressing D_n in terms of n. It is easier to apply the method of iteration to this new recurrence relation than to the one in Example 9.4 because the new one relates D_n to D_{n-1} rather than to both D_{n-1} and D_{n-2}. It produces the following terms:

$$\begin{aligned}
D_2 &= 1\\
D_3 &= 3(1) - 1 = 3 - 1\\
D_4 &= 4(3-1) + 1 = 4(3) - 4 + 1\\
D_5 &= 5[4(3) - 4 + 1] - 1 = 5(4)(3) - 5(4) + 5 - 1\\
D_6 &= 6[5(4)(3) - 5(4) + 5 - 1] + 1\\
&= 6(5)(4)(3) - 6(5)(4) + 6(5) - 6 + 1.
\end{aligned}$$

Note that

$$\begin{aligned}
D_6 &= P(6,4) - P(6,3) + P(6,2) - P(6,1) + P(6,0)\\
&= \frac{6!}{2!} - \frac{6!}{3!} + \frac{6!}{4!} - \frac{6!}{5!} + \frac{6!}{6!}\\
&= 6!\left(\frac{1}{2!} - \frac{1}{3!} + \frac{1}{4!} - \frac{1}{5!} + \frac{1}{6!}\right).
\end{aligned}$$

Thus we conjecture that

$$D_n = n!\left[\frac{1}{2!} - \frac{1}{3!} + \cdots + (-1)^n\frac{1}{n!}\right].$$

We leave as an exercise the verification that this formula is correct. ■

The method of iteration depends upon recognizing a pattern formed by successive terms. In practice this may be very difficult, or even impossible, to do. Nevertheless, the method of iteration can often be used to find a formula for the general term of a sequence defined by a recurrence relation, especially in problems where the recurrence relation is simple. Furthermore, the method of iteration is not limited to recurrence relations of a particular form. In the next section, we will use the method of iteration to find formulas for two very common types of recurrence relations.

EXERCISES 9.2

1. Prove by mathematical induction that $n^2 - n + 2$ is a solution to the recurrence relation $r_n = r_{n-1} + 2(n-1)$ for $n \geq 2$ with the initial condition $r_1 = 2$.
2. Prove by mathematical induction that $4(2^n) + 3$ is a solution to the recurrence relation $s_n = 2s_{n-1} - 3$ for $n \geq 1$ with the initial condition $s_0 = 7$.
3. Use the principle of mathematical induction to prove that $4^n - 3^n + 1$ is a solution to the recurrence relation $s_n = 7s_{n-1} - 12s_{n-2} + 6$ for $n \geq 2$ with the initial conditions $s_0 = 1$, $s_1 = 2$.

4. Prove by mathematical induction that $3^n(3+n)$ is a solution to the recurrence relation $s_n = 3s_{n-1} + 3^n$ for $n \geq 1$ with the initial condition $s_0 = 3$.

5. In Example 9.11, prove by mathematical induction that

$$D_n = n!\left[\frac{1}{2!} - \frac{1}{3!} + \cdots + (-1)^n \frac{1}{n!}\right] \qquad \text{for } n \geq 2.$$

6. Prove by mathematical induction that

$$\frac{1}{n}C(2n-2, n-1)$$

is a solution to the recurrence relation

$$s_n = \frac{4n-6}{n}s_{n-1} \qquad \text{for } n \geq 1$$

with the initial condition $s_1 = 1$.

7. Prove by mathematical induction that $C(2n+2, 3)$ is a solution to the recurrence relation $s_n = s_{n-1} + 4n^2$ for $n \geq 2$ with the initial condition $s_1 = 4$.

8. Prove by mathematical induction that $C(2n+1, 3)$ is a solution to the recurrence relation $s_n = s_{n-1} + (2n-1)^2$ for $n \geq 2$ with the initial condition $s_1 = 1$.

9. Compute $2^2 + 4^2 + 6^2 + \cdots + (2n)^2$.

10. Compute $1^2 + 3^2 + 5^2 + \cdots + (2n-1)^2$.

In Exercises 11–24, use the method of iteration to find a formula expressing s_n as a function of n for the given recurrence relation and initial conditions.

11. $s_n = s_{n-1} + 4$, $s_0 = 9$

12. $s_n = -2s_{n-1}$, $s_0 = 3$

13. $s_n = 3s_{n-1}$, $s_0 = 5$

14. $s_n = s_{n-1} - 2$, $s_0 = 7$

15. $s_n = -s_{n-1}$, $s_0 = 6$

16. $s_n = -s_{n-1} + 10$, $s_0 = -4$

17. $s_n = 5s_{n-1} + 3$, $s_0 = 1$

18. $s_n = 5 - 3s_{n-1}$, $s_0 = 2$

19. $s_n = s_{n-1} + 4(n-3)$, $s_0 = 10$

20. $s_n = -s_{n-1} + (-1)^n$, $s_0 = 6$

21. $s_n = -s_{n-1} + a^n$, $s_0 = 1, a \neq -1$

22. $s_n = s_{n-1} + 2n + 4$, $s_0 = 5$

23. $s_n = ns_{n-1} + 1$, $s_0 = 3$

24. $s_n = 4s_{n-2} + 1$, $s_0 = 1$, $s_1 = \frac{7}{3}$

25. Suppose that a high school had 1000 students enrolled at the beginning of the 1995 school term. The trend in enrollment over the previous 20 years was that the number s_n of students beginning a school year was 5% less than that of the previous year.

(a) Find a recurrence relation and initial conditions representing this situation, assuming that the enrollment trend continues.

(b) Find a formula expressing s_n as a function of n.

(c) If the enrollment trend continues, what number of students does the formula predict for the beginning of the school year 2005?

26. Zebra mussels are fresh water mollusks that attack underwater structures. Suppose that the volume of mussels in a confined area grows at a rate of 0.2% per day.

(a) If there are now 10 cubic feet of mussels in a lock on the Illinois River at Peoria, Illinois, develop a recurrence relation and initial conditions that represent the volume m_n of the mussel colony n days hence.

(b) Develop a formula expressing m_n as a function of n.

27. The following figure shows that 4 toothpicks are needed to make a 1×1 square, 12 toothpicks are needed to make a 2×2 square composed of four 1×1 squares, and 24 toothpicks are needed to make a 3×3 square composed of nine 1×1 squares:

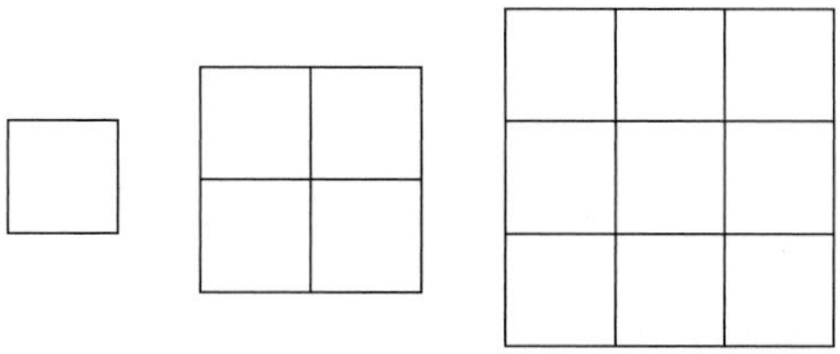

How many toothpicks are needed to make an $n \times n$ square composed of 1×1 squares?

28. (a) A rabbit breeder has one male–female pair of newborn rabbits. After reaching two months of age, these rabbits and their offspring breed two other male–female pairs each month. Write a recurrence relation and initial conditions for r_n, the number of pairs of rabbits after n months. Assume that no rabbits die during the n months.

(b) Show that the recurrence relation in (a) satisfies $r_n = 2r_{n-1} + (-1)^n$ for $n \geq 2$, and find a formula expressing r_n as a function of n. (*Hint:* Use a procedure like that in Example 9.11.)

29. Consider the sequences of n terms in which each term is -1, 0, or 1. Let s_n denote the number of such sequences in which no term of 0 occurs after a term of 1. Find a formula expressing s_n as a function of n.

30. Consider the sequences of n terms in which each term is -1, 0, or 1. Let s_n denote the number of such sequences that contain an even number of 1s. Find a formula expressing s_n as a function of n.

31. For some positive integer n, draw n lines in the Euclidean plane so that every pair of lines intersects and no three lines have a point in common. Determine r_n, the number of regions into which these lines divide the plane.

32. Let a_n denote the number of additions performed in evaluating the determinant of an $n \times n$ matrix by the cofactor expansion method. Find a formula expressing a_n as a function of n.

33. Suppose that the Towers of Hanoi game is played with $2n$ disks, two each of n different sizes. A disk may be moved on top of a disk of the same size or larger, but never on top of a smaller disk. Find a formula for the minimum number of moves required to transfer all the disks from one spoke to another.

34. Suppose that in the Towers of Hanoi game a disk can be moved only from one spoke to an *adjacent* spoke. Let m_n denote the minimal number of moves required to move n disks from the leftmost spoke to the rightmost spoke. Find a formula expressing m_n as a function of n.

9.3 LINEAR DIFFERENCE EQUATIONS WITH CONSTANT COEFFICIENTS

The simplest type of recurrence relation gives s_n as a function of s_{n-1} for $n \geq 1$. We call an equation of the form

$$s_n = as_{n-1} + b,$$

where a and b are constants and $a \neq 0$, a **first-order linear difference equation with constant coefficients**. For example, the following recurrence relations are all first-order linear difference equations with constant coefficients:

$$s_n = 3s_{n-1} - 1, \qquad s_n = s_{n-1} + 7, \qquad \text{and} \qquad s_n = 5s_{n-1}.$$

Recurrence relations of this type occur frequently in applications, especially in the analysis of financial transactions. The recurrence relations in Examples 9.2, 9.6, and 9.7 are first-order linear difference equations with constant coefficients.

Since a first-order linear difference equation with constant coefficients expresses s_n in terms of s_{n-1}, a sequence defined by such a difference equation is completely determined if a single term is known. We will use the method of iteration to find an explicit formula for this type of equation that expresses s_n as a function of n and s_0.

Consider the first-order linear difference equation with constant coefficients $s_n = as_{n-1} + b$ that has first term s_0. The first few terms of the sequence defined by this equation are

$$\begin{aligned}
s_0 &= s_0, \\
s_1 &= as_0 + b, \\
s_2 &= as_1 + b = a(as_0 + b) + b = a^2 s_0 + ab + b, \\
s_3 &= as_2 + b = a(a^2 s_0 + ab + b) + b = a^3 s_0 + a^2 b + ab + b, \\
s_4 &= as_3 + b = a(a^3 s_0 + a^2 b + ab + b) + b \\
&= a^4 s_0 + a^3 b + a^2 b + ab + b.
\end{aligned}$$

It appears that

$$\begin{aligned}
s_n &= a^n s_0 + a^{n-1} b + a^{n-2} b + \cdots + a^2 b + ab + b \\
&= a^n s_0 + (a^{n-1} + a^{n-2} + \cdots + a^2 + a + 1)b.
\end{aligned}$$

If $a = 1$, the expression in parentheses equals n; otherwise it can be simplified by using the identity from Example 2.51:

$$1 + x + x^2 + \cdots + x^n = \frac{x^{n+1} - 1}{x - 1}.$$

Applying this identity to the earlier expression for s_n, we obtain

$$\begin{aligned} s_n &= a^n s_0 + \left(\frac{a^n - 1}{a - 1}\right) b \\ &= a^n s_0 + a^n \left(\frac{b}{a-1}\right) - \left(\frac{b}{a-1}\right) \\ &= a^n (s_0 + c) - c, \end{aligned}$$

where

$$c = \frac{b}{a-1}.$$

We will state this result as Theorem 9.1, leaving a formal proof by mathematical induction as an exercise.

Theorem 9.1 The general term of the first-order linear difference equation with constant coefficients $s_n = as_{n-1} + b$ that has initial value s_0 satisfies

$$s_n = \begin{cases} a^n(s_0 + c) - c & \text{if } a \neq 1 \\ s_0 + nb & \text{if } a = 1, \end{cases}$$

where

$$c = \frac{b}{a-1}.$$

Example 9.12 Find a formula for s_n if $s_n = 3s_{n-1} - 1$ for $n \geq 1$ and $s_0 = 2$. Here, $a = 3$ and $b = -1$ in the notation of Theorem 9.1. Thus

$$c = \frac{b}{a-1} = \frac{-1}{3-1} = \frac{-1}{2},$$

so

$$\begin{aligned} s_n &= a^n(s_0 + c) - c \\ &= 3^n \left[2 + \frac{-1}{2}\right] - \frac{-1}{2} \\ &= \frac{3}{2}(3^n) + \frac{1}{2} \\ &= \frac{1}{2}(3^{n+1} + 1). \end{aligned}$$

Substituting $n = 0, 1, 2, 3, 4$, and 5 into this formula gives

$$s_0 = 2, \quad s_1 = 5, \quad s_2 = 14, \quad s_3 = 41, \quad s_4 = 122, \quad \text{and} \quad s_5 = 365,$$

which are easily checked by using the recurrence relation

$$s_n = 3s_{n-1} - 1 \qquad \text{for } n \geq 1$$

and the initial condition $s_0 = 2$. ■

Example 9.13 Find a formula for b_n, the unpaid balance after n months of the Thompsons' mortgage in Example 9.7.

We saw in Example 9.7 that b_n satisfies the recurrence relation

$$b_n = 1.01b_{n-1} - 1800 \qquad \text{for } n \geq 1$$

and the initial condition $b_0 = 175{,}000$. Since this recurrence relation is a first-order linear difference equation with constant coefficients, Theorem 9.1 can be used to find a formula expressing b_n as a function of n and b_0. In the notation of Theorem 9.1, we have $a = 1.01$ and $b = -1800$. Hence

$$c = \frac{b}{a-1} = \frac{-1800}{1.01 - 1} = \frac{-1800}{0.01} = -180{,}000.$$

Thus the desired formula for b_n is

$$\begin{aligned} b_n &= a^n(b_0 + c) - c \\ &= (1.01)^n[175{,}000 + (-180{,}000)] - (-180{,}000) \\ &= -5000(1.01)^n + 180{,}000. \end{aligned}$$

For example, the balance of the loan after 20 years (240 months) of payments is

$$\begin{aligned} b_{240} &= -5000(1.01)^{240} + 180{,}000 \approx -54{,}462.77 + 180{,}000 \\ &= 125{,}537.23. \end{aligned}$$

Thus the Thompsons will still owe \$125,537.23 after 20 years. ■

Example 9.14 A lumber company owns 7000 birch trees. Each year the company plans to harvest 12 percent of its trees and plant 600 new ones.

(a) How many trees will there be after 10 years?

(b) How many trees will there be in the long run?

Let s_n denote the number of trees after n years. During year n, 12 percent of the trees existing in year $n-1$ will be harvested; this number is $0.12s_{n-1}$. Since 600 additional trees will be planted during year n, the number of trees after n years is described by the equation

$$s_n = s_{n-1} - 0.12s_{n-1} + 600,$$

that is,

$$s_n = 0.88s_{n-1} + 600.$$

This is a first-order linear difference equation with constant coefficients $a = 0.88$ and $b = 600$. We are interested in the solution of this equation that satisfies the initial condition $s_0 = 7000$. In the notation of Theorem 9.1,

$$c = \frac{b}{a-1} = \frac{600}{0.88-1} = \frac{600}{-0.12} = -5000.$$

Hence a formula expressing s_n in terms of n is

$$\begin{aligned} s_n &= a^n(s_0 + c) - c \\ &= (0.88)^n(7000 - 5000) + 5000 \\ &= 2000(0.88)^n + 5000. \end{aligned}$$

(a) Therefore, after 10 years, the number of trees will be

$$s_{10} = 2000(0.88)^{10} + 5000 \approx 5557.$$

(b) As n increases, the quantity $(0.88)^n$ decreases to zero. Hence the formula

$$s_n = 2000(0.88^n) + 5000$$

implies that the number of trees approaches 5000. (Note that as the number of trees approaches 5000, the number of trees being harvested each year approaches the number of new trees being planted.) ■

Second-Order Homogeneous Linear Difference Equations

Certainly, one of the most famous sequences in mathematics is the sequence of Fibonacci numbers, named after Leonardo Fibonacci of Pisa (c. 1170–1250), the most famous European mathematician of the Middle Ages. The sequence first appeared in the following problem in his text *Liber Abaci*:

> A man has one male–female pair of rabbits in a hutch entirely surrounded by a wall. We wish to know how many pairs of rabbits can be bred from this pair in one year, if the nature of these rabbits is such that every month they breed one other male–female pair which begin to breed in the second month after their birth. Assume that no rabbits die during the year.

The diagram in Figure 9.8, with the letter "M" denoting a mature pair and the letter "I" denoting an immature pair, shows the pattern of reproduction described in the problem.

We see that the number of pairs at the beginning of any month equals the number at the beginning of the previous month plus the number of new pairs, which equals the number of pairs two months earlier. Thus the number of pairs at the beginning of month n satisfies the recurrence relation

$$F_n = F_{n-1} + F_{n-2} \qquad \text{for } n \geq 3$$

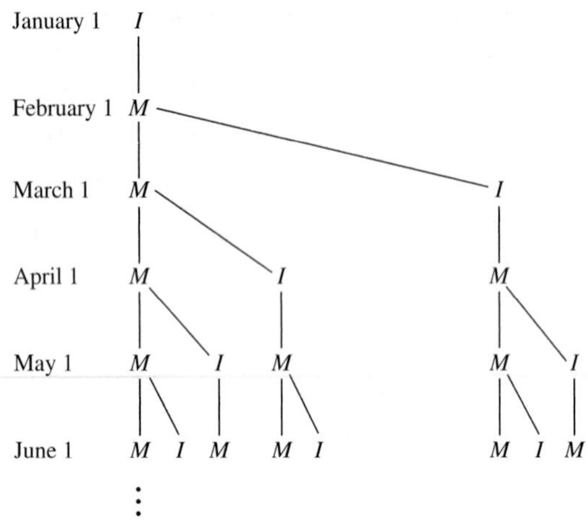

FIGURE 9.8

with the initial conditions $F_1 = F_2 = 1$. It is easily checked that there will be $F_{13} = 233$ pairs of rabbits in the hutch one year later!

The recurrence relation

$$F_n = F_{n-1} + F_{n-2} \qquad \text{for } n \geq 3$$

is called the **Fibonacci recurrence**. It appears in a wide variety of applications, often where it is least expected. Recall, for instance, that the Fibonacci recurrence occurred in Example 9.3.

A recurrence relation of the form

$$s_n = as_{n-1} + bs_{n-2},$$

where a and b are constants and $b \neq 0$, is called a **second-order homogeneous linear difference equation with constant coefficients**. The word "homogeneous" indicates that there is no constant term in the recurrence relation. The Fibonacci recurrence is an example of a second-order homogeneous linear difference equation with constant coefficients. Since this type of recurrence relation occurs frequently in applications, it is useful to have a formula expressing s_n as a function of n for a sequence defined by such a recurrence.

Theorem 9.2 Consider the second-order homogeneous linear difference equation with constant coefficients

$$s_n = as_{n-1} + bs_{n-2} \qquad \text{for } n \geq 2$$

that has initial values s_0 and s_1. Let r_1 and r_2 denote the roots of the equation

$$x^2 = ax + b.$$

Then

(a) If $r_1 \neq r_2$, there exist constants c_1 and c_2 such that $s_n = c_1 r_1^n + c_2 r_2^n$ for $n = 0, 1, 2, \ldots$.

(b) If $r_1 = r_2 = r$, there exist constants c_1 and c_2 such that $s_n = (c_1 + nc_2)r^n$ for $n = 0, 1, 2, \ldots$.

Proof. (a) If $s_n = c_1 r_1^n + c_2 r_2^n$ for $n = 0, 1, 2, \ldots$, then, for $n = 0$ and $n = 1$, we must have

$$\begin{aligned} s_0 &= c_1 + c_2 \\ s_1 &= c_1 r_1 + c_2 r_2. \end{aligned}$$

Multiplying the first equation by r_2 and subtracting the second yields

$$\begin{aligned} r_2 s_0 - s_1 &= c_1 r_2 - c_1 r_1 \\ &= c_1(r_2 - r_1). \end{aligned}$$

Hence, since $r_1 \neq r_2$, we have

$$\frac{r_2 s_0 - s_1}{r_2 - r_1} = c_1.$$

Thus

$$\begin{aligned} c_2 &= s_0 - c_1 \\ &= s_0 - \frac{r_2 s_0 - s_1}{r_2 - r_1} \\ &= \frac{s_0(r_2 - r_1)}{r_2 - r_1} - \frac{r_2 s_0 - s_1}{r_2 - r_1} \\ &= \frac{s_1 - r_1 s_0}{r_2 - r_1}. \end{aligned}$$

We leave it to the reader to show that for these values of c_1 and c_2, the expression $c_1 r_1^n + c_2 r_2^n$ with $n = 0$ and $n = 1$ yields the initial values s_0 and s_1.

This establishes the base for an induction proof. Assume now that, for $n = 0, 1, \ldots, k$, we have $s_n = c_1 r_1^n + c_2 r_2^n$. Since r_1 and r_2 are roots of the equation $x^2 = ax + b$, we have $ar_1 + b = r_1^2$ and $ar_2 + b = r_2^2$. Thus

$$\begin{aligned} s_{k+1} &= as_k + bs_{k-1} \\ &= a\left[c_1 r_1^k + c_2 r_2^k\right] + b\left[c_1 r_1^{k-1} + c_2 r_2^{k-1}\right] \\ &= c_1\left[ar_1^k + br_1^{k-1}\right] + c_2\left[ar_2^k + br_2^{k-1}\right] \\ &= c_1 r_1^{k-1}(ar_1 + b) + c_2 r_2^{k-1}(ar_2 + b) \\ &= c_1 r_1^{k-1}\left(r_1^2\right) + c_2 r_2^{k-1}\left(r_2^2\right) \\ &= c_1 r_1^{k+1} + c_2 r_2^{k+1}. \end{aligned}$$

Therefore $s_n = c_1 r_1^n + c_2 r_2^n$ for $n = k + 1$. It follows from the strong principle of mathematical induction that $s_n = c_1 r_1^n + c_2 r_2^n$ for all nonnegative integers n.

(b) Note that since $b \neq 0$ in the equation $x^2 = ax + b$ (because the given recurrence relation is of the second order), we must have $r \neq 0$. Therefore, if $s_n = (c_1 + nc_2)r^n$ for $n = 0, 1, 2, \ldots$, then taking $n = 0$ and $n = 1$ gives

$$\begin{aligned} s_0 &= c_1 \\ s_1 &= c_1 r + c_2 r. \end{aligned}$$

Hence we see that

$$c_1 = s_0 \qquad \text{and} \qquad c_2 = \frac{s_1 - s_0 r}{r}.$$

Again, we leave it to the reader to show that for these values of c_1 and c_2, the expression $(c_1 + nc_2)r^n$ with $n = 0$ and $n = 1$ yields the initial values s_0 and s_1. Assume that for $n = 0, 1, \ldots, k$ we have $s_n = (c_1 + nc_2)r^n$. Since

$$x^2 - ax - b = (x - r)^2 = x^2 - 2rx + r^2,$$

we have

$$a = 2r \qquad \text{and} \qquad b = -r^2.$$

Thus

$$\begin{aligned} s_{k+1} &= as_k + bs_{k-1} \\ &= a(c_1 + kc_2)r^k + b[c_1 + (k-1)c_2]r^{k-1} \\ &= 2r(c_1 + kc_2)r^k + (-r^2)[c_1 + (k-1)c_2]r^{k-1} \\ &= (2c_1 + 2kc_2)r^{k+1} - [c_1 + (k-1)c_2]r^{k+1} \\ &= [c_1 + (k+1)c_2]r^{k+1}. \end{aligned}$$

Therefore $s_n = (c_1 + nc_2)r^n$ for all nonnegative integers n by the strong principle of mathematical induction. ■

The quadratic equation $x^2 = ax + b$ in Theorem 9.2 is called the **auxiliary equation** of the recurrence relation $s_n = as_{n-1} + bs_{n-2}$.

Note that the proof of Theorem 9.2 actually produces the constants c_1 and c_2 that occur in the formula expressing $s_n = as_{n-1} + bs_{n-2}$ as a function of n. Rather than memorizing the formulas for these constants, however, we will obtain the values of c_1 and c_2 by solving a system of linear equations, as was done in the proof of Theorem 9.2. Examples 9.15 and 9.16 demonstrate this technique.

Example 9.15 Find a formula for s_n if s_n satisfies the recurrence relation

$$s_n = -s_{n-1} + 6s_{n-2} \qquad \text{for } n \geq 2$$

and the initial conditions $s_0 = 7$ and $s_1 = 4$.

The given recurrence relation is a second-order homogeneous linear difference equation with constant coefficients; its auxiliary equation is

$$x^2 = -x + 6.$$

Rewriting this equation in the form

$$x^2 + x - 6 = 0$$

and factoring gives

$$(x + 3)(x - 2) = 0.$$

Thus the roots of the auxiliary equation are -3 and 2. Because these roots are distinct, we use Theorem 9.2(a) to obtain a formula for s_n. Hence there are constants c_1 and c_2 such that $s_n = c_1(-3)^n + c_2(2)^n$. To determine these constants, we use the initial conditions $s_0 = 7$ and $s_1 = 4$. For $n = 0$, we have

$$7 = s_0 = c_1(-3)^0 + c_2(2)^0 = c_1 + c_2.$$

Likewise, for $n = 1$, we have

$$4 = s_1 = c_1(-3)^1 + c_2(2)^1 = -3c_1 + 2c_2.$$

Therefore the values of c_1 and c_2 satisfy the system of linear equations

$$\begin{aligned} c_1 + c_2 &= 7 \\ -3c_1 + 2c_2 &= 4. \end{aligned}$$

A simple calculation gives $c_1 = 2$ and $c_2 = 5$. Hence the terms of the sequence defined by the given recurrence relation and initial conditions satisfy

$$s_n = c_1(-3)^n + c_2(2)^n = 2(-3)^n + 5(2)^n. \quad \blacksquare$$

Example 9.16 Find a formula for s_n if s_n satisfies the recurrence relation

$$s_n = 6s_{n-1} - 9s_{n-1} \qquad \text{for } n \geq 2$$

and the initial conditions $s_0 = -2$ and $s_1 = 6$.

The given recurrence relation is a second-order homogeneous linear difference equation with constant coefficients; its auxiliary equation is

$$x^2 = 6x - 9.$$

Rewriting this equation and factoring gives

$$x^2 - 6x + 9 = (x - 3)^2 = 0.$$

In this case, the roots of the auxiliary equation are equal, and so we use Theorem 9.2(b) to obtain a formula for s_n. According to this theorem, there are constants c_1 and c_2 such that $s_n = (c_1 + nc_2)3^n$. To determine these constants, we make use of the initial conditions $s_0 = -2$ and $s_1 = 6$. For $n = 0$, we have

$$-2 = s_0 = (c_1 + 0c_2)3^0 = c_1.$$

Likewise, for $n = 1$, we have

$$6 = s_1 = (c_1 + 1c_2)3^1 = 3c_1 + 3c_2.$$

Therefore the values of c_1 and c_2 satisfy the system of linear equations

$$\begin{aligned} c_1 \qquad &= -2 \\ 3c_1 + 3c_2 &= 6. \end{aligned}$$

Clearly, $c_1 = -2$ and $c_2 = 4$. Hence the terms of the sequence defined by the given recurrence relation and initial conditions satisfy

$$s_n = (c_1 + nc_2)3^n = (-2 + 4n)3^n. \quad \blacksquare$$

In the next example, we use Theorem 9.2 to find a formula for the Fibonacci numbers.

Example 9.17 Find a formula expressing the nth Fibonacci number F_n as a function of n.

Recall that the recurrence relation satisfied by the Fibonacci numbers is

$$F_n = F_{n-1} + F_{n-2} \qquad \text{for } n \geq 3,$$

a second-order homogeneous linear difference equation with constant coefficients. Its auxiliary equation is

$$x^2 = x + 1.$$

Rewriting this equation in the form $x^2 - x - 1 = 0$ and applying the quadratic formula, we find that there are two distinct roots:

$$r_1 = \frac{1 + \sqrt{5}}{2} \qquad \text{and} \qquad r_2 = \frac{1 - \sqrt{5}}{2}.$$

Hence Theorem 9.2(a) guarantees that there are constants c_1 and c_2 such that

$$F_n = c_1 \left(\frac{1 + \sqrt{5}}{2} \right)^n + c_2 \left(\frac{1 - \sqrt{5}}{2} \right)^n.$$

To determine the values of c_1 and c_2, we use the initial values F_1 and F_2 to obtain

$$1 = c_1 \left(\frac{1 + \sqrt{5}}{2} \right)^1 + c_2 \left(\frac{1 - \sqrt{5}}{2} \right)^1 \quad \text{and}$$

$$1 = c_1 \left(\frac{1 + \sqrt{5}}{2} \right)^2 + c_2 \left(\frac{1 - \sqrt{5}}{2} \right)^2.$$

Solving this system of two equations for c_1 and c_2, we get $c_1 = \frac{1}{\sqrt{5}}$ and $c_2 = \frac{-1}{\sqrt{5}}$.

Substituting these values into the formula for F_n gives

$$F_n = \frac{1}{\sqrt{5}}\left(\frac{1+\sqrt{5}}{2}\right)^n - \frac{1}{\sqrt{5}}\left(\frac{1-\sqrt{5}}{2}\right)^n. \quad \blacksquare$$

Our final example provides a solution to a problem known as *the gambler's ruin*.

Example 9.18 Douglas and Jennifer have agreed to bet one dollar on each flip of a fair coin and to continue playing until one of them wins all of the other's money. What is the probability that Douglas will win all of Jennifer's money if Douglas starts with a dollars and Jennifer starts with b dollars?

To analyze this game, we will obtain a recurrence relation and initial conditions for d_n, the probability that Douglas will win all of Jennifer's money if he currently has n dollars. Let $t = a + b$, the total amount of money available to the players. Note that $d_0 = 0$ because Douglas has no money left, and $d_t = 1$ because Douglas has all of the money. Moreover, if $1 \le n \le t - 1$, then Douglas has a .5 probability of winning one dollar on the next flip (raising the amount of money he has to $n + 1$ dollars) and a .5 probability of losing one dollar on the next flip (reducing the amount of money he has to $n - 1$ dollars). Hence

$$d_n = .5d_{n+1} + .5d_{n-1} \qquad \text{for } 1 \le n \le t - 1.$$

Multiplying this equation by 2 and rearranging terms gives us

$$d_{n+1} = 2d_n - d_{n-1} \qquad \text{for } 1 \le n \le t - 1.$$

Replacing n by $n - 1$ in this equation yields the second-order homogeneous linear difference equation

$$d_n = 2d_{n-1} - d_{n-2} \qquad \text{for } 2 \le n \le t.$$

The auxiliary equation is

$$x^2 = 2x - 1,$$

which has $r = 1$ as a double root. Therefore Theorem 9.2(b) guarantees that there exist constants c_1 and c_2 such that $d_n = (c_1 + nc_2)1^n$. Using the initial conditions $d_0 = 0$ and $d_t = 1$, we see that

$$d_0 = 0 = c_1 + 0c_2$$
$$d_t = 1 = c_1 + tc_2.$$

The solution of this system of linear equations is $c_1 = 0$ and $c_2 = \frac{1}{t}$. Hence

$$d_n = \left(0 + \frac{n}{t}\right)(1)^n = \frac{n}{t} = \frac{n}{a+b}.$$

Therefore, at the start (when Douglas has a dollars), the probability of Douglas's winning the game is

$$\frac{a}{a+b},$$

and the probability of Jennifer's winning the game is

$$1 - \frac{a}{a+b} = \frac{b}{a+b}.$$

Hence a player's probability of winning is proportional to the amount of money with which he or she starts the game. ■

EXERCISES 9.3

In Exercises 1–24, find an explicit formula for s_n if $s_0, s_1, s_2, \ldots$ is a sequence satisfying the given recurrence relation and initial conditions.

1. $s_n = s_{n-1} + 3$, $s_0 = 2$

2. $s_n = 5s_{n-1} - 4$, $s_0 = 1$

3. $s_n = 4s_{n-1}$, $s_0 = 5$

4. $s_n = 1.5s_{n-1} - 1$, $s_0 = 4$

5. $s_n = -s_{n-1} + 6$, $s_0 = -4$

6. $s_n = s_{n-1} - 10$, $s_0 = 32$

7. $s_n = 3s_{n-1} - 8$, $s_0 = 3$

8. $s_n = -2s_{n-1}$, $s_0 = -5$

9. $s_n = s_{n-1} - 5$, $s_0 = 100$

10. $s_n = -s_{n-1} + 7$, $s_0 = 1$

11. $s_n = -2s_{n-1} - 9$, $s_0 = 7$

12. $s_n = 10s_{n-1} - 45$, $s_0 = 2$

13. $s_n = s_{n-1} + 2s_{n-2}$, $s_0 = 9$, $s_1 = 0$

14. $s_n = -2s_{n-1} - s_{n-2}$, $s_0 = 3$, $s_1 = 1$

15. $s_n = 8s_{n-1} - 16s_{n-2}$, $s_0 = 6$, $s_1 = 20$

16. $s_n = 4s_{n-2}$, $s_0 = -1$, $s_1 = -14$

17. $s_n = 9s_{n-2}$, $s_0 = 1$, $s_1 = 9$

18. $s_n = 6s_{n-1} - 9s_{n-2}$, $s_0 = 1$, $s_1 = 9$

19. $s_n = -4s_{n-1} - 4s_{n-2}$, $s_0 = -4$, $s_1 = 2$

20. $s_n = -8s_{n-1} - 15s_{n-2}$, $s_0 = 2$, $s_1 = 2$

21. $s_n = 10s_{n-1} - 25s_{n-2}$, $s_0 = -7$, $s_1 = -15$

22. $s_n = 10s_{n-1} - 24s_{n-2}$, $s_0 = 1$, $s_1 = 0$

23. $s_n = -5s_{n-1} - 4s_{n-2}$, $s_0 = 3$, $s_1 = 15$

24. $s_n = 4s_{n-1} - 4s_{n-2}$, $s_0 = -3$, $s_1 = 4$

25. In order to combat hypertension, Mr. Lorenzo is to take a capsule containing 25 mg of a drug each morning after awaking. During the next 24 hours, 20 percent of the amount of the drug in the body is eliminated.

(a) Write a difference equation and initial conditions giving the amount of the drug in Mr. Lorenzo's body immediately after he takes the nth capsule.

(b) How much of the drug will there be in Mr. Lorenzo's body immediately after he takes the eighth capsule?

(c) To what level will this drug eventually accumulate in his body?

26. In Mayville, 90 percent of the existing dog licenses are reissued each year, and 1200 new licenses are issued. In 2000 there were 15,000 dog licenses issued.

(a) Write a difference equation and initial conditions describing the number of dog licenses Mayville will issue n years after 2000.

(b) How many dog licenses will Mayville issue in 2009?

(c) If the present trend continues, how many dog licenses can Mayville expect to issue after many years?

27. Michelle has just opened a savings account with an initial deposit of \$1000. From the money she earns from her part-time job, Michelle will add \$100 to her savings account at the end of each month. If the account compounds interest monthly at the rate of 0.5% per month, how much will it be worth two years from now?

28. Suppose that a corporate executive deposits \$2000 per year for 35 years into an individual retirement account. If interest is compounded annually at the rate of 8%, how much will the account be worth after the last deposit?

29. The Johnson family is considering the purchase of a house costing \$159,000. They will make a \$32,000 down payment and take a 30-year mortgage for the remainder of the cost. The mortgage compounds interest monthly at the rate of 0.9% per month. How much will the Johnsons' monthly payment be under these conditions?

30. An automobile advertisement states that a new automobile can be purchased for \$175 per month. If payments are to be made for 60 months and interest is charged at the rate of 1.075% per month compounded monthly, how much does this car cost?

31. Write a recurrence relation and initial conditions for the number s_n of sequences of 1s and 2s having a sum of n. Use these to obtain a formula expressing s_n as a function of n.

32. Suppose that a bank, in order to promote long-term saving by its customers, authorizes a new savings account that pays 6% interest on money during the first year it is in the account and 8.16% interest on money that is in the account for more than one year. Interest is to be compounded annually. If you deposit \$1100 into such an account and allow the interest to accumulate, how much will the account be worth in n years?

33. Let $v_1, v_2, \ldots, v_n (n \geq 3)$ be the vertices of a cycle (as defined in Section 4.2), and let c_n denote the number of distinguishable ways to color these vertices with the colors red, yellow, blue, and green so that no adjacent vertices have the same color. Determine a formula expressing c_n as a function of n.

34. Use Theorem 9.1 to find a formula for

$$s_n = s_0 + s_0 r + s_0 r^2 + \cdots + s_0 r^n,$$

the sum of the first $n+1$ terms of a geometric progression with first term s_0 and common ratio $r \neq 1$. (*Hint:* The sequence $s_0, s_1, s_2 \ldots$ satisfies the first-order linear difference equation $s_n = rs_{n-1} + s_0$ with initial term s_0.)

35. Prove by mathematical induction that if $s_0, s_1, s_2, \ldots$ is a sequence satisfying the first-order linear difference equation $s_n = as_{n-1} + b$, then

$$s_n = \begin{cases} a^n(s_0 + c) - c & \text{if } a \neq 1 \\ s_0 + nb & \text{if } a = 1 \end{cases}$$

for all nonnegative integers n, where

$$c = \frac{b}{a-1}.$$

36. Let $s_0, s_1, s_2, \ldots$ be a sequence satisfying the first-order linear difference equation $s_n = as_{n-1} + b$ for $n \geq 1$, and define t_n for $n \geq 0$ by $t_n = s_0 + s_1 + s_2 + \cdots + s_n$. Prove by mathematical induction that

$$t_n = \begin{cases} \left(\dfrac{a^{n+1} - 1}{a - 1}\right) s_0 + b\left[\dfrac{a^{n+1} - (n+1)a + n}{(a-1)^2}\right] & \text{if } a \neq 1 \\[2ex] (n+1)\left(s_0 + \dfrac{nb}{2}\right) & \text{if } a = 1. \end{cases}$$

37. Show that if

$$c_1 = \frac{r_2 s_0 - s_1}{r_2 - r_1} \quad \text{and} \quad c_2 = \frac{s_1 - r_1 s_0}{r_2 - r_1},$$

then the values of $c_1 r_1^n + c_2 r_2^n$ for $n = 0$ and $n = 1$ equal s_0 and s_1, respectively, in the proof of Theorem 9.2(a).

38. Show that if

$$c_1 = s_0 \quad \text{and} \quad c_2 = \frac{s_1 - s_0 r}{r},$$

then the values of $(c_1 + nc_2)r^n$ for $n = 0$ and $n = 1$ equal s_0 and s_1, respectively, in the proof of Theorem 9.2(b).

9.4* ANALYZING THE EFFICIENCY OF ALGORITHMS WITH RECURRENCE RELATIONS

An important use of recurrence relations is in the analysis of the complexity of algorithms. In this section we will discuss the use of recurrence relations in determining the complexity of algorithms for searching and sorting, two fundamental processes of computer science. In order to keep the discussion simple, we will assume that the list of items to be searched or sorted consists of real numbers, but the algorithms we present can be used with any objects subject to a suitable order relation (for example, names and alphabetical order).

To illustrate this use, we will analyze the complexity of the following algorithm for checking whether a particular target value is present in an unsorted list. The algorithm proceeds in a natural manner, comparing each item in the list with the target value. It stops if a match is found or if the entire list has been searched.

Sequential Search Algorithm

This algorithm searches a list of n items $a_1, a_2, \ldots, a_n$ for a given target value t. If $t = a_k$ for some index k, then the algorithm gives the first such index k. Otherwise the algorithm gives $k = 0$.

Step 1 (initialize the starting point)
Set $j = 1$.
Step 2 (look for a match)
while $j \le n$ and $a_j \ne t$
  Step 2.1 (move to the next element)
  Replace j with $j + 1$.
endwhile

Step 3 (was there a match?)
if $j \le n$
Set $k = j$.
otherwise
Set $k = 0$.
endif

How efficient is the sequential search algorithm? To answer this question, we will count the maximum number of times that the target value can be compared with an item in the given list. Thus we will determine an upper bound for the number of comparisons made in using the algorithm. Such an analysis is called a *worst-case* analysis.

Suppose that the list to be searched contains n items and that we count the maximum number of comparisons between the target value and an item in the list. In the worst case, we will have to search the entire list to determine if the target value is present, and this requires comparing the target value with every item in the list. Thus n comparisons are required in the worst case, and so the sequential search algorithm is of order at most n.

Another way to count the maximum number of comparisons made by the sequential search algorithm is with a recurrence relation. For a list of n items, let c_n denote the maximum number of comparisons between the target value and an item in the list. To determine if the target is contained in a list of 0 items requires no comparisons; hence $c_0 = 0$. To search a list of n items, we compare the target value with the first item in the list and then, in the worst case, still have to search a list of $n - 1$ items. Thus we see that c_n satisfies the recurrence relation

$$c_n = c_{n-1} + 1 \qquad \text{for } n \ge 1$$

and the initial condition $c_0 = 0$. A formula for the solution to this first-order linear difference equation can be obtained from Theorem 9.1:

$$c_n = n \qquad \text{for } n \ge 0.$$

Example 9.19 We will determine the number b_n of comparisons of items that a bubble sort performs in arranging n numbers $a_1, a_2, \ldots, a_n$ in nondecreasing order.

Recall from Section 1.4 that in performing a bubble sort, the first iteration requires comparing a_{n-1} with a_n, then comparing a_{n-2} with a_{n-1}, and so forth until a_2 is compared with a_1. After each such comparison, we interchange the values of a_k and a_{k+1}, if necessary, so that $a_k \le a_{k+1}$. Thus, when the first iteration is complete, a_1 is the smallest number in the original list. A second iteration is then performed on the smaller list $a_2, a_3, \ldots, a_n$ in order to make a_2 the second-smallest number in the original list. After $n-1$ such iterations, the original list will be in nondecreasing order.

Since the first iteration of a bubble sort requires comparing a_k with a_{k+1} for $k = n-1, n-2, \ldots, 1$, we need $n-1$ comparisons to determine the smallest item in the original list. After this first iteration is complete, the bubble sort algorithm is continued to arrange the smaller list $a_2, a_3, \ldots, a_n$ in nondecreasing order. But the number of comparisons of items needed to do this is just b_{n-1}.

Thus we see that b_n satisfies the recurrence relation

$$b_n = b_{n-1} + (n-1) \qquad \text{for } n \geq 2.$$

Since no comparisons are needed to sort a list containing only 1 item, an initial condition for this recurrence relation is $b_1 = 0$. This is the same recurrence relation and initial condition encountered in Example 9.1 and analyzed in Example 9.8. Therefore, for $n \geq 1$, we have

$$b_n = \frac{n^2 - n}{2},$$

so the bubble sort algorithm has order at most n^2. ■

Divide-and-Conquer Algorithms

A special type of recurrence relation occurs in the analysis of a class of algorithms called divide-and-conquer algorithms. A **divide-and-conquer algorithm** is one in which a problem is split into several smaller problems of the same type. These smaller problems are each split into the same number of smaller problems, and so forth, until the problems become so small that they can be readily solved. The resulting solutions of the small problems are then reassembled to give a solution to the original problem.

Example 9.20 We will describe a divide-and-conquer approach for determining the largest item in a list of $n = 2^k$ items $a_1, a_2, \ldots, a_n$. First we divide the original list into two sublists

$$a_1, a_2, \ldots, a_r \qquad \text{and} \qquad a_{r+1}, a_{r+2}, \ldots, a_n,$$

where we take $r = \frac{n}{2}$ in order to make the two sublists of equal size. Then we find the largest item u in the first sublist and the largest item v in the second sublist. The larger of u and v is the largest item in the original list.

Thus we have reduced the original problem from one involving a list of 2^k items to two problems, each involving a list of 2^{k-1} items. To solve these problems, we subdivide each of the sublists in half, obtaining four sublists with 2^{k-2} items each. Continuing in this manner, we obtain 2^k sublists after k such subdivisions, and each of these sublists contains a single item. Since the largest item in a list with one item is obvious, we can ultimately determine the largest item in the sublists. ■

If, in Example 9.20, the number n is not a power of two, at some stage of the subdivision process we will have a list containing an odd number of items, say,

$$a_1, a_2, \ldots, a_m.$$

In order to obtain sublists of roughly equal size, we divide the list into

$$a_1, a_2, \ldots, a_r \qquad \text{and} \qquad a_{r+1}, a_{r+2}, \ldots, a_m,$$

where

$$r = \frac{m-1}{2}.$$

This puts r items in the first sublist and $m - r = r + 1$ items in the second sublist. Note that the number r is just the smallest positive integer less than $\frac{m}{2}$. We call the greatest integer less than or equal to a real number x the **floor** of x and denote it by $\lfloor x \rfloor$.

Example 9.21 Find the floors of the following numbers:

$$312.5, \quad \tfrac{10}{3}, \quad 7, \quad -3.6, \quad \text{and} \quad \sqrt{1000}.$$

Since the floor of a number is the greatest integer less than or equal to that number, we have

$$\lfloor 312.5 \rfloor = 312, \quad \left\lfloor \tfrac{10}{3} \right\rfloor = 3, \quad \lfloor 7 \rfloor = 7, \quad \lfloor -3.6 \rfloor = -4,$$

and since $31 < \sqrt{1000} < 32$, it follows that

$$\lfloor \sqrt{1000} \rfloor = 31. \quad ■$$

In Example 2.55 we used a divide-and-conquer approach to search for an unknown integer among the numbers 1, 2, ..., 64. We will now present an algorithm that formalizes this searching process.

Binary Search Algorithm

This algorithm searches a sorted list of n items $a_1 \le a_2 \le \cdots \le a_n$ for a given target value t. If $t = a_k$ for some index k, then the algorithm gives one such index k;

otherwise the algorithm gives $k = 0$. In the algorithm, b and e are the beginning and ending indices of the sublist of $a_1, a_2, \ldots, a_n$ currently being searched.

Step 1 (initialize the starting point)
Set $b = 1$ and $e = n$.

Step 2 (look for a match)
repeat
Step 2.1 (determine the middle value of the sublist)
Set $m = \left\lfloor \frac{1}{2}(b + e) \right\rfloor$.
Step 2.2 (determine the new sublist boundaries)
Step 2.2.1 (is t before a_m?)
if $t < a_m$
Replace e with $m - 1$.
endif
Step 2.2.2 (is t after a_m?)
if $t > a_m$
Replace b with $m + 1$.
endif
until either $a_m = t$ or $b > e$

Step 3 (was there a match?)
if $a_m = t$
Set $k = m$.
otherwise
Set $k = 0$.
endif

The next example illustrates the use of the binary search algorithm when the target number is not contained in the given list.

Example 9.22 We will apply the binary search algorithm to determine whether the target number 253 is contained in the list of the 500 even integers from 2 through 1000. (Thus $a_i = 2i$ for $1 \le i \le 500$.)

In step 2.1 our first value of m is

$$m = \left\lfloor \frac{1}{2}(1 + 500) \right\rfloor = \left\lfloor \frac{1}{2}(501) \right\rfloor = \lfloor 250.5 \rfloor = 250.$$

Comparing $a_m = 500$ with $t = 253$, we find that $t \ne a_m$, and, in fact, $t < a_m$. Therefore we change e to 249; b remains 1. The following table exhibits the working of the algorithm as a sequence of questions and answers:

b	e	m	a_m	Is $a_m = 253$?
1	500	$\left\lfloor \frac{1}{2}(1+500) \right\rfloor = 250$	500	no; greater
1	249	$\left\lfloor \frac{1}{2}(1+249) \right\rfloor = 125$	250	no; less
126	249	$\left\lfloor \frac{1}{2}(126+249) \right\rfloor = 187$	374	no; greater
126	186	$\left\lfloor \frac{1}{2}(126+186) \right\rfloor = 156$	312	no; greater
126	155	$\left\lfloor \frac{1}{2}(126+155) \right\rfloor = 140$	280	no; greater
126	139	$\left\lfloor \frac{1}{2}(126+139) \right\rfloor = 132$	264	no; greater
126	131	$\left\lfloor \frac{1}{2}(126+131) \right\rfloor = 128$	256	no; greater
126	127	$\left\lfloor \frac{1}{2}(126+127) \right\rfloor = 126$	252	no; less
127	127	$\left\lfloor \frac{1}{2}(127+127) \right\rfloor = 127$	254	no; greater
127	126			

Since $b > e$ in the last line of the table, we find that the target number is not in the given list. ■

We will analyze the complexity of the binary search algorithm by counting the maximum number of comparisons c_n performed by the algorithm in searching a list of n items. To simplify the analysis, we will assume that $n = 2^r$ for some nonnegative integer r. In step 2.2 we compare the target number to a middle item in the list, a_m. In the worst case, $t \neq a_m$, and we must search one of the two sublists $a_1 \leq a_2 \leq \cdots \leq a_{m-1}$ or $a_{m+1} \leq a_{m+2} \leq \cdots \leq a_n$. Since the longer of these sublists contains $n/2$ items, we see that c_n satisfies the recurrence relation

$$c_n = c_{n/2} + 1 \qquad \text{for } n = 2^r \geq 2.$$

Since one comparison is needed to search a list containing a single item, an initial condition for this recurrence relation is $c_1 = 1$. We leave it to the reader to verify that a formula for the general term of a sequence defined by this recurrence relation is

$$c_n = 1 + \log_2 n \qquad \text{for } n = 2^r \geq 1.$$

In the general case where n is any positive integer, it can be shown that c_n satisfies the recurrence relation

$$c_n = c_{\lfloor n/2 \rfloor} + 1 \qquad \text{for } n \geq 2$$

and the initial condition $c_1 = 1$. In this case, a formula for the general term of a sequence defined by this recurrence relation is

$$c_n = 1 + \lfloor \log_2 n \rfloor \qquad \text{for } n \geq 1.$$

(See Exercise 39.) Thus the binary search algorithm is of order at most $\log_2 n$, whereas the sequential search algorithm is of order at most n. Since

$$\log_2 n < n$$

for all positive integers n, the binary search algorithm is more efficient than the sequential search algorithm when searching a sorted list.

The recurrence relation just obtained is typical of that which results from the analysis of a divide-and-conquer algorithm. More generally, if a divide-and-conquer algorithm subdivides a problem into p smaller problems, a complexity analysis usually leads to a recurrence relation of the form

$$c_n = kc_{\lfloor n/p \rfloor} + f(n),$$

where k is a constant and f is some function of n.

We will now describe an efficient sorting algorithm called a **merge sort** that uses a divide-and-conquer approach. The sorting is accomplished by merging two sorted lists into one larger sorted list, as illustrated in Example 2.56. We first present a formal description of this merging process.

Merging Algorithm

This algorithm merges two sorted lists

$$A: a_1 \leq a_2 \leq \cdots \leq a_m \qquad \text{and} \qquad B: b_1 \leq b_2 \leq \cdots \leq b_n$$

into a single sorted list

$$C: c_1 \leq c_2 \leq \cdots \leq c_{m+n}.$$

Step 1 (initialization)
Set $i = 1$, $j = 1$, and $k = 1$.

Step 2 (construct C until either A or B is used up)
repeat
Step 2.1 (find the next item in C)
if $a_i < b_j$
(a) Set $c_k = a_i$.
(b) Replace i with $i + 1$.
(c) Replace k with $k + 1$.
otherwise
(a) Set $c_k = b_j$.
(b) Replace j with $j + 1$.
(c) Replace k with $k + 1$.
endif
until $i > m$ or $j > n$

Step 3 (copy end of A onto C if necessary)
while $i \le m$
(a) Set $c_k = a_i$.
(b) Replace i with $i + 1$.
(c) Replace k with $k + 1$.
endwhile

Step 4 (copy end of B onto C if necessary)
while $j \le n$
(a) Set $c_k = b_j$.
(b) Replace j with $j + 1$.
(c) Replace k with $k + 1$.
endwhile

We will now use the merging algorithm to sort a list of n items. We begin by regarding the original list as n sublists containing exactly one item. These sublists are necessarily sorted, and we merge them together in pairs. We continue merging in this manner until all the sublists are combined into a single list.

Merge Sort Algorithm

This algorithm sorts a list of n items $a_1, a_2, \ldots, a_n$ into nondecreasing order. In the algorithm, k denotes the number of sublists currently being processed.

Step 1 (initialization)
(a) Regard each item a_i as a one-item list.
(b) Set $k = n$.

Step 2 (merge sublists)
while $k > 1$
if k is even
Step 2.1 (merge an even number of sublists)
(a) Use the merging algorithm to merge sublist 1 with sublist 2, sublist 3 with sublist 4, ..., sublist $k - 1$ with sublist k.
(b) Set $k = n/2$.
otherwise
Step 2.2 (merge an odd number of sublists)
(a) Use the merging algorithm to merge sublist 1 with sublist 2, sublist 3 with sublist 4, ..., sublist $k - 2$ with sublist $k - 1$, and sublist k with the empty list.
(b) Set $k = \dfrac{(n+1)}{2}$.
endif
endwhile

The two examples that follow illustrate the working of the merge sort algorithm.

Example 9.23 We will use the merge sort algorithm to sort the list

$$(19, 14, 11, 18, 30, 17, 6)$$

into nondecreasing order. In step 1 we regard the original list as seven one-item sublists

$$(19), (14), (11), (18), (30), (17), (6).$$

Since each sublist contains only one item, each sublist is in nondecreasing order. There are precisely seven sublists, and so we go to step 2 and apply the merging algorithm to the first and second sublists, the third and fourth sublists, and the fifth and sixth sublists. The seventh sublist is merged with the empty list (the list having no items), so is unchanged. At this point, we have the following four sublists, each in nondecreasing order:

$$(14, 19), (11, 18), (17, 30), (6).$$

Again we go to step 2, where we apply the merging algorithm to the first and second and to the third and fourth of these sublists. This produces the following two ordered lists:

$$(11, 14, 18, 19), (6, 17, 30).$$

When we merge these two lists together, we obtain a single list, and so in step 2 we stop. The resulting list,

$$(6, 11, 14, 17, 18, 19, 30),$$

is the original list in nondecreasing order. ■

Example 9.24 The sorting performed in Example 9.23 can be illustrated with a tree diagram. Each level of the diagram shows one iteration of step 2 of the merge sort algorithm (the application of the merging algorithm to the existing sublists).

The final row in Figure 9.9 contains the output of the merge sort algorithm. ■

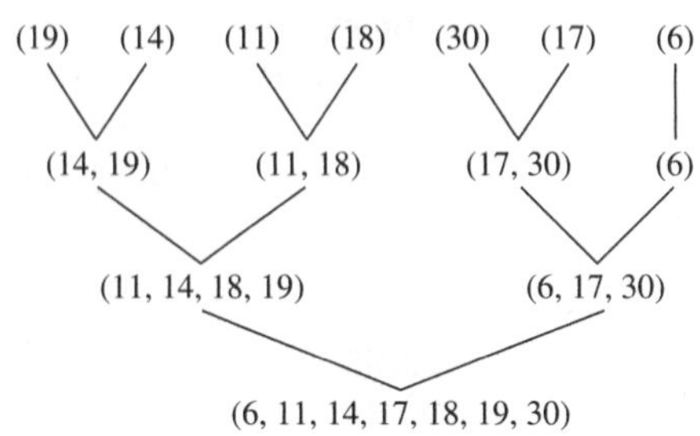

FIGURE 9.9

We will analyze the complexity of the merge sort algorithm by counting the maximum number of comparisons c_n performed by the algorithm when the list being sorted contains $n = 2^r$ items for some positive integer r. As we have seen, the merge sort algorithm works by successively applying the merging algorithm to combine two sublists. Recall from Theorem 2.9 that the merging algorithm requires at most

$$\left\lfloor \frac{n}{2} \right\rfloor + \left\lfloor \frac{n}{2} \right\rfloor - 1 = n - 1$$

comparisons to merge two lists of length $\left\lfloor \frac{n}{2} \right\rfloor$. Thus c_n satisfies the recurrence relation

$$c_n = 2c_{\lfloor n/2 \rfloor} + (n - 1) \qquad \text{for } n = 2^r \geq 2.$$

Since no comparisons are needed to sort a list containing a single item, an initial condition for this recurrence relation is $c_1 = 0$. A formula for the general term of a sequence defined by this recurrence relation can be shown to be

$$c_n = 1 + n(\log_2 n - 1) \qquad \text{for } n = 2^r \geq 1.$$

(See Exercise 37.) Hence the merge sort algorithm is of order at most $n \log_2 n$.

The Efficiency of Sorting Algorithms

In this section we have seen that the bubble sort algorithm has order at most n^2, whereas the merge sort algorithm has order at most $n \log_2 n$. Consequently, the merge sort algorithm is the more efficient algorithm for large values of n. But is it possible to obtain an even more efficient algorithm?

Suppose that we must sort a list of n distinct items $a_1, a_2, \ldots, a_n$ into nondecreasing order. In any sorting algorithm, we compare pairs of items a_i and a_j to determine whether $a_i \leq a_j$ or $a_i > a_j$. Such a comparison results in one of two possible outcomes. Some comparisons, of course, will not provide *new* information; for example, there is no reason to repeat a previous comparison or to compare a_1 with a_3 if we know that $a_1 \leq a_2$ and $a_2 \leq a_3$. But no matter how we make k comparisons, there will be at most 2^k different patterns for the pieces of information that we might receive. As a result, we can distinguish at most 2^k different orderings with k comparisons. If we are to sort the list, we must obtain enough information to distinguish among the $n!$ different sequences in which the items may be listed. Therefore, to sort the list, we must make at least k comparisons, where

$$2^k \geq n!.$$

It can be shown that $n! \geq n^{n/2}$. (See Exercises 33–36.) Hence we must have

$$\begin{aligned} 2^k &\geq n^{n/2} \\ \log_2(2^k) &\geq \log_2(n^{n/2}) \\ k &\geq \frac{n}{2} \log_2 n. \end{aligned}$$

From this inequality, we see that any sorting algorithm must have complexity at least $cn \log_2 n$ for some constant c. It follows that, up to a constant factor, the merge sort algorithm is as efficient as a sorting algorithm can be.

EXERCISES 9.4

In Exercises 1–8, find the floor of the given number.

1. 243

2. -34.5

3. $\frac{33}{7}$

4. 28.963

5. 0.871

6. -2487

7. $\frac{-(-34+2)}{2}$

8. $\frac{-343}{26}$

In Exercises 9–12, make a table, as in Example 9.22, showing the working of the binary search algorithm.

9. $t = 83, n = 100, a_i = i \quad \text{for } i = 1, 2, \ldots, 100$

10. $t = 17, n = 125, a_i = i \quad \text{for } i = 1, 2, \ldots, 125$

11. $t = 400, n = 300, a_i = 3i \quad \text{for } i = 1, 2, \ldots, 300$

12. $t = 305, n = 100, a_i = 2i + 100 \quad \text{for } i = 1, 2, \ldots, 100$

In Exercises 13–18, draw a diagram, as in Example 9.24, illustrating how the merge sort algorithm sorts the given numbers into nondecreasing order.

13. 19, 56, 87, 42

14. 42, 87, 56, 13

15. 13, 89, 56, 45, 62, 75, 68

16. 34, 67, 23, 54, 92, 18, 34, 54, 47

17. 95, 87, 15, 42, 56, 54, 16, 23, 73, 39

18. 34, 81, 46, 2, 53, 5, 4, 8, 26, 1, 0, 45, 35

19. Suppose that one sorting algorithm requires $\frac{n^2}{2}$ comparisons and a second requires $n \log_2 n$ comparisons. How large must n be for the second algorithm to be more efficient?

20. Explain how the merging algorithm treats equal items appearing in the two lists being merged.

In Exercises 21–23, prove the statement true for all real numbers x and y.

21. $\lfloor x \rfloor \le x < \lfloor x \rfloor + 1$.

22. $\lfloor x \rfloor + \lfloor y \rfloor \le \lfloor x + y \rfloor$. Show by example that equality need not hold.

23. $\lfloor x \rfloor + \lfloor y \rfloor \ge \lfloor x + y \rfloor - 1$. Show by example that equality need not hold.

24. If x is any real number and n is an integer, show that $\lfloor x + n \rfloor = \lfloor x \rfloor + n$.

25. Write a recurrence relation and initial conditions for the number e_n of elementary operations performed by the algorithm for evaluating x^n in Section 1.4. Regard an elementary operation as the addition, subtraction, multiplication, division, or comparison of two numbers.

26. Write a recurrence relation and initial conditions for the number e_n of elementary operations performed by the polynomial evaluation algorithm in Section 1.4. Regard an elementary operation as the addition, subtraction, multiplication, division, or comparison of two numbers.

27. Write a recurrence relation and initial conditions for the number e_n of elementary operations performed by Horner's polynomial evaluation algorithm in Section 1.4. Regard an elementary operation as the addition, subtraction, multiplication, division, or comparison of two numbers.

28. Write a recurrence relation and initial conditions for the number c_n of comparisons performed by the divide-and-conquer algorithm in Example 9.20.

29. In Exercise 25 find a formula expressing e_n as a function of n.

30. In Exercise 26 find a formula expressing e_n as a function of n.

31. In Exercise 27 find a formula expressing e_n as a function of n.

32. In Exercise 28 find a formula expressing c_n as a function of n.

33. Prove that, for any integer k such that $1 \le k \le n$, we have $k(n+1-k) \ge n$. (*Hint:* $(n-k)(k-1) \ge 0$.)

34. Prove that if n is an even positive integer, then $n! \ge n^{n/2}$. (*Hint:* Use Exercise 33.)

35. Prove that $\frac{n+1}{2} \ge \sqrt{n}$ for all positive integers n. (*Hint:* Show that $(n+1)^2 \ge (2\sqrt{n})^2$.)

36. Prove that if n is an odd positive integer, then $n! \ge n^{n/2}$. (*Hint:* Use Exercises 33 and 35.)

37. Prove that, for all positive integers of the form $n = 2^k$, $1 + n(\log_2 n - 1)$ is a solution to the recurrence relation $s_n = 2s_{\lfloor n/2 \rfloor} + (n-1)$ for $n \ge 2$ with the initial condition $s_1 = 0$.

38. Prove that, for all positive integers r,

$$\left\lfloor \log_2 \left\lfloor \frac{r+1}{2} \right\rfloor \right\rfloor = \left\lfloor \log_2 \left(\frac{r+1}{2} \right) \right\rfloor.$$

39. Let c and k be constants. Prove that, for all positive integers n, $k + c\lfloor \log_2 n \rfloor$ is a solution to the recurrence relation $s_n = s_{\lfloor n/2 \rfloor} + c$ for $n \ge 2$ with the initial condition $s_1 = k$. (*Hint:* Use the strong principle of mathematical induction and Exercise 38.)

40. The following divide-and-conquer algorithm for sorting is due to R.C. Bose and R.J. Nelson. (See [1] in the suggested readings.) For simplicity, we state it for lists of $n = 2^k$ items, where k is a nonnegative integer.

To sort a list of 2^k items, divide it into two sublists, each containing 2^{k-1} items. Sort each of the sublists, and then merge the two sorted sublists by using the following divide-and-conquer algorithm. To merge sorted lists A and B, subdivide each list into two (sorted) sublists A_1, A_2 and B_1, B_2, respectively, of equal length. Merge A_1 and B_1 into list C, and A_2 and B_2 into list D. Subdivide lists C and D into sublists C_1, C_2 and D_1, D_2, respectively, of equal length. Then merge C_2 and D_1 into list E. The final sorted list is C_1, E, D_2.

(a) Let m_k denote the number of comparisons needed to merge two lists, each containing 2^k items, by the procedure just described. Write a recurrence relation and initial conditions for m_k.

(b) Find a formula expressing m_k as a function of k.

(c) Write a recurrence relation and initial conditions for b_k, the number of comparisons needed to sort a list of 2^k items by the Bose–Nelson algorithm.

(d) Use the method of iteration to find a formula expressing b_k as a function of k.

9.5 COUNTING WITH GENERATING FUNCTIONS

We saw in Section 8.1 that the numbers $C(n, r)$ appear as the coefficients in the expansion of the binomial $(x + y)^n$. For example, we have

$$(1 + x)^n = C(n, 0) + C(n, 1)x + \cdots + C(n, r)x^r + \cdots + C(n, n)x^n.$$

Taking $n = 5$ yields

$$\begin{aligned}(1 + x)^5 &= C(5, 0) + C(5, 1)x + C(5, 2)x^2 + C(5, 3)x^3 + C(5, 4)x^4 + C(5, 5)x^5 \\ &= 1 + 5x + 10x^2 + 10x^3 + 5x^4 + x^5. \qquad (9.1)\end{aligned}$$

Thus the coefficient of x^r in the expansion of $(1 + x)^5$ is exactly the number of ways of choosing r objects from a set of 5 objects. This makes sense, since, as we saw in Chapter 8, the coefficient of x^r in $(1 + x)^5$ is just the number of ways of choosing the x instead of the 1 from exactly r factors of the product

$$(1 + x)(1 + x)(1 + x)(1 + x)(1 + x).$$

Example 9.25 A boy is allowed to choose two items from a basket containing an apple, an orange, a pear, a banana, and a plum. How many ways can this be done?

Since the boy is to choose 2 from a set of 5 items, the number of ways is 10, the coefficient of x^2 in (9.1). Of course, the expression (9.1) also reveals the number of ways the boy can choose any other number of items. It is suggestive to replace

$$(1 + x)(1 + x)(1 + x)(1 + x)(1 + x)$$

by

(0 apples + 1 apple)(0 oranges + 1 orange)(0 pears + 1 pear)(0 bananas + 1 banana)(0 plums + 1 plum)

in order to see the connection between choosing r fruits and the coefficient of x^r in the polynomial $(1 + x)^5$. ■

Example 9.26 A boy is allowed to choose two items from a basket containing two apples, an orange, a pear, and a banana. How many ways can this be done if we consider the two apples to be identical?

Instead of attempting a count by the methods of Chapter 8, we will look for a polynomial similar to that of (9.1) such that the coefficient of x^r gives the number of ways of choosing r items. One that does the job is

$$\underset{\text{apple}}{(1 + x + x^2)}\ \underset{\text{orange}}{(1 + x)}\ \underset{\text{pear}}{(1 + x)}\ \underset{\text{banana}}{(1 + x)}. \qquad (9.2)$$

It may help to think of the expression

(0 apples + 1 apple + 2 apples)(0 oranges + 1 orange)(0 pears + 1 pear)(0 bananas + 1 banana)

to understand (9.2). The boy may choose 0, 1, or 2 apples and 0 or 1 oranges, pears, and bananas, for a total of two items. The number of ways of doing this is exactly the coefficient of x^2 in (9.2). By computing

$$(1 + x + x^2)(1 + x)(1 + x)(1 + x) = 1 + 4x + 7x^2 + 7x^3 + 4x^4 + x^5, \quad (9.3)$$

we see that the number of ways is 7. As a check, we list these as follows:

number of apples	2	1	1	1	0	0	0
number of oranges	0	1	0	0	1	1	0
number of pears	0	0	1	0	1	0	1
number of bananas	0	0	0	1	0	1	1

For example, the third column corresponds to forming x^2 by choosing x from the first factor, 1 from the second factor, x from the third factor, and 1 from the fourth factor.

Of course, (9.3) tells much more than the fact that the boy can choose two items from the basket in 7 ways. From it, we can deduce that the boy can choose no items in 1 way, one item in 4 ways, two items in 7 ways, three items in 7 ways, etc.

Notice that in (9.2) distinguishable items (say, an orange and pear) give rise to different factors, while indistinguishable items (the two apples) are included in the same factor. ■

Generating Functions

In Examples 9.25 and 9.26, we found polynomials with the property that the coefficient of x^r gives us the number of elements in a set whose definition depends on r in some way. In Example 9.25, the coefficients of our polynomial count the ways of choosing r items from a basket containing one each of five different fruits, and in Example 9.26 they count the ways of choosing r items from a basket containing two apples and one each of three other fruits.

In general, consider an infinite sequence of numbers

$$a_0, a_1, a_2, \ldots,$$

where, for some integer n, we have $a_{n+1} = a_{n+2} = \cdots = 0$. We say that the polynomial

$$a_0 + a_1 x + a_2 x^2 + a_3 x^3 + \cdots + a_n x^n$$

is the **generating function** for the sequence. For example, if we define a_r to be the number of ways of choosing r items from a basket containing two apples, one orange, one pear, and one banana, then, since there are only 5 fruit in the basket, we have $a_6 = a_7 = \cdots = 0$, and the generating function for the sequence $\{a_r\}$ is

$$(1 + x + x^2)(1 + x)(1 + x)(1 + x) = 1 + 4x + 7x^2 + 7x^3 + 4x^4 + x^5,$$

according to Example 9.26.

Example 9.27 Each of r people wants to order a Danish pastry from a bakery. Unfortunately, the bakery only has 3 cheese, 2 apricot, and 4 raspberry pastries left. We want a generating function for $\{d_r\}$, where d_r is the number of fillable orders for the r pastries. In particular, what is d_7?

The generating function is

$$\underbrace{(1+x+x^2+x^3)}_{\text{cheese}}\ \underbrace{(1+x+x^2)}_{\text{apricot}}\ \underbrace{(1+x+x^2+x^3+x^4)}_{\text{raspberry}},$$

since we must pick 0 through 3 cheese Danishes, 0 through 2 apricot Danishes, and 0 through 4 raspberry Danishes, for a total of r Danishes. It is appropriate that this is a polynomial of degree 9, because the bakery only has 9 Danishes, so, clearly, $d_r = 0$ for $r > 9$. A tedious computation shows that our polynomial equals

$$1 + 3x + 6x^2 + 9x^3 + 11x^4 + 11x^5 + 9x^6 + 6x^7 + 3x^8 + x^9,$$

so there are exactly 6 fillable orders for seven pastries. As a check, we list them as follows:

number of cheese	3	3	3	2	2	1
number of apricot	2	1	0	2	1	2
number of raspberry	2*	3	4*	3	4*	4*

For example, column 4 corresponds to forming x^7 by choosing x^2 from the first factor, x^2 from the second factor, and x^3 from the third factor.

Now suppose that raspberry pastries only come two to a box, and so the bakery will sell them in multiples of two. Then the generating function for the number of fillable orders for r pastries becomes

$$\underbrace{(1+x+x^2+x^3)}_{\text{cheese}}\ \underbrace{(1+x+x^2)}_{\text{apricot}}\ \underbrace{(1+x^2+x^4)}_{\text{raspberry}},$$

since only 0, 2, or 4 raspberry can be bought. Multiplying this out gives

$$1 + 2x + 4x^2 + 5x^3 + 6x^4 + 6x^5 + 5x^6 + 4x^7 + 2x^8 + x^9.$$

For example, since the coefficient of x^7 is 4, there are 4 ways to order 7 pastries with the new restriction. These ways are marked with an asterisk in the previous table. ■

Formal Power Series

In Example 9.27, at most 9 pastries could be ordered. In some situations, however, the number of choices is effectively unlimited.

Example 9.28 Now suppose that a multinational corporation builds a large apricot Danish factory next to the bakery. Thus, for all practical purposes, the supply of apricot pastries has become unlimited. Unfortunately, still only 3 cheese and 4 raspberry pastries are available, and the raspberry must be bought two at a time. We would like a generating function for the number of ways of buying r pastries.

Since any number of apricot pastries can be supplied, the natural thing with which to replace the factor $(1 + x + x^2)$ in

$$\underset{\text{cheese}}{(1 + x + x^2 + x^3)}\ \underset{\text{apricot}}{(1 + x + x^2)}\ \underset{\text{raspberry}}{(1 + x^2 + x^4)}$$

seems to be

$$(1 + x + x^2 + x^3 + \cdots), \tag{9.4}$$

where the powers of x go on forever. Of course, there is a problem with this expression, since it indicates that infinitely many quantities are to be added. As long as we never substitute a specific number for x, however, this problem does not arise. We can treat the expression of (9.4) in a formal way, combining it with similar expressions by using the usual rules for adding and multiplying polynomials. For example, we would compute

$$\begin{aligned}
&(1 + 2x + 5x^3) + (1 + x + x^2 + x^3 + \cdots) \\
&\quad = (1 + 1) + (2 + 1)x + (0 + 1)x^2 + (5 + 1)x^3 + (0 + 1)x^4 + \cdots \\
&\quad = 2 + 3x + x^2 + 6x^3 + x^4 + x^5 + \cdots
\end{aligned}$$

and

$$\begin{aligned}
(1 + x + x^2 + x^3)&(1 + x + x^2 + x^3 + \cdots) \\
&= 1(1 + x + x^2 + x^3 + \cdots) \\
&\quad + x(1 + x + x^2 + x^3 + \cdots) \\
&\quad + x^2(1 + x + x^2 + x^3 + \cdots) \\
&\quad + x^3(1 + x + x^2 + x^3 + \cdots) \\
&= 1 + x + x^2 + x^3 + x^4 + x^5 + \cdots \\
&\quad + x + x^2 + x^3 + x^4 + x^5 + \cdots \\
&\quad + x^2 + x^3 + x^4 + x^5 + \cdots \\
&\quad + x^3 + x^4 + x^5 + \cdots \\
&= 1 + 2x + 3x^2 + 4x^3 + 4x^4 + 4x^5 + \cdots .
\end{aligned} \tag{9.5}$$

If we allow the expression (9.4), then the generating function we desire is

$$F = \underset{\text{cheese}}{(1 + x + x^2 + x^3)}\ \underset{\text{apricot}}{(1 + x + x^2 + x^3 + \cdots)}\ \underset{\text{raspberry}}{(1 + x^2 + x^4)}.$$

We have already multiplied out the first two factors to get (9.5). Thus we have

$$\begin{aligned}
F &= (1 + 2x + 3x^2 + 4x^3 + 4x^4 + 4x^5 + \cdots)(1 + x^2 + x^4) \\
&= (1 + 2x + 3x^2 + 4x^3 + 4x^4 + 4x^5 + \cdots)1 \\
&\quad + (1 + 2x + 3x^2 + 4x^3 + 4x^4 + 4x^5 + \cdots)x^2 \\
&\quad + (1 + 2x + 3x^2 + 4x^3 + 4x^4 + 4x^5 + \cdots)x^4 \\
&= 1 + 2x + 3x^2 + 4x^3 + 4x^4 + 4x^5 + 4x^6 + 4x^7 + 4x^8 + \cdots \\
&\qquad + x^2 + 2x^3 + 3x^4 + 4x^5 + 4x^6 + 4x^7 + 4x^8 + \cdots \\
&\qquad\qquad + x^4 + 2x^5 + 3x^6 + 4x^7 + 4x^8 + \cdots \\
&= 1 + 2x + 4x^2 + 6x^3 + 8x^4 + 10x^5 + 11x^6 + 12x^7 + 12x^8 + \cdots,
\end{aligned}$$

where the coefficient of x^r is 12 for $r \geq 7$. In particular, there are now 12 ways of choosing 7 pastries, since the coefficient of x^7 is 12. We list them as follows:

cheese	0	0	0	1	1	1	2	2	2	3	3	3
apricot	3	5	7	2	4	6	1	3	5	0	2	4
raspberry	4	2	0	4	2	0	4	2	0	4	2	0

■

In light of Example 9.28, we extend the definition of a **generating function** given earlier to be an expression of the form

$$a_0 + a_1x + a_2x^2 + a_3x^3 + \cdots,$$

where we now allow infinitely many of the coefficients a_r to be nonzero. Such an expression is called a **formal power series**. We add and multiply generating functions just like polynomials, so that

$$\begin{aligned}
&(a_0 + a_1x + a_2x^2 + a_3x^3 + \cdots) + (b_0 + b_1x + b_2x^2 + b_3x^3 + \cdots) \\
&\quad = (a_0 + b_0) + (a_1 + b_1)x + (a_2 + b_2)x^2 + (a_3 + b_3)x^3 + \cdots
\end{aligned}$$

and

$$\begin{aligned}
&(a_0 + a_1x + a_2x^2 + a_3x^3 + \cdots)(b_0 + b_1x + b_2x^2 + b_3x^3 + \cdots) \\
&\quad = a_0b_0 + (a_0b_1 + a_1b_0)x + (a_0b_2 + a_1b_1 + a_2b_0)x^2 + \cdots.
\end{aligned}$$

Example 9.29 At a restaurant in a ski area, a grilled cheese sandwich costs \$2 and a bowl of noodle soup costs \$3. Let a_r be the number of ways of ordering r dollars worth of grilled cheese sandwiches and bowls of noodle soup. We will find a generating function for the sequence $\{a_r\}$.

The desired generating function is

$$\underbrace{(1 + x^2 + x^4 + x^6 + \cdots)}_{\text{grilled cheese}}\ \underbrace{(1 + x^3 + x^6 + x^9 + \cdots)}_{\text{noodle soup}}.$$

Choosing a term from the first factor determines whether we spend \$0, \$2, \$4, etc., on grilled cheese sandwiches; likewise, the term from the second factor corresponds to the number of bowls of soup. Notice that

$$\begin{aligned}(1 + x^2 + x^4 + x^6 + \cdots)(1 + x^3 + x^6 + x^9 + \cdots)\\ = 1 + x^2 + x^3 + x^4 + x^5 + 2x^6 + x^7 + 2x^8 + 2x^9 + x^{10} + \cdots .\end{aligned}$$

For example, since $a_8 = 2$, there are exactly two ways of spending \$8. These are to buy four grilled cheese sandwiches and no soup, or else one grilled cheese sandwich and two bowls of soup. ■

Example 9.30 A woman has a large supply of 1-, 2-, and 3-cent stamps. (All the 1-cent stamps are identical, etc.) Find a generating function for $\{a_r\}$, where a_r is the number of ways she can arrange exactly 3 of these stamps in a row on an envelope so that their total value is r cents. What if any number of stamps can be used?

Since the first stamp will be worth 1, 2, or 3 cents, and likewise for the second and third stamp, the generating function for $\{a_r\}$ is

$$\begin{aligned}(x + x^2 + x^3)(x + x^2 + x^3)(x + x^2 + x^3) = (x + x^2 + x^3)^3\\ = x^3 + 3x^4 + 6x^5 + 7x^6 + 6x^7 + 3x^8 + x^9.\end{aligned}$$

For example, the 6 ways to total 5 are 113, 131, 311, 122, 212, and 221.

In the same way, if 4 stamps are to be used, the corresponding generating function is $(x + x^2 + x^3)^4$. If either 3 *or* 4 stamps are allowed, then the appropriate generating function is

$$(x + x^2 + x^3)^3 + (x + x^2 + x^3)^4,$$

since the coefficient of x^r in this expression will be the sum of the number of ways of totaling r cents with an arrangement of 3 or 4 stamps.

What if we wish to count *all* arrangements of stamps totaling r cents, no matter how many stamps are used? Since we wish to allow 0, 1, 2, ... stamps to be used, the corresponding generating function is

$$\begin{aligned}&1 + (x + x^2 + x^3) + (x + x^2 + x^3)^2 + (x + x^2 + x^3)^3 + \cdots\\ &= 1 + x + x^2 + x^3\\ &\quad + x^2 + 2x^3 + 3x^4 + 2x^5 + x^6\\ &\quad\quad + x^3 + 3x^4 + 6x^5 + 7x^6 + 6x^7 + 3x^8 + x^9\\ &\quad\quad \vdots\\ &= 1 + x + 2x^2 + 4x^3 + 7x^4 + \cdots .\end{aligned}$$

For example, the 4 ways of arranging 3 cents worth of stamps are as 3, 12, 21, and 111. ■

EXERCISES 9.5

Consider the generating functions

$$A = 1 + x + x^2, \qquad B = 1 + 2x + 4x^4 + x^5,$$
$$C = 1 - x^2 + x^4, \qquad D = 1 + x + x^2 + x^3 + \cdots,$$
$$E = 1 + x^3 + x^6 + x^9 + \cdots, \qquad F = 1 - x + x^2 - x^3 + x^4 - \cdots.$$

In Exercises 1–12, write each indicated expression in the form

$$a_0 + a_1x + a_2x^2 + a_3x^3 + \cdots.$$

If the expression is a polynomial, then compute it completely; otherwise compute it through the x^7 term.

1. $A + B$	**2.** $B + C$	**3.** AB	**4.** AC
5. $B + D$	**6.** $C + F$	**7.** AD	**8.** CF
9. EC	**10.** DC	**11.** DE	**12.** FD

In Exercises 13–22, give a generating function for the sequence $\{a_r\}$, and then write it in the form $a_0 + a_1x + a_2x^2 + a_3x^3 + \cdots$ through the x^6 term.

13. Let a_r be the number of ways of taking r drinks from a refrigerator containing 3 Cokes and 5 Pepsis.

14. Let a_r be the number of ways of choosing r cars from a rental agency that has a Buick, a Dodge, a Honda, and a Volkswagen available.

15. Let a_r be the number of ways of choosing r jellybeans from a basket containing 3 licorice, 4 strawberry, and 2 lemon jellybeans.

16. Let a_r be the number of ways of buying r batteries from a store that has 3 C batteries, 4 D batteries, and 6 AA batteries, if the AA batteries are only sold in sets of two.

17. Let a_r be the number of ways of buying r chicken parts from a grocery that has 4 wings, 3 breasts, and 5 drumsticks, if the drumsticks are wrapped in a package of 2 and a package of 3, and the packages cannot be broken up.

18. Let a_r be the number of ways of spending r dollars on posters, if four identical \$1 and three identical \$2 posters are available.

19. Let a_r be the number of ways of ordering r glasses of liquid, if 3 glasses of milk and an unlimited supply of water are available.

20. Let a_r be the number of ways of collecting r ounces of clams and mussels from a beach, if a clam weighs 3 ounces and a mussel 2 ounces.

21. Let a_r be the number of ways of choosing r oak and maple leaves for a scrapbook, if the book must contain at least 4 oak leaves and at least 2 maple leaves.

22. Let a_r be the number of ways of buying r baseball cards, if 1 Mickey Mantle, 1 Stan Musial, 1 Willie Mays, and an unlimited supply of Pete Rose cards are available.

In Exercises 23–26, find the generating function for $\{a_r\}$.

23. Let a_r be the number of ways of choosing r books from seven different mathematics books and five identical copies of *Peyton Place*.

24. Let a_r be the number of ways of spending r dollars on three different \$7 books (there is only one copy of each) and an unlimited number of identical \$9 books.

25. Let a_r be the number of ways of catching r pounds of bluegill, catfish, and bass, if a bluegill weighs 1 pound, a catfish weighs 3 pounds, and a bass weighs 4 pounds.

26. Let a_r be the number of solutions to $a + b = r$, where a and b are elements of the set $\{1, 2, 4, 8, \ldots\}$.

In Exercises 27–32, suppose the indicated generating function F is written as $a_0 + a_1x + a_2x^2 + a_3x^3 + \cdots$. Find a formula for a_r in terms of r.

27. $F = (1 + x + x^2 + x^3 + \cdots)^2$

28. $F = (1 + x + x^2 + x^3 + \cdots)(1 - x)$

29. $F = (1 + x + x^2 + x^3 + \cdots)(1 + x)$

30. $F = (1 + x + x^2 + x^3 + \cdots)(1 - x + x^2 - x^3 + \cdots)$

31. $F = (1 - x + x^2 - x^3 + \cdots)(1 + x)$

32. $F = (1 + x + x^2 + x^3 + \cdots)^3$

33. Let a_r be the number of solutions to $p + q = r$, where p and q are prime numbers. Find a generating function for $\{a_r\}$, and write it through the x^{10} term. It is an unproved conjecture (called the **Goldbach conjecture**) that $a_r > 0$ whenever r is even and greater than 2.

34. Let a_r be the number of solutions to $2^k + p = r$, where k is a nonnegative integer and p is a prime number. Express a generating function for $\{a_r\}$, and write it out through the x^{10} term. What is the smallest $r > 2$ such that $a_r = 0$?

35. Let a_r be the number of solutions to $a^2 + b^2 + c^2 + d^2 = r$, where a, b, c, and d are nonnegative integers. Express a generating function for $\{a_r\}$, and write it out through the x^{10} term. (It can be proved that $a_r > 0$ for all $r \geq 0$.)

9.6 THE ALGEBRA OF GENERATING FUNCTIONS

We saw in the previous section how generating functions, even those with infinitely many terms, can be added and multiplied just like polynomials. With these definitions, generating functions obey the same algebraic laws as polynomials. Examples are the associative and commutative laws of addition and multiplication and the distributive law. The generating function

$$0 = 0 + 0x + 0x^2 + 0x^3 + \cdots$$

takes the role of additive identity; that is,

$$0 + G = G + 0 = G$$

for every generating function G. Likewise, the generating function

$$1 = 1 + 0x + 0x^2 + 0x^3 + \cdots$$

is the multiplicative identity, so that

$$1G = G1 = G$$

for every generating function G.

We define the subtraction of generating functions by

$$(a_0 + a_1x + a_2x^2 + a_3x^3 + \cdots) - (b_0 + b_1x + b_2x^2 + b_3x^3 + \cdots)$$
$$= (a_0 - b_0) + (a_1 - b_1)x + (a_2 - b_2)x^2 + (a_3 - b_3)x^3 + \cdots .$$

Division presents more of a problem. The key is the existence of *inverses*; that is, given a generating function G, we would like to find another generating function G^{-1} such that $GG^{-1} = 1$, the multiplicative identity. Such an inverse often exists; for example,

$$(1 - x)(1 + x + x^2 + x^3 + \cdots) = 1 + x + x^2 + x^3 + \cdots - x - x^2 - x^3 - \cdots = 1.$$

Thus

$$(1 + x + x^2 + x^3 + \cdots)^{-1} = 1 - x$$

and

$$(1 - x)^{-1} = 1 + x + x^2 + x^3 + \cdots .$$

In fact, a similar computation shows that

$$(1 - G)(1 + G + G^2 + G^3 + \cdots) = 1 \tag{9.6}$$

for any generating function G having 0 as its constant term. For example, setting $G = -x$ in (9.6) gives

$$(1 + x)(1 - x + x^2 - x^3 + \cdots) = 1,$$

and taking $G = 2x$ in (9.6) gives

$$(1 - 2x)(1 + 2x + 4x^2 + 8x^3 + \cdots) = 1.$$

In Example 9.30 we found that the generating function for the number of ways of arranging a sequence of 1-, 2- and 3-cent stamps totaling r cents is

$$1 + (x + x^2 + x^3) + (x + x^2 + x^3)^2 + (x + x^2 + x^3)^3 + \cdots .$$

By taking $G = x + x^2 + x^3$ in (9.6), we can write this as

$$(1 - x - x^2 - x^3)^{-1}.$$

It turns out that all that is needed for the inverse of a generating function $a_0 + a_1x + a_2x^2 + a_3x^3 + \cdots$ to exist is that $a_0 \neq 0$.

Theorem 9.3 Suppose that

$$G = a_0 + a_1x + a_2x^2 + a_3x^3 + \cdots ,$$

where $a_0 \neq 0$. Then there is a unique generating function H such that $GH = 1$.

Proof. We are interested in a generating function

$$H = b_0 + b_1x + b_2x^2 + b_3x^3 + \cdots$$

such that

$$\begin{aligned} GH &= (a_0 + a_1x + a_2x^2 + a_3x^3 + \cdots)(b_0 + b_1x + b_2x^2 + b_3x^3 + \cdots) \\ &= a_0b_0 + (a_0b_1 + a_1b_0)x + (a_0b_2 + a_1b_1 + a_2b_0)x^2 + \cdots \\ &= 1. \end{aligned}$$

This leads to the following equations:

$$\begin{aligned} a_0b_0 &= 1 \\ a_0b_1 + a_1b_0 &= 0 \\ a_0b_2 + a_1b_1 + a_2b_0 &= 0 \\ &\vdots \end{aligned}$$

The first equation is true if and only if $b_0 = a_0^{-1}$, and a_0^{-1} exists since $a_0 \neq 0$. Substituting this value of b_0 into the second equation determines b_1 uniquely. Likewise, the third equation can be solved for b_2 after our previously determined values of b_0 and b_1 are substituted in it. Continuing in this way, we see that a unique sequence $b_0, b_1, b_2, b_3, \ldots$ is determined such that

$$(a_0 + a_1x + a_2x^2 + a_3x^3 + \cdots)(b_0 + b_1x + b_2x^2 + b_3x^3 + \cdots) = 1. \quad ■$$

Example 9.31 Let us try to find the inverse of the generating function

$$1 + 2x + 3x^2 + 4x^3 + \cdots.$$

We wish to determine a sequence $\{b_r\}$ such that

$$(1 + 2x + 3x^2 + 4x^3 + \cdots)(b_0 + b_1x + b_2x^2 + b_3x^3 + \cdots) = 1.$$

Equating the constant terms on both sides of this equation gives

$$1b_0 = 1,$$

so $b_0 = 1$. Likewise, the coefficients of x must be the same on both sides, so

$$1b_1 + 2b_0 = b_1 + 2 = 0.$$

This implies that $b_1 = -2$. Equating the coefficients of x^2 yields

$$1b_2 + 2b_1 + 3b_0 = b_2 - 4 + 3 = 0,$$

so $b_2 = 1$. The next equation is

$$1b_3 + 2b_2 + 3b_1 + 4b_0 = b_3 + 2 - 6 + 4 = 0,$$

so $b_3 = 0$. The reader should check that $b_4 = 0$ also. In fact, it can be proved that the rest of the coefficients b_r are all 0, so

$$(1 + 2x + 3x^2 + 4x^3 + \cdots)^{-1} = 1 - 2x + x^2.$$

The details are left for Exercise 33.

The same result could have been reached by another route if we assume that generating functions satisfy some familiar laws for exponents. According to Exercise 27 of Section 9.5, we have

$$(1 + x + x^2 + x^3 + \cdots)^2 = 1 + 2x + 3x^2 + 4x^3 + \cdots .$$

Thus

$$\begin{aligned}(1 + 2x + 3x^2 + 4x^3 + \cdots)^{-1} &= [(1 + x + x^2 + x^3 + \cdots)^2]^{-1} \\ &= [(1 + x + x^2 + x^3 + \cdots)^{-1}]^2 \\ &= [1 - x]^2 \\ &= 1 - 2x + x^2,\end{aligned}$$

where the next-to-last equality comes from (9.6). ■

Generating functions are an extremely flexible tool for studying combinatorial sequences, and we will only be able to touch on a few of their applications here. Given a recurrence relation, it is often possible to use it to construct the generating function for the corresponding sequence. This is illustrated by the next example.

Example 9.32 Consider the sequence $\{m_r\}$ of Example 9.2, which concerned the Towers of Hanoi game. The number m_r is the minimal number of moves needed to transfer a stack of r disks to an empty peg. We found that $m_1 = 1$ and $m_r = 2m_{r-1} + 1$ for $r \geq 2$. In fact, if we define m_0 to be 0, then our recurrence relation holds for $r \geq 1$.

Let us define M to be the generating function for $\{m_r\}$, so that

$$M = m_0 + m_1x + m_2x^2 + m_3x^3 + \cdots .$$

Then, since $m_0 = 0$ and $m_r = 2m_{r-1} + 1$ for $r \geq 1$, we have

$$\begin{aligned}M &= 0 + (2m_0 + 1)x + (2m_1 + 1)x^2 + (2m_2 + 1)x^3 + \cdots \\ &= 2m_0x + 1x + 2m_1x^2 + 1x^2 + 2m_2x^3 + 1x^3 + \cdots \\ &= 2x(m_0 + m_1x + m_2x^2 + m_3x^3 + \cdots) + x + x^2 + x^3 + \cdots \\ &= 2xM + x(1 + x + x^2 + \cdots).\end{aligned}$$

Then

$$M - 2xM = x(1 + x + x^2 + \cdots),$$

or

$$M(1 - 2x) = x(1 + x + x^2 + \cdots) = x(1 - x)^{-1},$$

where the last equation follows from setting $G = x$ in (9.6). Thus we have

$$M = \frac{x}{(1-2x)(1-x)},$$

where we have indicated the inverses by the usual fraction notation.

In order to get a formula for the coefficients of M, we will express the fraction on the right in the form

$$\frac{a}{1-2x} + \frac{b}{1-x},$$

where a and b are constants.[1] We have

$$\begin{aligned}\frac{x}{(1-2x)(1-x)} &= \frac{a}{1-2x} + \frac{b}{1-x} = \frac{a(1-x)+b(1-2x)}{(1-2x)(1-x)} \\ &= \frac{(a+b)+(-a-2b)x}{(1-2x)(1-x)},\end{aligned}$$

and therefore, by equating coefficients in the numerators, we get $a + b = 0$ and $-a - 2b = 1$. These equations are easily seen to have the solution $a = 1$ and $b = -1$. Thus

$$M = \frac{x}{(1-2x)(1-x)} = \frac{1}{1-2x} - \frac{1}{1-x}.$$

But then (9.6) yields

$$\begin{aligned}M &= (1-2x)^{-1} - (1-x)^{-1} \\ &= (1+2x+4x^2+8x^3+\cdots) - (1+x+x^2+x^3+\cdots) \\ &= (1-1)+(2-1)x+(4-1)x^2+(8-1)x^3+\cdots.\end{aligned}$$

Thus we see that m_r, the coefficient of x^r, is $2^r - 1$. This agrees with what we found in Section 9.2. ■

The method of Example 9.32 can be used on any first-order linear difference equation. In fact, an alternative proof of Theorem 9.1 can be based on it; the details are left for the exercises. Generating functions can also be applied to higher order recurrences, as in the next example.

Example 9.33 Let us consider the recurrence relation

$$s_0 = 0, \qquad s_1 = 1, \qquad s_n = 2s_{n-1} - s_{n-2} \qquad \text{for } n \geq 2.$$

[1]Calculus students may recognize the method of partial fractions.

If S is the generating function for this sequence, then

$$\begin{aligned} S &= s_0 + s_1x + s_2x^2 + s_3x^3 + \cdots \\ &= s_0 + s_1x + (2s_1 - s_0)x^2 + (2s_2 - s_1)x^3 + \cdots \\ &= 0 + x + 2x(s_1x + s_2x^2 + \cdots) - x^2(s_0 + s_1x + \cdots) \\ &= x + 2x(S - s_0) - x^2S \\ &= x + 2xS - x^2S. \end{aligned}$$

Thus we have

$$\begin{aligned} S - 2xS + x^2S &= x, \\ S(1 - 2x + x^2) &= x, \end{aligned}$$

so

$$S = x(1 - 2x + x^2)^{-1}.$$

But in Example 9.31 we found that the generating functions

$$1 - 2x + x^2 \qquad \text{and} \qquad 1 + 2x + 3x^2 + 4x^3 + \cdots$$

are inverses of each other. Thus

$$\begin{aligned} S &= x(1 + 2x + 3x^2 + 4x^3 + \cdots) \\ &= x + 2x^2 + 3x^3 + 4x^4 + \cdots, \end{aligned}$$

from which we see that $s_r = r$ for all nonnegative integers r. ■

Example 9.34 Let us use a generating function to find a formula for s_n, where $s_0 = s_1 = 1$, and $s_n = -s_{n-1} + 6s_{n-2}$ for $n \geq 2$. If S is the generating function for $\{s_n\}$, then

$$\begin{aligned} S &= s_0 + s_1x + s_2x^2 + s_3x^3 + \cdots \\ &= 1 + x + (-s_1 + 6s_0)x^2 + (-s_2 + 6s_1)x^3 + \cdots \\ &= 1 + x - x(s_1x + s_2x^2 + \cdots) + 6x^2(s_0 + s_1x + \cdots) \\ &= 1 + x - x(S - s_0) + 6x^2S \\ &= 1 + x - x(S - 1) + 6x^2S \\ &= 1 + 2x - xS + 6x^2S. \end{aligned}$$

Thus

$$\begin{aligned} S + xS - 6x^2S &= 1 + 2x, \\ S(1 + x - 6x^2) &= 1 + 2x, \\ S(1 + 3x)(1 - 2x) &= 1 + 2x, \end{aligned}$$

so

$$S = \frac{1 + 2x}{(1 + 3x)(1 - 2x)}.$$

We will attempt to find constants a and b so that the last fraction has the form

$$\frac{a}{1+3x}+\frac{b}{1-2x}.$$

This gives

$$\frac{1+2x}{(1+3x)(1-2x)}=\frac{a}{1+3x}+\frac{b}{1-2x}=\frac{a(1-2x)+b(1+3x)}{(1+3x)(1-2x)},$$

so $a+b=1$ and $-2a+3b=2$. Solving these equations simultaneously yields

$$a=\frac{1}{5} \quad \text{and} \quad b=\frac{4}{5}.$$

Now we use (9.6) to write

$$\begin{aligned} S &= \frac{1}{5}\frac{1}{1+3x}+\frac{4}{5}\frac{1}{1-2x} \\ &= \frac{1}{5}(1-3x+9x^2-27x^3+\cdots)+\frac{4}{5}(1+2x+4x^2+8x^3+\cdots). \end{aligned}$$

Picking off the coefficient of x^n tells us that

$$s_n=\frac{1}{5}(-3)^n+\frac{4}{5}(2)^n.$$

For example, we have

$$s_0=\frac{1}{5}+\frac{4}{5}=1, \quad s_1=\frac{1}{5}(-3)+\frac{4}{5}(2)=-\frac{3}{5}+\frac{8}{5}=1,$$

and

$$s_2=\frac{1}{5}(9)+\frac{4}{5}(4)=\frac{9}{5}+\frac{16}{5}=5. \quad \blacksquare$$

The result of Example 9.34 could also have been found by applying Theorem 9.2. In fact, a proof of Theorem 9.2 using generating functions is sketched in the exercises at the end of this section.

Example 9.35 An embassy communicates with its home country in code words consisting of a string of n decimal digits. In order to catch errors in transmission, it is agreed that the total number of 3s and 7s in each word should always be odd. How many code words are possible?

Let s_n be the number of allowable words of length n. We can get a recurrence relation for s_n as follows. Consider a word W of length $n+1$ counted by s_{n+1}. It either ends in a 3 or a 7, or neither. If it ends in a 3 or a 7, then the word W^* of length n formed by deleting the last digit from W must have an *even* number of 3s and 7s. Since there are 10^n strings of n decimal digits, the number of such words W^* is $10^n - s_n$. Thus the number of possible words W of this form is $2(10^n - s_n)$, since the last digit of W can be 3 or 7.

Now suppose that the allowable word W ends in a digit other than 3 or 7. Then deleting its last digit leaves a word counted by s_n. Since there are 8 possibilities for the last digit of W, the number of allowable words of this form is $8s_n$.

By combining the results of the last two paragraphs, we see that

$$s_{n+1} = 2(10^n - s_n) + 8s_n = 2 \cdot 10^n + 6s_n$$

for $n \geq 1$. Clearly, $s_0 = 0$, since the empty string cannot have an odd number of 3s and 7s. Using this relation allows us to compute the following table:

n	s_n
0	0
1	$2 \cdot 10^0 + 6 \cdot 0 = 2$
2	$2 \cdot 10^1 + 6 \cdot 2 = 32$
3	$2 \cdot 10^2 + 6 \cdot 32 = 392$

For example, s_2 counts the number of 2-digit strings with exactly one 3 or 7. Since we can use either a 3 or 7, since this can be either the first or second digit, and since there are 8 choices for the remaining digit, the number of such strings is $2 \cdot 2 \cdot 8 = 32$.

Now we will use generating functions to get an explicit formula for s_n. Let S be the generating function for $\{s_n\}$, so that

$$S = s_0 + s_1x + s_2x^2 + s_3x^3 + \cdots .$$

Then we have

$$\begin{aligned} S &= s_0 + (2 \cdot 10^0 + 6s_0)x + (2 \cdot 10^1 + 6s_1)x^2 + (2 \cdot 10^2 + 6s_2)x^3 + \cdots \\ &= s_0 + 2x(10^0 + 10^1x + 10^2x^2 + \cdots) + 6x(s_0 + s_1x + s_2x^2 + s_3x^3 + \cdots) \\ &= 0 + 2x(1 + 10x + (10x)^2 + \cdots) + 6xS \\ &= 2x(1 - 10x)^{-1} + 6xS. \end{aligned}$$

Solving for S yields

$$S(1 - 6x) = 2x(1 - 10x)^{-1},$$

or

$$S = \frac{2x}{(1 - 6x)(1 - 10x)}.$$

We will find constants a and b such that

$$\frac{2x}{(1 - 6x)(1 - 10x)} = \frac{a}{1 - 6x} + \frac{b}{1 - 10x} = \frac{a(1 - 10x) + b(1 - 6x)}{(1 - 6x)(1 - 10x)}.$$

Equating numerators gives the equations $a + b = 0$ and $-10a - 6b = 2$. We easily find that $a = -\frac{1}{2}$ and $b = \frac{1}{2}$. Thus

$$\begin{aligned} S &= -\frac{1}{2}(1-6x)^{-1} + \frac{1}{2}(1-10x)^{-1} \\ &= \frac{1}{2}[(1-10x)^{-1} - (1-6x)^{-1}] \\ &= \frac{1}{2}[(1+10x+100x^2+\cdots) - (1+6x+36x^2+\cdots)], \end{aligned}$$

from which we see that the coefficient of x^r in S is

$$s_r = \frac{10^r - 6^r}{2}.$$

For example, $s_2 = \frac{100-36}{2} = 32$, and $s_3 = \frac{1000-216}{2} = 392$. These values agree with our earlier computations. ■

EXERCISES 9.6

In Exercises 1–10, find the inverse of the given generating function.

1. $1 - 3x$

2. $1 - 5x$

3. $1 + 2x + 4x^2 + 8x^3 + \cdots$

4. $1 - 3x + 9x^2 - 27x^3 + \cdots$

5. $1 + x^2$

6. $1 + 2x^3$

7. $1 - x - x^2$

8. $1 + x + x^3$

9. $2 + 6x$

10. $\frac{1}{3} + x^4$

In Exercises 11–20, let S be the generating function of the sequence $\{s_n\}$. Find an equation satisfied by S (as in Examples 9.32 through 9.35), and solve for S.

11. $s_0 = 1$, and $s_n = 2s_{n-1} + 1$ for $n \geq 1$

12. $s_0 = 3$, and $s_n = -s_{n-1} + 2$ for $n \geq 1$

13. $s_0 = 1$, $s_1 = 1$, and $s_n = 2s_{n-1} - s_{n-2}$ for $n \geq 2$

14. $s_0 = 2$, $s_1 = 1$, and $s_n = s_{n-1} - 3s_{n-2}$ for $n \geq 2$

15. $s_0 = -1$, $s_1 = 0$, and $s_n = -s_{n-1} + 2s_{n-2}$ for $n \geq 2$

16. $s_0 = 0$, $s_1 = -2$, and $s_n = 3s_{n-1} + s_{n-2}$ for $n \geq 2$

17. $s_0 = -2$, $s_1 = 1$, and $s_n = s_{n-1} + 3s_{n-2} + 2$ for $n \geq 2$

18. $s_0 = -3$, $s_1 = 2$, and $s_n = 4s_{n-1} - 5s_{n-2} - 1$ for $n \geq 2$

19. $s_0 = 2$, $s_1 = -1$, $s_2 = 1$, and $s_n = s_{n-1} - 3s_{n-2} + s_{n-3}$ for $n \geq 3$

20. $s_0 = 1$, $s_1 = 1$, $s_2 = 5$, and $s_n = 2s_{n-1} + s_{n-2} - s_{n-3}$ for $n \geq 3$

In Exercises 21–26, find constants a and b such that the given equations are identities in x.

21. $\dfrac{x}{(1-x)(1+2x)} = \dfrac{a}{1-x} + \dfrac{b}{1+2x}$

22. $\dfrac{2}{(1+x)(1+3x)} = \dfrac{a}{1+x} + \dfrac{b}{1+3x}$

23. $\dfrac{1+3x}{(1+2x)(1-x)} = \dfrac{a}{1+2x} + \dfrac{b}{1-x}$

24. $\dfrac{1-x}{(1+2x)(1-3x)} = \dfrac{a}{1+2x} + \dfrac{b}{1-3x}$

25. $\dfrac{1+x}{(1+2x)^2} = \dfrac{a}{1+2x} + \dfrac{b}{(1+2x)^2}$

26. $\dfrac{3-x}{(1-x)^2} = \dfrac{a}{1-x} + \dfrac{b}{(1-x)^2}$

In Exercises 27–32, give a formula for s_n if $\{s_n\}$ has the given generating function S.

27. $S = \dfrac{1}{1-2x} + \dfrac{1}{1+x}$

28. $S = \dfrac{3}{1-x} - \dfrac{1}{1+3x}$

29. $S = \dfrac{-1}{1-2x} + \dfrac{4}{1+5x}$

30. $S = \dfrac{3}{1-x} + \dfrac{2}{1+2x}$

31. $S = \dfrac{2}{1-3x^2}$

32. $S = \dfrac{2}{1-x} + \dfrac{1}{1+x^2}$

33. Suppose that $b_0 = 1$, $b_1 = -2$, $b_2 = 1$, and

$$b_n + 2b_{n-1} + 3b_{n-2} + \cdots + (n+1)b_0 = 0$$

for all $n \geq 1$. Prove by mathematical induction that $b_n = 0$ for $n \geq 3$.

In Exercises 34–36, assume that s_0 is given and that $s_n = as_{n-1} + b$ for $n \geq 1$, where a and b are constants and $a \neq 1$.

34. Show that if S is the generating function for $\{s_n\}$, then

$$S = s_0 + axS + bx(1-x)^{-1}.$$

35. Show that

$$\frac{s_0 + (-s_0 + b)x}{(1-ax)(1-x)} = \frac{k_1}{1-ax} + \frac{k_2}{1-x},$$

where

$$k_1 = s_0 + \frac{b}{a-1} \quad \text{and} \quad k_2 = \frac{-b}{a-1}.$$

36. Show that

$$s_n = \left(s_0 + \frac{b}{a-1}\right)a^n - \frac{b}{a-1} \quad \text{for} \quad n \geq 0.$$

In Exercises 37–43, consider the second-order homogeneous difference equation

$$s_n = as_{n-1} + bs_{n-2},$$

where s_0 and s_1 are given. Assume that $x^2 - ax - b = (x - r_1)(x - r_2)$.

37. Show that $r_1 + r_2 = a$, $r_1 r_2 = -b$, and $1 - ax - bx^2 = (1 - r_1 x)(1 - r_2 x)$.

38. Show that if S is the generating function for $\{s_n\}$, then

$$S = s_0 + s_1 x + ax(S - s_0) + bx^2 S.$$

In Exercises 39 and 40, assume that $r_1 \neq r_2$.

39. Show that there exist constants c_1 and c_2 such that

$$\frac{s_0 + (s_1 + as_0)x}{(1 - r_1 x)(1 - r_2 x)} = \frac{c_1}{1 - r_1 x} + \frac{c_2}{1 - r_2 x}.$$

40. Show that $s_n = c_1 r_1^n + c_2 r_2^n$ for $n \geq 0$, where c_1 and c_2 are as in Exercise 39.

In Exercises 41–43, assume that $r_1 = r_2 = r \neq 0$.

41. Show that there exist constants k_1 and k_2 such that

$$\frac{s_0 + (s_1 + as_0)x}{(1 - rx)^2} = \frac{k_1}{1 - rx} + \frac{k_2}{(1 - rx)^2}.$$

42. Show that $s_n = k_1 r^n + k_2 (n + 1) r^n$ for $n \geq 0$, where k_1 and k_2 are as in Exercise 41.

43. Show that there exist constants c_1 and c_2 such that $s_n = c_1 r^n + n c_2 r^n$ for $n \geq 0$.

HISTORICAL NOTES

Recursion has been used from Greek times. Its formal development, however, dates back only to the past two and one-half centuries.

Archimedes had two relationships that involved recursion. If a_n and A_n are, respectively, the areas of the polygons with n sides inscribed in and circumscribed about a circle, then

$$a_{2n} = \sqrt{a_n A_n} \qquad \text{and} \qquad A_{2n} = \frac{2A_n a_{2n}}{A_n + a_{2n}}.$$

In a like manner, if p_n and P_n are the perimeters of the regular polygons inscribed in and circumscribed about a circle, then

$$p_{2n} = \sqrt{p_n P_{2n}} \qquad \text{and} \qquad P_{2n} = \frac{2P_n p_n}{P_n + p_n}.$$

Starting with a regular hexagon, Archimedes developed reasonable estimates for the value of π [73].

In his *Liber Abaci* of 1202, Leonardo of Pisa (c. 1175–1250), known as Fibonacci, provided the first systematic introduction for Europeans to the Arabic notation for numerals and their algorithms for arithmetic. As part of the text, Fibonacci presented his famous recursion problem dealing with generations of

Edouard Lucas

Abraham De Moivre

rabbits. While he did not develop any of the many relationships stemming from the recursion, the pattern was named in his honor by the French mathematician Edouard Lucas in the late 1800s.

The formula in Example 9.17 for the Fibonacci numbers was not derived until 1718, when Abraham De Moivre (1667–1754) obtained the result with an approach using a generating function. Extending the general techniques, Leonhard Euler (1707–1783) advanced the study of the partitions of integers in his 1748 two-volume opus *Introductio in Analysin Infinitorum*. Pierre Simon Laplace (1749–1827) also published a significant amount of work on generating functions and their applications in his 1754 work *The Calculus of Generating Functions*. The mathematical analysis of the Tower of Hanoi puzzle and its general closed-form solution via generating functions is credited to Lucas in his 1884 work *Récréations Mathématiques* [74].

SUPPLEMENTARY EXERCISES

In Exercises 1–5, determine s_5 for a sequence $s_0, s_1, s_2, \ldots$ satisfying the given recurrence relation and initial conditions.

1. $s_n = 3s_{n-1} + n^2$ for $n \geq 1$, $s_0 = 2$

2. $s_n = (-1)^n + s_{n-1}$ for $n \geq 1$, $s_0 = 1$

3. $s_n = 2ns_{n-1}$ for $n \geq 1$, $s_0 = 1$

4. $s_n = 3(s_{n-1} + s_{n-2})$ for $n \geq 2$, $s_0 = 1$, $s_1 = 2$

5. $s_n = ns_{n-1} - s_{n-2}$ for $n \geq 2$, $s_0 = 1$, $s_1 = 1$

6. Suppose that \$20,000 is deposited into an account with an annual interest rate of 8% compounded quarterly. Each quarter, there is a withdrawal of \$200 immediately after interest is credited to the account. Write a recurrence relation and initial conditions for v_n, the value of the account n quarters after the initial deposit.

7. A data processing position pays a starting salary of \$16,000 and offers yearly raises of \$500 plus a 4% cost of living adjustment on the present year's salary. Write a recurrence relation and initial conditions for s_n, the salary during year n.

8. An ecology group bought a printing press for \$18,000 to print leaflets. If the resale value of the press decreases by 12% of its current value each year, write a recurrence relation and initial conditions for v_n, the resale value of the press n years after its original purchase.

9. Write a recurrence relation and initial conditions for the number c_n of n-symbol codewords composed of dots and dashes with no two consecutive dashes.

10. Suppose that, at the beginning of an experiment, there are 500 cells in a sample and the number of cells is increasing at the rate of 150% per hour. Write a recurrence relation and initial conditions for c_n, the number of cells in the sample n hours after the start of the experiment.

11. Suppose that the efficiency e_n of a worker on an assembly line processing n units per minute is equal to the efficiency of the same worker processing $n - 1$ units per minute, minus an incremental loss for the nth unit. Assume that the incremental loss is inversely proportional to n^2. Write a recurrence relation describing the efficiency of the worker.

12. Twenty years ago, an individual invested an inheritance in an account that pays 8% interest compounded quarterly. If the present value of the account is \$75,569.31, what was the initial investment?

13. Prove by mathematical induction that $2n^2 + 2n$ is a solution to the recurrence relation $s_n = s_{n-1} + 4n$ for $n \geq 1$ with the initial condition $s_0 = 0$.

14. Prove by mathematical induction that $2^{n-1} + 2$ is a solution to the recurrence relation $s_n = 2s_{n-1} - 2$ for $n \geq 2$ with $s_1 = 3$.

15. Prove by mathematical induction that $(n + 1)! - 1$ is a solution to the recurrence relation $s_n = s_{n-1} + n \cdot n!$ for $n \geq 1$ with $s_0 = 0$.

16. Prove that $2^n + 2(3^n) + n - 7$ is a solution to the recurrence relation $s_n = 5s_{n-1} - 6s_{n-2} + 2n - 21$ for $n \geq 2$ with the initial conditions $s_0 = -4$, $s_1 = 2$.

Find an explicit formula for s_n if $s_0, s_1, s_2, \ldots$ is a sequence that satisfies the recurrence relation and initial conditions given in Exercises 17–20.

17. $s_n = 3s_{n-1} - 12$ for $n \geq 1$, $s_0 = 5$

18. $s_n = s_{n-1} + 7$ for $n \geq 1$, $s_0 = 2$

19. $s_n = 4s_{n-1} - 4s_{n-2}$ for $n \geq 2$, $s_0 = 4$, $s_1 = 6$

20. $s_n = 7s_{n-1} - 10s_{n-2}$ for $n \geq 2$, $s_0 = -2$, $s_1 = -1$

A ***Lucas sequence*** *is a second-order homogeneous linear difference equation with constant coefficients that is similar to the Fibonacci sequence. The general Lucas sequence can be defined as*

$$L_n = \begin{cases} p & \text{for } n = 1 \\ q & \text{for } n = 2 \\ L_{n-1} + L_{n-2} & \text{for } n \geq 3, \end{cases}$$

where p and q are integers.

21. Find the first 10 terms of the Lucas sequence with initial conditions $L_1 = 3$ and $L_2 = 4$.

22. Compute (to three decimal place accuracy) the quotient $\frac{L_{i+1}}{L_i}$ for the values obtained in Exercise 21. Compare the resulting quotients with the value of $\frac{1+\sqrt{5}}{2}$, the golden ratio.

23. Prove that if L_n is a Lucas sequence with initial conditions $L_1 = p$ and $L_2 = q$, then $L_n = qF_{n-1} + pF_{n-2}$ for all $n \geq 3$.

24. Solve the system of recurrence relations

$$s_n = 8s_{n-1} - 9t_{n-1}$$
$$t_n = 6s_{n-1} - 7t_{n-1}$$

with the initial conditions $s_0 = 4$, $t_0 = 1$. (*Hint:* Substitute $s_n = 3u_n + v_n$ and $t_n = 2u_n + v_n$, and solve for u_n and v_n.)

25. Let k be a positive integer and $a_1, a_2, \ldots, a_k$ be real numbers such that $a_k \neq 0$. We call the polynomial equation $x^k = a_1 x^{k-1} + a_2 x^{k-2} + \cdots + a_k$ the **auxiliary equation** of the recurrence relation

$$s_n = a_1 s_{n-1} + a_2 s_{n-2} + \cdots + a_k s_{n-k}. \tag{9.7}$$

Prove that r^n is a solution of (9.7) if and only if r is a root of the auxiliary equation.

26. Prove that if u_n and v_n satisfy (9.7) for $n \geq k$, then, for any constants b and c, $bu_n + cv_n$ also satisfies (9.7) for $n \geq k$.

27. Find an explicit formula for s_n if $s_0, s_1, s_2, \ldots$ is a sequence that satisfies $s_n = 3s_{n-1} + 10s_{n-2} - 24s_{n-3}$ for $n \geq 3$ and the initial conditions $s_0 = -4$, $s_1 = -9$, $s_2 = 13$. (*Hint:* Proceed as in Theorem 9.2(a), using the results of Exercises 25 and 26.)

28. Find an explicit formula for s_n if $s_0, s_1, s_2, \ldots$ is a sequence that satisfies $s_n = 6s_{n-1} - 12s_{n-2} + 8s_{n-3}$ for $n \geq 3$ and the initial conditions $s_0 = 5$, $s_1 = 6$, $s_2 = -20$. (*Hint:* Proceed as in Theorem 9.2(b), using the results of Exercises 25 and 26.)

29. Find an explicit formula for s_n if $s_0, s_1, s_2, \ldots$ is a sequence that satisfies $s_n = 3s_{n-2} + 2s_{n-3}$ for $n \geq 3$ and the initial conditions $s_0 = 4$, $s_1 = 4$, $s_2 = -3$. (*Hint:* Proceed as in Theorem 9.2(b), using the results of Exercises 25 and 26.)

30. A **linear inhomogeneous difference equation with constant coefficients** is a recurrence relation of the form

$$s_n = a_1 s_{n-1} + a_2 s_{n-2} + \cdots + a_k s_{n-k} + f(n), \tag{9.8}$$

where f is a nonzero function. Prove that if u_n satisfies (9.8) for $n \geq k$, then every solution of (9.8) has the form $u_n + v_n$, where v_n satisfies (9.7) for $n \geq k$. (*Hint:* If w_n is a solution of (9.8) for $n \geq k$, consider $w_n - u_n$.)

31. **(a)** Find values of a and b so that $an + b$ satisfies the recurrence relation $s_n = s_{n-1} + 6s_{n-2} + 6n - 1$ for $n \geq 2$.

(b) Use Exercise 30 to find an explicit formula for s_n if $s_0, s_1, s_2, \ldots$ is a sequence satisfying the recurrence relation $s_n = s_{n-1} + 6s_{n-2} + 6n - 1$ for $n \geq 2$ and the initial conditions $s_0 = -6$, $s_1 = 10$.

32. As in Exercise 31, find an explicit formula for s_n if $s_0, s_1, s_2, \ldots$ is a sequence satisfying the recurrence relation $s_n = 5s_{n-1} + 6s_{n-2} + 10n - 37$ for $n \geq 2$ and the initial conditions $s_0 = 7$, $s_1 = 3$.

33. Explain as in Example 9.22 the operation of the binary search algorithm to search the list 2, 4, 6, 8 for the number 6.

34. Explain as in Example 9.22 the operation of the binary search algorithm to search the list 2, 4, 6, 8 for the number 7.

Determine the number of comparisons needed by the merging algorithm to merge the lists in Exercises 35–38.

35. (45, 57) and (59, 87)

36. (45, 59) and (57, 87)

37. (1, 3, 5, 7) and (2, 4, 6, 8)

38. [(1) and (2, 3)] and (4)

39. Give an example of ordered lists $a_1, a_2, \ldots, a_m$ and $b_1, b_2, \ldots, b_n$ whose merging by the merging algorithm requires the minimum number of comparisons. Assume that $m \leq n$.

40. Give an example of ordered lists $a_1, a_2, \ldots, a_m$ and $b_1, b_2, \ldots, b_n$ whose merging by the merging algorithm requires the maximum number of comparisons. Assume that $m \leq n$.

In Exercises 41–44, let S be the generating function of the sequence $\{s_n\}$. Find an equation satisfied by S as in Section 9.6, and solve for S.

41. $s_0 = 1$ and $s_n = 2s_{n-1}$ for $n \geq 1$

42. $s_0 = 1$ and $s_n = s_{n-1} + 2$ for $n \geq 1$

43. $s_0 = 1$, $s_1 = 1$, and $s_n = -2s_{n-1} - s_{n-2}$ for $n \geq 2$

44. $s_0 = 0$, $s_1 = 1$, and $s_n = s_{n-2}$ for $n \geq 2$

In Exercises 45–48, give a formula for s_n if $\{s_n\}$ has the given generating function S.

45. $S = \dfrac{5}{1+2x}$

46. $S = \dfrac{-4}{1+6x}$

47. $S = \dfrac{1}{1-2x} + \dfrac{2}{1-3x}$

48. $S = \dfrac{1}{1-2x} + \dfrac{5}{1-x}$

In Exercises 49–55, give a generating function for the sequence $\{a_n\}$.

49. Let a_r be the number of ways to select r balls from 3 red balls, 2 green balls, and 5 white balls.

50. In a soft serve ice-cream shop, Great Northern Delites are made with one candy bar flavor chosen from Heath, Snickers, or Butterfinger. Let a_r be the number of possible orders for r Great Northern Delites.

51. In a store giveaway, 6 individual winners were identified. Each winner received at least 2 prizes, but no more than 4 prizes, and r identical prizes were awarded. Let a_r be the number of ways the prizes could have been distributed among the winners.

52. Let a_r be the number of ways of selecting r pastries from a cabinet containing three each of cherry-filled Bismarcks, lemon-filled Bismarcks, vanilla long-johns, chocolate long-johns, vanilla twists, chocolate twists, bearclaws, and apple fritters.

53. Let a_r be the number of ways r cents worth of postage can be placed on a letter using only 5-, 12-, and 25-cent stamps. The positions of the stamps on the letter do not matter.

54. Sample packages of chocolate and licorice are being made. Each package must contain r pieces of candy, including at least 3 pieces of chocolate and at most 2 pieces of licorice. Let a_r be the number of different ways to fill a package.

55. Let a_r be the number of ways to pay for an item costing r cents with pennies, nickels, and dimes.

COMPUTER PROJECTS

Write a computer program having the specified input and output.

1. Given a recurrence relation $s_n = a_1 s_{n-1} + a_2 s_{n-2} + \cdots + a_k s_{n-k}$ and initial values $s_0, s_1, \ldots, s_{k-1}$, compute a specified term of the sequence defined by these conditions.

2. Given a positive integer n, list the moves necessary to win the Towers of Hanoi game with n disks in the fewest possible moves.

3. Given a positive integer n, compute the nth Catalan number. (See Example 9.5.)

4. Given a positive integer n, list all the sequences in which the numbers $1, 2, \ldots, n$ can leave a stack if they enter it in sequence. (See Example 9.5.)
5. Given positive integers a and b, simulate 500 trials of the game in Example 9.18.
6. Given a list of n integers and a target integer t, find the first occurrence of t in the list by using the sequential search algorithm in Section 9.4.
7. Given a list of n integers and a target integer t, find an occurrence of t in the list by using the binary search algorithm in Section 9.4.
8. Given a list of n real numbers, sort the list into nondecreasing order by using the merge sort algorithm in Section 9.4.
9. Given a list of 2^k real numbers for some nonnegative integer k, sort the list by using the Bose–Nelson algorithm described in Exercise 40 of Section 9.4.
10. Give $f(x) = a_0 + a_1x + \cdots + a_nx^n$ and $g(x) = b_0 + b_1x + \cdots + b_nx^n$ and a nonnegative integer k, compute the coefficient of x^k in the polynomial $f(x)g(x)$.

SUGGESTED READINGS

1. Bose, R.C. and R.J. Nelson. "A Sorting Problem," *J. Assoc. Computing Machinery*, vol. 9 (1962): 282–296.
2. Goldberg, Samuel. *Introduction to Difference Equations*. New York: Wiley, 1958.
3. Horowitz, Ellis and Sartaj Sahni. *Fundamentals of Computer Algorithms*. New York: Freeman, 1984.
4. Levy, H. and F. Lessman. *Finite Difference Equations*. New York: Dover, 1992.
5. Ryser, Herbert John. *Combinatorial Mathematics*. Washington, DC: Mathematical Association of America, 1963.
6. Stanat, Donald F. and David F. McAllister. *Discrete Mathematics in Computer Science*. Englewood Cliffs, NJ: Prentice-Hall, 1977.

CHAPTER

10 Combinatorial Circuits and Finite State Machines

Today, tiny electronic devices called **microprocessors** are found in such diverse places as automobiles, digital watches, missiles, electronic games, compact disc players, and toasters. These devices control the larger machines in which they are embedded by responding to a variety of inputs according to a preset pattern. How they do this is determined by the circuits they contain. This chapter will provide an introduction to the logic of such circuits.

10.1 LOGICAL GATES

The sensitive electronic equipment in the control room of a recording studio needs to be protected from both high temperatures and excess humidity. An air conditioner is provided that must go on whenever either the temperature exceeds 80° or the humidity exceeds 50 percent. What is required is a control device that has two inputs—one coming from a thermostat and one from a humidistat—and one output going to the air conditioner. It must perform the function of turning on the air conditioner if it gets a yes signal from either of the input devices, as summarized in the following table:

Temperature > 80°?	*Humidity > 50%?*	*Air conditioner on?*
no	no	no
no	yes	yes
yes	no	yes
yes	yes	yes

We will follow the usual custom of using x and y to label our two inputs, and 1 and 0 to stand for the input or output signals yes and no, respectively. Thus

x and y can assume only the values 0 and 1; such variables are called **Boolean variables**. These conventions give our table a somewhat simpler form:

x	y	*Output*
0	0	0
0	1	1
1	0	1
1	1	1

The required device is an example of a **logical gate**, and the particular one whose working we have just described is called an **OR-gate**, since its output is 1 whenever either x *or* y is 1. We will denote the output of an OR-gate with inputs x and y by $x \vee y$, so that

$$x \vee y = \begin{cases} 1 & \text{if } x = 1 \quad \text{or} \quad y = 1 \\ 0 & \text{otherwise.} \end{cases}$$

We will not delve into the internal workings of the devices we call logical gates, but merely describe how they function. A logical gate is an electronic device that has either one or two inputs and a single output. These inputs and output are in one of two states, which we denote by 0 and 1. For example, the two states might be a low and high voltage.

Logical gates are represented graphically by standard symbols established by the Institute of Electrical and Electronics Engineers. The symbol for an OR-gate is shown in Figure 10.1.

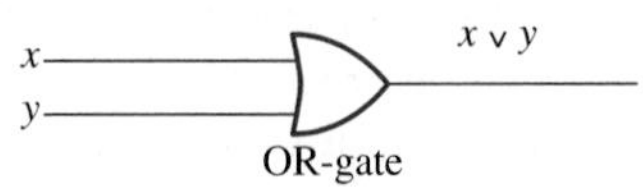

FIGURE 10.1

We will study only two other logical gates, the **AND-gate** and the **NOT-gate**. Their symbols are shown in Figure 10.2. Notice that the symbols for the OR-gate and AND-gate are quite similar, so care must be taken to distinguish between them.

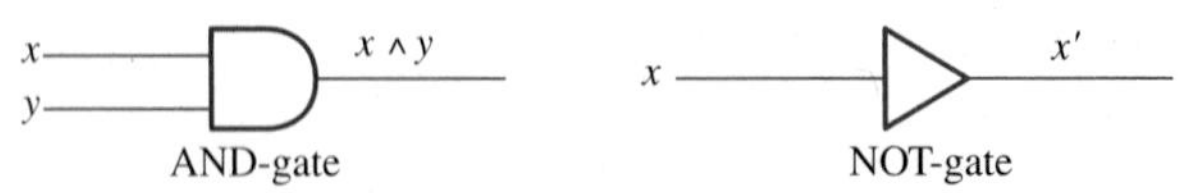

FIGURE 10.2

The output of an AND-gate with inputs x and y is 1 only when both x *and* y are 1. This output is denoted by $x \wedge y$, so that the values of $x \wedge y$ are given by the following table, which, as in logic, is called a **truth table**:

x	y	$x \wedge y$
0	0	0
0	1	0
1	0	0
1	1	1

Example 10.1 An ink-jet printer attached to a personal computer will print only when the "on-line" button on its case has been pressed *and* a paper sensor tells it that there is paper in the printer. We can represent this as an AND-gate, as in Figure 10.3. ■

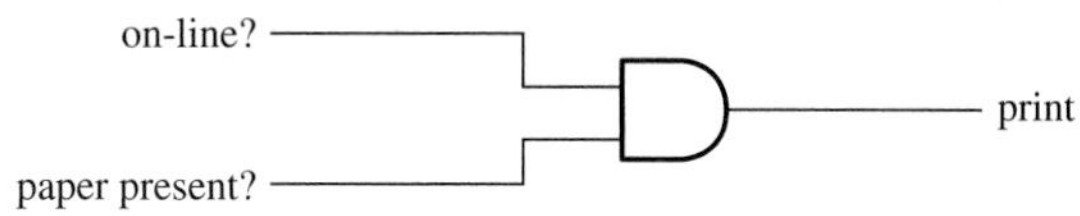

FIGURE 10.3

The other logical gate we will consider is the NOT-gate, which has only a single input. Its output is always exactly the opposite from its input. If the input is x, then the output of a NOT-gate, which we will denote by x', is as follows:

x	x'
0	1
1	0

Example 10.2 A rental truck is equipped with a governor. If the speedometer exceeds 70 miles per hour, the ignition of the truck is cut off. We can describe this with a NOT-gate, as in Figure 10.4. ■

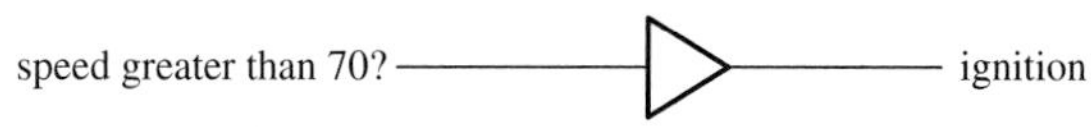

FIGURE 10.4

The reader familiar with logic will notice the similarity between the three gates we have described and the logical operators "or," "and," and "not." Although other logical gates may be defined, by appropriately combining the three gates we have introduced, we can simulate any logical gate that has no more than two inputs.

Example 10.3 A home gas furnace is attached to two thermostats, one in the living area of the house and the other in the chamber where the furnace heats air to be circulated. If the first thermostat senses that the temperature in the house is below 68°, it sends a signal to the furnace to turn on. On the other hand, if the thermostat in the heating chamber becomes hotter than 150°, it sends a message to the furnace to turn off. This signal is for reasons of safety and should be obeyed no matter what message the house thermostat is sending.

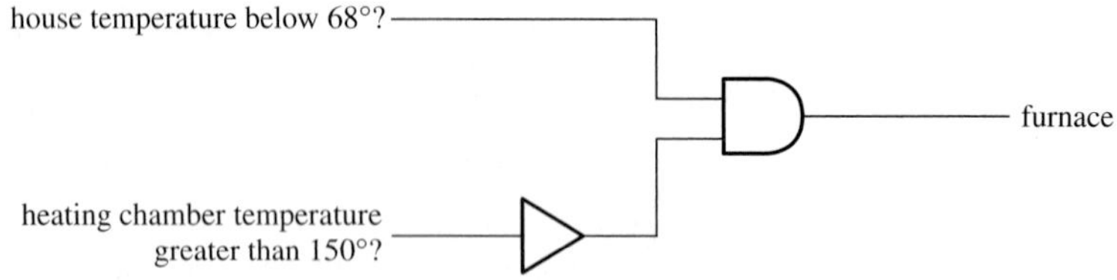

FIGURE 10.5

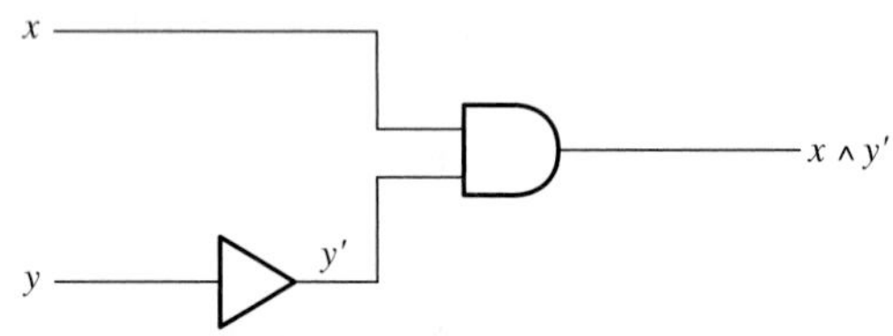

FIGURE 10.6

One arrangement of gates giving the desired output is shown in Figure 10.5. It is easier to check the effect of this arrangement if we denote the signals from the two thermostats by x and y, as in Figure 10.6. We can compute the value of $x \wedge y'$ for the possible values of x and y by means of a truth table:

x	y	y'	$x \wedge y'$
0 (room ok)	0 (chamber ok)	1	0 (furnace off)
0 (room ok)	1 (chamber hot)	0	0 (furnace off)
1 (room cold)	0 (chamber ok)	1	1 (furnace on)
1 (room cold)	1 (chamber hot)	0	0 (furnace off)

Notice that the furnace will run only when the house is cold and the heating chamber is not too hot. ■

Figure 10.6 shows an example of combining logical gates to produce what is called a **combinatorial circuit**, which we will usually refer to simply as a "circuit." More than two independent inputs are allowed, and an input may feed into more than one gate. A more complicated example is shown in Figure 10.7, in which the inputs are denoted by x, y, and z.

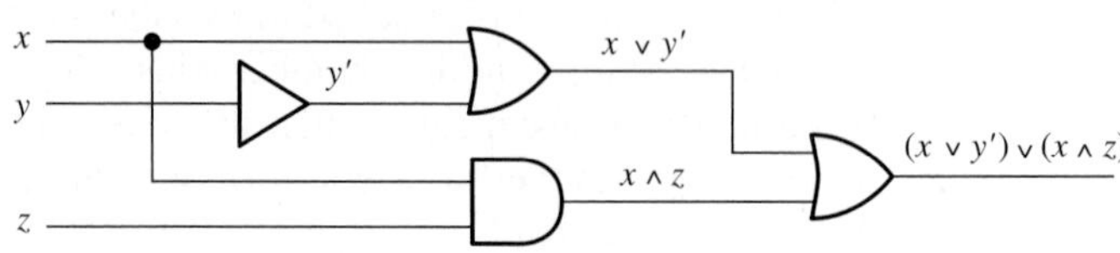

FIGURE 10.7

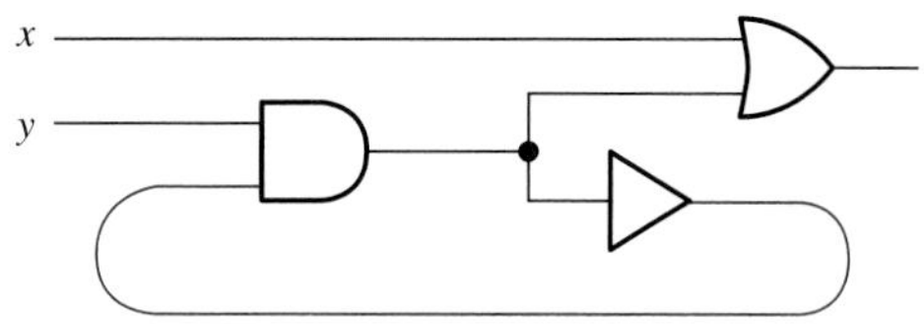

FIGURE 10.8

We will consider only circuits that have a single output, and we will not allow circuits such as the one shown in Figure 10.8, in which the output of the NOT-gate doubles back to be an input for the previous AND-gate. (We leave the precise formulation of this condition for the exercises.) In Figure 10.7 the input x splits at the heavy dot. In order to simplify our diagrams we may instead label more than one original input with the same variable. Thus Figure 10.9 is simply another way to draw Figure 10.7.

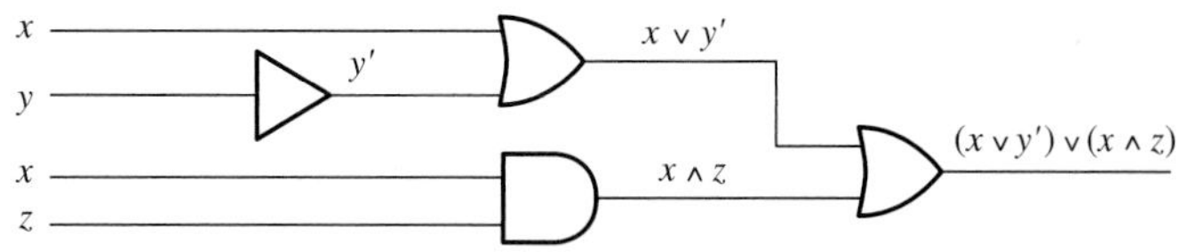

FIGURE 10.9

The effect of complicated circuits can be computed by successively evaluating the output of each gate for all possible values of the input variables, as in the following truth table:

x	y	z	y'	$x \vee y'$	$x \wedge z$	$(x \vee y') \vee (x \wedge z)$
0	0	0	1	1	0	1
0	0	1	1	1	0	1
0	1	0	0	0	0	0
0	1	1	0	0	0	0
1	0	0	1	1	0	1
1	0	1	1	1	1	1
1	1	0	0	1	0	1
1	1	1	0	1	1	1

The strings of symbols heading the columns of our table are examples of Boolean expressions. In general, given a finite set of Boolean variables, by a **Boolean expression** we mean any of these Boolean variables, either of the constants 0 and 1 (which represent variables with the constant value 0 or 1, respectively), and any subsequently-formed expressions of the type

$$B \vee C, \qquad B \wedge C, \qquad \text{or} \qquad B',$$

where B and C are Boolean expressions.

Example 10.4 Which of the following are Boolean expressions for the set of Boolean variables x, y, z?

$$x \vee (y \wedge (x \wedge z')') \qquad 1 \wedge y \qquad z$$
$$(x \wedge' z) \vee y \qquad \vee y' \wedge 0$$

The first three are Boolean expressions, but the last two are not, since neither $\wedge'$ nor $\vee y'$ makes sense. ■

Just as a combinatorial circuit leads to a Boolean expression, each Boolean expression corresponds to a circuit, which can be found by taking the expression apart from the outside. Consider the first expression of Example 10.4, which is $x \vee (y \wedge (x \wedge z')')$. This corresponds to a circuit with an OR-gate, having inputs x and $y \wedge (x \wedge z')'$, as in Figure 10.10. By continuing to work backward in this way, we find the circuit shown in Figure 10.11.

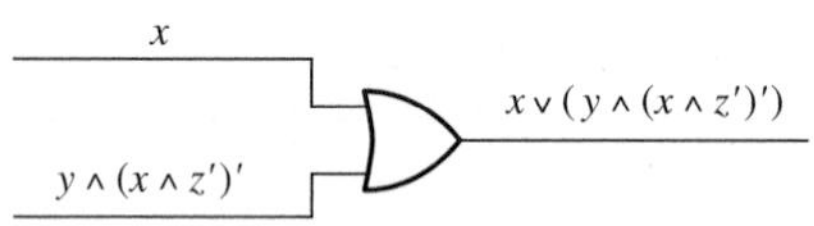

FIGURE 10.10

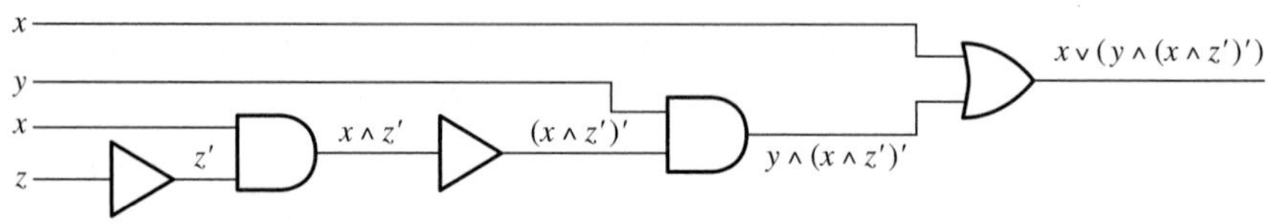

FIGURE 10.11

It may be that different circuits produce the same output for each combination of values of the input variables. For example, if we examine the table on page 533, in which we analyzed the effect of the circuit in Figure 10.9, we may notice that the output is 1 exactly when x is 1 or y is 0. Thus the circuit has exactly the same effect as that corresponding to $x \vee y'$, shown in Figure 10.12. Since this circuit is much simpler, manufacturing it rather than the circuit of Figure 10.9 will be cheaper. A simpler circuit will also usually run faster. Some integrated circuits contain more than 100,000 logical gates in an area of one square centimeter, and so their efficient use is very important.

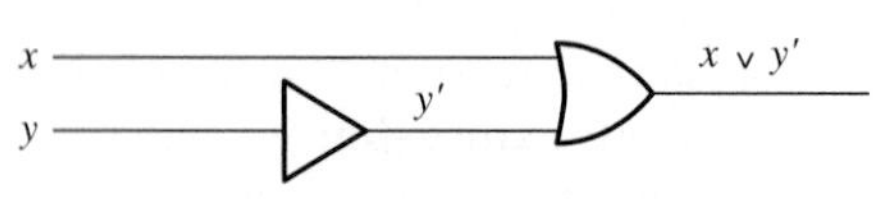

FIGURE 10.12

Circuits that give the same output for all possible values of their input variables are said to be **equivalent**, as are their corresponding Boolean expressions.

Thus $(x \vee y') \vee (x \wedge z)$ is equivalent to $x \vee y'$, as can be confirmed by comparing the following truth table with the one for $(x \vee y') \vee (x \wedge z)$ shown on page 533.

x	y	z	y'	$x \vee y'$
0	0	0	1	1
0	0	1	1	1
0	1	0	0	0
0	1	1	0	0
1	0	0	1	1
1	0	1	1	1
1	1	0	0	1
1	1	1	0	1

Since the circuits corresponding to equivalent Boolean expressions have exactly the same effect, we will write an equal sign between such expressions. For example, we will write

$$(x \vee y') \vee (x \wedge z) = x \vee y',$$

since the truth tables of the two expressions are the same. In subsequent sections, we will study how we can reduce Boolean expressions to simpler equivalent expressions to improve circuit design.

EXERCISES 10.1

In Exercises 1–8, write the Boolean expression associated with each circuit.

1.
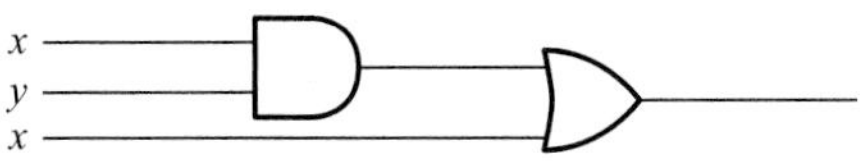

2.
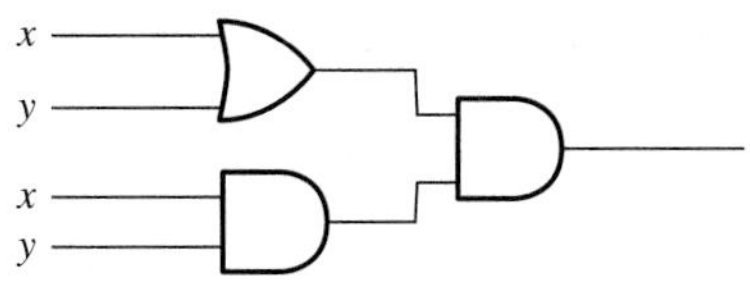

3.
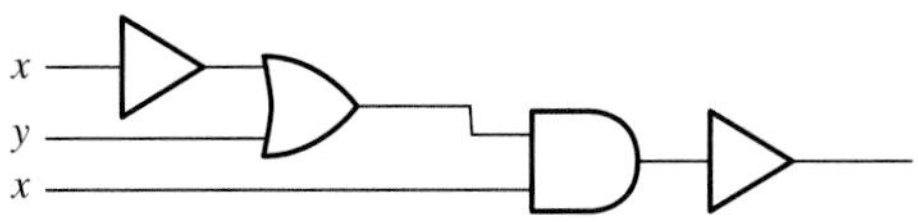

4.
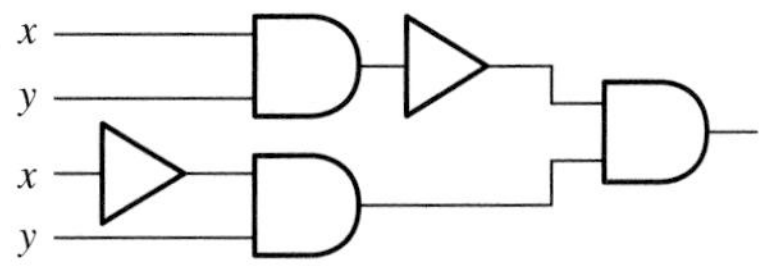

5.
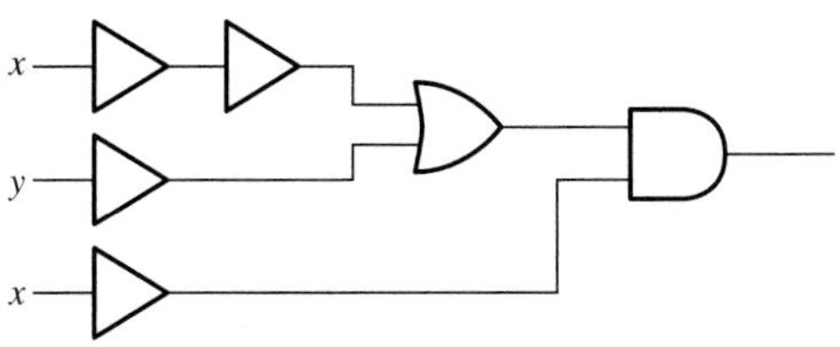

6.
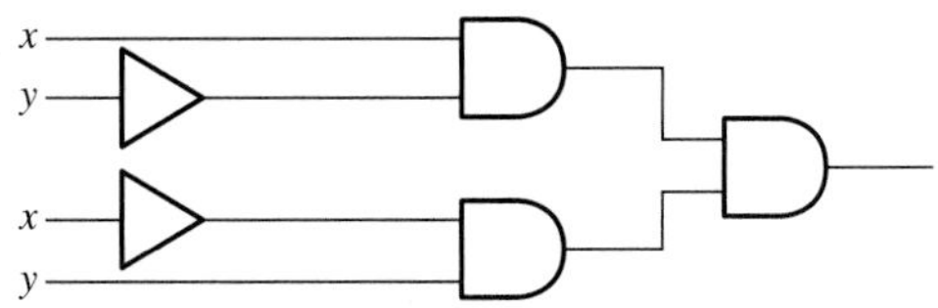

7. 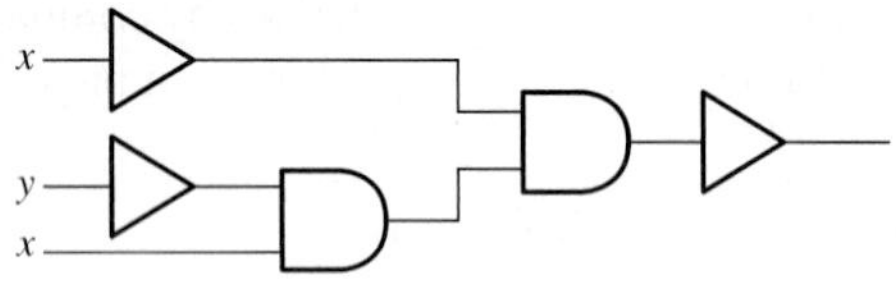

8. 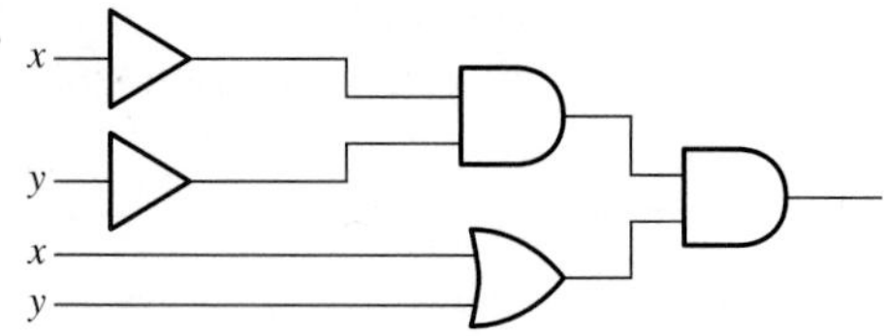

In Exercises 9–14, draw a circuit representing the given Boolean expression.

9. $(x \wedge y) \vee (x' \vee y)$

10. $(x' \wedge y) \vee [x \wedge (y \wedge z)]$

11. $[(x \wedge y') \vee (x' \wedge y')] \vee [x' \wedge (y \vee z)]$

12. $(w \wedge x) \vee [(x \vee y') \wedge (w' \vee x')]$

13. $(y' \wedge z') \vee [(w \wedge x') \wedge y']'$

14. $[x \wedge (y \wedge z)] \wedge [(x' \wedge y') \vee (z \wedge w')]$

In Exercises 15–18, give the output value for the Boolean expression with the given input values.

15. $(x \vee y) \wedge (x' \vee z)$ for $x = 1, y = 1, z = 0$

16. $[(x \wedge y) \vee z] \wedge [x \vee (y' \wedge z)]$ for $x = 0, y = 1, z = 1$

17. $[x \wedge (y \wedge z)]'$ for $x = 0, y = 1, z = 0$

18. $[(x \wedge (y \wedge z')) \vee ((x \wedge y) \wedge z)] \wedge (x \vee z')$ for $x = 0, y = 1, z = 0$

In Exercises 19–22, construct a truth table for the circuit shown.

19. 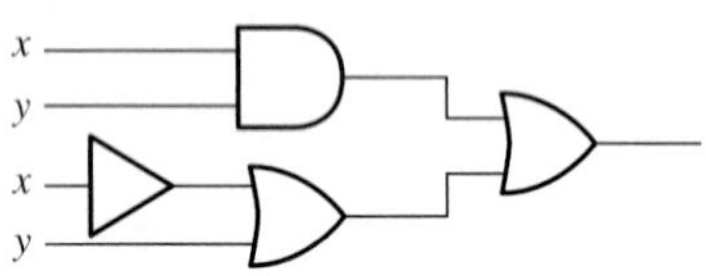

20. 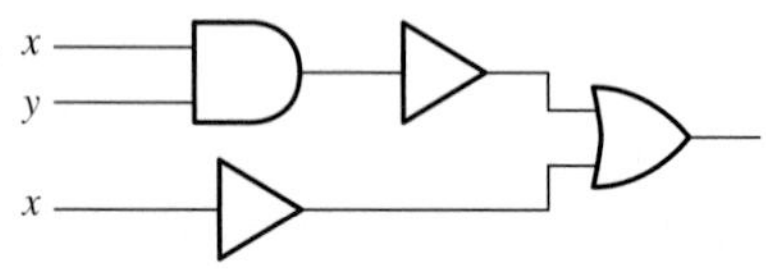

21. 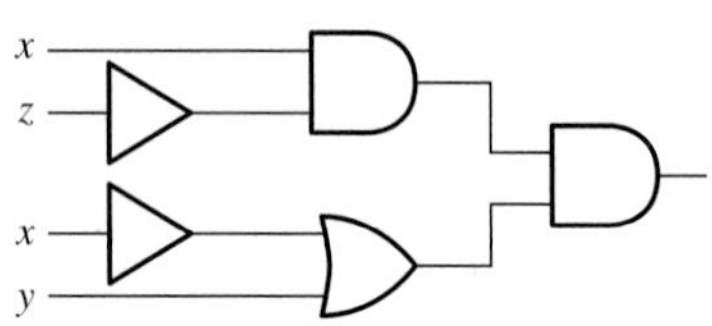

22. 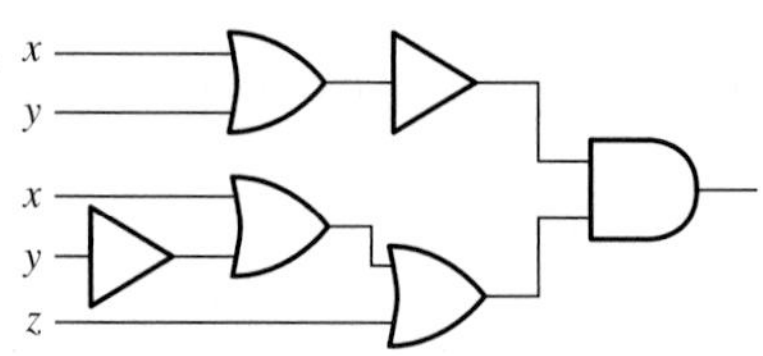

In Exercises 23–28, construct a truth table for the given Boolean expression.

23. $x \wedge (y \vee x')$

24. $(x \vee y')' \vee x$

25. $(x \wedge y) \vee (x' \wedge y')'$

26. $x \vee (x' \wedge y)$

27. $(x \vee y') \vee (x \wedge z')$

28. $[(x \wedge y) \wedge z] \vee [x \wedge (y \wedge z')]$

In Exercises 29–36, use truth tables to determine which pairs of circuits are equivalent.

29. 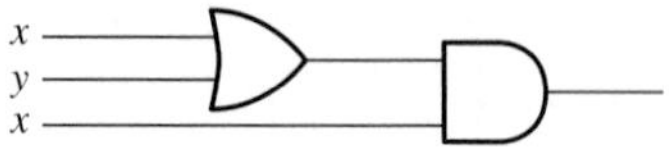

and

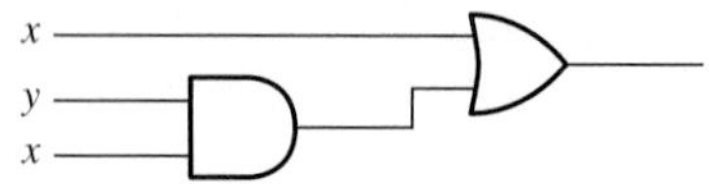

30. 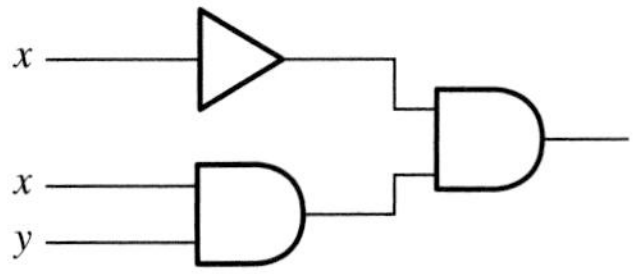

and
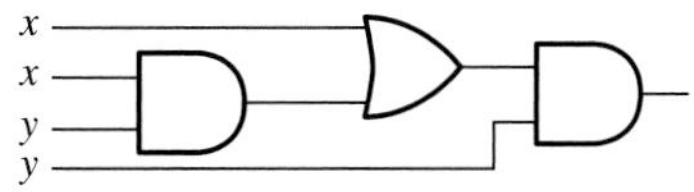

31. 

and
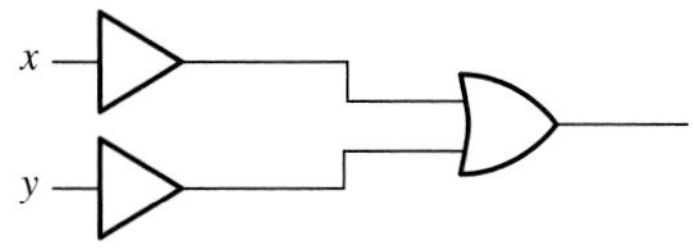

32. 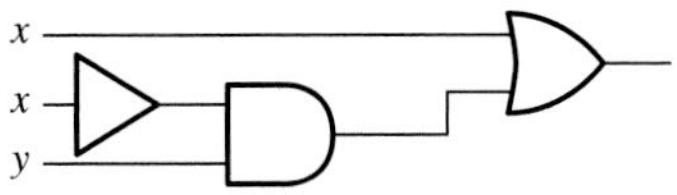

and
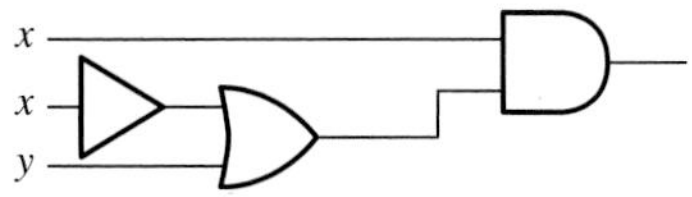

33. 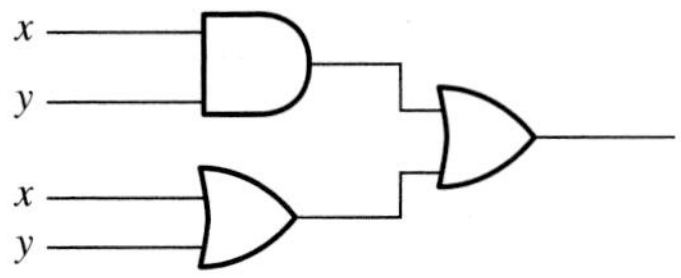

and
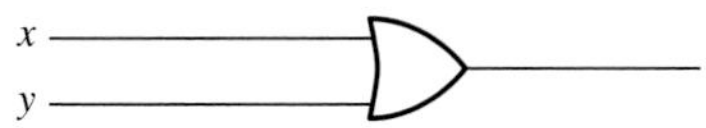

34. 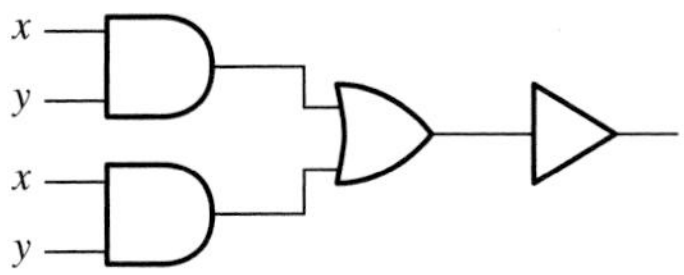

and
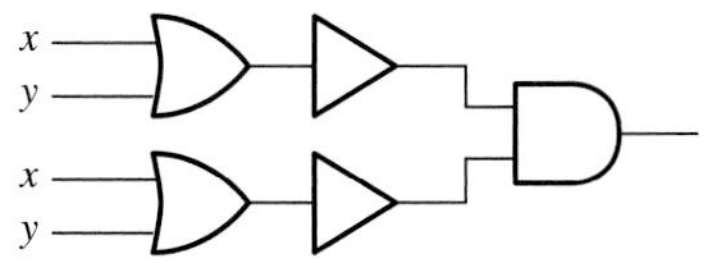

35. 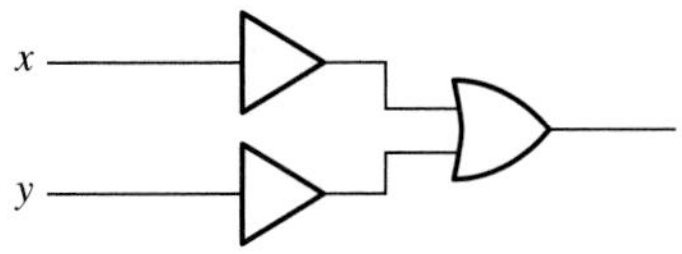

and
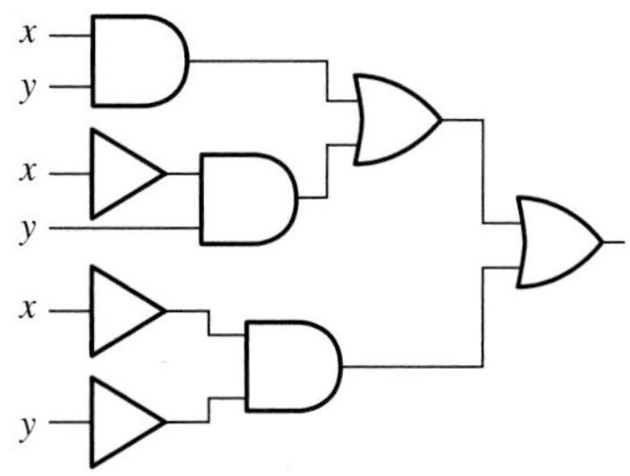

36. 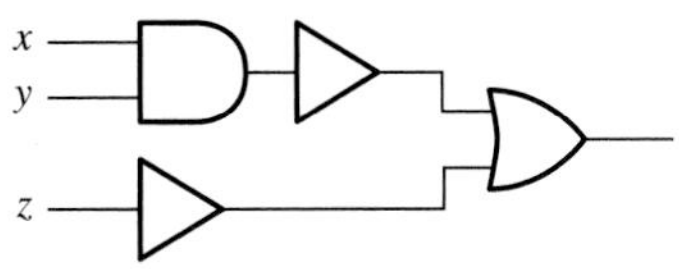

and
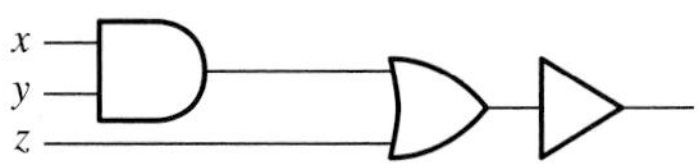

In Exercises 37–42, use truth tables to determine whether the Boolean expressions given are equivalent.

37. $x \vee (x \wedge y)$ and x

38. $x \wedge (x' \wedge y)$ and $x \wedge y$

39. $[(x \vee y) \wedge (x' \vee y)] \wedge (y \vee z)$ and $(x \vee y) \wedge (x' \vee z)$

40. $(x \wedge (y \wedge z)) \vee [x' \vee ((x \wedge y) \wedge z')]$ and $x' \vee y$

41. $y' \wedge (y \vee z')$ and $y' \wedge x'$

42. $x \wedge [w \wedge (y \vee z)]$ and $(x \wedge w) \wedge (y \vee z)$

43. A home security alarm is designed to alert the police department if a window signal is heard or if a door is opened without someone first throwing a safety switch. Draw a circuit for this situation, describing the meaning of your input variables.

44. The seatbelt buzzer for the driver's side of an automobile will sound if the belt is not buckled, the weight sensor indicates someone is in the seat, and the key is in the ignition. Draw a circuit for this situation, describing the meaning of your input variables.

45. Prove that the equivalence of Boolean expressions using a fixed finite set of Boolean variables is an equivalence relation as defined in Chapter 2.

46. Define the directed graph associated with a combinatorial circuit. State a condition on this directed graph that excludes circuits similar to that shown in Figure 10.8.

47. What is the output of the illegal circuit shown for $x = 0$ and 1?

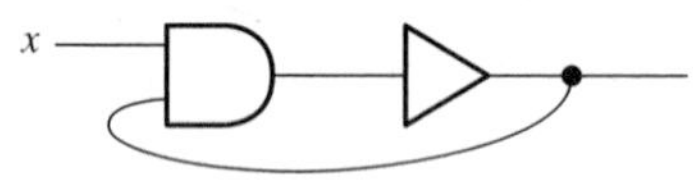

10.2 CREATING COMBINATORIAL CIRCUITS

In Section 10.1, we saw how each combinatorial circuit corresponds to a Boolean expression and observed that sometimes we could simplify a circuit by finding a simpler equivalent Boolean expression. One way to simplify Boolean expressions is by using standard identities, much in the way that the algebraic expression $(a+b)^2 - b(b-3a)$ can be reduced to $a(a+5b)$ by using the rules of algebra. Some of these identities for Boolean expressions are listed in Theorem 10.1.

Theorem 10.1 For any Boolean expressions X, Y, and Z,

(a) $X \wedge Y = Y \wedge X$ and $X \vee Y = Y \vee X$

(b) $(X \wedge Y) \wedge Z = X \wedge (Y \wedge Z)$ and $(X \vee Y) \vee Z = X \vee (Y \vee Z)$

(c) $X \wedge (Y \vee Z) = (X \wedge Y) \vee (X \wedge Z)$ and
$X \vee (Y \wedge Z) = (X \vee Y) \wedge (X \vee Z)$

(d) $X \vee (X \wedge Y) = X \wedge (X \vee Y) = X$

(e) $X \vee X = X \wedge X = X$

(f) $X \vee X' = 1$ and $X \wedge X' = 0$

(g) $X \vee 0 = X \wedge 1 = X$

(h) $X \wedge 0 = 0$ and $X \vee 1 = 1$

(i) $(X')' = X$, $0' = 1$, and $1' = 0$

(j) $(X \vee Y)' = X' \wedge Y'$ and $(X \wedge Y)' = X' \vee Y'$.

Many of these identities have the same form as familiar algebraic rules. For example, rule (a) says that the operations $\vee$ and $\wedge$ are commutative, and rule (b) is an associative law for these operations. In spite of rule (b), $x \vee (y \wedge z)$ is not equivalent to $(x \vee y) \wedge z$.

Rule (c) gives two distributive laws. For example, if, in the first equality of rule (c), we substitute multiplication for $\wedge$ and addition for $\vee$, we get

$$X(Y + Z) = (XY) + (XZ),$$

which is the distributive law of ordinary algebra. Making the same substitution in the second equality, however, produces

$$X + (YZ) = (X + Y)(X + Z),$$

which is not an identity of ordinary algebra. Thus these rules must be used with care; one should not jump to conclusions about how Boolean expressions may be manipulated on the basis of rules for other algebraic systems.

The equalities of rule (j) are known as **De Morgan's laws**; compare them with the rules for the complements of set unions and intersections in Theorem 2.2. The validity of all these identities can be proved by computing the truth tables for the expressions that are claimed to be equivalent.

Example 10.5 Prove rule (d) in Theorem 10.1.

We compute truth tables for the expressions $X \vee (X \wedge Y)$ and $X \wedge (X \vee Y)$ as follows:

X	Y	$X \wedge Y$	$X \vee (X \wedge Y)$	$X \vee Y$	$X \wedge (X \vee Y)$
0	0	0	0	0	0
0	1	0	0	1	0
1	0	0	1	1	1
1	1	1	1	1	1

Since the first, fourth, and sixth columns of this table are identical, rule (d) is proved. ■

As an example of the use of our rules, we will prove that the expressions $(x \vee y') \vee (x \wedge z)$ and $x \vee y'$ are equivalent without computing, as we did in Section 10.1, the truth table of each expression. We will start with the more complicated expression and use our rules to simplify it:

$$\begin{aligned}
(x \vee y') \vee (x \wedge z) &= (y' \vee x) \vee (x \wedge z) && \text{(rule (a))} \\
&= y' \vee (x \vee (x \wedge z)) && \text{(rule (b))} \\
&= y' \vee x && \text{(rule (d))} \\
&= x \vee y' && \text{(rule (a)).}
\end{aligned}$$

Example 10.6 Simplify the expression $x \vee (y \wedge (x \wedge z')')$, which corresponds to the circuit shown in Figure 10.11. We have

$$
\begin{aligned}
x \vee (y \wedge (x \wedge z')') &= x \vee (y \wedge (x' \vee z'')) && \text{(rule (j))}\\
&= x \vee (y \wedge (x' \vee z)) && \text{(rule (i))}\\
&= (x \vee y) \wedge (x \vee (x' \vee z)) && \text{(rule (c))}\\
&= (x \vee y) \wedge ((x \vee x') \vee z) && \text{(rule (b))}\\
&= (x \vee y) \wedge (1 \vee z) && \text{(rule (f))}\\
&= (x \vee y) \wedge 1 && \text{(rule (h))}\\
&= x \vee y && \text{(rule (g)).}
\end{aligned}
$$

We see that the complex circuit of Figure 10.11 can be replaced by a circuit having only one gate. ■

Because of rule (b) in Theorem 10.1, we can use expressions such as $X \vee Y \vee Z$ without ambiguity, since the result is the same no matter whether we calculate $X \vee Y$ or $Y \vee Z$ first. In terms of circuits, this means that the two circuits in Figure 10.13 are equivalent. Thus we will use the diagram of Figure 10.14 to represent either of the circuits in Figure 10.13; its output is 1 when any of X, Y, or Z is 1.

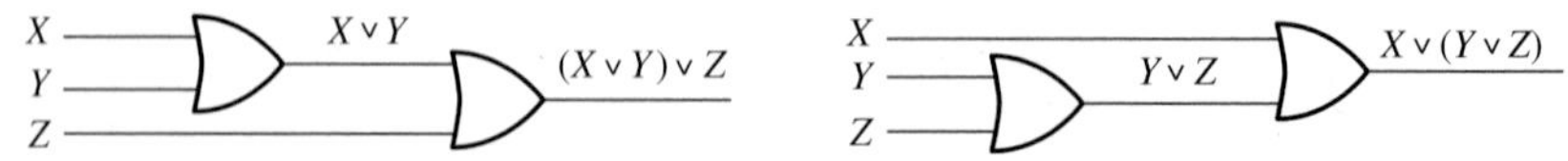

FIGURE 10.13

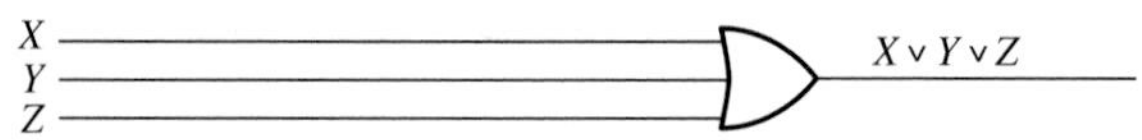

FIGURE 10.14

We use the same convention for more than three inputs. For example, the circuit shown in Figure 10.15 represents any of the equivalent circuits corresponding to a Boolean expression formed by putting parentheses in $W \wedge X \wedge Y \wedge Z$; one such expression is $(W \wedge X) \wedge (Y \wedge Z)$, another is $((W \wedge X) \wedge Y) \wedge Z$.

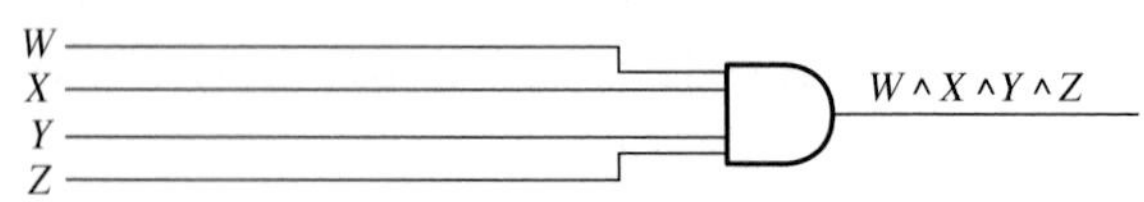

FIGURE 10.15

Of course, before we can simplify a circuit we must *have* a circuit. Thus we must consider the problem of finding a circuit that will accomplish the particular job we have in mind. Whether the circuit we find is simple or complicated is of secondary importance. There is always the possibility of simplifying a complicated circuit by reducing its corresponding Boolean expression.

As an example, we will consider the three-person finance committee of a state senate. The committee must vote on all revenue bills, and of course two or three yes votes are necessary for a bill to clear the committee. We will design a circuit that will take the three senators' votes as inputs and yield whether the bill passes or not as output. (Ours will be a scaled-down version of the electronic voting devices used in some legislatures.) If we denote yes votes and the passage of a bill by 1, we desire a circuit with the following truth table:

x	y	z	*Pass?*
0	0	0	0
0	0	1	0
0	1	0	0
*0	1	1	1
1	0	0	0
*1	0	1	1
*1	1	0	1
*1	1	1	1

We have marked the rows that have 1s in the output column because these rows will be used to construct a Boolean expression with this truth table. Consider, for example, the fourth row of the table. Since there is a 1 in the output column in this row, when x is 0 and y and z are 1, our Boolean expression should have a value 1. But x is 0 if and only if x' is 1; so this row corresponds to the condition that x', y, and z all have value 1. This happens exactly when $x' \wedge y \wedge z$ has value 1. The other marked rows indicate that the output is 1 also when $x \wedge y' \wedge z$, $x \wedge y \wedge z'$, or $x \wedge y \wedge z$ have value 1. Thus we want an output of 1 exactly when the expression $(x' \wedge y \wedge z) \vee (x \wedge y' \wedge z) \vee (x \wedge y \wedge z') \vee (x \wedge y \wedge z)$ has value 1, and this is the Boolean expression we seek. The circuit corresponding to this expression is shown in Figure 10.16.

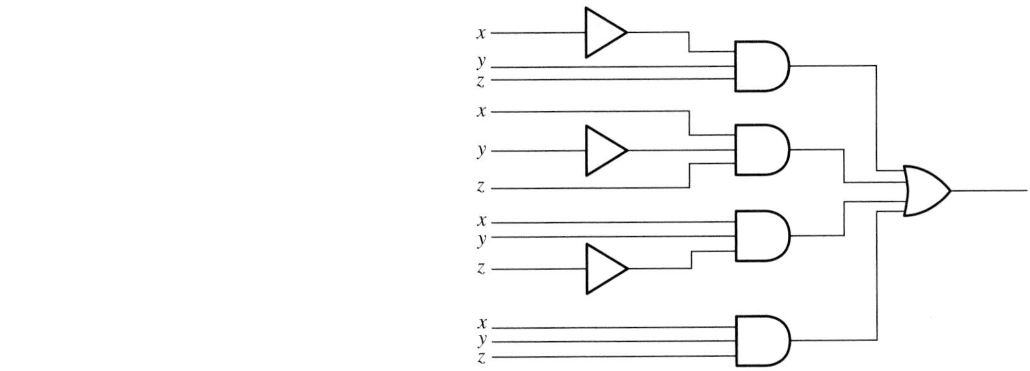

FIGURE 10.16

Notice that we have designed a crude arithmetic computer, since our circuit counts the number of yes votes and tells us whether there are 2 or more.

Now we summarize our method of finding a Boolean expression corresponding to a given truth table. Let us suppose the input variables are $x_1, x_2, \ldots, x_n$. If all outputs are 0, then the desired Boolean expression is 0. Otherwise we proceed as follows:

Step 1 Identify the rows of the truth table having output 1. For each such row, form the Boolean expression

$$y_1 \wedge y_2 \wedge \cdots \wedge y_n,$$

where y_i is taken to be x_i if there is a 1 in the x_i column, and y_i is taken to be x_i' if there is a 0 in the x_i column. The expressions thus formed are called **minterms**.

Step 2 If $B_1, B_2, \ldots, B_k$ are the minterms formed in step 1, form the expression

$$B_1 \vee B_2 \vee \cdots \vee B_k.$$

This Boolean expression has a truth table identical to the one with which we started.

Example 10.7 A garage light is to be controlled by three switches, one inside the kitchen to which the garage is attached, one at the garage door, and one at a back door to the garage. It should be possible to turn the light on or off with any of these switches, no matter what the positions of the other switches are. Design a circuit to make this possible.

The inputs are the three switches, which we will label 1 or 0 according to whether they are in an up or down position. We will design a circuit that turns the light on whenever the number of inputs equal to 1 is odd, since flipping any switch will change whether this number is odd or even. We want a circuit with the following truth table:

x	y	z	*Number of* 1s	*Output*
0	0	0	0	0
*0	0	1	1	1
*0	1	0	1	1
0	1	1	2	0
*1	0	0	1	1
1	0	1	2	0
1	1	0	2	0
*1	1	1	3	1

The rows having output 1 are marked, and the required Boolean expression is $(x' \wedge y' \wedge z) \vee (x' \wedge y \wedge z') \vee (x \wedge y' \wedge z') \vee (x \wedge y \wedge z)$. The corresponding circuit is shown in Figure 10.17. ■

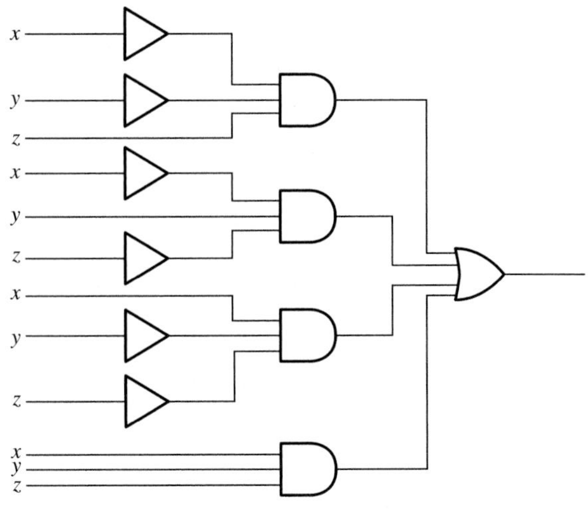

FIGURE 10.17

The Boolean expressions that our method produces tend to be complicated, and so correspond to complicated circuits. The circuit shown in Figure 10.17 is actually more complex than it appears, since if it were expressed using only our original three logical gates, each of the gates in the middle of the diagram with three inputs would have to be replaced by two standard 2-input AND-gates, and the gate on the right with four inputs would have to be replaced by three standard 2-input OR-gates. Thus the circuit of Figure 10.17 requires 6 NOT-gates, 8 AND-gates, and 3 OR-gates, for a total of 17 elementary gates. Although we might simplify the corresponding Boolean expression by using the rules given at the beginning of this section, it is not clear how to do this. In the next section, we will consider a method for simplifying Boolean expressions in an organized way.

EXERCISES 10.2

In Exercises 1–8, prove the equivalence by using truth tables.

1. $x \wedge y = y \wedge x$

2. $x \wedge (y \vee z) = (x \wedge y) \vee (x \wedge z)$

3. $x \vee x = x$

4. $(x')' = x$

5. $(x \wedge y)' = x' \vee y'$

6. $x' \wedge y' = (x \vee y)'$

7. $x \wedge x' = 0$

8. $x \wedge (y \wedge z) = (x \wedge y) \wedge z$

In Exercises 9–18, establish the validity of the equivalence by using Theorem 10.1. List by letter the rules you use, in order. Start with the expression on the left side.

9. $(x \wedge y) \vee (x \wedge y') = x$

10. $x \vee (x' \wedge y) = x \vee y$

11. $x \wedge (x' \vee y) = x \wedge y$

12. $[(x \wedge y) \vee (x \wedge y')] \vee [(x' \wedge y) \vee (x' \wedge y')] = 1$

13. $(x' \vee y)' \vee (x \wedge y') = x \wedge y'$

14. $[(x \vee y) \wedge (x' \vee y)] \wedge [(x \vee y') \wedge (x' \vee y')] = 0$

15. $(x \wedge y)' \vee z = x' \vee (y' \vee z)$

16. $((x \vee y) \wedge z)' = z' \vee (x' \wedge y')$

17. $(x \wedge y) \wedge [(x \wedge w) \vee (y \wedge z)] = (x \wedge y) \wedge (w \vee z)$

18. $(x \vee y)' \vee (x \wedge y)' = (x \wedge y)'$

In Exercises 19–22, show that the Boolean expressions are not equivalent.

19. $x \wedge (y \vee z)$ and $(x \wedge y) \vee z$

20. $(x \wedge y)'$ and $x' \wedge y'$

21. $(x \wedge y) \vee (x' \wedge z)$ and $(x \vee x') \wedge (y \vee z)$

22. $(1 \vee x) \vee x$ and x

In Exercises 23–28, find a Boolean expression of minterms that has the given truth table. Then draw the corresponding circuit.

23.

x	y	*Output*
0	0	0
0	1	1
1	0	1
1	1	0

24.

x	y	*Output*
0	0	0
0	1	1
1	0	0
1	1	1

25.

x	y	z	*Output*
0	0	0	0
0	0	1	0
0	1	0	0
0	1	1	1
1	0	0	0
1	0	1	1
1	1	0	1
1	1	1	0

26.

x	y	z	*Output*
0	0	0	1
0	0	1	0
0	1	0	0
0	1	1	1
1	0	0	1
1	0	1	0
1	1	0	0
1	1	1	1

27.

x	y	z	*Output*
0	0	0	0
0	0	1	0
0	1	0	0
0	1	1	1
1	0	0	0
1	0	1	1
1	1	0	0
1	1	1	1

28.

x	y	z	*Output*
0	0	0	1
0	0	1	0
0	1	0	0
0	1	1	0
1	0	0	0
1	0	1	1
1	1	0	0
1	1	1	1

In Exercises 29–34, give the number of AND-, OR-, and NOT-gates with one or two inputs it would take to represent the given circuits.

29.

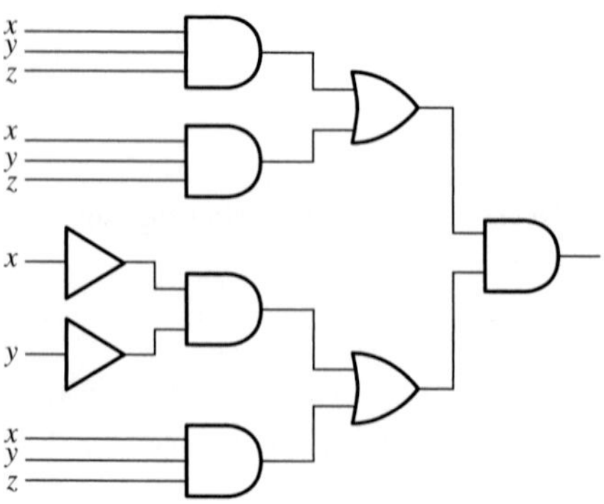

30.

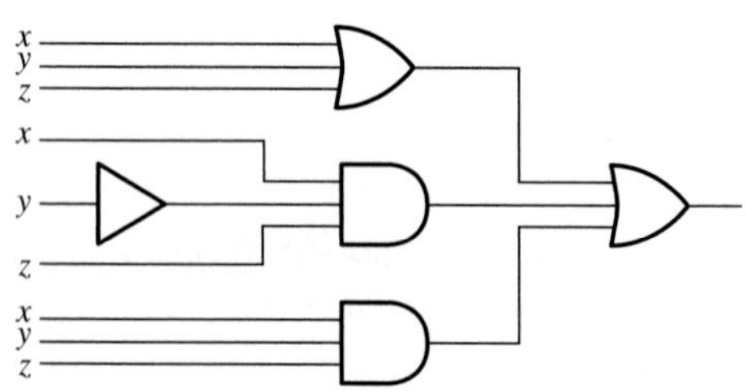

31.

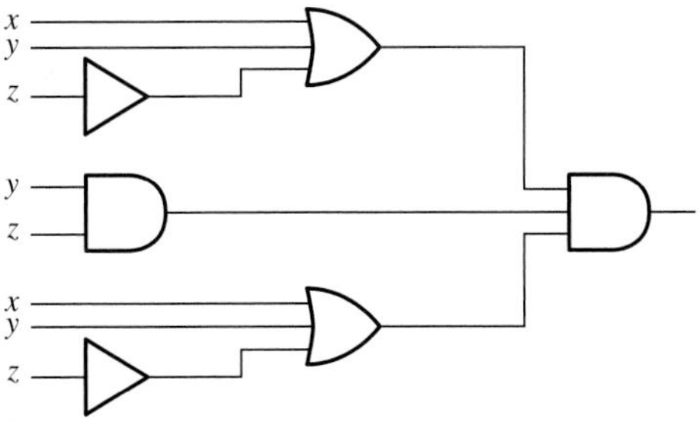

32.

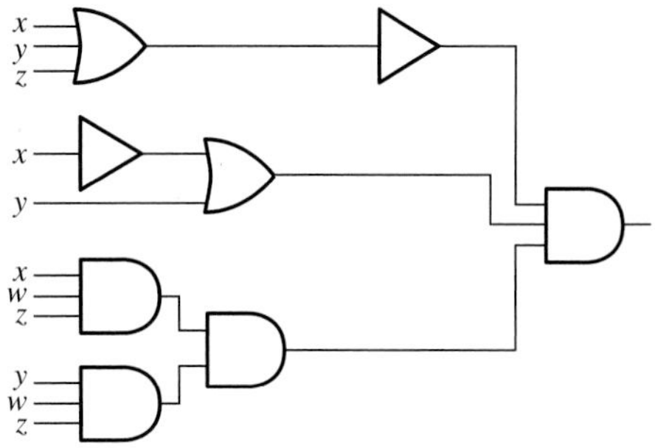

33.

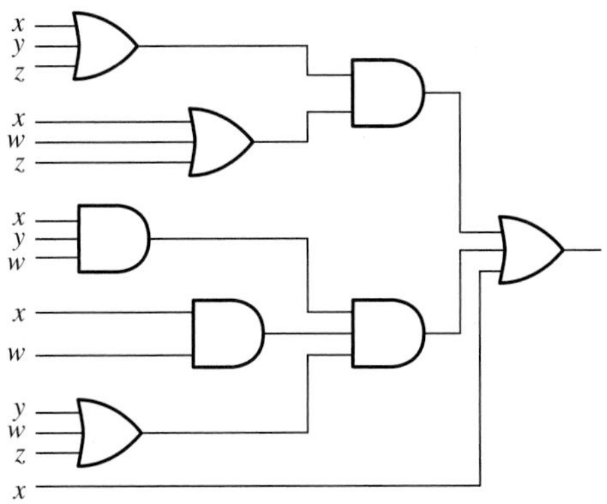

34.

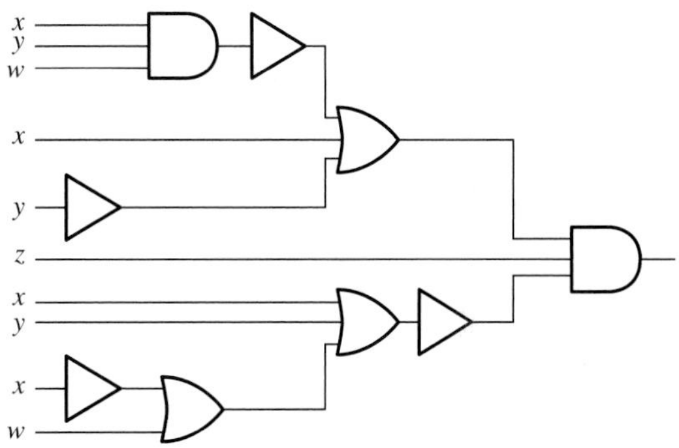

35. Suppose a company wishes to manufacture logical devices having inputs x and y and with output equivalent to the value of the logical statement $\sim (x \to y)$, where 1 corresponds to T and 0 to F. Using AND-, OR-, and NOT-gates, draw a circuit that will do this.

36. A security network for a three-guard patrol at a missile base is set up so that an alarm is sounded if guard one loses contact and at least one of the other two guards is not in contact, or if guard one and guard two are in contact but guard three loses contact. Find a Boolean expression that has value 1 exactly when the alarm sounds. Let the input 1 correspond to losing contact.

37. An inventory control system for a factory recognizes an error in an order if it contains part A and part B, but not part C; if it contains parts B or C, but not part D; or if it contains parts A and D. Find a Boolean expression in the variables a, b, c, and d that is 1 exactly when an error is recognized. Let a be 1 if part A is present, etc.

38. Which of the rules in Theorem 10.1 hold if X, Y, and Z stand for real numbers and we make the substitutions of multiplication for $\wedge$, addition for $\vee$, and $-X$ for X'?

39. Which of the rules in Theorem 10.1 hold if X, Y, and Z stand for subsets of a set U and we make the substitutions $\cap$ for $\wedge$, $\cup$ for $\vee$, $\bar{A}$ (the complement of A) for A', U for 1, and $\emptyset$ (the empty set) for 0?

We define a **Boolean algebra** to be a set B satisfying the following conditions:

(*i*) For each pair of elements a and b in B, there are defined unique elements $a \vee b$ and $a \wedge b$ in B.

(*ii*) If a and b are in B, then $a \vee b = b \vee a$ and $a \wedge b = b \wedge a$.

(*iii*) If a, b, and c are in B, then $a \vee (b \vee c) = (a \vee b) \vee c$ and $a \wedge (b \wedge c) = (a \wedge b) \wedge c$.

(*iv*) If a, b, and c are in B, then $a \vee (b \wedge c) = (a \vee b) \wedge (a \vee c)$ and $a \wedge (b \vee c) = (a \wedge b) \vee (a \wedge c)$.

(*v*) There exist distinct elements 0 and 1 in B such that if $a \in B$, then $a \vee 0 = a$ and $a \wedge 1 = a$.

(*vi*) If $a \in B$, there is defined a unique element $a' \in B$.

(*vii*) If $a \in B$, then $a \vee a' = 1$ and $a \wedge a' = 0$.

In Exercises 40–45, assume that B is a Boolean algebra. Exercises 41–45 show that the rules of Theorem 10.1 hold in any Boolean algebra.

40. Show that if a and b are in B and $a \vee b = 1$ and $a \wedge b = 0$, then $b = a'$. (*Hint:* Show that $b = b \wedge (a \vee a') = b \wedge a' = a' \wedge (a \vee b)$.)

41. Show that if $a \in B$, then $a \wedge 0 = 0$ and $a \vee 1 = 1$. (*Hint:* Compute $a \wedge (0 \vee a')$ and $a \vee (1 \wedge a')$ in two ways.)

42. Show that if a and b are in B, then $a \vee (a \wedge b) = a \wedge (a \vee b) = a$. (*Hint:* Compute $a \wedge (1 \vee b)$ and $a \vee (0 \wedge b)$ in two ways.)

43. Show that if $a \in B$, then $a \vee a = a \wedge a = a$. (*Hint:* Compute $(a \vee a) \wedge (a \vee a')$ and $(a \wedge a) \vee (a \wedge a')$ in two ways.)

44. Show that if $a \in B$, then $a'' = a$, $0' = 1$, and $1' = 0$. (*Hint:* Use Exercise 40.)

45. Show that if a and b are in B, then $(a \vee b)' = a' \wedge b'$ and $(a \wedge b)' = a' \vee b'$. (*Hint:* Use Exercise 40.)

10.3 KARNAUGH MAPS

In the previous section, we saw how to create a Boolean expression, and therefore a logical circuit, that corresponds to any given truth table. The circuits we created, however, were usually quite complicated. We will show how to create simpler circuits by, in effect, making a picture of the truth table. Of course, “simpler” has not been defined precisely, and in fact, various definitions might be appropriate. For compactness and economy of manufacture, we might want to consider one circuit better than another if it contains fewer gates. For speed of operation, however, we might prefer a circuit such that the maximal number of gates between any original input and the output is as small as possible. The method we will describe will lead to circuits that are in general much simpler than those we learned to create at the end of the previous section, although they will not necessarily be simplest by either of the criteria just suggested. We will only treat the cases of 2, 3, or 4 Boolean variables as inputs, although there are methods for dealing with more than 4 input variables. (See suggested reading [7].)

We will show how to produce a simple Boolean expression that has a prescribed truth table. A circuit can then be constructed from this expression. The truth table we start with may represent the desired output of a circuit we are designing, or it may have been computed from an existing circuit or Boolean expression that we wish to simplify.

To illustrate the technique, we will start with the following truth table:

x	y	*Output*
0	0	1
0	1	0
1	0	1
1	1	1

For this truth table, our previous method yields the Boolean expression $(x' \wedge y') \vee (x \wedge y') \vee (x \wedge y)$ and the circuit of Figure 10.18. To find a simpler circuit, we will represent our truth table graphically, as in Figure 10.19(a). Each cell in the grid shown corresponds to a row of the truth table, with the rows of the grid corresponding to x and x' and the columns to y and y'. For example, the top left cell corresponds to the row of the truth table with $x = 1$ and $y = 1$, and the 1 in that cell tells us that there is a 1 in the output column of this row. Since in the grid each row is labeled either x or x', each column either y or y', and there is either a 0 or 1 in each cell, from now on we will save time by omitting the labels x' and y' and 0s, as in Figure 10.19(b). This is called the **Karnaugh map** of the truth table.

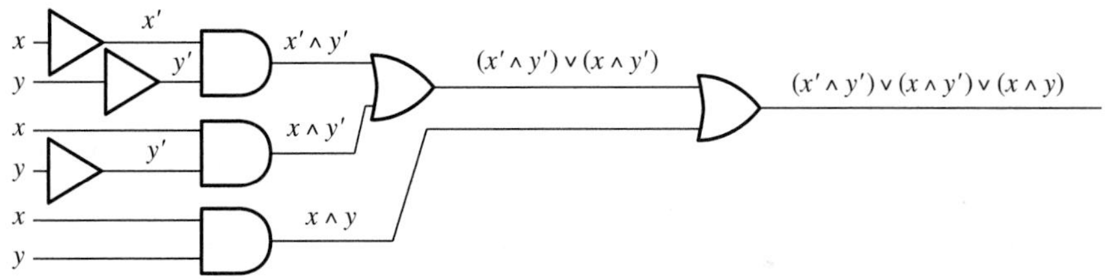

FIGURE 10.18

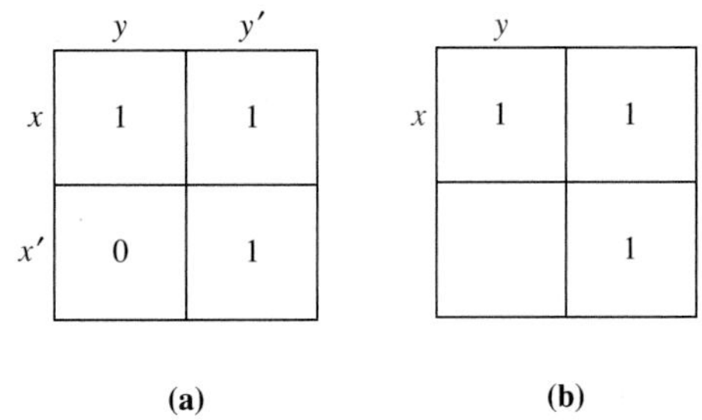

FIGURE 10.19

Each cell in the Karnaugh map corresponds to a minterm, as shown in Figure 10.20(a). Thus we can create a Boolean expression having the truth table we started with by joining with the symbol $\vee$ the minterms in cells containing a 1, as circled in Figure 10.20(b). This amounts to the method of the previous section and produces the Boolean expression

$$(x \wedge y) \vee (x \wedge y') \vee (x' \wedge y').$$

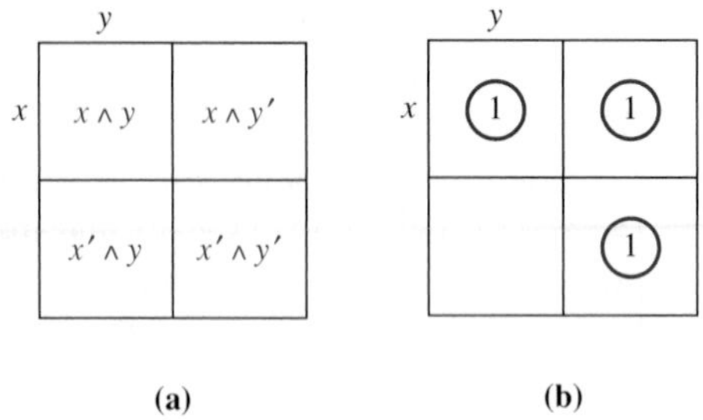

FIGURE 10.20

The key to our method is to notice that groups of adjacent cells may have even simpler Boolean expressions. For example, the two cells in the top row of the grid can be expressed simply as x. This can be confirmed by using Theorem 10.1 as follows:

$$\begin{aligned}(x \wedge y) \vee (x \wedge y') &= x \wedge (y \vee y') && \text{(rule (c))}\\ &= x \wedge 1 && \text{(rule (f))}\\ &= x && \text{(rule (g)).}\end{aligned}$$

Other such groups of two cells and the corresponding Boolean expressions are shown in Figures 10.21(a) and (b), where the ovals outline the cell groups named.

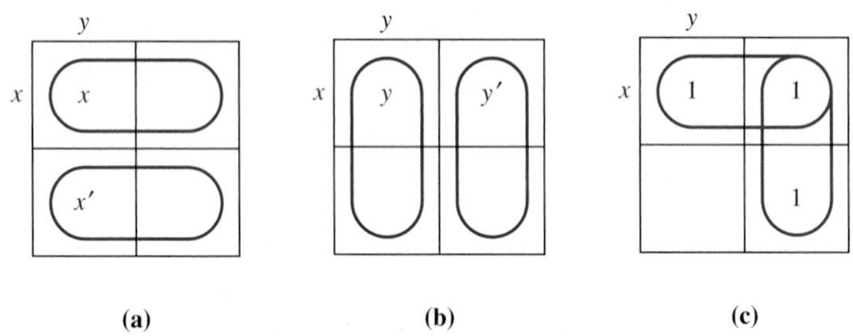

FIGURE 10.21

In Figure 10.21(c), we see that the three cells with 1s can be characterized as those cells in either the x oval or the y' oval, and so correspond to the Boolean expression $x \vee y'$. This is the simpler expression we have been seeking.

It is easily checked that $x \vee y'$ has the desired truth table. The corresponding circuit is shown in Figure 10.22. Comparison with the circuit of Figure 10.18 shows that it is simpler by any reasonable criterion.

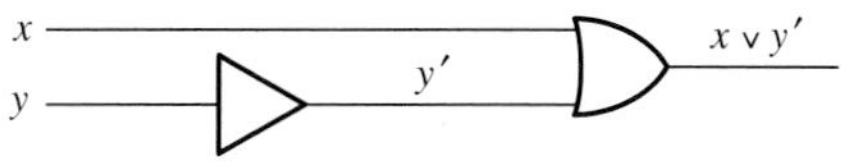

FIGURE 10.22

Since the case of two input variables is fairly straightforward, we shall proceed to three input variables, say, x, y, and z. The grid we will use is shown in Figure 10.23(a). Recall the convention that the unmarked second row corresponds to x'. Likewise, columns 3 and 4 correspond to y' and columns 1 and 4 to z'. The minterms for each cell are shown in Figure 10.23(b).

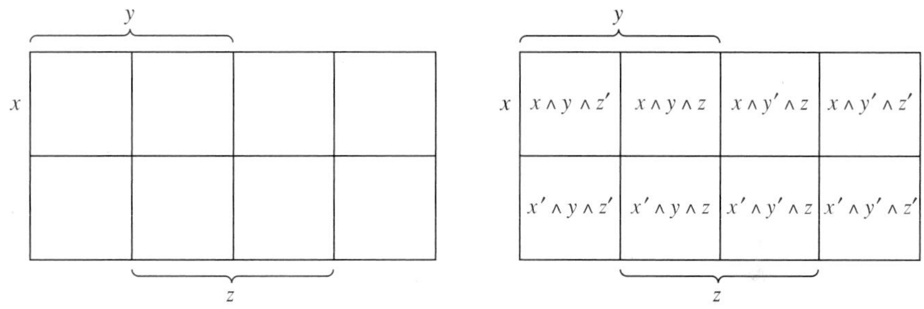

(a) **(b)**

FIGURE 10.23

We will make a somewhat technical definition. We define two cells to be **adjacent** in case the minterms to which they correspond differ in only a single variable. A pair of adjacent cells can be described by a Boolean expression with one variable fewer than a minterm. For example, the two cells in the second row and third and fourth columns correspond to

$$\begin{aligned}(x' \wedge y' \wedge z) \vee (x' \wedge y' \wedge z') &= (x' \wedge y') \wedge (z \vee z') \\ &= (x' \wedge y') \wedge 1 \\ &= x' \wedge y',\end{aligned}$$

where we have used rules (c), (f), and (g) of Theorem 10.1. Any two cells next to each other in a row or column are adjacent and have a 2-variable Boolean expression, as shown in Figure 10.24. There are also two pairs of adjacent cells that wrap around the sides of our grid; these are shown in Figure 10.25, along with their simplified Boolean expressions.

There are also groups of four cells with single-variable Boolean expressions. These are shown in Figure 10.26. The student should not try to memorize the Boolean expression for the groups of cells outlined in Figures 10.24, 10.25, and 10.26, but rather should study them to understand the principles behind them.

The method for constructing a simple Boolean expression corresponding to a truth table will be similar to the 2-variable case. We draw the Karnaugh map

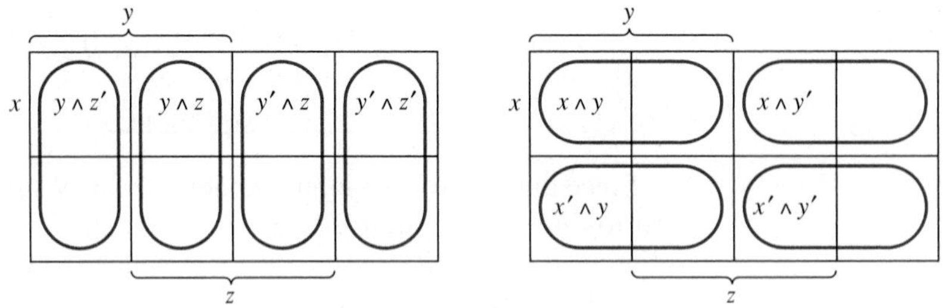

FIGURE 10.24

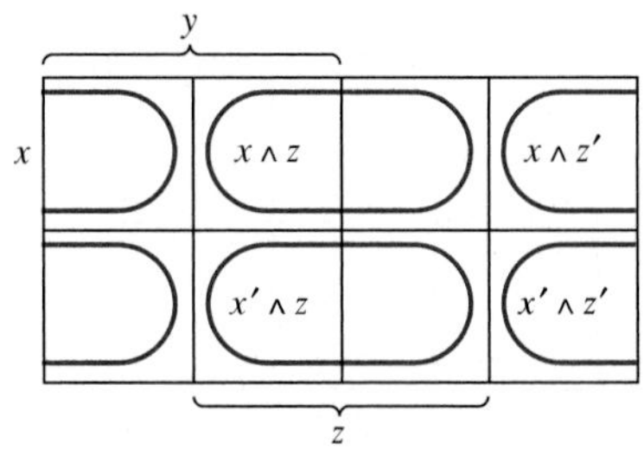

FIGURE 10.25

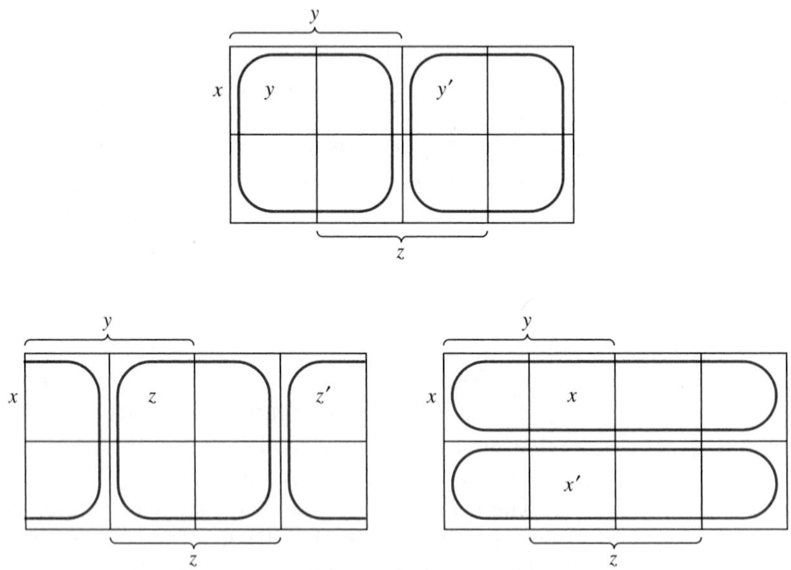

FIGURE 10.26

for the truth table, then enclose the cells containing 1s (and only those cells) in ovals corresponding to Boolean expressions. Since larger groups of cells have simpler Boolean expressions, we use them whenever possible, and we try not to use more groups than necessary. We then join these expressions by $\vee$ to form a Boolean expression with the required truth table.

Consider, for example, the two Karnaugh maps shown in Figure 10.27. The appropriate groups of cells are shown in Figure 10.28. The corresponding Boolean expressions are

$$x \vee (y' \wedge z) \qquad \text{and} \qquad (y \wedge z') \vee (x \wedge y) \vee (x' \wedge y' \wedge z),$$

respectively. Notice that the cell in the second row and third column of the second Karnaugh map is adjacent to no other cell with a 1, and so its 3-term minterm must be used.

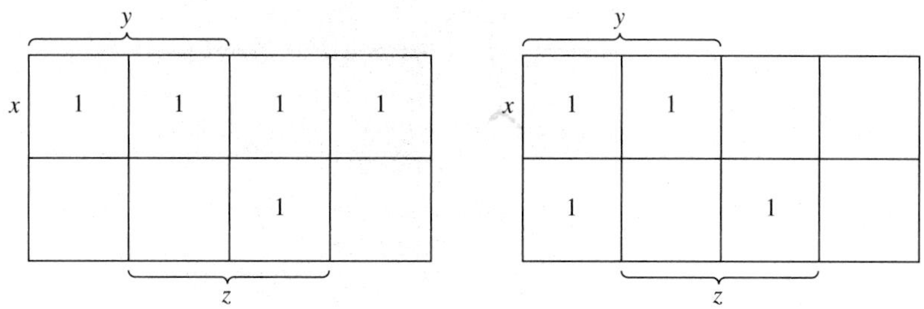

FIGURE 10.27

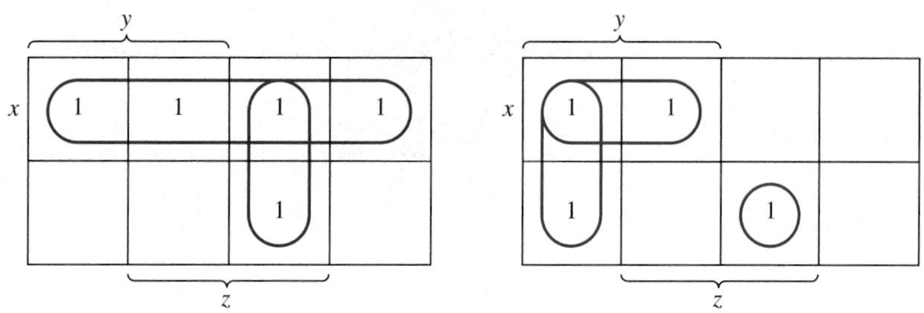

FIGURE 10.28

Example 10.8 Simplify the voting-machine circuit shown in Figure 10.16.

Since the machine is to give output 1 when at least two of x, y, and z are 1, the corresponding Karnaugh map is shown in Figure 10.29. Using the ovals indicated,

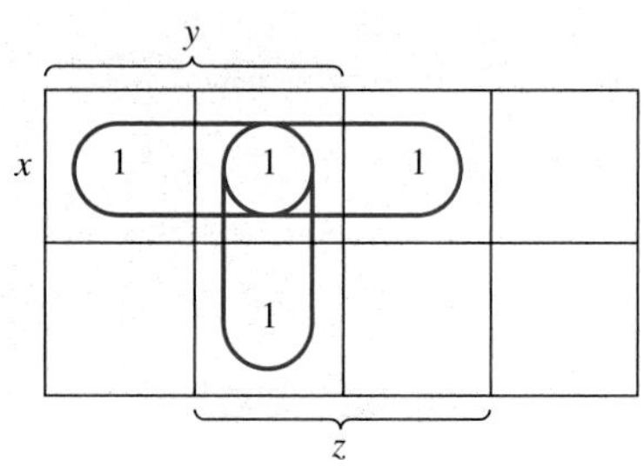

FIGURE 10.29

we write the Boolean expression $(x \wedge y) \vee (x \wedge z) \vee (y \wedge z)$. The corresponding circuit is shown in Figure 10.30. This circuit is considerably simpler than the one of Figure 10.16. In fact, if only gates with no more than two inputs are used, the previous circuit contains 14 while our new version has only 5. ■

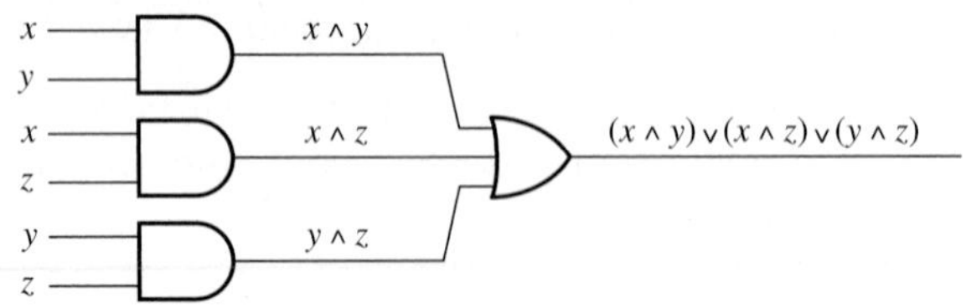

FIGURE 10.30

Example 10.9 Simplify the expression $x \vee (y \wedge (x \wedge z')')$ of Example 10.6.

We compute the following truth table:

x	y	z	z'	$x \wedge z'$	$(x \wedge z')'$	$y \wedge (x \wedge z')'$	$x \vee (y \wedge (x \wedge z')')$
0	0	0	1	0	1	0	0
0	0	1	0	0	1	0	0
0	1	0	1	0	1	1	1
0	1	1	0	0	1	1	1
1	0	0	1	1	0	0	1
1	0	1	0	0	1	0	1
1	1	0	1	1	0	0	1
1	1	1	0	0	1	1	1

This leads to the Karnaugh map of Figure 10.31. Using the indicated groups of cells produces the same Boolean expression $x \vee y$ that was derived in Example 10.6 by using the rules of Theorem 10.1. ■

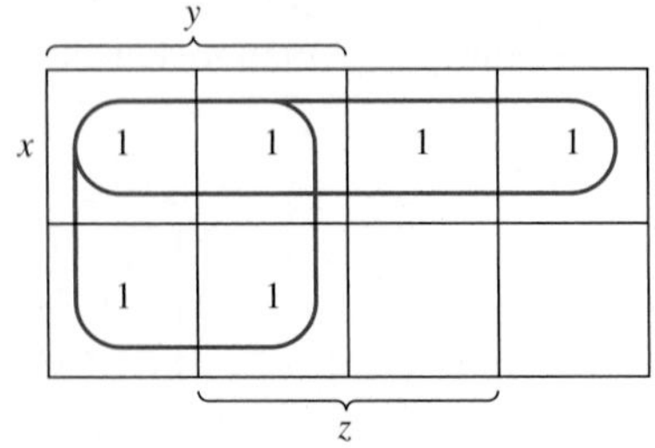

FIGURE 10.31

Finally, we consider Karnaugh maps for circuits with four inputs w, x, y, and z. We will use a 4-by-4-grid, labeled as in Figure 10.32(a). For example, the cell marked (1) corresponds to the minterm $w \wedge x' \wedge y \wedge z'$, and the cells marked (2) and (3) to the minterms $w \wedge x \wedge y' \wedge z$ and $w' \wedge x' \wedge y \wedge z$, respectively. Figure 10.32(b) shows various groups of two adjacent cells and their Boolean expressions. Of course, there are many more such groups. Examples of groups of four cells and their 2-variable Boolean expressions are shown in Figure 10.33. Notice that they can wrap around either horizontally or vertically. There are also 8-cell groups whose Boolean expressions have a single variable; some of these are shown in Figure 10.34.

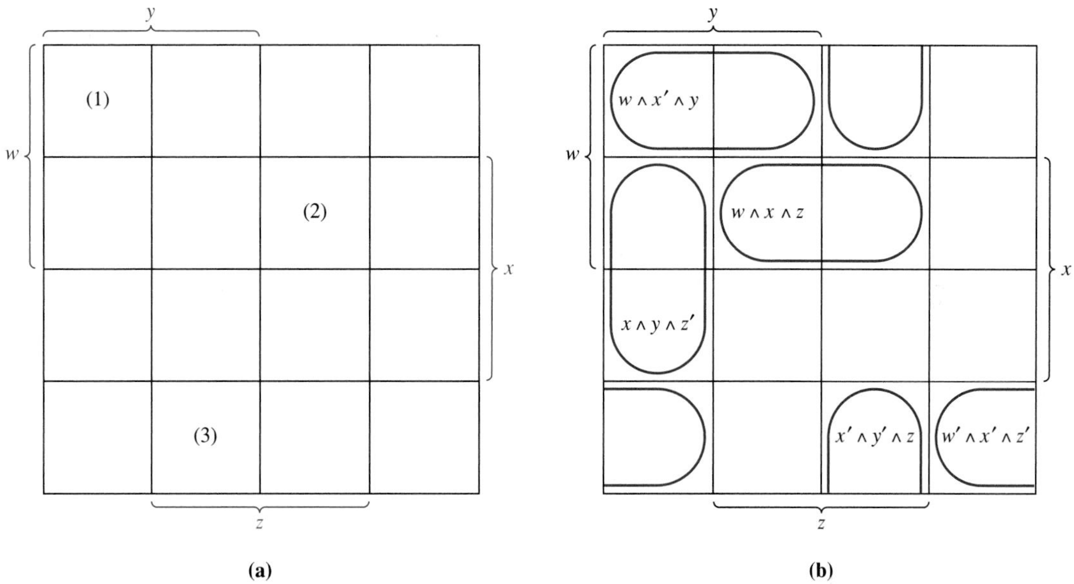

FIGURE 10.32

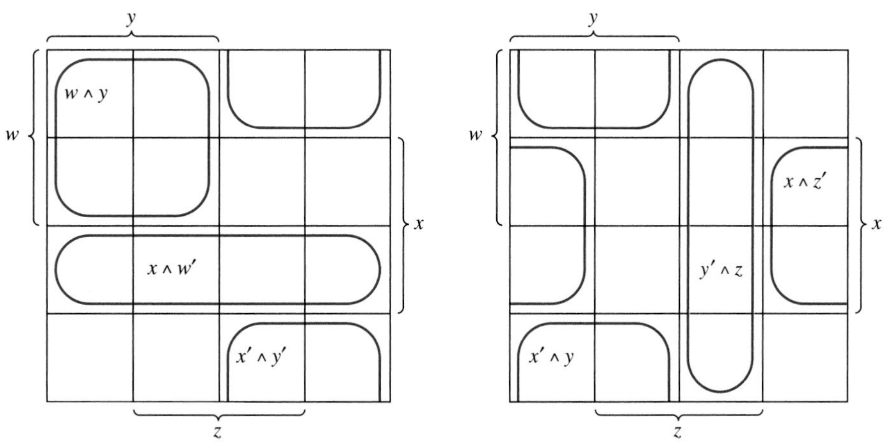

FIGURE 10.33

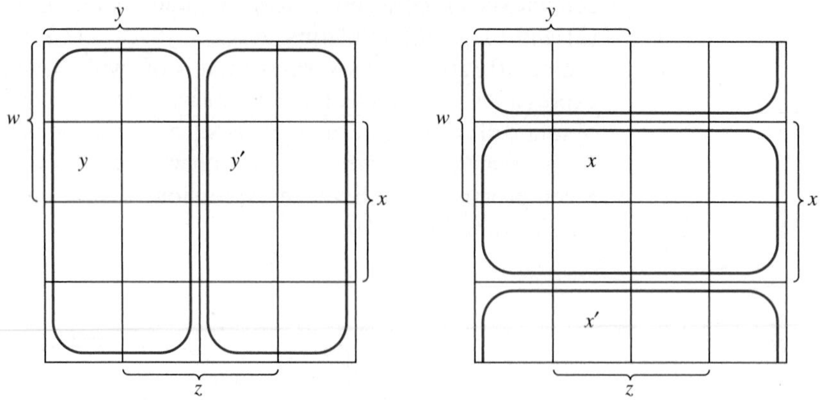

FIGURE 10.34

As before, given a truth table, we form its Karnaugh map and then enclose its 1s (and only its 1s) in rectangles of 1, 2, 4, or 8 cells that are as large as possible. The required Boolean expression is formed by joining the expressions for these rectangles with $\vee$.

Example 10.10 Find a circuit having the following truth table:

w	x	y	z	*Output*
0	0	0	0	1
0	0	0	1	0
0	0	1	0	1
0	0	1	1	1
0	1	0	0	1
0	1	0	1	1
0	1	1	0	1
0	1	1	1	0
1	0	0	0	1
1	0	0	1	0
1	0	1	0	1
1	0	1	1	0
1	1	0	0	1
1	1	0	1	1
1	1	1	0	1
1	1	1	1	1

The Karnaugh map for this table is shown in Figure 10.35. Using the rectangles of cells shown yields the expression

$$z' \vee (w \wedge x) \vee (x \wedge y') \vee (w' \wedge x' \wedge y).$$

Figure 10.36 shows the corresponding circuit. ■

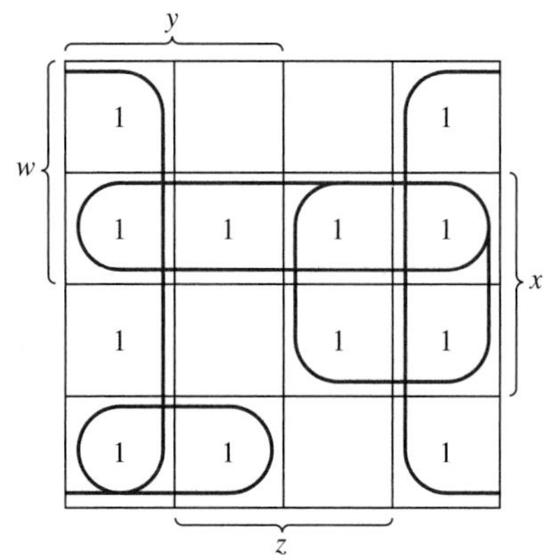

FIGURE 10.35

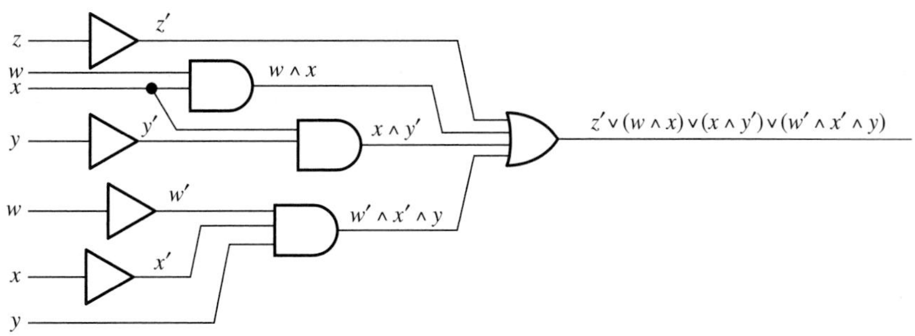

FIGURE 10.36

Example 10.11 Use Karnaugh maps to simplify the circuit of Figure 10.37(a).

We compute the Boolean expression

$$(w \wedge x \wedge y) \vee (w \wedge x \wedge z') \vee (w \wedge y' \wedge z) \vee (x' \wedge y' \wedge z)$$

for the circuit, as shown in Figure 10.37(b). The terms separated by $\vee$'s in this expression correspond to the four rectangles marked in Figure 10.38(a). The same cells can be enclosed by two rectangles, as shown in Figure 10.38(b). These yield the Boolean expression $(w \wedge x) \vee (x' \wedge y' \wedge z)$ and the circuit of Figure 10.39. ■

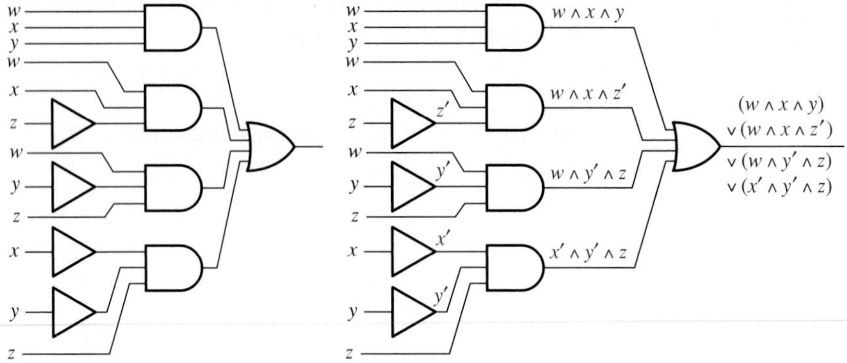

(a) **(b)**

FIGURE 10.37

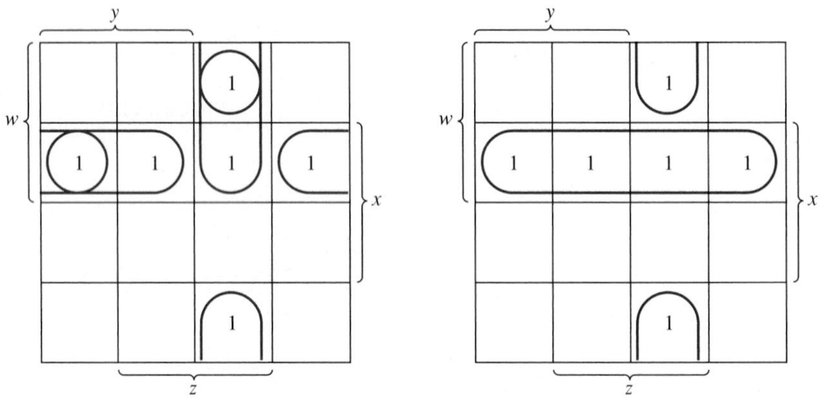

(a) **(b)**

FIGURE 10.38

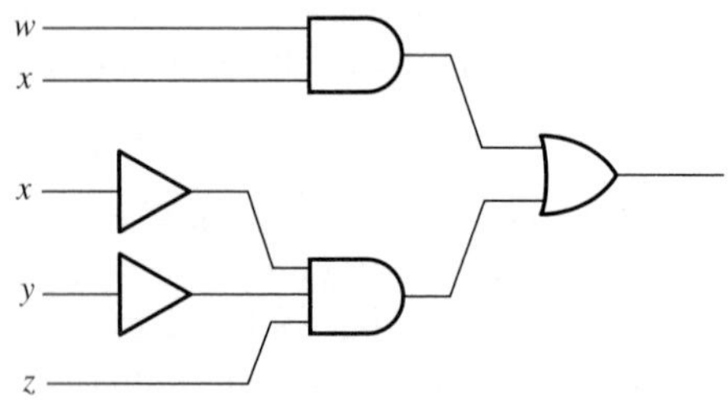

FIGURE 10.39

EXERCISES 10.3

In Exercises 1–6, find a Boolean expression of minterms that has the given truth table.

1.

x	y	*Output*
0	0	1
0	1	1
1	0	0
1	1	1

2.

x	y	z	*Output*
0	0	0	1
0	0	1	0
0	1	0	0
0	1	1	1
1	0	0	1
1	0	1	0
1	1	0	1
1	1	1	1

3.

x	y	z	*Output*
0	0	0	1
0	0	1	1
0	1	0	1
0	1	1	1
1	0	0	0
1	0	1	0
1	1	0	0
1	1	1	0

4.

x	y	z	*Output*
0	0	0	0
0	0	1	1
0	1	0	0
0	1	1	1
1	0	0	1
1	0	1	1
1	1	0	1
1	1	1	0

5.

w	x	y	z	*Output*
0	0	0	0	1
0	0	0	1	0
0	0	1	0	1
0	0	1	1	0
0	1	0	0	1
0	1	0	1	0
0	1	1	0	1
0	1	1	1	0
1	0	0	0	0
1	0	0	1	0
1	0	1	0	0
1	0	1	1	0
1	1	0	0	0
1	1	0	1	1
1	1	1	0	0
1	1	1	1	0

6.

w	x	y	z	*Output*
0	0	0	0	1
0	0	0	1	0
0	0	1	0	0
0	0	1	1	0
0	1	0	0	0
0	1	0	1	1
0	1	1	0	0
0	1	1	1	1
1	0	0	0	1
1	0	0	1	0
1	0	1	0	0
1	0	1	1	0
1	1	0	0	0
1	1	0	1	1
1	1	1	0	0
1	1	1	1	1

In Exercises 7–12, write the Boolean expression corresponding to the ovals in the Karnaugh map.

7.

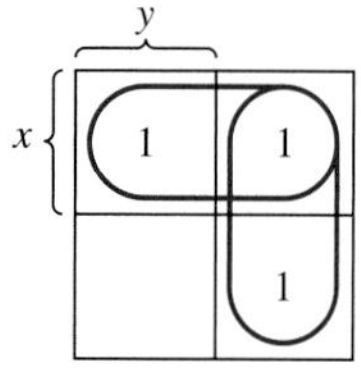

8.

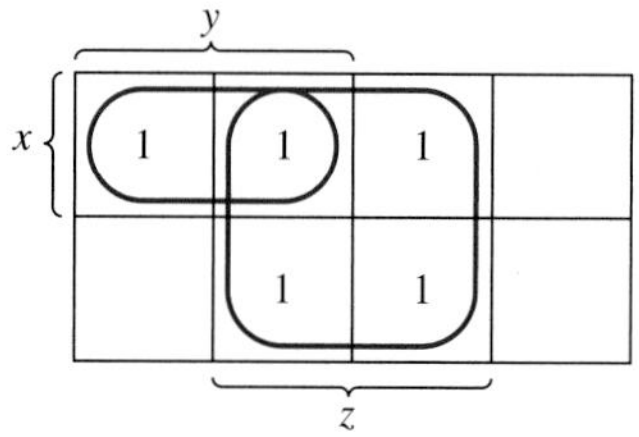

9.

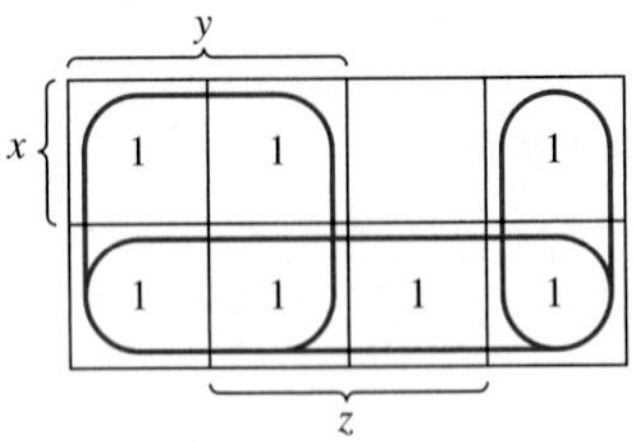

10.

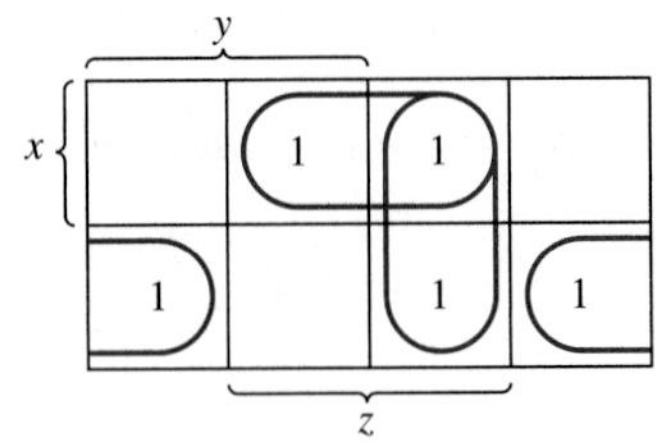

11.

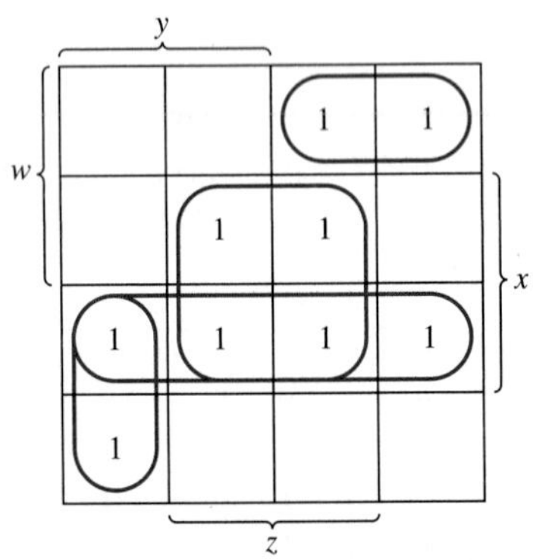

12.

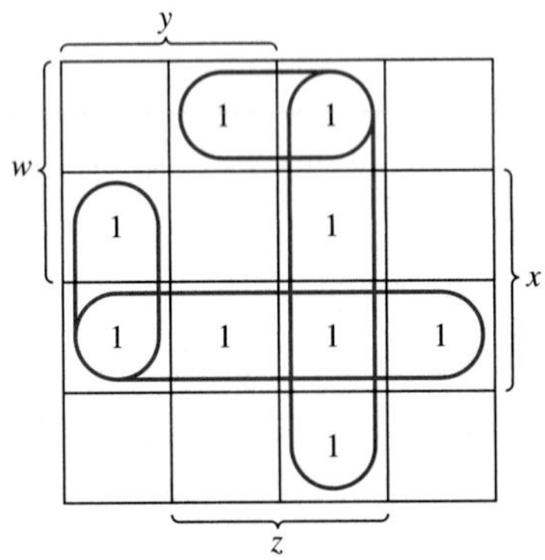

In Exercises 13–18, draw a Karnaugh map for the Boolean expression of the indicated exercise.

13. Exercise 1

14. Exercise 2

15. Exercise 3

16. Exercise 4

17. Exercise 5

18. Exercise 6

In Exercises 19–24, use the Karnaugh map method to simplify the Boolean expression in the indicated exercise. Then draw a circuit representing the simplified Boolean expression.

19. Exercise 1

20. Exercise 2

21. Exercise 3

22. Exercise 4

23. Exercise 5

24. Exercise 6

In Exercises 25–32, use the Karnaugh map method to simplify the expression.

25. $(x' \wedge y' \wedge z) \vee (x' \wedge y \wedge z) \vee (x \wedge y' \wedge z)$

26. $(x' \wedge y' \wedge z) \vee (x' \wedge y' \wedge z') \vee (x \wedge y \wedge z) \vee (x \wedge y' \wedge z')$

27. $(x' \wedge y' \wedge z) \vee (x' \wedge y \wedge z) \vee (x \wedge y' \wedge z')$

28. $[(x \vee y') \wedge (x' \wedge z')] \vee y$

29. $[x \wedge (y \vee z)] \vee (y' \wedge z')$

30. $(x \wedge y \wedge z) \vee (x \wedge y' \wedge z) \vee (x' \wedge y' \wedge z)$

31. $(w \wedge x \wedge y) \vee (w \wedge x \wedge z) \vee (w \wedge y' \wedge z') \vee (y' \wedge z')$

32. $(w' \wedge x' \wedge y') \vee (w' \wedge y' \wedge z) \vee (w \wedge y \wedge z) \vee (w \wedge x \wedge z') \vee (w \wedge y' \wedge z') \vee (w \wedge x' \wedge y \wedge z) \vee$ $(w' \wedge x \wedge y \wedge z')$

In Exercises 33 and 34, use Karnaugh maps to simplify the given circuit.

33.

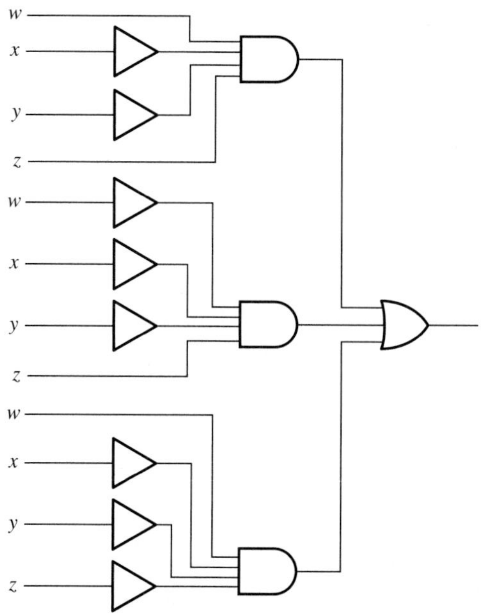

34.

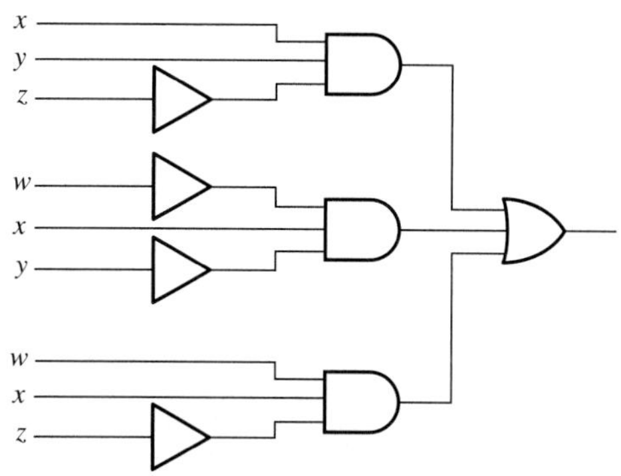

35. How many groups of two adjacent cells are there in a Karnaugh map grid for 4 Boolean variables?

36. How many 4-element square groups of adjacent cells are there in a Karnaugh map grid for 4 Boolean variables?

Although rule (b) of Theorem 10.1 suggests that we get equal expressions for any two ways we insert parentheses in $x_1 \vee x_2 \vee \cdots \vee x_n$, *we have not given a formal proof of this fact. (We will only treat* $\vee$; $\wedge$ *could be handled in the same way.) Define the expression* $x_1 \vee x_2 \vee \cdots \vee x_n$ *recursively as follows:*

$$x_1 \vee x_2 \vee \cdots \vee x_n = \begin{cases} x_1 & \text{if } n = 1 \\ (x_1 \vee x_2 \vee \cdots \vee x_{n-1}) \vee x_n & \text{for } n > 1. \end{cases}$$

37. Prove by induction on n that $(x_1 \vee x_2 \vee \cdots \vee x_m) \vee (y_1 \vee y_2 \vee \cdots \vee y_n) = x_1 \vee \cdots \vee x_m \vee y_1 \vee \cdots \vee y_n$ for any positive integers m and n.

38. Prove that any two expressions formed by inserting parentheses in the expression $x_1 \vee x_2 \vee \cdots \vee x_n$ are equal.

39. Prove by induction that $(x_1 \vee x_2 \vee \cdots \vee x_n)' = x_1' \wedge x_2' \wedge \cdots \wedge x_n'$ for all positive integers n, where the definition of the latter expression is similar to that for $\vee$.

40. Denote by q_n the number of ways of inserting $n-2$ sets of parentheses in $x_1 \vee x_2 \vee \cdots \vee x_n$ so that the order in which the $\vee$'s are applied is unambiguous. For example, $q_3 = 2$ counts the expressions $(x_1 \vee x_2) \vee x_3$ and $x_1 \vee (x_2 \vee x_3)$. Likewise, $q_4 = 5$. Show that

$$q_n = q_1 q_{n-1} + q_2 q_{n-2} + \cdots + q_{n-1} q_1 \text{ for } n > 1.$$

41. Let r_n be the number of ways of listing $x_1, x_2, \ldots, x_n$, joined by $\vee$'s in any order and with parentheses. For example, $r_1 = 1, r_2 = 2$ (the two expressions being $x_1 \vee x_2$ and $x_2 \vee x_1$), and $r_3 = 12$. Show that $r_{n+1} = (4n-2)r_n$ for all positive integers n.

42. Show that $r_n = \frac{(2n-2)!}{(n-1)!}$ and $q_n = \frac{(2n-2)!}{n!(n-1)!}$ for all positive integers n, where r_n and q_n are defined as in Exercises 40 and 41.

10.4 FINITE STATE MACHINES

In this section we will study devices, such as computers, that have not only inputs and outputs, but also a finite number of internal states. What the device does when presented with a given input will depend not only upon that input, but also upon the internal state that the device is in at the time. For example, if a person pushes the "PLAY" button on a CD player, what happens will depend on various things, such as whether or not the player is turned on, contains a CD, or is already playing.

The devices now considered will differ from those of the preceding sections in that output will depend not only on the immediate input, but also on the past history of inputs. Thus their action has the ability to change with *time*. Such devices are called *finite state machines*. Various formal definitions of a finite state machine may be given. We will study two types, one simple and the other somewhat more complicated. Our main concern will be to understand what such machines are and how they operate, rather than to construct finite state machines for specific tasks.

One simple example of a finite state machine is a newspaper vending machine. Such a vending machine has two states, locked and unlocked, which we will denote by L and U. We will consider a machine that only accepts quarters, the price of a paper. Two inputs are possible: to put a quarter into the machine (q) and to try to open and shut the door to get a paper (d). Putting in a quarter unlocks the machine, after which opening and shutting the door locks it again. Of course, putting a quarter into a machine that is already unlocked does not change the state of the machine, nor does trying to open the door of a locked machine.

There are various ways we can represent this machine. One way is to make a table showing how each input affects the state the machine is in.

		Present state	
		L	U
Input	q	U	U
	d	L	L

Here, the entries in the body of the table show the next state the machine enters, depending on the present state (column) and input (row). For example, the colored entry means that if the machine is in state L and the input is q, it changes to state U. Since this table gives a state for each ordered pair (i, s), where i is an input and s is a state, it describes a function with the Cartesian product $\{q, d\} \times \{U, L\}$ as its domain and the sets of states $\{U, L\}$ as its codomain. (The reader may want to review the concepts of Cartesian product and function in Sections 2.1 and 2.4.) Such a table is called the **state table** of the machine.

We can also represent our machine graphically, as in Figure 10.40. Here, the states L and U are shown as circles, and labeled arrows indicate the effect of each input when the machine is in each state. For example, the colored arrow indicates that a machine in state L with input q moves to state U. This diagram is called the **transition diagram** of the machine. (In the language of Section 4.5, the transition diagram is a directed multigraph.)

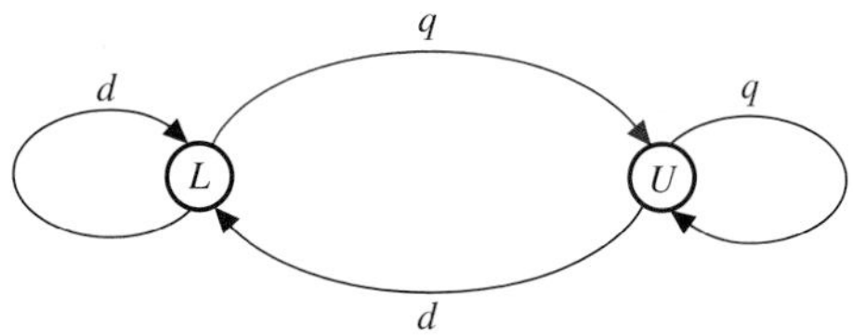

FIGURE 10.40

We will generally use the pictorial representation for finite state machines since our examples will be fairly simple. For a machine with many inputs and states, the picture may be so complicated that a state table is preferable.

A Parity Checking Machine

Before we give a formal definition of a finite state machine, we will give one more example. This example involves error-detecting codes, which are described in Sections 3.4–3.6, but you need not be familiar with these sections to understand it. Data sent between electronic devices are generally represented as a sequence of 0s and 1s. Some way of detecting errors in transmission is desirable. We will describe one simple means of doing this. Before a message is sent, the number of 1s in the message is counted. If this number is odd, a single 1 is added to the end of the message, and if it is even, a 0 is added. Thus all transmissions will contain an even number of 1s.

After a transmission is received, the 1s are counted again to determine whether there is an even or odd number of them. This is called a **parity check**. If there is an odd number of 1s, then some error must have occurred in transmission. In this case, a repeat of the message can be requested. Of course, if there are two or more errors in transmission, a parity check may not tell the receiver so. But if the transmission of each digit is reliable and the message is not too long, this may be far less likely than a single error. If the received transmission passes the parity check, its last digit is discarded to regain the original message.

Actually, it is not necessary to count the number of 1s in a message to tell whether this number is odd or even. Figure 10.41 represents a device that can be used to do this job. Here, the states are e (even) and o (odd), and the inputs are 0 and 1. The corresponding state table is as follows:

		State	
		e	o
Input	0	e	o
	1	o	e

We can use this device to determine whether the number of 1s in a string of 0s and 1s is even or odd by starting in state e and using each successive digit as a new input. For example, if the message 11010001 is used as input (reading from left to right), the machine starts in state e and moves to state o because the first input is 1. The second input is also 1, putting the machine back in state e, where it stays after the third input, 0. The way the machine moves from state to state is summarized in the following table:

Input:	Start	1	1	0	1	0	0	0	1
State:	e	o	e	e	o	o	o	o	e

If 11010001 is received, we would presume that no error occurred in transmission and that the original message was 1101000.

Two new symbolisms appear in Figure 10.41. One is the arrow pointing into state e. This indicates that we must start in state e for our device to work properly. The other is the double circle corresponding to state e. This indicates that this state is a desirable final state; otherwise some error has occurred in our example.

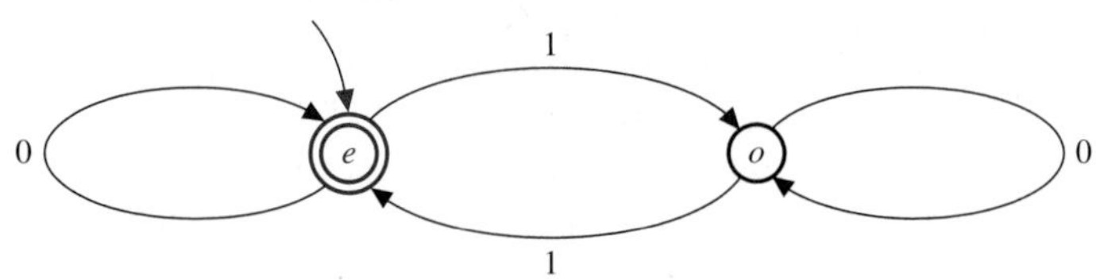

FIGURE 10.41

Now we formally define a **finite state machine** to consist of a finite set of states S, a finite set of inputs I, and a function f with $I \times S$ as its domain and S as its codomain such that if $i \in I$ and $s \in S$, then $f(i, s)$ is the state the machine moves to when it is in state s and is given input i. We may also, depending upon the application, specify an **initial state** s_0, as well as a subset S' of S. The elements of S', called **accepting states**, are the states we would like to end in.

Thus, our parity checking machine is a finite state machine with $S = \{e, o\}$, $I = \{0, 1\}$, $s_0 = e$, and $S' = \{e\}$. The function f is specified by

$$f(0, e) = e, \qquad f(0, o) = o,$$
$$f(1, e) = o, \qquad f(1, o) = e,$$

which corresponds to our previous state table.

A **string** is a finite sequence of inputs, such as 11010001 in our last example. Suppose, given the string $i_1 i_2 \ldots i_n$ and the initial state s_0, we successively compute $f(i_1, s_0) = s_1$, then $f(i_2, s_1) = s_2$, and so on, finally ending up with state s_n. This amounts to starting in the initial state, applying the inputs of the string from left to right, and ending up in state s_n. If s_n is in S', the set of accepting states, then we say that the string is **accepted**; otherwise it is **rejected**. In the parity check example, rejected transmissions contain some error, while accepted transmissions are presumed to be correct.

Example 10.12 Figure 10.42 shows a finite state machine with input set $I = \{0, 1\}$ that accepts a string precisely when it ends with the triple 100. Here, $S = \{A, B, C, D\}$, $s_0 = A$, $S' = \{D\}$, and the function f is as indicated by the labeled arrows in the diagram. For example, if the string 101010 is input, we move through the states $ABCBCBC$, and since C is not in S', the string is rejected. On the other hand, if 001100 is input, we move through the states $AAABBCD$, and the string is accepted because D is an accepting state.

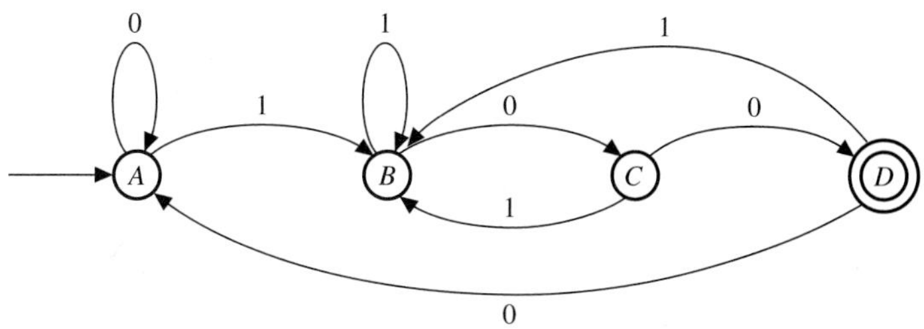

FIGURE 10.42

To see that the machine of Figure 10.42 does what we claim, the reader should first check that, no matter what state we are in, if the string 100 is input, we are taken to state D. This shows that all strings ending in 100 will be accepted by the machine. It remains to show that an accepted string must end in 100. Since we start in state A, an accepted string clearly must contain at least three digits. Since, when 1 is input, we move to state B no matter what the present state is, the accepted string must end in 0. Likewise, the reader should check that any string ending in 10 leaves the machine in state C. Thus our accepted string must end in two 0s. Finally, the reader should check that any string ending in 000 puts the machine in state A. Thus any accepted string must end in 100. ■

One important application of machines that accept certain strings and reject others is in compilers for computer languages. Before a program is run, each statement must be checked to see whether it conforms to the syntax of the language being used.

Finite State Machines with Output

Now we consider a slightly more complicated type of device. We start with an example more sophisticated than a newspaper vending machine, namely, a gum machine. Our gum machine accepts only quarters, which is the price of a pack of gum. Three varieties are available: Doublemint (denoted by D), Juicy Fruit (J), and Spearmint (S), which can be chosen by pressing buttons d, j, or s, respectively. The internal states of the machine are locked (denoted by L) and unlocked (U); and if the machine is unlocked, it will return any extra quarters put into it. The inputs are q (quarter), d, j, and s. A diagram showing some of the action of the machine is given in Figure 10.43(a). Figure 10.43(b) shows a more compact way of indicating multiple arrows going between the same two states; here, for example, the three arrows from U and L in Figure 10.43(a) have been replaced by a single arrow, and the corresponding inputs separated by commas.

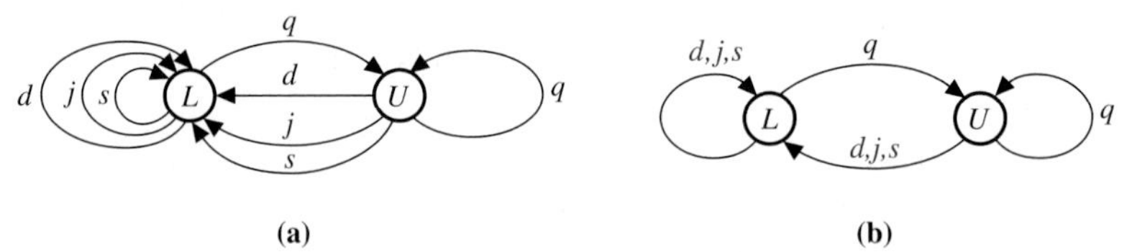

FIGURE 10.43

This diagram does not tell the whole story, however. Nowhere does it show that, if we press the d button on a machine in state U, we get a pack of Doublemint. Neither does it show that excess quarters are returned. We need to introduce the additional concept of **outputs** of the machine. In this example the possible outputs are D, J, S, Q (an excess quarter returned), and $\emptyset$, which we will use to stand for no output, as, for example, when a button is pressed while the machine is in state L.

Notice that the output may depend upon both the input and the state of the machine. The inputs d and j produce the distinct outputs D and J when the machine is in state U. Likewise, the input d produces the outputs $\emptyset$ or D, depending on whether the machine is in state L or U. Another function is involved here, having the Cartesian product of the set of inputs and the set of states as its domain and the set of outputs as its codomain. Since each arrow in our diagram stands for the result of an input being applied to a particular state, we can also label these arrows to show the corresponding outputs. This is done in Figure 10.44.

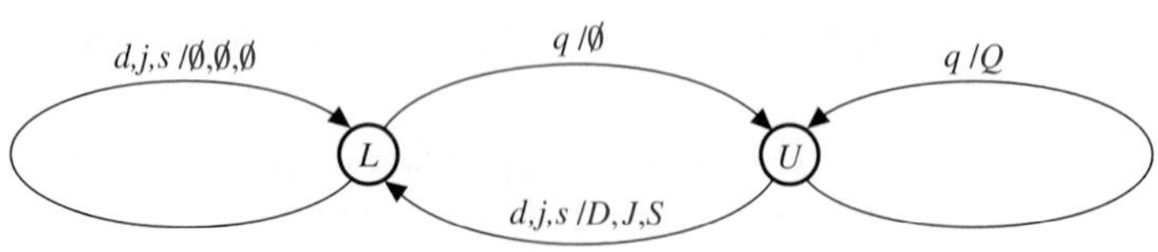

FIGURE 10.44

We will use slashes to separate the input and output labels on each arrow. Thus, in Figure 10.44, the $q/\emptyset$ on the arrow from L to U indicates that there is no output when we put a quarter in a locked machine; and the $d, j, s/D, J, S$ on the arrow from U to L indicates the outputs D, J, and S, respectively, when we push buttons d, j, and s on an unlocked machine.

We define a **finite state machine with output** to consist of finite sets S of states, I of inputs, and O of outputs, along with a function $f: I \times S \to S$ such that $f(i, s)$ is the state the machine goes to from state s when the input is i, and another function $g: I \times S \to O$ such that $g(i, s)$ is the output corresponding to input i when the machine is in state s. Depending on the application, we may again designate a particular state s_0 as the **initial state**.

In the gum machine example, we have $S = \{L, U\}$, $I = \{q, d, j, s\}$, and $O = \{D, J, S, Q, \emptyset\}$. The functions f and g are indicated in Figure 10.44, but they can also be described, as before, with tables:

		State	
		L	*U*
Input	*q*	*U*	*U*
	d	*L*	*L*
	j	*L*	*L*
	s	*L*	*L*

		State	
		L	*U*
Input	*q*	Ø	*Q*
	d	Ø	*D*
	j	Ø	*J*
	s	Ø	*S*

The first table, which gives the values of f, is still called the state table of the machine, while the second, which gives the values of g, is called the **output table**.

If a string of inputs is fed into a finite state machine with output, a corresponding sequence of outputs is produced, called the **output string**. This is illustrated in the next example.

Example 10.13 Figure 10.45 shows the transition diagram of a **unit delay** machine. This is a finite state machine with output in which $I = \{0, 1\}$, $S = \{A, B, C\}$, $O = \{0, 1\}$, and the initial state is A. Note that the first output is always 0, while any input of 0 puts the machine in state B, from which the next output will be 0. Likewise,

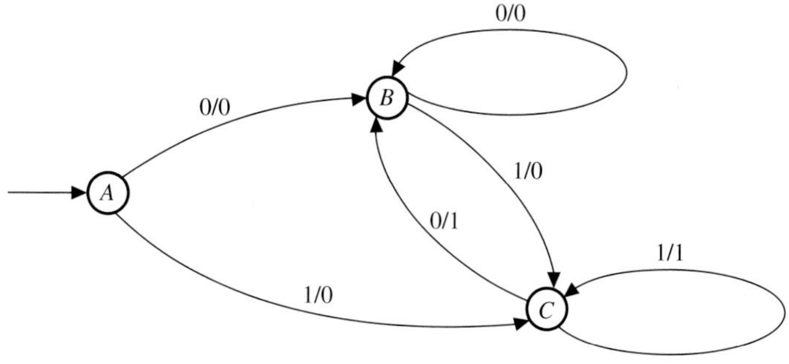

FIGURE 10.45

any input of 1 puts it in state C, from which the next output will be 1. Thus each output after the first is always the same as the input one step previously. An input string $i_1 i_2 \ldots i_n$ produces the output string $0 i_1 i_2 \ldots i_{n-1}$. For example, the input string 1100111 produces the output string 0110011. If it is desired to copy an entire input string, then a 0 must be appended to it before the string is input. ■

Example 10.14 If, in what follows, the initial state is A, draw the transition diagram for the finite state machine with output that has the following state and output tables, and describe what the machine does to an input string of x's and y's:

		State						State					
		A	B	C	D	E	F	A	B	C	D	E	F
Input	x	A	C	C	E	E	F	0	1	1	2	2	3
	y	B	B	D	D	F	F	1	1	2	2	3	3

The transition diagram is shown in Figure 10.46. Notice that, once an input x or y puts the machine into one of the states A, B, C, D, or E, the machine stays in that state until the input changes. The output is 0 or 1, according to whether the first input is x or y, and increases by one whenever the input changes from x to y. Thus the output at any time counts the number of groups of consecutive y's in the input string, up to three such groups. For example, the input string $xxyxxxyyyxx$ produces the output string 00111122222; and the last 2 counts the two groups of y's (y and yyy) in the input string. ■

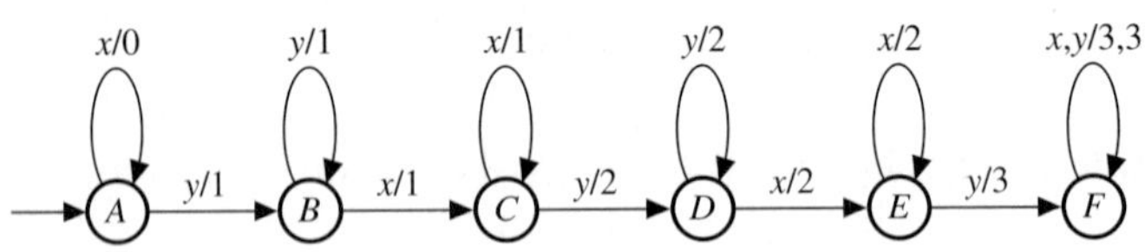

FIGURE 10.46

EXERCISES 10.4

In Exercises 1–6, draw the transition diagram for the finite state machine with the given state table.

1.

	A	B
0	A	A
1	A	B

2.

	A	B	C
0	B	C	A
1	A	C	B

3.

	x	y	z
0	y	z	z
1	x	x	y

Initial state x
Accepting state z

4.

	A	B
x	B	A
y	A	A
z	B	B

Initial state A
Accepting state A

5.

	A	B	C	D
a	B	A	D	C
b	C	C	A	A

Initial state B
Accepting states C, D

6.

	u	v	w
0	u	w	v
1	u	w	w
2	w	v	u

Accepting states u, v

In Exercises 7–10, give the state table for the finite state machine with the given transition diagram. List the initial and accepting states, if any.

7.

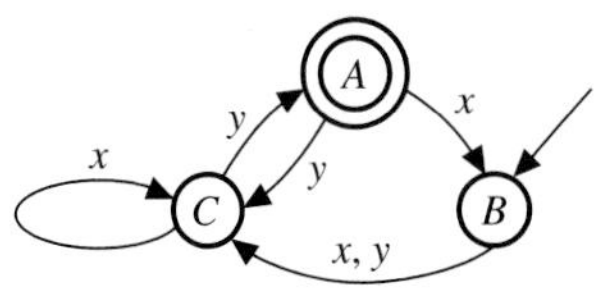

8.

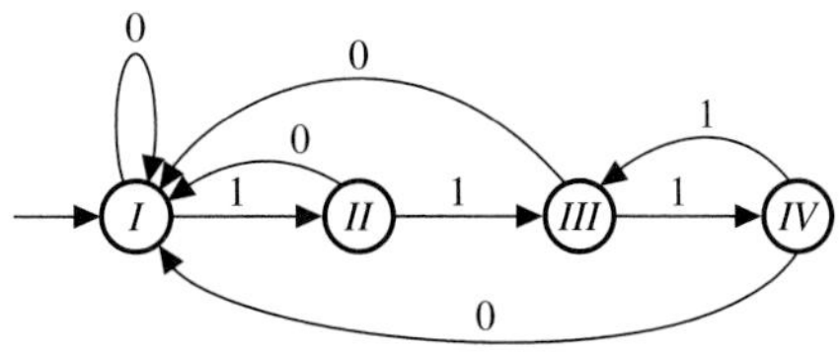

9.

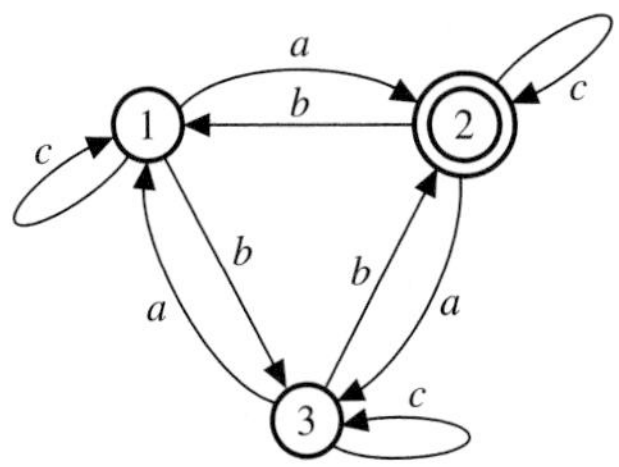

10.

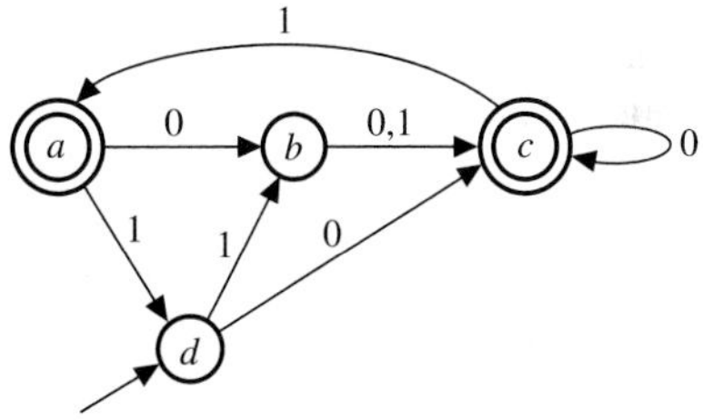

For the finite state machine and input string in Exercises 11–14, determine the state that the machine ends in if it starts at the initial state.

11. Input string 1011001, machine of Exercise 3

12. Input string $xyyzzx$, machine of Exercise 4

13. Input string $yxxxy$, machine of Exercise 7

14. Input string 0100011, machine of Exercise 8

In Exercises 15–18, tell whether the given input string would be accepted by the indicated finite state machine.

15. Input string $xyzxyzx$, machine of Exercise 4

16. Input string $aabbaba$, machine of Exercise 5

17. Input string $xyxxyy$, machine of Exercise 7

18. Input string 0011010, machine of Exercise 10

In Exercises 19–22, draw the transition diagram for the finite state machine with output whose state and output tables are given.

19.

	A	B	A	B
0	B	A	x	y
1	A	B	z	x

20.

	1	2	3	1	2	3
red	2	3	1	A	B	A
blue	1	1	3	A	A	B

21.

	A	B	A	B
0	A	A	x	y
1	B	B	w	x
2	A	B	y	w

Initial state A

22.

	00	01	10	11	00	01	10	11
A	11	10	01	00	1	−1	0	1
B	01	10	11	11	−1	0	1	−1

Initial state 10

In Exercises 23–26, give the state and output tables of the pictured finite state machine with output. Name the initial state, if any.

23.
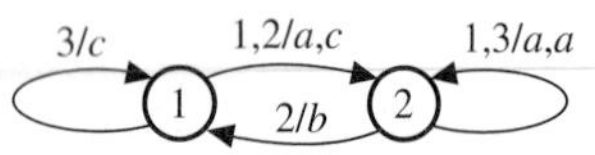

24.
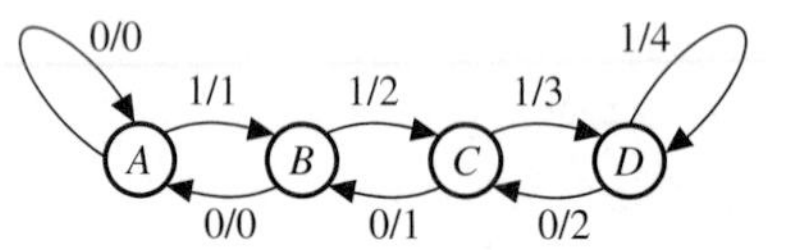

25.
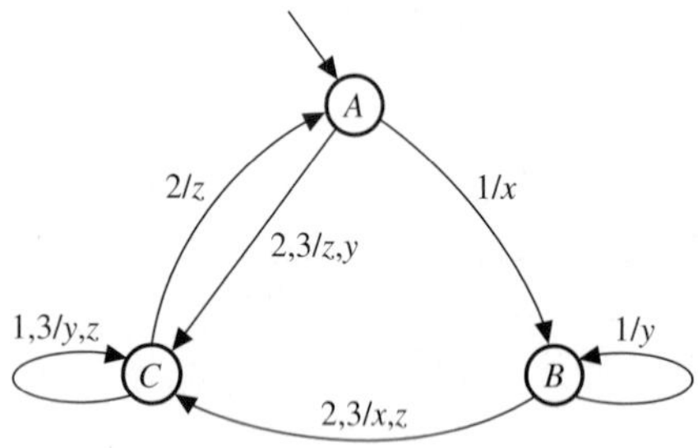

26.
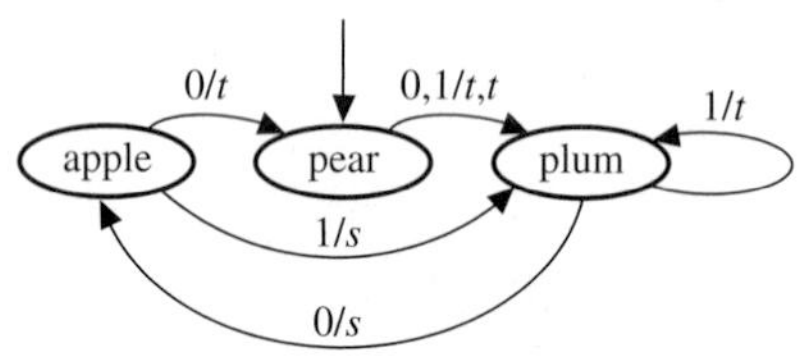

In Exercises 27–30, give the output string for the given input string and finite state machine with output.

27. Input string 2101211, machine of Exercise 21

28. Input string $BAABBB$, machine of Exercise 22

29. Input string 322113, machine of Exercise 25

30. Input string 10100110, machine of Exercise 26

In Exercises 31–34, describe which input strings of 0s and 1s are accepted by the pictured finite state machine.

31.
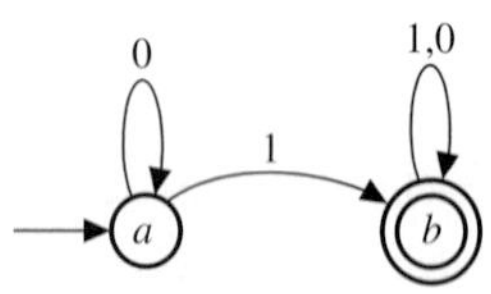

32.
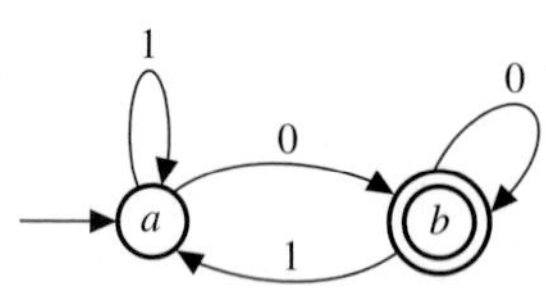

33.
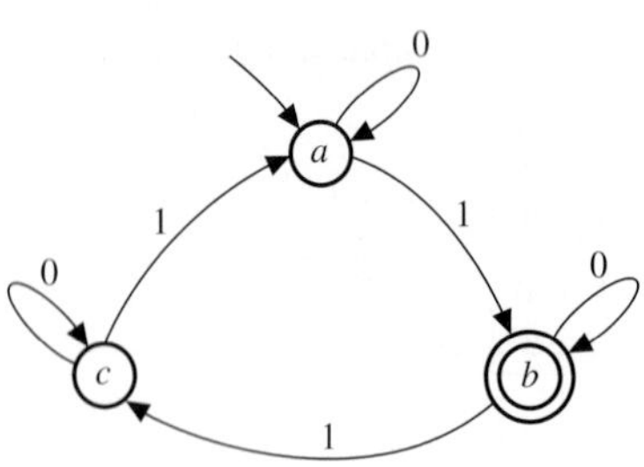

34.
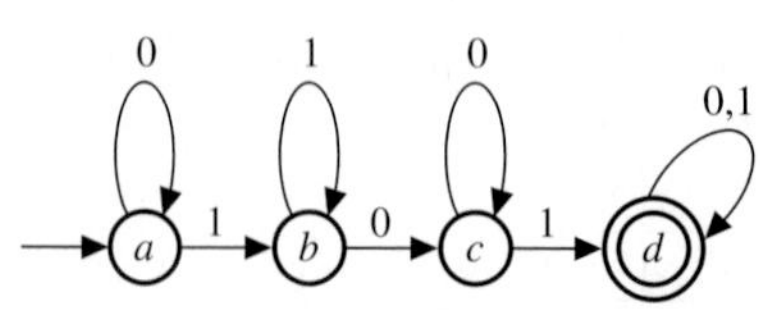

In Exercises 35–38, assume that the set of inputs is {0, 1}.

35. Design a finite state machine that accepts a string if and only if it ends with two 1s.

36. Design a finite state machine that accepts a string if and only if it does not contain two consecutive 0s.

37. Design a finite state machine with output such that, given an input string, its last output is the remainder when the number of 1s in the input string is divided by 3.

38. Design a finite state machine with output such that its output string contains as many 1s as there are pairs of consecutive 0s or 1s in the input string.

39. Let F and G be finite state machines. We say that F and G are **equivalent** if they have the same set of inputs and if, whenever a string is accepted by either of the machines, it is also accepted by the other. Let I and S be sets. Show that equivalence of finite state machines is an equivalence relation on the set of finite state machines having input sets and state sets that are subsets of I and S, respectively.

HISTORICAL NOTES

Claude Shannon

Willard Quine

Edward McCluskey

Gottfried Wilhelm Leibniz (1646–1716) was probably the first person to represent statements of logic algebraically. This freeing of logical symbols from representing specific interpretations allowed mathematicians and others to think about their forms abstractly. Augustus De Morgan (1806–1871) and others contributed to this formal model for deductive logic. George Boole's two texts, *The Mathematical Analysis of Logic* (1847) and *An Investigation of the Laws of Thought* (1854), detailed the results of his investigations. The structures developed by Boole were expanded by Charles Sanders Peirce (1839–1914) and Ernst Schröder (1841–1902) in the final third of the nineteenth century.

In 1869, the Englishman William Stanley Jevons (1835–1882) constructed a rudimentary machine capable of performing simple Boolean operations. His design was substantially improved by Peirce's student Allan Marquand (1853–1924) during the 1880s. Marquand's machine involved the use of circuits and electricity and required manually opening and closing circuits through a keyboard [77].

Despite these advances, Boolean algebra was still used mainly as a model for logical reasoning and formal algebraic structure. It was in the late 1930s that Claude Shannon (1916–2001) recognized the application of Boolean algebra to the design of switching circuitry and other applications. Almost immediately, machines employing two-state switches were developed, and other machines employing them and Boolean algebra constructs became central components in the emerging field of digital computing.

With this realization of Boolean statements and operations in mechanical form, there arose a need to minimize the number of switches or circuits for a given set of relations. Maurice Karnaugh (1924–) provided a method based on creating a map in 1953. Another method for minimization was the tabular approach developed by Willard Quine (1908–2000) during the period 1952–1955. This procedure was altered and improved by Edward McCluskey (1929–) in 1956 [76].

Finite state machines first appeared in the literature in the early 1950s with the works of G.H. Mealy, D.A. Huffman, and E.F. Moore.

SUPPLEMENTARY EXERCISES

1. Write a Boolean expression corresponding to the following circuit, and construct the corresponding truth table:

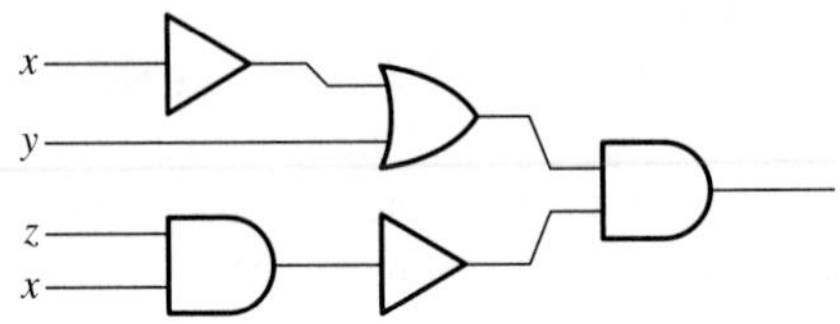

2. Draw a circuit representing the Boolean expression $[y \wedge (x' \vee z)] \vee (y \wedge z)'$, and construct the corresponding truth table.

3. Determine whether the two Boolean expressions in each of the following pairs are equivalent:

(a) $x \wedge (y \wedge z')'$ and $(x \wedge y') \vee (x \wedge z)$

(b) $x \wedge (y' \vee z)'$ and $(x \wedge y) \vee (x \wedge z')$

4. The lights on a private tennis court are to be controlled by either of two switches, one (labeled x) at the court and one (labeled y) at the house. If a third switch z at the house is thrown, however, then the switch at the court should no longer have any effect. Give the truth table modeling this situation. The output 1 means that the lights are on.

5. Establish the equivalences that follow by using Theorem 10.1. List by letter the rules you use. Start with the expression on the left side.

(a) $[(x \vee y) \wedge (x' \vee y)] \vee y' = 1$

(b) $x' \wedge (y \wedge z')' = (x \vee y)' \vee (x' \wedge z)$

6. Find a Boolean expression in minterms that has the following truth table:

x	y	z	*Output*
0	0	0	1
0	0	1	1
0	1	0	0
0	1	1	0
1	0	0	0
1	0	1	0
1	1	0	1
1	1	1	1

Now draw the corresponding circuit. How many gates with 1 or 2 inputs does the circuit represent?

7. Find a Boolean expression in minterms, and draw the corresponding circuit for the truth table of Exercise 4. How many gates with 1 or 2 inputs does this circuit represent?

8. Draw a Karnaugh map corresponding to each of the following truth tables:

(a)

x	y	*Output*
0	0	0
0	1	1
1	0	1
1	1	1

(b)

x	y	z	*Output*
0	0	0	1
0	0	1	0
0	1	0	0
0	1	1	1
1	0	0	1
1	0	1	0
1	1	0	0
1	1	1	1

(c)

w	x	y	z	*Output*
0	0	0	0	1
0	0	0	1	1
0	0	1	0	1
0	0	1	1	0
0	1	0	0	1
0	1	0	1	0
0	1	1	0	1
0	1	1	1	0
1	0	0	0	0
1	0	0	1	1
1	0	1	0	0
1	0	1	1	0
1	1	0	0	0
1	1	0	1	0
1	1	1	0	1
1	1	1	1	0

9. Write a Boolean expression corresponding to each of the following Karnaugh maps:

(a)

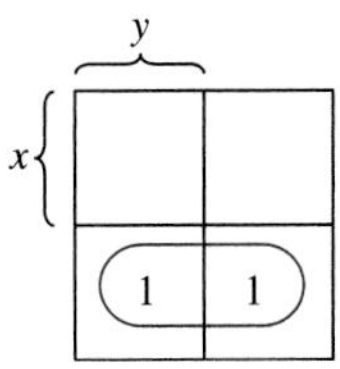

(b)

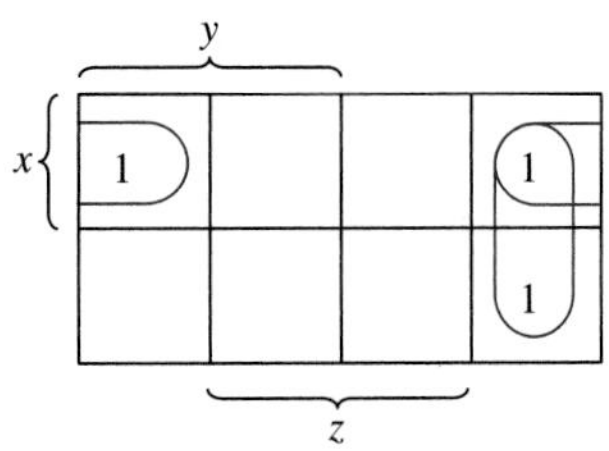

(c)

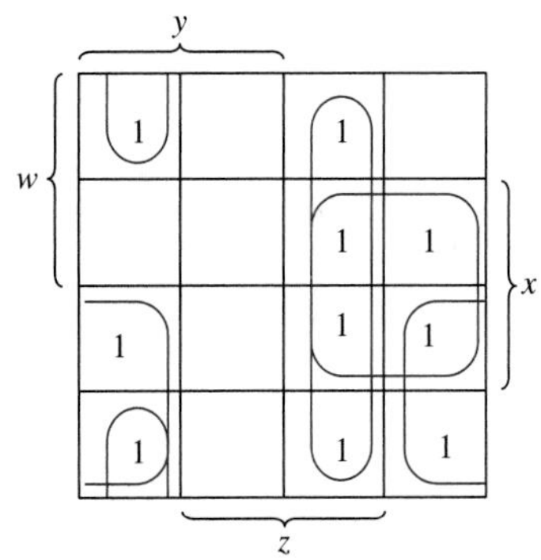

10. Use the Karnaugh map method to find a simple Boolean expression for each truth table in Exercise 8. Draw the corresponding circuit.

11. Use the Karnaugh map method to simplify the Boolean expression of Exercise 7, and draw the corresponding circuit. How many 1- or 2-input gates does the new circuit represent?

12. Use the Karnaugh map method to simplify each of the following expressions:

(a) $(x \wedge y' \wedge z) \vee (x' \wedge y \wedge z') \vee (x' \wedge y' \wedge z')$

(b) $(w \wedge x' \wedge y' \wedge z) \vee (w' \wedge x \wedge y \wedge z') \vee (w' \wedge x \wedge y') \vee (w' \wedge x' \wedge y' \wedge z') \vee (w' \wedge y' \wedge z)$

13. Use the Karnaugh map method to simplify the following circuit:

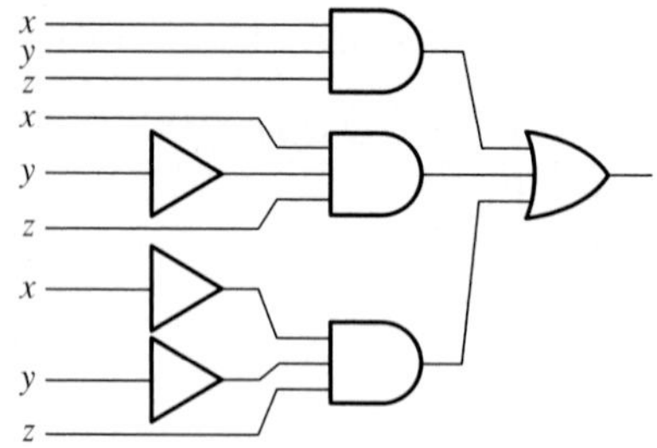

14. Draw the transition diagram for the finite state machine with the following state table:

	A	*B*	*C*
red	*B*	*B*	*B*
green	*A*	*B*	*C*
yellow	*B*	*C*	*A*

Initial state *A*

Accepting state *B*

15. What is the final state if the machine of the previous exercise has the following input string: green, red, green, red, yellow?

16. Give the state table for the finite state machine with the following transition diagram:

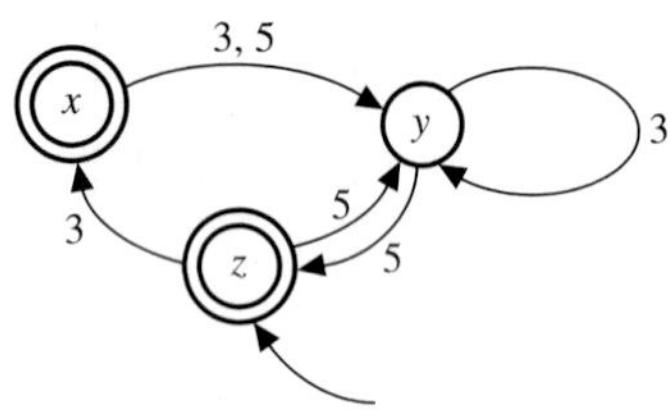

List the initial and accepting states.

17. What is the final state if the machine of the previous exercise has the input string 5, 5, 3, 3, 5, 5, 3?

18. Draw the transition diagram for the finite state machine with output having the following state and output tables:

	hot	*cold*	*hot*	*cold*
a	hot	cold	1	−1
b	cold	hot	0	1

Initial state cold

19. What is the output string if the machine of the previous exercise has input string *abaabba*?

20. Give the state and output tables for the finite state machine with output pictured as follows:

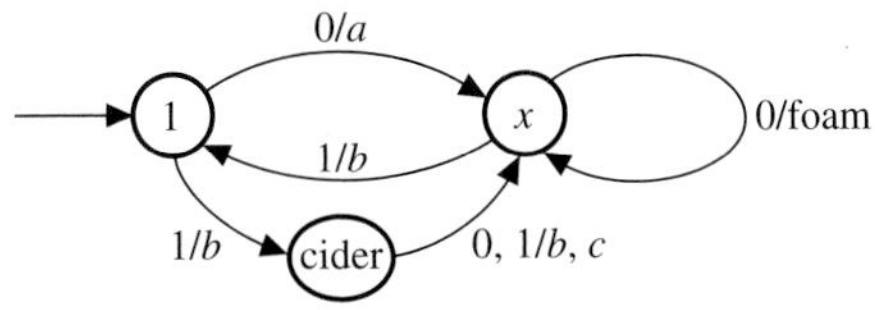

21. What is the output string if the machine of the previous exercise has input string 1001001?

22. Devise a finite state machine with inputs $I = \{0, 1\}$ that accepts a string $a_1 a_2 \ldots a_n$ exactly when $n \geq 2$ and $a_{n-1} \neq a_n$.

COMPUTER PROJECTS

Write a computer program having the specified input and output.

1. Given a triple (x, y, z), where each of x, y, and z is 0 or 1, output the corresponding value of $(x \vee z) \wedge (y \vee z')$.
2. Input a quintuple (A, B, x, y, z), where A and B are 2 or 3 and x, y, and z are 0 or 1. Here, 2 stands for $\wedge$ and 3 stands for $\vee$. Output the corresponding value of $(x\ A\ y)\ B\ z$.
3. Let a given string of eight 0s and 1s be interpreted as the rightmost column of a truth table with Boolean variables x, y, and z. Output the corresponding Boolean expression in minterms. For example, the input 11000000 would produce the output $(x' \wedge y' \wedge z') \vee (x' \wedge y' \wedge z)$.

Exercises 4–7 refer to exercises in Section 10.4.

4. Given a finite string of 0s and 1s, find the final state if the string is the input of the machine in Exercise 3.
5. Given a finite string of 0s and 1s, find the final state if the string is the input of the machine in Exercise 8.
6. Given an input string of 0s, 1s, and 2s, find the output string, using the machine of Exercise 21.
7. Given an input string of 1s, 2s, and 3s, find the output string, using the machine of Exercise 25.
8. Given a finite state machine with inputs $I = \{1, 2, 3\}$, states $S = \{1, 2, 3\}$, initial state 1, accepting states 1 and 3, and state table the 3×3 matrix A, determine whether a given input string is accepted or not.

SUGGESTED READINGS

1. Dornhoff, Larry L. and Franz E. Hohn. *Applied Modern Algebra.* New York: Macmillan, 1978.
2. Fisher, James L. *Application-Oriented Algebra.* New York: Crowell, 1977.
3. Friedman, Arthur D. and Premachandran R. Menon. *Theory & Design of Switching Circuits.* Rockville, MD: Computer Science Press, 1975.
4. Liu, C.L. *Elements of Discrete Mathematics*, 2d ed. New York: McGraw-Hill, 1985.
5. Stanat, Donald F. and David F. McAllister. *Discrete Mathematics in Computer Science.* Englewood Cliffs, NJ: Prentice-Hall, 1977.
6. Stone, Harold S. *Discrete Mathematical Structures and Their Applications.* Chicago: Science Research Associates, 1973.
7. Tremblay, J.P. and R. Manohar. *Discrete Mathematical Structures with Applications to Computer Science.* New York: McGraw-Hill, 1975.

A Appendix: An Introduction to Logic and Proof

It is essential that persons in such fields as mathematics, physics, and computer science understand the basic principles of logic so that they are able to recognize valid and invalid arguments. In Chapter 10 we explored an application of logic to the design of circuits such as those found in computers. In this appendix we present an informal introduction to logic and proof that provides a sufficient working knowledge of these subjects for students of computer science, mathematics, and the sciences.

A.1 STATEMENTS AND CONNECTIVES

One aspect of logic involves determining the truth or falsity of meaningful assertions. By a **statement**, we will mean any sentence that is either true or false, but not both. For example, each of the following is a statement:

(1) George Washington was the first president of the United States.
(2) Baltimore is the capital of Maryland.
(3) $6 + 3 = 9$.
(4) Texas has the largest area of any state in the United States.
(5) All dogs are animals.
(6) Some species of birds migrate.
(7) Every even integer greater than 2 is the sum of two primes.

In the sixth statement above, the word "some" appears. In logic, we interpret the word "some" to mean "at least one." Thus the sixth statement means that at least one species of birds migrates or that there is a species of birds that migrates.

The first, third, fifth, and sixth of the statements above are true, and the second and fourth are false. At this time, however, it is not known whether the seventh statement is true or false. (This statement is a famous unsolved mathematical problem called *Goldbach's conjecture*.) Nevertheless, it is a statement because it must be true or false, but not both.

On the other hand, the following sentences are *not* statements:

(1) Why should we study logic?
(2) Eat at the cafeteria.
(3) Enjoy your birthday!

The reason that these fail to be statements is that none of them can be judged to be true or false.

It is possible that a sentence is a statement and yet we are unable to ascertain its truth or falsity because of an ambiguity or lack of quantification. The following sentences are of this type:

(1) Yesterday it was cold.
(2) He thinks New York is a wonderful city.
(3) There is a number x such that $x^2 = 5$.
(4) Lucille is a brunette.

In order to decide whether the first sentence is true or false, we need to specify what we mean by the word "cold." Similarly, in the second sentence we need to know whose opinion is being considered in order to decide whether this sentence can be designated as true or false. Whether the third sentence is true or false depends upon what type of numbers are allowed as possible replacements, and the assignment of true or false to the last sentence depends upon which Lucille one might have in mind. Hereafter, we will not consider such ambiguous sentences as statements, due to their lack of specification, their quantification, or their lack of antecedents for critical words or variables needed to determine the truth or falsity of the sentence.

Example A.1 The following sentences are statements:

(a) On December 4, 1985, the temperature dropped below freezing in Miami, Florida.

(b) In the opinion of some citizens of Kuwait, George H. Bush is a hero.

(c) There is an integer x such that $x^2 = 5$.

(d) Singer Gloria Estefan is from Cuba. ■

We will be interested in studying the truth or falsity of statements formed with the use of the following words called **connectives**:

Connective	*Symbol*	*Name*
not	$\sim$	negation
and	$\wedge$	conjunction
or	$\vee$	disjunction
if ... then ...	$\rightarrow$	conditional
if and only if	$\leftrightarrow$	biconditional

The use of the connective "not" in logic is the same as in standard English; that is, its use denies the statement to which it applies. It is easy to form the negation of most simple statements, as we see in the following example:

Example A.2 Consider the following statements:

(a) Today is Friday.

(b) Los Angeles is not the capital of California.

(c) $3^2 = 9$.

(d) It is not true that I went to the movies today.

(e) The temperature is above 60° Fahrenheit.

The negations of the preceding statements are as follows:

(a) Today is not Friday.

(b) Los Angeles is the capital of California.

(c) $3^2 \neq 9$.

(d) It is true that I went to the movies today.

(e) The temperature is less than or equal to 60° Fahrenheit. ■

However, the negation of statements containing words such as "some," "all," and "every" requires more care. Consider, for instance, the following statement s:

s: Some bananas are blue.

Since "some" means "at least one," the negation of s is the statement

$\sim s$: No bananas are blue.

Likewise, the negation of the statement

t: Every banana is yellow.

is the statement

$\sim t$: Some bananas are not yellow.

As these examples suggest, a statement involving the word "some" is negated by changing "some" to "no," and a sentence involving the words "all," "each," or "every" is negated by changing these words to "some . . . not . . . ."

Example A.3 Negate each of the following statements:

(a) Some cowboys live in Wyoming.

(b) There are movie stars who are not famous.

(c) No integers are divisible by 5.

(d) All doctors are rich.
(e) Every college football player weighs at least 200 pounds.

The negations of these statements are as follows:

(a) No cowboys live in Wyoming.
(b) No movie stars are not famous. (Or, all movie stars are famous.)
(c) Some integers are divisible by 5.
(d) Some doctors are not rich.
(e) Some college football players do not weigh at least 200 pounds. ■

It is obvious that the negation of a true statement is false, and the negation of a false statement is true. We can record this information in the following table, called a **truth table**:

p	$\sim p$
T	F
F	T

Here, p denotes a statement and $\sim p$ denotes its negation. The letters T and F signify that the indicated statement is true or false, respectively.

The **conjunction** of two statements is formed by joining the statements with the word "and." For example, the conjunction of the statements

p: Today is Monday and q: I went to school

is the statement

$p \wedge q$: Today is Monday, and I went to school.

This statement is true only when both of the original statements p and q are true. Thus the truth table for the connective "and" is as follows:

p	q	$p \wedge q$
T	T	T
T	F	F
F	T	F
F	F	F

The **disjunction** of two statements is formed by joining the statements with the word "or." For example, the disjunction of the statements p and q is

$p \vee q$: Today is Monday, or I went to school.

This statement is true when at least one of the original statements is true. For example, the statement $p \vee q$ is true in each of the following cases:

(1) Today is not Monday, and I went to school.
(2) Today is Monday, and I did not go to school.
(3) Today is Monday, and I went to school.

Thus the truth table for the connective "or" is as follows:

p	q	$p \vee q$
T	T	T
T	F	T
F	T	T
F	F	F

The connectives "if . . . then . . ." and "if and only if" occur rather infrequently in ordinary discourse, but they are used very often in mathematics. A statement containing the connective "if . . . then . . ." is called a **conditional statement** or, more simply, a **conditional**. For example, suppose Mary is a student we know, and p and q are the statements

p: Mary was at the play Thursday night

and

q: Mary doesn't have an eight o'clock class on Friday morning.

Then the conditional $p \to q$ is the statement

$p \to q$: If Mary was at the play on Thursday night, then she doesn't have an eight o'clock class on Friday morning.

Another way of reading the statement "if p, then q" is "p implies q." In the conditional statement "if p, then q," statement p is called the **premise**, and statement q is called the **conclusion**.

It is important to note that conditional statements should not be interpreted in terms of cause and effect. Thus, when we say "if p, then q," we do not mean that the premise p causes the conclusion q, but only that when p is true, q must be true also.

In Normal, Illinois, there is a city ordinance designed to aid city street crews in removing snow. The ordinance states that if there is a snowfall of two or more inches, then cars may not be parked overnight on city streets. As applied to a particular day—say, December 15, 2004—this regulation is a conditional statement with premise

p: There is a snowfall of two or more inches on December 15, 2004

and conclusion

q: Cars are not parked overnight on city streets on December 15, 2004.

Let us consider under what circumstances the conditional statement $p \to q$ is false—that is, under what circumstances the ordinance has been violated. The ordinance is clearly violated if there is a snowfall of two or more inches and cars are parked overnight on city streets on December 15, 2004—that is, if p is true and q is false. Moreover, if there is a snowfall of two or more inches and cars are not parked overnight on the city streets on that day (that is, if both p and q are true), then the ordinance has been followed. If there is no snowfall of two or more inches (that is, if p is false), then the ordinance does not apply. Hence, in this case, the ordinance is not violated whether there are cars parked overnight on city streets or not. Thus the ordinance is violated only when the premise is true and the conclusion is false.

It may seem unnatural to regard a conditional statement $p \to q$ as being true whenever p is false. Indeed, it seems reasonable to regard a conditional statement as being not applicable when the premise is false. But then the conditional $p \to q$ would be neither true nor false when p is false, so $p \to q$ would no longer be a statement by our definition. For this reason, logicians consider a conditional statement to be true if its premise is false. Therefore the truth table for a conditional statement is as follows:

p	q	$p \to q$
T	T	T
T	F	F
F	T	T
F	F	T

The **biconditional** statement $p \leftrightarrow q$ means that $p \to q$ and $q \to p$. Thus a biconditional statement is the conjunction of two conditional statements. We read the biconditional statement $p \leftrightarrow q$ as "p if and only if q" or "p is necessary and sufficient for q." For instance, the following statements are biconditional statements:

> Mary was at the play Thursday night if and only if she doesn't have a class at eight o'clock on Friday morning.
>
> For John Snodgrass to drive his 1965 Mustang in the parade, it is necessary and sufficient that he buys a new muffler.

We can obtain the truth table for $p \leftrightarrow q$ from the tables for $p \to q$ and $q \to p$:

p	q	$p \to q$	$q \to p$	$(p \to q) \wedge (q \to p)$
T	T	T	T	T
T	F	F	T	F
F	T	T	F	F
F	F	T	T	T

Thus we see that the conditional statements $p \rightarrow q$ and $q \rightarrow p$ are both true only when p and q are both true or false. Hence the truth table for a biconditional statement is as follows:

p	q	$p \leftrightarrow q$
T	T	T
T	F	F
F	T	F
F	F	T

In the first table of the preceding paragraph, we can see that the conditional statements $p \rightarrow q$ and $q \rightarrow p$ do not always have the same truth values. Unfortunately, it is a common mistake to confuse these two conditionals and to assume that one is true if the other is. Although these two statements are different, they are obviously related because both involve the same p and q. We call the statement $q \rightarrow p$ the **converse** of $p \rightarrow q$. There are two other conditional statements that are related to the conditional $p \rightarrow q$. The statement $\sim p \rightarrow \sim q$ is called the **inverse** of $p \rightarrow q$, and the statement $\sim q \rightarrow \sim p$ is called the **contrapositive** of $p \rightarrow q$.

Example A.4 Form the converse, inverse, and contrapositive of the following statement about John Snodgrass: If John got a new muffler, then John drove his Mustang in the parade.

The given conditional statement is of the form $p \rightarrow q$, where p and q are the statements

p: John got a new muffler

and

q: John drove his Mustang in the parade.

The converse, inverse, and contrapositive of the given statement are as follows:

converse: If John drove his Mustang in the parade, then John got a new muffler.

inverse: If John didn't get a new muffler, then John didn't drive his Mustang in the parade.

contrapositive: If John didn't drive his Mustang in the parade, then John didn't get a new muffler. ■

Example A.5 Form the converse, inverse, and contrapositive of the statement

If it isn't raining today, then I am going to the beach.

The desired statements are as follows:

converse: If I am going to the beach today, then it isn't raining.

inverse: If it is raining today, then I am not going to the beach.

contrapositive: If I am not going to the beach today, then it is raining.

In this case, we must be careful not to read more into the given statement than it says. It is tempting to regard the given statement as a biconditional statement meaning that I am going to the beach today if it isn't raining and not going if it is. However, the given statement does *not* say that I am not going to the beach if it is raining. This is the inverse of the given statement. Likewise, one must constantly guard against assuming the truth or falsity of the converse of a conditional on the basis of the truth or falsity of the conditional itself. ■

It should be noted that the term "converse" is also commonly used in mathematics in a more complicated sense than that just defined. For example, consider the statement

A: If the 3 sides of a triangle are congruent, then the 3 angles of the triangle are congruent.

Most mathematicians would call the following statement the "converse" of statement A:

A^*: If the 3 angles of a triangle are congruent, then the 3 sides of the triangle are congruent.

Indeed, this seems to be consistent with the definition of "converse" given previously if we consider statement A to have the form $p \to q$, where p and q are as follows:

p: The 3 sides of a triangle are congruent.
q: The 3 angles of a triangle are congruent.

The problem with this interpretation is that p and q are not statements. For example, whether p is true or false depends on what triangle we are talking about.

Actually, the statement labeled A is an abbreviated, but common, way of expressing the following:

For all triangles T, if three sides of T are congruent, then the 3 angles of T are congruent.

This statement is of the form

$$B: \quad \text{For all } x \text{ in } S,\ p(x) \to q(x).$$

Here, x is a variable, S is some set (the set of all triangles in our example), and $p(x)$ and $q(x)$ are sentences that become statements when x is given any particular value in S. The conventional "converse" of statement B is thus the statement

$$B^*: \quad \text{For all } x \text{ in } S, q(x) \to p(x).$$

In our example with triangles, both statements A and its "converse" A^* are true in Euclidean geometry. On the other hand,

$$C: \quad \text{For all real numbers } x, \text{ if } x > 3, \text{ then } x > 2$$

is true, while

$$C^*: \quad \text{For all real numbers } x, \text{ if } x > 2, \text{ then } x > 3$$

is not. Note that both the statements

D: For all integers n, if n is even, then n is the square of an integer

and

D^*: For all integers n, if n is the square of an integer, then n is even

are false.

EXERCISES A.1

In Exercises 1–12, determine whether each sentence is a statement. If so, determine whether the statement is true or false.

1. Georgia is the southernmost state in the United States.
2. E.T., phone home.
3. If $x = 3$, then $x^2 = 9$.
4. Cats can fly.
5. What's the answer?
6. New York is the location of the United Nations building.
7. Five is an odd integer, and seven is an even integer.
8. Six is an even integer, or seven is an even integer.
9. Please be quiet until I am finished, or leave the room.
10. Nine is the largest prime number less than 10, and two is the smallest.
11. Five is a positive integer, or zero is a positive integer.
12. Go home and leave me alone.

Write the negations of the statements in Exercises 13–24.

13. $4 + 5 = 9$.
14. Christmas is celebrated on December 25.
15. California is not the largest state in the United States.
16. It has never snowed in Chicago.
17. All birds can fly.
18. Some people are rich.
19. There is a man who weighs 400 pounds.
20. Every millionaire pays taxes.

21. Some students do not pass calculus.

22. All residents of Chicago love the Cubs.

23. Everyone enjoys cherry pie.

24. There are no farmers in South Dakota.

For each of the given pairs of statements p and q in Exercises 25–32, write (a) the conjunction and (b) the disjunction. Then indicate which, if either, of these statements is true.

25. p: One is an even integer. q: Nine is a positive integer.

26. p: Oregon borders Canada. q: Egypt is in Asia.

27. p: The Atlantic is an ocean. q: The Nile is a river.

28. p: Cardinals are red. q: Robins are blue.

29. p: Birds have four legs. q: Rabbits have wings.

30. p: Oranges are fruit. q: Potatoes are vegetables.

31. p: Flutes are wind instruments. q: Timpani are string instruments.

32. p: Algebra is an English course. q: Accounting is a business course.

For each statement in Exercises 33–36, write (a) the converse, (b) the inverse, and (c) the contrapositive.

33. If this is Friday, then I will go to the movies.

34. If I complete this assignment, then I will take a break.

35. If Kennedy doesn't run for the Senate, then he will run for president.

36. If I get an A on the final exam, then I'll get a B for the course.

A.2 LOGICAL EQUIVALENCE

When analyzing a complicated statement involving connectives, we often find it useful to consider the simpler statements that form it. The truth or falsity of the complicated statement can then be determined by considering the truth or falsity of the simpler statements. Consider, for instance, the statement

> Fred Nitney starts at guard in tonight's game implies that Sam Smith scores fewer than 10 points if and only if Sam Smith scores fewer than 10 points or Fred Nitney doesn't start at guard.

This statement is formed from the two simpler statements

p: Fred Nitney starts at guard in tonight's game

and

q: Sam Smith scores fewer than 10 points in tonight's game.

We can write the given statement symbolically as $(p \to q) \leftrightarrow (q \vee \sim p)$.

Let us analyze the truth of this statement in terms of the truth of p and q. This analysis can be conveniently carried out in the following truth table, where each row corresponds to a different pair of truth values for p and q:

p	q	$p \to q$	$\sim p$	$q \vee \sim p$	$(p \to q) \leftrightarrow (q \vee \sim p)$
T	T	T	F	T	T
T	F	F	F	F	T
F	T	T	T	T	T
F	F	T	T	T	T

Thus we see that the original statement

> Fred Nitney starts at guard in tonight's game implies that Sam Smith scores fewer than 10 points if and only if Sam Smith scores fewer than 10 points or Fred Nitney doesn't start at guard

is always true, regardless of the truth or falsity of the statements Fred Nitney starts at guard in tonight's game and Sam Smith scores fewer than 10 points in tonight's game.

Example A.6 Assuming that $p, q,$ and r are statements, use a truth table to analyze the compound statement $p \vee [(p \wedge \sim q) \to r]$.

The following truth table shows that the statement $p \vee [(p \wedge \sim q) \to r]$ is always true:

p	q	r	$\sim q$	$p \wedge \sim q$	$[(p \wedge \sim q) \to r]$	$p \vee [(p \wedge \sim q) \to r]$
T	T	T	F	F	T	T
T	T	F	F	F	T	T
T	F	T	T	T	T	T
T	F	F	T	T	F	T
F	T	T	F	F	T	T
F	T	F	F	F	T	T
F	F	T	T	F	T	T
F	F	F	T	F	T	T

■

Compound statements that are true no matter what the truth values of their component statements, such as the one in Example A.6, are of special interest because of their use in constructing valid arguments. Such a statement is said to be a **tautology**. Likewise, it is possible for a compound statement to be false no matter what the truth values of its component statements; such a statement is called a **contradiction**. Obviously, the negation of a tautology is a contradiction, and vice versa.

Example A.7 As we can see in the following truth table, the statement $(p \wedge {\sim}q) \wedge ({\sim}p \vee q)$ is a contradiction:

p	q	$\sim p$	$\sim q$	$p \wedge {\sim}q$	${\sim}p \vee q$	$(p \wedge {\sim}q) \wedge ({\sim}p \vee q)$
T	T	F	F	F	T	F
T	F	F	T	T	F	F
F	T	T	F	F	T	F
F	F	T	T	F	T	F

Thus ${\sim}[(p \wedge {\sim}q) \wedge ({\sim}p \vee q)]$, the negation of the given statement, is a tautology. ■

Two compound statements are called **logically equivalent** if they have the same truth values for all possible truth values of their component statement variables. Thus two statements S and T are logically equivalent if and only if the biconditional $S \leftrightarrow T$ is a tautology. For example, we saw in the first truth table in this section that the biconditional $(p \to q) \leftrightarrow (q \vee {\sim}p)$ is a tautology. Therefore the statements $p \to q$ and $q \vee \sim p$ are logically equivalent.

Example A.8 Show that the compound statements ${\sim}(p \vee q)$ and $({\sim}p) \wedge ({\sim}q)$ are logically equivalent. (This fact is called *De Morgan's law.*)

In order to prove that the two statements are logically equivalent, it is sufficient to show that the columns in a truth table corresponding to these statements are identical. Since this is the case in the following truth table, we conclude that ${\sim}(p \vee q)$ and $({\sim}p) \wedge ({\sim}q)$ are logically equivalent:

p	q	$p \vee q$	${\sim}(p \vee q)$	${\sim}p$	${\sim}q$	$({\sim}p) \wedge ({\sim}q)$
T	T	T	F	F	F	F
T	F	T	F	F	T	F
F	T	T	F	T	F	F
F	F	F	T	T	T	T

■

In logical arguments, it is often necessary to simplify a complicated statement. In order for this simplification to result in a valid argument, it is essential that the replacement statement be logically equivalent to the original statement, for then the two statements always have the same truth values. Thus, because of the logical equivalences shown in Example A.8, we can replace either of the statements ${\sim}(p \vee q)$ or $({\sim}p) \wedge ({\sim}q)$ by the other without affecting the validity of an argument.

We will close this section by stating a theorem containing several important logical equivalences that occur frequently in mathematical arguments. The proof of this theorem will be left to the exercises. Note the similarity between parts (a) through (h) of this theorem and parts (a) through (c) of Theorem 2.1 and parts (a) and (b) of Theorem 2.2.

Theorem A.1 The following pairs of statements are logically equivalent:

(a) $p \wedge q$ and $q \wedge p$ (commutative law for conjunction)
(b) $p \vee q$ and $q \vee p$ (commutative law for disjunction)
(c) $(p \wedge q) \wedge r$ and $p \wedge (q \wedge r)$ (associative law for conjunction)
(d) $(p \vee q) \vee r$ and $p \vee (q \vee r)$ (associative law for disjunction)
(e) $p \vee (q \wedge r)$ and $(p \vee q) \wedge (p \vee r)$ (distributive law)
(f) $p \wedge (q \vee r)$ and $(p \wedge q) \vee (p \wedge r)$ (distributive law)
(g) $\sim(p \vee q)$ and $\sim p \wedge \sim q$ (De Morgan's law)
(h) $\sim(p \wedge q)$ and $\sim p \vee \sim q$ (De Morgan's law)
(i) $p \to q$ and $\sim q \to \sim p$ (law of the contrapositive)

EXERCISES A.2

In Exercises 1–10, construct a truth table for each compound statement.

1. $(p \vee q) \wedge [\sim(p \wedge q)]$

2. $(\sim p \vee q) \wedge (\sim q \wedge p)$

3. $(p \vee q) \to (\sim p \wedge q)$

4. $(\sim p \wedge q) \to (\sim q \vee p)$

5. $(p \to q) \to (p \vee r)$

6. $p \to (\sim q \vee r)$

7. $(\sim q \wedge r) \leftrightarrow (\sim p \vee q)$

8. $\sim[p \wedge (q \vee r)]$

9. $[(p \vee q) \wedge r] \to [(p \wedge r) \vee q]$

10. $(r \wedge \sim q) \leftrightarrow (q \vee p)$

In Exercises 11–16, show that the given statements are tautologies.

11. $\sim p \vee p$

12. $(p \to q) \vee (\sim q \wedge p)$

13. $(\sim p \wedge q) \to \sim(q \to p)$

14. $\sim(\sim p \wedge q) \to (\sim q \vee p)$

15. $\sim[((p \to q) \wedge (\sim q \vee r)) \wedge (\sim r \wedge p)]$

16. $[(p \wedge q) \to r] \to [\sim r \to (\sim p \vee \sim q)]$

In Exercises 17–24, show that the given pairs of statements are logically equivalent.

17. p and $\sim(\sim p)$

18. p and $p \vee (p \wedge q)$

19. $\sim(p \to q)$ and $\sim q \wedge (p \vee q)$

20. $p \leftrightarrow q$ and $(\sim p \vee q) \wedge (\sim q \vee p)$

21. $p \to (q \to r)$ and $(p \wedge q) \to r$

22. $(p \to q) \to r$ and $(p \vee r) \wedge (q \to r)$

23. $(p \vee q) \to r$ and $(p \to r) \wedge (q \to r)$

24. $p \to (q \vee r)$ and $(p \to q) \vee (p \to r)$

25. Prove Theorem A.1 parts (a) and (b).

26. Prove Theorem A.1 parts (c) and (d).

27. Prove Theorem A.1 parts (e) and (f).

28. Prove Theorem A.1 part (h).

29. Prove Theorem A.1 part (i).

30. The statement $[(p \to q) \wedge \sim q] \to \sim p$ is called **modus tollens**. Prove that modus tollens is a tautology.

31. The statement $[p \wedge (p \to q)] \to q$ is called **modus ponens**. Prove that modus ponens is a tautology.

32. The statement $[(p \vee q) \wedge \sim p] \to q$ is called the law of **disjunctive syllogism**. Prove that disjunctive syllogism is a tautology.

33. Define a new connective named "exclusive or" and denoted $\veebar$ by regarding $p \veebar q$ to be true if and only if exactly one of p or q is true.

(a) Write a truth table for "exclusive or."

(b) Show that $p \veebar q$ is logically equivalent to $\sim(p \leftrightarrow q)$.

34. The **Sheffer stroke** is a connective denoted | and defined by the following truth table:

p	q	$p\vert q$
T	T	F
T	F	T
F	T	T
F	F	T

The following parts prove that all of the basic connectives can be written by using only the Sheffer stroke:

(a) Show that $p|p$ is logically equivalent to $\sim p$.

(b) Show that $(p|p)|(q|q)$ is logically equivalent to $p \vee q$.

(c) Show that $(p|q)|(p|q)$ is logically equivalent to $p \wedge q$.

(d) Show that $p|(q|q)$ is logically equivalent to $p \to q$.

A.3 METHODS OF PROOF

Mathematics is probably the only human endeavor that places such a central emphasis on the use of logic and proof. Being able to think logically and to read proofs certainly increases mathematical understanding, but, more importantly, these skills enable us to apply mathematical ideas in new situations. In this section we will discuss basic methods of proof so that the reader will have a better understanding of the logical framework in which proofs are written.

A **theorem** is a mathematical statement that is true. Theorems are essentially conditional statements, although the wording of a theorem may obscure this fact. For instance, Theorem 1.3 is worded

A set with n elements has exactly 2^n subsets.

With this wording, the theorem does not seem to be a conditional statement; yet we can express this theorem as a conditional statement by writing

If S is a set with n elements, then S has exactly 2^n subsets.

When the theorem is expressed as a conditional statement, the premise and conclusion of the conditional statement are called the **hypothesis** and **conclusion** of the theorem.

By a **proof** of a theorem, we mean a logical argument that establishes the theorem to be true. The most natural form of proof is a **direct proof**. Suppose that we wish to prove the theorem $p \to q$. Since $p \to q$ is true whenever p is false, we need only show that whenever p is true, so is q. Therefore, *in a direct proof, we assume that the hypothesis of the theorem, p, is true and demonstrate that the conclusion, q, is true*. It then follows that $p \to q$ is true.

We will illustrate some types of proofs by proving certain elementary facts about integers that use the following two definitions:

(1) An integer n is called **even** if it can be written in the form $n = 2k$ for some integer k.
(2) An integer n is called **odd** if it can be written in the form $n = 2k + 1$ for some integer k.

We will also use the fact that every integer is either even or odd, but not both. The next theorem is proved by a direct proof.

Example A.9 Suppose that we wish to prove the following theorem: If n is an even integer, then n^2 is an even integer.

To prove this result by a direct proof, we assume the hypothesis and prove the conclusion. Accordingly, we assume that n is an even integer and prove that n^2 is even. Since n is even, $n = 2k$ for some integer k. Then

$$n^2 = (2k)^2 = 4k^2 = 2(2k^2).$$

If k is an integer, then so is $2k^2$. Hence n^2 can be expressed as 2 times the integer $2k^2$, and so n^2 is even. ■

Example A.10 Consider the following theorem: If x is a real number and $x^2 - 1 = 0$, then $x = -1$ or $x = 1$.

Since $x^2 - 1 = 0$, factoring gives $(x + 1)(x - 1) = 0$. But if the product of any two real numbers is 0, at least one of them must be 0. Hence $x + 1 = 0$ or $x - 1 = 0$. In the first case, $x = -1$; in the second, $x = 1$. Thus $x = -1$ or $x = 1$. ■

The argument in Example A.10 uses the **law of syllogism**, which states

$$[(p \to q) \wedge (q \to r)] \to (p \to r).$$

Suppose that x is some fixed real number, and let p, q, r, and s be the following statements:

$$\begin{aligned} p&: \quad x^2 - 1 = 0. \\ r&: \quad (x+1)(x-1) = 0. \\ s&: \quad x+1 = 0 \quad \text{or} \quad x - 1 = 0. \\ q&: \quad x = -1 \quad \text{or} \quad x = 1. \end{aligned}$$

Then the argument in Example A.10 shows that

$$(p \to r) \wedge (r \to s) \wedge (s \to q).$$

Hence, by two applications of the law of syllogism, we conclude that $p \to q$, proving the theorem.

Another type of proof is based on the law of the contrapositive, which asserts that the statements $p \to q$ and $\sim q \to \sim p$ are logically equivalent. To prove that $p \to q$ by this method, we give a direct proof of the statement $\sim q \to \sim p$ by assuming $\sim q$ and proving $\sim p$. The law of the contrapositive then allows us to conclude that $p \to q$ is also true.

Example A.11 We will prove the following theorem: If $x + y > 100$, then $x > 50$ or $y > 50$.

Suppose that x and y are some fixed real numbers. Then it suffices to show that $p \to q$, where p and q denote the following statements:

$$p: \quad x + y > 100 \qquad \text{and} \qquad q: \quad x > 50 \text{ or } y > 50.$$

We will establish the contrapositive of the desired result, which is $\sim q \to \sim p$. Consequently, we will assume that $\sim q$ is true and show that $\sim p$ is true. Using Example A.8, we see that $\sim q$ and $\sim p$ are the statements

$$\sim q: \quad x \leq 50 \text{ and } y \leq 50 \qquad \text{and} \qquad \sim p: \quad x + y \leq 100.$$

Suppose that $x \leq 50$ and $y \leq 50$. Then

$$x + y \leq 50 + 50 = 100.$$

Hence $\sim p$ is proved; that is, $\sim q \to \sim p$. It now follows from the law of the contrapositive that $p \to q$. ■

Example A.12 We will prove the following theorem: If n is an integer and n^2 is even, then n is even.

The contrapositive of this theorem is as follows: If an integer n is not even, then n^2 is not even. This statement can also be expressed in the form

If n is an odd integer, then n^2 is odd.

This statement can be proved by an argument like that in the proof of the theorem in Example A.9. Assume that n is an odd integer. Then $n = 2k + 1$ for some integer k. Then

$$n^2 = (2k+1)^2 = 4k^2 + 4k + 1 = 2(2k^2 + 2k) + 1.$$

Since k is an integer, so is $2k^2 + 2k$. Hence n^2 can be expressed as 2 times the integer $2k^2 + 2k$ plus 1, and so n^2 is odd. Thus we have proved that if n is an odd integer, then n^2 is odd. It follows that the contrapositive of this statement—if n is an integer and n^2 is even, then n is even—is also true. ■

A very different style of proof is a **proof by contradiction**. In this method of proof, we prove the theorem $p \to q$ by assuming that p and $\sim q$ are true and deducing a false statement r. Since $(p \wedge \sim q) \to r$ is true, but r is false, we can conclude that the premise $p \wedge \sim q$ of this conditional statement is false. But then its negation $\sim(p \wedge \sim q)$ is true, which is logically equivalent to the desired statement $p \to q$. (See Exercise 1.)

Example A.13 We will prove the following theorem: If n is the sum of the squares of two odd integers, then n is not a perfect square.

Proving this theorem by contradiction seems natural because the theorem expresses a negative idea (that n is *not* a perfect square). Thus, when we deny the conclusion, we obtain the positive statement that n is a perfect square.

Accordingly, we will use a proof by contradiction. Therefore we assume the hypothesis and deny the conclusion, and so we assume both that n is the sum of the squares of two odd integers and that n is a perfect square. Since n is a perfect square, we have $n = m^2$ for some integer m. But also, n is the sum of the squares of two odd integers. Since an odd integer is one more than an even integer, we can express n in the form

$$n = (2r+1)^2 + (2s+1)^2$$

for some integers r and s. It follows that

$$\begin{aligned} n &= (2r+1)^2 + (2s+1)^2 = (4r^2 + 4r + 1) + (4s^2 + 4s + 1) \\ &= 4(r^2 + s^2 + r + s) + 2, \end{aligned}$$

so that n is even. Thus $m^2 = n$ is even. We deduce from Example A.12 that m is even, and so $m = 2p$ for some integer p. Thus $n = m^2 = (2p)^2 = 4p^2$ is divisible by 4. But we saw previously that $n = 4(r^2 + s^2 + r + s) + 2$, which is not divisible by 4. Hence we have derived a false statement, namely, that n is both divisible by 4 and not divisible by 4. Thus assuming the hypothesis and denying the conclusion has led to a false statement. It follows that if the hypothesis is true, the conclusion must be true also. Consequently, the theorem has been proved. ■

Example A.14 Show that there is no rational number r such that $r^2 = 2$. (Recall that a rational number is one that can be written as the quotient of two integers.)

The theorem to be proved can be written as the following conditional statement: If r is a rational number, then $r^2 \neq 2$. Again, proving this theorem by contradiction seems natural because the theorem expresses a negative idea (that r^2 is *not* equal to 2). So if we deny the conclusion, we obtain the positive statement that there is a rational number r such that $r^2 = 2$.

Accordingly, we will use a proof by contradiction. Thus we assume the hypothesis and deny the conclusion, and so we assume that there is a rational number r such that $r^2 = 2$. Because r is a rational number, it can be expressed in the form $\frac{m}{n}$, where m and n are integers. Moreover, we may choose m and n to have no common factors greater than 1, so that the fraction m/n is in lowest terms. Then we have

$$\left(\frac{m}{n}\right)^2 = 2,$$

from which it follows that $m^2 = 2n^2$. Hence m^2 is an even integer. Thus, by Example A.12, m must be even; that is, $m = 2p$ for some integer p. Substituting this value for m in the equation $m^2 = 2n^2$ yields $4p^2 = 2n^2$, so that $2p^2 = n^2$. Hence n^2 is even, and it follows, as shown previously, that n is even. But then both m and n are even; that is, m and n have a common factor of 2. This fact contradicts our choice of m and n as having no common factor greater than 1, and so we deduce that the conclusion to our theorem must be true. Thus the theorem is proved. ■

We have discussed three basic methods of proof in this section, the direct proof, proof of the contrapositive, and proof by contradiction. There are other types of proofs as well. One method of proof that is quite important in discrete mathematics is proof by mathematical induction, which is discussed in Section 2.5. Another type of proof is a proof by cases, in which the theorem to be proved is subdivided into parts, each of which is proved separately. The next example demonstrates this technique.

Example A.15 Show that if n is an integer, then $n^3 - n$ is even.

Since every integer n is either even or odd, we will consider these two cases.

Case 1: n is even: Then $n = 2m$ for some integer m. Therefore

$$n^3 - n = (2m)^3 - 2m = 8m^3 - 2m = 2(4m^3 - m),$$

which is even.

Case 2: n is odd: Then $n = 2m + 1$ for some integer m. Hence

$$\begin{aligned} n^3 - n &= (2m+1)^3 - (2m+1) \\ &= (8m^3 + 12m^2 + 6m + 1) - (2m + 1) \\ &= 8m^3 + 12m^2 + 4m \\ &= 2(4m^3 + 6m^2 + 2m), \end{aligned}$$

which is even.

Because $n^3 - n$ is even in either case, we conclude that $n^3 - n$ is even for all integers n. ■

To close this section, we will briefly consider the problem of disproving a statement $p \to q$—that is, of showing that it is false. Because a conditional statement is false only when its premise is true and its conclusion is false, we must find an instance in which p is true and q is false. Such an instance is called a **counterexample** to the statement.

For example, consider the following statement: If an integer n is the sum of the squares of two even integers, then n is not a perfect square. To disprove this statement, we must find a counterexample—that is, an integer n that is the sum of the squares of two even integers and at the same time a perfect square itself. The equality $100 = 6^2 + 8^2$ shows that 100 is such a number. The existence of a single counterexample is enough to invalidate the statement, even though there are many values of n for which the statement holds (e.g., $40 = 2^2 + 6^2$). But the statement is false because it is not true for *all* integers that satisfy the hypothesis.

EXERCISES A.3

1. Prove that $\sim(p \wedge \sim q)$ is logically equivalent to $p \to q$.
2. Prove that the law of syllogism is a tautology.
3. Prove that if m is an integer and m^2 is odd, then m is odd. (*Hint:* Prove the contrapositive.)
4. Prove, as in Example A.14, that there is no rational number r such that $r^2 = 3$.

Prove the theorems in Exercises 5–12. Assume that all the symbols used in these exercises represent positive integers.

5. If a divides b, then ac divides bc for any c.
6. If ac divides bc, then a divides b.
7. If a divides b and b divides c, then a divides c.
8. If a divides b, then $a \le b$.
9. If p and q are primes and p divides q, then $p = q$.
10. If a divides b and a divides $b + 2$, then $a = 1$ or $a = 2$.
11. If xy is even, then x is even or y is even.
12. For all integers n greater than 10, $12(n - 2) < n^2 - n$.

Prove or disprove the results in Exercises 13–22. Assume that all the numbers mentioned in these exercises are integers.

13. The sum of two odd integers is odd.
14. The product of two odd integers is odd.
15. If $ac = bc$, then $a = b$.
16. If 3 divides xy, then 3 divides x or 3 divides y.

17. If 6 divides xy, then 6 divides x or 6 divides y.
18. If 3 divides x and 3 divides y, then 3 divides $ax + by$.
19. If a and b are odd, then $a^2 + b^2$ is even.
20. If a and b are odd, then $a^2 + b^2$ is not divisible by 4.
21. For all integers, n, n is odd if and only if 8 divides $n^2 - 1$.
22. The product of two integers is odd if and only if both of the integers are odd.

23. Prove or disprove: For every positive integer n, $n^2 + n + 41$ is prime.
24. Prove that, in any set of three consecutive odd positive integers other than 3, 5, and 7, at least one number is not prime.
25. Prove that, for each positive integer n, $n^2 - 2$ is not divisible by 3.
26. Prove that, for each positive integer n, $n^4 - n^2$ is divisible by 6.
27. Prove that if p is a prime positive integer, then $\log_{10} p$ is not expressible as the quotient of two integers.
28. Prove that there are infinitely many primes.

HISTORICAL NOTES

The study of logic and proof has played a central role in mathematics since the time of Thales of Miletus (ca. 580–500 B.C.), who is regarded as the first mathematician to offer deductive arguments. In the fifth century B.C., the Athenian Plato (429–348 B.C.) made a distinction between arithmetic (the theory of numbers) and logistic (the techniques of computation). In drawing this distinction, Plato discussed the differences of theory and application. He saw the essence of mathematics residing in the analytic method. One begins with givens, in the form of axioms or postulates, and works, step by step, to develop a line of reasoning resulting in a specific desired statement. Throughout his work, he elevated the role of theory in comparison to that of application. Plato's student Aristotle (384–322 B.C.) was the first to systematize deductive arguments into a system of principles. While he wrote little about mathematics directly, Aristotle's development of argument in his philosophical writings and his constant use of mathematical concepts in discussing argument forms left indelible marks on the subject [73, 74].

The German Gottfried Wilhelm Leibniz (1646–1716) presented his view of systematic argument forms in his 1666 book *De Arte Combinatoria*. While many of his scientific colleagues viewed his work as being "metaphysical," Leibniz worked from 1679 to 1690 developing a system built on undefined terms, axioms and postulates, logical rules, and derived statements. His goal was to develop a universal algebra for reasoning. It was not until the emergence of the work of the English mathematician George Boole (1815–1864) that such a system of thought became widely accepted. Boole's 1847 work *The Mathematical Analysis of Logic* and its 1854 extension *An Investigation of the Laws of Thought* ushered in a new focus on the nature of evidence, argument, and proof.

Augustus De Morgan

C.S. Peirce

These works were substantiated by Augustus De Morgan's (1806–1871) *Formal Logic; or, the Calculus of Inference, Necessary and Probable*. At the same time, the American Charles Sanders Peirce (1839–1914) argued for a separation of mathematics and logic and added emphasis on the role of quantifiers in logical arguments. Between 1880 and 1885, Peirce worked to develop a theory involving truth values and semantics for his logic. Peirce's work with quantification and the algebra of statements moved inference in a new direction. His algebra of logic was brought to new levels of rigor and formalism in German Ernst Schröder's (1841–1902) work *Vorlesungen über die Algebra der Logik* (1890–1905) and Alfred North Whitehead's (1861–1947) *A Treatise on Universal Algebra* (1898) [80].

At the same time, a number of mathematicians began to work to develop a system for rigorously expressing all of mathematics in terms of this new logical language. Between the appearances of his books *Foundation of Arithmetic* in 1879 and *The Fundamental Laws of Arithmetic* in 1903, the German Gottlieb Frege (1848–1925) tried to develop a more rigorous basis for mathematics. The Italian Giuseppe Peano's (1858–1932) seminal 1894 work *Formulaire de Mathématiques* laid out a view of arithmetic based on the undefined concepts of zero, number, and successor and the following fundamental axioms:

- Zero is a number.
- For any number n, its successor is a number.
- No number has zero as its successor.
- If two numbers m and n have the same successor, then $m = n$.
- If T is a set of numbers such that 0 is a member of T and the successor of n is in T whenever n is in T, then T is the set of all numbers.

In this, order (successor) and induction were tied to the development of number. These ideas were also used by Julius Wilhelm Richard Dedekind (1831–1916) and others to formulate a basis for arithmetic and, more generally, mathematics. The best known of these was the multi-volume work *Principia Mathematica* (1910–1913) developed by Bertrand Russell (1872–1970) and Alfred North Whitehead (1861–1947).

The Austrian-born mathematician Kurt Gödel's (1906–1978) 1931 paper showed that the axiomatic method has its limitations. His theorem, called Gödel's incompleteness theorem, showed that any set of axioms broad enough to contain the fundamentals of the positive integers must also contain a well-defined statement that can neither be proven true nor proven false within the system [80, 84].

SUPPLEMENTARY EXERCISES

In Exercises 1–8, determine whether each sentence is a statement. If so, tell whether it is true or false.

1. All integers are real numbers.

2. Each real number is an integer.

3. Tom is the smartest student in class.

4. The day before Thursday is Friday.

5. $n^2 \geq n!$ for all integers n.

6. Some rectangles are squares.

7. No square of an integer has a unit's digit of 7.

8. $n! + 1$ is a prime for all positive integers n.

Write the negation of each statement in Exercises 9–16. Indicate whether each negation is true or false.

9. No squares are triangles.

10. All isosceles triangles are equilateral.

11. Some scientists from the United States have received Nobel prizes.

12. Red is a primary color, and blue is not a primary color.

13. $2 + 2 > 4$, or 1 is a root of $x^5 + 1 = 0$.

14. It is not the case that $x^2 \geq 1$ for all integers x.

15. In circling the globe along a line of latitude, one must cross the equator twice, the North Pole, and the South Pole.

16. There are three solutions to the equation $x^3 - x = 0$.

For each pair of statements p and q in Exercises 17–20, write (a) their conjunction and (b) their disjunction. Indicate the truth value of each compound statement formed.

17. p: Squares have four sides. q: Triangles have three sides.

18. p: $2^3 = 8$ q: $3^2 = 8$

19. p: If $3 > 2$, then $3 \times 0 > 2 \times 0$. q: If $4 = 5$, then $5 = 9$.

20. p: Roses are animals. q: Tigers are plants.

For each statement in Exercises 21–24, write (a) the converse, (b) the inverse, and (c) the contrapositive. Indicate the truth value of each.

21. If $3 + 3 = 6$, then $3^2 = 6$.

22. If $(3 + 3)^2 = 18$, then $(3 + 3)^2 = 3^2 + 3^2$.

23. If $3^2 = 6$, then $3 \times 2 = 6$.

24. If $(3 + 3)^2 = 6^2$, then $3^2 + 2 \times 3 \times 3 + 3^2 = 6^2$.

In Exercises 25–28, construct a truth table for each compound statement.

25. $\sim[(p \vee q) \wedge \sim p] \wedge \sim p$

26. $\sim[(p \vee \sim q) \wedge p] \leftrightarrow (\sim p \wedge q)$

27. $[p \wedge (r \wedge (\sim p \vee q))] \rightarrow [(p \wedge r) \wedge (\sim p \vee q)]$

28. $[(p \rightarrow r) \wedge (p \rightarrow q)] \rightarrow [p \rightarrow (r \wedge q)]$

In Exercises 29–32, determine whether the given statements are tautologies.

29. $[p \vee (\sim p \wedge q)] \rightarrow (p \vee q)$

30. $(p \rightarrow q) \rightarrow [(p \vee q) \rightarrow q]$

31. $\{[(p \vee q) \wedge \sim p] \wedge q\} \rightarrow (q \wedge \sim q)$

32. $\sim\{[(p \vee q) \wedge (\sim p \vee r)] \vee \sim(q \wedge r)\}$

In Exercises 33–36, test whether the given statements are logically equivalent.

33. $[\sim p \wedge (\sim p \wedge q)] \vee [p \wedge (p \wedge \sim q)]$ and $(\sim p \wedge q) \vee (p \wedge \sim q)$

34. $(p \rightarrow q) \wedge (p \rightarrow r)$ and $q \rightarrow r$

35. $(p \wedge q \wedge r) \vee (p \wedge {\sim}q \wedge r) \vee ({\sim}p \wedge {\sim}q \wedge r) \vee ({\sim}p \wedge q \wedge r)$ and r

36. $[(p \vee q) \vee {\sim}r] \wedge [p \vee (q \vee r)]$ and $p \vee q$

In Exercises 37–44, prove or disprove the statement given. Assume that all the variables in these exercises represent positive integers.

37. If $n > 4$, then n can be written as the sum of two distinct primes.

38. If x is even and x does not have zero as a unit's digit, then x is not divisible by 5.

39. If x is even and x is a perfect square, then x is divisible by 4.

40. One or both of $6n + 1$ and $6n - 1$ are prime.

41. 10 divides $n^5 - n$.

42. If s is the sum of a positive integer n and its square n^2, then s is even.

43. If d is the difference of two consecutive cubes, then d is odd.

44. If k is the sum of a positive integer c and its cube c^3, then k is even.

SUGGESTED READINGS

1. Kenelly, John W. *Informal Logic*. Boston: Allyn and Bacon, 1967.
2. Lucas, John. *An Introduction to Abstract Mathematics*. Belmont, CA: Wadsworth, 1986.
3. Mendelson, Elliott. *Introduction to Mathematical Logic*. Princeton, NJ: Van Nostrand, 1964.
4. Polya, G. *How to Solve It*. 2d ed. Garden City, NY: Doubleday, 1957.
5. Solow, Daniel. *How to Read and Do Proofs*. New York: Wiley, 1982.

B

Appendix: Matrices

In our study of coding theory and graph theory, it is sometimes useful to represent an object as an array of 0s and 1s. Arrays of numbers are useful not only in representing graphs, but also in performing computations. In this appendix, we discuss the addition and multiplication of matrices, two operations that are used in Chapters 3–6.

An $m \times n$ **matrix** is a rectangular array of numbers in which there are m horizontal **rows** and n vertical **columns**. For example, if

$$A = \begin{bmatrix} 1 & -2 \\ 5 & 0 \\ 6 & 7 \end{bmatrix} \quad \text{and} \quad B = \begin{bmatrix} -1 & 2 & 1 \\ 3 & -4 & 0 \\ 5 & 9 & -1 \\ -7 & 8 & 2 \end{bmatrix},$$

then A is a 3×2 matrix and B is a 4×3 matrix. In this case, the third row of A and the second column of B are

$$\begin{bmatrix} 6 & 7 \end{bmatrix} \quad \text{and} \quad \begin{bmatrix} 2 \\ -4 \\ 9 \\ 8 \end{bmatrix},$$

respectively.

The numbers in a matrix are called its *entries*. More specifically, the number in row i and column j is called the i, j **entry** of the matrix. In the matrix B above, the 3, 2 entry is 9 and the 4, 1 entry is -7.

Two matrices A and B are called **equal** whenever A and B have the same number of rows and the same number of columns and the i, j entry of A equals the i, j entry of B for every possible choice of i and j. In other words, two matrices are equal when they have the same size and all pairs of corresponding entries are equal. As with real numbers, if matrices A and B are equal, then we write $A = B$; otherwise, we write $A \neq B$.

Example B.1 Consider the matrices

$$C = \begin{bmatrix} 1 & 2 \\ 3 & 4 \end{bmatrix}, \quad D = \begin{bmatrix} 1 & 2 & 0 \\ 3 & 4 & 0 \end{bmatrix}, \quad \text{and} \quad E = \begin{bmatrix} 1 & 3 \\ 2 & 4 \end{bmatrix}.$$

Because C and D are of different sizes (C is a 2×2 matrix, whereas D is a 2×3 matrix), we have $C \neq D$. Also, $C \neq E$ because the $1, 2$ entry of C, which is 2, does not equal the $1, 2$ entry of E, which is 3. However, if

$$F = \begin{bmatrix} (-1)^4 & \sqrt{4} \\ \sqrt[3]{27} & 2^2 \end{bmatrix},$$

then $C = F$. ■

MATRIX OPERATIONS

Matrices are useful for storing information. Moreover, there are operations that can be performed on matrices that correspond to natural ways to work with the data stored in them. We discuss here only the matrix operations used in this book, namely, matrix addition and multiplication.

Suppose that the mathematics department of a university maintains two computer labs. The lab for upper-level courses contains 25 computers and 3 printers, and the lab for lower-level courses contains 30 computers and 2 printers. One way to record this information is with the 2×2 matrix

$$\begin{matrix} & \text{Computer} & \text{Printer} \\ \text{Upper} & 25 & 3 \\ \text{Lower} & 30 & 2 \end{matrix} \qquad \begin{bmatrix} 25 & 3 \\ 30 & 2 \end{bmatrix} = M.$$

In addition, suppose that the computer science department also maintains two labs for upper- and lower-level courses, which have 50 computers and 10 printers and 28 computers and 4 printers, respectively. This information can be recorded in the 2×2 matrix

$$\begin{matrix} & \text{Computer} & \text{Printer} \\ \text{Upper} & 50 & 10 \\ \text{Lower} & 28 & 4 \end{matrix} \qquad \begin{bmatrix} 50 & 10 \\ 28 & 4 \end{bmatrix} = C.$$

Then the sum $M + C$ is the matrix

$$M + C = \begin{bmatrix} 25 & 3 \\ 30 & 2 \end{bmatrix} + \begin{bmatrix} 50 & 10 \\ 28 & 4 \end{bmatrix} = \begin{bmatrix} 25 + 50 & 3 + 10 \\ 30 + 28 & 2 + 4 \end{bmatrix} = \begin{bmatrix} 75 & 13 \\ 58 & 6 \end{bmatrix} \begin{matrix} \text{Upper} \\ \text{Lower} \end{matrix}$$

(columns of the result: Computer, Printer)

whose entries give the total number of computers and printers in the mathematics and computer science labs for each level of course.

In general, suppose that A and B are two $m \times n$ matrices. The **sum** of A and B, denoted $A + B$, is the $m \times n$ matrix in which the i, j entry equals the sum of the i, j entry of A and the i, j entry of B. In other words, matrices of the same size can be added by adding their corresponding entries. Observe that only matrices of the same size can be added, and the sum is the same size as the matrices being added.

Example B.2 Consider the matrices

$$A = \begin{bmatrix} 6 & 4 \\ 2 & 1 \\ 0 & -5 \end{bmatrix} \quad \text{and} \quad B = \begin{bmatrix} -2 & 6 \\ 3 & 0 \\ 7 & 4 \end{bmatrix}.$$

Since A and B are both 3×2 matrices, they can be added. Their sum is the 3×2 matrix

$$A + B = \begin{bmatrix} 6 & 4 \\ 2 & 1 \\ 0 & -5 \end{bmatrix} + \begin{bmatrix} -2 & 6 \\ 3 & 0 \\ 7 & 4 \end{bmatrix} = \begin{bmatrix} 6+(-2) & 4+6 \\ 2+3 & 1+0 \\ 0+7 & -5+4 \end{bmatrix} = \begin{bmatrix} 4 & 10 \\ 5 & 1 \\ 7 & -1 \end{bmatrix}. \quad \blacksquare$$

Unfortunately, the multiplication of matrices is more complicated than addition. Consider first the product of a $1 \times n$ matrix A and an $n \times 1$ matrix B. If

$$A = [a_1 \quad a_2 \quad \ldots \quad a_n] \quad \text{and} \quad B = \begin{bmatrix} b_1 \\ b_2 \\ \vdots \\ b_n \end{bmatrix},$$

then the **product** AB is the 1×1 matrix

$$AB = [a_1 \quad a_2 \quad \ldots \quad a_n] \begin{bmatrix} b_1 \\ b_2 \\ \vdots \\ b_n \end{bmatrix} = [a_1 b_1 + a_2 b_2 + \cdots + a_n b_n],$$

in which the single entry is the sum of the products of the corresponding entries of A and B.

Example B.3 Let

$$A = [1 \quad 2 \quad 3] \quad \text{and} \quad B = \begin{bmatrix} 7 \\ 8 \\ 9 \end{bmatrix}.$$

The product of the 1×3 matrix A and the 3×1 matrix B is

$$AB = [1 \quad 2 \quad 3]\begin{bmatrix} 7 \\ 8 \\ 9 \end{bmatrix} = [1(7) + 2(8) + 3(9)] = [50]. \quad ■$$

More generally, the product of an $m \times n$ matrix A and a $p \times q$ matrix B is defined whenever $n = p$—that is, whenever the number of columns in A equals the number of rows in B. In this case, the **product** AB is the $m \times q$ matrix whose i, j entry equals the sum of the products of the corresponding entries of row i of A and column j of B. Symbolically, if A is the $m \times n$ matrix whose i, j entry is a_{ij} and B is the $n \times q$ matrix whose i, j entry is b_{ij}, then AB is the $m \times q$ matrix whose i, j entry equals

$$a_{i1}b_{1j} + a_{i2}b_{2j} + \cdots + a_{in}b_{nj}.$$

Note that this value is the same as the entry in the 1×1 matrix obtained by multiplying row i of A by column j of B, as previously described.

Example B.4 Let

$$A = \begin{bmatrix} 1 & 2 & 3 \\ 4 & 5 & 6 \end{bmatrix} \quad \text{and} \quad B = \begin{bmatrix} 7 & 10 \\ 8 & 11 \\ 9 & 12 \end{bmatrix}.$$

Here, A is a 2×3 matrix and B is a 3×2 matrix, and so the product AB is defined and is a 2×2 matrix. Its 1, 1 entry is the sum of the products of the corresponding entries of row 1 of A and column 1 of B (as in Example B.3):

$$1(7) + 2(8) + 3(9) = 50.$$

Similarly, the 1, 2 entry of AB is the sum of the products of the corresponding entries of row 1 of A and column 2 of B:

$$1(10) + 2(11) + 3(12) = 68;$$

the 2, 1 entry of AB is the sum of the products of the corresponding entries of row 2 of A and column 1 of B:

$$4(7) + 5(8) + 6(9) = 122;$$

and the 2, 2 entry of AB is the sum of the products of the corresponding entries of row 2 of A and column 2 of B:

$$4(10) + 5(11) + 6(12) = 167.$$

Thus AB is the 2×2 matrix

$$AB = \begin{bmatrix} 1 & 2 & 3 \\ 4 & 5 & 6 \end{bmatrix} \begin{bmatrix} 7 & 10 \\ 8 & 11 \\ 9 & 12 \end{bmatrix}$$

$$= \begin{bmatrix} 1(7) + 2(8) + 3(9) & 1(10) + 2(11) + 3(12) \\ 4(7) + 5(8) + 6(9) & 4(10) + 5(11) + 6(12) \end{bmatrix}$$

$$= \begin{bmatrix} 50 & 68 \\ 122 & 167 \end{bmatrix}. \quad \blacksquare$$

For an application of matrix multiplication, we return to our example of the computer labs. Suppose that the mathematics department wishes to know the value of the equipment in its two labs. If each of its computers costs \$1000 and each of its printers costs \$200, then the value of the equipment in its lab for upper-level courses is

$$25(\$1000) + 3(\$200) = \$25{,}600,$$

and the value of the equipment in its lab for lower-level courses is

$$30(\$1000) + 2(\$200) = \$30{,}400.$$

Note that for

$$V = \begin{bmatrix} 1000 \\ 200 \end{bmatrix} \begin{matrix} \text{Computer} \\ \text{Printer} \end{matrix} \quad (\text{Value})$$

the product matrix

$$MV = \begin{bmatrix} 25 & 3 \\ 30 & 2 \end{bmatrix} \begin{bmatrix} 1000 \\ 200 \end{bmatrix} = \begin{bmatrix} 25(1000) + 3(200) \\ 30(1000) + 2(200) \end{bmatrix} = \begin{bmatrix} 25{,}600 \\ 30{,}400 \end{bmatrix} \begin{matrix} \text{Upper} \\ \text{Lower} \end{matrix} \quad (\text{Value})$$

gives the value of the equipment in each of the mathematics department's labs.

In this book, we often encounter the product of two $n \times n$ matrices. Note that such a product is defined and is another $n \times n$ matrix. In particular, if A is an $n \times n$ matrix, then the product AA is defined. As with real numbers, we denote this product as A^2. Since A^2 is also an $n \times n$ matrix, the product $A^3 = AA^2$ is defined and is another $n \times n$ matrix. In a similar fashion, we can define $A^{k+1} = AA^k$ for every positive integer k, and all of the matrices $A, A^2, A^3, \ldots$ are $n \times n$ matrices.

Two special matrices occur frequently in applications. The $m \times n$ **zero matrix** is the $m \times n$ matrix in which each entry is zero; and the $n \times n$ matrix whose i, j entry is 1 for $i = j$ and 0 otherwise is called the $n \times n$ **identity matrix** and denoted I_n. For example,

$$\begin{bmatrix} 0 & 0 & 0 \\ 0 & 0 & 0 \end{bmatrix} \quad \text{and} \quad \begin{bmatrix} 1 & 0 & 0 \\ 0 & 1 & 0 \\ 0 & 0 & 1 \end{bmatrix}$$

are the 2×3 zero matrix and the 3×3 identity matrix, respectively.

In general, the multiplication of matrices is not commutative; that is, AB need not equal BA. In Example B.4, for instance, AB is a 2×2 matrix and BA is a 3×3 matrix; so $AB \neq BA$. Moreover, even if both A and B are $n \times n$ matrices, it is possible that $AB \neq BA$. (See Exercise 25.) However, many of the other familiar properties of the addition and multiplication of real numbers remain true for matrices.

Theorem B.1 Let A, B, and C be $m \times n$ matrices, D be an $n \times r$ matrix, and P and Q be $r \times s$ matrices.

(a) $A + B = B + A$ *(commutative law of matrix addition)*

(b) $(A + B) + C = A + (B + C)$ *(associative law of matrix addition)*

(c) If O is the $m \times n$ zero matrix, then $A + O = A$.

(d) The product of any matrix and a zero matrix, if defined, is a zero matrix.

(e) $I_m A = A$ and $AI_n = A$.

(f) $(AD)P = A(DP)$ *(associative law of matrix multiplication)*

(g) $(A + B)D = AD + BD$ *(right distributive law)*

(h) $D(P + Q) = DP + DQ$ *(left distributive law)*

(i) Let e_i denote the $1 \times m$ matrix in which the only nonzero entry is the $1, i$ entry, which is 1. Then $e_i A$ equals row i of A.

Proof. We will prove only (f), (g), and (i); the rest are left as exercises.

(f) Note that the product matrix AD is defined and is an $m \times r$ matrix. Thus $(AD)P$ is defined and is an $m \times s$ matrix. Similarly, $A(DP)$ is an $m \times s$ matrix. Hence it suffices to show that the corresponding entries of $(AD)P$ and $A(DP)$ are equal. Let x_{ij}, y_{ij}, and z_{ij} denote the i, j entries of A, D, and P, respectively. The i, k entry of AD is

$$x_{i1}y_{1k} + x_{i2}y_{2k} + \cdots + x_{in}y_{nk},$$

so the i, j entry of $(AD)P$ is

$$(x_{i1}y_{11} + x_{i2}y_{21} + \cdots + x_{in}y_{n1})z_{1j} + (x_{i1}y_{12} + x_{i2}y_{22} + \cdots + x_{in}y_{n2})z_{2j} + \cdots + (x_{i1}y_{1r} + x_{i2}y_{2r} + \cdots + x_{in}y_{nr})z_{rj}.$$

Similarly, the k, j entry of DP is

$$y_{k1}z_{1j} + y_{k2}z_{2j} + \cdots + y_{kr}z_{rj},$$

so the i, j entry of $A(DP)$ is

$$x_{i1}(y_{11}z_{1j} + y_{12}z_{2j} + \cdots + y_{1r}z_{rj}) + x_{i2}(y_{21}z_{1j} + y_{22}z_{2j} + \cdots + y_{2r}z_{rj}) + \cdots + x_{in}(y_{n1}z_{1j} + y_{n2}z_{2j} + \cdots + y_{nr}z_{rj}).$$

After multiplying out these expressions, each term in the i, j entry of $(AD)P$ appears as a term in the i, j entry of $A(DP)$, and vice versa. Hence these entries are equal. It follows that $(AD)P = A(DP)$.

(g) Because both A and B are $m \times n$ matrices, so is $A + B$. Thus $(A + B)D$, AD, and BD are all $m \times r$ matrices, and so both $(A + B)D$ and $AD + BD$ are $m \times r$ matrices. Let a_{ij}, b_{ij}, and d_{ij} denote the i, j entries of A, B, and D, respectively. We will show that the i, j entries of $(A + B)D$ and $AD + BD$ are equal. Because the i, k entry of $A + B$ is $a_{ik} + b_{ik}$, the i, j entry of $(A + B)D$ is

$$(a_{i1} + b_{i1})d_{1j} + (a_{i2} + b_{i2})d_{2j} + \cdots + (a_{in} + b_{in})d_{nj}. \qquad \text{(B.1)}$$

Similarly, the i, j entries of AD and BD are

$$a_{i1}d_{1j} + a_{i2}d_{2j} + \cdots + a_{in}d_{nj} \qquad \text{and} \qquad b_{i1}d_{1j} + b_{i2}d_{2j} + \cdots + b_{in}d_{nj}.$$

Thus the i, j entry of $AD + BD$ is

$$(a_{i1}d_{1j} + a_{i2}d_{2j} + \cdots + a_{in}d_{nj}) + (b_{i1}d_{1j} + b_{i2}d_{2j} + \cdots + b_{in}d_{nj}). \qquad \text{(B.2)}$$

Clearly, expressions (B.1) and (B.2) are equal, and hence the i, j entries of $(A + B)D$ and $AD + BD$ are equal, completing the proof.

(i) Both e_iA and row i of A are $1 \times n$ matrices; so it suffices to show that their corresponding entries are equal. The i, j entry of e_iA equals the sum of the products of each entry of e_i and the corresponding entry from column j of A. But because e_i has only one nonzero entry (the ith), this sum equals $1 \cdot a_{ij} = a_{ij}$. Hence e_iA equals row i of A. ■

EXERCISES B.1

In Exercises 1–8, use the matrices

$$A = \begin{bmatrix} 1 & 2 & 3 \\ 0 & -1 & 2 \end{bmatrix}, \quad B = \begin{bmatrix} -2 & 0 \\ 1 & 3 \\ 4 & 1 \end{bmatrix}, \quad \text{and} \quad C = \begin{bmatrix} -1 & 0 & 2 \\ 3 & 1 & 4 \end{bmatrix}$$

to compute the given matrix if it is defined.

1. $A + B$ **2.** $B + A$ **3.** $C + A$ **4.** $A + C$

5. AB **6.** BA **7.** AC **8.** CA

In Exercises 9–16, use the matrices

$$A = \begin{bmatrix} 1 & 0 \\ -1 & 1 \end{bmatrix} \quad \text{and} \quad B = \begin{bmatrix} 3 & 2 \\ 1 & 5 \end{bmatrix}$$

to compute the given matrix if it is defined.

9. AB **10.** BA **11.** A^2 **12.** B^2

13. A^2B^2 **14.** $(AB)^2$ **15.** A^3 **16.** B^3

In Exercises 17–24, use the matrices

$$A = \begin{bmatrix} 1 & 1 & 1 \\ 1 & 1 & 1 \\ 1 & 1 & 1 \end{bmatrix} \quad \text{and} \quad B = \begin{bmatrix} 1 & 0 & 0 \\ 0 & -1 & 0 \\ 0 & 0 & 1 \end{bmatrix}$$

to compute the given matrix if it is defined.

17. AB **18.** BA **19.** A^2 **20.** B^2

21. A^2B^2 **22.** $(AB)^2$ **23.** A^3 **24.** B^3

25. Let

$$A = \begin{bmatrix} 1 & -1 \\ -1 & 1 \end{bmatrix} \quad \text{and} \quad B = \begin{bmatrix} 2 & 1 \\ 2 & 1 \end{bmatrix}.$$

(a) Compute AB and BA.

(b) This example illustrates two differences between the multiplication of matrices and the multiplication of real numbers. What are they?

26. Prove Theorem B.1(a). **27.** Prove Theorem B.1(b). **28.** Prove Theorem B.1(c).

29. Prove Theorem B.1(d) by establishing each of the following results:

(a) If A is an $m \times n$ matrix and O is the $n \times p$ zero matrix, then AO is the $m \times p$ zero matrix.

(b) If O is the $m \times n$ zero matrix and B is any $n \times p$ matrix, then OB is the $m \times p$ zero matrix.

30. Prove Theorem B.1(e). **31.** Prove Theorem B.1(h).

32. Prove that $I_n{}^m = I_n$ for any positive integers m and n.

HISTORICAL NOTES

The history of matrices traces back to Chinese mathematics in the period around 250 B.C. During that time, an unknown scribe wrote the *Chui-chang suan-shu*, which translates as *Nine Chapters on the Mathematical Art.* Like the Rhind papyrus from Egypt, the manuscript is a collection of worked problems, probably intended as a textbook for mathematics students. In considering the solution to the system of equations that would be written today as

$$\begin{aligned} 3x + 2y + z &= 39 \\ 2x + 3y + z &= 34 \\ x + 2y + 3z &= 26, \end{aligned}$$

the manuscript contains the following boxed array:

1	2	3
2	3	2
3	1	1
26	34	39

The solution of the system was then obtained by a series of operations on the columns of this rectangular array.

Girolamo Cardano

The use of such arrays to represent mathematical problems languished for some time after this. The Italian mathematician Girolamo Cardano (1501–1576) brought the methods back to Europe in 1545 in his *Ars Magna*. The Dutch mathematician Jan deWitt (1629–1672) used arrays in his *Elements of Curves* to represent transformations, but did not take the usage beyond that of representation. Gottfried Wilhelm Leibniz (1646–1716) was perhaps most responsible for turning the attention of European mathematicians to the use of arrays for recording information in problems and their solutions. During the period from 1700 to 1710, Leibniz's notes show that he experimented with over 50 array systems.

It was not until the middle of the 1800s that the topic of matrices moved beyond the writing of numbers in rectangular arrays to serve in the solution of equations. In 1848, James Joseph Sylvester (1814–1897) showed how arrays might be used to attack such problems more efficiently. In doing so, he called such an array of numbers a "matrix."

In 1858, the English mathematician Arthur Cayley (1821–1895) wrote a treatise on geometric transformations. In it, Cayley looked for a way to represent the transformation

$$T = \begin{cases} x' = ax + by \\ y' = cx + dy. \end{cases}$$

To do so, Cayley used a rectangular array reminiscent of that used by the Chinese, but without the rotation of the array. Cayley wrote his array between two sets of vertical bars:

$$\left\| \begin{matrix} a & b \\ c & d \end{matrix} \right\|.$$

In working with the arrays of coefficients, Cayley recognized that operations could be defined on these arrays, irrespective of the equations or transformations from which they were derived. He defined operations of addition and multiplication for these arrays and noted that the resulting mathematical system satisfied several of the same properties that characterized number systems, such as the associative properties and the distributive property of multiplication over addition. He also noted, however, that while addition was commutative, multiplication was not. Furthermore, he noted that the product of two matrices might be zero, although neither factor was the zero matrix. His 1858 paper, *Memoir on the Theory of*

Matrices, provided a framework for the later development of matrix theory. In it, he stated the famous Cayley-Hamilton theorem and illustrated its proof with a computational example.

The current bracket notation for matrices was first used by the English mathematician Cullis in 1913. His work also was the first to make significant use of the a_{ij} notation to represent the matrix entry in the ith row and jth column [73, 74, 75].

C Appendix: The Algorithms in this Book

*T*he algorithms in this book are written in a form that, while not corresponding to any particular computer language, is sufficiently structured so that they can be easily turned into programs. They are divided into steps, which are executed in order, subject to certain looping or branching instructions. Loops are begun by one of three special words: **while**, **repeat**, and **for**, printed in boldface type.

while . . . endwhile Loops

This construction has the following form:

> **while** *statement*
>   *some instructions*
> **endwhile**

Here, *statement* is checked, and if it is true, *some instructions* are executed. This is repeated until *statement* is false, at which point the algorithm resumes after the **endwhile**. An example follows.

Algorithm 1

Given a positive integer n, this algorithm computes the sum of the first n positive integers.

Step 1 Set $S = 0$ and $k = 1$.
Step 2 **while** $k \leq n$
    Replace S with $S + k$ and k with $k + 1$.
endwhile
Step 3 Print S.

The following table shows how the values of S and k change as the algorithm is applied to $n = 4$ (making such a table is often helpful in understanding a new algorithm):

S	k
0	1
$0+1=1$	$1+1=2$
$1+2=3$	$2+1=3$
$3+3=6$	$3+1=4$
$6+4=10$	$4+1=5$

Here, step 2 would be repeated until $k = 5$, making the statement $k \leq n$ false. Then step 3 would be executed, printing the value $S = 10$.

Note that the instructions between **while** and **endwhile** may not be executed even once. For example, if Algorithm 1 is applied with $n = 0$, nothing is ever done in step 2 and the value $S = 0$ is printed.

repeat ... until Loops

Another looping structure has the following form:

repeat
  some instructions
until *statement*

Here, *some instructions* are executed, then *statement* is checked, and, if found false, *some instructions* are executed again, etc. Only when *statement* is found true does the algorithm take up after the line containing **until**.

The following algorithm has the same effect as Algorithm 1 if n is any positive integer:

Algorithm 2

Given a positive integer n, this algorithm computes the sum of the first n positive integers.

Step 1 Set $S = 0$ and $k = 1$.
Step 2 **repeat**
  Replace S with $S + k$ and k with $k + 1$.
until $k > n$
Step 3 Print S.

Unlike a **while** . . . **endwhile** loop, the instructions between **repeat** and **until** are always executed at least once. Thus, if algorithm 2 were applied to $n = 0$, it would print the value $S = 1$.

for . . . endfor Loops

When algorithm 1 is run with $n = 4$, the instructions inside step 2 are executed for $k = 1, 2, 3$, and 4. Although writing the algorithm this way is useful to one who wants to count every elementary operation, most computer languages have a command like "for $k = 1$ to 4." In the algorithms in this book, such loops begin with **for** and end with **endfor**. We could rewrite algorithm 1, using this language, as follows:

Algorithm 3

Given a positive integer n, this algorithm computes the sum of the first n positive integers.

Step 1 Set $S = 0$.
Step 2 **for** $k = 1$ to n
        Replace S with $S + k$.
    endfor
Step 3 Print S.

In a **for** . . . **endfor** loop, the variable may be incremented by an amount d other than 1 by adding the words *by* d. For example, the following algorithm computes $1 + 2 + \cdots + n$ by adding the larger numbers first:

Algorithm 4

Given a positive integer n, this algorithm computes the sum of the first n positive integers.

Step 1 Set $S = 0$.
Step 2 **for** $k = n$ to 1 by -1
        Replace S with $S + k$.
    endfor
Step 3 Print S.

Branching

Branching in algorithms is accomplished by the **if** ... **otherwise** ... **endif** construction. This has the following form:

if *statement*
    some instructions
otherwise
    other instructions
endif

Here, *statement* is checked, and if it is found true, *some instructions* are executed. If *statement* is false, then *other instructions* are executed instead. In either case, the algorithm resumes after **endif**. If there are no *other instructions*, then the branching construction can be shortened to the following form:

if *statement*
    some instructions
endif

Algorithm 5 illustrates the **if** ... **otherwise** ... **endif** construction.

Algorithm 5

Given a real number x, this algorithm computes its absolute value.

Step 1 **if** $x \geq 0$
    Set $A = x$.
otherwise
    Set $A = -x$.
endif
Step 2 Print A.

In more complicated algorithms, looping and branching constructions are often nested. The example that follows has an **if** ... **otherwise** ... **endif** nested within a **while** ... **endwhile**. It concerns the *Collatz sequence*, in which a positive integer n is replaced by $n/2$ or $3n + 1$, according to whether it is even or odd, and this is repeated. It has been conjectured, but never proved, that eventually the number 1 is reached, no matter what positive integer n we start with. The following algorithm counts how many steps this takes for a given n:

Algorithm 6

This algorithm counts how many steps the Collatz sequence takes to reach 1 from a given positive integer n.

Step 1 Set $k = 0$ and $s = n$.
Step 2 **while** $s > 1$
  Step 2.1 Replace k with $k + 1$.
  Step 2.2 **if** s is even
    Replace s with $s/2$.
   otherwise
    Replace s with $3s + 1$.
   endif
 endwhile
Step 3 Print k.

The following table shows how k and s change when the algorithm is applied to $n = 3$:

k	s
0	3
1	$3 \cdot 3 + 1 = 10$
2	$10/2 = 5$
3	$3 \cdot 5 + 1 = 16$
4	$16/2 = 8$
5	$8/2 = 4$
6	$4/2 = 2$
7	$2/2 = 1$

Thus, when step 3 is executed, the value $k = 7$ is printed.

Recursive Algorithms

Section 5.5 presents several algorithms that are *recursive*, in the sense that they call themselves. The following is a simple example of such an algorithm:

Algorithm 7

Given a positive integer n, this algorithm prints out a sequence of integers $k_1, k_2, \ldots, k_t$ such that $n = k_1^2 + k_2^2 + \cdots + k_t^2$.

Step 1 Set $s = n$.
Step 2 **while** $s > 0$
  Step 2.1 Set $k = 1$.
  Step 2.2 **while** $(k+1)^2 \leq s$
    Replace k with $k + 1$.
  endwhile
  Step 2.3 Print k.
  Step 2.4 Apply Algorithm 7 with $n = s - k^2$.
endwhile

This algorithm finds the largest integer k such that $k^2 \leq n$, prints this integer, and then applies the same algorithm to $n - k^2$. The following table shows how s and k change, and what is printed out, for $n = 22$:

s	k	*printed*
22	1	
	2	
	3	
	4	4
$22 - 4^2 = 6$	1	
	2	2
$6 - 2^2 = 2$	1	1
$2 - 1^2 = 1$	1	1

Note that

$$22 = 4^2 + 2^2 + 1^2 + 1^2.$$

Bibliography

[1] Aho, Alfred, John Hopcroft, and Jeffrey Ullman. *Data Structures and Algorithms*. Reading, MA: Addison-Wesley, 1983.

[2] ______. *The Design and Analysis of Computer Algorithms*. Reading, MA: Addison-Wesley, 1974.

[3] Albertson, M.O. and J.P. Hutchinson. *Discrete Mathematics with Algorithms*. New York: Wiley, 1991.

[4] Althoen, Steven C. and Robert J. Bumcrot. *Introduction to Discrete Mathematics*. Boston: PWS-KENT, 1988.

[5] Anderson, Ian. *A First Course in Combinatorial Mathematics*. London: Oxford University Press, 1974.

[6] Barnier, William and Jean Chan. *Discrete Mathematics with Applications*. St. Paul, MN: West, 1989.

[7] Behzad, Mehdi, Gary Chartrand, and Linda Lesniak. *Graphs and Digraphs*. 3d ed. Boca Raton, FL: CRC Press, 1996.

[8] Benedicty, Mario and Frank R. Sledge. *Discrete Mathematical Structures*. San Diego: Harcourt Brace Jovanovich, 1987.

[9] Biggs, Norman. *Discrete Mathematics*. 3d ed. New York: Oxford University Press, 1993.

[10] Bogart, Kenneth. *Introductory Combinatorics*. 3d ed. San Diego: Academic Press, 1999.

[11] ______. *Discrete Mathematics*. Boston: Houghton Mifflin, 1988.

[12] Bondy, J.A. and U.S.R. Murty. *Graph Theory with Applications*. New York: Elsevier Science, 1979.

[13] Bradley, James. *Discrete Mathematics*. Reading, MA: Addison-Wesley, 1988.

[14] Brualdi, Richard A. *Introductory Combinatorics*. 3d ed. Upper Saddle River, NJ: Prentice Hall, 1999.

[15] Busacker, Robert, G. and Thomas L. Saaty. *Finite Graphs and Networks: An Introduction with Applications*. New York: McGraw-Hill, 1965.

[16] Chartrand, Gary. *Graphs as Mathematical Models*. Boston: Prindle, Weber & Schmidt, 1977.

[17] Cohen, Daniel I.A. *Basic Techniques of Combinatorial Theory*. New York: Wiley, 1978.

[18] Dierker, Paul F. and William L. Voxman. *Discrete Mathematics*. San Diego: Harcourt Brace Jovanovich, 1986.

[19] Doerr, Alan and Kenneth Levaseur. *Applied Discrete Structures for Computer Science*. 2d ed. New York: Macmillan, 1989.

[20] Epp, Susanna. *Discrete Mathematics with Applications*. Boston: PWS Publishers, 1990.

[21] Even, Shimon. *Graph Algorithms*. New York: Freeman, 1984.

[22] Finkbeiner, Daniel T., II and Wendell Lindstrom. *A Primer of Discrete Mathematics*. New York: Freeman, 1987.

[23] Fletcher, Peter, Hughes B. Hoyle III, and C. Wayne Patty. *Foundations of Discrete Mathematics*. Boston: PWS-KENT, 1991.

[24] Gerstein, Larry. *Discrete Mathematics and Algebraic Structures*. New York: Freeman, 1987.

[25] Gersting, Judith. *Mathematical Structures for Computer Science*. 2d ed. New York: Freeman, 1987.

[26] Grimaldi, Ralph. *Discrete and Combinatorial Mathematics*. 4th ed. Reading, MA: Addison-Wesley, 1998.

[27] Grossman, Jerrold W. *Discrete Mathematics*. New York: MacMillan, 1990.

[28] Harary, Frank. *Graph Theory*. Reading, MA: Addison-Wesley, 1994.

[29] Hillman, A., G.L. Alexanderson, and R.M. Grassl. *Discrete and Combinatorial Mathematics*. San Francisco: Dellen, 1987.

[30] Hirschfelder, R. and J. Hirschfelder. *Introduction to Discrete Mathematics*. Pacific Grove, CA: Brooks/Cole, 1991.

[31] Hu, T.C. *Combinatorial Algorithms*. Reading, MA: Addison-Wesley, 1982.

[32] Johnsonbaugh, Richard. *Discrete Mathematics*. 5th ed. Upper Saddle River, NJ: Prentice Hall, 2001.

[33] Kalmanson, Kenneth. *An Introduction to Discrete Mathematics*. Reading, MA: Addison-Wesley, 1986.

[34] Kincaid, David and E. Ward Cheney. *Introduction to Discrete Mathematics*. Pacific Grove, CA: Brooks/Cole, 1991.

[35] Knuth, Donald. *The Art of Computer Programming, Vol 1*. 3d ed. Reading, MA: Addison-Wesley, 1997.

[36] ———. *The Art of Computer Programming, Vol 2*. 3d ed. Reading, MA: Addison-Wesley, 1997.

[37] ______. *The Art of Computer Programming, Vol 3*. 2d ed. Reading, MA: Addison-Wesley, 1998.

[38] Kolman, Bernard and Robert Busby. *Discrete Mathematical Structures for Computer Science*. 2d ed. Upper Saddle River, NJ: Prentice Hall, 1987.

[39] ______. *Introductory Discrete Structures with Applications*. Englewood Cliffs, NJ: Prentice-Hall, 1987.

[40] Lawler, Eugene L. *Combinatorial Optimization: Networks and Matroids*. Mineola, NY: Dover, 1976.

[41] Liu, C.L. *Elements of Discrete Mathematics*. 2d ed. New York: McGraw-Hill, 1985.

[42] ______. *Introduction to Combinatorial Mathematics*. New York: McGraw-Hill, 1968.

[43] Maurer, Stephen B. and Anthony Ralston. *Discrete Algorithmic Mathematics*. 2d ed. Nantick, MA: A. K. Peters, 1998.

[44] McEliece, Robert, Robert Ash, and Carol Ash. *Introduction to Discrete Mathematics*. New York: McGraw-Hill, 1989.

[45] Minieka, Edward. *Optimization Algorithms for Networks and Graphs*. 2d ed. New York: Marcel Dekker, 1992.

[46] Molluzzo, John C. and Fred Buckley. *A First Course in Discrete Mathematics*. Prospect Heights, IL: Waveland Press, 1997.

[47] Mott, Joe, Abraham Kandel, and Theodore Baker. *Discrete Mathematics for Computer Scientists*. 2d ed. Englewood Cliffs, NJ: Prentice-Hall, 1985.

[48] Nicodemi, Olympia. *Discrete Mathematics*. St. Paul, MN: West, 1987.

[49] Niven, Ivan. *Mathematics of Choice*. Washington, DC: Mathematical Association of America, 1965.

[50] Norris, Fletcher. *Discrete Structures: An Introduction to Mathematics for Computer Science*. Englewood Cliffs, NJ: Prentice-Hall, 1985.

[51] Pfleeger, Shari Lawrence and David W. Straight. *Introduction to Discrete Structures*. Melbourne, FL: Kreiger, 1985.

[52] Polimeni, Albert D. and Joseph Straight. *Foundations of Discrete Mathematics*. 2d ed. Pacific Grove, CA: Brooks/Cole, 1985.

[53] Prather, Ronald E. *Elements of Discrete Mathematics*. Boston: Houghton Mifflin, 1986.

[54] Reingold, Edward M., Jurg Nievergelt, and Narsingh Deo. *Combinatorial Algorithms: Theory and Practice*. Englewood Cliffs, NJ: Prentice-Hall, 1977.

[55] Roberts, Fred S. *Discrete Mathematical Models*. Englewood Cliffs, NJ: Prentice-Hall, 1976.

[56] ———. *Applied Combinatorics*. Englewood Cliffs, NJ: Prentice-Hall, 1984.

[57] ———. *Graph Theory and Its Applications to Problems of Society*. Philadelphia: Society for Industrial and Applied Mathematics, 1978.

[58] Roman, Steven. *An Introduction to Discrete Mathematics*. 2d ed. Philadelphia: Saunders, 1986.

[59] Rosen, Kenneth H. *Discrete Mathematics and Its Applications*. 4th ed. New York: McGraw-Hill, 1998.

[60] Ross, Kenneth and Charles Wright. *Discrete Mathematics*. 2d ed. Englewood Cliffs, NJ: Prentice Hall, 1988.

[61] Ryser, Herbert John. *Combinatorial Mathematics*. Washington, DC: Mathematical Association of America, 1963.

[62] Shiflet, Angela. *Discrete Mathematics for Computer Science*. St. Paul, MN: West, 1987.

[63] Skvarcius, R. and W.B. Robinson. *Discrete Mathematics with Computer Science Applications*. Reading, MA: Addison-Wesley, 1986.

[64] Stanat, Donald and David McAllister. *Discrete Mathematics in Computer Science*. Englewood Cliffs, NJ: Prentice-Hall, 1977.

[65] Stanton, Dennis and Dennis White. *Constructive Combinatorics*. New York: Springer-Verlag, 1986.

[66] Stone, Harold. *Discrete Mathematical Structures and Their Applications*. Chicago: Science Research Associates, 1973.

[67] Townsend, Michael. *Discrete Mathematics: Applied Combinatorics and Graph Theory*. Reading, MA: Addison-Wesley, 1987.

[68] Tremblay, J.P. and R. Manohar. *Discrete Mathematical Structures with Applications to Computer Science*. New York: McGraw-Hill, 1975.

[69] Tucker, Alan. *Applied Combinatorics*. 3d ed. New York: Wiley, 1994.

[70] Wiitala, Stephen. *Discrete Mathematics: A Unified Approach*. New York: McGraw-Hill, 1987.

REFERENCES FOR THE HISTORICAL NOTES

[71] Biggs, N.L. "The Roots of Combinatorics." *Historica Mathematica*, vol. 6 (1979): 109–136.

[72] Biggs, N.L., E.K. Lloyd, and R.J. Wilson. *Graph Theory: 1736–1936*. Oxford, UK: Clarendon Press, 1986.

[73] Boyer, C.B., and U.C. Merzbach. *A History of Mathematics*. 2d ed. New York: John Wiley & Sons, 1989.

[74] Burton, D.M. *The History of Mathematics: An Introduction*. Dubuque, IA: William Brown, 1991.

[75] Cajori, F. *A History of Mathematical Notations* (2 vols). La Salle, IL: Open Court Publishing, 1928–29.

[76] Gardner, Martin. *Logic Machines and Diagrams*. New York: 1958.

[77] Goldstine, H. *The Computer from Pascal to von Neumann*. Princeton, NJ: Princeton University Press, 1972.

[78] Heath, T. *A History of Greek Mathematics* (vols 1–2). New York: Dover Publications, 1981.

[79] Heath, T. *Euclid's Elements* (vols 1–3). New York: Dover Publications, 1956.

[80] Houser, N. "Algebraic Logic from Boole to Schröder, 1840–1900." In I. Grattan-Guinness (ed.), *Companion Encyclopedia of the History and Philosophy of the Mathematical Sciences* (vol 2, pp. 600–616). London: Routledge, 1993.

[81] Kleiner, I. "Evolution of the Function Concept: A Brief Survey" *College Mathematics Journal*, vol. 20, no. 4 (1989): 282–300.

[82] Kline, M. *Mathematical Thought from Ancient to Modern Times*. New York: Oxford University Press, 1972.

[83] Perl, T. *Math Equals: Biographies of Women Mathematicians and Related Activities*. Menlo Park, CA: Addison Wesley, 1978.

[84] Rodríquez-Consuegra, F. A. "Mathematical Logic and Logicism from Peano to Quine, 1890–1940." In Ivor Grattan-Guinness (ed.), *Companion Encyclopedia of the History and Philosophy of the Mathematical Sciences* (vol 2, pp. 617–628). London: Routledge, 1993.

[85] Schreiber, P. "Algorithms and Algorithmic Thinking Through the Ages." In Ivor Grattan-Guinness (ed.), *Companion Encyclopedia of the History and Philosophy of the Mathematical Sciences* (vol 2, pp. 687–693). London: Routledge, 1993.

[86] Todhunter, I. *A History of the Mathematical Theory of Probability*. Cambridge, UK: Cambridge University Press, 1865.

[87] Wilson, R.J., and E.K. Lloyd. "Combinatorics." In Ivor Grattan-Guinness (ed.), *Companion Encyclopedia of the History and Philosophy of the Mathematical Sciences*. (vol 2, pp. 952–965). London: Routledge, 1993.

Answers to Odd-Numbered Exercises

CHAPTER 1

EXERCISES 1.1 *(page 8)*

1. 33; A-B-D-F-G or A-C-E-F-G **3.** 43; B-D-E-G **5.** 20.7; A-D-H-K **7.** 2.1; A-C-E-H-J

9. 23; A-D-F-G

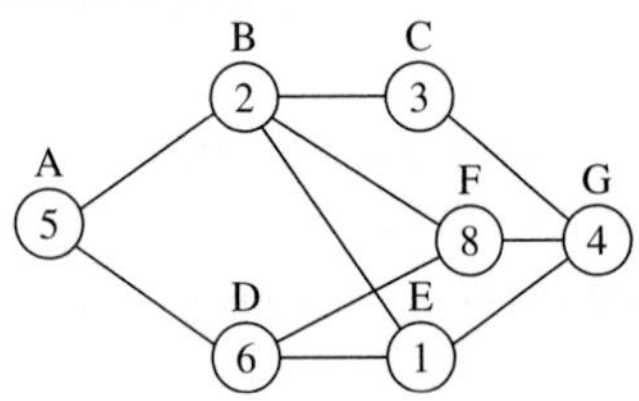

11. 24; B-C-F-G

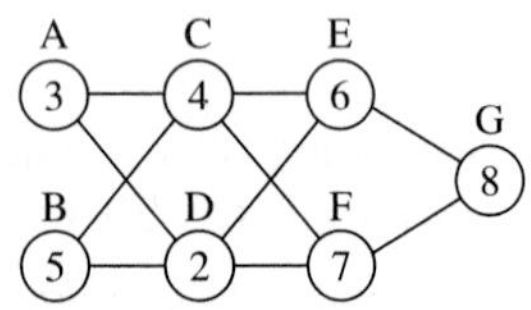

13. 15.7; D-I

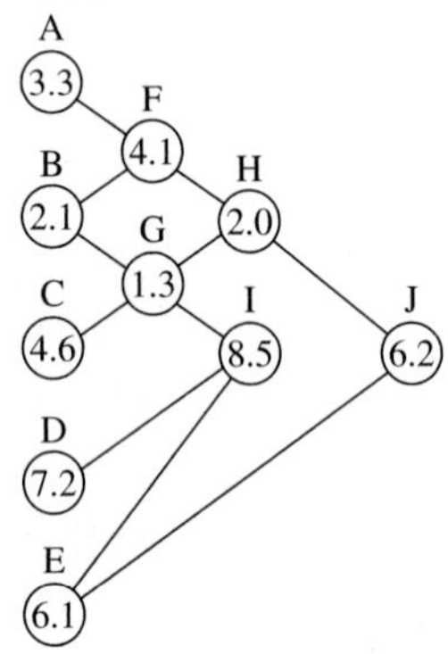

15. 0.29; E-F-C-D-I

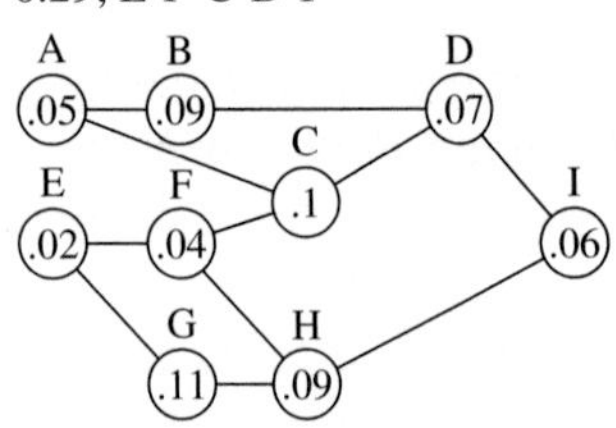

17. 27 minutes **19.** 15 days

EXERCISES 1.2 *(page 15)*

1. 120 **3.** 6720 **5.** 28 **7.** 840 **9.** 604,800 **11.** 25.2 **13.** 720 **15.** 56
17. 362,880 **19.** 720 **21.** 288 **23.** 210 **25.** 60 **27.** 20,118,067,200 **29.** 60

EXERCISES 1.3 *(page 22)*

1. F **3.** F **5.** F **7.** T **9.** F **11.** F **13.** F **15.** yes, 28 **17.** yes, 35
19. 1, 4, 5, 6, 7, 8, 9, 11, 12; 51 **21.** 128 **23.** 32 **25.** $n - m + 1$ **27.** 31 **29.** 12.7 days

EXERCISES 1.4 *(page 33)*

1. yes; 2 **3.** no **5.** no **7.** 3, 13; 5, 13 **9.** −7, 3, 11, 3; −1, 0, 5, 3 **11.** 110110
13. 001110

15.

k	j	a_1	a_2	a_3
3		1	0	1
2		1	1	1
	3	1	1	0

17.

k	j	a_1	a_2	a_3	a_4
4		1	1	0	1
3		1	1	1	1
	4	1	1	1	0

19.

a_1	a_2	a_3	a_4	j	k
13	56	87	42	1	3
13	56	42	87		2
13	42	56	87		1
13	42	56	87	2	3
13	42	56	87		2
13	42	56	87	3	3
13	42	56	87		

21.

a_1	a_2	a_3	a_4	a_5	j	k
6	33	20	200	9	1	4
6	33	20	9	200		3
6	33	9	20	200		2
6	9	33	20	200		1
6	9	33	20	200	2	4
6	9	33	20	200		3
6	9	20	33	200		2
6	9	20	33	200	3	4
6	9	20	33	200		3
6	9	20	33	200	4	4
6	9	20	33	200		

23. 58 minutes; 0.8 seconds **25.** 385,517 years; 6.4 seconds **27.** $3n+1$
29. $4n-2$ **31.** −7, 3, 11, 3

SUPPLEMENTARY EXERCISES *(page 37)*

1. 18; B-D-G-I **3.** 28 minutes **5.** 332,640 **7.** 990 **9.** F **11.** F **13.** T **15.** T
17. 32 **19.** 80 **21.** yes; 100 **23.** no **25.** 3, 9, 31, 88

27.

a_1	a_2	a_3	a_4	a_5	j	k
44	5	13	11	35	1	4
44	5	13	11	35		3
44	5	11	13	35		2
44	5	11	13	35		1
5	44	11	13	35	2	4
5	44	11	13	35		3
5	44	11	13	35		2
5	11	44	13	35	3	4
5	11	44	13	35		3
5	11	13	44	35	4	4
5	11	13	35	44		

29. 39 **31.** $4n-3$

CHAPTER 2

EXERCISES 2.1 *(page 46)*

1. $\{1, 2, 3, 4, 5, 6, 7, 8, 9\}$; $\{3, 5\}$; $\{2, 7, 8\}$; $\{1, 4, 6, 9\}$; $\{2, 7, 8\}$
3. $\{1, 2, 3, 4, 7, 8, 9\}$; $\emptyset$; $\{1, 2, 4, 8, 9\}$; $\{3, 5, 6, 7\}$; $\{1, 2, 4, 5, 6, 8, 9\}$
5. $\{(1, 7), (1, 8), (2, 7), (2, 8), (3, 7), (3, 8), (4, 7), (4, 8)\}$ **7.** $\{(a, x), (a, y), (a, z), (e, x), (e, y), (e, z)\}$
9.

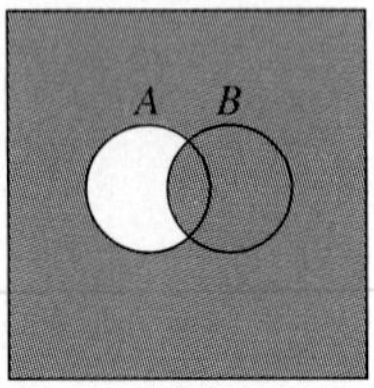

11.

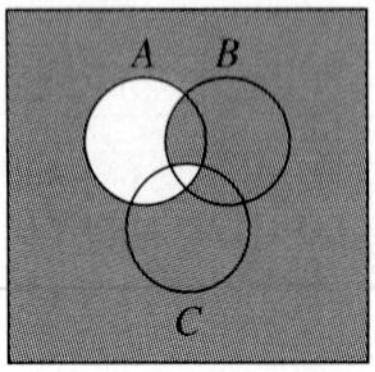

13. $A = \{1\}$, $B = \{2\}$, $C = \{1, 2\}$ **15.** $A = \{1, 2\}$, $B = \{1, 3\}$, $C = \{2, 3\}$ **17.** $\emptyset$ **19.** $A - B$
21. $B - A$ **23.** $A - B$ **25.** mn **27.** $B \subseteq A$ **39.** $A = \{1\}$, $B = \{2\}$, $C = \{3\}$, $D = \{4\}$

EXERCISES 2.2 *(page 52)*

1. symmetric and transitive **3.** reflexive, symmetric, and transitive **5.** reflexive and symmetric
7. reflexive, symmetric, and transitive **9.** reflexive, symmetric, and transitive **11.** reflexive and transitive
13. $[z]$ is the set of odd integers, 2 equivalence classes
15. $[z]$ is the set of integers greater than 1 that are divisible by 5 but not by any prime greater than 5, infinitely many equivalence classes
17. $[z]$ consists of all ordered pairs (x, y) that satisfy the equation $x^2 + y^2 = 5^2$, infinitely many equivalence classes
19. $\{(1, 1), (1, 5), (5, 1), (5, 5), (2, 2), (2, 4), (4, 2), (4, 4), (3, 3)\}$
23. There may be no element related to x; that is, xRy may not be true for any y.
25. 2^{n^2} **27.** $2^{n-1} - 1$ **29.** 15

EXERCISES 2.3 *(page 63)*

1. not reflexive **3.** a partial order **5.** a partial order **7.** not antisymmetric
9.

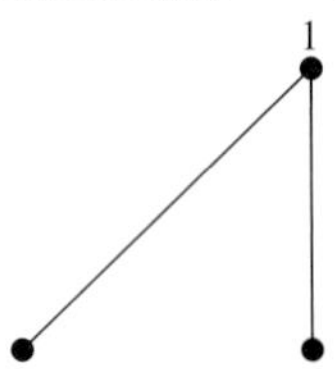

11.

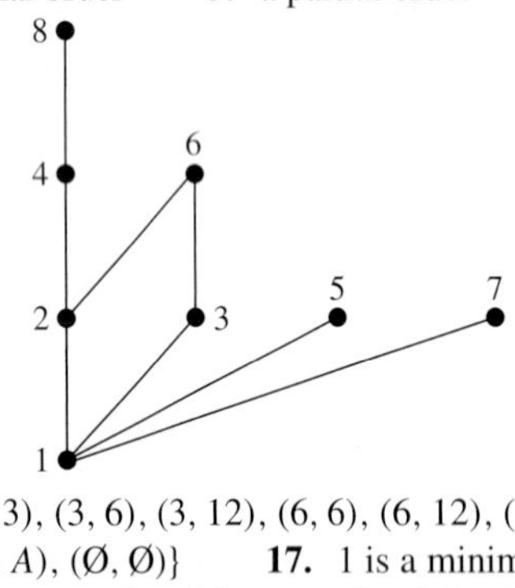

13. $\{(2, 2), (2, 6), (2, 12), (3, 3), (3, 6), (3, 12), (6, 6), (6, 12), (12, 12)\}$
15. $\{(2, 2), (x, x), (x, A), (A, A), (\emptyset, \emptyset)\}$ **17.** 1 is a minimal element; 4, 5, and 6 are maximal elements.
19. 2, 3, and 4 are minimal elements; 1 and 2 are maximal elements. **21.** $R \cup \{(1, 2), (1, 4)\}$
23. For $A_1 = \emptyset$, $A_2 = \{1\}$, $A_3 = \{2\}$, $A_4 = \{3\}$, $A_5 = \{1, 2\}$, and $A_6 = \{1, 3\}$, define $A_i T A_j$ if and only if $i \leq j$.
25. no **27.** $\{1, 2, 4, 8, 16\}$ **29.** $S = \{2, 3, 4, 5, 6, 9, 15\}$ with xRy if and only if x divides y.

31. 

37. The lexicographic order is a total order on $S_1 \times S_2$.

41. $n!$

EXERCISES 2.4 *(page 74)*

1. function with domain X **3.** not a function with domain X **5.** function with domain X **7.** not a function with domain X **9.** not a function with domain X **11.** function with domain X **13.** 8 **15.** $\frac{1}{4}$ **17.** 2 **19.** -9 **21.** 3 **23.** 0 **25.** -4 **27.** -5 **29.** 5.21 **31.** -0.22 **33.** 0.62 **35.** 9.97 **37.** $8x + 11; 8x - 5$ **39.** $5(2^x) + 7; 2^{5x+7}$ **41.** $|x|(\log_2 |x|); |x \log_2 x|$ **43.** $x^2 - 2x + 1; x^2 - 1$ **45.** one-to-one; not onto **47.** one-to-one; onto **49.** onto; not one-to-one **51.** neither one-to-one nor onto **53.** $f^{-1}(x) = \dfrac{x}{5}$ **55.** $f^{-1}(x) = -x$ **57.** $f^{-1}(x) = x^3$ **59.** does not exist **61.** $Y = \{x \in X : x > 0\}; g^{-1}(x) = -1 + \log_2\left(\dfrac{x}{3}\right)$ **63.** n^m

EXERCISES 2.5 *(page 83)*

1. 1, 1, 2, 3, 5, 8, 13, 21, 34, 55 **3.** 3, 4, 7, 11, 18, 29, 47, 76

5. Let x_n denote the nth even positive integer. Then $x_n = \begin{cases} 2 & \text{if } n = 1 \\ x_{n-1} + 2 & \text{if } n \geq 2. \end{cases}$

7. No base for the induction was established.

9. The proof of the inductive step is faulty because $x - 1$ and $y - 1$ need not be positive integers.

27. $\begin{cases} \dfrac{s_0(r^{n+1} - 1)}{r - 1} & \text{if } r \neq 1 \\ s_0(n + 1) & \text{if } r = 1 \end{cases}$

EXERCISES 2.6 *(page 91)*

1. 21 **3.** 252 **5.** 330 **7.** 462 **9.** 1 **11.** $\dfrac{n(n-1)}{2}$ **13.** 64 **15.** 128 **17.** 256 **19.** 21 **21.** 792 **23.** 20 **25.** 120 **27.** $\dfrac{52!}{13!\,39!}$ **43.** $\dfrac{n(n-3)}{2}$

47. Mr. and Mrs. Lewis each shook n hands.

SUPPLEMENTARY EXERCISES *(page 95)*

1. {3} **3.** {5, 6} **5.** {6} **7.** {2} **9.**

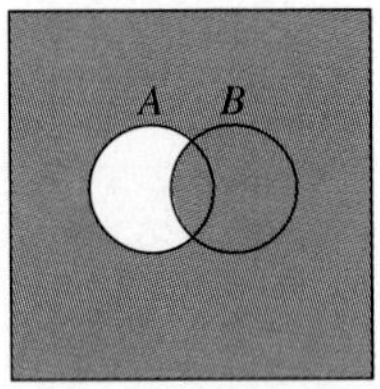

11.

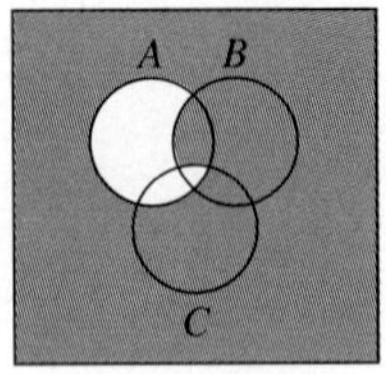

13. $f(g(x)) = 5 - 8x^2$, $g(f(x)) = -16x^2 - 48x - 35$ **15.** function with domain X
17. function with domain X **19.** one-to-one, not onto **21.** onto, not one-to-one **23.** does not exist
25. $f^{-1}(x) = \frac{1}{3}(x+6)$ **27.** 32 **29.** 5005 **31.** {1, 5}, {2, 6}, {3, 7}, {4, 8}
33. the sets $\{2n - 1, 2n\}$ for every integer n **35.** 512 **37.** 27 **39.** $R = \{(s, s): s \in S\}$
41. antisymmetric and transitive **43.** reflexive, antisymmetric, and transitive
45. A, B, C, D, E, F, G, H, I, J, K **47.** $R = \{(s, s): s \in S\}$ **49.** f must be one-to-one

CHAPTER 3

EXERCISES 3.1 *(page 105)*

1. $q = 7, r = 4$ **3.** $q = 0, r = 25$ **5.** $q = -9, r = 0$ **7.** $q = -9, r = 1$ **9.** $p \equiv q \pmod{m}$
11. $p \not\equiv q \pmod{m}$ **13.** $p \not\equiv q \pmod{m}$ **15.** $p \equiv q \pmod{m}$ **17.** [2] **19.** [4] **21.** [6]
23. [1] **25.** [2] **27.** [8] **29.** [4] **31.** [4] **33.** [2] **35.** [11] **37.** 8 P.M. **39.** 9
41. 7 **43.** **(a)** Wednesday **(b)** Saturday **45.** No, $10 \in A$, but $10 \notin B$.
47. Although 3 R 11 and 6 R 10 are true, both 9 R 21 and 18 R 110 are false.
49. The relation is not well defined if m does not divide n.

EXERCISES 3.2 *(page 111)*

1. −45, −15, −9, −5, −3, −1, 1, 3, 5, 9, 15, 45 **3.** −10, −5, −2, −1, 1, 2, 5, 10

5.

i	r_i
−1	715
0	312
1	91
2	39
3	13
4	0
[Print 13]	

7.

i	r_i
−1	247
0	117
1	13
2	0
[Print 13]	

9.

i	r_i
−1	76
0	123
1	76
2	47
3	29
4	18
5	11
6	7
7	4
8	3
9	1
10	0
[Print 1]	

11. 39

13.

i	q_i	r_i	x_i	y_i
−1		1479	1	0
0		272	0	1
1	5	119	1	−5
2	2	34	−2	11
3	3	17	7	−38
4	2	0	−16	87
[Print 17 ,7, −38]				

15.

i	q_i	r_i	x_i	y_i
−1		4050	1	0
0		1728	0	1
1	2	594	1	−2
2	2	540	−2	5
3	1	54	3	−7
4	10	0	−32	75
[Print 54, 3, −7]				

17.

i	q_i	r_i	x_i	y_i
−1		546	1	0
0		2022	0	1
1	0	546	1	0
2	3	384	−3	1
3	1	162	4	−1
4	2	60	−11	3
5	2	42	26	−7
6	1	18	−37	10
7	2	6	100	−27
8	3	0	−337	91
[Print 6, 100, −27]				

19. **(a)** no **(b)** yes **21.** **(a)** no **(b)** yes **23.** $x = 265, y = -1272$ **25.** $x = 267, y = 712$

EXERCISES 3.3 *(page 121)*

1. 32, 10, 16, 8 **3.** 35, 36, 14

5.

Q	R	r_1	r_2	p	e
			1	19	41
20	1	88	19	88	20
10	0	9		9	10
5	0	81		81	5
2	1	9	83	9	2
1	0	81		81	1
0	1	9	80	9	0
[Print 80]					

7.

Q	R	r_1	r_2	p	e
			1	11	73
36	1	121	11	121	36
18	0	55		55	18
9	0	33		33	9
4	1	154	176	154	4
2	0	154		154	2
1	0	154		154	1
0	1	154	176	154	0
[Print 176]					

9.

Q	R	r_1	r_2	p	e
			1	90	101
50	1	966	90	966	50
25	0	980		980	25
12	1	877	214	877	12
6	0	1035		1035	6
3	0	1125		1125	3
1	1	529	572	529	1
0	1	426	582	426	0
[Print 582]					

11. 64 **13.** 288 **15.** 5 **17.** 27 **19.** 65 **21.** 77 **23.** 8

EXERCISES 3.4 *(page 130)*

1. 1 **3.** 0 **5.** 0 **7.** 1 **9.** .0480 **11.** .0020 **13.** .9227 **15.** .0914 **17.** 2
19. 4 **21.** 2 **23.** 4 **25.** 1100 **27.** 00101 **29.** 100101 **31.** 11010110
33. **(a)** 7 **(b)** 3 **35.** **(a)** 14 **(b)** 9

EXERCISES 3.5 *(page 138)*

1. 32 **3.** 256 **5.** 10010011 **7.** 11010110 **9.** 9×6 **11.** 10×5 **13.** $\frac{2}{3}$
15. {00000, 10101, 01110, 11011} **17.** {000000, 100001, 010011, 001111, 110010, 101110, 011100, 111101}
19. {0000000, 1001001, 0100110, 0010101, 1101111, 1011100, 0110011, 1111010}

21. $\begin{bmatrix} 0 & 1 & 1 \\ 1 & 1 & 1 \\ 1 & 0 & 0 \\ 0 & 1 & 0 \\ 0 & 0 & 1 \end{bmatrix}$ **23.** $\begin{bmatrix} 1 & 0 & 1 \\ 1 & 1 & 0 \\ 0 & 1 & 1 \\ 1 & 0 & 0 \\ 0 & 1 & 0 \\ 0 & 0 & 1 \end{bmatrix}$ **25.** $\begin{bmatrix} 0 & 1 & 1 & 1 \\ 1 & 0 & 1 & 1 \\ 1 & 1 & 0 & 1 \\ 1 & 0 & 0 & 0 \\ 0 & 1 & 0 & 0 \\ 0 & 0 & 1 & 0 \\ 0 & 0 & 0 & 1 \end{bmatrix}$ **27.** $\begin{bmatrix} 1 & 0 & 1 & 0 & 1 \\ 0 & 1 & 0 & 1 & 0 \\ 0 & 1 & 1 & 1 & 0 \\ 1 & 0 & 0 & 0 & 0 \\ 0 & 1 & 0 & 0 & 0 \\ 0 & 0 & 1 & 0 & 0 \\ 0 & 0 & 0 & 1 & 0 \\ 0 & 0 & 0 & 0 & 1 \end{bmatrix}$

29. $\begin{bmatrix} 1 & 0 & 0 & 0 & 1 & 0 & 1 \\ 0 & 1 & 0 & 0 & 1 & 1 & 1 \\ 0 & 0 & 1 & 0 & 0 & 1 & 1 \\ 0 & 0 & 0 & 1 & 1 & 1 & 0 \end{bmatrix}$ **31.** no **33.** yes **35.** yes **37.** no **39.** 2^{k-n}

41. $\begin{bmatrix} 1 & 0 & 0 & 0 & 1 & 0 & 1 & 0 \\ 0 & 1 & 0 & 0 & 1 & 0 & 0 & 1 \\ 0 & 0 & 1 & 0 & 0 & 1 & 1 & 0 \\ 0 & 0 & 0 & 1 & 0 & 1 & 0 & 1 \end{bmatrix}$

EXERCISES 3.6 *(page 146)*

1. 0110, 11 **3.** 0000, 01 **5.** 0011, ?? **7.** 1001, 01 **9.** 0010, 101 **11.** 1110, ??
13. 0000, 011 **15.** 1010, 010 **17.** 0000, 001 **19.** 1011, ?? **21.** 0110, 110 **23.** 0000, 110
25. 0001, 100 **27.** 1101, 011 **29.** 11, ??, 11, 10 **31.** 12 **33.** 25 **35.** $k = 4, n = 7$
37. $k = 57, n = 63$

SUPPLEMENTARY EXERCISES *(page 149)*

1. false **3.** true **5.** [9] **7.** [1] **9.** [3] **11.** 10 **13.** 2 **15.** 14 **17.** 138
19. $55(770) - 24(1764) = 14$ **21.** $1(-9798) + 18(552) = 138$
23. **(a)** $x = -2070$, $y = 975$ **(b)** no solution **25.** 18, 4, 15, 32

27.

Q	R	r_1	r_2	p	e
			1	18	29
14	1	39	18	39	14
7	0	39	18	39	7
3	1	39	18	39	3
1	1	39	18	39	1
0	1	39	18	39	0
[Print 18]					

29. 1740 **31.** 50 **33.** Append 0.

35. approximately .000000000209 **37.** 2 **39.** The minimal Hamming distance between codewords is less than or equal to the minimum number of zeros in a codeword.
41. $C(n, s)$ **43.** 101010000 **45.** The codewords are 0000, 0100, 1011, and 1111.

47. $\begin{bmatrix} 1 & 1 & 1 & 1 \\ 0 & 0 & 1 & 1 \\ 1 & 0 & 1 & 0 \\ 1 & 1 & 0 & 0 \\ 0 & 1 & 0 & 1 \\ 1 & 0 & 0 & 0 \\ 0 & 1 & 0 & 0 \\ 0 & 0 & 1 & 0 \\ 0 & 0 & 0 & 1 \end{bmatrix}$ **49.** $\begin{bmatrix} 1 & 0 & 1 & 0 & 1 \\ 0 & 0 & 1 & 1 & 1 \\ 1 & 1 & 0 & 1 & 1 \\ 1 & 0 & 0 & 0 & 0 \\ 0 & 1 & 0 & 0 & 0 \\ 0 & 0 & 1 & 0 & 0 \\ 0 & 0 & 0 & 1 & 0 \\ 0 & 0 & 0 & 0 & 1 \end{bmatrix}$ **51.** $\begin{bmatrix} 1 & 0 & 1 & 0 & 1 \\ 0 & 1 & 0 & 1 & 1 \end{bmatrix}$ **53.** $[1 \; 1 \; 1 \; 1 \; 1 \; 1]$

55. 0000, 011 **57.** 0100, 101 **59.** 0000, 100 **61.** **(b)** 0101, 1110, 1101 **(c)** 2 **(d)** 1 **63.** 5
67. no

CHAPTER 4

EXERCISES 4.1 *(page 161)*

1. $\mathcal{V} = \{A, B, C, D\}$; $\mathcal{E} = \{\{A, B\}, \{A, C\}, \{B, C\}, \{B, D\}, \{C, D\}\}$ **3.** $\mathcal{V} = \{F, G, H\}, \emptyset$ **5.**

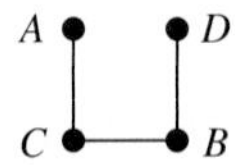

7.

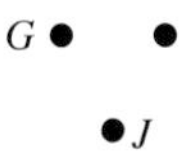

9. yes **11.** no **13.** no

15. *For a Female*

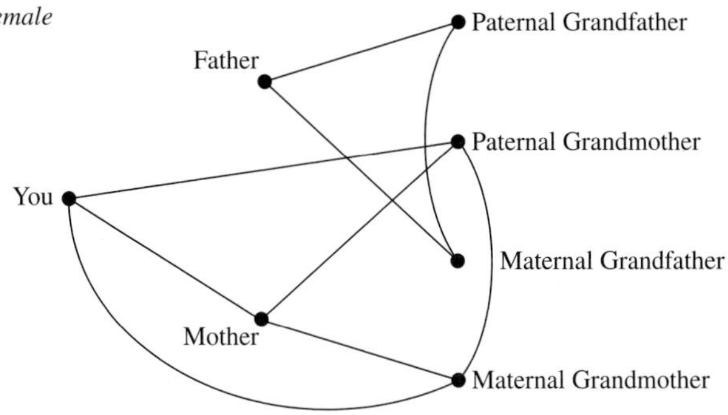

For a Male

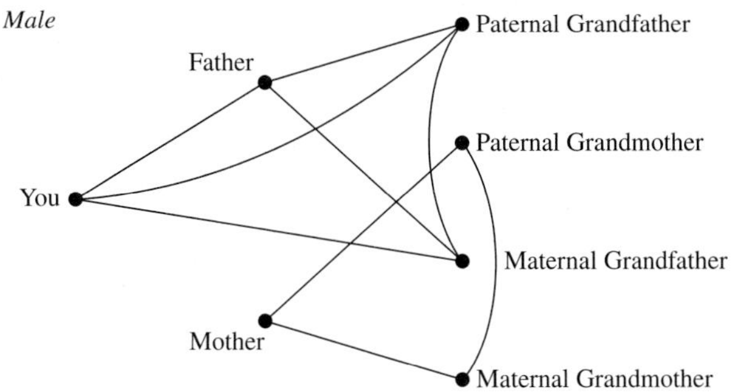

17. 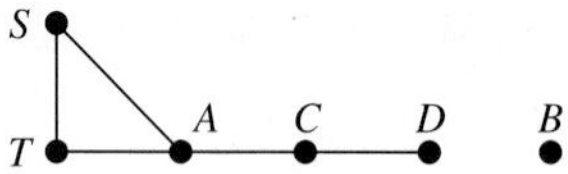

19. A: B, C, D, E (degree 4); B: A, C, F (degree 3)

21. **(a)** 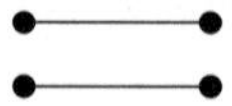 **(b)** 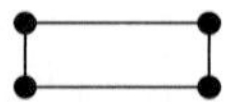

23. 3, 6, 10, $C(n, 2) = \dfrac{n(n-1)}{2}$

25. 10

27. $\begin{bmatrix} 0 & 1 & 0 & 0 & 1 \\ 1 & 0 & 1 & 0 & 0 \\ 0 & 1 & 0 & 1 & 0 \\ 0 & 0 & 1 & 0 & 1 \\ 1 & 0 & 0 & 1 & 0 \end{bmatrix}$ V_1: V_2, V_5; V_2: V_1, V_3; V_3: V_2, V_4; V_4: V_3, V_5; V_5: V_1, V_4

29. $\begin{bmatrix} 0 & 1 & 0 \\ 1 & 0 & 0 \\ 0 & 0 & 0 \end{bmatrix}$ V_1: V_2; V_2: V_1; V_3: none

31. 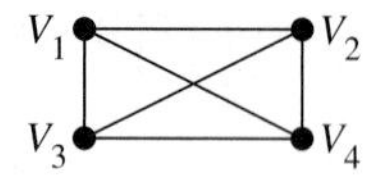

33. 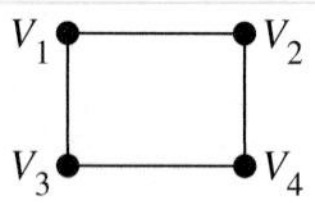

35. yes

37. No, diagonal entries are nonzero.

39. no

41. no

43. **(a)** yes **(b)** No, the first graph has two vertices of degree 2. **(c)** yes

45. 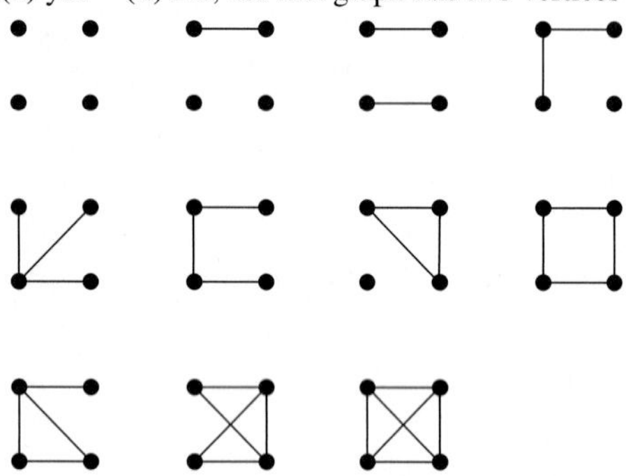

47. 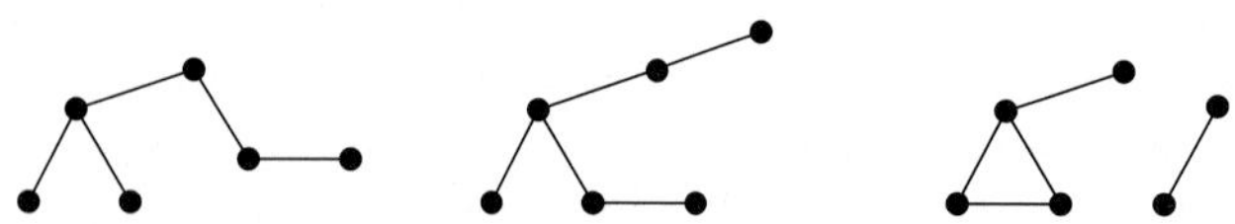

49. $\dfrac{n}{2}$ if n is even and $\dfrac{n+1}{2}$ if n is odd.

51. the least integer greater than or equal to $\dfrac{1+\sqrt{8m+1}}{2}$

53. Each shook 3 hands.

EXERCISES 4.2 *(page 176)*

1. a graph **3.** not a graph **5.** parallel edges: none; loops: a, c

7. parallel edges: a, b, c, d; loops: none

9. **(a)** c length 1; a, c length 2; a, c, b length 3 **(b)** c length 1 **(c)** all have c as a simple path **(d)** a length 1; b length 1

11. **(a)** 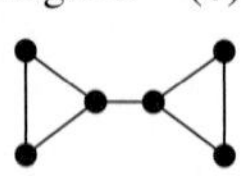 **(b)** 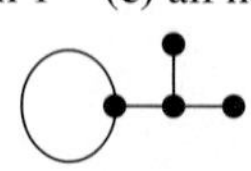 **(c)** 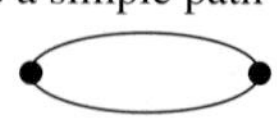

13. yes

15. no **17.** yes **19.** no **21.** yes, $d, a, b, c, e, g, k, m, h, i, j, f$ **23.** no **25.** no **27.** no

29. no **31.** no **33.** no **35.** yes **37.** yes **39.** c, h, i, d, a and g, k, f, b, e, j, n, m

41. (a) 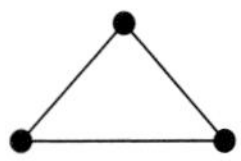 (b)  (c) 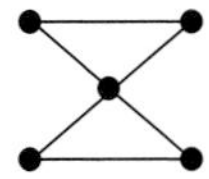 (d) 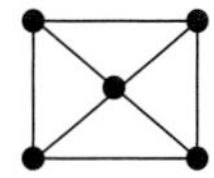 (e) 

43.

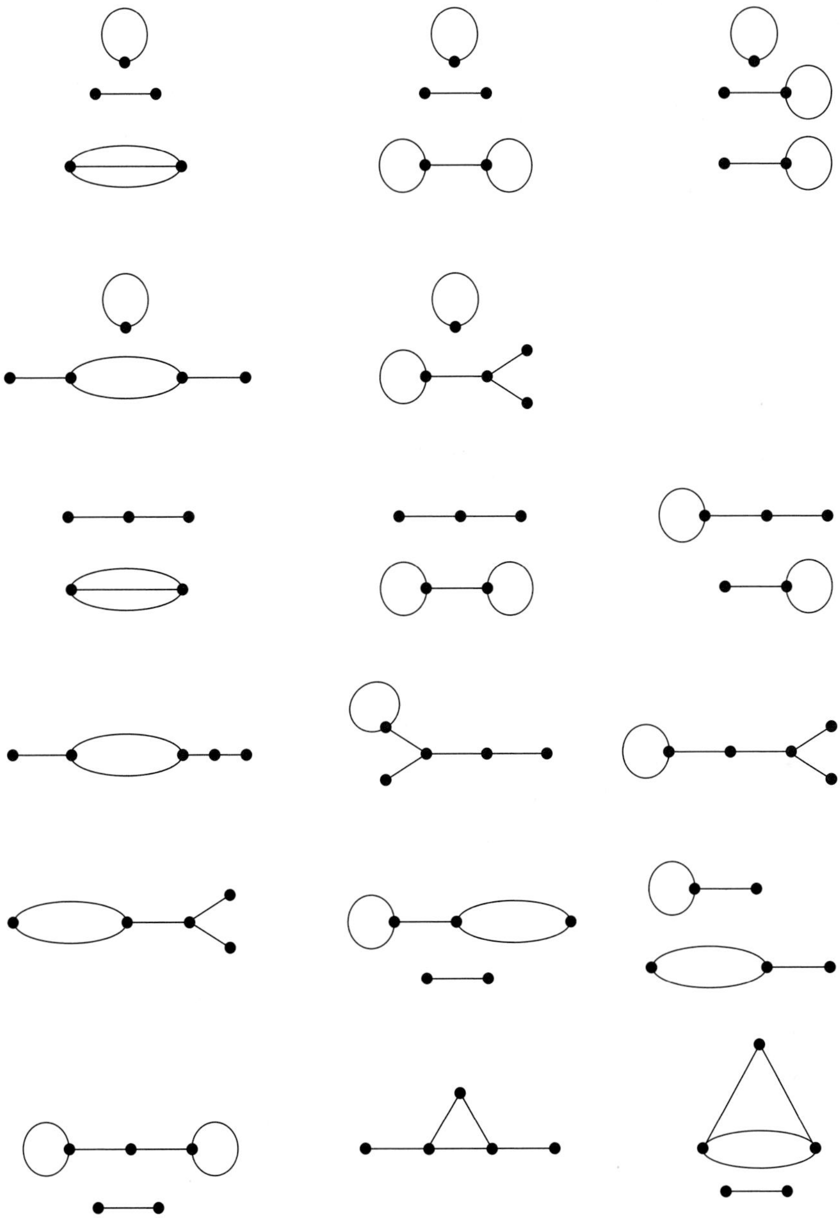

45.

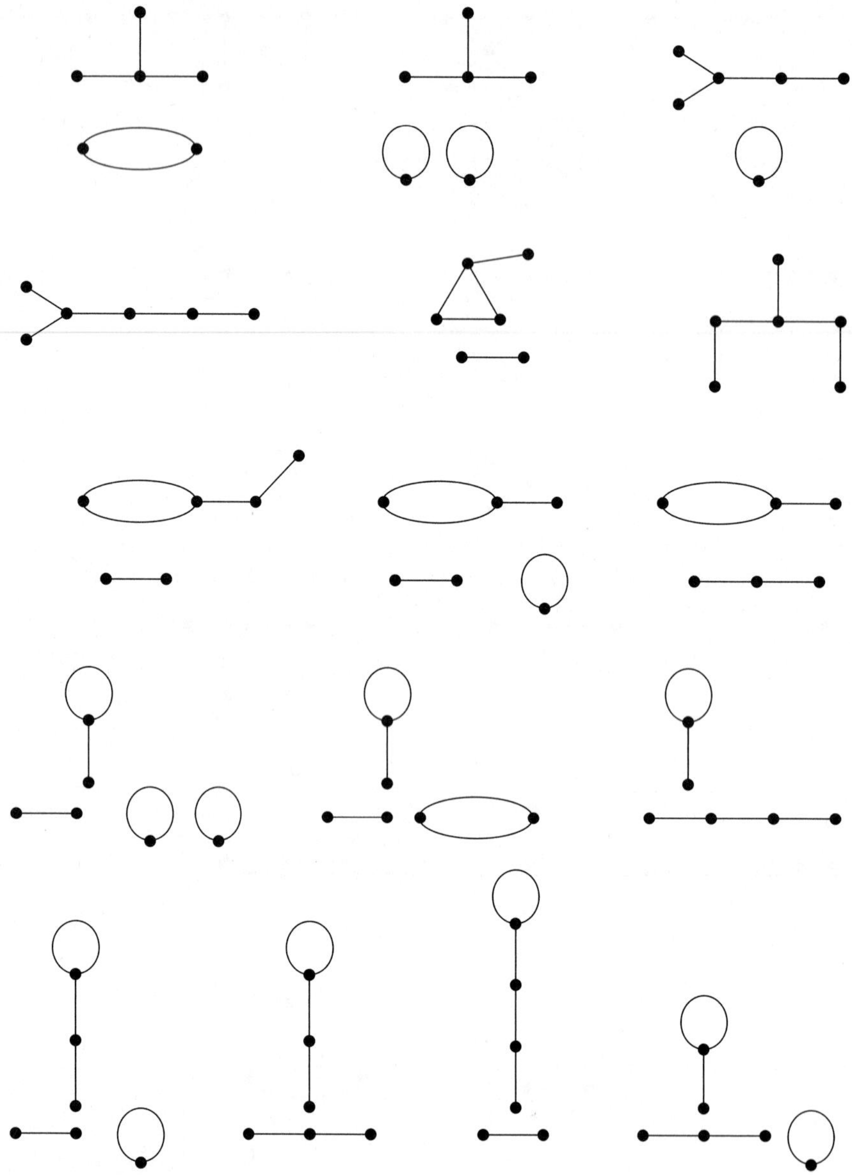

47. yes **49.** yes **51.** No, the second graph has a cycle of length 3.
53. m and n are both even.
59. **(a)** $\{A, C, G, E\}, \{B, D, H, F\}$ **(b)** $\{I, J, K, L\}$ **(c)** $\{M, O, Q\}, \{N\}, \{P, R, S, T\}$

EXERCISES 4.3 *(page 190)*

1. 5; S, D, G, E, F, T

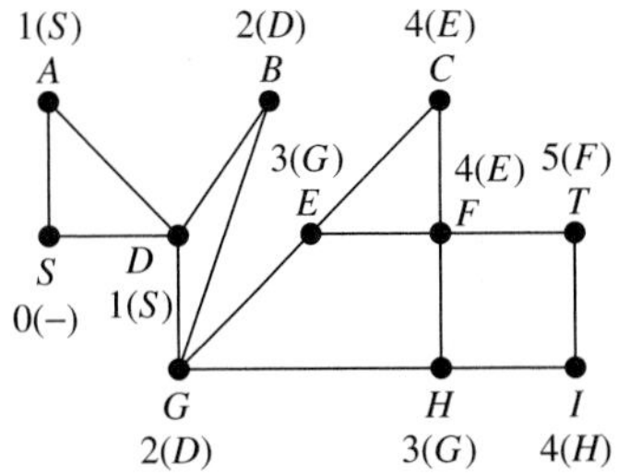

3. 7; S, A, F, G, B, C, O, T

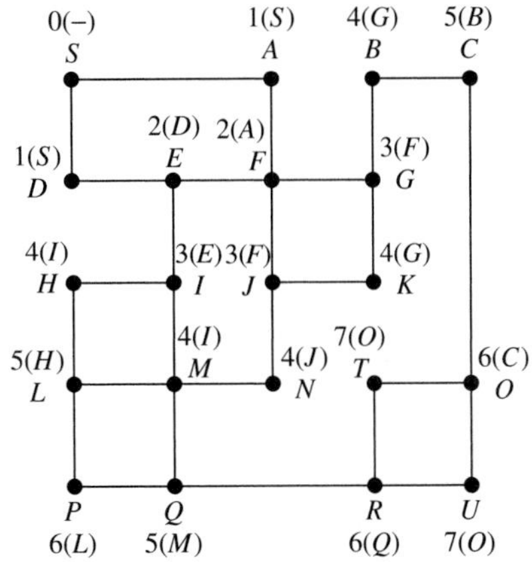

5. A: 8; B: 9; C: 3; D: 5; E: 4; F: 6; G: 7; H: 5; I: 6; shortest path to A: S, C, E, F, G, A

7. A: 8; B: 6; C: 3; D: 5; E: 2; F: 3; G: 6; H: 1; shortest path to A: S, C, D, G, A

9. $S, E, F, K, L, G, A, G, L, M, T$; Find shortest path from S to A and then from A to T.

11. S, F, A, C, D, E, T; Find shortest path from S to A and then from A to T.

13. From V_1 to V_2: 1, 2, 7, 20; from V_2 to V_3: 1, 2, 7, 20

15. From V_1 to V_1: 0, 4, 8, 34; from V_4 to V_3: 0, 2, 4, 18

17. the number of paths from V_i to V_j with length at most 3

EXERCISES 4.4 *(page 199)*

1. 3 **3.** 3 **5.** 3 **7.** 2 **9.** There are no edges. **11.** **(a)**

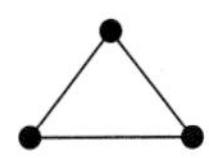

(b)

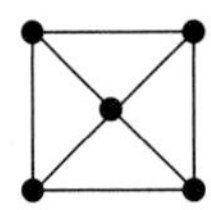

13. This algorithm uses two colors (red and blue) to color a graph having no cycles of odd length. In the algorithm, $\mathcal{L}$ denotes the set of labeled vertices (those that have been colored).

Step 1 (initialization)
  Let $\mathcal{L} = \emptyset$.
Step 2 (color all the vertices in another component)
  repeat
    Step 2.1 (color some vertex in an uncolored component)
      (a) Select a vertex S not in $\mathcal{L}$.
      (b) Assign S the label 0, and color S red.
      (c) Include S in $\mathcal{L}$.
      (d) Set $k = 0$.
    Step 2.2 (color the other vertices in this component)
      repeat
        Step 2.2.1 (increase the label)
          Replace k with $k + 1$.
        Step 2.2.2 (enlarge the labeling)
          while $\mathcal{L}$ contains a vertex V with label $k - 1$ that
            is adjacent to a vertex W not in $\mathcal{L}$

(a) Assign the label k to W.
(b) Color W red if k is even and blue if k is odd.
(c) Include W in $\mathcal{L}$.
endwhile
until no vertex in $\mathcal{L}$ is adjacent to a vertex not in $\mathcal{L}$
until every vertex is in $\mathcal{L}$

15. 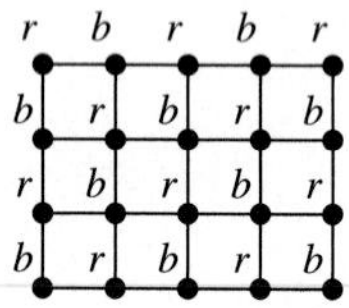

17. 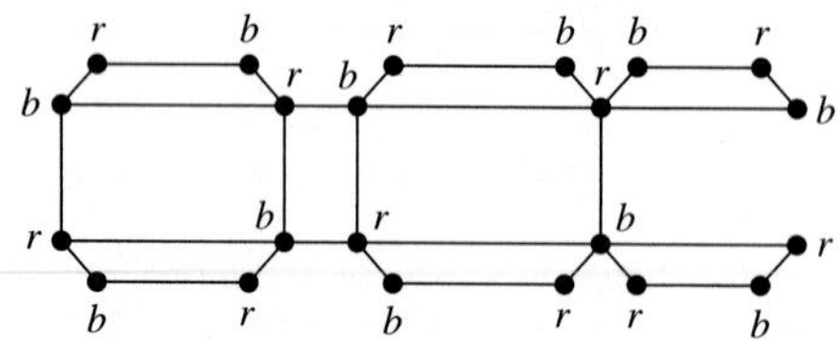

19. 5 **21.** n^n

23. 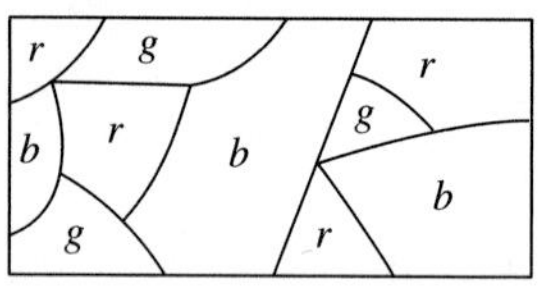

25. 

27. Three separate meeting times are needed, with finance and agriculture meeting at the same time, and likewise for budget and labor. **29.** 3

EXERCISES 4.5 *(page 212)*

1. vertices: A, B, C, D
directed edges: $(A, B), (B, D), (C, A), (C, D)$

3. vertices: A, B, C, D
directed edges: $(A, B), (A, C), (B, A), (C, D), (D, C)$

5. 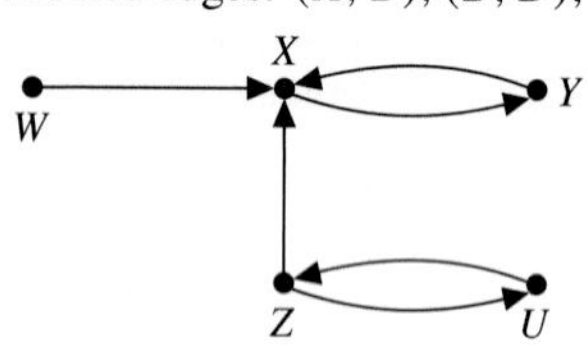

7. 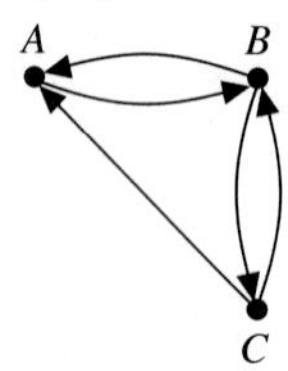

9. 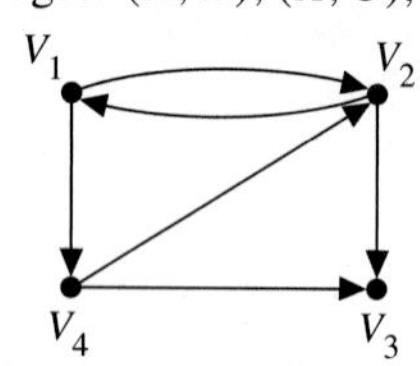

11. 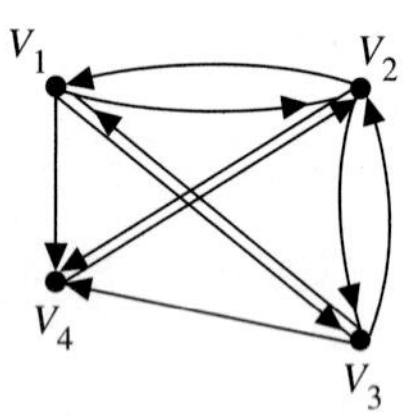

13. to A: B and C, from A: B and D, indegree 2, outdegree 2

15. to A: $B, C, D,$ and E, from A: none, indegree 4, outdegree 0

17. **(a)** A, B (1); A, C, B (2); A, D, B (2); A, D, C, B (3); A, C, D, B (3) **(b)** A, B, A (2); C, D, C (2); A, C, B, A (3); A, D, B, A (3); A, C, D, B, A (4); A, D, C, B, A (4)

19. $\begin{bmatrix} 0 & 1 & 0 & 1 \\ 1 & 0 & 1 & 0 \\ 1 & 1 & 0 & 0 \\ 0 & 1 & 0 & 0 \end{bmatrix}$ $A: B, D$; $B: A, C$; $C: A, B$; $D: B$

21. $\begin{bmatrix} 0 & 0 & 0 & 0 & 0 \\ 1 & 0 & 0 & 0 & 0 \\ 1 & 1 & 0 & 0 & 0 \\ 1 & 1 & 0 & 0 & 0 \\ 1 & 0 & 0 & 0 & 0 \end{bmatrix}$ A: none; $B: A$; $C: A, B$; $D: A, B$; $E: A$

23.

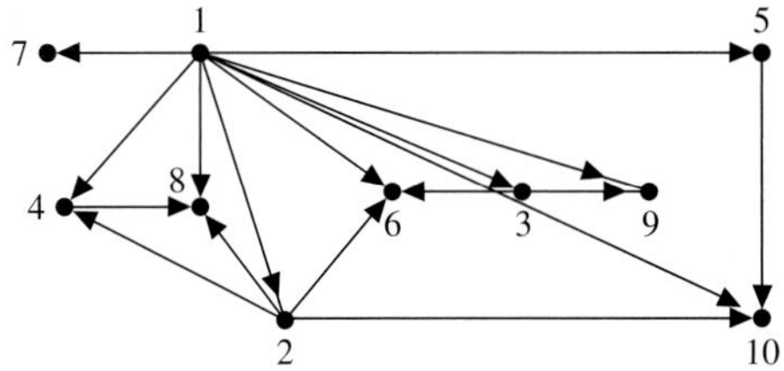

25.

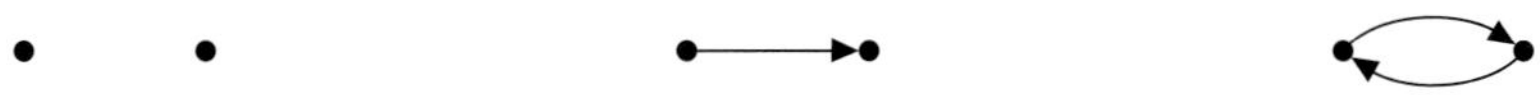

27.

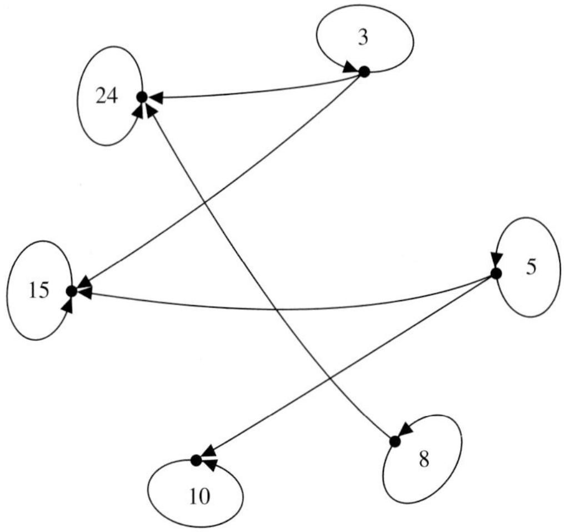

29. There is a directed edge from each vertex to itself.

31. If there is a directed edge from A to B and one from B to A, then $A = B$.

33.

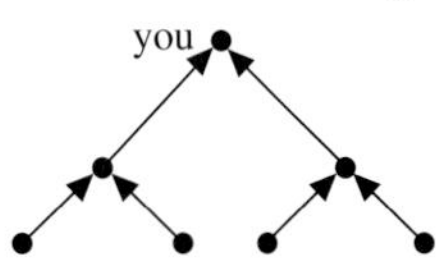

(A, B) is a directed edge in the directed graph of Exercise 32 if and only if (B, A) is a directed edge in the directed graph of Exercise 33.

35. yes **37.**

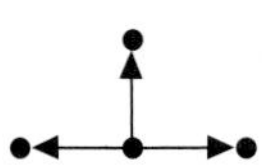

39. no **41.**

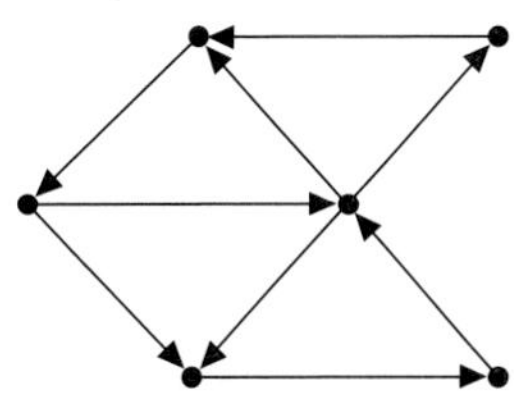

43. Any vertex can be reached from any other along the directed Hamiltonian cycle.

45. Euler circuit: $a, c, g, i, j, k, h, f, e, d, b$ **47.** Euler path: $j, g, f, n, o, k, h, i, d, a, b, c, e, m$

49. Neither—there is a vertex with indegree 3 and outdegree 1. **53.** only 1; d, b, c

55. Cookies, ice cream, eclairs, pie, pudding is the only ranking.
57. B and C have a maximum score of 3. There is a directed path of length 1 from B to A, D, and E and one of length 2 to C. There is a directed path of length 1 from C to B, D, and E and one of length 2 to A.
59. The only ranking is Bears, Vikings, Packers, Lions.
61. This algorithm determines the distance and a shortest directed path in a directed graph from vertex S to every other vertex for which such a path exists. In the algorithm, $\mathcal{L}$ denotes the set of labeled vertices, and the *predecessor* of vertex A is a vertex in $\mathcal{L}$ that is used in labeling A.

Step 1 (label S)
(a) Assign S the label 0, and let S have no predecessor.
(b) Set $\mathcal{L} = \{S\}$ and $k = 0$.
Step 2 (label vertices)
repeat
Step 2.1 (increase the label)
Replace k with $k + 1$.
Step 2.2 (enlarge labeling)
while $\mathcal{L}$ contains a vertex V with label $k - 1$ and with a directed edge to a vertex W not in $\mathcal{L}$
(a) Assign the label k to W.
(b) Assign V to be the predecessor of W.
(c) Include W in $\mathcal{L}$.
endwhile
until no vertex in $\mathcal{L}$ has a directed edge to a vertex not in $\mathcal{L}$
Step 3 (construct a directed shortest path to T)
if a vertex T is in $\mathcal{L}$
The length of a directed path from S to T is the label on T. A shortest directed path from S to T is formed by taking in reverse order T, the predecessor of T, the predecessor of the predecessor of T, and so forth, until S is reached.
otherwise
There is no directed path from S to T.
endif

63. $S, B, G, N, H, C, D, I, Q, J, K, T$; length 11 **65.** $S, A, F, G, M, N, V, W, O, I, D, T$; length 11
67. 5 to A, 10 to B, 4 to C, 3 to D, 5 to E, 2 to F, 4 to G; S, F, G, A
69. 7 to A, 11 to B, 5 to C, 6 to D, 14 to E, 2 to F, 5 to G, 6 to H, 8 to I; S, F, G, H, A
73. from V_1 to V_4: 0, 2, 1, 4; from V_4 to V_1: 1, 0, 2, 2 **75.** the number of directed edges; a directed path of length n
77. **(a)** no **(b)** no

SUPPLEMENTARY EXERCISES *(page 220)*

1.

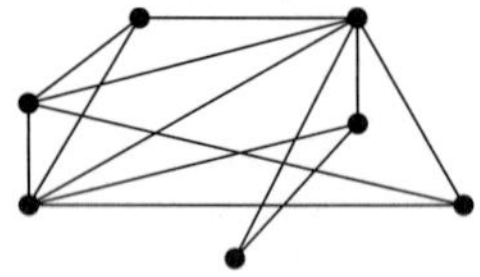

3.

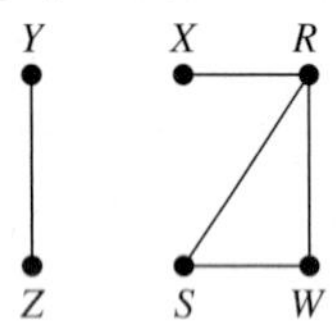

5. No, the first graph has a vertex of degree 2.

7.

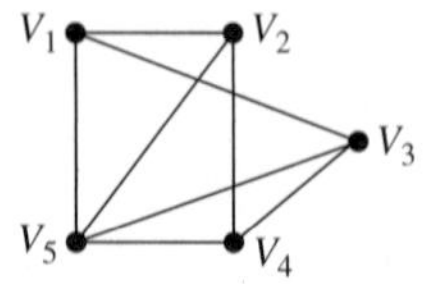

9. no

11. Yes, tear down any bridge and connect the other two land masses by a new bridge.
13. yes; $a, b, d, h, j, i, g, f, e, c$ **15.** yes **17.** 6; S, D, H, E, F, J, T **19.** 11 to A, 13 to B, 2 to C, 7 to D, 3 to E, 4 to F, 6 to G, 8 to H, 12 to I, 5 to J, 9 to K, 13 to L; S, C, D, H, A; S, C, E, J, K, B
21. from V_1 to V_2: 0, 1, 1, 4; from V_1 to V_4: 0, 1, 1, 4 **23.**

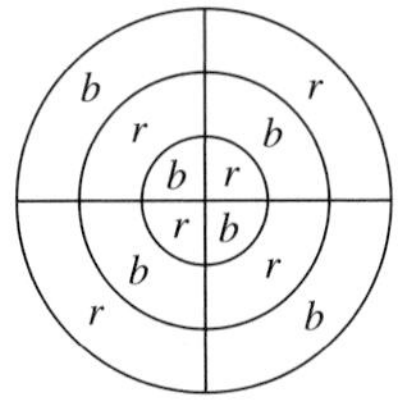

27. 4

29.

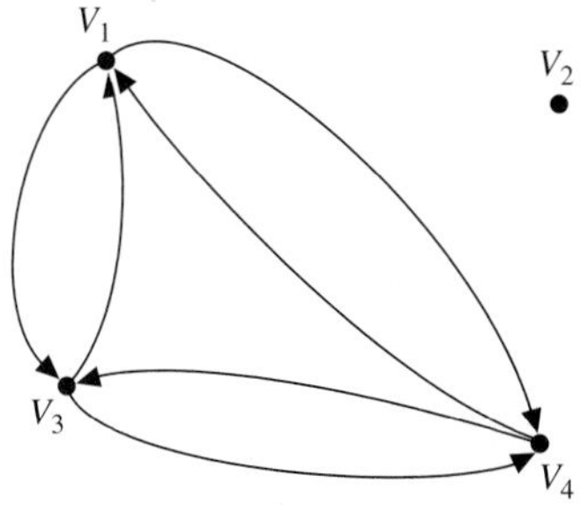

31.

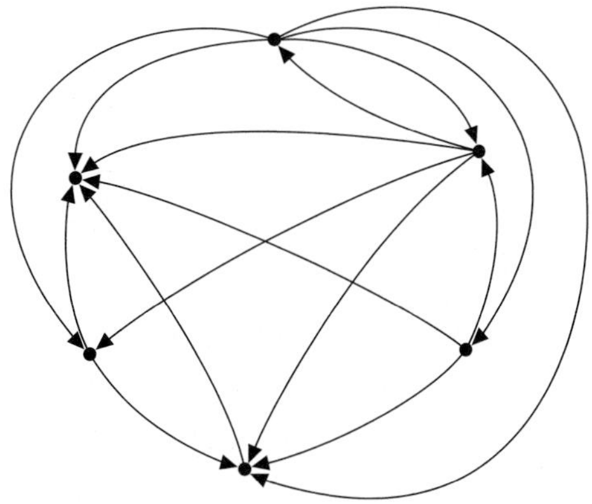

33. 0000100111101011 **37.** from V_1 to V_4: 1, 0, 2, 4; from V_2 to V_5: 0, 1, 1, 4
39. It is reflexive when there is a directed loop at each vertex. It is symmetric if, whenever there is a directed edge from A to B, then there is a directed edge from B to A. It is transitive if, whenever there is a directed edge from A to B and a directed edge from B to C, then there is a directed edge from A to C.

CHAPTER 5

EXERCISES 5.1 *(page 234)*

1. yes **3.** no **5.** no **7.** yes **9.** 16 **11.** Connect Lincoln to each other town, using only 6 lines.
13. 12 **15.**

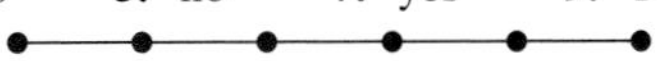

17. $n + 1$ **21.** 1, 2
23.

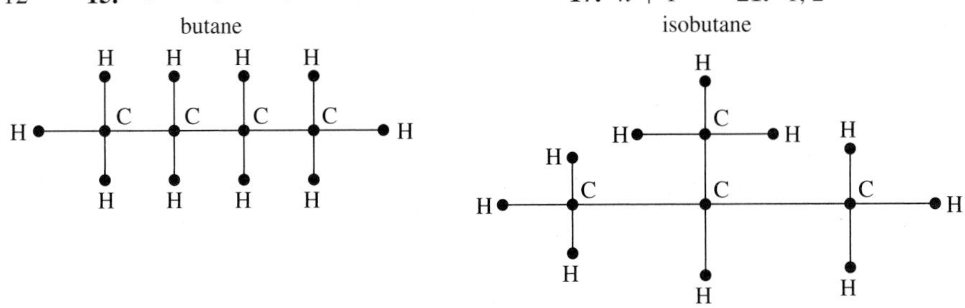

25. no

27.

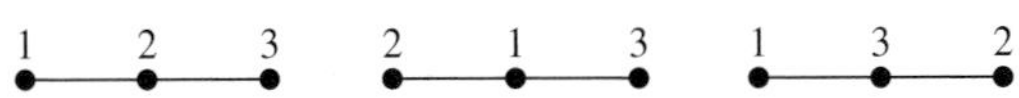

29. 2; 1; 3 **31.** 2, 2, 5, 5, 5

33.

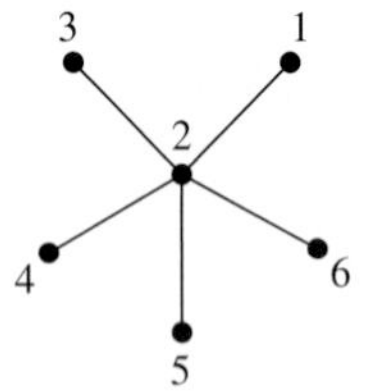

35.

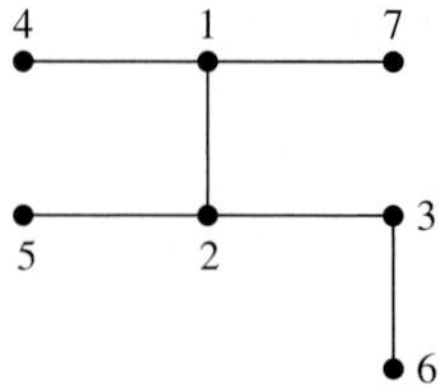

37.

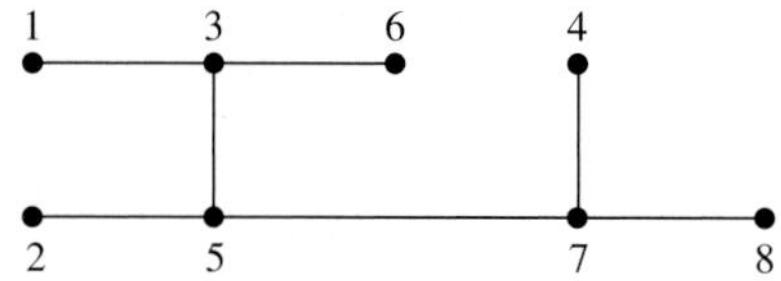

EXERCISES 5.2 *(page 248)*

1.

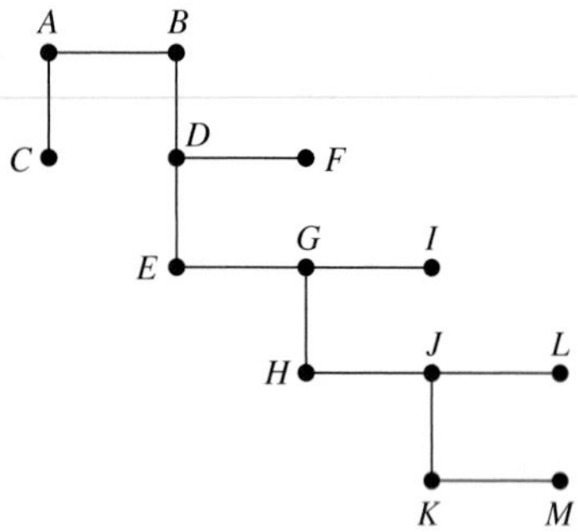

3.

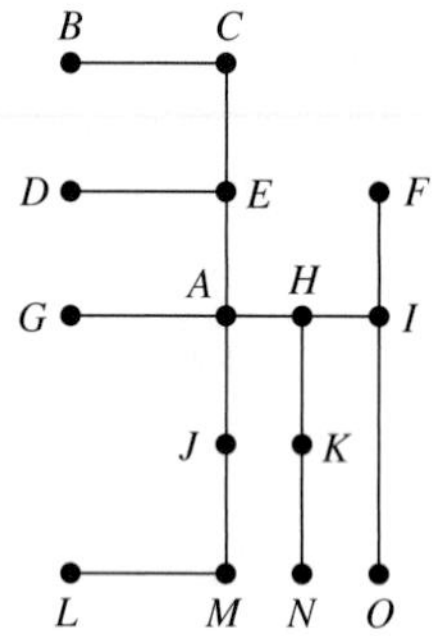

5.

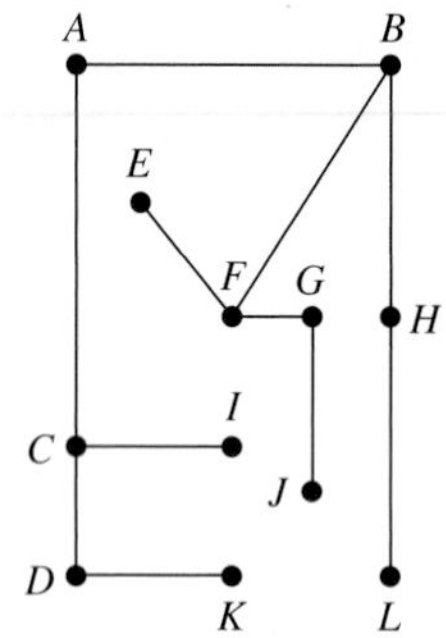

7.

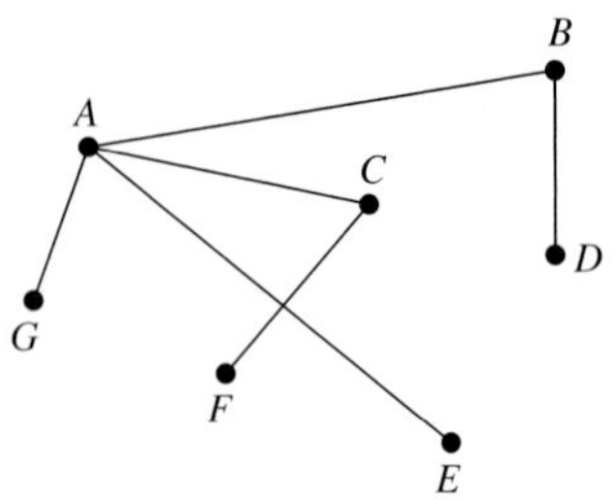

9. no

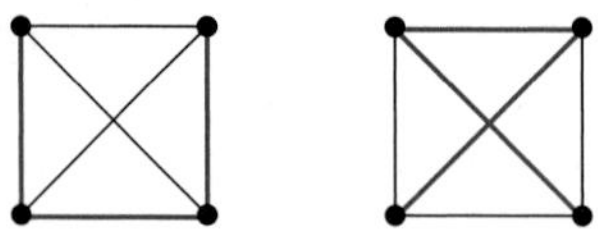

13.

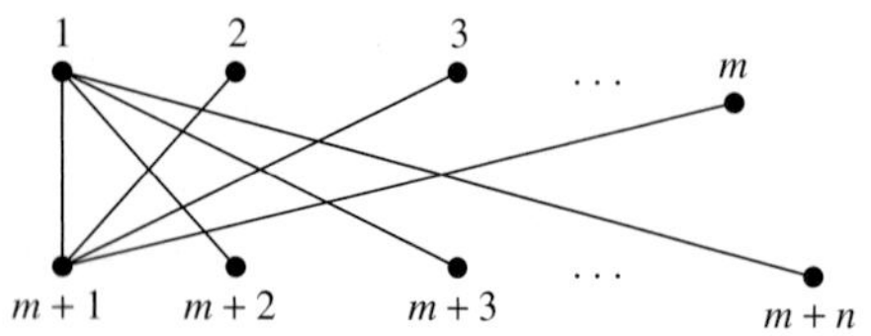

15. yes

17. a, c, g, f; 9 **19.** c, a, d, e, k, f, i, j; 21 **21.** g, f, c, a; 9 **23.** k, e, f, i, j, d, c, a; 21
25. d, e, b, c; 18 **27.** m, j, g, h, e, n, a, b; 31 **29.** {1, 5}, {5, 6}, {6, 4}, {4, 2}, {2, 7}, {6, 3}

31.

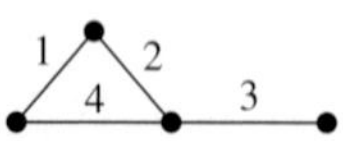

33. b, c, d, e, k, f, i, j

35.

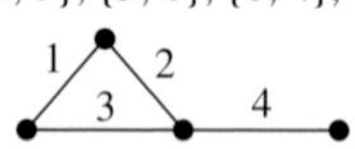

37. If the package cost 26 cents to mail, the greedy algorithm would use one 22-cent stamp and four 1-cent stamps. However, two 13-cent stamps would also do the job with fewer stamps.

39. $i, m, d, f, g, b, c, n, a$ **41.** $k, f, j, c, e, g, b, d, q, i, o$ **43.** b, k, e, f, i, c, j, d

EXERCISES 5.3 *(page 263)*

1. A, C, F, B, D, E, G, H **3.** $A, B, E, C, D, H, J, I, G, F$ **5.** $A, C, E, B, F, J, D, G, H, I$

7. 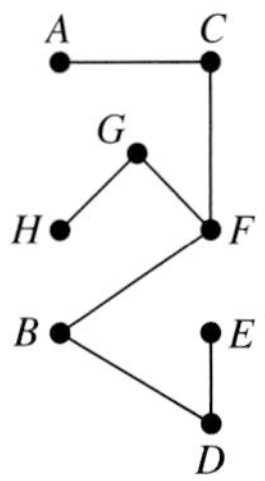

9. 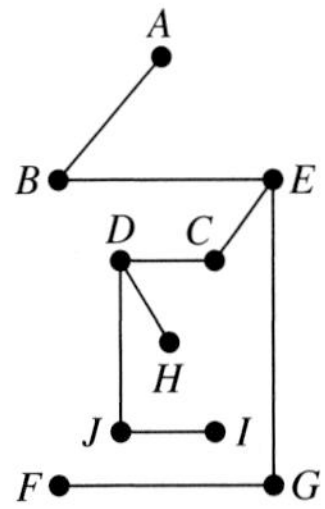

11. 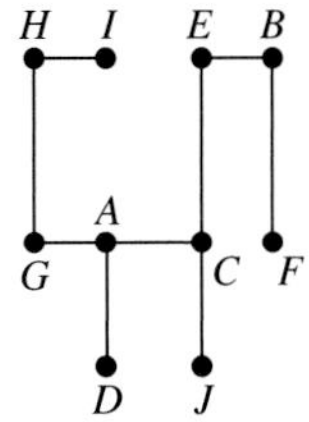

13. $\{A, H\}, \{F, E\}, \{B, E\}, \{G, C\}, \{H, F\}$ **15.** $\{A, E\}, \{B, F\}, \{C, H\}, \{C, I\}$ **17.** $\{A, I\}, \{F, C\}$

19. There is no bridge. **21.** There is a bridge, $\{B, E\}$.

23. 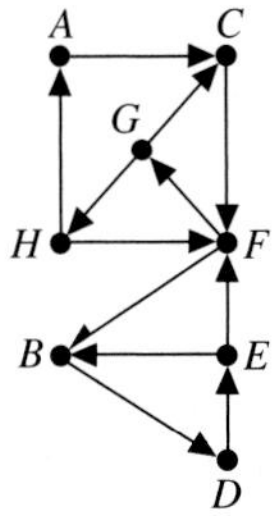

25. 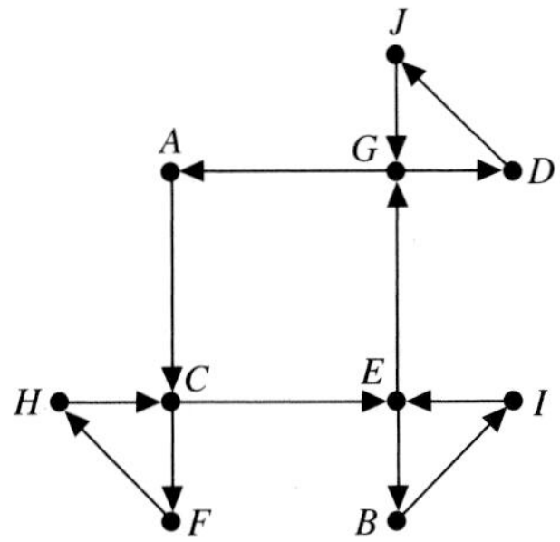

27. 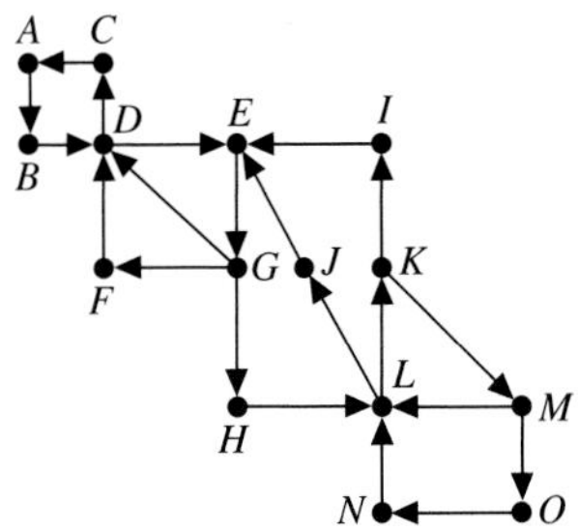

31. 2 **33.** $(n-1)!$ **39.**

Q				
			Q	
	Q			
				Q
		Q		

41. 12131231

EXERCISES 5.4 *(page 271)*

1. yes **3.** no **5.** no

7. no **9.**  **11.** 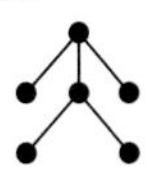 **13.** 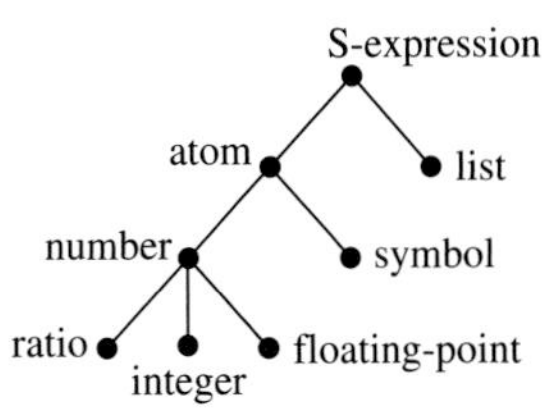

15. A vertex would have indegree greater than 1.

17. This algorithm directs the edges of a tree with a vertex labeled R to transform it into a rooted tree with root R.

Step 1 (label the root)
Label R.

Step 2 (direct the edges)
while there are unlabeled vertices
(a) Find an unlabeled vertex V that is adjacent to a labeled vertex U.
(b) Direct the edge joining U and V from U to V.
(c) Label V.
endwhile

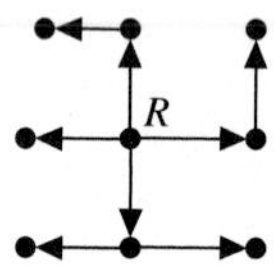

19. There is only one way.

21. **(a)** A **(b)** A, B, C, D, H, I **(c)** J, K, L, E, F, G **(d)** C **(e)** D, E, F **(f)** H, I, J, K, L **(g)** A, B, D

23. **(a)** E **(b)** E, A, D, I, J **(c)** B, K, G, F, H, C **(d)** D **(e)** none **(f)** G **(g)** E, J

25.

27.

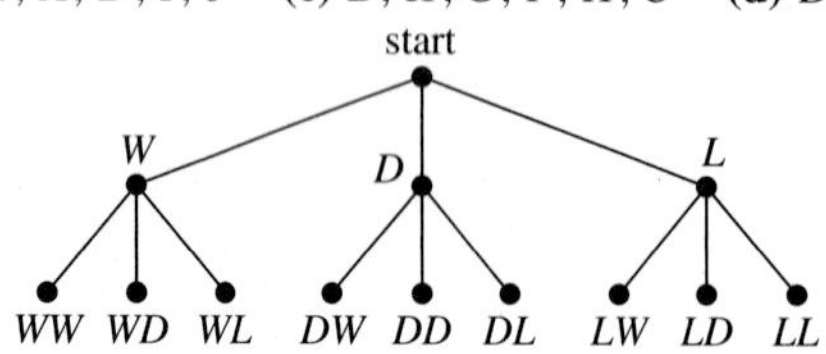

29.

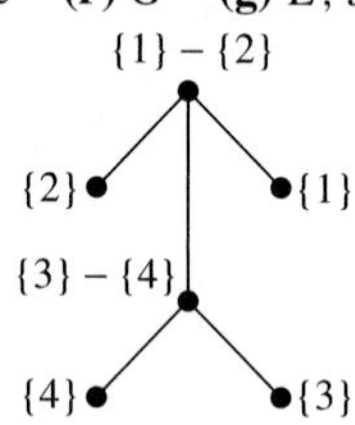

31.

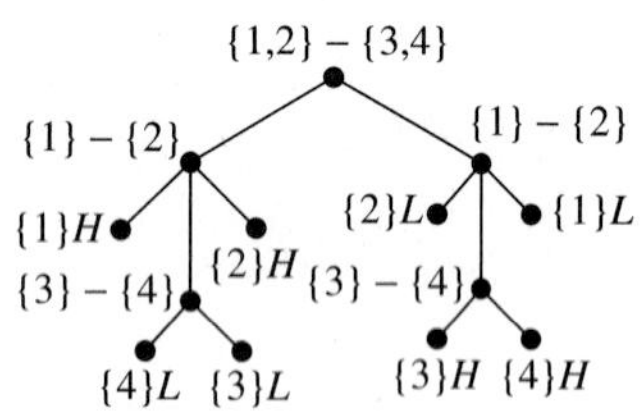

33. 0 **35.** 2 **37.** 1

EXERCISES 5.5 *(page 284)*

1.

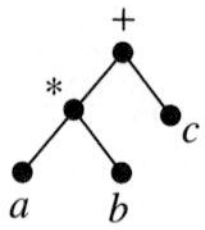

3.

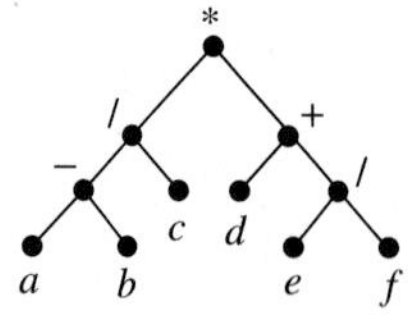

5.

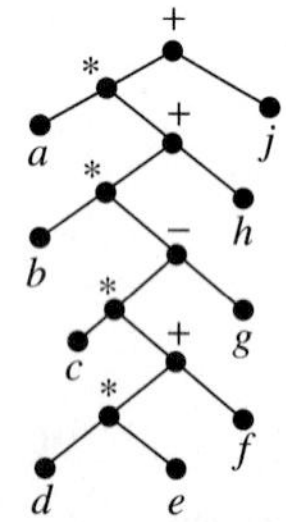

7.

9.

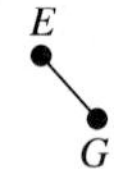

11.

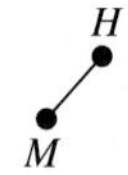

13. A, B, C **15.** A, B, D, F, C, E, G **17.** $A, B, D, G, L, E, H, M, I, N, C, F, J, O, P, K, Q$

19. B, C, A **21.** F, D, B, G, E, C, A **23.** $L, G, D, M, H, N, I, E, B, O, P, J, Q, K, F, C, A$

25. B, A, C **27.** D, F, B, A, E, G, C **29.** $L, G, D, B, M, H, E, N, I, A, C, O, J, P, F, K, Q$
31. $+ * a\,b\,c$ **33.** $*/-a\,b\,c + d\,/\,e\,f$ **35.** $+ * a + *b - *c + *d\,e\,f\,g\,h\,j$ **37.** $a\,b * c +$
39. $a\,b - c\,/\,d\,e\,f\,/ + *$ **41.** $a\,b\,c\,d\,e * f + *g - *h + *j +$ **43.** 13 **45.** 13 **47.** −2
49. 39 **51.**

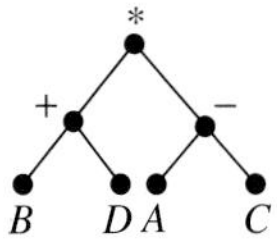

53.

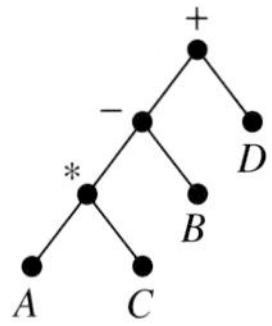

55.

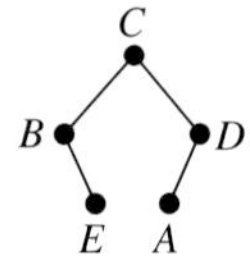

57.

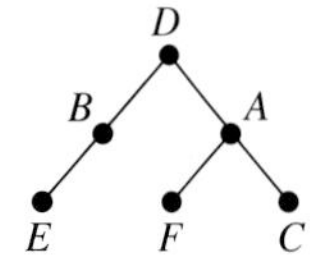

59.

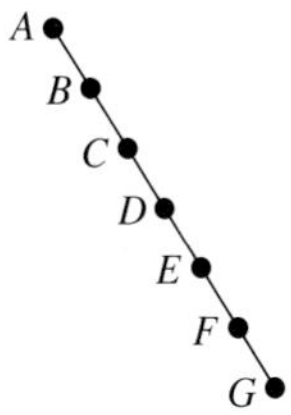

61.

63.

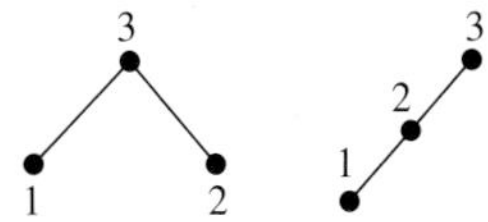

EXERCISES 5.6 *(page 302)*

1. no **3.** no **5.** no **7.** $a = 1, b = 1, c = 1$ **9.**

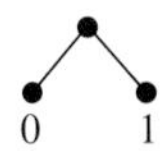

11.

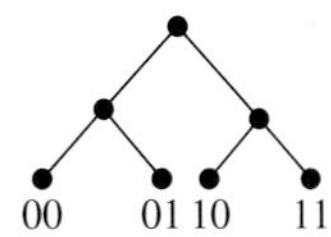

13.

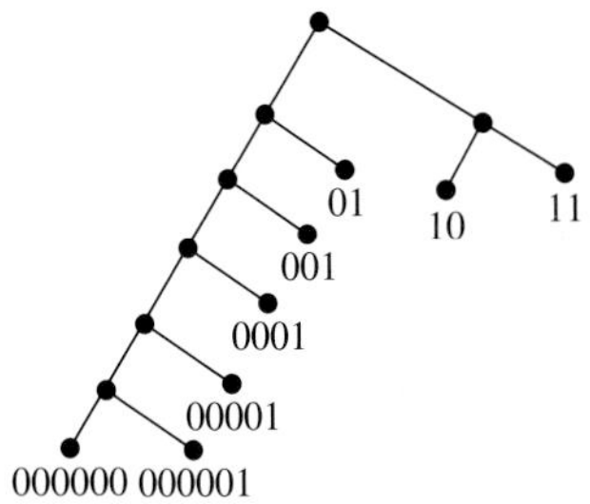

15.

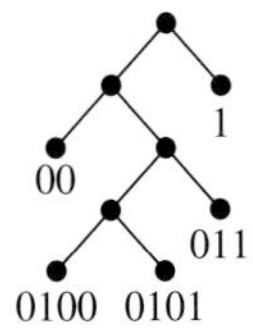

17.

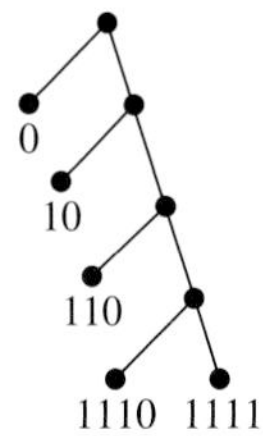

19. BATMAN **21.** TONTO **23.** GOGO **25.** THEHATS **27.** DOG **29.** QUIET
31.

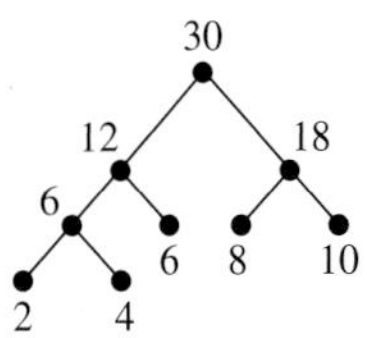

33.

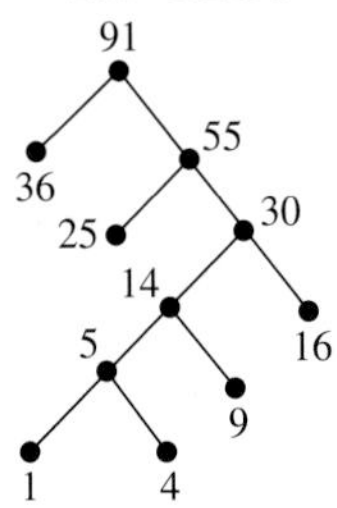

35. 277 **37.** 955

39. R: 0, I: 11, H: 101, V: 100
41. 1: 100, 2: 1111, 3: 1110, 4: 01001, 5: 0101, 6: 00, 7: 01000, 8: 011, 9: 110, 10: 101

49.

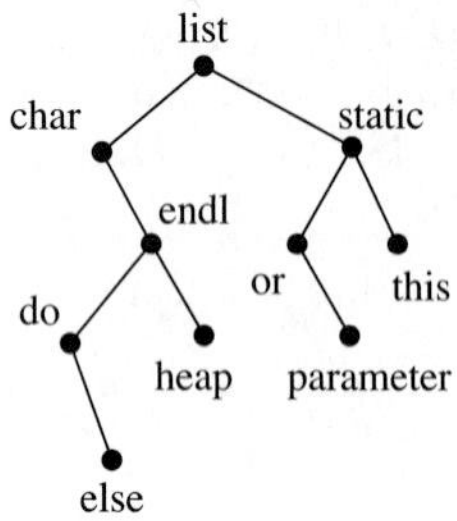

51.

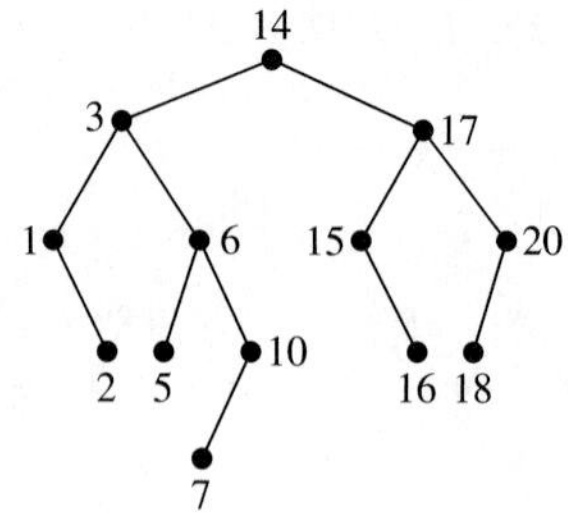

53.

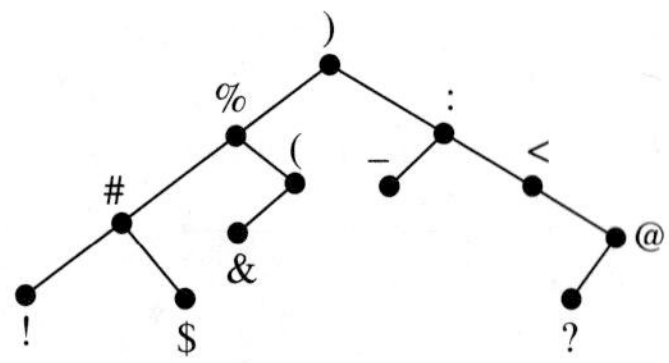

55.

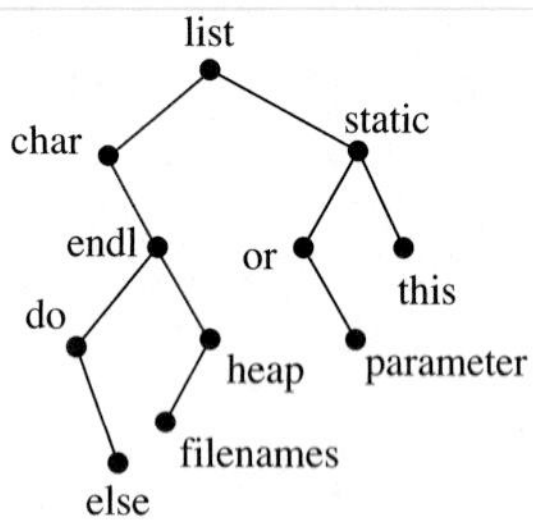

57.

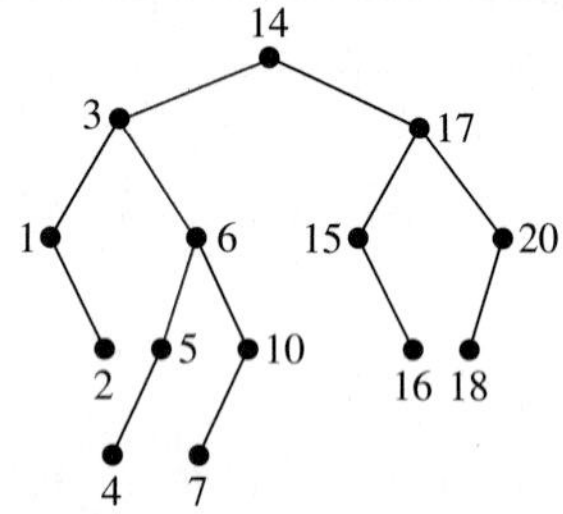

59.

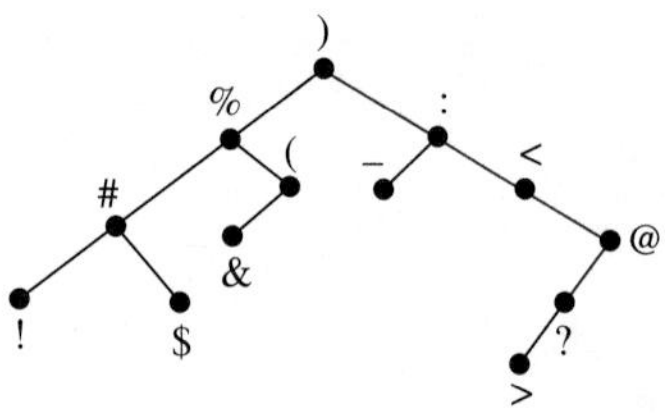

61. no

63.

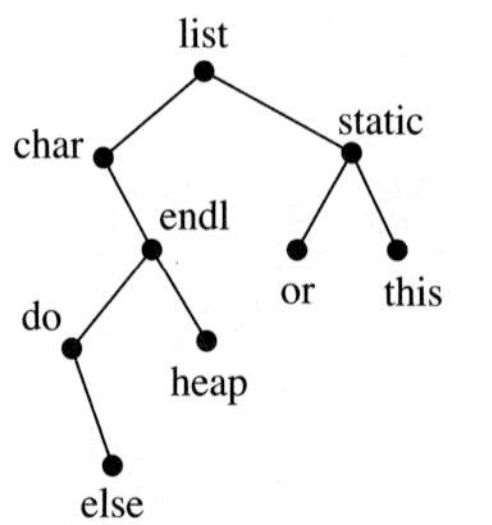

65.

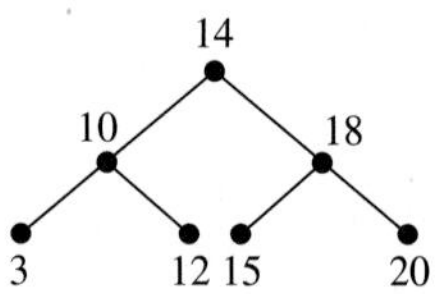

67.

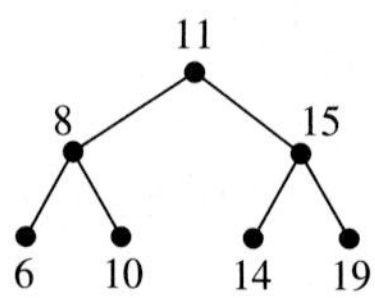

69.

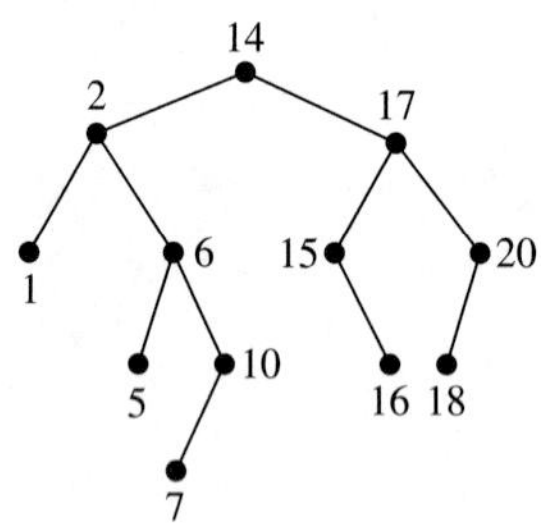

71.

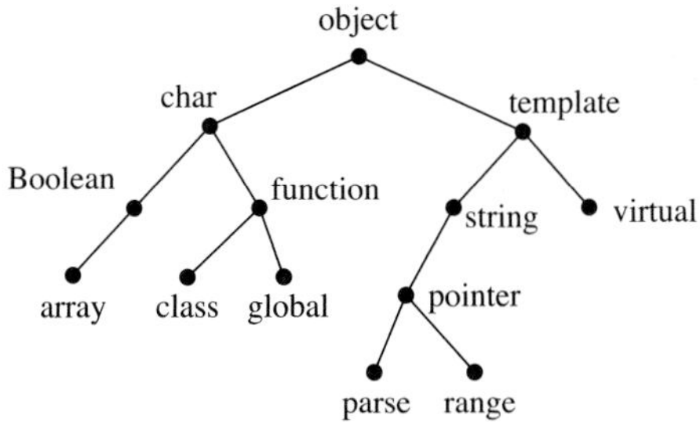

73.

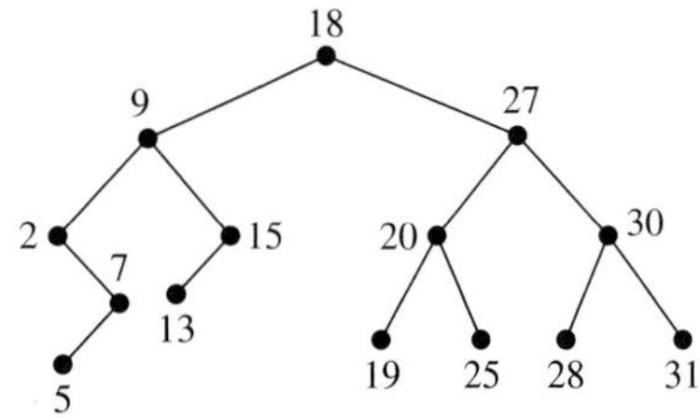

SUPPLEMENTARY EXERCISES *(page 308)*

1. 6 **3.** 1 **9.**

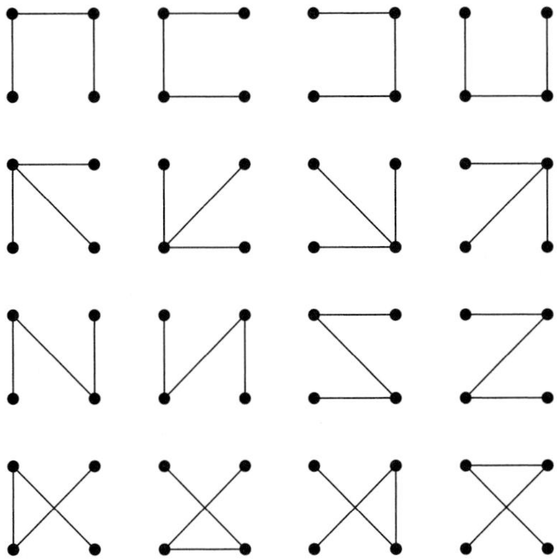

11. 10 **13.** no **17.** yes

19.

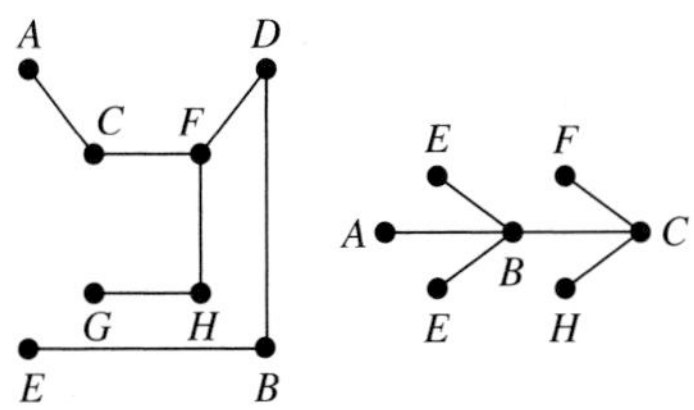

23.

Q					
			Q		
	Q				
				Q	
		Q			
					Q

25. yes **27.** $pq + 1$

39.

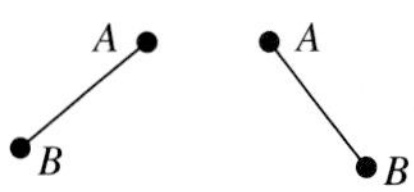

41. $* a * b c = * * a b c, a b c * * = a b * c *$ **43.** yes

45.

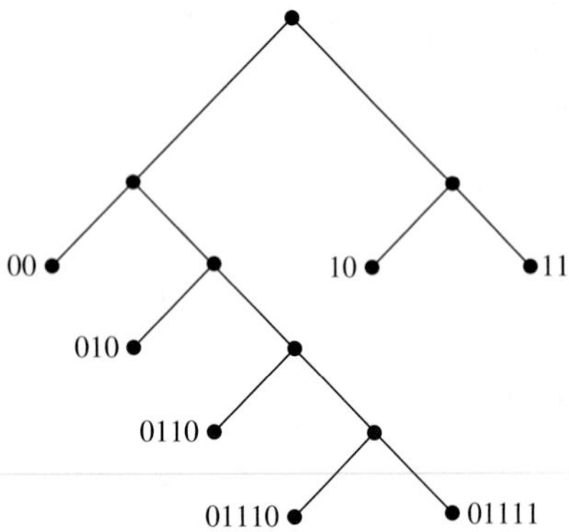

47.

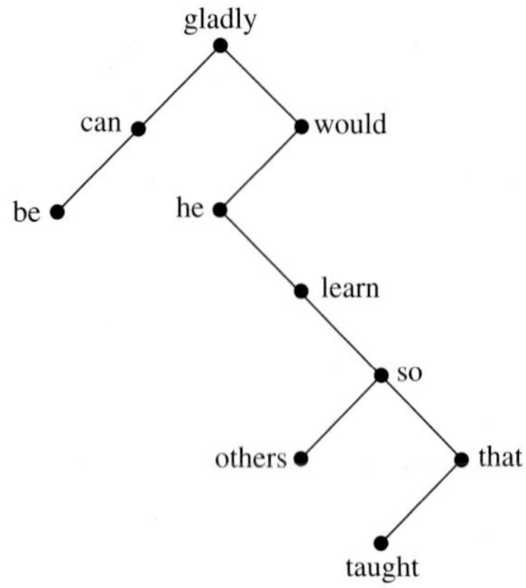

CHAPTER 6

EXERCISES 6.1 *(page 317)*

1. 2 **3.** 6 **5.** 0 **7.** yes **9.** no **11.** {3} **13.** {1, 2, 3, 4, 5} **15.** {1, 3, 4, 6}
17. $n!$ **19.** 0 **21.** Amy, Burt, Dan, and Edsel like only 3 flavors among them.
25. Timmack, Alfors, Tang, Ramirez, Washington, Jelinek, Rupp **29.** 2^n

EXERCISES 6.2 *(page 325)*

1. $\mathcal{V}_1 = \{1, 3, 6, 8, 9, 11, 13\}$, $\mathcal{V}_2 = \{2, 4, 5, 7, 10, 12\}$ **3.** no **5.** no
7. {{1, 2}, {3, 4}, {5, 6}, {7, 8}, {9, 10}, {12, 13}}, {{1, 4}, {3, 5}, {6, 7}}, {{1, 2}, {3, 4}}
9. {2, 4, 5, 7, 10, 12}, {1, 3, 6, 7}, {1, 2, 4}

11. $\begin{bmatrix} 1 & 0 & 0 & 1 & 0 & 0 \\ 1 & 0 & 1 & 0 & 0 & 0 \\ 0 & 1 & 1 & 0 & 0 & 1 \\ 1 & 0 & 1 & 1 & 0 & 0 \\ 0 & 0 & 0 & 1 & 1 & 0 \\ 0 & 0 & 1 & 0 & 0 & 0 \end{bmatrix}$ **13.** $\begin{bmatrix} 1 & 1 & 1 & 0 & 0 & 0 \\ 1 & 1 & 0 & 1 & 0 & 0 \\ 0 & 1 & 1 & 0 & 1 & 0 \\ 1 & 0 & 0 & 1 & 0 & 1 \\ 0 & 0 & 0 & 1 & 1 & 1 \\ 0 & 0 & 1 & 0 & 1 & 1 \end{bmatrix}$ **15.** $\begin{bmatrix} 1 & 1 & 0 & 0 & 0 & 0 \\ 0 & 0 & 1 & 0 & 0 & 0 \\ 1 & 1 & 0 & 1 & 0 & 0 \\ 0 & 0 & 0 & 0 & 1 & 0 \\ 0 & 1 & 0 & 0 & 0 & 1 \\ 0 & 0 & 1 & 0 & 0 & 0 \\ 0 & 0 & 0 & 0 & 1 & 0 \end{bmatrix}$

17. $\begin{bmatrix} * & 0 & 0 & 0 & 0 & 0 \\ 0 & 0 & * & 0 & 0 & 0 \\ 0 & * & 0 & 0 & 0 & 0 \\ 0 & 0 & 0 & * & 0 & 0 \\ 0 & 0 & 0 & 0 & * & 0 \\ 0 & 0 & 0 & 0 & 0 & 0 \end{bmatrix} \begin{bmatrix} 0 & 0 & 0 & * \\ * & 0 & 0 & 0 \\ 0 & * & 0 & 0 \\ 0 & 0 & 0 & 0 \\ 0 & 0 & 0 & 0 \end{bmatrix}$ main diagonal

19. row named 5, columns named 2, 6, 8, 10; rows named 3, 5, column named 8; all rows

21.

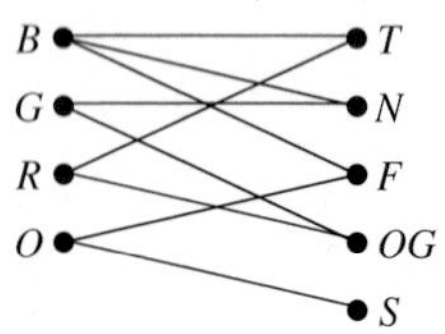

	T	N	F	OG	S
B	1^*	1	1	0	0
G	0	1^*	0	1	0
R	1	0	0	1^*	0
O	0	0	1	0	1^*

23.

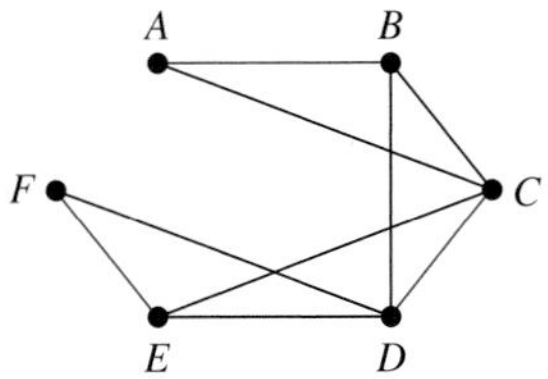

25.

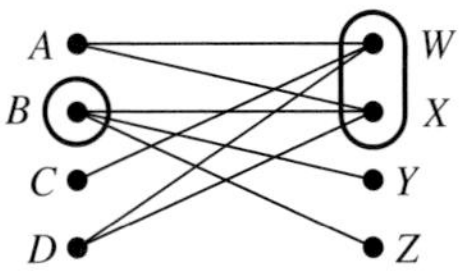

$$\begin{array}{c} \\ A \\ B \\ C \\ D \end{array}\begin{array}{cccc} W & X & Y & Z \\ \end{array}\begin{bmatrix} 1 & 1 & 0 & 0 \\ 0 & 1 & 1 & 1 \\ 1 & 0 & 0 & 0 \\ 1 & 1 & 0 & 0 \end{bmatrix}$$

EXERCISES 6.3 *(page 336)*

1. $\begin{bmatrix} 0 & 1^* & 0 & 1 \\ 1^* & 0 & 0 & 1 \\ 1 & 1 & 0 & 0 \\ 1 & 0 & 0 & 0 \end{bmatrix}\begin{array}{l} D\surd \\ D\surd \\ \\ \\ \end{array}$
$\quad 2 \quad 1 \quad \#\surd \quad \#\surd$

3. $3A, 2A, 2D$ **5.** $1D, 2A, 3C, 4B$ **7.** $1D, 2E, 3B, 4A$

9. $1D, 2A, 3B, 5C$ **11.** $1B, 3A, 4D$ **13.** $\{1, B\}, \{2, C\}, \{3, A\}$
15. $\{1, A\}, \{2, D\}, \{3, E\}, \{4, B\}, \{5, C\}$ **17.** B, C, A, D **19.** W, Z, Y, X
21. carrot, banana, egg, apple **23.** Constantine to 1, Egmont to 2, Fungo to 3, Drury to 4, Arabella to 5

EXERCISES 6.4 *(page 345)*

1. row 2, columns 2 and 4 **3.** row 3, columns 1, 3, 4 **5.** $\{2, A, C\}$ **7.** $\{B, C, D, E\}$ **9.** impossible
11. $\{1, 3, 5, 6\}$ **13.** 7 hours **15.** $\{1, 4, 5, 6, 7, 8, B\}$

EXERCISES 6.5 *(page 353)*

1. 13 **3.** 13 **5.** 18 **7.** 11 **9.** 16 **11.** 28 **13.** Addams to Chicago, Hart to Las Vegas, Young to New York, Herriman to Los Angeles **15.** The Hungarian algorithm must be applied to a square matrix.

SUPPLEMENTARY EXERCISES *(page 355)*

1. (a) 60 **(b)** 36 **(c)** 0 **3. (a)** no **(b)** yes; $\mathcal{V}_1 = \{1, 2, 5, 6, 7, 8, 11, 12\}$, $\mathcal{V}_2 = \{3, 4, 9, 10, 13, 14, 15, 16\}$
5. (a) $\{2, 4, 6, 7, 9, 11\}$ **(b)** $\{1, 2, 5, 6, 7, 8, 11, 12\}$ **7.** $\{1, 2\}, \{3, 4\}, \{6, 7\}, \{8, 5\}$

9. $\begin{bmatrix} 1^* & 0 & 0 & 0 & 1 \\ 0 & 0 & 1 & 1^* & 0 \\ 0 & 0 & 1^* & 0 & 0 \\ 1 & 1^* & 0 & 1 & 1 \\ 0 & 0 & 1 & 0 & 0 \end{bmatrix}$; the 1st and 4th rows and 3rd and 4th columns

11. w, z, v, x, y **13.** 5 hours
15. One way is as follows: Adam, Studebakers; Beth, Hupmobiles; Cal, Packards; Danielle, Hudsons.

CHAPTER 7

EXERCISES 7.1 *(page 367)*

1. a network with source A and sink E **3.** not a network because arc (C, B) has negative capacity
5. a network with source D and sink B **7.** not a flow because 5 comes into D and 6 comes out of D
9. a flow with value 3 **11.** not a flow because 2 comes into D and 3 comes out of D
13. not a cut because vertex C is not in $\mathcal{S}$ or $\mathcal{T}$ **15.** a cut with capacity 40 **17.** a cut with capacity 34

19.

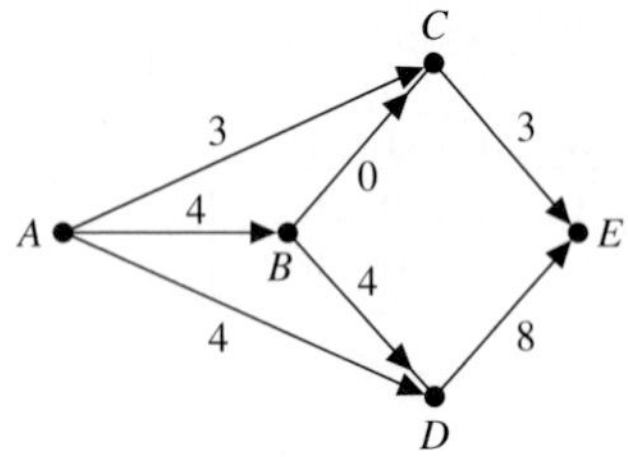

21.

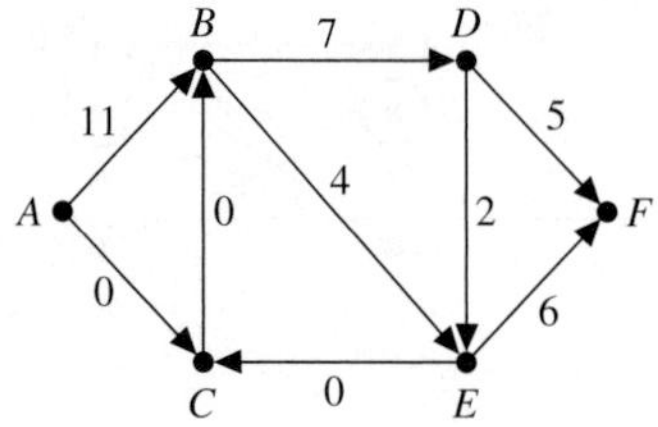

23.

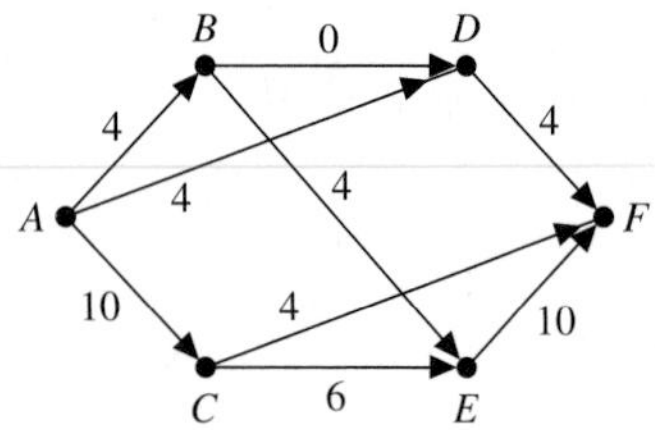

25. $\{A, B, C, D\}, \{E\}$ **27.** $\{A, B, C\}, \{D, E, F\}$

29. $\{A, C\}, \{B, D, E, F\}$ **31.**

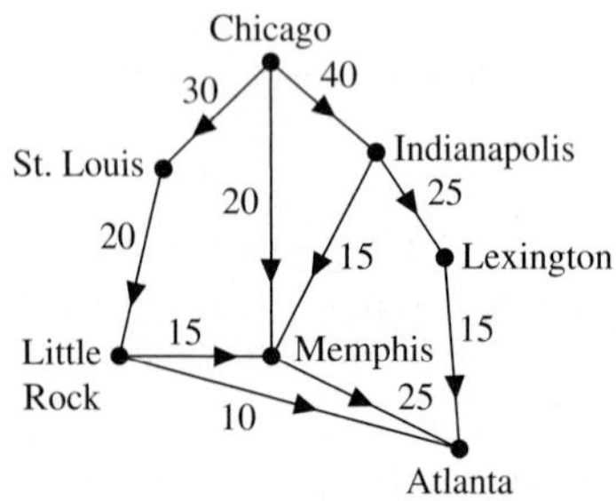

33. $f(\mathcal{U}, \mathcal{V}) = 15$ and $f(\mathcal{V}, \mathcal{U}) = 3$

35. For the flow in Exercise 10, take $\mathcal{U} = \{D\}$, $\mathcal{V}_1 = \{A, B, C\}$, and $\mathcal{V}_2 = \{B, E, F\}$.

EXERCISES 7.2 *(page 378)*

1. 1 **3.** 2 **5.**

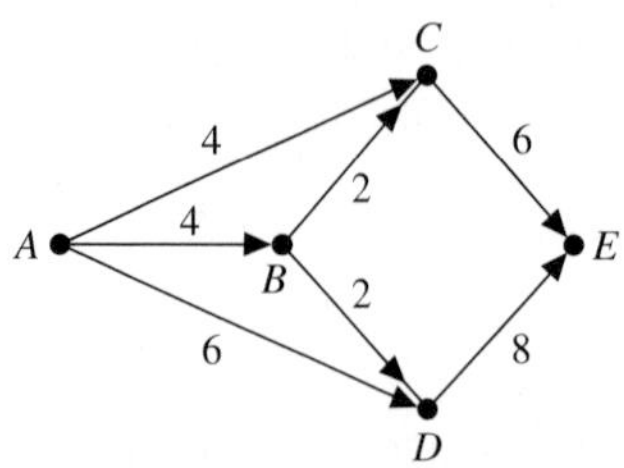

7.

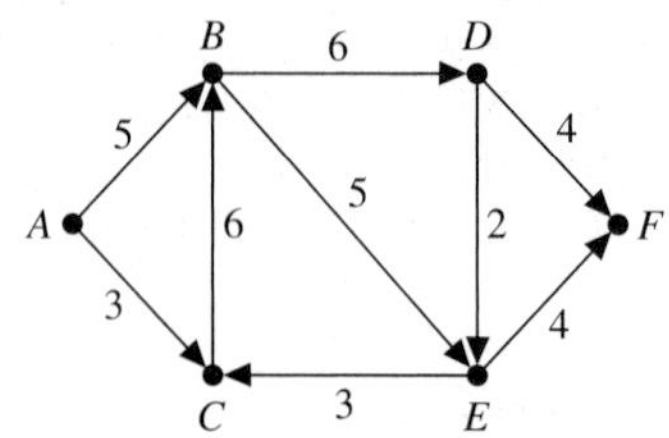

9. Increase the flow by 3 along A, C, E. **11.** The given flow is maximal. **13.** Increase the flow by 2 along A, D, B, E, F. **15.** Increase the flow by 2 along A, B, D, C, F, E, G. **17.** The given flow is maximal.

19.

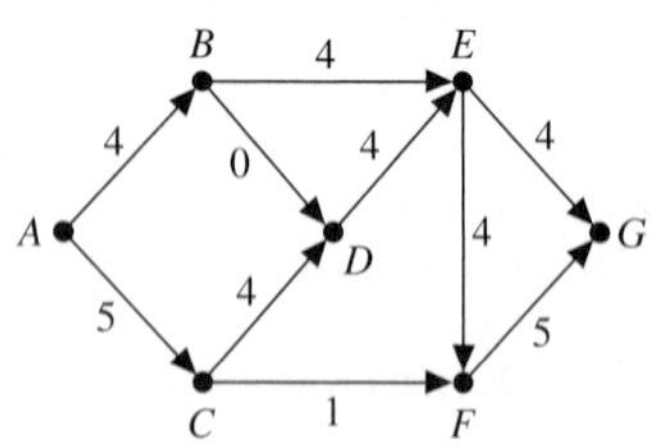

21.

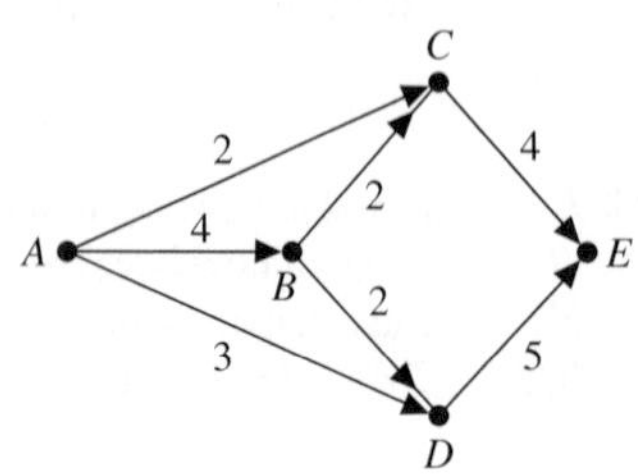

23.

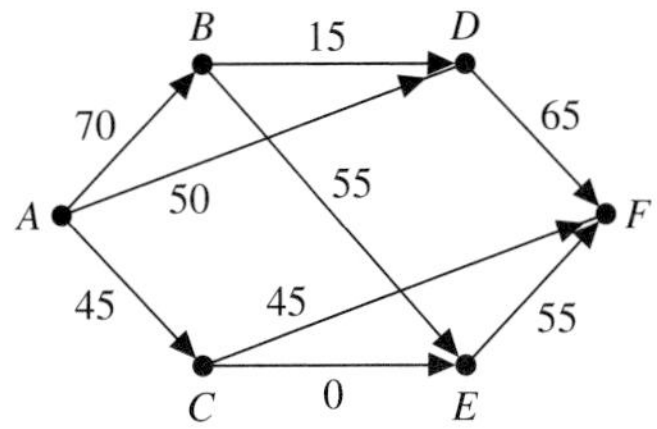

25.

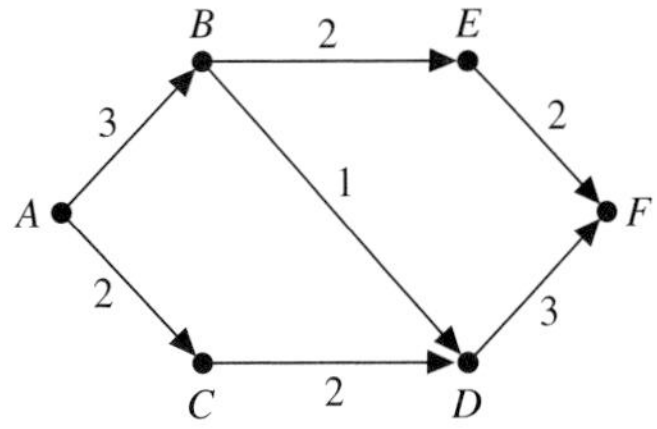

27.

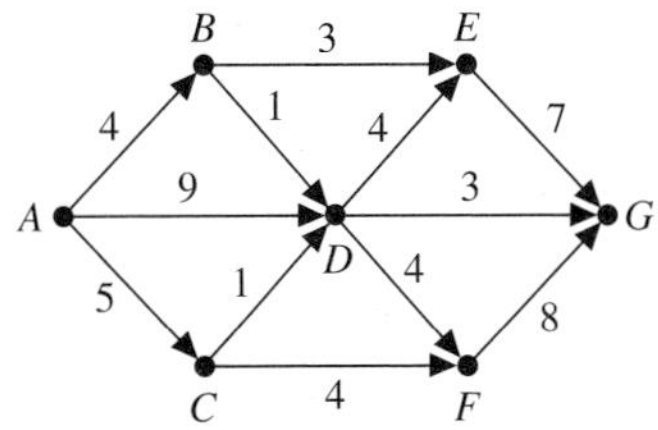

29.

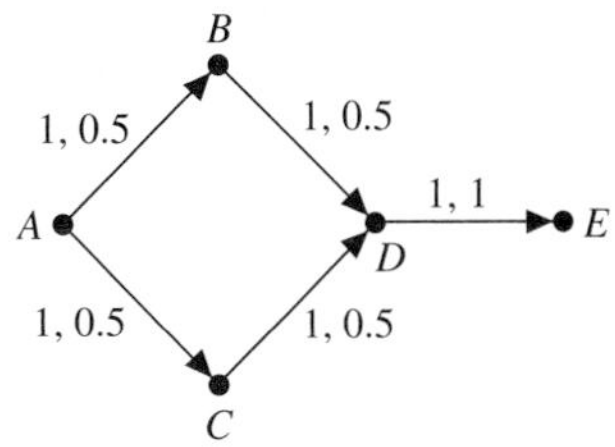

EXERCISES 7.3 *(page 386)*

1. 21 **3.** 28 **5.** $\{A, B, C, E\}, \{D, F\}$ **7.** $\{A, B, D\}, \{C, E, F\}$ **9.** $\{A, B, C, D\}, \{E\}$

11. $\{A, C, F\}, \{B, D, E, G\}$ **13.**

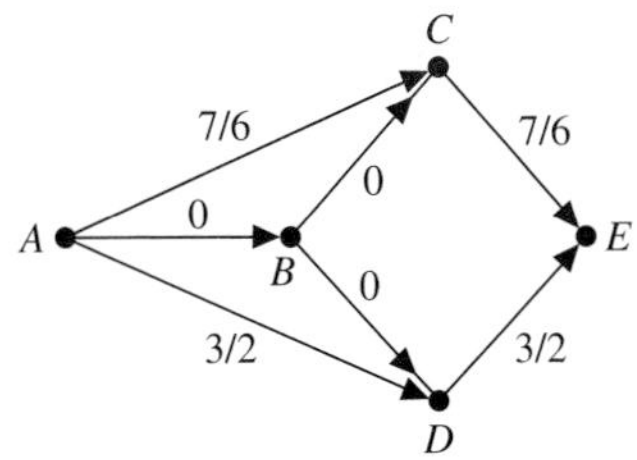

17. 2^{n-2} **19.** $\{(S, A), (F, T)\}$ **23.**

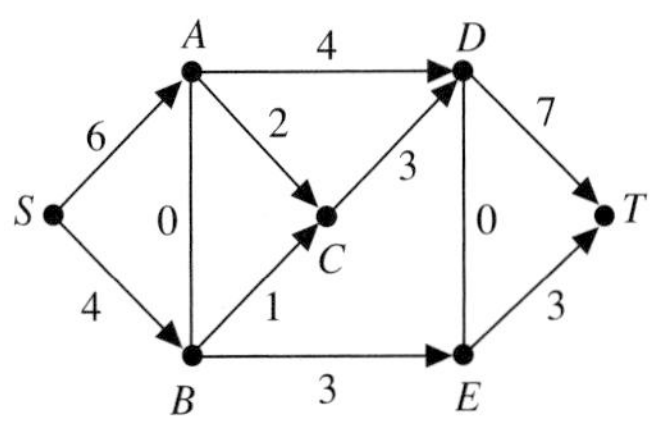

EXERCISES 7.4 *(page 394)*

1. bipartite; $\mathcal{V}_1 = \{A, D, E\}$ and $\mathcal{V}_2 = \{B, C, F\}$ **3.** not bipartite

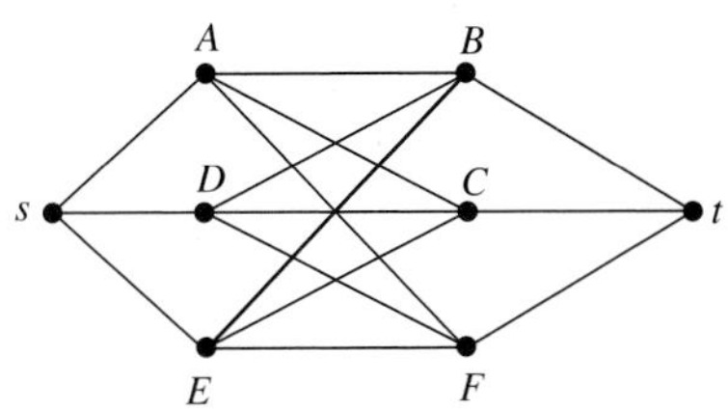

5. bipartite; $\mathcal{V}_1 = \{A, D\}$ and $\mathcal{V}_2 = \{B, C, E, F\}$ **7.** $\{(A, Y), (B, Z), (D, X)\}$

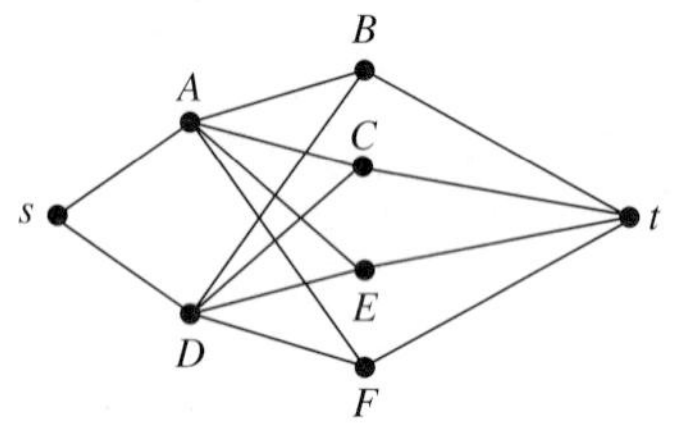

9. The given matching is a maximum matching. **11.** $\{(A, 1), (C, 3), (D, 2)\}$
13. $\{(a, A), (b, C), (c, B), (d, D)\}$ **15.** Andrew and Greta, Bob and Hannah, Dan and Flo, Ed and Iris
17. Craig files, Dianne distributes paychecks, Gale collates, Marilyn types, and Sharon helps students.
19. Create a bipartite graph $\mathcal{G}$ with vertices U_i that correspond to the sets $S_1, S_2, \ldots, S_n$, and vertices V_j that correspond to the elements in $S_1 \cup S_2 \cup \ldots \cup S_n$. Join U_i and V_j if and only if the element corresponding to V_j belongs to set S_i. Apply the flow augmentation algorithm to the network $\mathcal{N}$ associated with $\mathcal{G}$. Then $S_1, S_2, \ldots, S_n$ has a system of distinct representatives if and only if the value of a maximal flow in $\mathcal{N}$ is n.
21. No system of distinct representatives exists. **23.** No acceptable assignment exists.

SUPPLEMENTARY EXERCISES *(page 397)*

1. A minimal cut is $\{A\}, \{B, C, D, E, F\}$. A maximal flow is as follows:

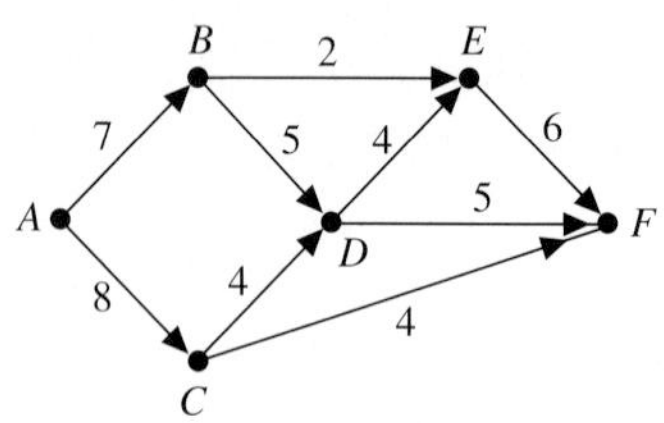

3. A minimal cut is $\{A, B, C\}, \{D, E, F, G\}$. A maximal flow is as follows:

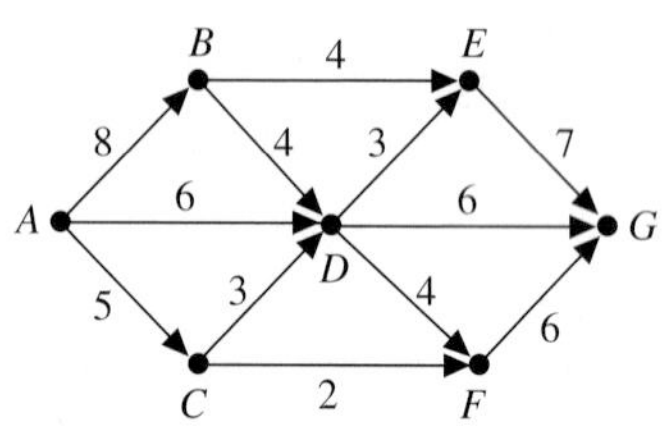

5. A minimal cut is $\{A, B, C, D, F\}, \{E, G, H\}$. A maximal flow is as follows:

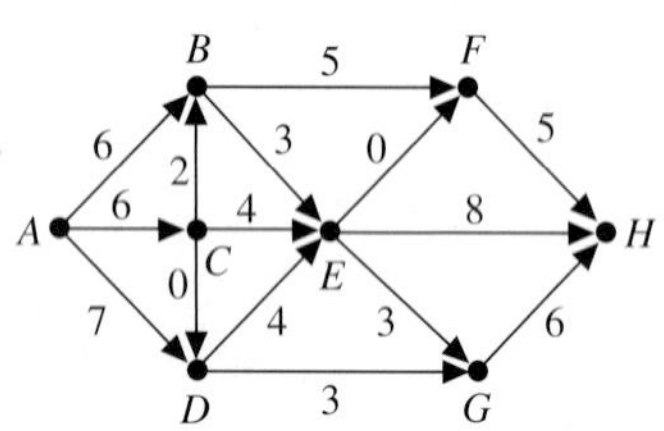

7. A minimal cut is $\{A, D\}, \{B, C, E, F, G, H, I, J, K\}$. A maximal flow is as follows:

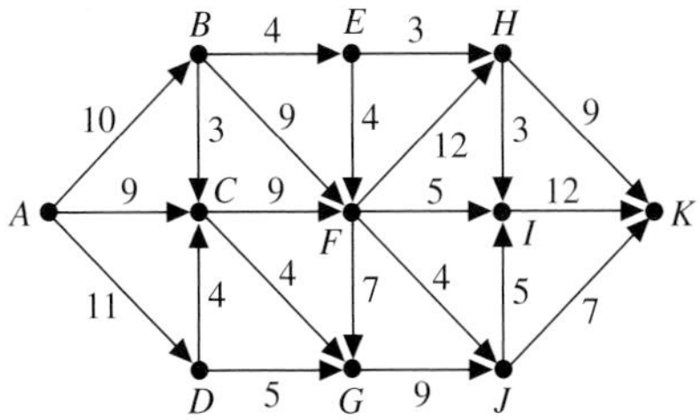

11.

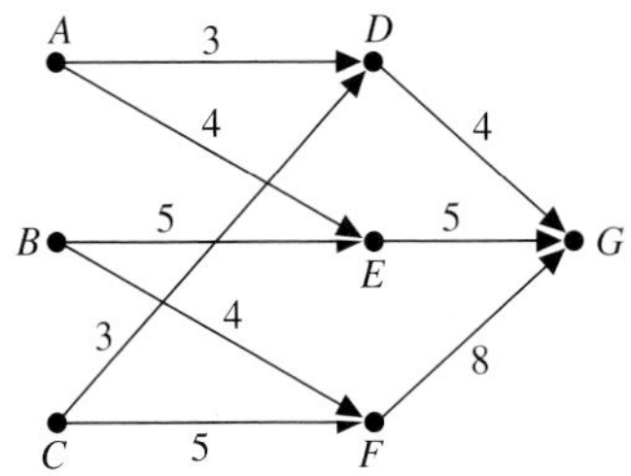

15.

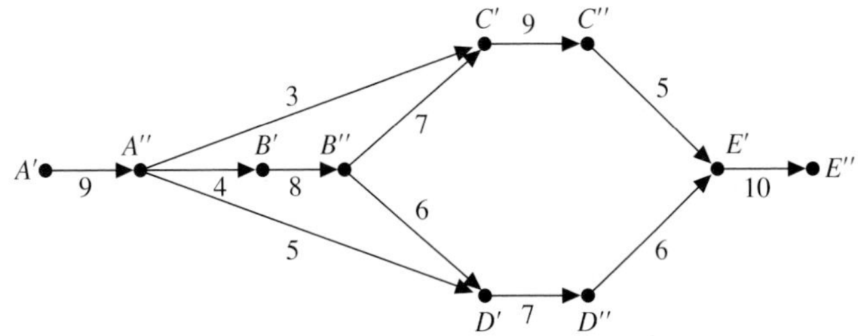

17.

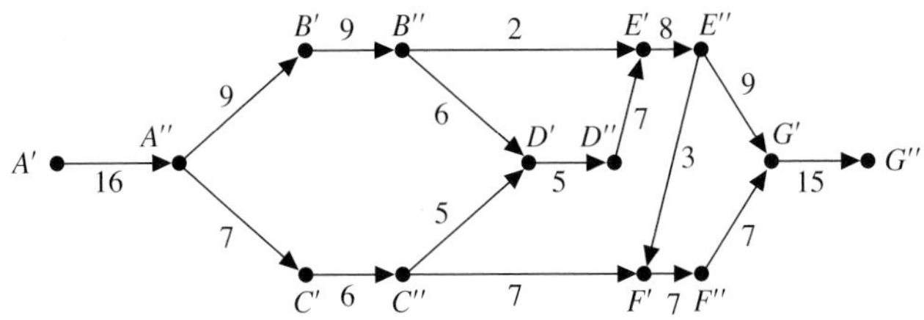

19.

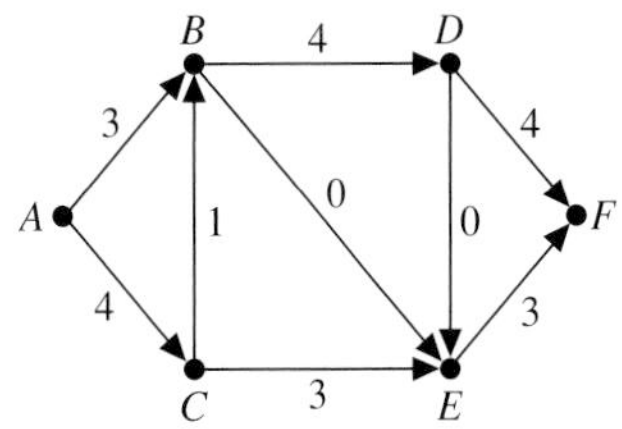

25.
$$\begin{bmatrix} 1 & 1 & 1 & 0 & 0 & 1 \\ 1 & 1 & 1 & 0 & 0 & 1 \\ 1 & 0 & 0 & 1 & 0 & 0 \\ 0 & 0 & 1 & 1 & 1 & 1 \end{bmatrix}$$

CHAPTER 8

EXERCISES 8.1 *(page 405)*

1. 10 **3.** 56 **5.** 6 **7.** −220 **9.** 3360 **11.** −262,440 **13.** 1, 6, 15, 20, 15, 6, 1
15. $x^6 + 6x^5y + 15x^4y^2 + 20x^3y^3 + 15x^2y^4 + 6xy^5 + y^6$ **17.** $81x^4 - 108x^3y + 54x^2y^2 - 12xy^3 + y^4$
19. 35 **21.** 252 **23.** 15

EXERCISES 8.2 *(page 413)*

1. 13 **3.** 5 **5.** 14 **7.** 45 **9.** 32 **11.** 2^n **13.** 168 **15.** 160
17. **(a)** 720 **(b)** 144 **(c)** 36 **(d)** 48 **19.** 3219 **21.** 70 **23.** 48 **25.** 36,504
27. **(a)** 720 **(b)** 360 **(c)** 240 **(d)** 576 **29.** 112 **31.** **(a)** 1296 **(b)** 360 **(c)** 60 **(d)** 240

EXERCISES 8.3 *(page 419)*

1. 20 **3.** 10 **5.** 12 **7.** 15,120 **9.** 5040 **11.** n **13.** 24 **15.** 360 **17.** 286 **19.** 210 **21.** 120 **23.** 84 **25.** 200 **27.** 10,584 **29.** **(a)** 495 **(b)** 5 **(c)** 72 **(d)** 54

EXERCISES 8.4 *(page 426)*

1. 1260 **3.** 210 **5.** 45 **7.** 1820 **9.** 63,063,000 **11.** 35 **13.** 165 **15.** 10 **17.** 140 **19.** 3,864,861 **21.** 4200 **23.** 462 **25.** 1050 **27.** 165 **29.** 6062 **31.** 220 **33.** 287,134,346

EXERCISES 8.5 *(page 432)*

1. $\frac{5}{6}$ **3.** $\frac{1}{32}$ **5.** $\frac{1}{18}$ **7.** $\frac{5}{16}$ **9.** $\frac{4}{33}$ **11.** $\frac{1}{210}$ **13.** $\frac{63}{125}$ **15.** $\frac{1}{105}$ **17.** $\frac{1}{120}$ **19.** $\frac{1}{4}$ **21.** $\frac{10{,}010}{59{,}049}$ **23.** $\frac{175}{429}$ **25.** $\frac{3}{28}$ **27.** $\frac{140}{2187}$ **29.** $\frac{5}{12}$ **31.** $\frac{352}{833}$ **33.** $\frac{88}{4165}$ **35.** $\frac{969}{2530}$

EXERCISES 8.6 *(page 442)*

1. 40 **3.** 160 **5.** 55 **7.** 27 **9.** $\frac{1}{64}$ **11.** 7 **13.** $k^4 - 4k^3 + 6k^2 - 3k$ **15.** 61 **17.** 231 **19.** 13,824 **21.** 685,464 **23.** 35 **27.** $D_k = k!\left[\frac{1}{0!} - \frac{1}{1!} + \cdots + (-1)^k \frac{1}{k!}\right]$ **29.** $S(n, 0) = 0$, $S(n, 1) = 1$, $S(n, 2) = 2^{n-1} - 1$, $S(n, n-2) = C(n, 3) + 3 \cdot C(n, 4)$, $S(n, n-1) = C(n, 2)$, and $S(n, n) = 1$ **31.** $S(n, 1) + S(n, 2) + \cdots + S(n, n)$

EXERCISES 8.7 *(page 451)*

1. $p < q$ **3.** $p < q$ **5.** $p > q$ **7.** (2, 1, 4, 3, 6, 5) **9.** (2, 1, 5, 3, 4, 6) **11.** (5, 6, 4, 1, 2, 3) **13.** none **15.** (5, 3, 1, 2, 4, 6) **17.** (6, 4, 1, 2, 3, 5) **19.** (1, 2, 3, 4); (1, 2, 4, 3); (1, 3, 2, 4); (1, 3, 4, 2); (1, 4, 2, 3); (1, 4, 3, 2); (2, 1, 3, 4); (2, 1, 4, 3); (2, 3, 1, 4); (2, 3, 4, 1); (2, 4, 1, 3); (2, 4, 3, 1); (3, 1, 2, 4); (3, 1, 4, 2); (3, 2, 1, 4); (3, 2, 4, 1); (3, 4, 1, 2); (3, 4, 2, 1); (4, 1, 2, 3); (4, 1, 3, 2); (4, 2, 1, 3); (4, 2, 3, 1); (4, 3, 1, 2); (4, 3, 2, 1) **21.** {1, 3, 5, 8, 9} **23.** {2, 3, 6, 7, 8} **25.** {3, 4, 5, 7, 9} **27.** {3, 6, 7, 8, 9} **29.** {4, 6, 7, 8, 9} **31.** none

SUPPLEMENTARY EXERCISES *(page 453)*

1. 36 **3.** 3024 **5.** $x^6 - 6x^5 + 15x^4 - 20x^3 + 15x^2 - 6x + 1$ **7.** $32x^5 + 240x^4y + 720x^3y^2 + 1080x^2y^3 + 810xy^4 + 243y^5$ **9.** (8, 2, 4, 1, 3, 5, 6, 7) **11.** 24310, 19448, 12376, 6188, 2380, 680, 136, 17, 1 **13.** 18,564 **15.** 21 **17.** 960 **19.** 240,240 **21.** $\frac{29}{59}$ **23.** 27,720 **25.** 43,243,200 **27.** 1820 **29.** 14 **31.** **(a)** 2520 **(b)** $\frac{1}{7}$ **(c)** $\frac{10}{21}$ **33.** **(a)** 657,720 **(b)** 142,506 **35.** 63 **37.** $\frac{10{,}005}{191{,}216}$ **39.** 6188 **41.** .3439 **43.** 514,594,080 **45.** 63,063,000 **47.** 630,630 **49.** no **51.** $\frac{105{,}293}{3{,}811{,}606}$ **55.** $C(n+1, 3)$

CHAPTER 9

EXERCISES 9.1 *(page 468)*

1. 126 **3.** 331 **5.** −63 **7.** 56 **9.** 31 **11.** 3
13. $t_n = 1.0525t_{n-1}$ for $n \geq 1$, $t_0 = 28{,}000$ **15.** $r_n = r_{n-1} + 6$ for $n \geq 1$, $r_0 = 24$
17. $b_n = 1.015b_{n-1} - 25$ for $n \geq 1$, $b_0 = 280$ **19.** $w_n = 0.15w_{n-1} + 2.0$ for $n \geq 1$, $w_0 = 1.7$
21. $s_n = 2s_{n-1} + 3s_{n-2} + s_{n-3}$ for $n \geq 4$, $s_1 = 2$, $s_2 = 7$, $s_3 = 21$ **23.** $a_n = na_{n-1}$ for $n \geq 2$, $a_1 = 1$
25. $s_n = s_{n-1} + (n-1)$ for $n \geq 1$, $s_0 = 0$
27. $s_n = s_{n-1} + s_{n-2} + s_{n-5}$ for $n \geq 6$, $s_1 = 1$, $s_2 = 2$, $s_3 = 3$, $s_4 = 5$, $s_5 = 9$; There are 128 sequences for a drink costing 50 cents
29. $c_n = (2n-1)c_{n-1}$ for $n \geq 2$, $c_1 = 1$ **31.** $r_n = r_{n-1} + 2(n-1)$ for $n \geq 2$, $r_1 = 2$
33. $s_n = s_{n-1} + s_{n-2} + s_{n-3}$ for $n \geq 4$, $s_1 = 2$, $s_2 = 4$, $s_3 = 7$
35. $c_n = c_0c_{n-1} + c_1c_{n-2} + \cdots + c_{n-1}c_0$ for $n \geq 2$, $c_0 = 1$, $c_1 = 1$
37. $s_n = 2s_{n-1} - s_{n-2} + s_{n-3}$ for $n \geq 4$, $s_1 = 2$, $s_2 = 4$, $s_3 = 7$; $s_6 = 37$

EXERCISES 9.2 *(page 479)*

9. $\frac{2}{3}n(2n+1)(n+1) = C(2n+2, 3)$ **11.** $s_n = 9 + 4n$ **13.** $s_n = 5(3)^n$ **15.** $s_n = 6(-1)^n$
17. $s_n = 1.75(5)^n - 0.75$ **19.** $s_n = 2(n^2 - 5n + 5)$ **21.** $s_n = \dfrac{a^{n+1} + (-1)^n}{a+1}$
23. $s_n = n!\left(2 + \frac{1}{0!} + \frac{1}{1!} + \cdots + \frac{1}{n!}\right)$
25. **(a)** $s_n = 0.95s_{n-1}$ for $n \geq 1$, $s_0 = 1000$ **(b)** $s_n = 1000(0.95)^n$ for $n \geq 0$ **(c)** 599
27. $s_n = 2n^2 + 2n$ **29.** $s_n = 2^{n-1}(n+2)$ **31.** $r_n = \dfrac{n^2+n+2}{2}$ **33.** $m_n = 2^{n+1} - 2$

EXERCISES 9.3 *(page 492)*

1. $s_n = 2 + 3n$ **3.** $s_n = 5(4^n)$ **5.** $s_n = 3 - 7(-1)^n$ **7.** $s_n = 4 - 3^n$ **9.** $s_n = 100 - 5n$
11. $s_n = 10(-2)^n - 3$ **13.** $s_n = 6(-1)^n + 3(2)^n$ **15.** $s_n = (6-n)4^n$ **17.** $s_n = 2(3)^n - (-3)^n$
19. $s_n = (3n-4)(-2)^n$ **21.** $s_n = (4n-7)5^n$ **23.** $s_n = 9(-1)^n - 6(-4)^n$
25. **(a)** $d_n = 0.80d_{n-1} + 25$ for $n \geq 1$, $d_0 = 0$ **(b)** $d_8 = 104.02848$ mg **(c)** 125 mg
27. About \$3670.36 **29.** \$1190.30 **31.** $s_n = s_{n-1} + s_{n-2}$ for $n \geq 3$, $s_1 = 1$, $s_2 = 2$

$$s_n = \frac{5+\sqrt{5}}{10}\left(\frac{1+\sqrt{5}}{2}\right)^n + \frac{5-\sqrt{5}}{10}\left(\frac{1-\sqrt{5}}{2}\right)^n$$

33. $c_n = 3^n + 3(-1)^n$

EXERCISES 9.4 *(page 504)*

1. 243 **3.** 4 **5.** 0 **7.** 16

9.

b	e	m	a_m	Is $a_m = t$?
1	100	$\left\lfloor \frac{101}{2} \right\rfloor = 50$	50	no, less
51	100	$\left\lfloor \frac{151}{2} \right\rfloor = 75$	75	no, less
76	100	$\left\lfloor \frac{176}{2} \right\rfloor = 88$	88	no, greater
76	87	$\left\lfloor \frac{163}{2} \right\rfloor = 81$	81	no, less
82	87	$\left\lfloor \frac{169}{2} \right\rfloor = 84$	84	no, greater
82	83	$\left\lfloor \frac{165}{2} \right\rfloor = 82$	82	no, less
83	83	83	83	yes

11.

b	e	m	a_m	Is $a_m = t$?
1	300	$\left\lfloor \frac{301}{2} \right\rfloor = 150$	450	no, greater
1	149	$\left\lfloor \frac{150}{2} \right\rfloor = 75$	225	no, less
76	149	$\left\lfloor \frac{225}{2} \right\rfloor = 112$	336	no, less
113	149	$\left\lfloor \frac{262}{2} \right\rfloor = 131$	393	no, less
132	149	$\left\lfloor \frac{281}{2} \right\rfloor = 140$	420	no, greater
132	139	$\left\lfloor \frac{271}{2} \right\rfloor = 135$	405	no, greater
132	134	$\left\lfloor \frac{266}{2} \right\rfloor = 133$	399	no, less
134	134	$\left\lfloor \frac{268}{2} \right\rfloor = 134$	402	no, greater
134	133			

Since $b > e$, the target t is not in the list.

13.

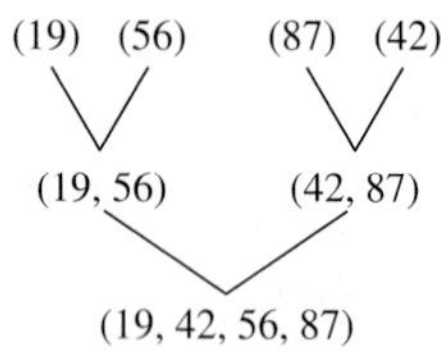

15.

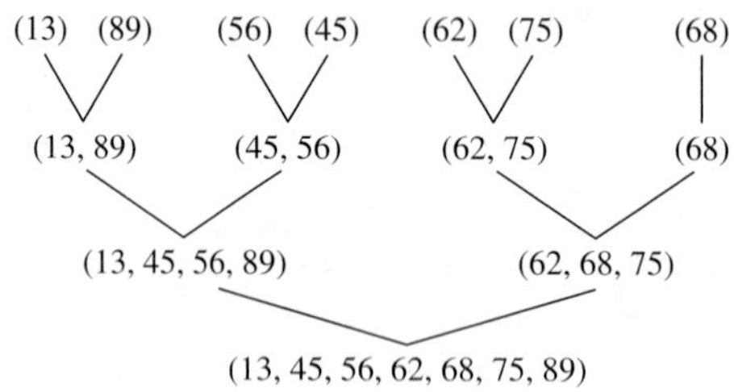

17.

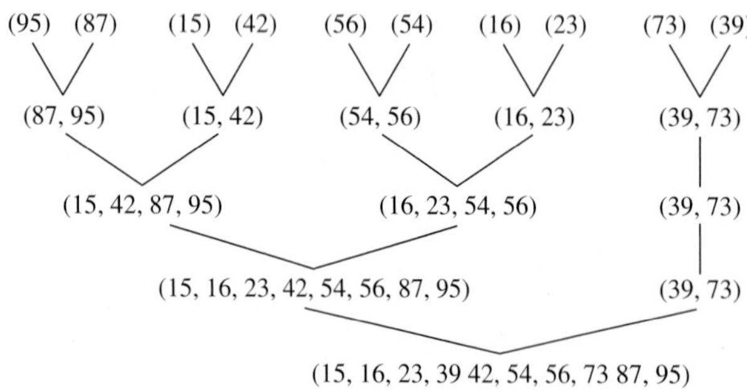

19. $n > 4$ **25.** $e_n = e_{n-1} + 3$ for $n \geq 2$, $e_1 = 1$ **27.** $e_n = e_{n-1} + 5$ for $n \geq 1$, $e_0 = 1$
29. $e_n = 3n - 2$ for $n \geq 1$ **31.** $e_n = 5n + 1$ for $n \geq 0$

EXERCISES 9.5 *(page 512)*

1. $2 + 3x + x^2 + 4x^4 + x^5$ **3.** $1 + 3x + 3x^2 + 2x^3 + 4x^4 + 5x^5 + 5x^6 + x^7$
5. $2 + 3x + x^2 + x^3 + 5x^4 + 2x^5 + x^6 + x^7 + \cdots$ **7.** $1 + 2x + 3x^2 + 3x^3 + 3x^4 + 3x^5 + 3x^6 + 3x^7 + \cdots$
9. $1 - x^2 + x^3 + x^4 - x^5 + x^6 + x^7 - \cdots$ **11.** $1 + x + x^2 + 2x^3 + 2x^4 + 2x^5 + 3x^6 + 3x^7 + \cdots$
13. $(1 + x + x^2 + x^3)(1 + x + x^2 + x^3 + x^4 + x^5) = 1 + 2x + 3x^2 + 4x^3 + 4x^4 + 4x^5 + 3x^6 + \cdots$
15. $(1 + x + x^2 + x^3)(1 + x + x^2 + x^3 + x^4)(1 + x + x^2) = 1 + 3x + 6x^2 + 9x^3 + 11x^4 + 11x^5 + 9x^6 + \cdots$
17. $(1 + x + x^2 + x^3 + x^4)(1 + x + x^2 + x^3)(1 + x^2)(1 + x^3)$
$= 1 + 2x + 4x^2 + 7x^3 + 9x^4 + 11x^5 + 12x^6 + \cdots$
19. $(1 + x + x^2 + x^3)(1 + x + x^2 + \cdots) = 1 + 2x + 3x^2 + 4x^3 + 4x^4 + 4x^5 + 4x^6 + \cdots$
21. $(x^4 + x^5 + \cdots)(x^2 + x^3 + \cdots) = x^6 + \cdots$ **23.** $(1 + x)^7(1 + x + x^2 + x^3 + x^4 + x^5)$
25. $(1 + x + x^2 + \cdots)(1 + x^3 + x^6 + \cdots)(1 + x^4 + x^8 + \cdots)$ **27.** $a_r = r + 1$
29. $a_0 = 1$ and $a_r = 2$ if $r > 0$ **31.** $a_0 = 1$ and $a_r = 0$ if $r > 0$
33. $(x^2 + x^3 + x^5 + x^7 + x^{11} + \cdots)^2 = x^4 + 2x^5 + x^6 + 2x^7 + 2x^8 + 2x^9 + 3x^{10} + \cdots$
35. $(1 + x + x^4 + x^9 + \cdots)^4$
$= 1 + 4x + 6x^2 + 4x^3 + 5x^4 + 12x^5 + 12x^6 + 4x^7 + 6x^8 + 16x^9 + 18x^{10} + \cdots$

EXERCISES 9.6 *(page 521)*

1. $1 + 3x + 9x^2 + \cdots$ **3.** $1 - 2x$ **5.** $1 - x^2 + x^4 - x^6 + \cdots$ **7.** $1 + (x + x^2) + (x + x^2)^2 + \cdots$

9. $\frac{1}{2} - \frac{3}{2}x + \frac{9}{2}x^2 - \frac{27}{2}x^3 + \cdots$ **11.** $S = 2xS + 1 + x + x^2 + \cdots$; $S = (1 - 2x)^{-1}(1 - x)^{-1}$ **13.** $S = 1 + x + 2x(S - 1) - x^2S$; $S = (1 - x)S = (1 - x)^{-1}$

15. $S = -1 - x(S + 1) + 2x^2S$; $S = -(1 + x)(1 - x)^{-1}(1 + 2x)^{-1}$ **17.** $S = -2 + x + x(S + 2) + 3x^2S + 2x^2(1 + x + x^2 + \cdots)$; $S = (1 - x)^{-1}(1 - x - 3x^2)^{-1}(-2 + 5x - x^2)$

19. $S = 2 - x + x^2 + x(S - 2 + x) - 3x^2(S - 2) + x^3S$; $S = (2 - 3x + 8x^2)(1 - x + 3x^2 - x^3)^{-1}$ **21.** $a = \frac{1}{3}, b = -\frac{1}{3}$ **23.** $a = -\frac{1}{3}, b = \frac{4}{3}$

25. $a = \frac{1}{2}, b = \frac{1}{2}$ **27.** $s_n = 2^n + 1$ **29.** $s_n = -2^n + 4(-5)^n$
31. $s_n = 2(3^{n/2})$ if n is even and $s_n = 0$ if n is odd

SUPPLEMENTARY EXERCISES *(page 524)*

1. 829 **3.** 3840 **5.** 33 **7.** $s_n = 1.04s_{n-1} + 500$ for $n \geq 2$ and $s_1 = 16{,}000$
9. $c_n = c_{n-1} + c_{n-2}$ for $n \geq 3$, $c_1 = 2$, and $c_2 = 3$ **11.** $e_n = e_{n-1} - k/n^2$ for $n \geq 1$, where k is a constant
17. $s_n = 6 - 3^n$ for $n \geq 0$ **19.** $s_n = (4 - n)2^n$ for $n \geq 0$ **21.** 3, 4, 7, 11, 18, 29, 47, 76, 123, 199
27. $s_n = 2(4^n) + (-3)^n - 7(2^n)$ for $n \geq 0$ **29.** $s_n = 2^n + (3 - 5n)(-1)^n$ for $n \geq 0$
31. **(a)** $a = -1, b = -2$ **(b)** $s_n = 3^n - 5(-2)^n - n - 2$ for $n \geq 0$
33.

b	e	m	a_m	Is $a_m = 6$?
1	4	$\left\lfloor \frac{(1+4)}{2} \right\rfloor = 2$	4	no, less
3	4	$\left\lfloor \frac{(3+4)}{2} \right\rfloor = 3$	6	yes

35. 2 **37.** 7

39. $1, 2, \ldots, m$ and $m + 1, m + 2, \ldots, m + n$ **41.** $S = 1 + 2xS$; $S = (1 - 2x)^{-1}$
43. $S = 1 + x - x^2S - 2x(S - 1)$; $S = \frac{1 + 3x}{(1 + x)^2}$ **45.** $s_n = 5(-2)^n$ **47.** $s_n = 2^n + 2(3^n)$

49. $(1 + x + x^2 + x^3)(1 + x + x^2)(1 + x + x^2 + x^3 + x^4 + x^5)$
51. $(x^2 + x^3 + x^4)^6$ **53.** $(1 - x^5)^{-1}(1 - x^{12})^{-1}(1 - x^{25})^{-1}$ **55.** $(1 - x)^{-1}(1 - x^5)^{-1}(1 - x^{10})^{-1}$

CHAPTER 10

EXERCISES 10.1 *(page 535)*

1. $(x \wedge y) \vee x$ **3.** $((x' \vee y) \wedge x)'$ **5.** $(x'' \vee y') \wedge x'$ **7.** $(x' \wedge (y' \wedge x))'$

9.

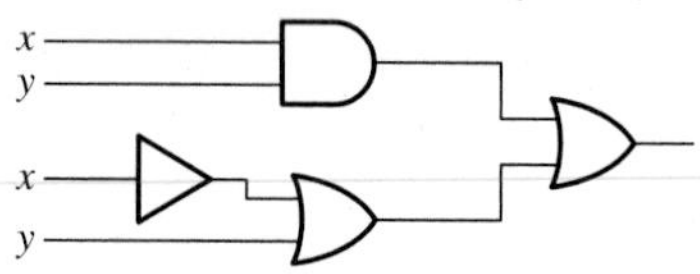

11.

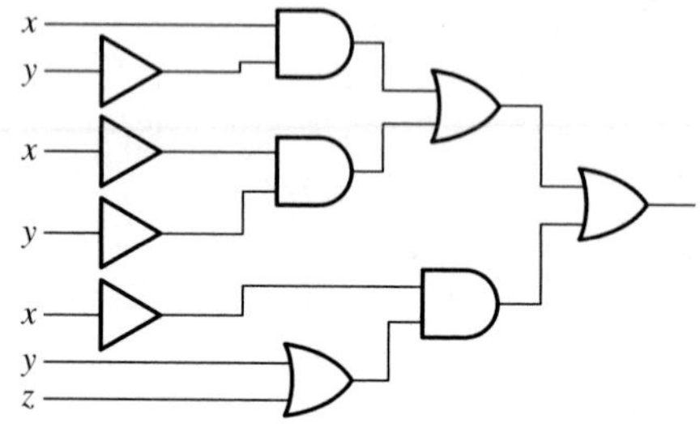

13.

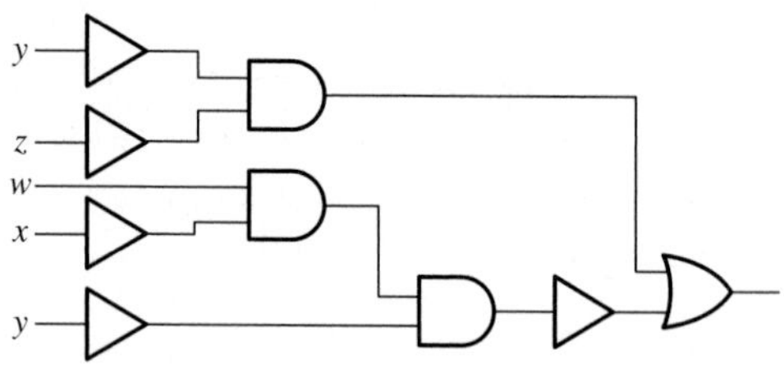

15. 0 **17.** 1

19.

x	y	*Output*
0	0	1
0	1	1
1	0	0
1	1	1

21.

x	y	z	*Output*
0	0	0	0
0	0	1	0
0	1	0	0
0	1	1	0
1	0	0	0
1	0	1	0
1	1	0	1
1	1	1	0

23.

x	y	*Output*
0	0	0
0	1	0
1	0	0
1	1	1

25.

x	y	*Output*
0	0	0
0	1	1
1	0	1
1	1	1

27.

x	y	z	*Output*
0	0	0	1
0	0	1	1
0	1	0	0
0	1	1	0
1	0	0	1
1	0	1	1
1	1	0	1
1	1	1	1

29. equivalent **31.** equivalent **33.** equivalent

35. not equivalent **37.** equivalent **39.** not equivalent **41.** not equivalent

43.

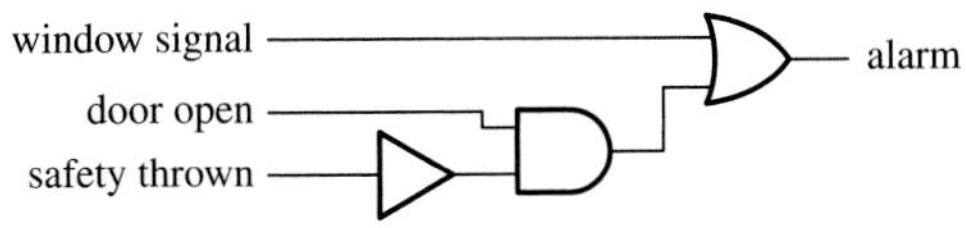

47. 1, undefined

EXERCISES 10.2 *(page 543)*

9. (c), (f), (g) **11.** (c), (f), (a), (g) **13.** (j), (i), (e) **15.** (j), (b)

17. (c), (a), (b), (b), (b), (b), (e), (e), (b), (b), (a), (c)

19. When $x = y = 0$, $z = 1$, the first is 0 and the second is 1.

21. When $x = z = 0$, $y = 1$, the first is 0 and the second is 1.

23. $(x' \wedge y) \vee (x \wedge y')$

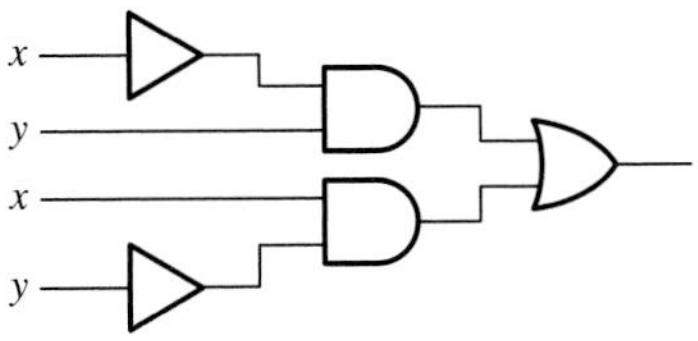

25. $(x' \wedge y \wedge z) \vee (x \wedge y' \wedge z) \vee (x \wedge y \wedge z')$

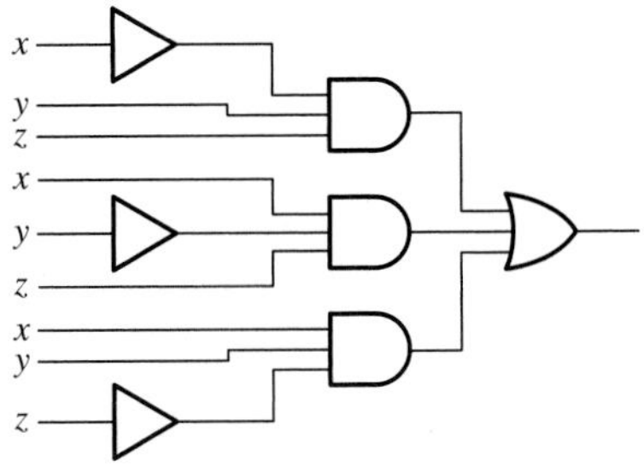

27. $(x' \wedge y \wedge z) \vee (x \wedge y' \wedge z) \vee (x \wedge y \wedge z)$

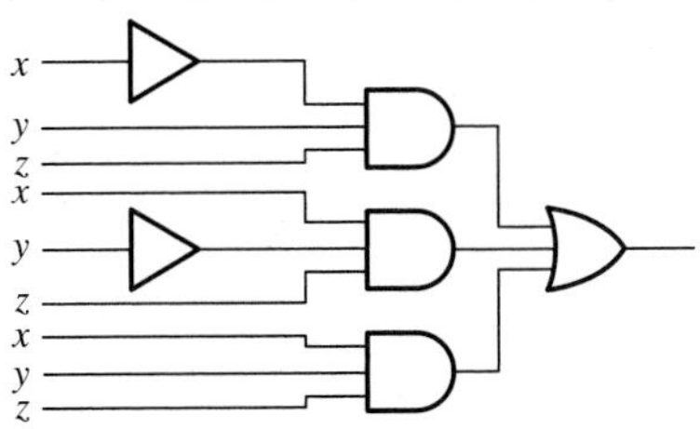

29. 12 **31.** 9 **33.** 14

35.

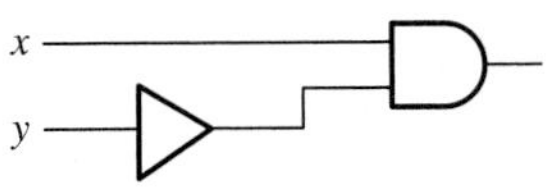

37. $(a \wedge b \wedge c') \vee ((b \wedge c) \wedge d') \vee (a \wedge d)$ **39.** all

EXERCISES 10.3 *(page 557)*

1. $(x' \wedge y') \vee (x' \wedge y) \vee (x \wedge y)$ **3.** $(x' \wedge y' \wedge z') \vee (x' \wedge y' \wedge z) \vee (x' \wedge y \wedge z') \vee (x' \wedge y \wedge z)$

5. $(w' \wedge x' \wedge y' \wedge z') \vee (w' \wedge x' \wedge y \wedge z') \vee (w' \wedge x \wedge y' \wedge z') \vee (w' \wedge x \wedge y \wedge z') \vee (w \wedge x \wedge y' \wedge z)$

7. $x \vee y'$ **9.** $y \vee x' \vee (y' \wedge z')$ **11.** $(x \wedge z) \vee (w' \wedge x) \vee (w \wedge x' \wedge y') \vee (w' \wedge y \wedge z')$

13.

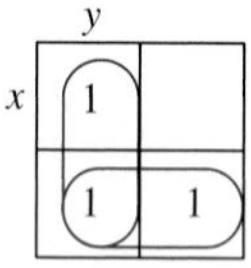

15.

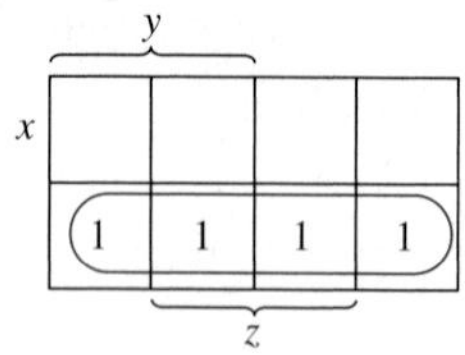

17.

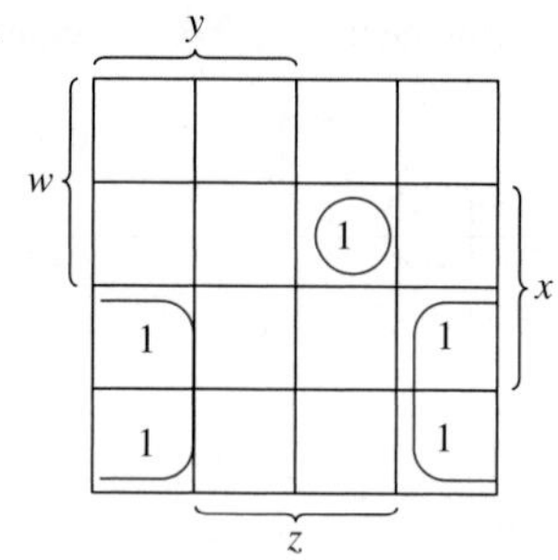

19. $x' \vee y$

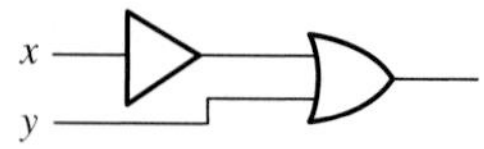

21. x'

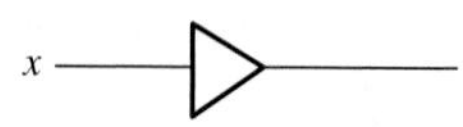

23. $(w' \wedge z') \vee (w \wedge x \wedge y' \wedge z)$

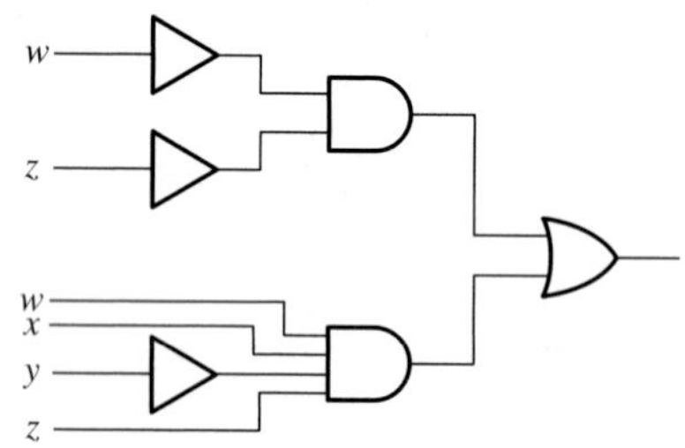

25. $(x' \wedge z) \vee (y' \wedge z)$ **27.** $(x' \wedge z) \vee (x \wedge y' \wedge z')$ **29.** $x \vee (y' \wedge z')$ **31.** $(w \wedge x) \vee (y' \wedge z')$

33.

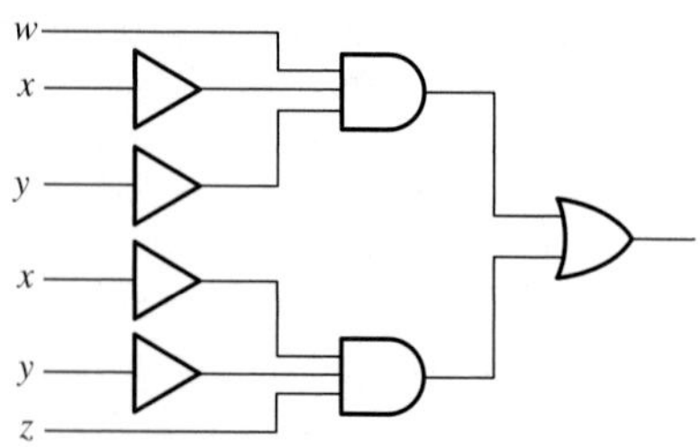

35. 32

EXERCISES 10.4 *(page 566)*

1.

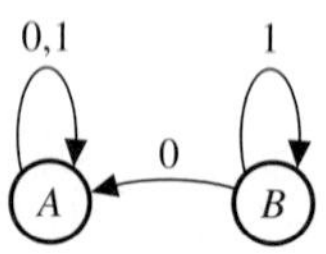

3.

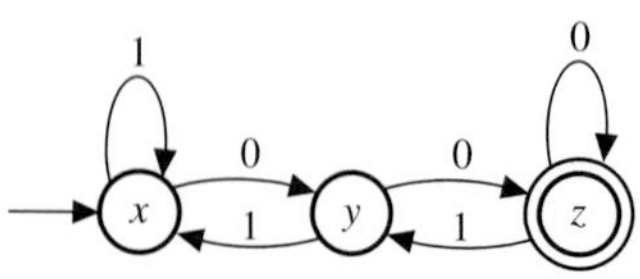

5.

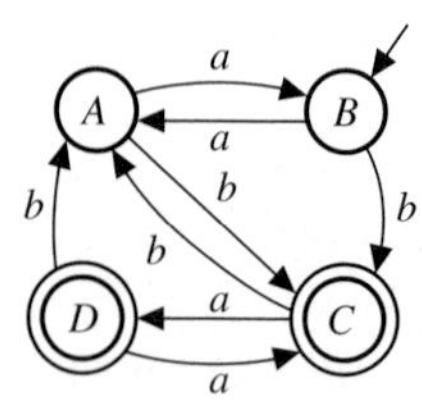

7.

	A	B	C
x	B	C	C
y	C	C	A

Initial state B
Accepting state A

9.

	1	2	3
a	2	3	1
b	3	1	2
c	1	2	3

Accepting state 2

11. y **13.** A **15.** yes **17.** no

19.

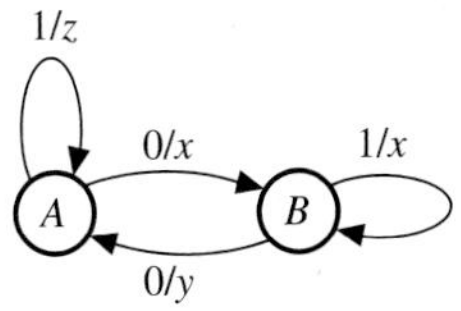

21.

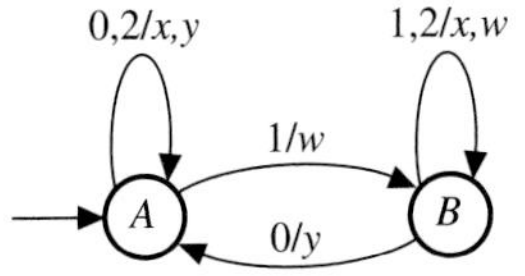

23.

	1	2	1	2
1	2	2	*a*	*a*
2	2	1	*c*	*b*
3	1	2	*c*	*a*

25.

	A	*B*	*C*	*A*	*B*	*C*
1	*B*	*B*	*C*	*x*	*y*	*y*
2	*C*	*C*	*A*	*z*	*x*	*z*
3	*C*	*C*	*C*	*y*	*z*	*z*

27. *ywywwxx* **29.** *yzzyyz* **31.** All strings containing a 1

33. All strings containing exactly n 1s, where $n \equiv 1 \pmod 3$

35.

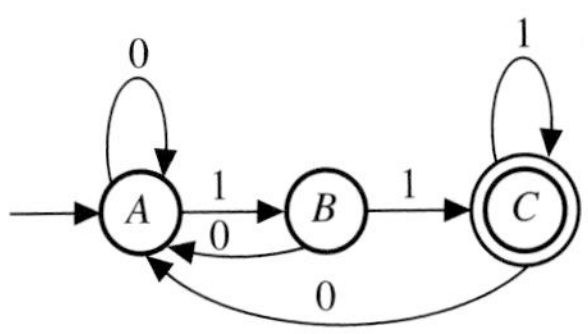

37.

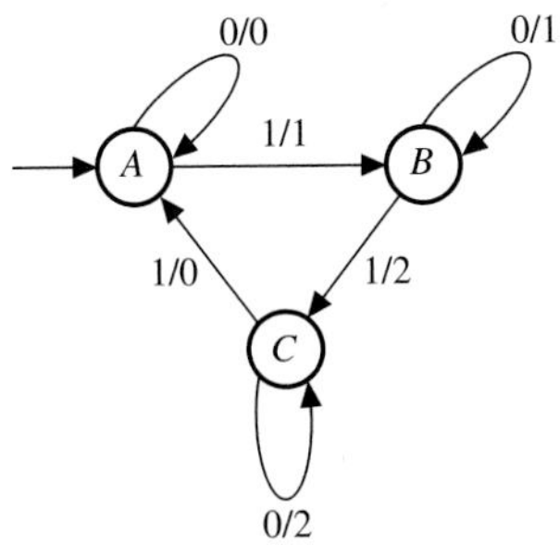

SUPPLEMENTARY EXERCISES *(page 570)*

1. $(x' \vee y) \wedge (z \wedge x)'$

x	y	z	$(x' \vee y) \wedge (z \wedge x)'$
0	0	0	1
0	0	1	1
0	1	0	1
0	1	1	1
1	0	0	0
1	0	1	0
1	1	0	1
1	1	1	0

3. **(a)** yes **(b)** no

5. **(a)**
$$\begin{aligned}[(x \vee y) \wedge (x' \vee y)] \vee y' &= [(y \vee x) \wedge (y \vee x')] \vee y' && \text{(a)}\\ &= [y \vee (x \wedge x')] \vee y' && \text{(c)}\\ &= (y \vee 0) \vee y' && \text{(f)}\\ &= y \vee y' && \text{(g)}\\ &= 1 && \text{(f)}\end{aligned}$$

(b)
$$\begin{aligned}x' \wedge (y \wedge z')' &= x' \wedge (y' \vee z'') && \text{(j)}\\ &= x' \wedge (y' \vee z) && \text{(i)}\\ &= (x' \wedge y') \vee (x' \wedge z) && \text{(c)}\\ &= (x \vee y)' \vee (x' \wedge z) && \text{(j)}\end{aligned}$$

7. $(x' \wedge y \wedge z') \vee (x' \wedge y \wedge z) \vee (x \wedge y' \wedge z') \vee (x \wedge y \wedge z)$

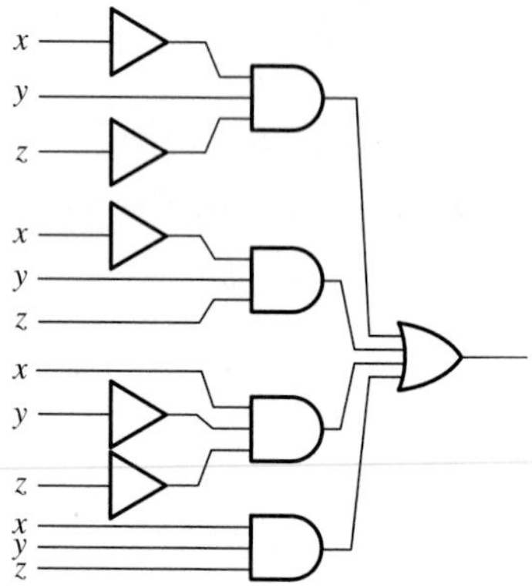

16 gates

9. **(a)** x' **(b)** $(x \wedge z') \vee (y' \wedge z')$ **(c)** $(x' \wedge y \wedge z') \vee (y' \wedge z) \vee (x \wedge y') \vee (w' \wedge z')$

11. $(x' \wedge y) \vee (y \wedge z) \vee (x \wedge y' \wedge z')$

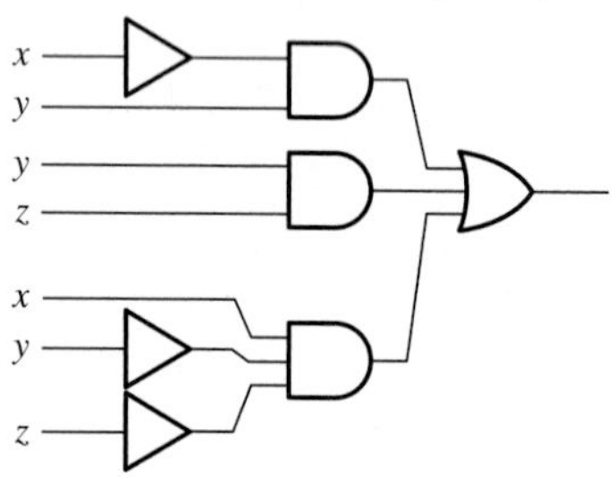

9 gates

13.

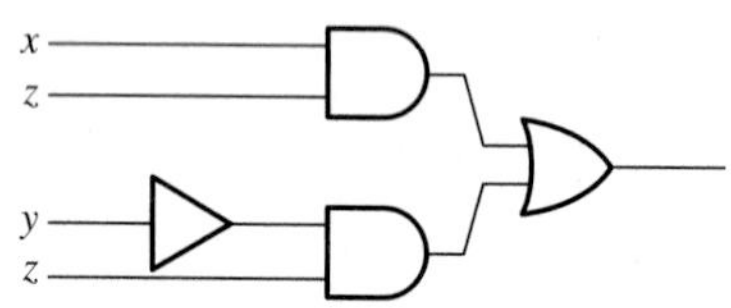

15. C **17.** y **19.** $-1, 1, 1, 1, 0, 1, 1$ **21.** b, b, foam, b, a, foam, b

APPENDIX A

EXERCISES A.1 *(page 582)*

1. false statement **3.** true statement **5.** not a statement **7.** false statement **9.** not a statement
11. true statement **13.** $4 + 5 \neq 9$ **15.** California is the largest state in the United States.
17. Some birds cannot fly. **19.** No man weighs 400 pounds.
21. No students do not pass calculus. (All students pass calculus.) **23.** Someone does not enjoy cherry pie.
25. **(a)** One is an even integer and nine is a positive integer. (false) **(b)** One is an even integer or nine is a positive integer. (true)
27. **(a)** The Atlantic is an ocean and the Nile is a river. (true) **(b)** The Atlantic is an ocean or the Nile is a river. (true)
29. **(a)** Birds have four legs and rabbits have wings. (false) **(b)** Birds have four legs or rabbits have wings. (false)
31. **(a)** Flutes are wind instruments and timpani are string instruments. (false) **(b)** Flutes are wind instruments or timpani are string instruments. (true)
33. **(a)** If I go to the movies, then this is Friday. **(b)** If this isn't Friday, then I won't go to the movies. **(c)** If I don't go to the movies, then this isn't Friday.

35. **(a)** If Kennedy runs for president, then he won't run for the Senate. **(b)** If Kennedy runs for the Senate, then he won't run for president. **(c)** If Kennedy doesn't run for president, then he is running for the Senate.

EXERCISES A.2 *(page 586)*

Note: Only the last columns of truth tables are given.

1.

p	q	$(p \vee q) \wedge [\sim(p \wedge q)]$
T	T	F
T	F	T
F	T	T
F	F	F

3.

p	q	$(p \vee q) \rightarrow (\sim p \wedge q)$
T	T	F
T	F	F
F	T	T
F	F	T

5.

p	q	r	$(p \rightarrow q) \rightarrow (p \vee r)$
T	T	T	T
T	T	F	T
T	F	T	T
T	F	F	T
F	T	T	T
F	T	F	F
F	F	T	T
F	F	F	F

7.

p	q	r	$(\sim q \wedge r) \leftrightarrow (\sim p \vee q)$
T	T	T	F
T	T	F	F
T	F	T	F
T	F	F	T
F	T	T	F
F	T	F	F
F	F	T	T
F	F	F	F

9.

p	q	r	$[(p \vee q) \wedge r] \rightarrow [(p \wedge r) \vee q]$
T	T	T	T
T	T	F	T
T	F	T	T
T	F	F	T
F	T	T	T
F	T	F	T
F	F	T	T
F	F	F	T

17.

p	$\sim p$	$\sim(\sim p)$
T	F	T
F	T	F

19. If the truth table is arranged as in Exercise 1, then the column corresponding to the given statement is F, T, F, F.
21. If the truth table is arranged as in Exercise 5, then the column corresponding to the given statements is T, F, T, T, T, T, T, T.
23. If the truth table is arranged as in Exercise 5, then the column corresponding to the given statements is T, F, T, F, T, F, T, T.

25.

p	q	$p \wedge q$	$q \wedge p$	$p \vee q$	$q \vee p$
T	T	T	T	T	T
T	F	F	F	T	T
F	T	F	F	T	T
F	F	F	F	F	F

27. **(e)** If the truth table is arranged as in Exercise 5, then the column corresponding to both statements is T, T, T, T, T, F, F, F.
(f) If the truth table is arranged as in Exercise 5, then the column corresponding to both statements is T, T, T, F, F, F, F, F. **29.** If the truth table is arranged as in Exercise 1, then the column corresponding to both statements is T, F, T, T.

33. **(a)**

p	q	$p \veebar q$
T	T	F
T	F	T
F	T	T
F	F	F

(b) If the truth table is arranged as in part (a), then the column corresponding to both statements is F, T, T, F.

EXERCISES A.3 *(page 592)*

1.

p	q	$\sim q$	$p \wedge (\sim q)$	$\sim[p \wedge (\sim q)]$	$p \to q$
T	T	F	F	T	T
T	F	T	T	F	F
F	T	F	F	T	T
F	F	T	F	T	T

13. The statement is false. For example, $3 + 5 = 8$.

15. The statement is false. For example, if $a = 3$, $b = 2$, and $c = 0$, then $ac = bc$, but $a \neq b$.
17. The statement is false. For example, if $x = 4$ and $y = 9$, then 6 divides xy, but 6 does not divide either x or y.
19. The statement is true. **21.** The statement is true.
23. The statement is false. For example, if $n = 41$, then

$$n^2 + n + 41 = 41^2 + 41 + 41 = 41(43).$$

SUPPLEMENTARY EXERCISES *(page 594)*

1. true statement **3.** not a statement **5.** false statement **7.** true statement
9. There exists a square that is a triangle. (false) **11.** No scientist from the United States has received a Nobel prize. (false) **13.** $2 + 2 \leq 4$, and 1 is not a root of $x^5 + 1 = 0$. (true)
15. In circling the globe along a line of latitude, one must not cross the equator exactly twice or not cross the North Pole or not cross the South Pole. (true)
17. **(a)** Squares have four sides and triangles have three sides. (true) **(b)** Squares have four sides or triangles have three sides. (true) **19.** **(a)** If $3 > 2$, then $3 \times 0 > 2 \times 0$, and if $4 = 5$, then $5 = 9$. (false) **(b)** If $3 > 2$, then $3 \times 0 > 2 \times 0$, or if $4 = 5$, then $5 = 9$. (true) **21.** **(a)** If $3^2 = 6$, then $3 + 3 = 6$. (true) **(b)** If $3 + 3 \neq 6$, then $3^2 \neq 6$. (true) **(c)** If $3^2 \neq 6$, then $3 + 3 \neq 6$. (false) **23.** **(a)** If $3 \times 2 = 6$, then $3^2 = 6$. (false) **(b)** If $3^2 \neq 6$, then $3 \times 2 \neq 6$. (false) **(c)** If $3 \times 2 \neq 6$, then $3^2 \neq 6$. (true)

25.

p	q	$\sim[(p \vee q) \wedge \sim p] \wedge \sim p$
T	T	F
T	F	F
F	T	F
F	F	T

27.

p	q	r	$[p \wedge (r \wedge (\sim p \vee q))] \to [(p \wedge r) \wedge (\sim p \vee q)]$
T	T	T	T
T	T	F	T
T	F	T	T
T	F	F	T
F	T	T	T
F	T	F	T
F	F	T	T
F	F	F	T

29. yes **31.** no **33.** yes **35.** yes **37.** 6 cannot **39.** The statement is true.
41. The statement is true. **43.** The statement is true.

APPENDIX B

EXERCISES B.1 *(page 603)*

1. not defined **3.** $\begin{bmatrix} 0 & 2 & 5 \\ 3 & 0 & 6 \end{bmatrix}$ **5.** $\begin{bmatrix} 12 & 9 \\ 7 & -1 \end{bmatrix}$ **7.** not defined **9.** $\begin{bmatrix} 3 & 2 \\ -2 & 3 \end{bmatrix}$

11. $\begin{bmatrix} 1 & 0 \\ -2 & 1 \end{bmatrix}$ **13.** $\begin{bmatrix} 11 & 16 \\ -14 & -5 \end{bmatrix}$ **15.** $\begin{bmatrix} 1 & 0 \\ -3 & 1 \end{bmatrix}$ **17.** $\begin{bmatrix} 1 & -1 & 1 \\ 1 & -1 & 1 \\ 1 & -1 & 1 \end{bmatrix}$ **19.** $\begin{bmatrix} 3 & 3 & 3 \\ 3 & 3 & 3 \\ 3 & 3 & 3 \end{bmatrix}$

21. $\begin{bmatrix} 3 & 3 & 3 \\ 3 & 3 & 3 \\ 3 & 3 & 3 \end{bmatrix}$ **23.** $\begin{bmatrix} 9 & 9 & 9 \\ 9 & 9 & 9 \\ 9 & 9 & 9 \end{bmatrix}$

25. $AB = \begin{bmatrix} 0 & 0 \\ 0 & 0 \end{bmatrix}$, $BA = \begin{bmatrix} 1 & -1 \\ 1 & -1 \end{bmatrix}$ Matrix multiplication is not commutative, and the product of two nonzero matrices can be a zero matrix.

Photo Credits

pp. 35, 36, 93, 94, and 148 (Vigenère) University of St. Andrews, MacTutor Archive; p. 148 (Rivest) Courtesy of Ronald Rivest; p. 148 (Shamir) The Weizmann Institute of Science; p. 148 (Adleman) Courtesy of Leonard Adleman; p. 149 Bell Laboratories; p. 219 University of St. Andrews, MacTutor Archive; p. 220 (Dijkstra) Courtesy of Dr. Hamilton Richards; p. 220 (König) Courtesy of János Bolyai Mathematical Society; pp. 307 and 354; University of St. Andrews, MacTutor Archive; p. 397 Mathematical Association of America; pp. 452, 453, 524, and 569 University of St. Andrews, MacTutor Archive; p. 569 (McCluskey) Courtesy of Edward McCluskey; pp. 594 and 605 University of St. Andrews, MacTutor Archive

Index

DISCRETE MATHEMATICS TIMELINE

c. 300 B.C. | 0 | 1200 | 1500 | 1600 | 1700

■ c. 300 B.C.
Euclid published the Euclidean Algorithm in his *Elements*.

■ 1202
Fibonacci introduced the Fibonacci numbers in his *Liber Abaci*.

■ 1575
Francesco Maurolico wrote his *Arithmeticorum*, which employed the principle of mathematical induction in proofs.

■ 1654
Blaise Pascal's famous triangle appeared in *Arithmetical Triangle*.

■ 1654
Blaise Pascal and Pierre Fermat corresponded on the fundamentals of probability theory.

■ 1666
Gottfried Leibnitz published his ideas on combinatorics in *Diertatio de Arte Combinatoria*.

■ 1713
James Bernoulli's *Ars Conjectandi* introduced several ideas related to permutations and combinations.

■ 1718
Abraham De Moivre wrote about the principle of inclusion–exclusion in his *Doctrine of Chances*.

■ 1720
Abraham De Moivre began work on generating functions.

■ 1735
Leonhard Euler introduced functional notation.

■ 1736
Leonhard Euler published a paper on the Königsberg Bridge problem.

■ 1750
Leonhard Euler proposed his formula for the number of edges, vertices, and faces of polyhedra.

c. 300 B.C. | 0 | 1200 | 1500 | 1600 | 1700